AF876701

PROGRESS IN BRAIN RESEARCH

VOLUME 165

COMPUTATIONAL NEUROSCIENCE: THEORETICAL INSIGHTS INTO BRAIN FUNCTION

Other volumes in PROGRESS IN BRAIN RESEARCH

PROGRESS IN BRAIN RESEARCH

VOLUME 165

COMPUTATIONAL NEUROSCIENCE: THEORETICAL INSIGHTS INTO BRAIN FUNCTION

EDITED BY

PAUL CISEK
TREVOR DREW
JOHN F. KALASKA
Department of Physiology, University of Montréal, Montréal, QC, Canada

ELSEVIER

AMSTERDAM – BOSTON – HEIDELBERG – LONDON – NEW YORK – OXFORD
PARIS – SAN DIEGO – SAN FRANCISCO – SINGAPORE – SYDNEY – TOKYO

Elsevier
Radarweg 29, PO Box 211, 1000 AE Amsterdam, The Netherlands
Linacre House, Jordan Hill, Oxford OX2 8DP, UK

First edition 2007

Copyright © 2007 Elsevier B.V. All rights reserved

No part of this publication may be reproduced, stored in a retrieval system or transmitted in any form or by any means electronic, mechanical, photocopying, recording or otherwise without the prior written permission of the publisher

Permissions may be sought directly from Elsevier's Science & Technology Rights Department in Oxford, UK: phone (+44) (0) 1865 843830; fax (+44) (0) 1865 853333; e-mail: permissions@elsevier.com. Alternatively you can submit your request online by visiting the Elsevier web site at http://www.elsevier.com/locate/permissions, and selecting *Obtaining permission to use Elsevier material*

Notice
No responsibility is assumed by the publisher for any injury and/or damage to persons or property as a matter of products liability, negligence or otherwise, or from any use or operation of any methods, products, instructions or ideas contained in the material herein. Because of rapid advances in the medical sciences, in particular, independent verification of diagnoses and drug dosages should be made

Library of Congress Cataloging-in-Publication Data
A catalog record for this book is available from the Library of Congress

British Library Cataloguing in Publication Data
A catalogue record for this book is available from the British Library

ISBN: 978-0-444-52823-0 (this volume)
ISSN: 0079-6123 (Series)

For information on all Elsevier publications
visit our website at books.elsevier.com

Printed and bound in The Netherlands

07 08 09 10 11 10 9 8 7 6 5 4 3 2 1

Working together to grow
libraries in developing countries
www.elsevier.com | www.bookaid.org | www.sabre.org
ELSEVIER BOOK AID International Sabre Foundation

List of Contributors

L.F. Abbott, Department of Physiology and Cellular Biophysics, Center for Neurobiology and Behavior, Columbia University College of Physicians and Surgeons, New York, NY 10032-2695, USA

A.P.L. Abdala, Department of Physiology, School of Medical Sciences, University of Bristol, Bristol BS8 1TD, UK

D.E. Angelaki, Department of Anatomy and Neurobiology, Box 8108, Washington University School of Medicine, 660 South Euclid Avenue, St. Louis, MO 63110, USA

J. Beck, Department of Brain and Cognitive Sciences, University of Rochester, Rochester, NY, USA

Y. Bengio, Department IRO, Université de Montréal, P.O. Box 6128, Downtown Branch, Montreal, QC H3C 3J7, Canada

C. Cadieu, Redwood Center for Theoretical Neuroscience and Helen Wills Neuroscience Institute, University of California, Berkeley, CA, USA

C.E. Carr, Department of Biology, University of Maryland, College Park, MD 20742, USA

P. Cisek, Groupe de Recherche sur le Système Nerveux Central, Département de Physiologie, Université de Montréal, Montréal, QC H3C 3J7, Canada

C.M. Colbert, Biology and Biochemistry, University of Houston, Houston, TX, USA

E.P. Cook, Department of Physiology, McGill University, 3655 Sir William Osler, Montreal, QC H3G 1Y6, Canada

P. Dario, CRIM Laboratory, Scuola Superiore Sant'Anna, Viale Rinaldo Piaggio 34, 56025 Pontedera (Pisa), Italy

A.G. Feldman, Center for Interdisciplinary Research in Rehabilitation (CRIR), Rehabilitation Institute of Montreal, and Jewish Rehabilitation Hospital, Laval, 6300 Darlington, Montreal, QC H3S 2J4, Canada

M.S. Fine, Department of Biomedical Engineering, Washington University, 1 Brookings Dr., St Louis, MO 63130, USA

D.W. Franklin, Kobe Advanced ICT Research Center, NiCT, *and* ATR Computational Neuroscience Laboratories, 2-2-2 Hikaridai, Keihanna Science City, Kyoto 619-0288, Japan

W.J. Freeman, Department of Molecular and Cell Biology, University of California, Berkeley, CA 94720-3206, USA

T. Gisiger, Récepteurs et Cognition, Institut Pasteur, 25 rue du Docteur Roux, 75015 Paris Cedex 15, France

S. Giszter, Neurobiology and Anatomy, Drexel University College of Medicine, 2900 Queen Lane, Philadelphia, PA 19129, USA

V. Goussev, Center for Interdisciplinary Research in Rehabilitation (CRIR), Rehabilitation Institute of Montreal, and Jewish Rehabilitation Hospital, Laval, 6300 Darlington, Montreal, QC H3S 2J4, Canada

R. Grashow, Volen Center MS 013, Brandeis University, 415 South St., Waltham, MA 02454-9110, USA

A.M. Green, Départment de Physiologie, Université de Montréal, 2960 Chemin de la Tour, Rm 2140, Montréal, QC H3T 1J4, USA

S. Grillner, Nobel Institute for Neurophysiology, Department of Neuroscience, Karolinska Institutet, Retzius väg 8, SE-171 77 Stockholm, Sweden

S. Grossberg, Department of Cognitive and Neural Systems, Center for Adaptive Systems, and Center for Excellence for Learning in Education, Science and Technology, Boston University, 677 Beacon Street, Boston, MA 02215, USA

J.A. Guest, Biology and Biochemistry, University of Houston, Houston, TX, USA
C. Hart, Neurobiology and Anatomy, Drexel University College of Medicine, 2900 Queen Lane, Philadelphia, PA 19129, USA
M. Hawken, Center for Neural Science, New York University, 4 Washington Place, New York, NY 10003, USA
M.R. Hinder, Perception and Motor Systems Laboratory, School of Human Movement Studies, University of Queensland, Brisbane, Queensland 4072, Australia
G.E. Hinton, Department of Computer Science, University of Toronto, 10 Kings College Road, Toronto, M5S 3G4 Canada
A. Ijspeert, School of Computer and Communication Sciences, Ecole Polytechnique Fédérale de Lausanne (EPFL), Station 14, CH-1015 Lausanne, Switzerland
J.F. Kalaska, GRSNC, Département de Physiologie, Faculté de Médecine, Pavillon Paul-G. Desmarais, Université de Montréal, C.P. 6128, Succursale Centre-ville, Montréal, QC H3C 3J7, Canada
M. Kerszberg, Université Pierre et Marie Curie, Modélisation Dynamique des Systèmes Intégrés UMR CNRS 7138—Systématique, Adaptation, évolution, 7 Quai Saint Bernard, 75252 Paris Cedex 05, France
U. Knoblich, Center for Biological and Computational Learning, McGovern Institute for Brain Research, Computer Science and Artificial Intelligence Laboratory, Brain and Cognitive Sciences Department, Massachusetts Institute of Technology, 43 Vassar Street #46-5155B, Cambridge, MA 02139, USA
C. Koch, Division of Biology, California Institute of Technology, MC 216-76, Pasadena, CA 91125, USA
J.H. Kotaleski, Computational Biology and Neurocomputing, School of Computer Science and Communication, Royal Institute of Technology, SE 10044 Stockholm, Sweden
M. Kouh, Center for Biological and Computational Learning, McGovern Institute for Brain Research, Computer Science and Artificial Intelligence Laboratory, Brain and Cognitive Sciences Department, Massachusetts Institute of Technology, 43 Vassar Street #46-5155B, Cambridge, MA 02139, USA
A. Kozlov, Computational Biology and Neurocomputing, School of Computer Science and Communication, Royal Institute of Technology, SE 10044 Stockholm, Sweden
J.W. Krakauer, The Motor Performance Laboratory, Department of Neurology, Columbia University College of Physicians and Surgeons, New York, NY 10032, USA
G. Kreiman, Department of Ophthalmology and Neuroscience, Children's Hospital Boston, Harvard Medical School and Center for Brain Science, Harvard University
N.I. Krouchev, GRSNC, Département de Physiologie, Faculté de Médecine, Pavillon Paul-G. Desmarais, Université de Montréal, C.P. 6128, Succursale Centre-ville, Montréal, QC H3C 3J7, Canada
I. Kurtzer, Centre for Neuroscience Studies, Queen's University, Kingston, ON K7L 3N6, Canada
A. Lansner, Computational Biology and Neurocomputing, School of Computer Science and Communication, Royal Institute of Technology, SE 10044 Stockholm, Sweden
P.E. Latham, Gatsby Computational Neuroscience Unit, London WC1N 3AR, UK
M.F. Levin, Center for Interdisciplinary Research in Rehabilitation, Rehabilitation Institute of Montreal and Jewish Rehabilitation Hospital, Laval, QC, Canada
J. Lewi, Georgia Institute of Technology, Atlanta, GA, USA
Y. Liang, Biology and Biochemistry, University of Houston, Houston, TX, USA
W.J. Ma, Department of Brain and Cognitive Sciences, University of Rochester, Rochester, NY 14627, USA
K.M. MacLeod, Department of Biology, University of Maryland, College Park, MD 20742, USA
L. Maler, Department of Cell and Molecular Medicine and Center for Neural Dynamics, University of Ottawa, 451 Smyth Rd, Ottawa, ON K1H 8M5, Canada
E. Marder, Volen Center MS 013, Brandeis University, 415 South St., Waltham, MA 02454-9110, USA
S.N. Markin, Department of Neurobiology and Anatomy, Drexel University College of Medicine, 2900 Queen Lane, Philadelphia, PA 19129, USA

N.Y. Masse, Department of Physiology, McGill University, 3655 Sir William Osler, Montreal, QC H3G 1Y6, Canada
D.A. McCrea, Spinal Cord Research Centre and Department of Physiology, University of Manitoba, 730 William Avenue, Winnipeg, MB R3E 3J7, Canada
A. Menciassi, CRIM Laboratory, Scuola Superiore Sant'Anna, Viale Rinaldo Piaggio 34, 56025 Pontedera (Pisa), Italy
T. Mergner, Neurological University Clinic, Neurocenter, Breisacher Street 64, 79106 Freiburg, Germany
T.E. Milner, School of Kinesiology, Simon Fraser University, Burnaby, BC V5A 1S6, Canada
P. Mohajerian, Computer Science and Neuroscience, University of Southern California, Los Angeles, CA 90089-2905, USA
L. Paninski, Department of Statistics and Center for Theoretical Neuroscience, Columbia University, New York, NY 10027, USA
V. Patil, Neurobiology and Anatomy, Drexel University College of Medicine, 2900 Queen Lane, Philadelphia, PA 19129, USA
J.F.R. Paton, Department of Physiology, School of Medical Sciences, University of Bristol, Bristol BS8 1TD, UK
J. Pillow, Gatsby Computational Neuroscience Unit, University College London, Alexandra House, 17 Queen Square, London WC1N 3AR, UK
T. Poggio, Center for Biological and Computational Learning, McGovern Institute for Brain Research, Computer Science and Artificial Intelligence Laboratory, Brain and Cognitive Sciences Department, Massachusetts Institute of Technology, 43 Vassar Street #46-5155B, Cambridge, MA 02139, USA
A. Pouget, Department of Brain and Cognitive Sciences, University of Rochester, Rochester, NY 14627, USA
A. Prochazka, Centre for Neuroscience, 507 HMRC University of Alberta, Edmonton, AB T6G 2S2, Canada
R. Rohrkemper, Physics Department, Institute of Neuroinformatics, Swiss Federal Institute of Technology, Zürich CH-8057, Switzerland
I.A. Rybak, Department of Neurobiology and Anatomy, Drexel University College of Medicine, Philadelphia, PA 19129, USA
A. Sangole, Center for Interdisciplinary Research in Rehabilitation, Rehabilitation Institute of Montreal and Jewish Rehabilitation Hospital, Laval, QC, Canada
S. Schaal, Computer Science and Neuroscience, University of Southern California, Los Angeles, CA 90089-2905, USA
S.H. Scott, Centre for Neuroscience Studies, Department of Anatomy and Cell Biology, Queen's University, Botterell Hall, Kingston, ON K7L 3N6, Canada
T. Serre, Center for Biological and Computational Learning, McGovern Institute for Brain Research, Computer Science and Artificial Intelligence Laboratory, Brain and Cognitive Sciences Department, Massachusetts Institute of Technology, 43 Vassar Street #46-5155B, Cambridge, MA 02139, USA
R. Shadmehr, Laboratory for Computational Motor Control, Department of Biomedical Engineering, Johns Hopkins School of Medicine, Baltimore, MD 21205, USA
R. Shapley, Center for Neural Science, New York University, 4 Washington Place, New York, NY 10003, USA
J.C. Smith, Cellular and Systems Neurobiology Section, National Institute of Neurological Disorders and Stroke, National Institutes of Health, Bethesda, MD 20892-4455, USA
C. Stefanini, CRIM Laboratory, Scuola Superiore Sant'Anna, Viale Rinaldo Piaggio 34, 56025 Pontedera (Pisa), Italy
J.A. Taylor, Department of Biomedical Engineering, Washington University, 1 Brookings Dr., St Louis, MO 63130, USA

K.A. Thoroughman, Department of Biomedical Engineering, Washington University, 1 Brookings Dr., St Louis, MO 63130, USA
L.H. Ting, The Wallace H. Coulter Department of Biomedical Engineering at Georgia Tech and Emory University, 313 Ferst Drive, Atlanta, GA 30332-0535, USA
A.-E. Tobin, Volen Center MS 013, Brandeis University, 415 South St., Waltham, MA 02454-9110, USA
E. Torres, Division of Biology, California Institute of Technology, Pasadena, CA 91125, USA
J.Z. Tsien, Center for Systems Neurobiology, Departments of Pharmacology and Biomedical Engineering, Boston University, Boston, MA 02118, USA
D. Tweed, Departments of Physiology and Medicine, University of Toronto, 1 King's College Circle, Toronto, ON M5S 1A8, Canada
D.B. Walther, Beckman Institute for Advanced Science and Technology, University of Illinois at Urbana-Champaign, 405 N. Mathews Ave., Urbana, IL 61801, USA
A.C. Wilhelm, Department of Physiology, McGill University, 3655 Sir William Osler, Montreal, QC H3G 1Y6, Canada
D. Xing, Center for Neural Science, New York University, 4 Washington Place, New York, NY 10003, USA
S. Yakovenko, Departément de Physiologie, Université de Montréal, Pavillon Paul-G. Desmarais, Universite de Montreal, C.P. 6128, Succ. Centre-ville, Montreal, QC H3C 3J7, Canada
D. Zipser, Department of Cognitive Science, UCSD 0515, 9500 Gilman Drive, San Diego, CA 92093, USA

Preface

In recent years, computational approaches have become an increasingly prominent and influential part of neuroscience research. From the cellular mechanisms of synaptic transmission and the generation of action potentials, to interactions among networks of neurons, to the high-level processes of perception and memory, computational models provide new sources of insight into the complex machinery which underlies our behaviour. These models are not merely mathematical surrogates for experimental data. More importantly, they help us to clarify our understanding of a particular nervous system process or function, and to guide the design of our experiments by obliging us to express our hypotheses in a language of mathematical formalisms. A mathematical model is an explicit hypothesis, in which we must incorporate all of our beliefs and assumptions in a rigorous and coherent conceptual framework that is subject to falsification and modification. Furthermore, a successful computational model is a rich source of predictions for future experiments. Even a simplified computational model can offer insights that unify phenomena across different levels of analysis, linking cells to networks and networks to behaviour. Over the last few decades, more and more experimental data have been interpreted from computational perspectives, new courses and graduate programs have been developed to teach computational neuroscience methods and a multitude of interdisciplinary conferences and symposia have been organized to bring mathematical theorists and experimental neuroscientists together.

This book is the result of one such symposium, held at the Université de Montréal on May 8–9, 2006 (see: http://www.grsnc.umontreal.ca/XXVIIIs). It was organized by the Groupe de Recherche sur le Système Nerveux Central (GRSNC) as one of a series of annual international symposia held on a different topic each year. This was the first symposium in that annual series that focused on computational neuroscience, and it included presentations by some of the pioneers of computational neuroscience as well as prominent experimental neuroscientists whose research is increasingly integrated with computational modelling. The symposium was a resounding success, and it made clear to us that computational models have become a major and very exciting aspect of neuroscience research. Many of the participants at that meeting have contributed chapters to this book, including symposium speakers and poster presenters. In addition, we invited a number of other well-known computational neuroscientists, who could not participate in the symposium itself, to also submit chapters.

Of course, a collection of 34 chapters cannot cover more than a fraction of the vast range of computational approaches which exist. We have done our best to include work pertaining to a variety of neural systems, at many different levels of analysis, from the cellular to the behavioural, from approaches intimately tied with neural data to more abstract algorithms of machine learning. The result is a collection which includes models of signal transduction along dendrites, circuit models of visual processing, computational analyses of vestibular processing, theories of motor control and learning, machine algorithms for pattern recognition, as well as many other topics. We asked all of our contributors to address their chapters to a broad audience of neuroscientists, psychologists, and mathematicians, and to focus on the broad theoretical issues which tie these fields together.

The conference, and this book, would not have been possible without the generous support of the GRSNC, the Canadian Institute of Advanced Research (CIAR), Institute of Neuroscience, Mental Health and Addiction (INMHA) of the Canadian Institutes of Health Research (CIHR), the Fonds de la

Recherche en Santé Québec (FRSQ), and the Université de Montréal. We gratefully acknowledge these sponsors as well as our contributing authors who dedicated their time to present their perspectives on the computational principles which underlie our sensations, thoughts, and actions.

Paul Cisek

Trevor Drew

John F. Kalaska

Contents

P. Cisek, T. Drew & J.F. Kalaska (Eds.)
Progress in Brain Research, Vol. 165
ISSN 0079-6123
Copyright © 2007 Elsevier B.V. All rights reserved

CHAPTER 1

The neuronal transfer function: contributions from voltage- and time-dependent mechanisms

Erik P. Cook[1,*], Aude C. Wilhelm[1], Jennifer A. Guest[2], Yong Liang[2], Nicolas Y. Masse[1] and Costa M. Colbert[2]

[1]*Department of Physiology, McGill University, 3655 Sir William Osler, Montreal, QC H3G 1Y6, Canada*
[2]*Biology and Biochemistry, University of Houston, Houston, TX, USA*

Abstract: The discovery that an array of voltage- and time-dependent channels is present in both the dendrites and soma of neurons has led to a variety of models for single-neuron computation. Most of these models, however, are based on experimental techniques that use simplified inputs of either single synaptic events or brief current injections. In this study, we used a more complex time-varying input to mimic the continuous barrage of synaptic input that neurons are likely to receive in vivo. Using dual whole-cell recordings of CA1 pyramidal neurons, we injected long-duration white-noise current into the dendrites. The amplitude variance of this stimulus was adjusted to produce either low subthreshold or high suprathreshold fluctuations of the somatic membrane potential. Somatic action potentials were produced in the high variance input condition. Applying a rigorous system-identification approach, we discovered that the neuronal input/output function was extremely well described by a model containing a linear bandpass filter followed by a nonlinear static-gain. Using computer models, we found that a range of voltage-dependent channel properties can readily account for the experimentally observed filtering in the neuronal input/output function. In addition, the bandpass signal processing of the neuronal input/output function was determined by the time-dependence of the channels. A simple active channel, however, could not account for the experimentally observed change in gain. These results suggest that nonlinear voltage- and time-dependent channels contribute to the linear filtering of the neuronal input/output function and that channel kinetics shape temporal signal processing in dendrites.

Keywords: dendrite; integration; hippocampus; CA1; channel; system-identification; white noise

The neuronal input/output function

What are the rules that single neurons use to process synaptic input? Put another way, what is the neuronal input/output function? Revealing the answer to this question is central to the larger task of understanding information processing in the brain. The past two decades of research have significantly increased our knowledge of how neurons integrate synaptic input, including the finding that dendrites contain nonlinear voltage- and time-dependent mechanisms (for review, see Johnston et al., 1996). However, there is still no consensus on the precise structure of the rules for synaptic integration.

*Corresponding author. Tel.: +1 514 398 7691;
Fax: +1 514 398 8241; E-mail: erik.cook@mcgill.ca

DOI: 10.1016/S0079-6123(06)65001-2

Early theoretical models of neuronal computation described the neuronal input/output function as a static summation of the synaptic inputs (McCulloch and Pitts, 1943). Rall later proposed that cable theory could account for the passive electrotonic properties of dendritic processing (Rall, 1959). This passive theory of dendritic integration has been extremely useful because it encompasses both the spatial and temporal aspects of the neuronal input/output function using a single quantitative framework. For example, the passive model predicts that the temporal characteristics of dendrites are described by a lowpass filter with a cutoff frequency that is inversely related to the distance from the soma.

The recent discovery that dendrites contain a rich collection of time- and voltage-dependent channels has renewed and intensified the study of dendritic signal processing at the electrophysiological level (for reviews, see Hausser et al., 2000; Magee, 2000; Segev and London, 2000; Reyes, 2001; London and Hausser, 2005). The central goal of this effort has been to understand how these active mechanisms augment the passive properties of dendrites. These studies, however, have produced somewhat conflicting results as to whether dendrites integrate synaptic inputs in a linear or nonlinear fashion (Urban and Barrionuevo, 1998; Cash and Yuste, 1999; Nettleton and Spain, 2000; Larkum et al., 2001; Wei et al., 2001; Tamas et al., 2002; Williams and Stuart, 2002). The focus of past electrophysiological studies has also been to identify the conditions in which dendrites initiate action potentials (Stuart et al., 1997; Golding and Spruston, 1998; Larkum and Zhu, 2002; Ariav et al., 2003; Gasparini et al., 2004; Womack and Khodakhah, 2004), to understand how dendrites spatially and temporally integrate inputs (Magee, 1999; Polsky et al., 2004; Williams, 2004; Gasparini and Magee, 2006; Nevian et al., 2007), and to reveal the extent of local dendritic computation (Mel, 1993; Hausser and Mel, 2003; Williams and Stuart, 2003).

Although these past studies have shed light on many aspects of single-neuron computation, most studies have focused on quiescent neurons in vitro. A common experimental technique is to observe how dendrites process brief "single-shock" inputs, either a single EPSP or the equivalent dendritic current injection, applied with no background activity present (but see Larkum et al., 2001; Oviedo and Reyes, 2002; Ulrich, 2002; Oviedo and Reyes, 2005; Gasparini and Magee, 2006). Based on the average spike rate of central neurons, it is unlikely that dendrites receive single synaptic inputs in isolation. A more likely scenario is that dendrites receive constant time-varying excitatory and inhibitory synaptic input that together produces random fluctuations in the membrane potential (Ferster and Jagadeesh, 1992; Destexhe and Pare, 1999; Chance et al., 2002; Destexhe et al., 2003; Williams, 2004). The challenge is to incorporate this type of temporally varying input into our study of the neuronal input/output function. Fortunately, system-identification theory provides us with several useful tools for addressing this question.

Using a white-noise input to reveal the neuronal input/output function

The field of system-identification theory has developed rigorous methods for describing the input/output relationships of unknown systems (for reviews, see Marmarelis and Marmarelis, 1978; Sakai, 1992; Westwick and Kearney, 2003) and has been used to describe the relationship between external sensory inputs and neuronal responses in a variety of brain areas (for reviews, see Chichilnisky, 2001; Wu et al., 2006). A prominent tool in system-identification is the use of a "white-noise" stimulus to characterize the system. Such an input theoretically contains all temporal correlations and power at all frequencies. If the unknown system is linear, or slightly nonlinear, it is a straightforward process to extract a description of the system by correlating the output with the random input stimulus. If the unknown system is highly nonlinear, however, this approach is much more difficult.

One difficulty of describing the input/output function of a single neuron is that we lack precise statistical descriptions of the inputs neurons receive over time. Given that a typical pyramidal neuron has over ten thousand synaptic contacts, one might reasonably estimate that an input arrives on the dendrites every millisecond or less, producing membrane fluctuations that are constantly varying in time. Thus, using a white-noise input has two advantages: (1) it affords the use of quantitative methods for identifying the dendrite input/output function and (2) it may represent a stimulus that is statistically closer to the type of input dendrites receive in vivo.

We applied a system-identification approach to reveal the input/output function of hippocampal CA1 pyramidal neurons in vitro (Fig. 1). We used standard techniques to perform dual whole-cell patch clamp recordings in brain slices (Colbert and Pan, 2002). More specifically, we injected 50 s of white-noise current (I_d) into the dendrites with one

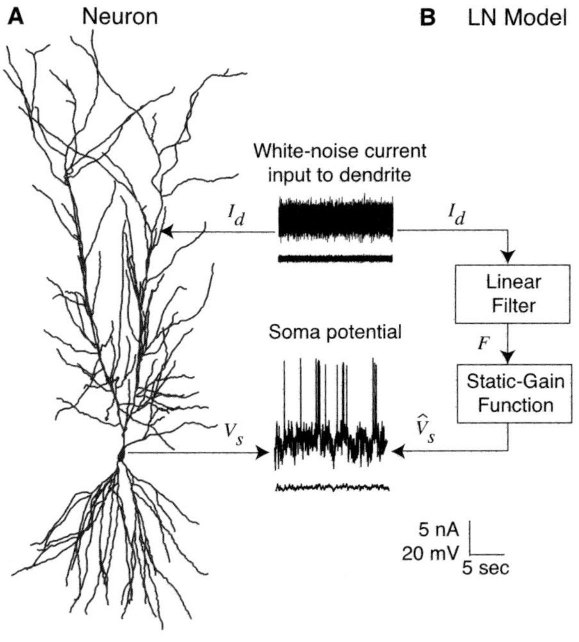

Fig. 1. Using a system-identification approach to characterize the dendrite-to-soma input/output function. (A) Fifty seconds of zero-mean Gaussian distributed random current (I_d) was injected into the proximal apical dendrites of CA1 pyramidal neurons and the membrane potential (V_s) was recorded at the soma. The variance of the injected current was switched between low (bottom traces) and high (top traces) on alternate trials. Action potentials were produced with the high-variance input. (B) An LN model was fit to the somatic potential. The input to the model was the injected current and the output of the model was the predicted soma potential ($\hat{V}_S$). The LN model was composed of a linear filter that was convolved with the input current followed by a static-gain function. The output of the linear filter, F (arbitrary units), was scaled by the static-gain function to produce the predicted somatic potential. The static-gain function was modeled as a quadratic function of F.

electrode and measured the membrane potential at the soma (V_s) with a second electrode. The amplitude distribution of the injected current was Gaussian with zero mean. Electrode separation ranged from 125 to 210 μm with the dendrite electrode placed on the main proximal apical dendritic branch. Figure 1 illustrates a short segment of the white-noise stimulus and the corresponding somatic membrane potentials.

To examine how the input/output function changed with different input conditions, we alternately changed the variance of the input current between low and high values. The low-variance input produced small subthreshold fluctuations in the somatic membrane potential. In contrast, the high-variance input produced large fluctuations that caused the neurons to fire action potentials with an average rate of 0.9 spikes/s. This rate of firing was chosen because it is similar to the average firing rate of CA1 hippocampal neurons in vivo (Markus et al., 1995; Yoganarasimha et al., 2006). Thus, we examined the dendrite-to-soma input/output function under physiologically reasonable subthreshold and suprathreshold operating regimes.

The LN model

Figure 1 illustrates our approach for describing the input/output function of the neuron using an LN model (Hunter and Korenberg, 1986). This is a functional model that provides an intuitive description of the system under study and has been particularly useful for capturing temporal processing in the retina in response to random visual inputs (for reviews, see Meister and Berry, 1999; Chichilnisky, 2001) and the processing of current injected at the soma of neurons (Bryant and Segundo, 1976; Poliakov et al., 1997; Binder et al., 1999; Slee et al., 2005). The LN model is a cascade of two processing stages: The first stage is a filter (the "L" stage) that linearly convolves the input current I_d. The output of the linear filter, F, is the input to the nonlinear second stage (the "N" stage) that converts the output of the linear filter into the predicted somatic potentials ($\hat{V}_S$). This second stage is static and can be viewed as capturing the gain of the system. The two stages of the LN model are represented mathematically as

$$F = H^* I_d$$
$$\hat{V}_S = G(F) \quad (1)$$

where H is a linear filter, * the convolution operator, and G a quadratic static-gain function.

Having two stages of processing is an important aspect of the model because it allows us to separate temporal processing from gain control. The linear filter describes the temporal processing while the nonlinear static-gain captures amplitude-dependent changes in gain. Thus, this functional model permits us to describe the neuronal input/output function using quantitatively precise terms such as filtering and gain control. In contrast, highly detailed biophysical models of single neurons, with their large number of nonlinear free parameters, are less likely to provide such a functionally clear description of single-neuron computation.

It is important to note that we did not seek to describe the production of action potentials in the dendrite-to-soma input/output function. Action potentials are extremely nonlinear events and would not be captured by the LN model. We instead focused on explaining the subthreshold fluctuations of the somatic voltage potential. Thus, action potentials were removed from the somatic potential before the data were analyzed. This was accomplished by linearly interpolating the somatic potential from 1 ms before the occurrence of the action potential to either 5 or 10 ms after the action potential. Because action potentials make up a very small part of the 50 s of data (typically less than 2%), our results were not qualitatively affected when the spikes were left in place during the analysis.

The LN model accounts for the dendrite-to-soma input/output function

Using standard techniques, we fit the LN model to reproduce the recorded somatic potential in response to the injected dendritic current (Hunter and Korenberg, 1986). We wanted to know how the low and high variance input conditions affected the components of the LN model. Therefore, these conditions were fit separately. An example of the LN model's ability to account for the neuronal input/output function is shown in Fig. 2. For this neuron, the LN model's predicted somatic membrane voltage ($\hat{V}_S$, dashed line) almost perfectly overlapped the neuron's actual somatic potential (V_s, thick gray line) for both input conditions (Fig. 2A and B). The LN model was able to fully describe the somatic potentials in response to the random input current with very little error. Computing the Pearson's correlation coefficient over the entire 50 s of data, the LN model accounted for greater than 97% of the variance of this neuron's somatic potential.

Repeating this experiment in 11 CA1 neurons, the LN model accounted for practically all of the somatic membrane potential (average $R^2 > 0.97$). Both the low and high variance input conditions were captured equally well by the LN model. Thus, the LN model is a functional model that describes the neuronal input/output function over a range of input regimes from low-variance subthreshold to high-variance suprathreshold stimulation.

Gain but not filtering adapts to the input variance

The LN model's linear filters and nonlinear static-gain functions are shown for our example neuron in Fig. 2C and D. The impulse-response function of the linear filters (Fig. 2C) for both the low (solid line) and high (dashed line) variance inputs had pronounced negativities corresponding to a band-pass in the 1–10 Hz frequency range (inset). Although the two input conditions were significantly different, the filters for the low- and high-variance inputs were very similar. Across our population of neurons, we found no systematic change in the linear filters as the input variance was varied between low and high levels. Therefore, the temporal processing performed by CA1 pyramidal neurons on inputs arriving at the proximal apical dendrites does not change with the input variance.

In contrast to the filtering properties of CA1 neurons, the static-gain function changed as a function of input variance. Figure 2D illustrates the static-gain function for both input conditions. In this plot, the resting membrane potential corresponds to 0 mV and the units for the output of the linear filter (F) are arbitrary. The static-gain function for the

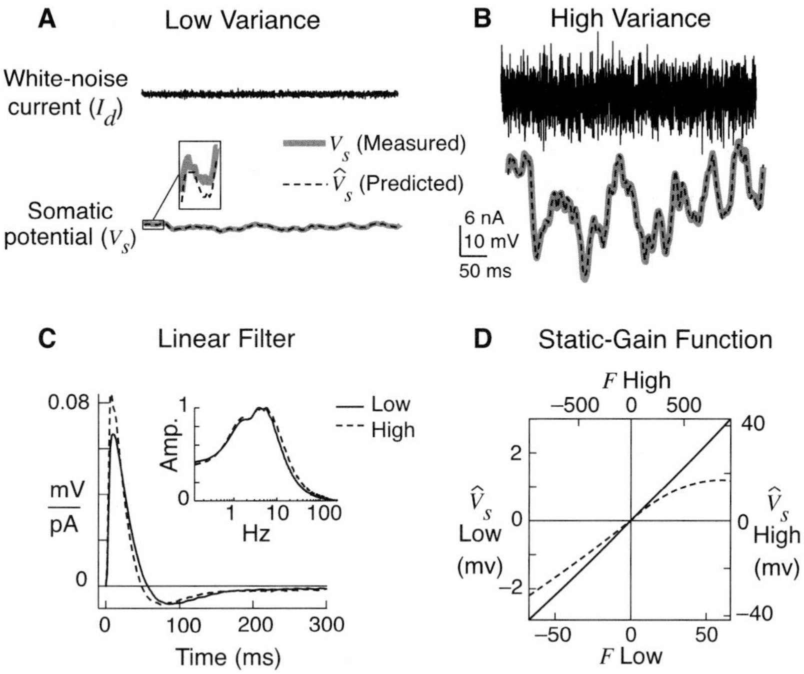

Fig. 2. The dendrite-to-soma input/output function of a CA1 neuron is well described by the LN model. (A) Example of 500 ms of the input current and somatic potential for the low-variance input. The predicted somatic membrane potential of the LN model ($\hat{V}_S$, dashed line) overlaps the recorded somatic potential (V_S, thick gray line). (B) Example of the LN model's fit to the high-variance input. Action potentials were removed from the recorded somatic potential before fitting the LN model to the data. (C) The impulse-response function of the linear filters for the optimized LN model corresponding to the low (solid line) and high (dashed line) variance inputs. Inset is the frequency response of the filters. (D) Static-gain function for the optimized LN model plotted for the low (solid line) and high (dashed line) variance inputs. The axes for the high variance input were appropriately scaled so that the slope of both static-gain functions could be compared.

low-variance input was a straight line indicating that the neuronal input/output function was linear. For the high-variance input, however, the static-gain function demonstrated two important nonlinearities. First, the static-gain function showed a compressive nonlinearity at depolarized potentials. Thus, at large depolarizing potentials, there was a reduction in the gain of the input/output relationship. Second, there was a general reduction in slope of the static-gain function for high-variance input compared with the low-variance slope, indicating an overall reduction in gain. Thus, for this neuron, increasing the variance of the input reduced the gain of the input/output function at rest that was further reduced for depolarizing potentials.

Across our population of 11 neurons, we found that increasing the variance of the input reduced the gain of CA1 neurons by an average of 16% at the resting membrane potential. This reduction in gain also increased with both hyperpolarized and depolarized potentials. Adapting to the variance of an input is an important form of gain control because it ensures that the input stays within the operating range of the neuron. Although a 16% reduction may seem small in comparison to the large change in the input-variance, there are many instances where small changes in neuronal activity are related to significant changes in behavior. For visual cortical neurons, it has been shown that small changes in spike activity (<5%) are correlated with pronounced changes in perceptual abilities (Britten et al., 1996; Dodd et al., 2001; Cook and Maunsell, 2002; Uka and DeAngelis, 2004; Purushothaman and Bradley, 2005). Thus, even small modulations of neuronal activity can have large effects on behavior.

Voltage- and time-dependent properties that underlie neuronal bandpass filtering

The above experimental results suggest that the dendrite-to-soma input/output relationship is well described as a linear filter followed by an adapting static-gain function. We wanted to know the biophysical components that produce the filtering and gain control. To address this, we used the computer program NEURON (Hines and Carnevale, 1997) to simulate a multi-compartment "ball & stick" model neuron (Fig. 3A).

We applied the random stimulus that we used in the experimental recordings to the dendrite of the passive model and then fit the data with the LN model to describe its input/output function. As would be expected from Rall's passive theory of dendrites, the estimated filters and gain functions were identical for the low and high variance input conditions (Fig. 3B). In addition, the filters from the passive model's impulse-response function had no negativity and thus were not bandpass (inset) and the static-gain function was linear (Fig. 3C). Thus, the passive properties of dendrites in the compartmental model do not produce the same characteristics of the experimentally observed dendrite-to-soma input/output function.

We wanted to know what type of voltage- and time-dependent channels might account for our experimental observations. Active channels come in a variety of classes. Instead of focusing on one particular class, we used the freedom of computer simulations to construct a hypothetical channel. Using a generic channel, referred to as I_x, we systematically varied channel parameters to investigate how the voltage- and time-dependent properties affected temporal filtering and gain control in the ball & stick model. Our theoretical channel was based on the classic Hodgkin and Huxley formulation (Hodgkin and Huxley, 1952) that incorporated a voltage- and time-dependent activation variable, $n(v, t)$. This activation variable had sigmoidal voltage-dependent steady-state activation with first-order kinetics.

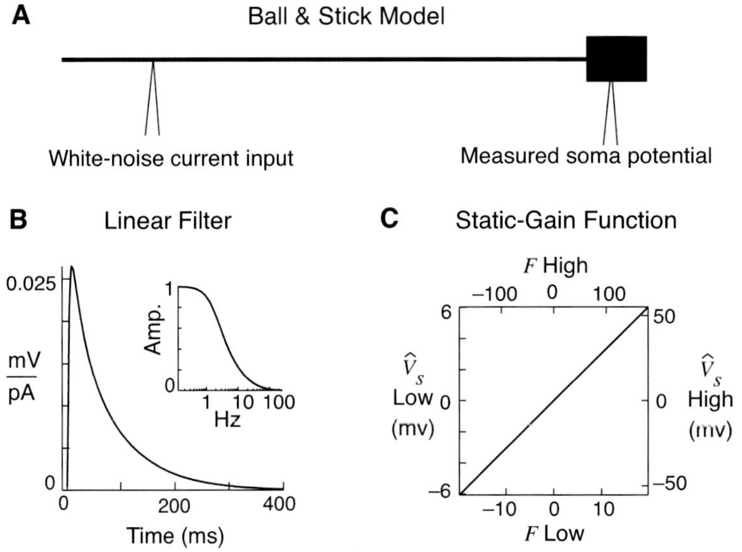

Fig. 3. Dendrite-to-soma input/output function of a passive neuron model. (A) The passive model had 20 dendrite compartments with a total length of 2000 μm and a diameter that tapered distally from 3 to 1 μm. The soma was a single 20 × 20 μm compartment. The passive parameters of the model were $R_m = 40{,}000\,\Omega\,cm^2$, $C_m = 2\,\mu F/cm^2$, and $R_a = 150\,\Omega\,cm$. (B) The optimized filters of the LN model were fit to the passive model. Filters for the low- and high-variance input were identical. (C) Static-gain functions of the optimized LN model were linear and had the same slope for both input conditions.

Mathematically, our hypothetical channel is described as

$$I_x = \bar{g}_x \cdot n(v,t) \cdot (v - E_{rev})$$
$$n_\infty = 1 - \frac{1}{1 + e^{-\beta(v - v_{1/2})}}$$
$$\frac{dn}{dt} = \frac{(n_\infty - n)}{\tau} \quad (2)$$

where n_∞ is the steady-state activation based on a sigmoid centered at $v_{1/2}$ with a slope of $1/\beta$, $\bar{g}_x$ the maximal conductance, τ the time constant of activation, and E_{rev} the reversal potential of the channel.

We first examined the effects of varying the steady-state voltage activation curve on the input/output function of the model. Voltage-dependent channels can have either depolarizing or hyperpolarizing activation curves. We inserted a uniform density of our I_x current throughout the dendrites and left the soma compartment passive. We set the parameters of I_x to have decreasing activation with depolarizing voltage (Fig. 4A) and stimulated the model with our low- and high-variance dendritic current injection. Fitting the LN model to the results of the simulation resulted in a bandpass filter and linear static-gain (Fig. 4A). The LN model accounted for greater than 98% of the somatic membrane potential and thus represented an excellent description of the input/output relationship of the compartmental model. It is worth mentioning that the simulated properties of I_x resembled the prominent dendritic current

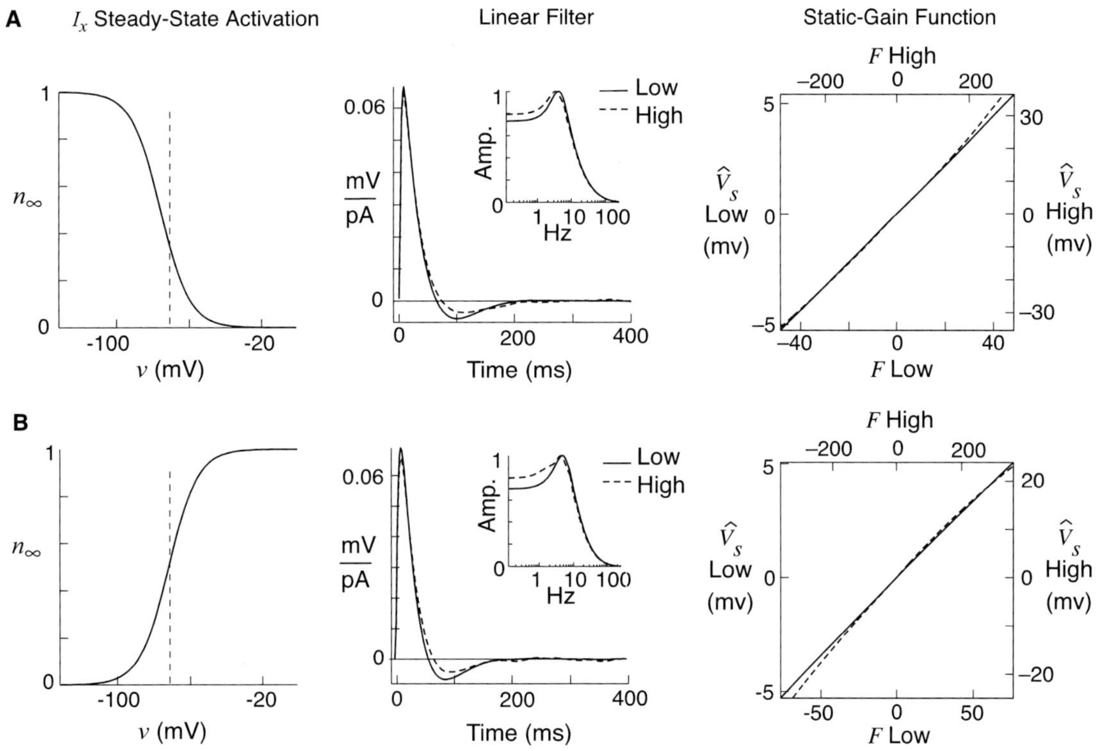

Fig. 4. The direction of steady-state voltage activation has little effect on bandpass features of the LN model. This figure shows the LN models that describe the dendrite-to-soma input/output function of the compartmental model containing the dendritic channel I_x. Two different steady-state activation curves were used for I_x. (A) Hyperpolarizing steady-state voltage activation of I_x produced bandpass features in the LN model (i.e., biphasic impulse-response function) but did not produce a reduction in gain between the low (solid line) and high (dashed line) variance input conditions. (B) Depolarizing steady-state voltage activation of I_x also produced bandpass features with no reduction in gain. In all simulations, I_x had a τ of 50 ms. The vertical dashed line in the activation plots indicates the resting membrane potential of -65 mV.

I_h (Magee, 1998). Thus, a simple voltage- and time-dependent channel can account for the bandpass filtering observed in our experimental data.

To see how the activation properties of the channel affected the input/output function, we reversed the activation curve of our hypothetical channel to have an increasing activation with depolarized potentials (Fig. 4B). The other parameters were the same except that the reversal potential of I_x was changed and the activation curve was shifted slightly to maintain stability. Injecting the low- and high-variance input current and fitting the LN model to the somatic potential, we found that this active current also produced a bandpass input/output function. Interestingly, there was still a lack of change in gain with input variance as can be seen in the static-gain function. Similar results were also observed when the slope of the activation curves was varied (data not shown).

From these simulations we can draw two conclusions. First, it appears that a variety of voltage dependencies can produce the bandpass filtering observed in neurons. Of course, this is only true when the membrane potential falls within the voltage activation range of the channel. In other words, a voltage-dependent channel that is always open or closed would not produce bandpass filtering. Second, a simple voltage-dependent mechanism does not seem to account for the experimentally observed change in gain between the low and high variance input conditions (compare the static-gain functions in Figs. 2D and 4).

Next, we examined the effect of the time dependencies of our theoretical channel on the neuronal input/output function. In the above simulations, we held the τ of I_x fixed at 50 ms. By varying τ we found that the time dependencies of the channel greatly affected the filtering properties. A shorter τ of 8 ms produced a model with an input/output function that exhibited less bandpass filtering that was shifted to higher frequencies (Fig. 5A). The shorter τ, however, created a slight increase in gain

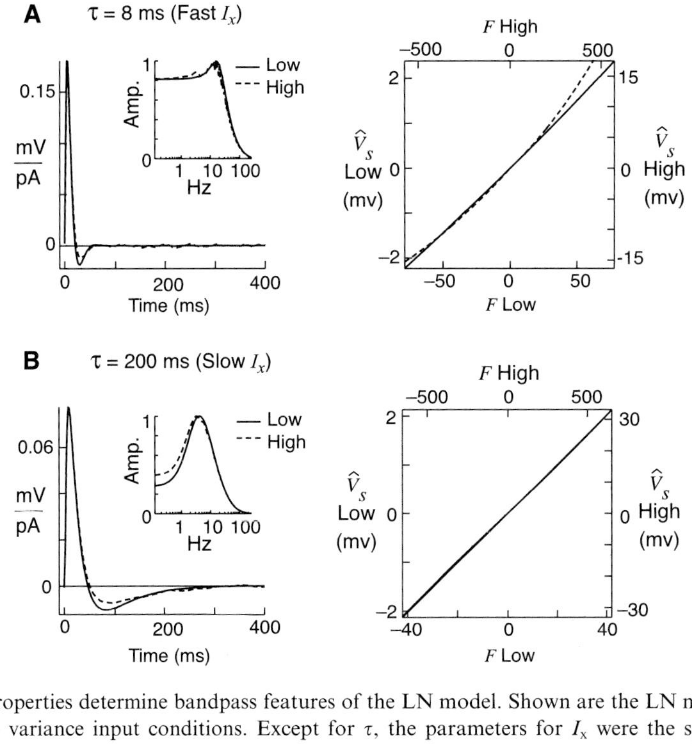

Fig. 5. Temporal channel properties determine bandpass features of the LN model. Shown are the LN models for both the low (solid line) and high (dashed line) variance input conditions. Except for τ, the parameters for I_x were the same as in Fig. 4A. (A) Fast activation of I_x ($\tau = 8$ ms) moved the bandpass to higher frequencies, but did not produce a reduction in gain with increased input variance. (B) Slow activation of I_x ($\tau = 200$ ms) increased the bandpass property of the filter and moved it toward lower frequencies with no reduction in gain.

for the high-variance input compared with the low-variance input, which is opposite to the gain change observed experimentally. In comparison, increasing τ to 200 ms had the opposite effect of enhancing the bandpass filtering of the model (Fig. 5B). Compared with a τ of 50 ms (Fig. 4A), the slower channel also moved the bandpass region to a lower frequency range. However, increasing τ produced no change in the gain of the neuron from the low-variance to the high-variance condition.

Discussion

Determining how neurons integrate synaptic input is critical for revealing the mechanisms underlying higher brain function. A precise description of the dendrite-to-soma input/output function is an important step. We found that the dendrite-to-soma input/output function of CA1 pyramidal neurons is well described by a simple functional LN model that combines linear filtering with static nonlinear gain control. The fact that the LN model accounted for over 97% of the somatic potential variance during a relatively long random input cannot be overemphasized. Even when producing action potentials during the high-variance input, the neuronal input/output function was well described by the LN model. The combination of bandpass filtering characteristics and nonlinear gain changes suggests that the input/output function cannot be explained by passive cellular properties, but requires active membrane mechanisms.

The advantages of characterizing the neuronal input/output relationship using a functional LN model are many. This model allows us to describe neuronal processing using the well-defined signal processing concepts of linear filtering and gain control. Although useful in understanding the biophysical aspects of neurons, a realistic compartmental model of a neuron would not allow such a clear description of the dendrite-to-soma input/output function. As demonstrated by our modeling of a hypothetical voltage-dependent conductance, I_x, different channel parameters can produce the same qualitative input/output characteristics of a compartmental neuron model.

That a simple functional model accounted so well for the dendrite-to-soma processing was initially surprising given that dendrites contain a wide range of nonlinear voltage- and time-dependent channels (Johnston et al., 1996). However, our subsequent computer simulations using a compartmental model indicate that nonlinear channels can underlie the linear temporal dynamics observed experimentally. The bandpass filtering produced by our theoretical voltage- and time-dependent channel is a result of a complex interaction between the passive filtering properties of the membrane and the temporal dynamics of the channel (for review, see Hutcheon and Yarom, 2000). Although the steady-state activation curve also influenced the bandpass filtering, we found that channel kinetics had the greatest effect on the temporal filtering of the model.

It is significant that the dendrite-to-soma input/output relationship contains a prominent bandpass in the theta frequency range. Neuronal networks in the hippocampus have prominent theta oscillations that are correlated with specific cognitive and behavioral states. Hippocampal theta oscillations occur during active exploration of the environment, during REM sleep, and may underlie memory-related processes (for reviews, see Buzsaki, 2002; Lengyel et al., 2005). Thus, the bandpass dynamics of the dendrite-to-soma input/output function may contribute directly to network-level oscillations in the hippocampus and other brain areas such as the neocortex (Ulrich, 2002).

Adaptation of gain is an important signal-processing mechanism because it ensures that the amplitude of the stimulus is maintained within the dynamic range of the system (for review, see Salinas and Thier, 2000). Information theory provides a basis for the popular idea that the brain adapts to the statistical properties of the signals encoded for efficient representation (e.g., Barlow, 1961; Atick, 1992; Bialek and Rieke, 1992; Hosoya et al., 2005). For example, the spike activity of neurons in the visual system has repeatedly been shown to adapt to the variance (or contrast) of a visual stimulus (e.g., Maffei et al., 1973; Movshon and Lennie, 1979; Albrecht et al., 1984; Fairhall et al., 2001; Kim and Rieke, 2001; Baccus and Meister, 2002). We found a similar change in gain to the variance of the injected current, suggesting that the intrinsic properties of dendrites may provide part of the foundation for gain adaptation observed at the circuit and systems level. Recent studies have reported similar changes in the gain of signals injected into the soma of cortical neurons in vitro. It has been proposed that this regulation of gain may be due to either intrinsic channel mechanisms (Sanchez-Vives et al., 2000), changes in background synaptic activity (Chance et al., 2002; Rauch et al., 2003; Shu et al., 2003), or both (Higgs et al., 2006). Because of the importance of maintaining the optimal level of activity in the brain, it is not surprising that there may exist multiple mechanisms for regulating gain.

With our computer simulations, however, we were not able to link the properties of our simple theoretical channel to the experimentally observed adaptation of the static-gain function. Although we observed changes in gain that occurred between the low- and high-variance input conditions, these were in the wrong direction (compare Fig. 2D and 5A). In addition, the model did not produce the compressive reduction in gain observed at depolarized potentials with the high-variance input. This suggests that the experimentally observed change in the static-gain function may be due to other mechanisms such as an increase in intracellular Ca^{2+} during the high-variance input. Another possibility is that the reduction in gain with increased input variance may arise from the interaction of many different channel types and mechanisms.

The theoretical channel model in Fig. 4A is based closely on the voltage-dependent current, I_h. This channel is expressed throughout the dendrites and has been shown to affect the temporal integration of synaptic inputs (Magee, 1999). Using a "chirp" sinusoidal stimulus, Ulrich showed that I_h plays a role in dendrite-to-soma bandpass filtering in neocortical neurons (Ulrich, 2002). Our preliminary experiments conducted in the presence of pharmacological blockers suggest that I_h may have a similar role in hippocampal pyramidal cells. However, dendrites contain many other voltage-dependent mechanisms and understanding how they work together to shape the dendrite-to-soma input/output function is an important topic for future studies.

Acknowledgments

Supported by the Canada Foundation for Innovation and operating grants from the Canadian Institutes of Health Research and the Natural Sciences and Engineering Research Council of Canada (EPC).

References

Albrecht, D.G., Farrar, S.B. and Hamilton, D.B. (1984) Spatial contrast adaptation characteristics of neurones recorded in the cat's visual cortex. J. Physiol., 347: 713–739.

Ariav, G., Polsky, A. and Schiller, J. (2003) Submillisecond precision of the input-output transformation function mediated by fast sodium dendritic spikes in basal dendrites of CA1 pyramidal neurons. J. Neurosci., 23: 7750–7758.

Atick, J.J. (1992) Could information theory provide an ecological theory of sensory processing? Network, 3: 213–251.

Baccus, S.A. and Meister, M. (2002) Fast and slow contrast adaptation in retinal circuitry. Neuron, 36: 909–919.

Barlow, H.B. (1961) In: RosenBlith W.A. (Ed.), Sensory Communication. MIT Press, Cambridge, MA, pp. 217–234.

Bialek, W. and Rieke, F. (1992) Reliability and information transmission in spiking neurons. Trends Neurosci., 15: 428–434.

Binder, M.D., Poliakov, A.V. and Powers, R.K. (1999) Functional identification of the input-output transforms of mammalian motoneurones. J. Physiol. Paris, 93: 29–42.

Britten, K.H., Newsome, W.T., Shadlen, M.N., Celebrini, S. and Movshon, J.A. (1996) A relationship between behavioral choice and the visual responses of neurons in macaque MT. Vis. Neurosci., 13: 87–100.

Bryant, H.L. and Segundo, J.P. (1976) Spike initiation by transmembrane current: a white-noise analysis. J. Physiol., 260: 279–314.

Buzsaki, G. (2002) Theta oscillations in the hippocampus. Neuron, 33: 325–340.

Cash, S. and Yuste, R. (1999) Linear summation of excitatory inputs by CA1 pyramidal neurons. Neuron, 22: 383–394.

Chance, F.S., Abbott, L.F. and Reyes, A.D. (2002) Gain modulation from background synaptic input. Neuron, 35: 773–782.

Chichilnisky, E.J. (2001) A simple white noise analysis of neuronal light responses. Network, 12: 199–213.

Colbert, C.M. and Pan, E. (2002) Ion channel properties underlying axonal action potential initiation in pyramidal neurons. Nat. Neurosci., 5: 533–538.

Cook, E.P. and Maunsell, J.H. (2002) Dynamics of neuronal responses in macaque MT and VIP during motion detection. Nat. Neurosci., 5: 985–994.

Destexhe, A. and Pare, D. (1999) Impact of network activity on the integrative properties of neocortical pyramidal neurons in vivo. J. Neurophysiol., 81: 1531–1547.

Destexhe, A., Rudolph, M. and Pare, D. (2003) The high-conductance state of neocortical neurons in vivo. Nat. Rev. Neurosci., 4: 739–751.

Dodd, J.V., Krug, K., Cumming, B.G. and Parker, A.J. (2001) Perceptually bistable three-dimensional figures evoke high choice probabilities in cortical area MT. J. Neurosci., 21: 4809–4821.

Fairhall, A.L., Lewen, G.D., Bialek, W. and de Ruyter Van Steveninck, R.R. (2001) Efficiency and ambiguity in an adaptive neural code. Nature, 412: 787–792.

Ferster, D. and Jagadeesh, B. (1992) EPSP-IPSP interactions in cat visual cortex studied with in vivo whole-cell patch recording. J. Neurosci., 12: 1262–1274.

Gasparini, S. and Magee, J.C. (2006) State-dependent dendritic computation in hippocampal CA1 pyramidal neurons. J. Neurosci., 26: 2088–2100.

Gasparini, S., Migliore, M. and Magee, J.C. (2004) On the initiation and propagation of dendritic spikes in CA1 pyramidal neurons. J. Neurosci., 24: 11046–11056.

Golding, N.L. and Spruston, N. (1998) Dendritic sodium spikes are variable triggers of axonal action potentials in hippocampal CA1 pyramidal neurons. Neuron, 21: 1189–1200.

Hausser, M. and Mel, B. (2003) Dendrites: bug or feature? Curr. Opin. Neurobiol., 13: 372–383.

Hausser, M., Spruston, N. and Stuart, G.J. (2000) Diversity and dynamics of dendritic signaling. Science, 290: 739–744.

Higgs, M.H., Slee, S.J. and Spain, W.J. (2006) Diversity of gain modulation by noise in neocortical neurons: regulation by the slow after-hyperpolarization conductance. J. Neurosci., 26: 8787–8799.

Hines, M.L. and Carnevale, N.T. (1997) The NEURON simulation environment. Neural Comput., 9: 1179–1209.

Hodgkin, A.L. and Huxley, A.F. (1952) A quantitative description of membrane current and its application to conduction and excitation in nerve. J. Physiol., 117: 500–544.

Hosoya, T., Baccus, S.A. and Meister, M. (2005) Dynamic predictive coding by the retina. Nature, 436: 71–77.

Hunter, I.W. and Korenberg, M.J. (1986) The identification of nonlinear biological systems: Wiener and Hammerstein cascade models. Biol. Cybern., 55: 135–144.

Hutcheon, B. and Yarom, Y. (2000) Resonance, oscillation and the intrinsic frequency preferences of neurons. Trends Neurosci., 23: 216–222.

Johnston, D., Magee, J.C., Colbert, C.M. and Cristie, B.R. (1996) Active properties of neuronal dendrites. Annu. Rev. Neurosci., 19: 165–186.

Kim, K.J. and Rieke, F. (2001) Temporal contrast adaptation in the input and output signals of salamander retinal ganglion cells. J. Neurosci., 21: 287–299.

Larkum, M.E. and Zhu, J.J. (2002) Signaling of layer 1 and whisker-evoked Ca^{2+} and Na^{+} action potentials in distal and terminal dendrites of rat neocortical pyramidal neurons in vitro and in vivo. J. Neurosci., 22: 6991–7005.

Larkum, M.E., Zhu, J.J. and Sakmann, B. (2001) Dendritic mechanisms underlying the coupling of the dendritic with the axonal action potential initiation zone of adult rat layer 5 pyramidal neurons. J. Physiol., 533: 447–466.

Lengyel, M., Huhn, Z. and Erdi, P. (2005) Computational theories on the function of theta oscillations. Biol. Cybern., 92: 393–408.

London, M. and Hausser, M. (2005) Dendritic computation. Annu. Rev. Neurosci., 28: 503–532.

Maffei, L., Fiorentini, A. and Bisti, S. (1973) Neural correlate of perceptual adaptation to gratings. Science, 182: 1036–1038.

Magee, J.C. (1998) Dendritic hyperpolarization-activated currents modify the integrative properties of hippocampal CA1 pyramidal neurons. J. Neurosci., 18: 7613–7624.

Magee, J.C. (1999) Dendritic I_h normalizes temporal summation in hippocampal CA1 neurons. Nat. Neurosci., 2: 508–514.

Magee, J.C. (2000) Dendritic integration of excitatory synaptic input. Nat. Rev. Neurosci., 1: 181–190.

Markus, E.J., Qin, Y.L., Leonard, B., Skaggs, W.E., McNaughton, B.L. and Barnes, C.A. (1995) Interactions between location and task affect the spatial and directional firing of hippocampal neurons. J. Neurosci., 15: 7079–7094.

Marmarelis, P.Z. and Marmarelis, V.Z. (1978) Analysis of Physiological Systems: The White Noise Approach. Plenum Press, New York, NY.

McCulloch, W.S. and Pitts, W. (1943) A logical calculus of ideas immanent in nervous activity. Bull. Math Biophys., 5: 115–133.

Meister, M. and Berry II, M.J. (1999) The neural code of the retina. Neuron, 22: 435–450.

Mel, B.W. (1993) Synaptic integration in an excitable dendritic tree. J. Neurophysiol., 70: 1086–1101.

Movshon, J.A. and Lennie, P. (1979) Pattern-selective adaptation in visual cortical neurones. Nature, 278: 850–852.

Nettleton, J.S. and Spain, W.J. (2000) Linear to supralinear summation of AMPA-mediated EPSPs in neocortical pyramidal neurons. J. Neurophysiol., 83: 3310–3322.

Nevian, T., Larkum, M.E., Polsky, A. and Schiller, J. (2007) Properties of basal dendrites of layer 5 pyramidal neurons: a direct patch-clamp recording study. Nat. Neurosci., 10: 206–214.

Oviedo, H. and Reyes, A.D. (2002) Boosting of neuronal firing evoked with asynchronous and synchronous inputs to the dendrite. Nat. Neurosci., 5: 261–266.

Oviedo, H. and Reyes, A.D. (2005) Variation of input-output properties along the somatodendritic axis of pyramidal neurons. J. Neurosci., 25: 4985–4995.

Poliakov, A.V., Powers, R.K. and Binder, M.D. (1997) Functional identification of the input-output transforms of motoneurones in the rat and cat. J. Physiol., 504(Pt 2): 401–424.

Polsky, A., Mel, B.W. and Schiller, J. (2004) Computational subunits in thin dendrites of pyramidal cells. Nat. Neurosci., 7: 621–627.

Purushothaman, G. and Bradley, D.C. (2005) Neural population code for fine perceptual decisions in area MT. Nat. Neurosci., 8: 99–106.

Rall, W. (1959) Branching dendritic trees and motoneuron membrane resistivity. Exp. Neurol., 1: 491–527.

Rauch, A., La Camera, G., Luscher, H.R., Senn, W. and Fusi, S. (2003) Neocortical pyramidal cells respond as integrate-and-fire neurons to in vivo-like input currents. J. Neurophysiol., 90: 1598–1612.

Reyes, A. (2001) Influence of dendritic conductances on the input-output properties of neurons. Annu. Rev. Neurosci., 24: 653–675.

Sakai, H.M. (1992) White-noise analysis in neurophysiology. Physiol. Rev., 72: 491–505.

Salinas, E. and Thier, P. (2000) Gain modulation: a major computational principle of the central nervous system. Neuron, 27: 15–21.

Sanchez-Vives, M.V., Nowak, L.G. and McCormick, D.A. (2000) Membrane mechanisms underlying contrast adaptation in cat area 17 in vivo. J. Neurosci., 20: 4267–4285.

Segev, I. and London, M. (2000) Untangling dendrites with quantitative models. Science, 290: 744–750.

Shu, Y., Hasenstaub, A., Badoual, M., Bal, T. and McCormick, D.A. (2003) Barrages of synaptic activity control the gain and sensitivity of cortical neurons. J. Neurosci., 23: 10388–10401.

Slee, S.J., Higgs, M.H., Fairhall, A.L. and Spain, W.J. (2005) Two-dimensional time coding in the auditory brainstem. J. Neurosci., 25: 9978–9988.

Stuart, G., Schiller, J. and Sakmann, B. (1997) Action potential initiation and propagation in rat neocortical pyramidal neurons. J. Physiol., 505(Pt 3): 617–632.

Tamas, G., Szabadics, J. and Somogyi, P. (2002) Cell type- and subcellular position-dependent summation of unitary postsynaptic potentials in neocortical neurons. J. Neurosci., 22: 740–747.

Uka, T. and DeAngelis, G.C. (2004) Contribution of area MT to stereoscopic depth perception: choice-related response modulations reflect task strategy. Neuron, 42: 297–310.

Ulrich, D. (2002) Dendritic resonance in rat neocortical pyramidal cells. J. Neurophysiol., 87: 2753–2759.

Urban, N.N. and Barrionuevo, G. (1998) Active summation of excitatory postsynaptic potentials in hippocampal CA3 pyramidal neurons. Proc. Natl. Acad. Sci. U.S.A., 95: 11450–11455.

Wei, D.S., Mei, Y.A., Bagal, A., Kao, J.P., Thompson, S.M. and Tang, C.M. (2001) Compartmentalized and binary behavior of terminal dendrites in hippocampal pyramidal neurons. Science, 293: 2272–2275.

Westwick, D.T. and Kearney, R.E. (2003) Identification of Nonlinear Physiological Systems. IEEE Press, Piscataway, NJ.

Williams, S.R. (2004) Spatial compartmentalization and functional impact of conductance in pyramidal neurons. Nat. Neurosci., 7: 961–967.

Williams, S.R. and Stuart, G.J. (2002) Dependence of EPSP efficacy on synapse location in neocortical pyramidal neurons. Science, 295: 1907–1910.

Williams, S.R. and Stuart, G.J. (2003) Role of dendritic synapse location in the control of action potential output. Trends Neurosci., 26: 147–154.

Womack, M.D. and Khodakhah, K. (2004) Dendritic control of spontaneous bursting in cerebellar Purkinje cells. J. Neurosci., 24: 3511–3521.

Wu, M.C., David, S.V. and Gallant, J.L. (2006) Complete functional characterization of sensory neurons by system identification. Annu. Rev. Neurosci., 29: 477–505.

Yoganarasimha, D., Yu, X. and Knierim, J.J. (2006) Head direction cell representations maintain internal coherence during conflicting proximal and distal cue rotations: comparison with hippocampal place cells. J. Neurosci., 26: 622–631.

P. Cisek, T. Drew & J.F. Kalaska (Eds.)
Progress in Brain Research, Vol. 165
ISSN 0079-6123
Copyright © 2007 Elsevier B.V. All rights reserved

CHAPTER 2

A simple growth model constructs critical avalanche networks

L.F. Abbott[1,]* and R. Rohrkemper[2]

[1]*Department of Physiology and Cellular Biophysics, Center for Neurobiology and Behavior, Columbia University College of Physicians and Surgeons, New York, NY 10032-2695, USA*
[2]*Physics Department, Institute of Neuroinformatics, Swiss Federal Institute of Technology, Zürich CH-8057, Switzerland*

Abstract: Neurons recorded from electrode arrays show a remarkable scaling property in their bursts of spontaneous activity, referred to as "avalanches" (Beggs and Plenz, 2003, 2004). Such scaling suggests a critical property in the coupling of these circuits. We show that similar scaling laws can arise in a simple model for the growth of neuronal processes. In the model (Van Ooyen and Van Pelt, 1994, 1996), the spatial range of the processes extending from each neuron is represented by a circle that grows or shrinks as a function of the average intracellular calcium concentration. Neurons interact when the circles corresponding to their processes intersect, with a strength proportional to the area of overlap.

Keywords: network activity; homeostasis; plasticity; network development

Introduction

Theoretical (also known as computational) neuroscience seeks to use mathematical analysis and computer simulation to link the anatomical and physiological properties of neural circuits to behavioral and cognitive functions. Often, researchers working in this field have a general principle of circuit design or a computational mechanism in mind when they start to work on a project. For the project to be described here, the general issue concerns the connectivity of neural circuits. For all but the smallest of neural circuits, we typically do not have a circuit diagram of synaptic connectivity or a list of synaptic strengths. How can we model a circuit when we are ignorant of such basic facts about its structure? One answer is to approach the problem statistically, put in as much as we know and essentially average over the rest. Another approach, and the one that inspires this work, is to hope that we can uncover properties of a neural circuit from basic principles of synapse formation and plasticity. In other words, if we knew the rules by which neural circuits develop, maintain themselves, and change in response to activity, we could work out their architecture on the basis of that knowledge. To this end, we need to uncover the basic rules and principles by which neural circuits construct themselves.

When neurons are removed from the brain and grown in culture, they change from disassociated neurons into reconnected networks or, in the case of slice cultures, from brain slices to essentially two-dimensional neural circuits. These re-development processes provide an excellent opportunity for exploring basic principles of circuit formation. Using slice cultures from rat cortex (and also acute slices),

*Corresponding author. Tel.: +1 212-543-5070;
Fax: +1 212-543-5797; E-mail: lfa2103@columbia.edu

DOI: 10.1016/S0079-6123(06)65002-4

Beggs and Plenz (2003, 2004) uncovered an intriguing property of networks of neurons developed in this way. By growing neural circuits on electrode arrays, they were able to record activity over long periods of time and accumulate a lot of data on the statistical properties of the activity patterns that arise spontaneously in such networks. Of particular interest are the observations of scaling behavior and criticality. These results provide the inspiration for the model we construct and study here.

The networks recorded by Beggs and Plenz (2003, 2004) are often silent, but silent periods are punctuated by spontaneous bursts of activity observed on variable numbers of electrodes for different periods of time. Beggs and Plenz called these bursts avalanches. To define and parameterize neural avalanches, they divided time into bins of size t_{bin} through a procedure that selects an optimal size. Here, we simply use $t_{\text{bin}} = 10\,\text{ms}$, typical of the values they used. An avalanche is defined as an event in which activity is observed on at least one electrode for a contiguous sequence of time bins, bracketed before and after by at least one bin of silence on all electrodes. We use an identical definition here, except that electrode activity is replaced by neuronal activity, because our model has no electrodes and we can easily monitor each neuron we simulate.

The results of Beggs and Plenz (2003, 2004) of particular importance for our study are histograms characterizing both the durations and sizes of the avalanches they recorded. Duration was determined by counting the number of consecutive bins within an avalanche. Size was measured either in terms of the number of electrodes on which activity was recorded during an avalanche, or by a measure of the total signal seen on all electrodes during the course of an avalanche. In our modeling work, we measure the size of an avalanche by counting the total number of action potentials generated during its time course.

The histograms of duration and size constructed from the data revealed a fascinating property (Beggs and Plenz, 2003, 2004; Fig. 1); both were of a power-law form. The number of events of a given size fell as the size to the $-3/2$ power, and the number of events of a given duration fell as the duration to the -2 power. Power-law distributions are interesting because they contain no natural scale. For example, in this context we might expect the typical size of a neuronal dendritic tree or axonal arbor (around $100\,\mu\text{m}$) to set the spatial scale for avalanches. Similarly, we might expect a typical membrane time constant of around 10 ms to set the scale for avalanche durations. If this were true, the distributions should be exponential rather than power-law. Power-law distributions indicate that these networks can, at least occasionally, produce activity patterns that are much larger and much long-lasting that we would have expected. This is what makes power-law distributions so interesting. Another intriguing feature is that power-law behavior typically arises in systems

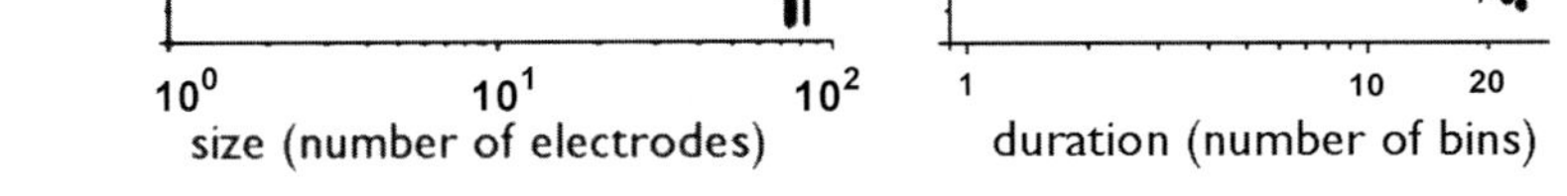

Fig. 1. Results of Beggs and Plenz on avalanche distributions. Left: probability of avalanches of different spatial sizes. The dashed line corresponds to a $-3/2$ power. Right: probability of avalanches of different durations. The dashed line corresponds to a -2 power. (Adapted with permission from Beggs and Plenz, 2004).

when they are critical, meaning that they are close to a transition in behavior. Thus, power laws arise when systems are specially configured.

Beggs and Plenz (2003, 2004) went on to note that the powers they observed, $-3/2$ and -2, are the same as those that arise in a very simple model (Zapperi et al., 1995). In this model, each neuron connects to n other neurons and, if it fires an action potential, causes each of its targets to fire with probability p. If $p<1/n$, activity in this model tends to die out, and if $p>1/n$ it tends to blow up. If $p = 1/n$, on the other hand, this simple model produces distributions with the same power-law dependence and same powers as those observed in the data. The condition $p = 1/n$ implies that every neuron that fires an action potential causes, on average, one other neuron to fire. This is critical in the sense discussed above that smaller values of p tend to induce patterns of activity that die out over time, and larger values of p tend to produce exploding bursts of activity. Thus, the results from these array recordings lead to the puzzle of how networks develop and maintain patterns of connectivity that satisfy this criticality condition. Do neurons somehow count the number of other neurons they project to and adjust the strengths of their synapses in inverse proportion to this number? If so, what would be the biophysical substrate for such a computation and adjustment (Teramae and Fukai, 2007)?

To address these questions, we made use of a model of neuronal circuit growth due to Van Ooyen and Van Pelt (1994, 1996). The model is simple, but here simplicity is exactly the point. We ask, in place of the above questions, whether a simple, biophysically plausible mechanism could account for the power-law behavior seen in the avalanche histograms without requiring any counting of synapses or criticality calculations. We are not proposing that the model we present is realistic, but rather use it to show that adjusting a network to be critical may not be as difficult as it would first appear.

The model

Following the work of (Van Ooyen and Van Pelt (1994, 1996); for reviews, see Van Ooyen, 2001, 2003), our model consists of N neurons positioned at random locations within a square region. The length and width of this square defines 1 unit of length. We can think of each location as the position of the soma of a neuron. The axonal and dendritic processes of each neuron are characterized by a circle drawn around its location. The size of this circle represents the extent of the processes projecting from the centrally located soma. Neurons interact synaptically when the circles representing their processes overlap, and the strength of the coupling is proportional to the area of overlap between these two circles. This is reasonable because synapses form in areas where neuronal processes intersect, and more intersections are likely to result in more synapses. All synaptic connections are excitatory.

The critical component of the model is the growth rule that determines how the process-defining circles expand or contract as a function of neuronal activity. The rule is simple: high levels of activity, which signify excessively strong excitation, cause the neuronal circle to contract, and low levels of activity, signifying insufficient excitation, cause it to grow. The initial sizes of the circles are chosen randomly and uniformly over the range from 0 to 0.05, in the units defined by the size of the square "plating" region.

Each neuron in the model is characterized by a firing rate and a radius, which is the radius of the circle defining the extent of its processes. Neuronal activity is generated by a Poisson spiking model on the basis of a computed firing rate. The firing rate for neuron i, where $i = 1, 2, 3, \ldots, N$, relaxes exponentially to a background rate r_0 with a time constant τ_r according to

$$\tau_r \frac{\mathrm{d}r_i}{\mathrm{d}t} = r_0 - r_i. \qquad (1)$$

We took $r_0 = 0.1$ Hz and $\tau_r = 5$ ms. The low background firing rate of 0.1 Hz is important to prevent the network from simply remaining silent. At every time step Δt, neuron i fires an action potential with probability $r_i \Delta t$. We took $\Delta t = 1$ ms. After a neuron fires an action potential, it is held in a refractory state in which it cannot fire for 20 ms.

Whenever another neuron, neuron j, fires an action potential, the firing rate of neuron i is incremented by

$$r_i \rightarrow r_i + gA_{ij} \tag{2}$$

where A_{ij} is the overlap area between the two circles characterizing the processes of neurons i and j. In our simulations, the constant g, which sets the scale of synaptic strength in the model, is set to $g = 500\,\text{Hz}$. This number is large because the overlap areas between the neurons are quite small in the units we are using.

The average level of activity of neuron i is monitored by a variable C_i that represents the internal calcium concentration in that neuron. C_i decays to zero exponentially,

$$\tau_C \frac{dC_i}{dt} = -C_i \tag{3}$$

and is incremented by one unit ($C_i \rightarrow C_i + 1$) whenever neuron i fires an action potential. This step size defines the unit of calcium concentration. The value of the time constant τ_C is not critical in what follows, but we took it to be 100 ms.

Two features make calcium a useful indicator of neuronal activity. First, resting calcium concentrations inside neurons are very small, but calcium enters the cell whenever the neuron fires an action potential. Because of this, the calcium concentration acts as an integrator of the action potential response and, for this reason, imaging calcium concentrations is a common way to monitor neuronal activity. Second, many molecules in a neuron are sensitive to the internal calcium concentrations, so this indicator can activate numerous biochemical cascades, including those responsible for growth.

The remaining equation in the model is the one that determines the contraction or growth of the radius a_i characterizing neuron i. This is

$$\frac{da_i}{dt} = k(C_{\text{target}} - C_i) \tag{4}$$

where k determines the rate of growth. We used a variety of values for k, but growth was always slow on the time scale of neuronal activity. We often started a run with a larger value of k ($k = 0.02\,\text{s}^{-1}$) to speed up growth, but as an equilibrium state was reached we lowered this to $k = 0.002\,\text{s}^{-1}$. The parameter C_{target} plays the dominant role in determining the behavior of the model. This sets a target level of calcium, and therefore a target level of activity, for the neurons. If activity is low so that $C_i < C_{\text{target}}$, the above equation causes the processes from neuron i to grow (a_i increases) leading to more excitatory connections with other neurons and hence more activity. If activity is high so that $C_i > C_{\text{target}}$, the processes will retract (a_i decreases) lowering the amount of excitation reaching neuron i. In this way, each neuron grows or contracts in an attempt to maintain the target level of calcium concentration ($C_i = C_{\text{target}}$), which implies a certain target level of activity. We discuss the value of C_{target} more fully below, but $C_{\text{target}} = 0.08$ was used to obtain the results in the figures we show.

Results

The left panel of Fig. 2 shows a typical configuration at the beginning of a run. In this case, 100 neurons have been located randomly with various radii, also chosen randomly. At this initial point, many of the neurons are disconnected or, at most, connected together in small clusters. Each neuron has a spontaneous firing rate of 0.1 Hz, even when isolated, so this network exhibits activity, but at a low level. Fig. 2 (left) shows a typical initial state of the model, but the results of running a model simulation are independent of the initial state unless a highly unlikely initial configuration (such as many neurons at the same position) limits the possibilities for developing connections through growth. The target calcium level we use, $C_{\text{target}} = 0.08$, is larger than the average calcium level attained by the neurons in this initial configuration. Thus, when the simulation starts, the neurons (the circles in Fig. 2, left) grow larger.

As the neurons grow, they begin to form more and stronger connections, which causes the level of activity in the network to increase. Growth continues until the neurons are active enough to bring their average calcium concentrations near to the value C_{target}. At this point, the average rate of growth of the network goes to zero, but there are still small adjustments in the sizes of individual neurons. As neurons adjust their own radii, and

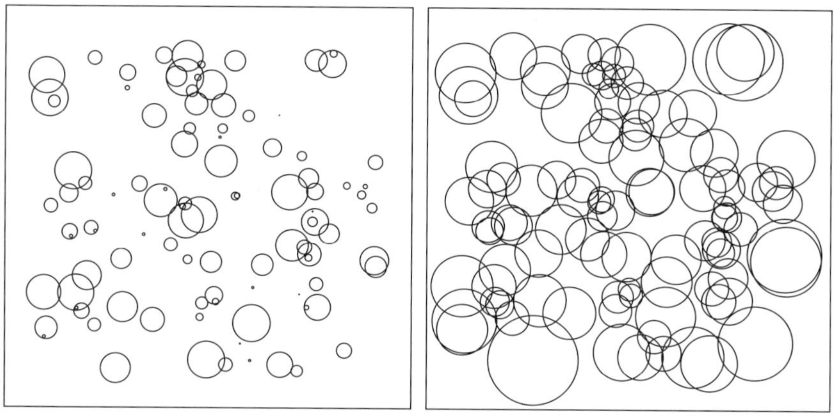

Fig. 2. Configuration of the model network before (left) and after (right) activity-dependent growth. Each circle represents the extent of the processes for one neuron. Neurons with overlapping circles are connected. Initially (left), the neurons are either uncoupled or coupled in small clusters. At equilibrium (right), the network is highly connected.

react to the adjustments of their neighbors, they eventually achieve a quasi-equilibrium point in which their time-averaged calcium concentrations remain close to C_{target}, with small fluctuations in their radii over time. From this point on, the network will remain in the particular configuration it has achieved indefinitely. This growth process has been described previously (Van Ooyen and Van Pelt, 1994, 1996; Abbott and Jensen, 1997). Our only modification on the original growth model of Van Ooyen and Van Pelt (1994, 1996) was to add Poisson spikes to their firing-rate model. The right panel of Fig. 2 shows the equilibrium configuration that arose from the initial configuration shown in the left panel.

The size of the small fluctuations in neuronal size about the equilibrium configuration is determined by the magnitude of the growth rate, k. Because growth processes are much slower than the processes generating activity in a network, we chose k to be as small as we could without requiring undue amounts of computer time to achieve equilibrium. The results we report are insensitive to the exact value of k.

Once the network has achieved an equilibrium configuration, we analyze its patterns of activity using the same approach as Beggs and Plenz (2003, 2004). In other words, we constructed histograms of the duration and total number of action potentials in periods of activity that were bracketed by 10 ms time bins in which no activity was observed. To assure that the resulting histograms reflect the dynamics of the network and not of the growth process, we shut off growth (set $k = 0$) while we accumulated data for the histograms, although for the small growth rate we use, this did not make any noticeable difference to the results.

Histograms of the durations and number of action potentials for the avalanches seen in the model at equilibrium are shown in Fig. 3. These are log-log plots, and the straight lines drawn indicate $-3/2$ (Fig. 3, left) and -2 (Fig. 3, right) power-law dependences. Over the range shown, the histograms follow the power-law dependences of a critical cascade model. As in the data (Beggs and Plenz, 2003, 2004), there are deviations for large, rare events due to finite-size effects.

Changing the initial size of the circles representing the neuronal processes in these simulations has no effect, because the growth rule simply expands small circles or shrinks large circles until they are in the equilibrium range. The model is, however, sensitive to the value of the target calcium concentration. The most sensitive result is the exponent of the power function describing the distribution of spike counts, as shown in the left panels of Figs. 1 and 3. The exponent for the distribution of durations is less sensitive. Fig. 4 shows how the spike count distribution exponent depends on C_{target} over a range of values from 0.04 to 1.2, with the value used for the previous figures, 0.08, in the middle of this range.

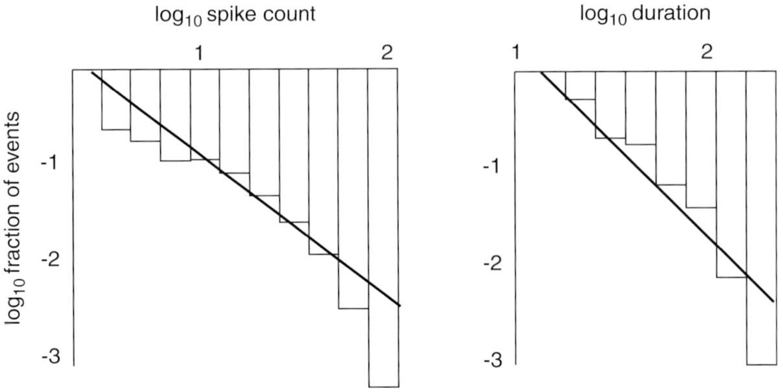

Fig. 3. Histograms of the fraction of avalanches with different numbers of spikes (left) and different durations (right). The plots are log-log and the lines indicate $-3/2$ (left) and -2 (right) powers.

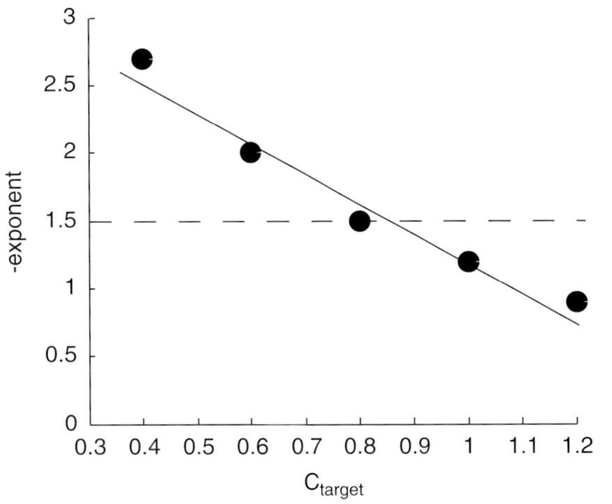

Fig. 4. Value of minus the exponent of the power function describing the spike count distribution as a function of the target calcium concentration. The value seen in the experiments, indicated by the dashed line, is 1.5, corresponding to $C_{\text{target}} = 0.08$. The solid line is drawn only to guide the eye.

Discussion

In our network model, the spontaneous level of activity for each neuron, 0.1 Hz, is insufficient to allow the internal calcium concentration to approach the target level we set. Therefore, disconnected neurons grow, and they can only reach an equilibrium size if they "borrow" activity from other neurons. Even the activity in small clusters is insufficient to halt growth. However, the target calcium concentration was set so that all-to-all connections or excessive large-scale firing over the entire network would produce internal calcium concentrations that exceed the target level and therefore induce process withdrawal. Therefore, the network is forced to find a middle ground in which individual neurons share activity in variable-sized groups, drawing excitation from both nearby and faraway neurons. This is what provides the potential for critical, power-law behavior.

The power-laws shown in Figs. 3 and 4 occur over a range of values of C_{target}, but they are not an inevitable consequence in the model. Values of C_{target} significantly higher than those we have used lead to an essentially flat distribution (over the finite range) of event sizes and durations. Smaller values lead to a shortage of large, long-lasting events.

The model we have considered warrants studying in more depth, and it can be extended in a number of ways. Obviously, inhibitory neurons should be added. In addition, it would be of interest to provide each neuron with two circles, one representing the extent of dendritic outgrowth and the other axonal. Separate growth rules would be needed for the two circles in this case. Finally, the axonal projections could be given both local extension, represented by a circle around the somatic location, and distal projections, represented by additional circles located away from the soma.

The fact that a simple growth rule can generate circuits with critical, power-law behavior suggests that it could be the basis for developing interesting network models. We have only explored uncontrolled spontaneous activity, but the fact that this can occur over such a large range of sizes and durations makes the functional implications of these networks quite intriguing. If we can learn to grow circuits like this in which we can control the size and time scale of the activity, this could form a basis for building functional circuits that go beyond spontaneous activity to perform useful tasks.

Acknowledgments

Research supported by the National Science Foundation (IBN-0235463) and by an NIH Director's Pioneer Award, part of the NIH Roadmap for Medical Research, through grant number 5-DP1-OD114-02. We thank Tim Vogels and Joe Monaco for valuable input.

References

Abbott, L.F. and Jensen, O. (1997) Self-organizing circuits of model neurons. In: Bower J. (Ed.), Computational Neuroscience, Trends in Research 1997. Plenum, NY, pp. 227–230.

Beggs, J.M. and Plenz, D. (2003) Neuronal avalanches in neocortical circuits. J. Neurosci., 23: 11167–11177.

Beggs, J.M. and Plenz, D. (2004) Neuronal avalanches are diverse and precise activity patterns that are stable for many hours in cortical slice cultures. J. Neurosci., 24: 5216–5229.

Teramae, J.n. and Fukai, T. (2007) Local cortical circuit model inferred from power-law distributed neuronal avalanches. J. Comput. Neurosci., 22: 301–312.

Van Ooyen, A. (2001) Competition in the development of nerve connections: a review of models. Network, 12: R1–R47.

Van Ooyen, A. (Ed.). (2003) Modeling Neural Development. MIT Press, Cambridge, MA.

Van Ooyen, A. and Van Pelt, J. (1994) Activity-dependent outgrowth of neurons and overshoot phenomena in developing neural networks. J. Theor. Biol., 167: 27–43.

Van Ooyen, A. and Van Pelt, J. (1996) Complex periodic behaviour in a neural network model with activity-dependent neurite outgrowth. J. Theor. Biol., 179: 229–242.

Zapperi, S., Baekgaard, I.K. and Stanley, H.E. (1995) Self-organized branching processes: mean-field theory for avalanches. Phys. Rev. Lett., 75: 4071–4074.

P. Cisek, T. Drew & J.F. Kalaska (Eds.)
Progress in Brain Research, Vol. 165
ISSN 0079-6123
Copyright © 2007 Elsevier B.V. All rights reserved

CHAPTER 3

The dynamics of visual responses in the primary visual cortex

Robert Shapley*, Michael Hawken and Dajun Xing

Center for Neural Science, New York University, 4 Washington Place, New York, NY 10003, USA

Abstract: There is a transformation in behavior in the visual system of cats and primates, from neurons in the Lateral Geniculate Nucleus (LGN) that are not tuned for orientation to orientation-tuned cells in primary visual cortex (V1). The visual stimuli that excite V1 can be well controlled, and the thalamic inputs to V1 from the LGN have been measured precisely. Much has been learned about basic principles of cortical neurophysiology on account of the intense investigation of the transformation between LGN and V1. Here we present a discussion of different models for visual cortex and orientation selectivity, and then discuss our own experimental findings about the dynamics of orientation selectivity. We consider what these theoretical analyses and experimental results imply about cerebral cortical function. The conclusion is that there is a very important role for intracortical interactions, especially cortico-cortical inhibition, in producing neurons in the visual cortex highly selective for orientation.

Keywords: V1 cortex; orientation selectivity; computational model; untuned suppression; tuned suppression; dynamics

Introduction

Orientation tuning, as an emergent property in visual cortex, must be an important clue to how the cortex works and why it is built the way it is. There is a transformation in behavior, from neurons in the Lateral Geniculate Nucleus (LGN) that are not tuned for orientation to orientation-tuned cells in V1 cortex (for example, in cat area 17, Hubel and Wiesel, 1962; in monkey V1, Hubel and Wiesel, 1968; Schiller et al., 1976; De Valois et al., 1982). We have learned about basic principles of cortical neurophysiology from the intense investigation and constructive disagreements about the mechanisms of the orientation transformation between LGN and V1 as discussed below. Here we will present our own findings about the dynamics of orientation selectivity, and contrast our results and conclusions with others. Our results suggest that intracortical interactions, especially cortico-cortical inhibition, play an important role in producing highly selective neurons in the cortex.

Theories of orientation selectivity

The rationale of our experiments came from considering different models or theories for visual cortical function, so it makes sense to begin with theory. There are two poles of thought about theoretical solutions for the problem of orientation selectivity: feedforward filtering on the one hand, and attractor states where networks develop

*Corresponding author. Tel.: +1 212 9987614;
Fax: +1 212 9954860; E-mail: shapley@cns.nyu.edu

DOI: 10.1016/S0079-6123(06)65003-6

"bumps of activity" in the orientation domain as a response to weakly oriented input on the other (Ben-Yishai et al., 1995). Our own view based on our experimental work, and also on recent theoretical work (Troyer et al., 1998; Chance et al., 1999; McLaughlin et al., 2000; Tao et al., 2004; Marino et al., 2005) is that the major cause of orientation selectivity in V1 is recurrent network filtering. We believe that feedforward excitation induces an orientation preference in V1 neurons but that cortico-cortical inhibitory interactions within the V1 network are needed to make V1 neurons highly selective for orientation.

Feedforward model of orientation selectivity

The first model offered chronologically, and first discussed here, is the feedforward model that is descended from the pioneering work of Hubel and Wiesel (1962). The HW model has the great virtue of being explicit and calculable. It involves the addition of signals from LGN cells that are aligned in a row along the long axis of the receptive field of the orientation-selective neuron, as in Fig. 1. Such connectivity is likely the basis of orientation *preference* (the preferred orientation) but whether or not feedforward connectivity can account for orientation *selectivity* (how much bigger the preferred response is than responses to nonpreferred orientations) is a more difficult question. There is some support for a feedforward neural architecture based on studies that have determined the pattern of LGN input to V1 cells. In the ferret visual cortex Chapman et al. (1991) inhibited cortical activity with Muscimol, a GABA agonist, and observed the spatial pattern of LGN inputs to a small zone of V1. Reid and Alonso (1995) did dual recordings in LGN and cat V1 and mapped the overlapping receptive fields of cortical cells and their LGN inputs. The experiment on cooling of cat V1 to block cortical activity by Ferster et al. (1996) is somewhat analogous to the Chapman et al. (1991) study with the technical improvement of intracellular recording of synaptic current in V1 cells; it was interpreted to mean that there is substantial orientation tuning of the collective thalamic input to a cortical neuron, consistent with the HW feedforward model. In spite of all this evidence, there is general agreement that the HW model predicts rather weak orientation selectivity, and therefore does not account for the visual properties of those V1 cells that are highly selective (Sompolinsky and Shapley, 1997; Troyer et al., 1998; McLaughlin et al., 2000).

The reason for the shortfall of orientation selectivity in the HW model has been discussed before. LGN cells have a low spontaneous rate but are quite responsive to visual stimuli. An LGN cell's firing rate during visual stimulation by an optimal grating pattern has a sharp peak at one temporal phase and dips to zero spikes/s at the opposite temporal phase. Such nonlinear behavior depends on stimulus contrast; at very low stimulus contrast the LGN cells' minimum firing rate may not go down as low as zero spikes/s. But at most

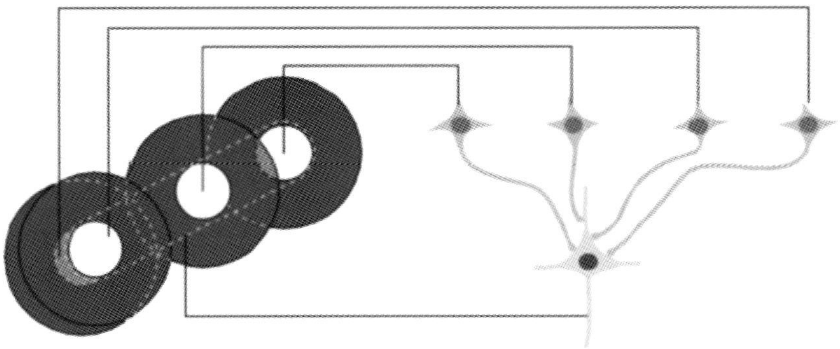

Fig. 1. Classic feedforward model from LGN to simple cells in V1 cortex. Adapted with permission from Hubel and Wiesel (1962). Four LGN cells are drawn as converging onto a single V1 cell. The circular LGN receptive fields aligned in a row on the left side of the diagram make the receptive field of the cortical cell elongated.

stimulus contrasts used in experiments on cortex (that is contrast >0.1) the LGN cells' firing rate will hit zero on the downswing. This clipping of the spike rate at zero spikes/s makes the LGN cells act like nonlinear excitatory subunits as inputs to their cortical targets (Palmer and Davis, 1981; Tolhurst and Dean, 1990; Shapley, 1994). Since the HW model simply adds up the LGN sources, its summation of the clipped LGN inputs results in a nonzero response at 90° from the optimal orientation. Computational simulations of feedforward models with estimates of LGN convergent input derived from the work of Reid and Alonso (1995) support this analysis (Sompolinsky and Shapley, 1997; McLaughlin et al., 2000). An example is given in Fig. 2, which shows a computation of the summed excitatory synaptic input from an HW model onto a cortical cell (cf. Sompolinsky and Shapley, 1997). Such a model produces a substantial LGN input to a cortical cell at 90° from the preferred orientation, as seen in the figure. However, highly selective V1 cells respond little or not at all at 90° from peak orientation. Therefore, feedforward convergence can be only a part of the story of cortical orientation selectivity.

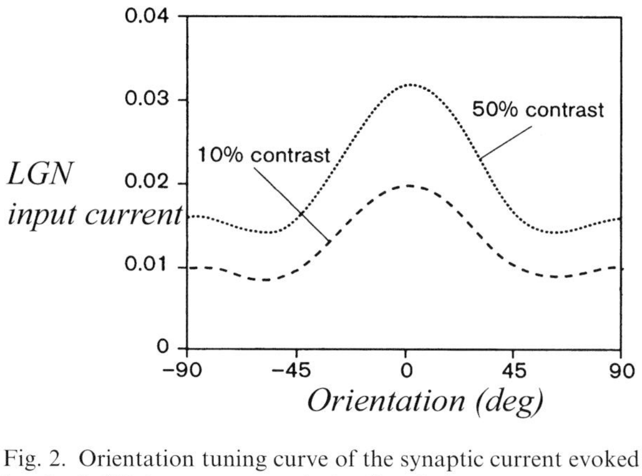

Fig. 2. Orientation tuning curve of the synaptic current evoked by the LGN input to a cortical cell, relative to spontaneous levels of LGN input calculated from a feedforward model (Sompolinsky and Shapley, 1997). In this model, the LGN afferents formed an ON–OFF–ON receptive field. Each subregion had an aspect ratio of 2. A total of 24 OFF-center cells comprised the OFF subfield, while 12 ON cells comprised each ON subregion, in the model. The pattern of wiring was based on the experimental results of Reid and Alonso (1995).

It might be supposed that one could rescue the feedforward model by setting the spike threshold just high enough that the off-peak LGN input would be sub-threshold (Carandini and Ferster, 2000). However, this strategy will only work for one contrast. One can infer this from Fig. 2. If one adds a threshold that makes the 10% contrast curve highly selective, the 50% contrast curve will have a very broadly tuned response. This has been pointed out often before (cf. Ben-Yishai et al., 1995; Sompolinsky and Shapley, 1997; Troyer et al., 1998). To understand cortical orientation selectivity we must answer the theoretical question: how does V1 reduce large feedforward responses at orientations far from the preferred orientation, like those illustrated in Fig. 2? The important experimental issue therefore is, what is the global shape of the orientation tuning curve? This focuses attention on global measures of orientation selectivity like circular variance (Ringach et al., 2002) or 1 minus circular variance, sometimes called the orientation selectivity index (Dragoi et al., 2000). Kang et al. (2004) showed that global measures like circular variance or orientation selectivity index are equivalent to informational measures of discriminability of widely separated orientations, an important function for visual perception.

Models with cortical inhibition and excitation

There is a well-known addition to the HW model that would increase the orientation selectivity greatly. One can obtain increased orientation selectivity by adding *inhibition* that is more broadly tuned for orientation than excitation. The inhibition can be either spatial-phase-specific, so-called push–pull inhibition (Palmer and Davis, 1981; Ferster, 1988, 1992; Tolhurst and Dean, 1990; Troyer et al., 1998), or some other kind of cross-orientation inhibition (Bonds, 1989; Ben-Yishai et al., 1995; Somers et al., 1995; McLaughlin et al., 2000). What matters for explaining orientation selectivity is not the phase specificity of the inhibition but the breadth of tuning. Thalamo-cortical synapses are thought to be purely excitatory (Freund et al., 1989; Callaway, 1998), so the inhibition must come through cortical interneurons rather than directly from the thalamic afferents. Experiments about intracortical inhibition in V1 have given mixed results. Initially, Sillito's (1975) and Sillito et al. (1980) experiments with bicuculline, a GABA antagonist, suggested that intracortical inhibition is necessary for orientation tuning. However, the interpretation of these results is moot because of possible ceiling effects. Subsequent experiments of Nelson et al. (1994) blocking inhibition intracellularly were interpreted to mean that inhibition onto a single neuron is not necessary for that neuron to be orientation tuned. There is some question about this interpretation because in the Nelson experiments the blocked cells were hyperpolarized, mimicking the effect of sustained inhibition. Somewhat later, an important role for intracortical inhibition was indicated by pharmacological experiments (Allison et al., 1995; Sato et al., 1996; Crook et al., 1998).

There are several models that explain cortical orientation selectivity in terms of broadly tuned inhibition and more narrowly tuned excitation. One such theory of orientation tuning in cat cortex (Troyer et al., 1998) explains orientation selectivity in V1 in terms of "push–pull," that is spatial-phase-specific, inhibition (Palmer and Davis, 1981; Ferster, 1988, 1992; Tolhurst and Dean, 1990). However, the phase specificity is not the main reason the Troyer et al. model generates orientation selectivity. The mechanism for sharpening of orientation tuning in the Troyer et al. (1998) model is cortico-cortical inhibition that is broadly tuned for orientation. In the Troyer et al. model there is broadly tuned LGN convergent excitation as in the HW model, and then more broadly tuned inhibition that cancels out the wide angle responses but that leaves the tuning curve around the peak orientation relatively unchanged. In having broadly tuned inhibition and more narrowly tuned excitation, this particular model resembles many other cortico-cortical interaction models for orientation selectivity (Somers et al., 1995; Ben-Yishai et al., 1995; McLaughlin et al., 2000).

More recently, our colleagues David McLaughlin and Michael Shelley and their colleagues (McLaughlin et al., 2000; Wielaard et al., 2001; Shelley et al., 2002) designed a realistic network model for macaque V1. They constructed a large-scale model (16,000 neurons) of four hypercolumns in layer $4c\alpha$ of macaque V1 incorporating known facts about the physiology and anatomy. This model accounts for many visual properties of V1 neurons, among them orientation selectivity. One innovation in this model is its realism: the spatial strength of connections between neurons is taken to be the spatial density of synaptic connections revealed by anatomical investigations of cortex (e.g., Lund, 1988; Callaway, 1998). This model causes significant sharpening of orientation selectivity of V1 neurons compared to their feedforward LGN input. The mechanism of sharpening of orientation tuning is, as in the Troyer et al. (1998) model, broadly tuned inhibition. The big difference between this model and that of Troyer et al. (1998) is that in the McLaughlin et al. model the inhibitory conductance input to a cell is phase-insensitive (and not push–pull). This is a consequence of the realistic simulation of cortical anatomy: because inhibition onto a model cell is a sum from many inhibitory neurons and each cortical inhibitory cell has a fixed phase preference that is different from that of other inhibitory neurons. This view of the nonselective nature of local cortico-cortical inhibitory interactions is supported by the measured phase insensitivity of synaptic inhibitory conductance in V1 neurons (Borg-Graham et al., 1998; Anderson et al., 2000, discussed in Wielaard et al., 2001). Another distinguishing feature of the large-scale model of McLaughlin et al. (2000) is that it provides a mechanism for diversity in orientation selectivity that has been observed (Ringach et al., 2002).

Others have suggested that cortico-cortical excitatory interactions play a crucial role in orientation selectivity. Somers et al. (1995) presented an elaborate computational model for orientation tuning that includes both recurrent cortical excitation and inhibition as crucial elements. Douglas et al. (1995) argued for the importance of recurrent excitation in cortical circuits, reinforcing the message of Douglas and Martin (1991) on the "canonical microcircuit" of V1 cortex. A third paper in this genre was Ben-Yishai et al. (1995). Ben-Yishai et al. offered an analytical model from which they make several qualitative and quantitative predictions. One of their theoretical results is that if recurrent feedback is strong enough, one will observe a "marginal phase" state in which V1 behaves like a set of attractors for orientation. The attractor states in recurrent excitatory models are discussed not only in Ben-Yishai et al. (1995), but also in Tsodyks et al. (1999). The concept is that the tuning of very weakly orientation-tuned feed-forward signals can be massively sharpened by strong recurrent excitatory feedback. In such a network, the neurons will respond to any visual signal by relaxing into a state of activity governed by the pattern of cortico-cortical feedback. A similar idea was proposed in Adorjan et al. (1999). Our motivation was to try to decide between the different cortical models by performing and analyzing experiments on cortical orientation dynamics.

Cortical orientation dynamics

In an attempt to provide data to test models of orientation selectivity, we used a reverse correlation method developed originally by Dario Ringach. The idea was to measure the time evolution of orientation selectivity extracellularly in single V1 neurons, with a technique that drove most cortical neurons above threshold. The technique is illustrated in Fig. 3. The input image sequence is a stimulus "movie" that runs for 15–30 min. Grating patterns of orientations drawn randomly from a set of equally spaced orientations around the clock (usually in 10^0 steps) are presented for a fixed time (17 ms = 1 frame at a 60 Hz refresh rate in the early experiments reported in Ringach et al., 1997, and 20 ms = 2 frames at 100 Hz refresh rate in the more recent experiments reported in Ringach et al., 2003; Xing et al., 2005). Each orientation is presented at eight spatial phases and the response is phase averaged. For each fixed time interval between a spike and a preceding stimulus, the probability distribution for orientation is calculated by incrementing the orientation bin corresponding to the orientation that precedes each of the N spikes, and then dividing the bin counts by N. N is usually of the order of 5000 spikes. This is done for each value of time interval between spike and stimulus to create a sequence of orientation tuning curves, one for each time interval — an "orientation selectivity movie."

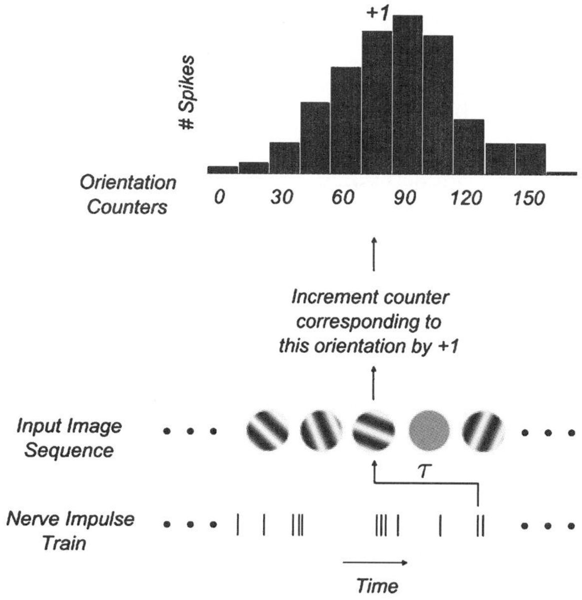

Fig. 3. Reverse correlation in the orientation domain. The input image sequence runs for 15–30 min. Grating patterns of orientations drawn randomly from a set of equally spaced orientations in the interval [0°, 180°] (usually in 10° angle steps) are presented for 20 ms each (2 frames at 100 Hz frame rate). Each orientation is presented at eight spatial phases; response is phase averaged. For each time offset, the probability distribution for orientation is calculated by incrementing the orientation bin corresponding to the orientation that precedes each of the N spikes, and then dividing the bin counts by N. N is usually of the order of 5000 spikes. This is done for each time offset τ to create an "orientation selectivity movie." In these experiments an additional pattern is added — a blank stimulus at the mean luminance of the grating patterns. This allows us to create a baseline with which the responses at different angles can be compared. Adapted with permission from Shapley et al. (2003).

In more recent experiments on orientation dynamics (Ringach et al., 2003; Xing et al., 2005), we used a refined technique that allowed us to uncover the mechanisms of orientation selectivity. As shown in Fig. 3, an additional pattern is added to the sequence — a blank stimulus at the mean luminance of the grating patterns. This allows us for the first time to measure untuned excitation and inhibition because, with this new technique, one can estimate whether the effect of one of the oriented patterns is greater or less than that of the blank pattern. If the probability of producing a spike by a pattern of orientation θ is greater than that of a blank, we view as evidence that a pattern of orientation θ produces net excitation, while if the probability of producing a spike by a pattern of orientation θ is less than that of a blank, we take this as an indication of inhibition. Specifically, we take $R(\theta, \tau) = \log[p(\theta, \tau)/p(\text{Blank}, \tau)]$. If the probability that angle θ evokes a spike is greater than that of a blank screen, then the sign of R is +. If the probability that angle θ evokes a spike is less than that of a blank screen, then the sign of R is −. If all angles evoke a response above the response to a blank, then $R(\theta)$ will have a positive value for all θ. A visual neuron equally well excited by stimuli of all orientation angles would produce a constant, positive $R(\theta)$.

The shape of the orientation tuning curve $R(\theta, \tau)$ changes with time, τ, and this dynamic behavior has a number of important properties that are revealed in Fig. 4 for a representative V1 neuron. The black curve is a graph of $R(\theta, \tau)$ at the time offset τ_{peak} when the orientation modulation depth, that is the difference between R_{max} and R_{min}, reaches its maximum value. The red and blue curves are graphs of $R(\theta, \tau)$ at the two times bracketing τ_{peak} at which the orientation modulation depth is half the maximum value; the red curve is at the development time τ_{dev}, the earlier of the two times when the modulation depth first rises from zero to half maximum, and the blue curve is at the declining time τ_{dec} when the response has declined back down to half maximum from maximum. One striking feature of these curves is that the dynamic tuning curve at the earlier time, $R(\theta, \tau_{dev})$, has a large positive pedestal of response, a sign of untuned or very broadly tuned excitation early in the response. This is just what one might predict from the analysis of feedforward models (see Fig. 2), if indeed the earliest response measurable were predominantly feedforward excitation. But then, as the response evolves in time, the maximum value of $R(\theta, \tau)$ at the preferred orientation grows only a little, while the responses at nonpreferred orientations decline substantially. Thus, Fig. 4 demonstrates that the maximum orientation modulation depth occurs at a time when inhibition has suppressed nonpreferred responses. Because such inhibition suppresses all responses far from the preferred orientation, we infer that this is untuned inhibition. It is also reasonable to infer that tuned excitation near the preferred orientation counteracts the untuned inhibition to maintain the peak value of $R(\theta, \tau)$.

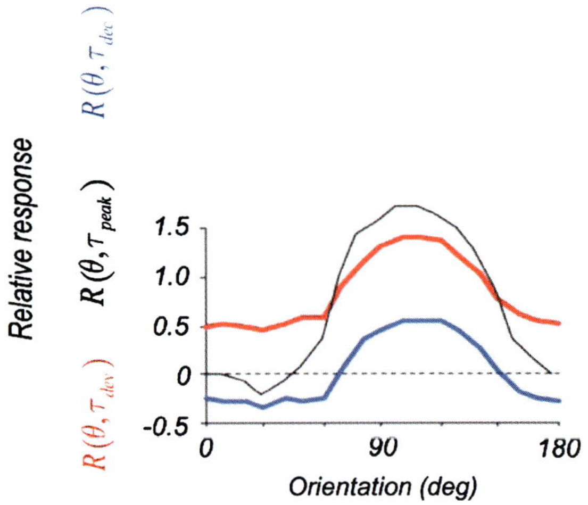

Fig. 4. Dynamics of orientation tuning in a representative V1 neuron. The black curve is a graph of $R(\theta, \tau)$ at the time offset τ_{peak} when the orientation modulation depth reaches its maximum value. The red and blue curves are graphs of $R(\theta, \tau)$ at the two times before and after τ_{peak} at which orientation modulation is half maximal: the red curve is at τ_{dev}, the earlier of the two times, and the blue curve is for τ_{dec}, the later time. Adapted with permission from Shapley et al. (2003).

While bandwidth often has been the focus of interest in previous research, it is rather the global shape of the tuning curve at all orientations that differentiates between different theoretical mechanisms. One simple way to study the global shape of the tuning curve is to compare the response at the preferred orientation with the response at orthogonal-to-preferred. Therefore, we studied $R(\theta_{pref}, \tau)$ and $R(\theta_{ortho}, \tau)$ in a population of V1 neurons because these features of the dynamical tuning curves are related to the overall shape of the tuning curve and lead to insight about the role of inhibition in the time evolution of orientation selectivity. The average behaviors of $R(\theta_{pref}, \tau)$, $R(\theta_{ortho}, \tau)$ averaged over a population of 101 neurons are depicted in Fig. 5. An important feature is the positive sign of $R(\theta_{pref}, \tau)$ and $R(\theta_{ortho}, \tau)$ early in the response, indicating that, on average, V1 cells tended to respond to *all* orientations early in the response. This is a feature that is consistent with the idea that at early times feedforward input as in Fig. 2 controls the response. Another important feature of the data

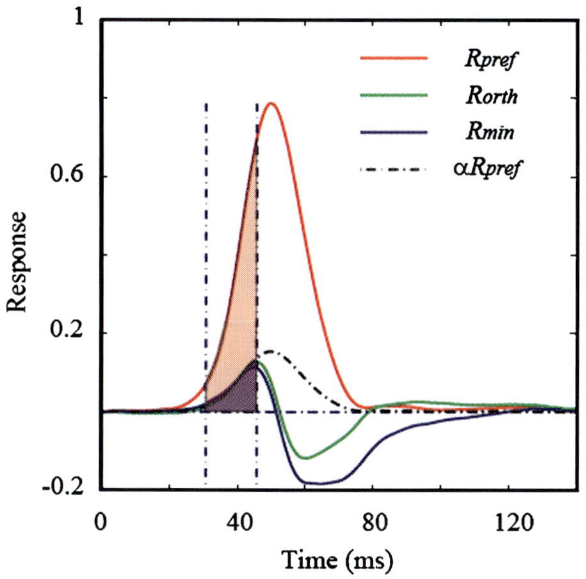

Fig. 5. Time course of the population-averaged (101 cells) response to preferred orientation (R_{pref}, red curve), to orthogonal orientation (R_{orth}, green curve) and to the orientation where the response was minimum ($R_{\min}$, blue curve) in responses to stimuli of large size. Black dash-dot curve (aR_{pref}) is the rescaled R_{pref}. The time course of each cell's responses was shifted so that its half-max rise time of R_{pref} is at 41 ms. Adapted with permission from Xing et al. (2005).

is that the time course of $R(\theta_{\text{ortho}}, \tau)$ was different from $R(\theta_{\text{pref}}, \tau)$. Note especially in the time courses in Fig. 5 the downward turn of $R(\theta_{\text{ortho}}, \tau)$ just before $R(\theta_{\text{pref}}, \tau)$ reached its peak value. Eventually $R(\theta_{\text{ortho}}, \tau)$ declined to negative values meaning that later in the response orientations far from the preferred orientation were suppressive not excitatory. If the entire response were dominated by feedforward input, one would expect that preferred and orthogonal responses would have the same time course simply scaled by the relative sensitivity. Therefore, the results in Fig. 5 qualitatively rule out an explanation of the time evolution of orientation selectivity in terms of feedforward inputs alone.

The results about the population averages in Fig. 5 support the hypothesis that there is untuned suppression generated in the cortex that is rapid, but still somewhat delayed with respect to the early excitatory input. The untuned suppression contributes to the amount of orientation selectivity at the time when the neuron is most selective. These results could be explained with a theory in which feedforward excitation drives the early weakly selective response. Evidence in favor of weakly selective excitation was obtained by Xing et al. (2005) when they analyzed the orientation dynamics into a sum of excitation and untuned and tuned suppression. The orientation tuning of the excitatory term is shown in Fig. 6 where it is compared to the predicted broad tuning curve for

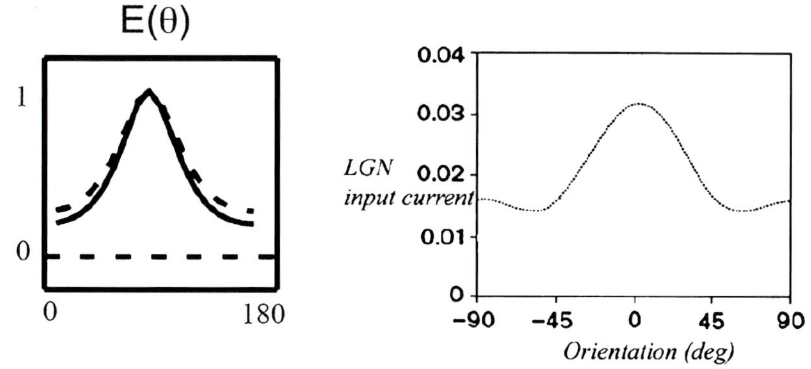

Fig. 6. The time course of measured excitation compared with the prediction of feedforward LGN input current. The orientation dependence of tuned excitation is plotted in the left hand panel, redrawn from Xing et al. (2005). Note especially the broad tuning with nonzero response at orthogonal-to-preferred. The dashed curve is for responses to stimulus of optimal size; the solid curve is for a large stimulus of 2–4 × the diameter of an optimal stimulus. The right panel reproduces the theoretical prediction of Fig. 2.

feedforward input, from Fig. 2. Sharpening of this broadly tuned input occurs when, with a very short delay, relatively rapid intracortical inhibition reduces the response at all orientations, acting like an untuned suppression. Data from intracellular recording in V1 indicate that a wide variety of patterns of cortico-cortical inhibition may influence orientation selectivity (Monier et al., 2003).

Discussion: inhibition and selectivity

The data in Figs. 4 and 5 from the orientation dynamics experiments demonstrate that early excitation in V1 is very broadly tuned for orientation, just as predicted for models of feedforward convergence like the HW model (see Fig. 2). Indeed in simulations of the dynamics experiments with a large-scale network model of V1, McLaughlin et al. demonstrated that feedforward excitation generates dynamical orientation tuning curves with very high circular variance, meaning poor selectivity, at all time offsets between stimulus and spike (see McLaughlin et al., 2000, Fig. 2). Therefore, to us, an important question about orientation selectivity in V1 is, as we have stated it above, how does the cortex suppress the feedforward excitation far from the preferred orientation? Our experimental results show that untuned inhibition in the cortex answers the question for those V1 neurons that are highly selective for orientation. The inhibitory signals must be fairly rapid, though not quite as fast in arrival at the V1 neuron as the earliest excitatory signals. Also, inhibition appears to persist longer than excitation, as illustrated in Fig. 5. A more comprehensive and detailed analysis of the dynamics of orientation selectivity, and in particular of untuned suppression, can be found in Xing et al. (2005).

Additional compelling evidence for the important role of inhibition in orientation selectivity has come from experiments on intracellular recording from neurons in cat V1 (Borg-Graham et al., 1998; Monier et al., 2003). Furthermore, the very elegant pharmacological experiments in macaque V1 cortex by Sato et al. (1996) established that when cortical inhibition was weakened by pharmacological competitive inhibitors, neuronal orientation selectivity was reduced because the response to off-peak orientations grew stronger relative to the peak response (cf. especially Fig. 8 in Sato et al., 1996). This is further support for the idea that the feedforward excitatory input is very broadly tuned in orientation, and that cortical inhibition suppresses the responses far from the preferred orientation. As presented earlier, the importance of broadly tuned cortical inhibition has been suggested also in computational models of the cortex (Troyer et al., 1998; McLaughlin et al., 2000; Wielaard et al., 2001).

Untuned suppression and cortical inhibition

To judge whether or not cortico-cortical inhibition is the source of untuned suppression requires more detailed considerations. When we stimulated a cell with a stimulus of optimal size (0.45° radius on average in our data), we most likely activated a compact region of V1 (Van Essen et al., 1984; Tootell et al., 1988). This region in V1 cortex corresponds to the cell's local neighborhood (Angelucci et al., 2002). That we see a strong untuned suppression even with a stimulus of optimal size suggests that the untuned suppression mainly comes from the center mechanism and the local circuitry within a cortical hypercolumn. This is consistent with the recent anatomical findings (Angelucci et al., 2002; Marino et al., 2005) that a V1 cell gets most of its inhibitory synaptic input from a local area in the cortex of approximate diameter of 100–250 μm. Untuned suppression exists in all layers as well as in simple and complex cell groups (Xing et al., 2005). This suggests that untuned suppression is a general mechanism in primary visual cortex (Ringach et al., 2002; Shapley et al., 2003; Xing et al., 2005). Broadly tuned cortico-cortical inhibition that arises locally in the cortical circuitry is the likely source of the untuned suppression we have measured (Troyer et al., 1998; McLaughlin et al., 2000; Tao et al., 2004). There are other candidate mechanisms for untuned suppression in V1, for instance synaptic depression at the thalamo-cortical synapses, as proposed by Carandini et al. (2002). The fact that untuned suppression is stronger in layer 4B and layer 5 than in the main thalamo-recipient layers (layer 4C and layer 6) suggests that the untuned suppression is mainly from cortico-cortical effects instead of from thalamic-cortical effects (Xing et al., 2005). Furthermore, the untuned suppression we measured had short persistence (Xing et al., 2005), while rapid synaptic depression has 200–600 ms recovery time (Abbott et al., 1997). So the time course of untuned suppression is unlike what has been assumed for synaptic depression (e.g., Carandini et al., 2002). A likely possibility is that fast cortical inhibition is the source of the untuned suppression.

Cortico-cortical excitation and selectivity

There is a possibility that tuned cortico-cortical excitation may contribute also to enhancement of orientation selectivity by boosting the response only around the preferred orientation. The possibility that cortico-cortical excitation could enhance orientation selectivity was suggested previously in theories of V1 (Ben-Yishai et al., 1995; Somers et al., 1995). However, we did not observe a substantial sharpening of the excitatory input during the time evolution of orientation selectivity (Xing et al., 2005). Therefore, the orientation dynamics data suggest that the role of tuned cortical excitation is less than that of untuned inhibition in generating selectivity in V1.

Comparison with other studies

In the Introduction we reviewed previous experiments that were taken to support a completely different point of view, namely that the pattern of feedforward thalamic input is enough to determine orientation selectivity. Our results as a whole are not consistent with this viewpoint. There are in the literature two studies with dynamical stimuli that have been interpreted as supporting the feedforward theory. Gillespie et al. (2001), recording intracellularly in cat V1, reported that the bandwidth of orientation tuning curves did not change with time in their dynamic experiments. As stated above, we think that examining bandwidth misses the point that the crucial question in orientation selectivity is how the orthogonal response is suppressed by the cortex. Interestingly, Gillespie et al. (2001, Figs. 2h, and 3b, f, j) do report a change in the intracellular baseline with time that reinforces our observations on the dynamic growth of inhibition. Therefore, our interpretation of the results of Gillespie et al. (2001) is that they support the concept that inhibition plays an important role in enhancing orientation selectivity, by untuned inhibition.

In a study that purports to assign a dominant role to feedforward connections in orientation, Mazer et al. (2002) recorded extracellularly in V1 of awake macaques, and used a reverse correlation technique very similar to the one we introduced in 1997 (Ringach et al., 1997). However, unlike the results we have presented here, Mazer et al.'s results were interpreted to indicate that the orientation tuning curves measured dynamically did not change shape with time. Because they did not have a baseline stimulus, as we did with the blank stimuli in the stimulus sequence, Mazer et al. (2002) could not measure the presence of untuned suppression, or broadly tuned excitation either. Therefore, their conclusions about the time course of orientation dynamics were not well supported by the data they had available.

Diversity

The diversity of orientation selectivity is interesting. Others have also reported data that indicate wide diversity of orientation tuning in cat V1 (Dragoi et al., 2000) and in ferret V1 (Chapman and Stryker, 1993) when the orientation tuning curves were analyzed with global measures of selectivity like those we have employed. There is a need for understanding what are the functional consequences for visual perception of the wide diversity of orientation tuning that is observed. This question was considered by Kang et al. (2004) in a paper that applied a new technique for measuring information transmission by populations of neurons. Kang et al. concluded that diversity of orientation selectivity could make the cortical population better at discriminations of different orientation differences. It is also plausible that the visual cortex is not only designed for tasks like orientation discrimination, and that diversity of orientation selectivity may be a result of specializations of neurons in other stimulus dimensions besides orientation.

Orientation selectivity and cortical circuits

Our view of V1 is that it is a nonlinear dynamical system and one of its tasks is to find local stimulus features in the neural image of the visual scene relayed to V1 from the eye through the LGN. Different sources of excitation drive the activity in V1 cells: local thalamo-cortical projections, local-circuit cortico-cortical excitation, long-distance horizontal V1 axons, and also feedback. Different sources of intracortical inhibition contribute to the selectivity of V1 neurons: local-circuit inhibition, inhibition mediated by signals from long-distance intrinsic V1 horizontal connections (Gilbert and Wiesel, 1983; Rockland and Lund, 1983; Crook et al., 1998; Roerig and Chen, 2002), and feedback from extra-striate cortex (Angelucci et al., 2002) that drives inhibitory interneurons in the local circuit. While feedforward excitation must play a role in giving V1 cells preferences for particular orientations, intracortical inhibition makes some V1 cells highly selective for their preferred orientation over all others.

Acknowledgments

We thank the US National Eye Institute for support of our research through grants EY01472 and EY8300.

References

Abbott, L.F., Varela, J.A., Sen, K. and Nelson, S.B. (1997) Synaptic depression and cortical gain control. Science, 275: 220–224.

Adorjan, P., Levitt, J.B., Lund, J.S. and Obermayer, K. (1999) A model for the intracortical origin of orientation preference and tuning in macaque striate cortex. Vis. Neurosci., 16: 303–318.

Allison, J.D., Casagrande, V.A. and Bonds, A.B. (1995) Dynamic differentiation of GABAA-sensitive influences on orientation selectivity of complex cells in the cat striate cortex. Exp. Brain Res., 104: 81–88.

Anderson, J.S., Carandini, M. and Ferster, D. (2000) Orientation tuning of input conductance, excitation, and inhibition in cat primary visual cortex. J. Neurophysiol., 84: 909–926.

Angelucci, A., Levitt, J.B., Walton, E.J., Hupe, J.M., Bullier, J. and Lund, J.S. (2002) Circuits for local and global signal integration in primary visual cortex. J. Neurosci., 22: 8633–8646.

Ben-Yishai, R., Bar-Or, R.L. and Sompolinsky, H. (1995) Theory of orientation tuning in visual cortex. Proc. Natl. Acad. Sci. U.S.A., 92: 3844–3848.

Bonds, A.B. (1989) Role of inhibition in the specification of orientation selectivity of cells in the cat striate cortex. Vis. Neurosci., 2: 41–55.

Borg-Graham, L.J., Monier, C. and Fregnac, Y. (1998) Visual input evokes transient and strong shunting inhibition in visual cortical neurons. Nature, 393: 369–373.

Callaway, E.M. (1998) Local circuits in primary visual cortex of the macaque monkey. Ann. Rev. Neurosci., 21: 47–74.

Carandini, M. and Ferster, D. (2000) Membrane potential and firing rate in cat primary visual cortex. J. Neurosci., 20: 470–484.

Carandini, M., Heeger, D.J. and Senn, W. (2002) A synaptic explanation of suppression in visual cortex. J. Neurosci., 22: 10053–10065.

Chance, F.S., Nelson, S.B. and Abbott, L.F. (1999) Complex cells as cortically amplified simple cells. Nat. Neurosci., 2: 277–282.

Chapman, B. and Stryker, M.P. (1993) Development of orientation selectivity in ferret visual cortex and effects of deprivation. J. Neurosci., 13: 5251–5262.

Chapman, B., Zahs, K.R. and Stryker, M.P. (1991) Relation of cortical cell orientation selectivity to alignment of receptive fields of the geniculocortical afferents that arborize within a single orientation column in ferret visual cortex. J. Neurosci., 11: 1347–1358.

Crook, J.M., Kisvarday, Z.F. and Eysel, U.T. (1998) Evidence for a contribution of lateral inhibition to orientation tuning and direction selectivity in cat visual cortex: reversible inactivation of functionally characterized sites combined with neuroanatomical tracing techniques. Eur. J. Neurosci., 10: 2056–2075.

De Valois, R.L., Yund, E.W. and Hepler, N. (1982) The orientation and direction selectivity of cells in macaque visual cortex. Vision Res., 22: 531–544.

Douglas, R.J., Koch, C., Mahowald, M., Martin, K.A. and Suarez, H.H. (1995) Recurrent excitation in neocortical circuits. Science, 269: 981–985.

Douglas, R.J. and Martin, K.A. (1991) A functional microcircuit for cat visual cortex. J. Physiol., 440: 735–769.

Dragoi, V., Sharma, J. and Sur, M. (2000) Adaptation-induced plasticity of orientation tuning in adult visual cortex. Neuron, 28: 287–298.

Ferster, D. (1988) Spatially opponent excitation and inhibition in simple cells of the cat visual cortex. J. Neurosci., 8: 1172–1180.

Ferster, D. (1992) The synaptic inputs to simple cells of the cat visual cortex. Prog. Brain Res., 90: 423–441.

Ferster, D., Chung, S. and Wheat, H. (1996) Orientation selectivity of thalamic input to simple cells of cat visual cortex. Nature, 380: 249–252.

Freund, T.F., Martin, K.A., Soltesz, I., Somogyi, P. and Whitteridge, D. (1989) Arborisation pattern and postsynaptic targets of physiologically identified thalamocortical afferents in striate cortex of the macaque monkey. J. Comp. Neurol., 289: 315–336.

Gilbert, C.D. and Wiesel, T.N. (1983) Clustered intrinsic connections in cat visual cortex. J. Neurosci., 3: 1116–1133.

Gillespie, D.C., Lampl, I., Anderson, J.S. and Ferster, D. (2001) Dynamics of the orientation-tuned membrane potential response in cat primary visual cortex. Nat. Neurosci., 4: 1014–1019.

Hubel, D.H. and Wiesel, T.N. (1962) Receptive fields, binocular interaction and functional architecture in the cat's visual cortex. J. Physiol., 160: 106–154.

Hubel, D.H. and Wiesel, T.N. (1968) Receptive fields and functional architecture of monkey striate cortex. J. Physiol., 195: 215–243.

Kang, K., Shapley, R.M. and Sompolinsky, H. (2004) Information tuning of populations of neurons in primary visual cortex. J. Neurosci., 24: 3726–3735.

Lund, J.S. (1988) Anatomical organization of macaque monkey striate visual cortex. Ann. Rev. Neurosci., 11: 253–288.

Marino, J., Schummers, J., Lyon, D.C., Schwabe, L., Beck, O., Wiesing, P., Obermayer, K. and Sur, M. (2005) Invariant computations in local cortical networks with balanced excitation and inhibition. Nat. Neurosci., 8: 194–201.

Mazer, J.A., Vinje, W.E., McDermott, J., Schiller, P.H. and Gallant, J.L. (2002) Spatial frequency and orientation tuning dynamics in area V1. Proc. Natl. Acad. Sci. U.S.A., 99: 1645–1650.

McLaughlin, D., Shapley, R., Shelley, M. and Wielaard, J. (2000) A neuronal network model of sharpening and dynamics of orientation tuning in an input layer of macaque primary visual cortex. Proc. Natl. Acad. Sci. U.S.A., 97: 8087–8092.

Monier, C., Chavane, F., Baudot, P., Graham, L.J. and Fregnac, Y. (2003) Orientation and direction selectivity of synaptic inputs in visual cortical neurons: a diversity of combinations produces spike tuning. Neuron, 37: 663–680.

Nelson, S., Toth, L., Sheth, B. and Sur, M. (1994) Orientation selectivity of cortical neurons during intracellularblockade of inhibition. Science, 265: 774–777.

Palmer, L.A. and Davis, T.L. (1981) Receptive-field structure in cat striate cortex. J. Neurophysiol., 46: 260–276.

Reid, R.C. and Alonso, J.M. (1995) Specificity of monosynaptic connections from thalamus to visual cortex. Nature, 378: 281–284.

Ringach, D., Hawken, M. and Shapley, R. (1997) The dynamics of orientation tuning in the macaque monkey striate cortex. Nature, 387: 281–284.

Ringach, D.L., Hawken, M.J. and Shapley, R. (2003) Dynamics of orientation tuning in macaque V1: the role of global and tuned suppression. J. Neurophysiol., 90: 342–352.

Ringach, D.L., Shapley, R.M. and Hawken, M.J. (2002) Orientation selectivity in macaque v1: diversity and laminar dependence. J. Neurosci., 22: 5639–5651.

Rockland, K.S. and Lund, J.S. (1983) Intrinsic laminar lattice connections in primate visual cortex. J. Comp. Neurol., 216: 303–318.

Roerig, B. and Chen, B. (2002) Relationships of local inhibitory and excitatory circuits to orientation preference maps in ferret visual cortex. Cereb. Cortex, 12: 187–198.

Sato, H., Katsuyama, N., Tamura, H., Hata, Y. and Tsumoto, T. (1996) Mechanisms underlying orientation selectivity of neurons in the primary visual cortex of the macaque. J. Physiol., 494: 757–771.

Schiller, P.H., Finlay, B.L. and Volman, S.F. (1976) Quantitative studies of single-cell properties in monkey striate cortex. II. Orientation specificity and ocular dominance. J. Neurophysiol., 39: 1320–1333.

Shapley, R., Hawken, M. and Ringach, D.L. (2003) Dynamics of orientation selectivity in macaque V1 cortex, and the importance of cortical inhibition. Neuron, 38: 689–699.

Shapley, R.M. (1994) Linearity and non-linearity in cortical receptive fields. In: Higher Order Processing in the Visual System, Ciba Symposium 184, pp. 71–87. Wiley, Chichester.

Shelley, M., McLaughlin, D., Shapley, R. and Wielaard, J. (2002) States of high conductance in a large-scale model of the visual cortex. J. Comput. Neurosci., 13: 93–109.

Sillito, A.M. (1975) The contribution of inhibitory mechanisms to the receptive field properties of neurones in the striate cortex of the cat. J. Physiol., 250: 305–329.

Sillito, A.M., Kemp, J.A., Milson, J.A. and Berardi, N. (1980) A re-evaluation of the mechanisms underlying simple cell orientation selectivity. Brain Res., 194: 517–520.

Somers, D.C., Nelson, S.B. and Sur, M. (1995) An emergent model of orientation selectivity in cat visual cortical simple cells. J. Neurosci., 15: 5448–5465.

Sompolinsky, H. and Shapley, R. (1997) New perspectives on the mechanisms for orientation selectivity. Curr. Opin. Neurobiol., 7: 514–522.

Tao, L., Shelley, M., McLaughlin, D. and Shapley, R. (2004) An egalitarian network model for the emergence of simple and complex cells in visual cortex. Proc. Natl. Acad. Sci. U.S.A., 101: 366–371.

Tolhurst, D.J. and Dean, A.F. (1990) The effects of contrast on the linearity of spatial summation of simple cells in the cat's striate cortex. Exp. Brain Res., 79: 582–588.

Tootell, R.B., Switkes, E., Silverman, M.S. and Hamilton, S.L. (1988) Functional anatomy of macaque striate cortex II. Retinotopic organization. J. Neurosci., 8: 1531–1568.

Troyer, T.W., Krukowski, A.E., Priebe, N.J. and Miller, K.D. (1998) Contrast-invariant orientation tuning in cat visual cortex: thalamocortical input tuning and correlation-based intracortical connectivity. J. Neurosci., 18: 5908–5927.

Tsodyks, M., Kenet, T., Grinvald, A. and Arieli, A. (1999) Linking spontaneous activity of single cortical neurons and the underlying functional architecture. Science, 286: 1943–1946.

Van Essen, D.C., Newsome, W.T. and Maunsell, J.H. (1984) The visual field representation in striate cortex of the macaque monkey: asymmetries, anisotropies, and individual variability. Vision Res., 24: 429–448.

Wielaard, J., Shelley, M., McLaughlin, D.M. and Shapley, R.M. (2001) How simple cells are made in a nonlinear network model of the visual cortex. J. Neurosci., 21: 5203–5211.

Xing, D., Shapley, R.M., Hawken, M.J. and Ringach, D.L. (2005) The effect of stimulus size on the dynamics of orientation selectivity in macaque V1. J. Neurophysiol., 94: 799–812.

P. Cisek, T. Drew & J.F. Kalaska (Eds.)
Progress in Brain Research, Vol. 165
ISSN 0079-6123
Copyright © 2007 Elsevier B.V. All rights reserved

CHAPTER 4

A quantitative theory of immediate visual recognition

Thomas Serre*, Gabriel Kreiman, Minjoon Kouh, Charles Cadieu, Ulf Knoblich and Tomaso Poggio

Center for Biological and Computational Learning, McGovern Institute for Brain Research, Computer Science and Artificial Intelligence Laboratory, Brain and Cognitive Sciences Department, Massachusetts Institute of Technology, 43 Vassar Street #46-5155B, Cambridge, MA 02139, USA

Abstract: Human and non-human primates excel at visual recognition tasks. The primate visual system exhibits a strong degree of selectivity while at the same time being robust to changes in the input image. We have developed a quantitative theory to account for the computations performed by the feedforward path in the ventral stream of the primate visual cortex. Here we review recent predictions by a model instantiating the theory about physiological observations in higher visual areas. We also show that the model can perform recognition tasks on datasets of complex natural images at a level comparable to psychophysical measurements on human observers during rapid categorization tasks. In sum, the evidence suggests that the theory may provide a framework to explain the first 100–150 ms of visual object recognition. The model also constitutes a vivid example of how computational models can interact with experimental observations in order to advance our understanding of a complex phenomenon. We conclude by suggesting a number of open questions, predictions, and specific experiments for visual physiology and psychophysics.

Keywords: visual object recognition; hierarchical models; ventral stream; feedforward

Introduction

The primate visual system rapidly and effortlessly recognizes a large number of diverse objects in cluttered, natural scenes. In particular, it can easily categorize images or parts of them, for instance as an office scene or a face within that scene, and identify a specific object. This remarkable ability is evolutionarily important since it allows us to distinguish friend from foe and identify food targets in complex, crowded scenes. Despite the ease with which we see, visual recognition — one of the key issues addressed in computer vision — is quite difficult for computers. The problem of object recognition is even more difficult from the point of view of neuroscience, since it involves several levels of understanding from the information processing or computational level to circuits and biophysical mechanisms. After decades of work in different brain areas ranging from the retina to higher cortical areas, the emerging picture of how cortex performs object recognition is becoming too complex for any simple qualitative "mental" model.

A quantitative, computational theory can provide a much-needed framework for summarizing and integrating existing data and for planning, coordinating, and interpreting new experiments. Models are powerful tools in basic research, integrating knowledge across several levels of analysis — from molecular to synaptic, cellular, systems and to complex visual behavior. In this paper, we

*Corresponding author. Tel.: +1 617 253 0548; Fax: +1 617 253 2964; E-mail: serre@mit.edu

DOI: 10.1016/S0079-6123(06)65004-8

describe a quantitative theory of object recognition in primate visual cortex that (1) bridges several levels of understanding from biophysics to physiology and behavior and (2) achieves human level performance in rapid recognition of complex natural images. The theory is restricted to the feedforward path of the ventral stream and therefore to the first 100–150 ms of visual recognition; it does not describe top-down influences, though it should be, in principle, capable of incorporating them.

In contrast to other models that address the computations in any one given brain area (such as primary visual cortex) or attempt to explain a particular phenomenon (such as contrast adaptation or a specific visual illusion), we describe here a large-scale neurobiological model that attempts to describe the basic processes across multiple brain areas. One of the initial key ideas in this and many other models of visual processing (Fukushima, 1980; Perrett and Oram, 1993; Mel, 1997; Wallis and Rolls, 1997; Riesenhuber and Poggio, 1999) come from the pioneering physiological studies and models of Hubel and Wiesel (1962).

Following their work on striate cortex, they proposed a hierarchical model of cortical organization. They described a hierarchy of cells within the primary visual cortex: at the bottom of the hierarchy, the radially symmetric cells behave similarly to cells in the thalamus and respond best to small spots of light. Second, the simple cells which do not respond well to spots of light require bar-like (or edge-like) stimuli at a particular orientation, position, and phase (i.e., white bar on a black background or dark bar on a white background). In turn, complex cells are also selective for bars at a particular orientation but they are insensitive to both the location and the phase of the bar within their receptive fields. At the top of the hierarchy, hypercomplex cells not only respond to bars in a position and phase invariant way like complex cells, but also are selective for bars of a particular length (beyond a certain length their response starts to decrease). Hubel and Wiesel suggested that such increasingly complex and invariant object representations could be progressively built by integrating convergent inputs from lower levels. For instance, position invariance at the complex cell level could be obtained by pooling over simple cells at the same preferred orientation but at slightly different positions. The main contribution from this and other models of visual processing (Fukushima, 1980; Perrett and Oram, 1993; Mel, 1997; Wallis and Rolls, 1997; Riesenhuber and Poggio, 1999) has been to extend the notion of hierarchy beyond V1 to extrastriate areas and show how this can explain the tuning properties of neurons in higher areas of the ventral stream of the visual cortex.

A number of biologically inspired algorithms have been described (Fukushima, 1980; LeCun et al., 1998; Ullman et al., 2002; Wersing and Koerner, 2003), i.e., systems which are only qualitatively constrained by the anatomy and physiology of the visual cortex. However, there have been very few neurobiologically plausible models (Olshausen et al., 1993; Perrett and Oram, 1993; Mel, 1997; Wallis and Rolls, 1997; Riesenhuber and Poggio, 1999; Thorpe, 2002; Amit and Mascaro, 2003) that try to address a generic, high-level computational function such as object recognition by summarizing and integrating a large body of data from different levels of understanding. What should a general theory of biological object recognition be able to explain? It should be constrained to match data from anatomy and physiology at different stages of the ventral stream as well as human performance in complex visual tasks such as object recognition. The theory we propose may well be incorrect. Yet it represents a set of claims and ideas that deserve to be either falsified or further developed and refined.

The scope of the current theory is limited to "immediate recognition," i.e., to the first 100–150 ms of the flow of information in the ventral stream. This is behaviorally equivalent to considering "rapid categorization" tasks for which presentation times are fast and back-projections are likely to be inactive (Lamme and Roelfsema, 2000). For such tasks, presentation times do not allow sufficient time for eye movements or shifts of attention (Potter, 1975). Furthermore, EEG studies (Thorpe et al., 1996) provide evidence that the human visual system is able to solve an object detection task — determining whether a natural scene contains an animal or not — within 150 ms.

Extensive evidence shows that the responses of inferior temporal (IT) cortex neurons begin 80–100 ms after onset of the visual stimulus (Perrett et al., 1992). Furthermore, the neural responses at the IT level are tuned to the stimulus essentially from response onset (Keysers et al., 2001). Recent data (Hung et al., 2005) show that the activity of small neuronal populations in IT (~100 randomly selected cells) over very short time intervals from response onset (as small as 12.5 ms) contains surprisingly accurate and robust information supporting visual object categorization and identification tasks. Finally, rapid detection tasks, e.g., animal vs. non-animal (Thorpe et al., 1996), can be carried out without top-down attention (Li et al., 2002). We emphasize that none of these rules out the use of local feedback — which is in fact used by the circuits we propose for the two main operations postulated by the theory (see section on "A quantitative framework for the ventral stream") — but suggests a hierarchical forward architecture as the core architecture underlying "immediate recognition."

We start by presenting the theory in section "A quantitative framework for the ventral stream:" we describe the architecture of a model implementing the theory, its two key operations, and its learning stages. We briefly review the evidence about the agreement of the model with single cell recordings in visual cortical areas (V1, V2, V4) and describe in more detail how the final output of the model compares to the responses in IT cortex during a decoding task that attempts to identify or categorize objects (section on "Comparison with physiological observations"). In section "Performance on natural images," we further extend the approach to natural images and show that the model performs surprisingly well in complex recognition tasks and is competitive with some of the best computer vision systems. As an ultimate and more stringent test of the theory, we show that the model predicts the level of performance of human observers on a rapid categorization task. The final section discusses the state of the theory, its limitations, a number of open questions including critical experiments, and its extension to include top-down effects and cortical back-projections.

A quantitative framework for the ventral stream

Organization of the ventral stream of visual cortex

Object recognition in cortex is thought to be mediated by the ventral visual pathway (Ungerleider and Haxby, 1994). Information from the retina is conveyed to the lateral geniculate nucleus in the thalamus and then to primary visual cortex, V1. Area V1 projects to visual areas V2 and V4, and V4 in turn projects to IT, which is the last exclusively visual area along the ventral stream (Felleman and van Essen, 1991). Based on physiological and lesion experiments in monkeys, IT has been postulated to play a central role in object recognition (Schwartz et al., 1983). It is also a major source of input to prefrontal cortex (PFC) that is involved in linking perception to memory and action (Miller, 2000).

Neurons along the ventral stream (Perrett and Oram, 1993; Logothetis and Sheinberg, 1996; Tanaka, 1996) show an increase in receptive field size as well as in the complexity of their preferred stimuli (Kobatake and Tanaka, 1994). Hubel and Wiesel (1962) first described *simple cells* in V1 with small receptive fields that respond preferentially to oriented bars. At the top of the ventral stream, IT cells are tuned to complex stimuli such as faces and other objects (Gross et al., 1972; Desimone et al., 1984; Perrett et al., 1992).

A hallmark of the cells in IT is the robustness of their firing over stimulus transformations such as scale and position changes (Perrett and Oram, 1993; Logothetis et al., 1995; Logothetis and Sheinberg, 1996; Tanaka, 1996). In addition, as other studies have shown, most neurons show specificity for a certain object view or lighting condition (Hietanen et al., 1992; Perrett and Oram, 1993; Logothetis et al., 1995; Booth and Rolls, 1998) while other neurons are view-invariant and in agreement with earlier predictions (Poggio and Edelman, 1990). Whereas view-invariant recognition requires visual experience of the specific novel object, significant position and scale invariance seems to be immediately present in the view-tuned neurons (Logothetis et al., 1995) without the need of visual experience for views *of the specific object* at different positions and scales (see also Hung et al., 2005).

In summary, the accumulated evidence points to four, mostly accepted, properties of the feedforward path of the ventral stream architecture: (a) a hierarchical build-up of invariances first to position and scale and then to viewpoint and other transformations; (b) an increasing selectivity, originating from inputs from previous layers and areas, with a parallel increase in both the size of the receptive fields and in the complexity of the optimal stimulus; (c) a basic feedforward processing of information (for "immediate recognition" tasks); and (d) plasticity and learning probably at all stages with a time scale that decreases from V1 to IT and PFC.

Architecture and model implementation

The physiological data summarized in the previous section, together with computational considerations on image invariances, lead to a theory that summarizes and extends several previously existing neurobiological models (Hubel and Wiesel, 1962; Poggio and Edelman, 1990; Perrett and Oram, 1993; Mel, 1997; Wallis and Rolls, 1997; Riesenhuber and Poggio, 1999) and biologically motivated computer vision approaches (Fukushima, 1980; LeCun et al., 1998; Ullman et al., 2002). The theory maintains that: One of the main functions of the ventral stream pathway is to achieve an exquisite trade-off between selectivity and invariance at the level of shape-tuned and invariant cells in IT from which many recognition tasks can be readily accomplished; the key computational issue in object recognition is to be able to finely discriminate between different objects and object classes while at the same time being tolerant to object transformations such as scaling, translation, illumination, viewpoint changes, changes in context and clutter, non-rigid transformations (such as a change of facial expression) and, for the case of categorization, also to shape variations within a class.

The underlying architecture is hierarchical, with a series of stages that gradually increase invariance to object transformations and tuning to more specific and complex *features*.

There exist at least two main functional types of units, *simple* and *complex*, which represent the result of two main operations to achieve selectivity (*S* layer) and invariance (*C* layer). The two corresponding operations are a (bell-shaped) Gaussian-like TUNING of the simple units and a MAX-like operation for invariance to position, scale, and clutter (to a certain degree) of the complex units.

Two basic operations for selectivity and invariance

The *simple S* units perform a TUNING operation over their afferents to build object-selectivity. The *S* units receive convergent inputs from retinotopically organized units tuned to *different preferred stimuli* and combine these *subunits* with a bell-shaped tuning function, thus increasing object selectivity and the complexity of the preferred stimulus. Neurons with a Gaussian-like bell-shaped tuning are prevalent across cortex. For instance, simple cells in V1 exhibit a Gaussian tuning around their preferred orientation; cells in AIT are typically tuned around a particular view of their preferred object. From the computational point of view, Gaussian-like tuning profiles may be the key in the generalization ability of the cortex. Indeed, networks that combine the activity of several units tuned with a Gaussian profile to different training examples have proved to be a powerful learning scheme (Poggio and Edelman, 1990).

The *complex C* units perform a MAX-like operation over their afferents to gain invariance to several object transformations. The complex *C* units receive convergent inputs from retinotopically organized *S* units tuned to the *same preferred stimulus* but at slightly different positions and scales and combine these subunits with a MAX-like operation, thereby introducing tolerance to scale and translation. The existence of a MAX operation in visual cortex was proposed by Riesenhuber and Poggio (1999) from theoretical arguments [and limited experimental evidence (Sato, 1989)] and was later supported experimentally in both V4 (Gawne and Martin, 2002) and V1 at the complex cell level (Lampl et al., 2004).

A gradual increase in both selectivity and invariance, to 2D transformations, as observed along the ventral stream and as obtained in the model by interleaving the two key operations, is critical for avoiding both a combinatorial explosion in the number of units and the binding problem between features. Below we shortly give idealized mathematical expressions for the operations.

Idealized mathematical descriptions of the two operations: In the following, we denote by y the response of a unit (simple or complex). The set of inputs to the cell (i.e., pre-synaptic units) are denoted with subscripts $j = 1, \ldots, N$. When presented with a pattern of activity $\mathbf{x} = (x_1, \ldots, x_N)$ as input, an idealized and static description of a complex unit response y is given by:

$$y = \max_{j=1, \ldots, N} x_j \qquad (1)$$

As mentioned above, for a complex cell, the inputs x_j are retinotopically organized (selected from an $m \times m$ grid of afferents with the same selectivity). For instance, in the case of a V1-like complex cell tuned to a horizontal bar, all input subunits are tuned to a horizontal bar but at slightly different positions and scales. Similarly, an idealized description of a simple unit response is given by:

$$y = exp\left(-\frac{1}{2\sigma^2}\sum_{j=1}^{N}(w_j - x_j)^2\right) \qquad (2)$$

σ defines the sharpness of the TUNING of the unit around its preferred stimulus corresponding to the synaptic strengths $\mathbf{w} = (w_1, \ldots, w_N)$. As for complex cells, the subunits of the simple cells are also retinotopically organized (selected from an $m \times m$ grid of possible afferents). In contrast with complex cells, the subunits of a simple cell have different selectivities to increase the complexity of the preferred stimulus. For instance, for the S_2 units, the subunits are V1-like complex cells at different preferred orientations. The response of a simple unit is maximal when the current pattern of input $\mathbf{x}$ matches exactly the synaptic weights $\mathbf{w}$ (for instance the frontal view of a face) and decreases with a bell-shaped profile as the pattern of input becomes more dissimilar (e.g., for IT-like face-tuned units, as the face is rotated away from the preferred view).

Both of these mathematical descriptions are only meant to describe the response behavior of cells at a phenomenological level. Plausible biophysical circuits for the TUNING and MAX operations have been proposed based on feedforward and/or feedback shunting inhibition combined with normalization [see Serre et al. (2005) and references therein].

Building a dictionary of shape-components from V1 to IT

The overall architecture is sketched in Fig. 1 and reflects the general organization of the visual cortex in a series of layers from V1 to IT and PFC. Colors encode the tentative correspondences between the functional primitives of the theory (right) and the structural primitives of the ventral stream in the primate visual system (modified from Gross (1998) (left, modified from Gross, 1998). Below we give a brief description of a model instantiating the theory. The reader should refer to Serre (2006) for a more complete description of the architecture and detailed parameter values.

The first stage of simple units (S_1), corresponding to the classical simple cells of Hubel and Wiesel, represents the result of the first tuning operation. Each S_1 cell is tuned in a Gaussian-like way to a bar (a gabor) of one of four possible orientations. Each of the complex units in the second layer (C_1), corresponding to the classical complex cells of Hubel and Wiesel, receives, within a neighborhood, the outputs of a group of simple units in the first layer at slightly different positions and sizes but with the same preferred orientation. The operation is a nonlinear MAX-like operation [see Eq. (1)] that increases invariance to local changes in position and scale while maintaining feature specificity.

At the next simple cell layer (S_2), the units pool the activities of several complex units (C_1) with weights dictated by the unsupervised learning stage (see below), yielding selectivity to more complex patterns such as combinations of oriented lines. Simple units in higher layers (S_3 and S_4) combine more and more complex features with a Gaussian tuning function [see Eq. (2)], while the complex units (C_2 and C_3) pool their afferents through a MAX-like function [see Eq. (1)], providing increasing invariance to position and scale.

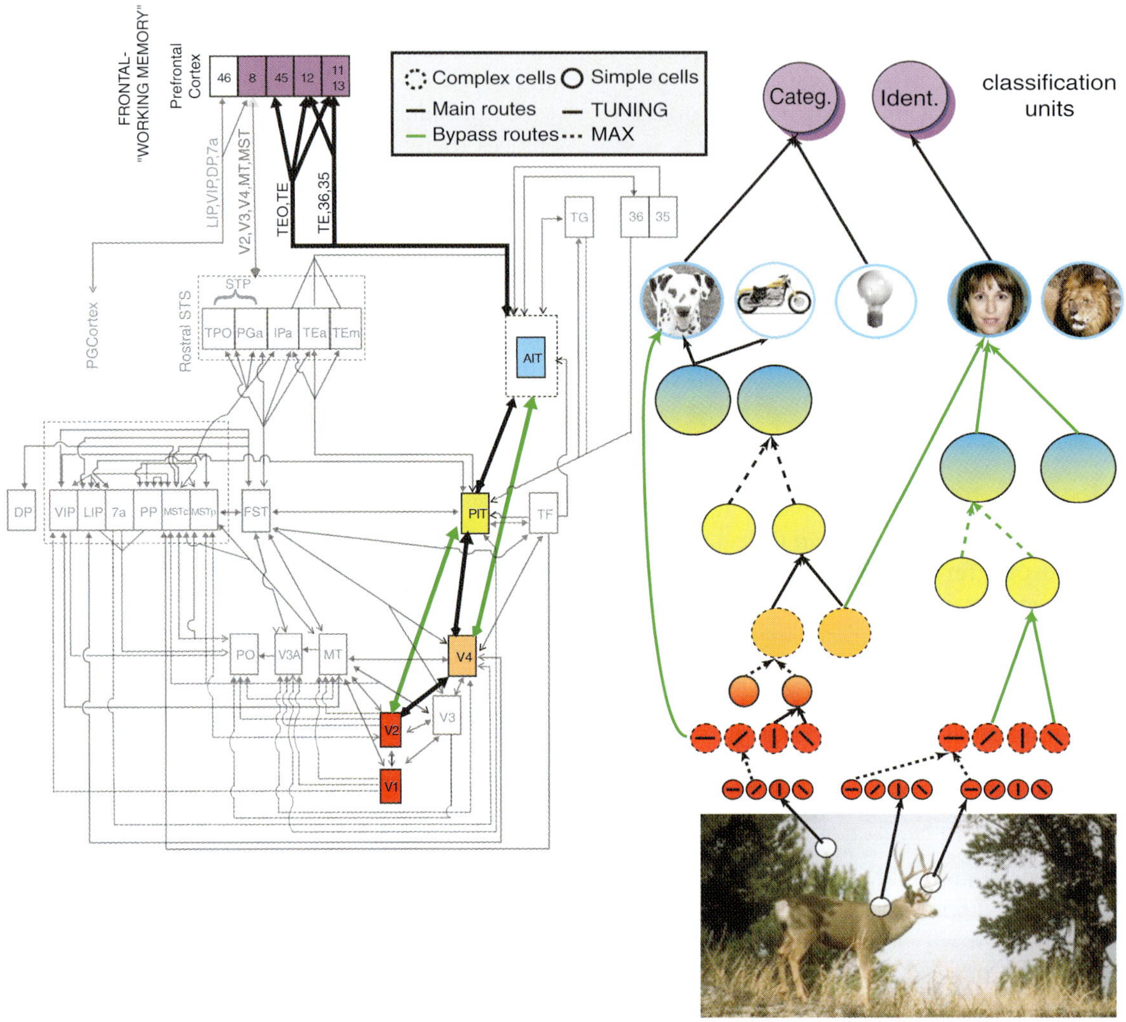

Fig. 1. Tentative mapping between structural primitives of the ventral stream in the primate visual system (modified from Gross (1998)) (left) and functional primitives of the theory. The model, which is feedforward (apart from local recurrent circuits), attempts to describe the initial stage of visual processing and immediate recognition, corresponding to the output of the top of the hierarchy and to the first 150 ms in visual recognition. Colors encode the tentative correspondences between model layers and brain areas. Stages of simple cells with Gaussian-like tuning (bold circles and arrows), which provide generalization (Poggio and Bizzi, 2004), are interleaved with layers of complex units (dashed circles and arrows), which perform a MAX-like operation on their inputs and provide invariance to position and scale (pooling over scales is not shown in the figure). Both operations may be performed by the same local recurrent circuits of lateral inhibition (see text). It is important to point out that the hierarchy is probably not as strict as depicted here. In addition there may be cells with relatively complex receptive fields already in V1. The main route from the feedforward ventral pathway is denoted with black arrows while the bypass route (Nakamura et al., 1993) is denoted with green arrows. Learning in the simple unit layers from V2/V4 up to IT (including the S_4 view-tuned units) is assumed to be stimulus-driven. It only depends on task-independent visual experience-dependent tuning of the units. Supervised learning occurs at the level of the circuits in PFC (two sets of possible circuits for two of the many different recognition tasks — identification and categorization — are indicated in the figure at the level of PFC). (Adapted with permission from Serre et al., 2007a, Fig. 1.)

In the model, the two layers alternate (see Riesenhuber and Poggio, 1999). Besides the main route that follows stages along the hierarchy of the ventral stream step-by-step, there are several routes which *bypass* some of the stages, e.g., direct projections from V2 to posterior IT (bypassing V4) and from V4 to anterior IT (bypassing posterior IT cortex). In the model, such *bypass* routes correspond, for instance, to the projections from the C_1 layer to the S_{2b} and then C_{2b} layers. Altogether the various layers in the architecture — from V1 to IT — create a large and redundant dictionary of features with different degrees of selectivity and invariance.

Although the present implementation follows the hierarchy of Fig. 1, the ventral stream's hierarchy may not be as strict. For instance there may be units with relatively complex receptive fields already in V1 (Mahon and DeValois, 2001; Victor et al., 2006). A mixture of cells with various levels of selectivity has also commonly been reported in V2, V4, and IT (Tanaka, 1996; Hegdé and van Essen, 2007). In addition, it is likely that the same stimulus-driven learning mechanisms implemented for the S_2 units and above operate also at the level of the S_1 units. This may generate S_1 units with TUNING not only for oriented bars but also for more complex patterns (e.g., corners), corresponding to the combination of LGN-like, center-surround subunits in specific geometrical arrangements. Indeed it may be advantageous for circuits in later stages (e.g., task-specific circuits in PFC) to have access not only to the highly invariant and selective units of AIT but also to less invariant and simpler units such as those in V2 and V4. Fine orientation discrimination tasks, for instance, may require information from lower levels of the hierarchy such as V1. There might also be high level recognition tasks that benefit from less invariant representations.

Learning

Unsupervised developmental-like learning from V1 to IT: Various lines of evidence suggest that visual experience, both during and after development, together with genetic factors, determine the connectivity and functional properties of cells in cortex. In this work, we assume that learning plays a key role in determining the wiring and the synaptic weights for the model units. We suggest that the TUNING properties of simple units at various levels in the hierarchy correspond to learning combinations of features that appear most frequently in natural images. This is roughly equivalent to learning a dictionary of image patterns that appear with high probability. The wiring of the S layers depends on learning correlations of features in the image that are present at the same time (i.e., for S_1 units, the bar-like arrangements of LGN inputs, for S_2 units, more complex arrangements of bar-like subunits, etc.).

The wiring of complex cells, on the other hand, may reflect learning from visual experience to associate frequent transformations in time, such as translation and scale, of specific complex features coded by simple cells. The wiring of the C layers could reflect learning correlations *across time*: e.g., at the C_1 level, learning that afferent S_1 units with the same orientation and neighboring locations should be wired together because such a pattern often changes smoothly in time (under translation) (Földiák, 1991). Thus, learning at the S and C levels involves learning correlations present in the visual world. At present it is still unclear whether these two types of learning require different types of synaptic learning rules or not.

In the present model we have only implemented learning at the level of the S units (beyond S_1). Connectivity at the C level was hardwired based on physiology data. The goal of this learning stage is to determine the selectivity of the S units, i.e., set the weight vector $\mathbf{w}$ (see Eq. (2)) of the units in layers S_2 and higher. More precisely, the goal is to define the basic types of units in each of the S layers, which constitute a dictionary of shape-components that reflect the statistics of natural images. This assumption follows the notion that the visual system, through visual experience and evolution, may be adapted to the statistics of its natural environment (Barlow, 1961). Details about the learning rule can be found in (Serre, 2006).

Supervised learning of the task-specific circuits from IT to PFC: For a given task, we assume that a particular program or routine is set up somewhere beyond IT (possibly in PFC (Freedman et al., 2002; Hung et al., 2005), but the exact locus may depend on the task). In a passive state (no specific visual task is set) there may be a default routine running (perhaps the routine: what is out there?). Here we think of a particular classification routine as a particular PFC-like unit that combines the activity of a few hundred S_4 units tuned to produce a high response to examples of the target object and low responses to distractors. While learning in the S layers is stimulus-driven, the PFC-like classification units are trained in a supervised way. The concept of a classifier that takes its inputs from a few broadly tuned example-based units is a learning scheme that is closely related to Radial Basis Function (RBF) networks (Poggio and Edelman, 1990), which are among the most powerful classifiers in terms of generalization ability. Computer simulations have shown the plausibility of this scheme for visual recognition and its quantitative consistency with many data from physiology and psychophysics (Poggio and Bizzi, 2004).

In the model, the response of a PFC-like *classification* unit with input weights $\mathbf{c} = (c_1, \ldots, c_n)$ is given by:

$$f(\mathbf{x}) = \sum_i c_i K(\mathbf{x^i}, \mathbf{x}) \qquad (3)$$

$$\text{where} \quad K(\mathbf{x^i}, \mathbf{x}) = exp\left(-\frac{1}{2\sigma^2} \sum_{j=1}^{n} (x_j^i - x_j)^2 \right)$$

$K(\mathbf{x^i},\mathbf{x})$ characterizes the activity of the ith S_4 unit, tuned to the training example $\mathbf{x^i}$, in response to the input image $\mathbf{x}$ and was obtained by replacing the weight vector $\mathbf{w}$ in Eq. (2) by the training example $\mathbf{x^i}$ (i.e., $\mathbf{w} = \mathbf{x^i}$). The superscript i indicates the index of the image in the training set and the subscript j indicates the index of the pre-synaptic unit. Supervised learning at this stage involves adjusting the synaptic weights $\mathbf{c}$ to minimize the overall classification error on the training set (see Serre, 2006).

Comparison with physiological observations

The quantitative implementation of the model, as described in the previous section, allows for direct comparisons between the responses of units in the model and electrophysiological recordings from neurons in the visual cortex. Here we illustrate this approach by directly comparing the model against recordings from the macaque monkey area V4 and IT cortex while the animal was passively viewing complex images.

Comparison of model units with physiological recordings in the ventral visual cortex

The model includes several layers that are meant to mimic visual areas V1, V2, V4, and IT cortex (Fig. 1). We directly compared the responses of the model units against electrophysiological recordings obtained throughout all these visual areas. The model is able to account for many physiological observations in early visual areas. For instance, at the level of V1, model units agree with the tuning properties of cortical cells in terms of frequency and orientation bandwidth, as well as peak frequency selectivity and receptive field sizes (see Serre and Riesenhuber, 2004). Also in V1, we observe that model units in the C_1 layer can explain responses of a subpopulation of complex cells obtained upon presenting two oriented bars within the receptive field (Lampl et al., 2004). At the level of V4, model C_2 units exhibit tuning for complex gratings (based on the recordings from Gallant et al., 1996), and curvature (based on Pasupathy and Connor, 2001), as well as interactions of multiple dots (based on Freiwald et al., 2005) or the simultaneous presentation of two-bar stimuli [based on Reynolds et al. (1999), see Serre et al. (2005) for details].

Here we focus on one comparison between C_2 units and the responses of V4 cells. Figure 2 shows the side-by-side comparison between a population of model C_2 units and V4 cell responses to the presentation of one-bar and two-bar stimuli. As in (Reynolds et al., 1999) model units were presented with either (1) a *reference* stimulus alone (an oriented bar at position 1, see Fig. 2A), (2) a *probe* stimulus alone (an oriented bar at position 2), or (3) both a reference and a probe stimulus simultaneously. We used stimuli of 16 different orientations for a total of $289 = (16+1)^2$ total stimulus

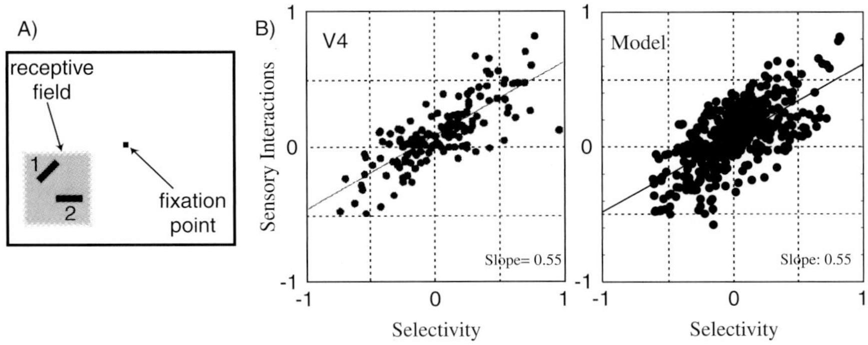

Fig. 2. A quantitative comparison between model C_2 units and V4 cells. (A) Stimulus configuration (adapted with permission from Reynolds et al., 1999, Fig. 1A): The stimulus in position 1 is denoted as the reference and the stimulus in position 2 as the probe. As in Reynolds et al. (1999) we computed a *selectivity* index (which indicates how selective a cell is to an isolated stimulus in position 1 vs. position 2 alone) and a *sensory interaction* index (which indicates how selective the cell is to the paired stimuli vs. the reference stimulus alone) (see text and Serre et al., 2005 for details). (B) Side-by-side comparison between V4 neurons (left, adapted with permission from Reynolds et al., 1999, Fig. 5) while the monkey attends away from the receptive field location and C_2 units (right). Consistent with the physiology, the addition of a second stimulus in the receptive field of the C_2 unit moves the response of the unit toward that of the second stimulus alone, i.e., the response to the clutter condition lies between the responses to the individual stimuli.

combinations for each unit [see Serre et al. (2005) for details]. Each unit's response was normalized by the maximal response of the unit across all conditions. As in Reynolds et al. (1999) we computed a *selectivity* index as the normalized response of the unit to the reference stimulus minus the normalized response of the unit to one of the probe stimuli. This index was computed for each of the probe stimuli, yielding 16 selectivity values for each model unit. This selectivity index ranges from -1 to $+1$, with negative values indicating that the reference stimulus elicited the stronger response, a value of 0 indicating identical responses to reference and probe, and positive values indicating that the probe stimulus elicited the strongest response. We also computed a *sensory interaction* index that corresponds to the normalized response to a pair of stimuli (the reference and a probe) minus the normalized response to the reference alone. The selectivity index also takes on values from -1 to $+1$. Negative values indicate that the response to the pair is smaller than the response to the reference stimulus alone (i.e., adding the probe stimulus suppresses the neuronal response). A value of 0 indicates that adding the probe stimulus has no effect on the neuron's response while positive values indicate that adding the probe increases the neuron's response.

As shown in Fig. 2B, model C_2 units and V4 cells behave very similarly to the presentation of two stimuli within their receptive field. Indeed the slope of the *selectivity* vs. *sensory interaction* indices is ~0.5 for both model units and cortical cells. That is, at the population level, presenting a preferred and a non-preferred stimulus together produces a neural response that falls between the neural responses to the two stimuli individually, sometimes close to an average.[1] We have found that such a "clutter effect" also happens higher up in the hierarchy at the level of IT (see Serre et al., 2005). Since normal vision operates with many objects appearing within the same receptive fields and embedded in complex textures (unlike the artificial experimental setups), understanding the behavior of neurons under clutter conditions is important and warrants more experiments (see later section "Performance on natural images" and section "A quantitative framework for the ventral stream").

In sum, the model can capture many aspects of the physiological responses of neurons along the

[1]We only compare the response of the model units to V4 neurons when the monkey is attending away from the receptive field location of the neuron. When the animal attends at the location of the receptive field the response to the pairs is shifted towards the response to the attended stimulus.

ventral visual stream from V1 to IT cortex (see also Serre et al., 2005).

Decoding object information from IT and model units

We recently used a linear statistical classifier to quantitatively show that we could accurately, rapidly, and robustly decode visual information about objects from the activity of small populations of neurons in anterior IT cortex (Hung et al., 2005). In collaboration with Chou Hung and James DiCarlo at MIT, we observed that a binary response from the neurons (using small bins of 12.5 ms to count spikes) was sufficient to encode information with high accuracy. The visual information, as measured by our classifiers, could in principle be decoded by the targets of IT cortex such as PFC to determine the class or identity of an object (Miller, 2000). Importantly, the population response generalized across object positions and scales. This scale and position invariance was evident even for novel objects that the animal never observed before (see also Logothetis et al., 1995). The observation that scale and position invariance occurs for novel objects strongly suggests that these two forms of invariance do not require multiple examples of each specific object. This should be contrasted with other forms of invariance, such as robustness to depth rotation, which requires multiple views in order to be able to generalize (Poggio and Edelman, 1990).

Read-out from C_{2b} units is similar to decoding from IT neurons

We examined the responses of the model units to the same set of 77 complex object images seen by the monkey. These objects were divided into eight possible categories. The model unit responses were divided into a training set and a test set. We used a one-versus-all approach, training eight binary classifiers, one for each category against the rest of the categories, and then taking the classifier prediction to be the maximum among the eight classifiers (for further details, see Hung et al., 2005; Serre et al., 2005). Similar observations were made when trying to identify each individual object by training 77 binary classifiers. For comparison, we also tried decoding object category from a random selection of model units from other layers of the model (see Fig. 1). The input to the classifier consisted of the responses of randomly selected model units and the labels of the object categories (or object identities for the identification task). Data from multiple units were concatenated assuming independence.

We observed that we could accurately read out the object category and identity from model units. In Fig. 3A, we compare the classification performance, for the categorization task described above, between the IT neurons and the C_{2b} model units. In agreement with the experimental data from IT, units from the C_{2b} stage of the model yielded a high level of performance (>70% for 100 units; where chance was 12.5%). We observed that the physiological observations were in agreement with the predictions made by the highest layers in the model (C_{2b}, S_4) but not by earlier stages (S_1 through S_2). As expected, the layers from S_1 through S_2 showed a weaker degree of scale and position invariance.

The classification performance of S_{2b} units (the input to C_{2b} units, see Fig. 1) was qualitatively close to the performance of local field potentials (LFPs) in IT cortex (Kreiman et al., 2006). The main components of LFPs are dendritic potentials and therefore LFPs are generally considered to represent the dendritic input and local processing within a cortical area (Mitzdorf, 1985; Logothetis et al., 2001). Thus, it is tempting to speculate that the S_{2b} responses in the model capture the type of information conveyed by LFPs in IT. However, care should be taken in this interpretation as the LFPs constitute an aggregate measure of the activity over many different types of neurons and large areas. Further investigation of the nature of the LFPs and their relation with the spiking responses could help unravel the transformations that take place across cortical layers.

The pattern of errors made by the classifier indicates that some groups were easier to discriminate than others. This was also evident in the correlation matrix of the population responses between all pairs of pictures (Hung et al., 2005; Serre et al., 2005). The units yielded similar responses to

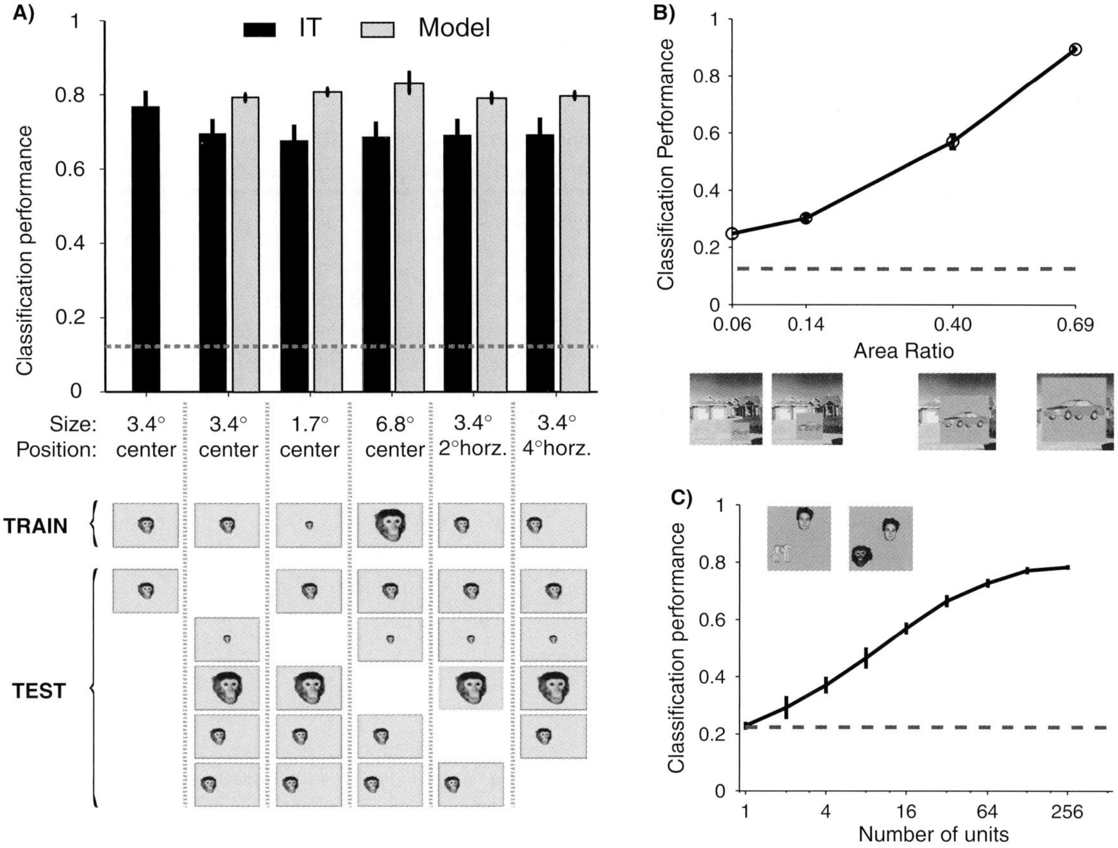

Fig. 3. (A) Classification performance based on the spiking activity from IT neurons (black) and C_{2b} units from the model (gray). The performance shown here is based on the categorization task where the classifier was trained based on the category of the object. A linear classifier was trained using the responses to the 77 objects at a single scale and position (shown for one object by "TRAIN"). The classifier performance was evaluated using shifted or scaled versions of the same 77 objects (shown for one object by "TEST"). During training, the classifier was never presented with the unit responses to the shifted or scaled objects. The left-most column shows the performance for training and testing on separate repetitions of the objects at the same standard position and scale (this is shown only for the IT neurons because there is no variability in the model which is deterministic). The second bar shows the performance after training on the standard position and scale (3.4°, center of gaze) and testing on the shifted and scaled images. The dashed horizontal line indicates chance performance (12.5%, one out of eight possible categories). Error bars show standard deviations over 20 random choices of the units used for training/testing. (B) Classification performance for reading out object category as a function of the relative size (area ratio) of object to background. Here the classifier was trained using the responses of 256 units to the objects presented in cluttered backgrounds. The classifier performance was evaluated using the same objects embedded in different backgrounds. The horizontal dashed line indicates chance performance obtained by randomly shuffling the object labels during training. (C) Classification performance for reading out object category in the presence of two objects. We exhaustively studied all possible pairs using the same 77 objects as in part A (see two examples on the upper left part of the figure). The classifier was trained with images containing two objects and the label corresponded to the category of one of them. During testing, the classifier's prediction was considered to be a hit if it correctly categorized either of the objects present in the image. The dashed line indicates chance performance obtained by randomly assigning object labels during training.

stimuli that looked alike at the pixel level. The performance of the classifier for categorization dropped significantly upon arbitrarily defining the categories as random groups of pictures.

We also tested the ability of the model to generalize to novel stimuli not included in the training set. The performance values shown in Fig. 3A are based on the responses of model units to single stimulus presentations that were not included in the classifier training and correspond to the results obtained using a linear classifier. Although the way in which the weights were learned (using a support vector machine classifier) is probably very different in biology (see Serre, 2006); once the weights are established, the linear classification boundary could very easily be implemented by neuronal hardware [see Eq. (3)]. Therefore, the recognition performance provides a lower bound to what a real downstream unit (e.g., in PFC) could, in theory, perform on a single trial given input consisting of a few spikes from the neurons in IT cortex. Overall, we observed that the population of C_{2b} model units yields a read-out performance level that is very similar to the one observed from a population of IT neurons.

Extrapolation to larger object sets

One of the remarkable aspects of primate visual recognition is the large number of different objects that can be identified. Although the exact limits are difficult to estimate, coarse estimates suggest that it is possible to visually recognize on the order of 10^4 different concepts (Biederman, 1987). The physiological recordings were necessarily limited to a small set of objects due to time constraints during a recording session. Here we show that this type of encoding can extrapolate to reading out object category in a set consisting of 787 objects divided into 20 categories (the physiological observations and the model results discussed above were based on 77 objects divided into 8 categories).

The population of C_{2b} units conveyed information that could be decoded to indicate an object's category across novel objects. The classifier was trained with objects from 20 possible categories presented at different random locations and the test set included novel objects never seen before by the classifier but belonging to the same categories. These results show that a relatively small neuronal population can in principle support object recognition over large object sets. Similar results were obtained in analogous computer vision experiments using an even larger set known as the *Caltech-101* object dataset (Serre et al., 2007b) where the model could perform object categorization among 101 categories. Other investigators have also used models that can extrapolate to large numbers of objects (Valiant, 2005) or suggested that neuronal populations in IT cortex can also extrapolate to many objects (Abbott et al., 1996; Hung et al., 2005).

The number of objects (or classes) that can be decoded at a given level of accuracy grows approximately as an exponential function of the number of units. Even allowing for a strong redundancy in the number of units coding each type of feature, these results suggest that networks of thousands of units could display a very large capacity. Of course the argument above relies on several assumptions that could well be wrong. However, at the very least, these observations suggest that there do not seem to be any obvious capacity limitations for hierarchical models to encode realistically large numbers of objects and categories.

Robustness in object recognition

Many biological sources of noise could affect the encoding of information. Among the most drastic sources of noise are synaptic failures and neuronal death. To model this, we considered the performance of the classifier after randomly deleting a substantial fraction of the units during testing. As shown for the experimental data in Hung et al. (2005), the classifier performance was very robust to this source of noise.

As discussed in the introduction, one of the main achievements of visual cortex is the balance of invariance *and* selectivity. Two particularly important forms of invariance are the robustness to changes in scale and position of the images. In order to analyze the degree of invariance to scale and position changes, we studied the responses of units at different stages of the model to scaled ($0.5\times$ and $2\times$) and translated (2° and 4°) versions of the images. The earlier stages of the model show a poor read-out performance under these transformations, but the performance of the C_{2b} stage is quite robust to these transformations as shown in Fig. 3A, in good agreement with the experimental data (Hung et al., 2005).

We also observed that the population response could extrapolate to novel objects within the same categories by training the classifier on the responses to 70% of the objects and testing its performance on the remaining 30% of the objects (Serre et al., 2005). This suggests another dimension of robustness, namely, the possibility of learning about a category from some exemplars and then extrapolating to novel objects within the same category.

The results shown above correspond to randomly selecting a given number of units to train and test the classifier. The brain could be wired in a very specific manner so that only the neurons highly specialized for a given task project to the neurons involved in decoding the information for that task. Preselecting the units (e.g., using those yielding the highest signal-to-noise ratio) yields similar results while using a significantly smaller number of units. Using a very specific set of neurons (instead of randomly pooling from the population and using more neurons for decoding) may show less robustness to neuronal death and spike failures. The bias toward using only a specific subset of neurons could be implemented through selection mechanisms including attention. For example, when searching for the car keys, the weights from some neurons could be adjusted so as to increase the signal-to-noise ratio for those keys. This may suggest that other concomitant recognition tasks would show weaker performance. In this case, the selection mechanisms take place before recognition by biasing specific populations for certain tasks.

Recognition in clutter

The decoding experiments described above as well as a large fraction of the studies reported in the literature, involve the use of well-delimited single objects on a uniform background. This is quite remote from natural vision where we typically encounter multiple objects embedded in different backgrounds, with potential occlusions, changes in illumination, etc.

Ultimately, we would like to be able to read out information from IT or from model units under natural vision scenarios in which an everyday image can be presented and we can extract from the population activity the same type and quality of information that a human observer can (in a flash). Here we show the degree of decoding robustness of objects that are embedded in complex backgrounds (see also section "Performance on natural images" describing the performance of the model in an animal vs. non-animal categorization task using objects embedded in complex backgrounds).

We presented the same 77 objects used in Fig. 3A overlayed on top of images containing complex background scenes (Fig. 3B). We did not attempt to make the resulting images realistic or meaningful in any way. While cognitive influences, memory, and expectations play a role in object recognition, these high-level effects are likely to be mediated by feedback biasing mechanisms that would indicate that a monitor is more likely to be found on an office desk than in the jungle. However, the model described here is purely feedforward and does not include any of these potential biasing mechanisms. We used four different relative sizes of object-to-background (ratio of object area to whole image area) ranging from 6% to 69%. The latter condition is very similar to the single object situation analyzed above, both perceptually and in terms of the performance of the classifier. The smaller relative size makes it difficult to detect the object at least in some cases when it is not salient (see also section "Performance on natural images").

The classifier was trained on all objects using 20% of the background scenes and performance was evaluated using the same objects presented on the remaining novel background scenes (we used a total of 98 complex background scenes with photographs of outdoor scenes). The population of C_{2b} units allowed us to perform both object recognition (Fig. 3B) and identification significantly above chance in spite of the background. Performance depended quite strongly on the relative image size (Fig. 3B). The largest size (69%) yielded results that were very close to the single isolated object results discussed above (cf. Fig. 3A). The small relative image size (6%) yielded comparatively lower results but the performance of C_{2b}

units was still significantly above chance levels both for categorization and identification.

Recognizing (and searching for) small objects embedded in a large complex scene (e.g., searching for the keys in your house), constitutes an example of a task that may require additional resources. These additional resources may involve serial attention that is likely to be dependent on feedback connections. Therefore, the model may suggest tasks and behaviors that require processes that are not predominantly feedforward.

Reading-out from images containing multiple objects

In order to further explore the mechanisms for representing information about an object's identity and category in natural scenes, we studied the ability to read out information from the model units upon presentation of more than one object. We presented two objects simultaneously in each image (Fig. 3C). During testing, the classifier was presented with images containing multiple objects. We asked two types of questions: (1) what is the most likely object in the image? and (2) what are all the objects present in the image?

Training was initially performed with single objects. Interestingly, we could also train the classifier using images containing multiple objects. In this case, for each image, the label was the identity (or category) of one of the objects (randomly chosen so that the overall training set had the same number of examples for each of the objects or object categories). This is arguably a more natural situation in which we learn about objects since we rarely see isolated objects. However, it is possible that attentional biases to some extent "isolate" an object (e.g., when learning about an object with an instructor that points to it).

In order to determine the most likely object present in the image (question 1, above), the classifier's prediction was considered to be a hit if it correctly predicted either one of the two objects presented during testing. The population of C_{2b} model units yielded very high performance reaching more than 90% both for categorization and identification with the single object training and reaching more than 80% with the multiple object training. Given that in each trial there are basically two possibilities to get a hit, the chance levels are higher than the ones reported in Fig. 3A. However, it is clear that the performance of the C_{2b} population response is significantly above chance indicating that accurate object information can be read-out even in the presence of another object. We also extended these observations to 3 objects and to 10 objects (Serre et al., 2005), obtaining qualitatively similar conclusions.

Ultimately, we would like to be able to understand an image in its entirety, including a description of all of its objects. Therefore, we asked a more difficult question by requiring the classifier to correctly predict all the objects (or all the object categories) present in the image. During perception, human observers generally assume that they can recognize and describe every object in an image during a glimpse. However, multiple psychophysics studies suggest that this is probably wrong. Perhaps one of the most striking demonstrations of this fallacy is the fact that sometimes we can be oblivious to large changes in the images (see Simons and Rensink, 2005). What is the capacity of the representation at-a-glance? There is no consensus answer to this question but some psychophysical studies suggest that only a handful of objects can be described in a brief glimpse of an image (on the order of five objects). After this first glance, eye movements and/or attentional shifts may be required to further describe an image. We continue here referring to this rapid vision scenario and we strive to explain our perceptual capabilities during the glance using the model. Thus, the goal is to be able to fully describe a set of about five objects that can be simultaneously presented in multiple backgrounds in a natural scenario.

For this purpose, we addressed our second question by taking the two most likely objects (or object categories) given by the two best classifier predictions (here the number of objects was hard-wired). A hit from the classifier output was defined as a perfect match between these predictions and the two objects present in the image. This task is much more difficult (compared to the task where the goal is to categorize or identify *any* of the objects in the image). The performance of the classifier was also much smaller than the one reported for the single-object predictions. However, performance was significantly above chance, reaching almost 40% for categorization (chance $= 0.0357$) and almost 8% for identification (chance $= 3.4 \times 10^{-4}$).

Similar results were obtained upon reading out the category or identity of all objects present in the image in the case of 3-object and 10-object images. Briefly, even in images containing 10 objects, it is possible to reliably identify one arbitrary object significantly above chance from the model units. However, the model performance in trying to describe all objects in the image drops drastically with multiple objects to very low levels for 4–5 objects.

In summary, these observations suggest that it is possible to recognize objects from the activity of small populations of IT-like model units under natural situations involving complex backgrounds and several objects. The observations also suggest that, in order to fully describe an image containing many objects, eye movements, feedback, or other additional mechanisms may be required.

Performance on natural images

For a theory of visual cortex to be successful, it should not only mimic the response properties of neurons and the behavioral response of the system to artificial stimuli like the ones typically used in physiology and psychophysics, but should also be able to perform complex categorization tasks in a real-world setting.

Comparison between the model and computer vision systems

We extensively tested the model on standard computer vision databases for comparison with several state-of-the-art AI systems (see Serre, 2006; Serre et al., 2007b, for details). Such real-world image datasets tend to be much more challenging than the typical ones used in a neuroscience lab. They usually involve different object categories and the systems that are evaluated have to cope with large variations in shape, contrast, clutter, pose, illumination, size, etc. Given the many specific biological constraints that the theory had to satisfy (e.g., using only biophysically plausible operations, receptive field sizes, range of invariances, etc.), it was not clear how well the model implementation described in section "A quantitative framework for the ventral stream" would perform in comparison to systems that have been heuristically engineered for these complex tasks.

Surprisingly we found that the model is capable of recognizing complex images (see Serre et al., 2007b). For instance, the model performs at a level comparable to some of the best existing systems on the *CalTech-101* image database of 101 object categories (Fei-Fei et al., 2004) with a recognition rate of ~55% [chance level <1%, see Serre et al. (2007b) and also the extension by Mutch and Lowe (2006); using only 30 training examples per object class].[2] Additionally, Bileschi and Wolf have developed an automated real-world Street Scene recognition system (Serre et al., 2007b) based in part on the model described in section "A quantitative framework for the ventral stream." The system is able to recognize seven different object categories (cars, bikes, pedestrians, skies, roads, buildings, and trees) from natural images of street scenes despite very large variations in shape (e.g., trees in summer and winter, SUVs as well as compact cars under any view point).

Comparison between the model and human observers

Finally, we tested whether the level of performance achieved by the model was sufficient to account for the level of performance of human observers. To test this hypothesis, in the same way as an experimental test of Newton's second law requires choosing a situation in which friction is negligible, we looked for an experimental paradigm in which recognition has to be fast and cortical back-projections are likely to be inactive. Ultra-rapid object categorization (Thorpe et al., 1996) likely depends only on feedforward processing (Thorpe

[2]These benchmark evaluations relied on an earlier partial implementation of the model which only included the bypass route from $S_1 \rightarrow C_{2b}$.

et al., 1996; Keysers et al., 2001; Thorpe and Fabre-Thorpe, 2001; Li et al., 2002; VanRullen and Koch, 2003) and thus satisfies our criterion. Here we used a backward masking paradigm (Bacon-Mace et al., 2005) in addition to the rapid stimulus presentation to try to efficiently block recurrent processing and cortical feedback loops (Enns and Di Lollo, 2000; Lamme and Roelfsema, 2000; Breitmeyer and Ogmen, 2006).

Human observers can discriminate a scene that contains a particular prominent object, such as an animal or a vehicle, after only 20 ms of exposure. Evoked response potential components related to either low-level features of the image categories (e.g., animal or vehicles) or to the image status (animal present or absent) are available at 80 and 150 ms respectively. These experimental results establish a lower bound on the latency of visual categorization decisions made by the human visual system, and suggest that categorical decisions can be implemented within a feedforward mechanism of information processing (Thorpe et al., 1996; Keysers et al., 2001; Thorpe and Fabre-Thorpe, 2001; Li et al., 2002; VanRullen and Koch, 2003).

Predicting human performance during a rapid categorization task

In collaboration with Aude Oliva at MIT, we tested human observers on a rapid animal vs. non-animal categorization task [see Serre et al. (2007a), for details]. The choice of the animal category was motivated by the fact that (1) it was used in the original paradigm by Thorpe et al. (1996) and (2) animal photos constitute a rich class of stimuli exhibiting large variations in texture, shape, size, etc. providing a difficult test for a computer vision system.

We used an image dataset that was collected by Antonio Torralba and Aude Oliva and consisted of a balanced set of 600 animal and 600 non-animal images (see Torralba and Oliva, 2003). The 600 animal images were selected from a commercially available database (Corel Photodisc) and grouped into four categories, each category corresponding to a different viewing-distance from the camera: *heads* (close-ups), *close-body* (animal body occupying the whole image), *medium-body* (animal

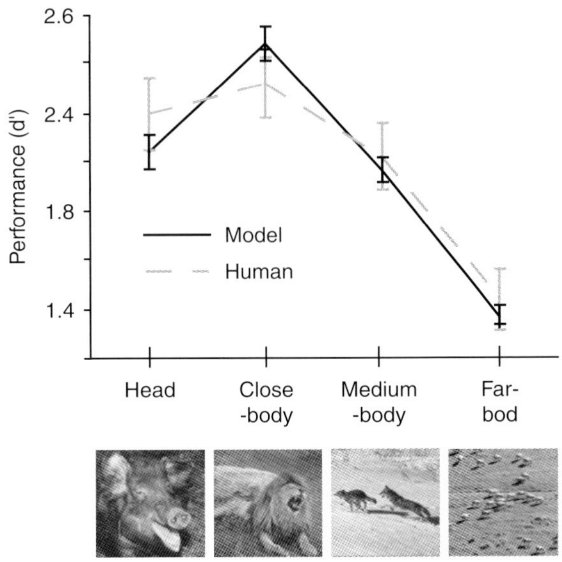

Fig. 4. Comparison between the model and human observers. Images showed either an animal embedded in a natural background or a natural scene without any animals. Images were flashed for 20 ms followed by a 30 ms blank and a 80 ms mask. Human observers or the model were queried to respond indicating whether an animal was present or not. The figure shows the accuracy as d' (the higher the value of the d', the higher the performance), for the model (black) and humans (gray) across 1200 animal and non-animal stimuli. The model is able to predict the level of performance of human observers (overall 82% for the model vs. 80% for human observers). For both the model and human observers the level of performance is highest on the close-body condition and drops gradually as the amount of clutter increases in the image from close-body to medium-body and far-body. (Adapted with permission from Serre et al., 2007a, Fig. 3A.)

in scene context), and *far-body* (small animal or groups of animals in larger context). One example from each group is shown in Fig. 4.

To make the task harder and prevent subjects from relying on low-level cues such as image-depth, the 600 distractor images were carefully selected to match each of the four viewing-distances. Distractor images were of two types (300 of each): artificial or natural scenes [see Serre et al. (2007a), for details].

During the experiment, images were briefly flashed for 20 ms, followed by an inter-stimulus interval (i.e., a blank screen) of 30 ms, followed by a mask (80 ms, 1/f noise). This is usually considered a long stimulus onset asynchrony ($SOA = 50$ ms) for which human observers are close to ceiling performance (Bacon-Mace et al., 2005). On the other hand, based on latencies in visual cortex, such an *SOA* should minimize the possibility of feedback and top-down effects in the task: we estimated from physiological data (see Serre et al., 2007a) that feedback signals from say, V4 to V1 or IT/PFC to V4, should not occur earlier than 40–60 ms after stimulus onset. Human observers ($n_h = 24$) were asked to respond as fast as they could to the presence or absence of an animal in the image by pressing either of the two keys.

Before we could evaluate the performance of the model, the task-specific circuits from IT to PFC (see section on "A quantitative framework for the ventral stream") had to be trained. These task-specific circuits correspond to a simple linear classifier that reads out the activity of a population of high level model units analogous to recordings from anterior IT cortex (see section on "Comparison with physiological observations"). The training for these task-specific circuits was done by using ($n_m = 20$) random splits of the 1200 stimuli into a training set of 600 images and a test set of 600 images. For each split, we learned the synaptic weights of the task-specific circuits of the model by minimizing the error on the training set (see Serre et al., 2007a) and evaluated the model performance on the test set. The reported performance corresponds to the average performance from the random runs.

The performance of the model and of human observers was very similar (see Fig. 4). As for the model, human observers performed best on "close-body" views and worst on "far-body" views. An intermediate level of performance was obtained for "head" and "medium-far" views. Overall no significant difference was found between the level of performance of the model and human subjects. Interestingly, the observed dependency between the level of performance and the amount of clutter in the images (which increases from the close-body to the far-body condition) for both human observers and the model seems consistent with the read-out experiment from IT neurons (for both the model and human observers) as described in section "Comparison with physiological observations."

Importantly, lower stages of the model (C_1 units) alone could not account for the results (see Serre et al., 2007a). Additionally, performing a V4 lesion in the model (i.e., leaving the bypass routes (C_{2b} units as the only source of inputs to the final classifier), see Fig. 1), also resulted in a significant loss in performance (this was true even after retraining the task-specific circuits thus accounting for a "recovery" period). This lesion experiment suggests that the large dictionary of shape-tuned units in the model (from V1 to IT) with different levels of complexity and invariance learned from natural images is the key in explaining the level of performance.

Beyond comparing levels of performance, we also performed an image-by-image comparison between the model and human observers. For this comparison, we defined an index of "animalness" for each individual image. For the model, this index was computed by calculating the percentage of times each image was classified as an animal (irrespective of its true label) for each random run ($n_m = 20$) during which it was presented as a test image. For human observers we computed the number of times each individual image was classified as an animal by each observer ($n_h = 24$). This index measures the confidence of either the model ($n_m = 20$) or human observers ($n_h = 24$) in the presence of an animal in the image. A percentage of 100% (correspondingly 0%) indicates a very high level of confidence in the presence (absence) of an animal. The level of correlation for the animalness index between the model and human observers was 0.71, 0.84, 0.71, and 0.60 for heads, close-body, medium-body, and far-body respectively ($p < 0.01$ for testing the hypothesis of no correlation against the alternative that there is a non-zero correlation). This suggests that the model and human observers tend to produce consistent responses on individual images.

Additionally, to further challenge the model, we looked at the effect of image orientation (90° and 180° in-the-plane rotation): Rousselet et al. (2003) previously suggested that the level of performance of human observers during a rapid categorization task tends to be robust to image rotation. We found that the model and human observers exhibited a similar degree of robustness (see Serre et al.,

2007a). Interestingly, the good performance of the model on rotated images was obtained without the need for retraining the model. This suggests that according to the dictionary of shape-tuned units from V1 to IT in the model (and presumably in visual cortex), an image of a rotated animal is more similar to an image of an upright animal than to distractors. In other words, a small image patch of a rotated animal is more similar to a patch of an upright animal than to a patch of image from a distractor.

Discussion: feedforward vs. feedback processing

As discussed earlier, an important assumption for the experiment described above is that with an *SOA* 50 ms, the mask leaves sufficient time to process the signal and estimate firing rates at each stage of the hierarchy (i.e., 20–50 ms, see Tovee et al., 1993; Rolls et al., 1999; Keysers et al., 2001; Thorpe and Fabre-Thorpe, 2001; Hung et al., 2005), yet selectively blocks top-down signals [e.g., from IT or PFC to V4 that we estimated to be around 40–60 ms, see Serre et al. (2007a) for a complete discussion]. The prediction is thus that the feedforward system should: (1) outperform human observers for very short *SOAs* (i.e., under 50 ms when there is not enough time to reliably perform local computations or estimate firing rates within visual areas), (2) mimic the level of performance of human observers for *SOAs* around 50 ms such that there is enough time to reliably estimate firing rates within visual areas but not enough time for back-projections from top-down to become active, and (3) underperform human observers for long *SOAs* (beyond 60 ms) such that feedbacks are active.

We thus tested the influence of the mask onset time on visual processing with four experimental conditions, i.e., when the mask followed the target image (a) without any delay (with an *SOA* of 20 ms), (b) with an *SOA* of 50 ms (corresponding to an inter-stimulus interval of 30 ms), (c) with an *SOAs* of 80 ms, or (d) never ("no-mask" condition). For all four conditions, the target presentation was fixed to 20 ms as before. As expected, the delay between the stimulus and the mask onset modulated the level of performance of the observers, improving gradually from the 20 ms *SOA* condition to the no-mask condition. The performance of the model was superior to the performance of human observers for the *SOA* of 20 ms. The model closely mimicked the level of performance of human observers for the 50 ms condition (see Fig. 4). The implication would be that, under these conditions, the present feedforward version of the model already provides a satisfactory description of information processing in the ventral stream of visual cortex. Human observers however outperformed the model for the 80 ms *SOA* and the no-mask condition.

Discussion

General remarks about the theory

We have developed a quantitative model of the feedforward pathway of the ventral stream in visual cortex — from cortical area V1 to V2 to V4 to IT and PFC — that captures its ability to learn visual tasks, such as identification and categorization of objects from images. The quantitative nature of the model has allowed us to directly compare its performance against experimental observations at different scales and also against current computer vision algorithms. In this paper we have focused our discussion on how the model can explain experimental results from visual object recognition within short times at two very different levels of analysis: human psychophysics and physiological recordings in IT cortex. The model certainly does not account for all possible aspects of visual perception or illusions (see also extensions, predictions, and future directions below). However, the success of the model in explaining experimental data across multiple scales and making quantitative predictions strongly suggests that the theory provides an important framework for the investigation of the feedforward path in visual cortex and the processes involved in immediate recognition.

An important component of a theory is that it should be falsifiable. In that spirit, we list some key experiments and findings here that could refute the present framework. First, a strong dissociation between experimental observations and model predictions would suggest that revisions need to be made to the model (e.g., psychophysical or physiological observations that cannot be explained or contradict predictions made by the model). Second, as stated in the introduction, the present framework relies entirely on a feedforward architecture from V1 to IT and PFC. Any evidence that feedback plays a key role *during the early stages* of immediate recognition should be considered as hard evidence suggesting that important revisions would need to be made in the main architecture of the model (Fig. 1).

A wish-list of experiments

Here we discuss some predictions from the theory and an accompanying "wish list" of experiments that could be done to test, refute, or validate those predictions. We try to focus on what we naively think are feasible experiments.

1. The distinction between simple and complex cells has been made only in primary visual cortex. Our theory and parsimony considerations suggest that a similar circuit is repeated throughout visual cortex. Therefore, *unbiased* recordings from neurons in higher visual areas may reveal the existence of two classes of neurons which could be distinguished by their degree of invariance to image transformations.
2. As the examples discussed in this manuscript illustrate, our theory can make quantitative predictions about the limits of immediate recognition at the behavioral level (section on "Performance on natural images") and also at the neuronal level (section on "Comparison with physiological observations"). The biggest challenges to recognition include conditions in which the objects are small relative to the whole image and the presence of multiple objects, background, or clutter. It would be interesting to compare these predictions to behavioral and physiological measurements. This could be achieved by adding extra conditions in the psychophysical experiment of section on "Performance on natural images" and by extending the read-out experiments from section "Comparison with physiological observations" to natural images and more complex recognition scenarios.
3. The theory suggests that immediate recognition may rely on a large dictionary of shape-components (i.e., common image-features) with different levels of complexity and invariance. This fits well with the concept of "unbound features" (Treisman and Gelade, 1980; Wolfe and Bennett, 1997) postulated by cognitive theories of pre-attentive vision. Importantly, the theory does not rely on any figure-ground segregation. This suggests that, at least for immediate recognition, recognition can work without an intermediate segmentation step. Furthermore, it also suggests that it is not necessary to define *objects* as fundamental units in visual recognition.
4. There is no specific computational role for a functional topography of units in the model. Thus, the strong degree of topography present throughout cortex, may arise from developmental reasons and physical constraints (a given axon may be more likely to target two adjacent neurons than two neurons that are far away; also, there may be a strong pressure to minimize wiring) as opposed to having a specific role in object recognition or the computations made in cortex.
5. The response of a given simple unit in the model can be described by Eq. (2). Thus, there are multiple *different* inputs that could activate a particular unit. This may explain the somewhat puzzling observations of why physiologists often find neurons that seem to respond to apparently dissimilar objects. Following this reasoning, it should be possible to generate an iso-response stimulus set, i.e., a series of stimuli that should elicit similar responses in a given unit even when the stimuli apparently look different or the shape of the iso-response stimulus set appear non-intuitive.
6. It is tempting to anthropomorphize the responses of units and neurons. This has been carried as far as to speak of a neuron's "preferences." The current theory suggests that an

input that gives rise to a high response from a neuron is at the same time simpler and more complex than this anthropomorphized account. It is simpler because it can be rigorously approximated by specific simple equations that control its output. It is more complex because these weight vectors and equations are not easily mapped to words such as "face neuron," "curvature," etc., and taken with the previous point, that visually dissimilar stimuli can give rise to similar responses, the attribution of a descriptive word may not be unique.

7. There are many tasks that may not require back-projections. The performance of the model may provide a reliable signature of whether a task can be accomplished during immediate recognition in the absence of feedback (e.g., the model performs well for immediate recognition of single objects on uncluttered backgrounds, but fails for attention-demanding tasks Li et al., 2002). As stated above, one of the main assumptions of the current model is the feedforward architecture. This suggests that the model may not perform well in situations that require multiple fixations, eye movements, and feedback mechanisms. Recent psychophysical work suggests that performance on dual tasks can provide a diagnostic tool for characterizing tasks that do or do not involve attention (Li et al., 2002). Can the model perform these dual tasks when psychophysics suggests that attention is or is not required? Are back-projections and feedback required?

In addition to the predictions listed above, we recently discussed other experiments and predictions that are based on a more detailed discussion of the biophysical circuits implementing the main operations in the model (see Serre et al., 2005).

Future directions

We end this article by reflecting on several of the open questions, unexplained phenomena, and missing components of the theory. Before we begin, we should note that visual recognition encompasses much more than what has been attempted and achieved with the current theory. A simple example may illustrate this point. In the animal categorization task discussed in the previous sections, humans make mistakes upon being pressed to respond promptly. Given 10 s and no mask, performance would be basically 100%. As stated several times, the goal here is to provide a framework to quantitatively think about the initial steps in vision, but it is clear that much remains to be understood beyond immediate recognition.

Open questions

How strict is the hierarchy and how precisely does it map into cells of different visual areas? For instance, are cells corresponding to S_2 units in V2 and C_2 units in V4 or are some cells corresponding to S_2 units already in V1? The theory is rather open about these possibilities: the mapping of Fig. 1 is just an educated guess. However, because of the increasing arborization of cells and the number of boutons from V1 to PFC (Elston, 2003), the number of subunits to the cells should increase and thus their potential size and complexity. In addition, C units should show more invariance from the bottom to the top of the hierarchy.

What is the nature of the cortical and subcortical connections (both feedforward and feedback) to and from the main areas of the ventral visual stream that are involved in the model? A more thorough characterization at the anatomical level of the circuits in visual cortex would lead to a more realistic architecture of the model by better constraining some of the parameters such as the size of the dictionary of shape-components or the number of inputs to units in different layers. This would also help refine and extend the existing literature on the organization of visual cortex (Felleman and van Essen, 1991). With the recent development of higher resolution tracers (e.g., PHA-L, biocytin, DBA), visualization has greatly improved and it is now possible to go beyond a general layout of interconnected structures and start addressing the finer organization of connections.

What are the precise biophysical mechanisms for the learning rule described in section "A quantitative framework for the ventral stream" and how can invariances be learned within the same framework? Possible synaptic mechanisms for learning should be described in biophysical detail. As suggested earlier, synaptic learning rules should allow for three types of learning: (1) the TUNING of the units at the S level by detecting correlations among subunits at the same time, (2) the invariance to position and scale at the C level by detecting correlations among subunits across time, and (3) the training of task-specific circuits (probably from IT to PFC) in a supervised fashion.

Is learning in areas below IT purely unsupervised and developmental-like as assumed in the theory? Or is there task- and/or object-specific learning in adults occurring below IT in V4, V2, or even V1?

Have we reached the limit of what feedforward architectures can achieve in terms of performance? In other words, is the somewhat better performance of humans on the animal vs. non-animal categorization task (see section on "Comparison between the model and human observers") over the model for *SOAs* longer than 80 ms due to feedback effects mediated by back-projections or can the model be improved to attain human performance in the absence of a mask? There could be several directions to follow in order to try to improve the model performance. One possibility would involve experimenting with the size of the dictionary of shape-components (that could be further reduced with feature selection techniques for instance). Another possibility would involve adding intermediate layers to the existing ones.

Are feedback loops always desirable? Is the performance on a specific task guaranteed to always increase when subjects are given more time? Or are there tasks for which blocking the effect of back-projections with rapid masked visual presentation increases the level of performance compared to longer presentation times?

Future extensions

Learning the tuning of the S_1 units: In the present implementation of the model the tuning of the simple cells in V1 is hardwired. It is likely that it could be determined through the same passive learning mechanisms postulated for the S_2, S_{2b}, and S_3 units (in V4 and PIT respectively), possibly with a slower time scale and constrained to LGN center-surround subunits. We would expect the automatic learning from natural images mostly of oriented receptive fields but also of more complex ones, including end-stopping units [as reported for instance in DeAngelis et al. (1992) in layer 6 of V1].

Dynamics of neuronal responses: The current implementation is completely static, for a given static image the model produces a single response in each unit. This clearly does not account for the intricate dynamics present in the brain and also precludes us from asking several questions about the encoding of visual information, learning, the relative timing across areas, etc. Perhaps the easiest way to solve this is by using simple single neuron models (such as an integrate-and-fire neuron) for the units in the model. This question is clearly related to the biophysics of the circuitry, i.e., what type of biological architectures and mechanisms can give rise to the global operations used by the model. A dynamical model would allow us to more realistically compare to experimental data. For example, the experiments described in section "Performance on natural images" compare the results in a categorization task between the model and human subjects. In the human psychophysics, the stimuli were masked briefly after stimulus presentation. A dynamical model would allow us to investigate the role and mechanisms responsible for masking. A dynamical model may also allow investigation of time-dependent phenomena as well as learning based on correlations across time.

Extensions of the model to other visual inputs: There are many aspects of vision that are not currently implemented in the model. These include color, stereo, motion, and time-varying stimuli. Initial work has been done to extend the model to the visual recognition of action and motions (Giese and Poggio, 2003; Sigala et al., 2005).

Color mechanisms from V1 to IT should be included. The present implementation only deals with gray level images (it has been shown that the addition of color information in rapid categorization tasks only leads to a mild increase in performance, see Delorme et al., 2000). More complex phenomena involving color such as color constancy and the influence of the background and integration in color perception should ultimately be explained.

Stereo mechanisms from V1 to IT should also be included. Stereo (in addition to motion) is likely to play an important role in the learning of invariances such as position and size invariance via a correlation-based rule such as the trace rule (Földiák, 1991).

Extensions of the anatomy of the model: Even staying within the feedforward skeleton outlined here, there are many connections that are known to exist in the brain that are not accounted for in the current model. The goal of the model is to extract the basic principles used in recognition and not to copy, neuron by neuron, the entire brain. However, certain connectivity patterns may have important computational consequences. For example, there are horizontal connections in the cortex that may be important in modulating and integrating information across areas beyond the receptive field.

Beyond a feedforward model: It has been known for many decades now that there are abundant back-projections in the brain. In the visual system, every area projects back to its input area (with the exception of the lateral geniculate nucleus in the thalamus that does not project back to the retina). Some of these connections (e.g., from V2 to V1), may play a role even during immediate recognition. However, a central assumption of the current model is that long-range backprojections (e.g., from area IT to V1) do not play a role during the first 100–150 ms of vision. Given enough time, humans make eye movements to scan an image and performance in many object recognition tasks can increase significantly over that obtained during fast presentation.

Visual illusions: A variety of visual illusions show striking effects that are often counterintuitive and require an explanation in terms of the neuronal circuits. While in some cases specific models have been proposed to explain one phenomenon or another, it would be interesting to explore how well the model (and thus feedforward vision) can account for those observations. A few simple examples include illusory contours (such as the Kanizsa triangle), long-range integration effects (such as the Cornsweet illusion), etc. More generally, it is likely that early Gestalt-like mechanisms — for detecting collinearity, symmetry, parallelism, etc. — exist in V1 or V2 or V4. They are not present in this version of the model. It is an open and interesting question how they could be added to it in a plausible way.

Acknowledgments

We would like to acknowledge Chou Hung and James DiCarlo for their contributions in the physiology experiments described in section "Comparison with physiological observations." We would like to acknowledge Aude Oliva for her contribution to the psychophysics experiments described in section "Performance on natural images." We thank Timothee Masquellier for useful comments on this manuscript. This research was sponsored by grants from: Office of Naval Research (DARPA) under Contract No. N00014-00-1-0907, McGovern fellowship (GK), National Science Foundation (ITR) under Contract No. IIS-0085836, National Science Foundation (KDI) under Contract No. DMS-9872936, and National Science Foundation under Contract No. IIS-9800032.

References

Abbott, L.F., Rolls, E.T. and Tovee, M.T. (1996) Representational capacity of face coding in monkeys. Cereb. Cortex, 6: 498–505.

Amit, Y. and Mascaro, M. (2003) An integrated network for invariant visual detection and recognition. Vision Res., 43(19): 2073–2088.

Bacon-Mace, N., Mace, M.J., Fabre-Thorpe, M. and Thorpe, S.J. (2005) The time course of visual processing: backward masking and natural scene categorisation. Vision Res., 45: 1459–1469.

Barlow, H.B. (1961) Possible principles underlying the transformation of sensory messages. In: Rosenblith W.A. (Ed.), Sensory Communication. MIT Press, Cambridge, MA, pp. 217–234.

Biederman, I. (1987) Recognition-by-components: a theory of human image understanding. Psychol. Rev., 94: 115–147.

Booth, M.C. and Rolls, E.T. (1998) View-invariant representations of familiar objects by neurons in the inferior temporal visual cortex. Cereb. Cortex, 8: 510–523.
Breitmeyer, B. and Ogmen, H. (2006) Visual Masking: Time Slices through Conscious and Unconscious Vision. Oxford University Press, UK.
DeAngelis, G.C., Robson, J.G., Ohzawa, I. and Freeman, R.D. (1992) Organization of suppression in receptive fields of neurons in cat visual cortex. J. Neurophysiol., 68(1): 144–163.
Delorme, A., Richard, G. and Fabre-Thorpe, M. (2000) Ultra-rapid categorisation of natural images does not rely on colour: a study in monkeys and humans. Vision Res., 40: 2187–2200.
Desimone, R., Albright, T.D., Gross, C.G. and Bruce, C. (1984) Stimulus-selective properties of inferior temporal neurons in the macaque. J. Neurosci., 4(8): 2051–2062.
Elston, G.N. (2003) Comparative studies of pyramidal neurons in visual cortex of monkeys. In: Kaas J.H. and Collins C. (Eds.), The Primate Visual System. CRC Press, Boca Raton, FL, pp. 365–385.
Enns, J.T. and Di Lollo, V. (2000) What's new in masking? Trends Cogn. Sci., 4(9): 345–351.
Fei-Fei, L., Fergus, R. and Perona, P. (2004) Learning generative visual models from few training examples: an incremental Bayesian approach tested on 101 object categories. In: Proc. IEEE CVPR, Workshop on generative-model based vision, Washington, DC.
Felleman, D.J. and van Essen, D.C. (1991) Distributed hierarchical processing in the primate cerebral cortex. Cereb. Cortex, 1: 1–47.
Földiák, P. (1991) Learning invariance from transformation sequences. Neural Comput., 3: 194–200.
Freedman, D.J., Riesenhuber, M., Poggio, T. and Miller, E.K. (2002) Visual categorization and the primate prefrontal cortex: neurophysiology and behavior. J. Neurophysiol., 88: 930–942.
Freiwald, W.A., Tsao, D.Y., Tootell, R.B.H. and Livingstone, M.S. (2005) Complex and dynamic receptive field structure in macaque cortical area V4d. J. Vis., 4(8): 184a.
Fukushima, K. (1980) Neocognitron: a self-organizing neural network model for a mechanism of pattern recognition unaffected by shift in position. Biol. Cybern., 36: 193–202.
Gallant, J.L., Connor, C.E., Rakshit, S., Lewis, J.W. and Van Essen, D.C. (1996) Neural responses to polar, hyperbolic, and Cartesian gratings in area V4 of the macaque monkey. J. Neurophysiol., 76: 2718–2739.
Gawne, T.J. and Martin, J.M. (2002) Responses of primate visual cortical V4 neurons to simultaneously presented stimuli. J. Neurophysiol., 88: 1128–1135.
Giese, M. and Poggio, T. (2003) Neural mechanisms for the recognition of biological movements and action. Nat. Rev. Neurosci., 4: 179–192.
Gross, C.G. (1998) Brain Vision and Memory: Tales in the History of Neuroscience. MIT Press, Cambridge, MA.
Gross, C.G., Rocha-Miranda, C.E. and Bender, D.B. (1972) Visual properties of neurons in inferotemporal cortex of the macaque. J. Neurophysiol., 35: 96–111.
Hegdé, J. and van Essen, D.C. (2007) A comparative study of shape representation in macaque visual areas V2 and V4. Cereb. Cortex., 17: 1100–1116.
Hietanen, J.K., Perrett, D.I., Oram, M.W., Benson, P.J. and Dittrich, W.H. (1992) The effects of lighting conditions on responses of cells selective for face views in the macaque temporal cortex. Exp. Brain Res., 89: 157–171.
Hubel, D.H. and Wiesel, T.N. (1962) Receptive fields, binocular interaction and functional architecture in the cat's visual cortex. J. Physiol., 160: 106–154.
Hung, C., Kreiman, G., Poggio, T. and DiCarlo, J. (2005) Fast read-out of object identity from macaque inferior temporal cortex. Science, 310: 863–866.
Keysers, C., Xiao, D.K., Földiák, P. and Perrett, D.I. (2001) The speed of sight. J. Cogn. Neurosci., 13: 90–101.
Kobatake, E. and Tanaka, K. (1994) Neuronal selectivities to complex object features in the ventral visual pathway of the macaque cerebral cortex. J. Neurophysiol., 71: 856–867.
Kreiman, G., Hung, C., Poggio, T. and DiCarlo, J. (2006) Object selectivity of local field potentials and spikes in the inferior temporal cortex of macaque monkeys. Neuron, 49: 433–445.
Lamme, V.A.F. and Roelfsema, P.R. (2000) The disctinct modes of vision offered by feedforward and recurrent processing. Trends Neurosci., 23: 571–579.
Lampl, I., Ferster, D., Poggio, T. and Riesenhuber, M. (2004) Intracellular measurements of spatial integration and the MAX operation in complex cells of the cat primary visual cortex. J. Neurophysiol., 92: 2704–2713.
LeCun, Y., Bottou, L., Bengio, Y. and Haffner, P. (1998) Gradient-based learning applied to document recognition. Proc. IEEE, 86(11): 2278–2324.
Li, F.F., VanRullen, R., Koch, C. and Perona, P. (2002) Rapid natural scene categorization in the near absence of attention. Proc. Natl. Acad. Sci. U.S.A., 99: 9596–9601.
Logothetis, N.K., Pauls, J., Augath, M., Trinath, T. and Oeltermann, A. (2001) Neurophysiological investigation of the basis of the fMRI signal. Nature, 412: 150–157.
Logothetis, N.K., Pauls, J. and Poggio, T. (1995) Shape representation in the inferior temporal cortex of monkeys. Curr. Biol., 5: 552–563.
Logothetis, N.K. and Sheinberg, D.L. (1996) Visual object recognition. Ann. Rev. Neurosci., 19: 577–621.
Mahon, L.E. and DeValois, R.L. (2001) Cartesian and non-Cartesian responses in LGN, V1, and V2 cells. Vis. Neurosci., 18: 973–981.
Mel, B.W. (1997) SEEMORE: combining color, shape, and texture histogramming in a neurally inspired approach to visual object recognition. Neural Comput., 9: 777–804.
Miller, E.K. (2000) The prefrontal cortex and cognitive control. Nat. Rev. Neurosci., 1: 59–65.
Mitzdorf, U. (1985) Current source-density method and application in cat cerebral cortex: investigation of evoked potentials and EEG phenomena. Physiol. Rev., 65: 37–99.
Mutch, J. and Lowe, D. (2006) Multiclass object recognition with sparse, localized features. In: Proc. IEEE Conf. Comput. Vis. Pattern Recognit., New York, NY.

Nakamura, H., Gattass, R., Desimone, R. and Ungerleider, L.G. (1993) The modular organization of projections from areas V1 and V2 to areas V4 and TEO in macaques. J. Neurosci., 13(9): 3681–3691.

Olshausen, B.A., Anderson, C.H. and Van Essen, D.C. (1993) A neurobiological model of visual attention and invariant pattern recognition based on dynamic routing of information. J. Neurosci., 13(11): 4700–4719.

Pasupathy, A. and Connor, C.E. (2001) Shape representation in area V4: position-specific tuning for boundary conformation. J. Neurophysiol., 86(5): 2505–2519.

Perrett, D.I., Hietanen, J.K., Oram, M.W. and Benson, P.J. (1992) Organization and functions of cells responsive to faces in the temporal cortex. Philos. Trans. R. Soc. B, 335: 23–30.

Perrett, D.I. and Oram, M. (1993) Neurophysiology of shape processing. Image Vis. Comput., 11: 317–333.

Poggio, T. and Bizzi, E. (2004) Generalization in vision and motor control. Nature, 431: 768–774.

Poggio, T. and Edelman, S. (1990) A network that learns to recognize 3D objects. Nature, 343: 263–266.

Potter, M.C. (1975) Meaning in visual search. Science, 187: 565–566.

Reynolds, J.H., Chelazzi, L. and Desimone, R. (1999) Competitive mechanisms subserve attention in macaque areas V2 and V4. J. Neurosci., 19: 1736–1753.

Riesenhuber, M. and Poggio, T. (1999) Hierarchical models of object recognition in cortex. Nat. Neurosci., 2: 1019–1025.

Rolls, E.T., Tovee, M.J. and Panzeri, S. (1999) The neurophysiology of backward visual masking: information analysis. J. Comp. Neurol., 11: 300–311.

Rousselet, G.A., Mace, M.J. and Fabre-Thorpe, M. (2003) Is it an animal? Is it a human face? Fast processing in upright and inverted natural scenes. J. Vis., 3: 440–455.

Sato, T. (1989) Interactions of visual stimuli in the receptive fields of inferior temporal neurons in awake macaques. Exp. Brain Res., 74(2): 263–271.

Schwartz, E.L., Desimone, R., Albright, T.D. and Gross, C.G. (1983) Shape recognition and inferior temporal neurons. Proc. Natl. Acad. Sci. U.S.A., 80(18): 5776–5778.

Serre, T. (2006) Learning a dictionary of shape-components in visual cortex: comparison with neurons, humans and machines. PhD thesis, Massachusetts Institute of Technology, Cambridge, MA, April 2006.

Serre, T., Kouh, M., Cadieu, C., Knoblich, U., Kreiman, G. and Poggio, T. (2005) A theory of object recognition: computations and circuits in the feedforward path of the ventral stream in primate visual cortex. AI Memo 2005-036/CBCL Memo 259, MIT, Cambridge, MA.

Serre, T., Oliva, A. and Poggio, T. (2007a) A feedforward architecture accounts for rapid categorization. Proc. Natl. Acad. Sci., 104(15): 6424–6429.

Serre, T. and Riesenhuber, M. (2004) Realistic modeling of simple and complex cell tuning in the HMAX model, and implications for invariant object recognition in cortex. AI Memo 2004-017/CBCL Memo 239, MIT, Cambridge, MA.

Serre, T., Wolf, L., Bileschi, S., Riesenhuber, M. and Poggio, T. (2007b) Robust object recognition with cortex-like mechanisms. IEEE Trans. Pattern Anal. Machine Intell., 29(3): 411–426.

Sigala, R., Serre, T., Poggio, T. and Giese, M. (2005) Learning features of intermediate complexity for the recognition of biological motion. In: Proc. Int. Conf. Artif. Neural Netw., Warsaw, Poland.

Simons, D.J. and Rensink, R.A. (2005) Change blindness: past, present and future. Trends Cogn. Sci., 9(1): 16–20.

Tanaka, K. (1996) Inferotemporal cortex and object vision. Ann. Rev. Neurosci., 19: 109–139.

Thorpe, S.J. (2002) Ultra-rapid scene categorisation with a wave of spikes. In: Proc. Biologically Motivated Comput. Vis., Tubingen, Germany.

Thorpe, S.J. and Fabre-Thorpe, M. (2001) Seeking categories in the brain. Science, 291: 260–263.

Thorpe, S.J., Fize, D. and Marlot, C. (1996) Speed of processing in the human visual system. Nature, 381: 520–522.

Torralba, A. and Oliva, A. (2003) Statistics of natural image categories. Netw Comput. Neural Syst., 14: 391–412.

Tovee, M.J., Rolls, E.T., Treves, A. and Bellis, R.P. (1993) Information encoding and the response of single neurons in the primate temporal visual cortex. J. Neurophysiol., 70: 640–654.

Treisman, A.M. and Gelade, G. (1980) A feature-integration theory of attention. Cogn. Psychol., 12: 97–136.

Ullman, S., Vidal-Naquet, M. and Sali, E. (2002) Visual features of intermdediate complexity and their use in classification. Nat. Neurosci., 5(7): 682–687.

Ungerleider, L.G. and Haxby, J.V. (1994) "What" and "where" in the human brain. Curr. Opin. Neurobiol., 4: 157–165.

Valiant, L.G. (2005) Memorization and association on a realistic neural model. Neural Comput., 17: 527–555.

VanRullen, R. and Koch, C. (2003) Visual selective behavior can be triggered by a feed-forward process. J. Comp. Neurol., 15: 209–217.

Victor, J.D., Mechler, F., Repucci, M.A., Purpura, K.P. and Sharpee, T. (2006) Responses of V1 neurons to two-dimensional hermite functions. J. Neurophysiol., 95: 379–400.

Wallis, G. and Rolls, E.T. (1997) A model of invariant object recognition in the visual system. Prog. Neurobiol., 51: 167–194.

Wersing, H. and Koerner, E. (2003) Learning optimized features for hierarchical models of invariant recognition. Neural Comput., 15(7): 1559–1588.

Wolfe, J.M. and Bennett, S.C. (1997) Preattentive object files: shapeless bundles of basic features. Vision Res., 37: 25–44.

P. Cisek, T. Drew & J.F. Kalaska (Eds.)
Progress in Brain Research, Vol. 165
ISSN 0079-6123
Copyright © 2007 Elsevier B.V. All rights reserved

CHAPTER 5

Attention in hierarchical models of object recognition

Dirk B. Walther[1,*] and Christof Koch[2]

[1]*Beckman Institute for Advanced Science and Technology, University of Illinois at Urbana-Champaign, 405 N. Mathews Ave., Urbana, IL 61801, USA*
[2]*Division of Biology, California Institute of Technology, MC 216-76, Pasadena, CA 91125, USA*

Abstract: Object recognition and visual attention are tightly linked processes in human perception. Over the last three decades, many models have been suggested to explain these two processes and their interactions, and in some cases these models appear to contradict each other. We suggest a unifying framework for object recognition and attention and review the existing modeling literature in this context. Furthermore, we demonstrate a proof-of-concept implementation for sharing complex features between recognition and attention as a mode of top-down attention to particular objects or object categories.

Keywords: attention; object recognition; modeling; object-based attention; bottom-up attention; top-down attention

> *"At first he'd most easily make out the shadows; and after that the phantoms of the human beings and the other things in water; and, later, the things themselves."* — Socrates describing the visual experience of a man exposed to the richness of the visual world outside his cave for the first time (Plato, The Republic).

Introduction

Vision is the sense we humans rely on most for our everyday activities. But what does it mean to see? When light reflected by an object hits the photoreceptors of the retina, electrical impulses are created by retinal ganglion cells and sent out to other parts of the brain. How do these electrical impulses give rise to the percepts of the visual world surrounding us?

Two important parts of visual perception are the recognition of objects and the gating of visual information by attention. By object recognition we mean our visual system's ability to infer the presence of a particular object or member of an object category from the retinal image. By attention we mean the process of selecting and gating visual information based on saliency in the image itself (bottom-up), and on prior knowledge about scenes or a particular task (top-down) (Desimone and Duncan, 1995; Itti and Koch, 2001).

The algorithms required for object recognition are governed by the complementary forces of *specificity* and *invariance*. The activation level of a particular cone in the retina has a very *specific* spatial location, but it is likely to have the exact same activation level for a wide range of objects. Activation of object-selective cells in inferior temporal (IT) cortex, on the other hand, *specifically* indicates the presence of a particular object (or a member of a particular object category). To a large extent, this representation is *invariant* to where the object is located (at least in areas near the fovea),

*Corresponding author. Tel.: +1-217-333-9961; Fax: +1-217-333-2922; E-mail: walther@uiuc.edu

DOI: 10.1016/S0079-6123(06)65005-X

which way it is oriented, if it appears to be large or small, or whether it is brightly illuminated or in the shadow.

The majority of models of object recognition have at their heart a hierarchy of processing steps that more or less gradually increase both *specificity* to the structure of the stimulus and *invariance* to translation, rotation, size, and illumination. There is less agreement about the details of these steps, the tuning of units at intermediate levels, and the representation of their spatial relations.

Gating by attention may occur at any level of processing. Typically, attention is modeled as the preferred processing of some visual information selected by spatial location and/or the encoded feature(s). We analyze the various modes of attention in more detail later on.

Virtually all models of object recognition in cortex start with filtering the incoming image with orientation-sensitive filters. They approximate the receptive fields of the simple and complex cells found by Hubel and Wiesel (1962) in cat striate cortex with Gabor filters, steerable filters, or other orientation-tuned filters.

Details of the subsequent steps of processing are much more contentious. One of the controversies is about whether the three-dimensional structure of objects is represented by an explicit description of its components and their spatial relations, or whether it is inferred by interpolation between several two-dimensional views. We will briefly review both approaches in the next section and then show how they can be described by a unifying framework in "A unifying framework for attention and object recognition (UNI)". This framework will also allow us to explain roles of attention in object recognition.

Hierarchical models of object recognition

The two main ideas for implementing recognition of three-dimensional objects are recognition by components and view-based recognition. In this section we survey models for both approaches and highlight their differences and the common elements.

Recognition by components

Marr postulated a primal sketch for inferring the presence of surfaces, which are then combined into a relief-like $2\frac{1}{2}$ d sketch and eventually into 3d models of objects, which can be indexed and later recalled and referenced (Marr and Nishihara, 1978; Marr, 1982). Building on Marr's work, Biederman (1987) suggested that the next processing step should be fitting simple 3d shapes (generalized cones and cylinders termed "geons") to the surfaces, and that spatial relationships between geons are encoded explicitly. Crucial to the correct recognition of 3d objects are non-accidental properties such as T-junctions and line intersections.

Experimental evidence for this recognition-by-components (RBC) model comes from a study demonstrating successful priming for object identity by degraded line drawings, even when the priming and the primed stimulus have no lines in common (Biederman and Cooper, 1991). A recent study supports these results by finding stronger fMRI adaptation to line drawings with local features deleted than for line drawings with entire components (geons) removed (Hayworth and Biederman, 2006). Demonstration of a computational implementation of the model, however, was limited to carefully selected line drawings (Hummel and Biederman, 1992). In fact, the biggest criticism of RBC points at the difficulty of fitting geons with a potentially large variety of parameters to images of natural objects (e.g., Edelman, 1997).

View-based recognition in HMAX

An alternative approach to recognition by components is the recognition of 3d objects by interpolating between 2d views of the objects at various rotations. Almost all models of view-based recognition trace their origins to the "Neocognitron", a hierarchical network developed by Fukushima (1980). The Neocognitron consists of alternating S and C layers, in an allusion to the simple and complex cells found by Hubel and Wiesel (1962) in cat visual cortex. The first layer of S units consists of Gabor-like edge detectors, and their output is pooled spatially by the corresponding C layer, leaving the shape tuning unaffected. The next S layer is responsible for recombining activations from the preceding layer into new, more complex patterns. The output is again pooled spatially by the next C layer and so forth. Typically, three such S and C layer sandwiches are stacked into a hierarchy, providing increasing complexity of the features and increasing invariance to stimulus translation as well as to slight deformations. The network performs well on the recognition of hand-written digits, which are inherently two-dimensional.

How does this help in recognizing 3d objects? Ullman (1979) showed that it is possible to infer the 3d structure of objects from as few as two planar views, given the correspondence of feature points between the views. Poggio and Edelman (1990) applied this idea to the recognition of 3d objects from a series of 2d views in a network that uses Generalized Radial Basis Functions to match the spatial coordinates of sets of feature points between the views. These results inspired a study of the tuning properties of cells in IT cortex of macaque monkeys by Logothetis et al. (1994), who found that IT cells show a high degree of invariance to size changes and translations of previously learned objects, but that tuning to rotations of these objects is fairly narrow, thereby supporting a view-based theory of the perception of 3d objects. Further support for view-based object recognition comes from behavioral studies of the ability to learn and recognize novel objects from different angles (Tarr and Bülthoff, 1995; Gauthier and Tarr, 1997).

In their HMAX model of object recognition, Riesenhuber and Poggio (1999) combined Fukushima's idea of gradual build-up of invariance and complexity with the insight of view-based recognition of 3d objects. In its original form, HMAX consists of a sequence of two sets of S and C layers and so-called view-tuned units (VTUs), which are fed by the C2 layer. The S1 layer consists of edge detectors using Gabor filters. Layer C1 pools over spatial location (and scale) of S1 activity using a maximum (*max*) operation. The *max* operation was chosen instead of a linear sum in order to retain feature specificity of the signal across the pooling step. Recombination of C1 activations into S2 activity is achieved by hard-wired connections of all possible combinations of orientation-specific C1 units in a 2×2 neighborhood. Layer C2 pools over the remaining spatial locations, so that the spatial receptive field of C2 units encompasses the entire visual field of the model. Patterns of C2 activity, learned from pre-labeled training examples, are associated with specific object views, and VTUs belonging to different aspects of the same object are pooled into object sensitive cells. See Fig. 1 (feed-forward connections only) for a schematic of HMAX.

The model was shown to successfully learn and recognize computer-generated images of 3d wire frame ("paperclip") stimuli, faces, and rendered cats and dogs (Riesenhuber and Poggio, 1999; Freedman et al., 2003). Serre et al. (2005a, b) endowed the model with a method for learning the connections from layer S1 to C2 from natural scene statistics and later added an additional set of S and C cells for better correspondence with functional areas of the primate brain. With these additions, the model (now frequently called the "standard model") is able to learn and recognize a large variety of real-world object categories in photographs (Serre et al., 2007a, b).

Other view-based models

The idea of view-based object recognition was also followed by others. Wallis and Rolls (1997), for instance, presented a hierarchical model of object recognition based on closely matching the receptive field properties of simple and complex cells. They report good performance of a computational implementation of their network for detecting "L", "T", and "+" stimuli as well as faces.

LeCun et al. (1998) refined the ideas of Fukushima in their back-propagation neural network for character recognition ("Le Net"). In his SEEMORE model of object recognition, Mel (1997) employed a rich set of low-level features, including oriented edge filters, color filters, and blobs. Edge filter responses were combined into contours and corners. A neural network trained on the output of all these filters was able to recognize simple objects

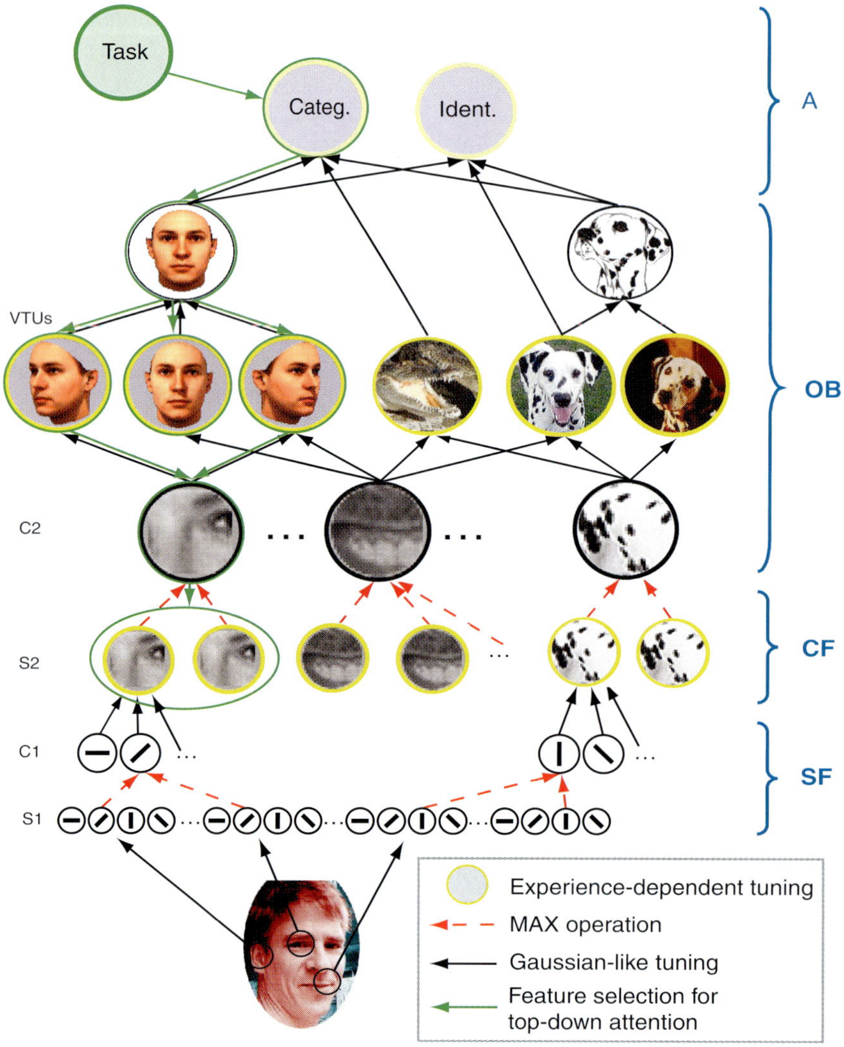

Fig. 1. The basic architecture of the model of object recognition and top-down attention in "Sharing features between object detection and top-down attention" (adapted with permission from Walther et al., 2005b and Serre et al., 2005a). In the feed-forward pass, orientation-selective S1 units filter the input image, followed by max-like pooling over space and scale in the C1 units. The S1 and C1 layers correspond to the "simple features map bundle" SF in Eq. (10). S2 units (complex features CF in Eq. (10)) respond according to the Gaussian distance of each image patch from a set of prototypes. After another step of max-pooling over space and scale, C2 units (OB layer in Eq. (10)) respond to the presence of particular features anywhere in the visual field. View-tuned units (VTUs) and, finally, units selective to individual objects or object categories are acquired from labeled training data (abstract object space $\mathbb{A}$ in Eq. (10)). By association with a particular object or object category, activity due to a given task could traverse down the hierarchy (green) to identify a small subset of features at the S2 level that are indicative for the particular object category.

in photographs with good performance and, as expected, decreasing performance for degraded images.

The model by Amit and Mascaro (2003) for combining object recognition and visual attention is also view-based, and it also employs features of increasing complexity. Translation invariant detection is achieved by pooling over the output of detectors at multiple locations using an *or* operation, the binary equivalent to Riesenhuber and Poggio's *max* operation. Basic units in their model consist of feature-location pairs, where location is measured with respect to the center of mass. Detection proceeds at many locations simultaneously, using hypercolumns with replica units that store copies of some image areas. Complex features are defined as combinations of orientations. There is a trade-off between accuracy and combinatorics: more complex features lead to a better detection algorithm, but more features are needed to represent all objects, i.e., the dimensionality of the feature space increases.

Further support for the suitability of a hierarchy with increasing feature complexity comes from a study by Ullman et al. (2002), who showed that features (in their case rectangular image patches) of intermediate complexity carry more information about object categories than features of low or high complexity (see also Ullman, 2007).

Representation of spatial relations

An important difference between the view-based and the component-based models is the way in which spatial relations between parts are encoded (see Table 1). Biederman (1987) describes a system, in which the relations between the detected components (geons) are encoded in an explicit way, either qualitatively (e.g., using spatial relations such as "above", "to the right" etc.) or quantitatively (e.g., "2 cm to the right").

In view-based models, spatial relations are represented implicitly by the design of the increasingly complex features. Their tuning with respect to the feature representations in the preceding layer inherently includes the spatial relations between these earlier, simpler features (Table 1).

There has been a long debate in the literature over those two contrary views of the representation of spatial relations, with experimental evidence presented for both (e.g., Biederman and Cooper, 1991; Bülthoff et al., 1995). At the current stage, component-based vision in both experiments and models appears to be limited to carefully prepared line drawings, while view-based models generalize well to photographs of real-world objects (Mutch and Lowe, 2006; Serre et al., 2007b; Ullman, 2007).

Regardless of the nature of the encoding of spatial relations, all models have in common the notion of increasing receptive field size and increasing complexity of tuning due to recombinations of simpler features. These traits can be summarized in a unifying formal framework. In the following section we describe such a framework, which we will use to explain roles of attention in object recognition later in this chapter.

A unifying framework for attention and object recognition (UNI)

In a very broad sense, object recognition is a correspondence between the space of all possible images and the abstract space of all possible objects, linking objects with the images in which they appear. Note that this correspondence is not a

Table 1. Comparison of recognition by components and view-based object recognition

	Recognition by components	View-based recognition
Early features	Orientations	Orientations
Intermediate features	3d geometric shapes (geons)	2d patterns of orientations
Object representation	3d assemblies of geons	Multiple 2d views
Spatial relations	Explicit representation	Implicit representation, given by the design of increasingly complex features
Translation and scale invariance	Given by explicit representation of spatial relations	Pooling over space and scale bands
Rotation invariance	Mental rotation of 3d objects	Established by interpolation between views
Key papers	Marr and Nishihara (1978); Biederman (1987); Biederman and Cooper (1991); Hummel and Biederman (1992); Hayworth and Biederman (2006)	Ullman (1979); Fukushima (1980); Logothetis et al. (1994); Bülthoff et al. (1995); Riesenhuber and Poggio (1999); Serre et al. (2007b)

function, since many images may be linked to the same object, and one image may contain several objects. For instance, an image of granny with a hat is associated the representation of grandmother as well as that of grandmother's hat. Likewise, many different images of grandma are associated with the same grandmother representation.

The specifics of this correspondence may depend on the individual observer's prior experience, the task the individual is involved in, the state of alertness, and many other factors. Here we attempt to break down object recognition into intermediate steps that are in approximate agreement with neurophysiological and psychophysical evidence. Subdividing object recognition in this way also allows for the injection of attentional biases at various stages of processing.

Some definitions

We model object recognition as a hierarchy of operations. Each level of the hierarchy consists of a bundle of retinotopic maps. By "map" we mean a retinotopically organized array with scalar values, encoding a particular feature. Bundles of maps encode several features, one map for each feature.

In the brain, these bundles of maps could be implemented in two general types of arrangements: as spatially separate maps, or as an array of hypercolumns, where each hypercolumn contains the activations of all maps at this location in retinotopic space. The particular type of arrangement does not matter for the computational principles outlined in the following section. In fact, a combination of the two types is often the most likely scenario.

As we move upward in the hierarchy, the complexity of the features encoded in the bundles of maps as well as the receptive field size will increase, until we arrive at object-selective units whose receptive field spans large parts of the visual field.

We start out by providing a formal way of encoding one such layer of the hierarchy. Then we will show how activity is transformed from layer to layer. Finally, we will construct a general formal framework for object recognition from stacking multiple layers in a hierarchy.

Bundles of maps

Let us start with the definition of a bundle of maps. Each map corresponds to a feature $k \in \{1, \ldots, K\}$ and assigns an activation level $m(x, y, k)$ to map coordinates (x, y). Thus, we can write a bundle of K maps as a set of quadruples:

$$\mathrm{M} = \{(x, y, k, m) | (x, y) \in \mathbb{N}^2, k \in \{1, \ldots, K\}, \quad m = m(x, y, k) \in \mathbb{R}\} \tag{1}$$

We can construct feature maps recursively, starting with the image. May the first layer (M) encode the image, and may k index the color channels for red, green, and blue. Then the second layer (M′) could encode center-surround color contrasts from the colors in the first layer, taking into account the structure of the features within a neighborhood of a given location in the first layer (e.g., Mel, 1997; Itti et al., 1998). This neighborhood is the spatial receptive field of the second layer. Let us describe this more formally.

Spatial receptive fields

The spatial receptive field can be described by an index function:

$$r(x, y, x_0, y_0) = \begin{cases} 1 & \text{if } (x, y) \text{ is part of} \\ & \text{the receptive field at } (x_0, y_0), \\ 0 & \text{otherwise} \end{cases} \tag{2}$$

The receptive field around (x_0, y_0) is the support of r with x_0 and y_0 fixed, i.e., the set of all pairs (x, y), for which $r(x, y, x_0, y_0)$ is larger than zero:

$$\begin{aligned} RF(x_0, y_0) &= \sup(r(., ., x_0, y_0)) \\ &= \{(x, y) \mid r(x, y, x_0, y_0) > 0\}. \end{aligned} \tag{3}$$

Typically, the receptive field is defined as a contiguous disk-shaped region with center (x_0, y_0). However, the definitions in Eqs. (2) and (3) are general enough to also allow for non-contiguous regions, e.g., for units that are selective to the co-occurrence of two stimuli, such as geons, at separate locations.

Feature recombination functions

The features in layer M′ can be modeled as a feature recombination function ϕ, which maps combinations of activations for all K features within the receptive field RF at (x_0, y_0) to the new features K' at (x_0, y_0):

$$\phi : \mathbb{R}^{K \cdot \|RF(x_0,y_0)\|} \to \mathbb{R}^{K'} \tag{4}$$

In an exclusively linear model, ϕ would be a matrix. For example, the columns of ϕ could contain Gabor convolution kernels for mapping local image regions to the respective filter responses. More generally, ϕ can be any kind of function, including non-linearities such as sigmoidal functions as used in back-propagation networks (e.g., LeCun et al., 1998).

Spatial and feature-based attentional modulation

We would like to have the ability to modulate activity according to spatial and feature-based attentional biases. We achieve this by introducing a spatial modulation function s and a feature modulation function f. The spatial modulation function assigns a non-negative real modulation factor to each coordinate pair (x, y):

$$s : \mathbb{N}^2 \to [0, \infty) \tag{5}$$

Locations with $s = 1$ are not modulated; locations with $s = 0$ are entirely suppressed; locations with $0 < s < 1$ are slightly suppressed; and locations with $s > 1$ are enhanced due to spatial attention.

Similarly, the feature modulation function assigns a non-negative modulation factor to each feature index:

$$f : \{1, \ldots, K\} \to [0, \infty) \tag{6}$$

Both s and f can be used in a binary mode with only the values 0 for *not selected* and 1 for *selected*, or they can be used for more fine-grained attentional modulation.

It may seem like an omission to reduce the effects of attention to gain modulation only. In fact, other effects such as increased baseline activity, shifting of tuning curves, or biasing competitive interactions have been suggested in models and demonstrated experimentally (Desimone and Duncan, 1995; Rees et al., 1997; McAdams and Maunsell, 1999). In our unifying framework, these effects can be implemented by gain modulation of the afferent connections, i.e., by modulating activity in the preceding layer.

Linking layers

With these definitions, the feature activation $m'(x', y', k')$ at location (x', y') for feature k' in M′ is given by:

$$m' = \phi_{k'}\left(\left\{m(x, y, k) \cdot s(x, y) \cdot f(k) | (x, y) \in RF(x', y'),\; k \in \{1, \ldots, K\}\right\}\right) \tag{7}$$

where $\phi_{k'}$ denotes the k'th component of the vector-valued function ϕ.

Finally, we can write the bundle of maps M′ as:

$$\mathrm{M}' = \left\{(x', y', k', m') | (x', y') \in \mathbb{N}^2,\; k' \in \{1, \ldots, K'\},\; m' \in \mathbb{R}\right\} \tag{8}$$

with $m'(x', y', k')$ as defined in Eq. (7).

Observe that we have used four functions to derive M′ from M: the receptive field index function r, the feature recombination function ϕ, the spatial modulation function s, and the feature modulation function f. To indicate this fact, we will use the following notation:

$$\mathrm{M} \xrightarrow{(r,\phi,s,f)} \mathrm{M}' \tag{9}$$

The recognition hierarchy

The definitions of a bundle of maps and of the mapping from one bundle to the next provide us with the building blocks for assembling the framework for hierarchical object recognition. We will build our framework with four basic layers: the *image* IM, a *simple features layer* SF, a *complex features layer* CF, and an *object layer* OB. An *abstract object layer* $\mathbb{A}$ on top of the hierarchy as a fifth layer is a placeholder for a variety of cognitive functions, such as the realization of the percept, i.e., the awareness that particular objects are present, or the answer to the question whether a certain object is contained in the scene, or committing the perceived objects to memory. The entire feed-forward hierarchy looks like this:

$$\begin{array}{l} \mathbb{A} \\ \Big\uparrow f_{ob} \\ \mathbf{OB} \\ \Big\uparrow (r_{\mathrm{ob}}, \phi_{\mathrm{ob}}, s_{\mathrm{cf}}, f_{\mathrm{cf}}) \\ \mathbf{CF} \\ \Big\uparrow (r_{\mathrm{cf}}, \phi_{\mathrm{cf}}, s_{\mathrm{sf}}, f_{\mathrm{sf}}) \\ \mathbf{SF} \\ \Big\uparrow (r_{\mathrm{sf}}, \phi_{\mathrm{sf}}, s_{\mathrm{im}}, f_{\mathrm{im}}) \\ \mathbf{IM} \end{array} \qquad (10)$$

Let us now discuss the details of each of these layers.

The recognition process starts out with an image, which can be expressed as a bundle of maps:

$$\mathrm{IM} = \{(x, y, k_{\mathrm{im}}, m_{\mathrm{im}}) | (x, y) \in \mathbb{N}^2, \\ k_{\mathrm{im}} \in \{1, \ldots, K_{\mathrm{im}}\}, m_{\mathrm{im}} \in \mathbb{R}\} \qquad (11)$$

k_{im} enumerates the color channels of the image, and $m_{\mathrm{im}}(x, y, k_{\mathrm{im}})$ is the pixel value of color channel k_{im} at location (x, y).

In the case of an RGB image, we would have $K_{\mathrm{im}} = 3$, and $m_{\mathrm{im}}(3,4,1)$ would be the value of the red channel at the location with coordinates $x = 3$ and $y = 4$. Depending on the color space of the image, K_{im} can have different values: $K_{\mathrm{im}} = 1$ (e.g., a gray-level image), $K_{\mathrm{im}} = 3$ (e.g., RGB, LMS, HSV, YUV, or CIELAB color spaces), $K_{\mathrm{im}} = 4$ (e.g., CMYK), or other values for any kind of color space.

When used in a binary mode, the spatial modulation function s_{im} allows for spatial selection of a sub-region of the image for further processing. This mechanism is frequently described as an attentional window sliding over the image (e.g., Olshausen et al., 1993; Rutishauser et al., 2004; Walther et al., 2005a), and it is sometimes seen as the equivalent of eye movements selecting parts of a complex scene (e.g., Rybak et al., 1998).

Color channels can be selected with the feature modulation function f_{im}. When using HSV color space, for instance, processing could be limited to the luminance (value) channel.

In the first processing step, simple features (SFs) are derived from the image pixels. Typically, SFs are modeled as convolutions with Gabor filters of various orientations spatial frequencies, and phases (e.g., Fukushima, 1980; LeCun et al., 1998; Riesenhuber and Poggio, 1999). Other possibilities include color center-surround contrasts at various scales (e.g., Mel, 1997; Itti et al., 1998) or texture detectors (e.g., Ullman et al., 2002). All these operations are consolidated in the feature recombination function ϕ_{sf}, and their spatial receptive field properties are encoded in r_{sf}.

These operations result in a bundle of maps SF with K_{sf} simple feature maps. K_{sf} can become quite large when considering Gabor filters with different spatial frequencies and phases at several orientations and spatial scales, for instance. The feature modulation function f_{sf} provides the mechanism for feature-based attention, allowing for modulation of activity or even restricting processing to only a few of these features. Spatial attention is expressed in the spatial modulation function s_{sf}.

Complex features are encoded in the bundle of maps CF. They can encompass a variety of structures, such as corners and line intersections, parts of objects (patches) of an intermediate size, or 3d geometric shapes (geons). Their construction from SFs is described by ϕ_{cf}, and the spatial receptive field properties are given by r_{cf}. The feature recombination function ϕ_{cf} can be hard-wired into the model architecture (e.g., Mel, 1997; Riesenhuber and Poggio, 1999), or it can be learned from image statistics (e.g., LeCun et al., 1998; Ullman et al., 2002; Serre et al., 2007b).

In some models of object recognition, several layers of increasingly complex features follow before objects are recognized or categorized (e.g., Fukushima, 1980, LeCun et al., 1998; Serre et al., 2007b). In a gross simplification we collapse all these intermediate complexity layers into the one complex feature layer CF. The activation in this layer can be modulated by spatial (s_{sf}) and feature-based attention (f_{cf}).

We assume that objects are recognized based on the information present in CF, activating object units OB according to the rules in ϕ_{ob}. The OB layer is functionally approximately equivalent to cells in the monkey IT cortex. While OB units respond very specifically to an object category, a particular object, or a particular view of an object, they have very little spatial resolution. This means that their receptive fields (r_{ob}) typically encompass large regions of the visual field, and that their feature activations $m_{ob}(x, y, k)$ respond specifically to the presence of a particular object anywhere within that receptive field. Several models replace the maps in OB with only one unit for each of the K_{ob} features.

Note that we start out with high spatial resolution (i.e., high spatial specificity) in IM, but with only one to four features (color channels). As information is processed in the hierarchy of Eq. (10), spatial specificity decreases, but the number of features, and therefore their specificity, increases. The OB layer, finally, is fairly insensitive to the location of objects in the visual field, but it contains a potentially very large number of object-specific maps — up to tens of thousands in a fully trained model, according to estimates of the number of visual categories (Biederman, 1987).

It should be pointed out that the formalism described so far is agnostic to the way spatial relations between object parts are encoded. The definition of the receptive field index function [Eq. (2)] and the feature recombination function [Eq. (4)] are sufficiently general to encompass both explicit encoding of spatial relations, as in the RBC model, and implicit encoding by increasingly complex features, as in view-based recognition (see "Representation of spatial relations"). In fact, once encoded in the feature recombination function ϕ, explicit and implicit encoding become almost indistinguishable.

The specifics of the mapping from the object-sensitive units of layer OB to the abstract object space $\mathbb{A}$ depend highly on task and context. A typical instantiation of $\mathbb{A}$ in computational models would be the look-up and report of an object label. In a behaving human or animal, this could be the behavioral response required by the task or situation. Other potential instantiations of $\mathbb{A}$ include committing the percept to memory or assigning an emotional value to the perceived objects. The feature modulation function f_{ob} allows for preferential perception of particular objects or object categories.

Mechanisms of attention

Now that we have mapped out this general formal framework, we use it to review a number of ways of integrating attention with object recognition. Modes of attention can be characterized by the origin of the attentional signal and by the way attention is deployed.

Bottom-up attention is derived only from low-level image properties, typically determining the salience of target regions by some sort of feature contrast. This mode of attention is fast, automatic, and task-independent. Top-down attention, on the other hand, is driven by a task or an intention. Its deployment is comparatively slow and volition-controlled. Most of the models surveyed in this section have provisions for both bottom-up and top-down attention.

The most common way to deploy attention is spatial. In this mode, an attentional "spotlight" selectively enhances processing at a particular location in the visual field (Posner, 1980; Treisman and Gelade, 1980). Occasionally, attention is compared to a zoom lens (Eriksen and James, 1986), adapting the size of the spotlight to the attended object.

In feature-based attention, processing of particular features is biased in a way that is optimal for detecting a known target. Feature-based attention is deployed either directly to the object recognition hierarchy, or it is deployed spatially by biasing the computation of a spatial saliency map.

Object-based attention captures a variety of effects that range from spatially limiting attention to the attended object to setting optimal feature biases for a particular search target.

The unifying framework (UNI) described in the previous section provides the means to model deployment of both spatial and feature-based attention via the spatial and feature modulation functions at each processing level. In this section we review various models of visual attention and show that most attention effects can be modeled within the UNI framework.

The saliency map

The saliency map is a retinotopic map whose activation strength is an indicator for how much a particular image region should attract attention based solely on bottom-up, image-based information. This notion of saliency was introduced by Koch and Ullman (1985).

A computational implementation of the model was given by Itti et al. (1998). In this implementation, several SFs are extracted from the input image at multiple scales in feature pyramids: red-green and blue-yellow color opponencies, luminance, and the four canonical orientations. Center-surround contrasts in these features are computed as differences between scales in the respective feature pyramids and, after normalization, stored in "feature maps". Feature maps can be written as a bundle of maps FM, and the center-surround and normalization operations for computing FM from the SFs as ϕ_{fm}.

In the original implementation by Itti et al. (1998) the number of feature maps is $K_{\mathrm{fm}} = 42$ (4 orientations, 2 color contrasts, 1 luminance contrast, all at 6 spatial scale combinations). In later versions of the model, features such as flicker, motion, and extended contours were added (Itti, 2005; Peters et al., 2005; Carmi and Itti, 2006).

In a series of normalization and across-scale pooling steps, feature maps are eventually combined into a saliency map, itself a bundle of one map SM with $K_{\mathrm{sm}} = 1$. We model the contribution of each feature to the saliency map with a feature modulation function f_{fm}.

A winner-take-all (WTA) network of integrate-and-fire neurons determines the most active location in the saliency map, which is then attended to. The attended location is inhibited (inhibition of return, IOR), and competition continues for the next most salient location. The succession of attended locations can be written as a bundle of maps LOC, where each map is 0 everywhere except for the attended location, where it is 1. LOC contains as many maps as there are successive fixations on the image.

In order to arrive at a spatial modulation function with spatially extended regions, Itti et al. (1998) convolve the maps in LOC with a disk-shaped kernel of fixed size, arriving at a bundle of binary spatial modulation maps $\mathrm{SI}_{\mathrm{disc}}$. The entire flow of processing can be summarized as:

$$\begin{array}{l} \mathbb{A} \\ \uparrow \\ \mathbf{OB} \\ \uparrow s_{\mathrm{cf}} \\ \mathbf{CF} \\ \uparrow s_{\mathrm{sf}} \\ \mathbf{SF} \xrightarrow{\phi_{\mathrm{fm}}} \mathbf{FM} \xrightarrow{(\phi_{\mathrm{sm}},\, f_{\mathrm{fm}})} \mathbf{SM} \xrightarrow{\mathrm{WTA}} \mathbf{LOC} \longrightarrow \mathbf{SI}_{\mathrm{disk}} \longrightarrow s_{\mathrm{xx}} \\ \uparrow s_{\mathrm{im}} \\ \mathbf{IM} \end{array} \qquad (12)$$

The individual spatial modulation maps $m_{\mathrm{disc}}(x, y, k)$ in $\mathrm{SI}_{\mathrm{disk}}$ can be applied at any of the processing stages in the object recognition hierarchy, written as s_{xx} in Eq. (12).

Miau et al. (2001) demonstrated the use of these maps as binary spatial modulation functions s_{im} for the first processing step from IM to SF. They implemented this deployment of spatial attention by cutting out rectangular sections of the image around the attended locations and limiting object recognition to these regions.

Building on the work of Itti et al. (1998), we have developed a mechanism for attending to salient proto-objects (Walther et al., 2002; Walther and Koch, 2006). This work was inspired by Rensink's coherence theory, in which low-level "proto-objects" are formed rapidly and in parallel across the visual field prior to attention. When focused attention accesses a proto-object, it becomes available to higher level perception as a coherent object, but it loses its coherence once attention is released (Rensink, 2000a, b).

In our version of the saliency model, feedback mechanisms within the saliency computations identify the feature map with the strongest contribution to the saliency at the attended location. Spreading of attention from the attended location over a contiguous region of high activation within that feature map yields the shape of the attended proto-object. This provides us with a first estimate of the size and extent of an attended object, object part, or group of objects.[1] In a modification of Eq. (12), our approach can be summarized as:

$$
\begin{array}{llllllll}
\mathbb{A} & & & & & & & \\
\uparrow & & & & & & & \\
\mathbf{OB} & & & & & & & \\
\uparrow s_{\mathrm{cf}} & & & & & & & \\
\mathbf{CF} & & & & & & & \\
\uparrow s_{\mathrm{sf}} & & & & & & & \\
\mathbf{SF} & \xrightarrow{\phi_{\mathrm{fm}}} & \mathbf{FM} & \xrightarrow{(\phi_{\mathrm{sm}},\, f_{\mathrm{fm}})} & \mathbf{SM} & \xrightarrow{\mathrm{WTA}} & \mathbf{LOC} & \\
\uparrow s_{\mathrm{im}} & & & \searrow & & \swarrow & & \\
\mathbf{IM} & & & & \mathrm{SI}_{\mathrm{proto}} & \longrightarrow & s_{\mathrm{xx}} &
\end{array} \qquad (13)
$$

We (Walther and Koch, 2006) have modeled the deployment of spatial attention to proto-objects at the level of s_{cf} in the HMAX model. In a task that required the successive recognition of two objects in an image, we varied the strength of the spatial modulation between 0% (no attentional modulation, corresponding to $s_{\mathrm{cf}} = 1$ at all locations) and 100% (binary modulation with total suppression of regions outside of the attended proto-object, corresponding to $s_{\mathrm{cf}} = 0$ there). We found that a modulation strength of 20% suffices for successful deployment of attention for the recognition of simple wire frame object, and 40% for the recognition of faces. These values are in good agreement with attentional modulation found in the response of neurons in area V4 of macaques (Spitzer et al., 1988; Connor et al., 1997; Luck et al., 1997; Reynolds et al., 2000; Chelazzi et al., 2001; McAdams and Maunsell, 2000). Deploying spatial attention at the level of s_{sf} yielded very similar results.

[1]Matlab code for this model is available online at http://www.saliencytoolbox.net.

Working with a non-biological object recognition algorithm (Lowe, 2004), we applied binary versions of the maps in $\mathrm{SI}_{\mathrm{proto}}$ to s_{im} and enabled learning and recognition of multiple objects in cluttered scenes (Rutishauser et al., 2004; Walther et al., 2005a). Using binary versions of the spatial modulation functions and applying them directly to the image instead of later steps is sensible in computer vision, because computationally expensive processing can be restricted to the support of s_{im}, i.e., those image regions (x, y), for which $s_{\mathrm{im}}(x, y) > 0$.

The idea of a saliency map was used by others as well. Milanese et al. (1994), for instance, describe a method for combining bottom-up and top-down information through relaxation in an associative memory. Frintrop et al. (2005) included depth information from a laser range finder in their version of a saliency map.

Other models of spatial attention

With their MORSEL model of object recognition and selective attention, Mozer (1991) and Mozer and Sitton (1998) implemented a connectionist network and tested it successfully with stylized letters in four standard positions in the display. Spatial attention is deployed as gain modulation of the activity of their first layer ("retina"), corresponding to the IM bundle of maps in our UNI framework.

In the "shifter circuit" model by Olshausen et al. (1993), spatial attention is deployed as gain modulation at various levels of the visual processing hierarchy. Modulation is controlled such that visual information is selectively "re-routed" from the initial layer to the final object-processing area, implemented as an associative shape memory (Hopfield, 1984). In combination with the Hopfield network (the $\mathbb{A}$ layer in the UNI framework), the model by Olshausen and colleagues is capable of detecting objects invariant to translation and scale.

The idea of dynamic re-routing of visual information was also used by Heinke and Humphreys (1997) in their SAIM model (selective attention for identification model). Instead of an associative Hopfield network, SAIM uses a "content layer" for matching the visual information in the focus of attention (FOA) with stored object templates. The model is able to predict a range of experimental results such as reaction times in detection tasks and behavior of the system when lesioned in a systematic way (Heinke and Humphreys, 2003).

The "Selective Tuning" (ST) model of visual attention by Tsotsos et al. (1995) tightly integrates visual attention and object detection (Rothenstein and Tsotsos, 2007). In the first feed-forward pass through this hierarchical system, features of increasing complexity compete locally in WTA networks. After top-down selection of a particular feature or feature combination, competition is biased in a feedback pass such that the stimulus with the selected property is enhanced, and the activity around it is suppressed (inhibitive surround). Once spatially isolated in this manner, the stimulus is processed in another feed-forward pass for ultimate detection or identification. In the UNI framework, this is equivalent to selectively tuning the spatial modulation functions s_{xx} for particular maps. The ST model has been demonstrated successfully for motion-defined shapes (Tsotsos et al., 2005).

Rybak et al. (1998) proposed a model for learning and recognizing objects as combinations of their parts, with the relations between the parts encoded in saccade vectors. Their attention window coincides with s_{im} in our UNI framework, their primary feature detection and invariant transformation modules to layers SF and CF, and their "what" structure to OB. All processing steps in the model of Rybak and colleagues contribute to the formation of the next saccade, i.e., the formation of the next spatial modulation function:

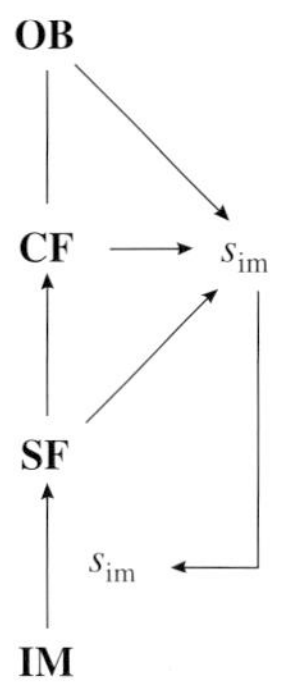

(14)

The model of Rybak and colleagues can memorize and recognize objects and faces in gray-level photographs.

In a different approach to deploying attention, Deco and Schürmann (2000) suggested a model for using spatial attention to selectively enhance resolution for object processing in order to iteratively test hypotheses about the object identity. In the UNI framework, this would amount to actually modifying the feature recombination functions ϕ_{xx} based on the task.

In their physiologically detailed model of visual learning and recognition, Deco and Rolls (2004) implemented attention as biased competition between spatial locations (Duncan and Humphreys, 1989). Depending on the task (spatial search versus object identification), spatial or feature modulation functions are adjusted throughout the hierarchy of visual processing layers in order to bias competition toward the target:

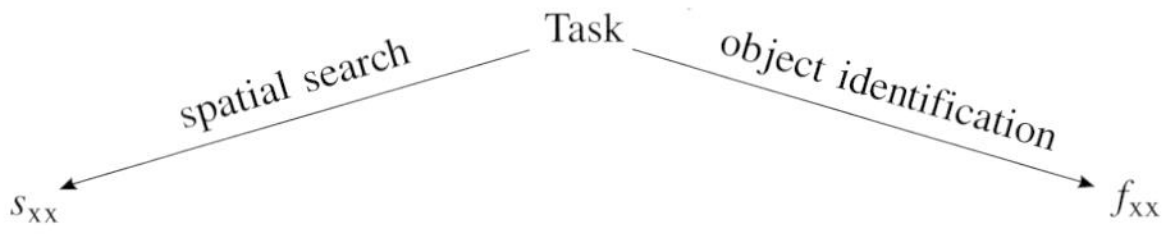

(15)

Feature-based attention

Many of these models of visual attention contain provisions for biasing particular features from the top down based on a task or intention. In the UNI framework, two basic approaches for feature-based attention are possible. First, features can be biased in the recognition hierarchy by using the mechanism of the feature modulation functions f_{xx} in Eq. (10). In this manner, red targets, for instance, could be processed preferentially throughout the visual field. Experimental evidence for this kind of biasing has been found for color and luminance in macaque area V4 (Motter, 1994), for motion in macaque MT (Treue and Martinez Trujillo, 1999), and for color and motion in human V4 and MT (Saenz et al., 2002).

This approach is followed by the ST model by Tsotsos and colleagues when running the model in top-down search mode. A particular property chosen at the top of the hierarchy is selectively enhanced throughout the hierarchy to find an object with this property (Tsotsos et al., 1995, 2005; Rothenstein and Tsotsos, 2007).

The other possibility for feature-based attention is to bias the computation of the saliency map for particular features or combinations of features. This can be achieved with the feature modulation function f_{fm} in Eq. (12). Feature-based attention of this kind is deployed by way of the spatial modulation functions that are derived from the saliency map (Eqs. (12) and (13)). This mode of attention is the essence of "Guided Search", a model proposed by Wolfe and colleagues to explain human behavior in visual search for targets that are defined by individual features or feature conjunctions (Wolfe et al., 1989; Wolfe, 1994; Cave, 1999).

Most models of attention and object recognition follow the second approach. Navalpakkam and Itti (2005), for instance, model the effects of task on visual attention by adjusting the weights for the combination of feature maps into the saliency map. The combinations of weights for particular targets are learned from training images. In the UNI framework, this corresponds to learning the feature modulation function f_{fm} in Eq. (13) from the SF map bundle:

$$\mathrm{SF} \rightarrow f_{fm} \tag{16}$$

Schill et al. (2001) proposed a method for learning features that maximize the gain of information in each saccade in a belief propagation network using orientations only. Their system is tested on 24,000 artificially created scenes which can be classified with a 80% hit rate.

In his model of dynamic interactions between prefrontal areas, IT, and V4, Hamker (2004) proposed a method for setting feature biases in order to guide spatial attention to target locations. This biologically detailed model was fitted to reproduce the neurophysiological data by Chelazzi et al. (1998) for a cued target selection task in rhesus monkeys (Hamker, 2003). Additionally, the model is capable of detecting objects in natural scenes (Hamker, 2005).

No matter how feature-based attention is deployed, there is always the question of how to choose the optimal modulation function for a particular task. Navalpakkam and Itti (2007) showed an elegant way of choosing the weights for a linear modulation function by maximizing the signal-to-noise ratio between search targets and distracters. Counterintuitively, it is sometimes optimal to enhance the activity of neurons tuned to an exaggerated property of the target instead of the perfectly tuned neuron.

Object-based attention

Experimental evidence suggests that attention can be tied to objects, object parts, or groups of objects (Duncan, 1984; Egly et al., 1994; Roelfsema et al., 1998). In particular, Egly et al. (1994) reported that attention spreads over objects defined by luminance contrast. In their study, an invalid spatial cue for the location of a briefly flashed target is still effective when cue and target are located on the same object, but not when they are on different objects. The effect has been replicated for objects defined by color (Mitchell et al., 2003; Reynolds et al., 2003) and illusory contours (Moore et al., 1998).

We have modeled this effect as spreading of attention over a contiguous region of high activity in the feature map that contributes most to the saliency of the attended location, thus obtaining an estimate for the size and shape of the attended objects (see Eq. (13)).

Rolls and Deco (2006) model object-based attention by shrinking the size of the receptive field of model IT neurons to match the size of the attended object. This idea of dynamically adjusting the size of the attended region to the attended object like a zoom lens was pioneered by Eriksen and James (1986). In a similar spirit, the shape of both the enhanced center-region and the suppressive surround in the ST model adapt to the shape of the attended object (Tsotsos et al., 1995).

Objects can also be attended to when they are not clearly separated. The model by Lee and Lee (2000) is able to learn optimal biases for top-down attention using back-propagation in a multilayer perceptron network. Their system can segment superimposed handwritten digits on the pixel level. Treating attention as a by-product of a recognition model based on Kalman filtering, the system by Rao (1998) can attend to spatially overlapping (occluded) objects on a pixel basis as well.

Object-based attention does not need to be spatial, however. O'Craven et al. (1999) showed that human subjects could attend selectively to faces and houses when both stimuli were semi-transparently superimposed. In these fMRI experiments the attended object could also be defined by its coherent motion. In the UNI framework, these results can be interpreted as feature-based attention by adjusting f_{sf} for a particular direction of motion or f_{ob} for a particular object category in Eq. (10).

If it is possible to effectively deploy top-down attention for particular objects as feature biases for the relatively unspecific SFs, then using complex, more object-specific features should make selection even easier. Object recognition as described earlier in this chapter provides us with a set of complex features and with a mapping from these features to the object layer and finally to the abstract object space:

$$\mathrm{CF} \overset{\phi_{ob}}{\rightarrow} \mathrm{OB} \rightarrow \mathbb{A} \qquad (17)$$

We propose a mode of object-specific attention that uses *feedback connections* from abstract object representations $\mathbb{A}$ via the object layer OB back to the complex features CF in order to infer suitable complex feature maps for the localization of a given object or object category. Both the complex features CF and the feature recombination function ϕ_{ob} are acquired when the system is trained for object detection. We suggest that these same representations can be used for top-down attention as well. As a proof of concept for this idea, we show in our simulations in the next section that it is possible to share complex features between object detection and attention.

Sharing features between object detection and top-down attention

In the hierarchy of processing steps for object recognition in Eq. (10), the structure of objects is encoded in the feature recombination functions ϕ_{xx}. While these functions are fairly generic at early stages (e.g., Gabor filters for ϕ_{sf}), at later stages they are more complex and more specific for particular object categories. Some models of object recognition use hard-wired features at these higher levels (e.g., Riesenhuber and Poggio, 1999; Amit and Mascaro, 2003), others learn these features from their visual input (e.g., LeCun et al., 1998; Serre et al., 2005a).

As mentioned in "View-based recognition in HMAX", Serre et al. (2005a) extended the HMAX model of Riesenhuber and Poggio (1999) by learning the structure of complex features, i.e., the details of ϕ_{cf} in Eq. (10), from large numbers of natural images. Once learned, these complex features are fixed and used as prototypes for feature recombination functions that resemble radial basis functions. With these functions, the model can be trained to categorize objects in photographs with high accuracy (Serre et al., 2007b).

Here we suggest a method of sharing these same complex feature representations between object detection and top-down attention. We propose that by cortical feedback connections, top-down processes can re-use these same features to bias attention to locations with a higher probability of containing the target object. We compare the performance of a computational implementation of such a model with pure bottom-up attention and, as a benchmark, with biasing for skin hue, which is known to work well as a top-down bias for faces.

Methods

The basic architecture of our model is shown in Fig. 1. Proto-types for the S2 features are randomly initiated from a set of training images. For the work presented in this chapter, we trained the model on detecting frontal views of human faces in photographs and investigated the suitability of the corresponding S2 features for top-down attention to faces.

For feature learning and training, we used 200 color images, each containing one face among clutter, and 200 distracter images without faces (see Fig. 2 for examples). For testing the recognition performance of the system, we used a separate test set of 201 face images and 2119 non-face distracter images. To evaluate top-down attention, we used a third set of 179 color images containing between 2 and 20 frontal views of faces (Fig. 2 third and fourth row). All images were obtained from the world wide web, and face images were labeled by hand, with the eyes, nose, and mouth of each face marked.[2]

During feature learning, 100 patches of size 6×6 pixels were extracted from the C1 maps for each presentation of a training image. Over five iterations of presenting the 200 training images in random order, 100 stable features were learned. Two separate sets of features were learned in this manner: set A was derived from patches that were extracted from any location in the training images (Fig. 2, top row); patch selection for set B was limited to regions around faces (Fig. 2, second row).

To evaluate feature sets A and B for top-down attention, the S2 maps were computed for the third set of 179 images containing multiple faces. These top-down feature maps were compared to the bottom-up saliency map of Itti et al. (1998) and to a skin hue detector for each of the images.

Skin hue is known to be an excellent indicator for the presence of faces in color images (Darrel et al., 2000). Here we use it as a benchmark. Since we want our model of skin hue to be independent of light intensity, we model it in intensity-normalized (r',g') color space. If (r, g, b) are the RGB values of a given color pixel, then we compute our (r',g') color coordinates as

$$r' = \frac{r}{r+g+b} \quad \text{and} \quad g' = \frac{g}{r+g+b} \qquad (18)$$

Note that it is not necessary to have a separate value for blue, because the blue content of the pixel can be inferred from r' and g' at any given light intensity $(r+g+b)$.

For the description of skin hue in this color space, we use a simple Gaussian model with mean (μ_r, μ_g), standard deviations (σ_r, σ_g), and correlation coefficient ρ. For a given color pixel with coordinates (r', g'), the model's hue response is given by

$$h(r', g') = \exp\left[-\frac{1}{2}\left(\frac{(r'-\mu_r)^2}{\sigma_r^2} + \frac{(g'-\mu_g)^2}{\sigma_g^2} - \frac{\rho(r'-\mu_r)(g'-\mu_g)}{\sigma_r\sigma_g}\right)\right] \qquad (19)$$

To estimate the parameters of the skin hue distribution, we used 1153 color photographs containing a total of 3947 faces from the world wide web[3] and fitted the hue distribution of the faces. The resulting parameters are shown in Table 2. The images used for estimating the skin hue model are separate from the sets of images used elsewhere in this section.

The images depict humans of many different ages and ethnicities, both female and male. There is a slight bias toward Caucasian males, reflecting a general bias of images of humans in the world wide web. We observed that the skin *hue* does not vary much between different ethnicities, while brightness of the skin shows much more variations. Figure 3 shows the hue of the training faces and the fitted distribution.

Results

After feature learning, the recognition model was trained to detect faces using the training images. Recognition performance on the test images was 98.9% with feature set A and 99.4% with set B (measured as area under the ROC curve).

For the purpose of testing the suitability of the features for top-down attention we use an analysis of fixations on faces based on the respective activation maps. The S2 feature maps for both feature sets, for the bottom-up saliency map, and for the skin hue bias map were computed for the 179 multiple-face images. Each of the four maps was treated like a saliency map, and the locations in the map were visited in order of decreasing saliency, neglecting spatial relations between the locations. While this procedure falls short of the full

[2]We would like to thank Xinpeng Huang and Thomas Serre for collecting and labeling the images. The image database is available online at http://web.mit.edu/serre/www/Resources.htm.

[3]We would like to thank Pietro Perona for making the images available.

Fig. 2. Examples for training stimuli for feature set A (top row), feature set B (second row), test stimuli with two or more faces (third and fourth row), and for non-face distracters (bottom row). See "Methods" for details.

simulation of a WTA network with IOR as described in Koch and Ullman (1985), it nevertheless provides a simple and consistent means of scanning the maps.

Table 2. Parameters of the distribution of skin hue in intensity-normalized (r',g') color space

Parameter	Value
μ_r	0.434904
μ_g	0.301983
σ_r	0.053375
σ_g	0.024349
ρ	0.5852

For each map we determined the number of "fixations" required to find a face and, once the FOA leaves the most salient face, how many fixations are required to attend to each subsequent face. The fraction of all faces that required one, two, three, or more than three fixations was determined for each map type and used as a measure for the quality of the respective mode of computing an attention signal.

The results are shown in Fig. 4 for the best features from sets A and B, for bottom-up attention, and for skin hue detection. Feature set B shows slightly higher performance than our benchmark skin hue detection, followed by feature set A and bottom-up attention. Results are significantly

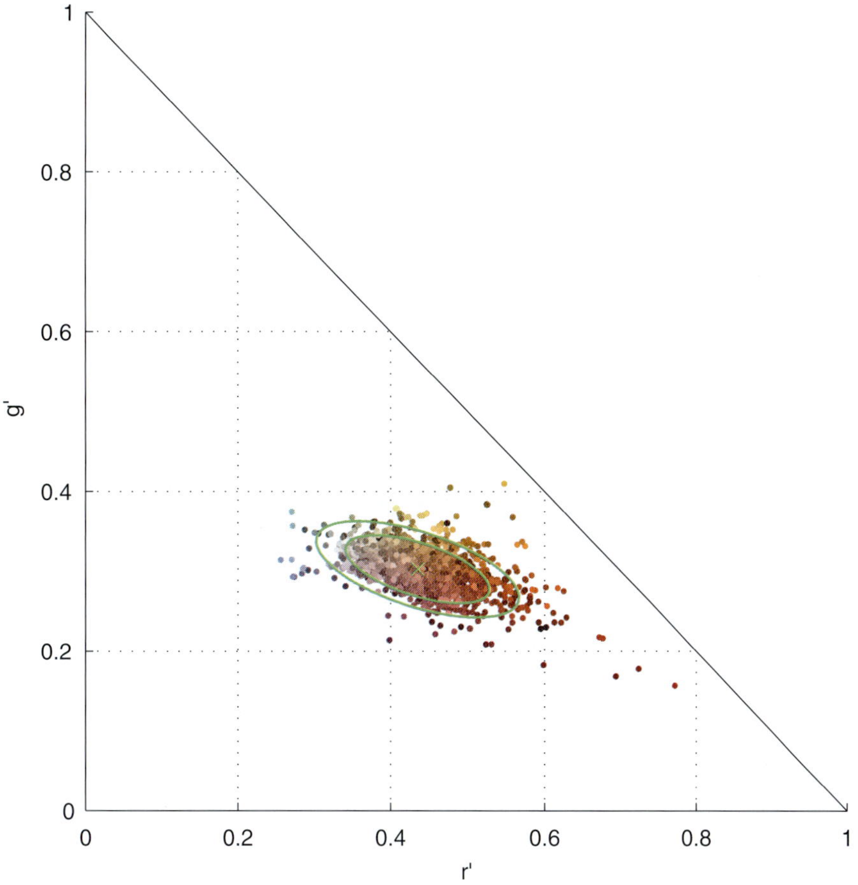

Fig. 3. The Gaussian model for skin hue. The individual training points are derived from 3974 faces in 1153 color photographs. Each dot represents the average hue for one face and is plotted in the color of the face. The green cross represents the mean (μ_r, μ_g), and the green ellipses the 1σ and 2σ intervals of the hue distribution.

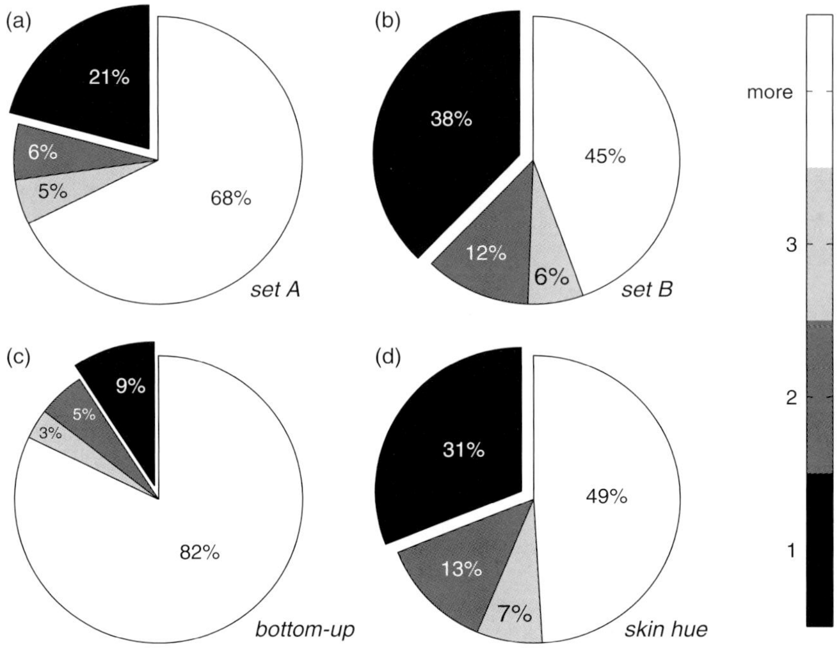

Fig. 4. Fractions of faces in test images requiring one, two, three, or more than three fixations to be found when using top-down feature sets A or B, bottom-up attention, or biasing for skin hue.

better when feature selection is restricted to faces (set B) compared to unrestricted feature selection (set A).

Top-down attention based on S2 features by far outperform bottom-up attention in our experiments. While bottom-up attention is well suited to identify salient regions in the absence of a specific task, it cannot be expected to localize a specific object category as well as feature detectors that are specialized for this category.

Discussion

We show that features learned for recognizing a particular object category may also serve for top-down attention to that object category. Object detection can be understood as a mapping from a set of features to an abstract object representation. When a task implies the importance of an object, the respective abstract object representation may be invoked, and feedback connections may reverse the mapping, allowing inference as to which features are useful to guide top-down attention to image locations that have a high probability of containing the target object.

The important question of how to combine several S2 feature maps optimally for the search for a specific target remains unanswered in this section. It is possible that feature combination strategies like the one that Navalpakkam and Itti (2007) applied to SFs are also viable for our more complex features. For now, however, this remains an open issue.

Note that this mode of top-down attention does not necessarily imply that the search for any object category can be done in parallel using an explicit map representation. Search for faces, for instance, has been found to be efficient (Hershler and Hochstein, 2005), although this result is disputed (VanRullen, 2006). We have merely shown a method for identifying features that can be used to search for an object category. The efficiency of the search will depend on the complexity of those features and, in particular, on the frequency of the same features for other object categories, which constitute the set of distracters in visual search.

To analyze this aspect further, it would be of interest to explore the overlap in the sets of features that are useful for multiple object categories. Torralba et al. (2004) have addressed this problem for multiple object categories as well as multiple views of objects in a machine vision context.

Conclusions

We have reviewed a number of models of object recognition and visual attention, showing that many seemingly disparate ideas can in fact be captured by a common formal framework. In this unifying framework for attention and object recognition we have explained recognition by components as well as view-based object recognition; we have covered many aspects of spatial, feature-based, and object-based attention, as well as the interactions between attention and object recognition. Furthermore, we have shown in a particular instantiation how complex features learned for the purpose of object detection can be shared with top-down attention.

In our review we have focused on the ideas for attention and object recognition that fit within our hierarchical framework. However, there are other ideas that are not captured by the UNI framework, either because they describe the neurobiological processes at a more detailed level than is provided by the UNI framework, or because they make use of information from sources that fall outside the framework, such as visual context.

Grossberg and Raizada (2000) and Raizada and Grossberg (2001), for instance, proposed a model of attention and visual grouping based on biologically realistic models of neurons and neural networks. Their model relies on grouping edges within a laminar cortical structure by synchronous firing, allowing it to extract real as well as illusory contours.

Recognizing individual objects is only part of visual perception. Objects are typically embedded in context with other objects or with the general layout of a scene (the "gist"). Interactions between object recognition and scene perception go both ways: gist provides a powerful top-down cue, restricting the possibilities for object identity; objects contribute to the general context and interpretation of a scene. Attention serves as a means of deploying this top-down information.

Some aspects of this interaction were modeled by Oliva et al. (2003) in a probabilistic framework. Oliva and colleagues incorporate context information into the spatial probability function for seeing certain objects (e.g., people) at particular locations. Comparison with human eye tracking results show improvement over purely bottom-up saliency-based attention.

Work on scene perception has been progressing at a rapid pace over the last few years (see, for instance, Bar, 2004; Oliva and Torralba, 2006; Fei-Fei et al., 2007; Walther and Fei-Fei, 2007), and integrating scene and object information into a general framework is an interesting challenge for years to come. Exploring and modeling the interactions between scene perception, object recognition, and visual attention will bring us closer to understanding the rich and varied experience afforded to us by visual perception.

Acknowledgments

Thomas Serre and Tomaso Poggio collaborated on parts of the work on sharing features between object detection and top-down attention. We would like to thank Karen F. Bernhardt-Walther, Kerstin Preuschoff, and Daniel Simons for helpful discussions and feedback on versions of the manuscript. This work was funded by DARPA, the NSF, the NIH, the NIMH, the ONR, the Keck Foundation, and a Beckman Postdoctoral Fellowship to D.B.W.

References

Amit, Y. and Mascaro, M. (2003) An integrated network for invariant visual detection and recognition. Vision Res., 43: 2073–2088.

Bar, M. (2004) Visual objects in context. Nat. Rev. Neurosci., 5: 617–629.

Biederman, I. (1987) Recognition-by-components: a theory of human image understanding. Psychol. Rev., 94: 115–147.

Biederman, I. and Cooper, E.E. (1991) Priming contour-deleted images: evidence for intermediate representations in visual object recognition. Cognit. Psychol., 23: 393–419.

Bülthoff, H.H., Edelman, S.Y. and Tarr, M.J. (1995) How are three-dimensional objects represented in the brain? Cereb. Cortex, 5: 247–260.

Carmi, R. and Itti, L. (2006) Visual causes versus correlates of attentional selection in dynamic scenes. Vision Res., 46: 4333–4345.

Cave, K.R. (1999) The FeatureGate model of visual selection. Psychol. Res., 62: 182–194.

Chelazzi, L., Duncan, J., Miller, E.K. and Desimone, R. (1998) Responses of neurons in inferior temporal cortex during memory-guided visual search. J. Neurophysiol., 80: 2918–2940.

Chelazzi, L., Miller, E.K., Duncan, J. and Desimone, R. (2001) Responses of neurons in macaque area V4 during memory-guided visual search. Cereb. Cortex, 11: 761–772.

Connor, C.E., Preddie, D.C., Gallant, J.L. and van Essen, D.C. (1997) Spatial attention effects in macaque area V4. J. Neurosci., 17: 3201–3214.

Darrel, T., Gordon, G., Harville, M. and Woodfill, J. (2000) Integrated person tracking using stereo, color, and pattern detection. Int. J. Comput. Vis., 37: 175–185.

Deco, G. and Rolls, E.T. (2004) A neurodynamical cortical model of visual attention and invariant object recognition. Vision Res., 44: 621–642.

Deco, G. and Schürmann, B. (2000) A hierarchical neural system with attentional top-down enhancement of the spatial resolution for object recognition. Vision Res., 40: 2845–2859.

Desimone, R. and Duncan, J. (1995) Neural mechanisms of selective visual-attention. Annu. Rev. Neurosci., 18: 193–222.

Duncan, J. (1984) Selective attention and the organization of visual information. J. Exp. Psychol. Gen., 113: 501–517.

Duncan, J. and Humphreys, G.W. (1989) Visual search and stimulus similarity. Psychol. Rev., 96: 433–458.

Edelman, S. (1997) Computational theories of object recognition. Trends Cognit. Sci., 1: 296–304.

Egly, R., Driver, J. and Rafal, R.D. (1994) Shifting visual attention between objects and locations: evidence from normal and parietal lesion subjects. J. Exp. Psychol. Gen., 123: 161–177.

Eriksen, C.W. and James St., J.D. (1986) Visual attention within and around the field of focal attention: a zoom lens model. Percept. Psychophys., 40: 225–240.

Fei-Fei, L., Iyer, A., Koch, C. and Perona, P. (2007) What do we perceive in a glance of a real-world scene? J. Vis., 7(1): 1–29.

Freedman, D.J., Riesenhuber, M., Poggio, T. and Miller, E.K. (2003) A comparison of primate prefrontal and inferior temporal cortices during visual categorization. J. Neurosci., 23: 5235–5246.

Frintrop, S., Rome, E., Nüchter, A. and Surmann, H. (2005) A bimodal laser-based attention system. Comput. Vis. Image Underst., 100: 124–151.

Fukushima, K. (1980) Neocognitron: a self-organizing neural network model for a mechanism of pattern recognition unaffected by shift in position. Biol. Cybern., 36: 193–202.

Gauthier, I. and Tarr, M.J. (1997) Becoming a "Greeble" expert: exploring mechanisms for face recognition. Vision Res., 37: 1673–1682.

Grossberg, S. and Raizada, R.D. (2000) Contrast-sensitive perceptual grouping and object-based attention in the laminar circuits of primary visual cortex. Vision Res., 40: 1413–1432.

Hamker, F.H. (2003) The reentry hypothesis: linking eye movements to visual perception. J. Vis., 3: 808–816.

Hamker, F.H. (2004) A dynamic model of how feature cues guide spatial attention. Vision Res., 44: 501–521.

Hamker, F.H. (2005) The emergence of attention by population-based inference and its role in distributed processing and cognitive control of vision. Comput. Vis. Image Underst., 100: 64–106.

Hayworth, K.J. and Biederman, I. (2006) Neural evidence for intermediate representations in object recognition. Vision Res., 46: 4024–4031.

Heinke, D. and Humphreys, G.W. (1997) SAIM: a model of visual attention and neglect. In: 7th international conference on artificial neural networks — ICANN 97, Springer Verlag, Lausanne, Switzerland, pp. 913–918.

Heinke, D. and Humphreys, G.W. (2003) Attention, spatial representation, and visual neglect: simulating emergent attention and spatial memory in the selective attention for identification model (SAIM). Psychol. Rev., 110: 29–87.

Hershler, O. and Hochstein, S. (2005) At first sight: a high-level pop out effect for faces. Vision Res., 45: 1707–1724.

Hopfield, J.J. (1984) Neurons with graded response have collective computational properties like those of two-state neurons. Proc. Natl. Acad. Sci. U.S.A., 81: 3088–3092.

Hubel, D.H. and Wiesel, T.N. (1962) Receptive fields, binocular interaction and functional architecture in the cat's visual cortex. J. Physiol., 160: 106–154.

Hummel, J.E. and Biederman, I. (1992) Dynamic binding in a neural network for shape recognition. Psychol. Rev., 99: 480–517.

Itti, L. (2005) Quantifying the contribution of low-level saliency to human eye movements in dynamic scenes. Vis. Cogn., 12: 1093–1123.

Itti, L. and Koch, C. (2001) Computational modelling of visual attention. Nat. Rev. Neurosci., 2: 194–203.

Itti, L., Koch, C. and Niebur, E. (1998) A model of saliency-based visual attention for rapid scene analysis. IEEE Trans. Pattern Anal. Machine Intell., 20: 1254–1259.

Koch, C. and Ullman, S. (1985) Shifts in selective visual-attention: towards the underlying neural circuitry. Hum. Neurobiol., 4: 219–227.

LeCun, Y., Bottou, L., Bengio, Y. and Haffner, P. (1998) Gradient-based learning applied to document recognition. Proc. IEEE, pp. 2278–2324.

Lee, S.I. and Lee, S.Y. (2000) Top-down attention control at feature space for robust pattern recognition. In: Biologically Motivated Computer Vision, Seoul, Korea.

Logothetis, N.K., Pauls, J., Bülthoff, H.H. and Poggio, T. (1994) View-dependent object recognition by monkeys. Curr. Biol., 4: 401–414.

Lowe, D.G. (2004) Distinctive image features from scale-invariant keypoints. Int. J. Comput. Vis., 60: 91–110.

Luck, S.J., Chelazzi, L., Hillyard, S.A. and Desimone, R. (1997) Neural mechanisms of spatial selective attention in areas V1,

V2, and V4 of macaque visual cortex. J. Neurophysiol., 77: 24–42.

Marr, D. (1982) Vision. W. H. Freeman and Company, New York.

Marr, D. and Nishihara, H.K. (1978) Representation and recognition of the spatial organization of three-dimensional shapes. Proc. R. Soc. Lond. B Biol. Sci., 200: 269–294.

McAdams, C.J. and Maunsell, J.H.R. (1999) Effects of attention on orientation-tuning functions of single neurons in macaque cortical area V4. J. Neurosci., 19: 431–441.

McAdams, C.J. and Maunsell, J.H.R. (2000) Attention to both space and feature modulates neuronal responses in macaque area V4. J. Neurophysiol., 83: 1751–1755.

Mel, B.W. (1997) SEEMORE: combining color, shape, and texture histogramming in a neurally inspired approach to visual object recognition. Neural Comput., 9: 777–804.

Miau, F., Papageorgiou, C. and Itti, L. (2001) Neuromorphic algorithms for computer vision and attention. In: SPIE 46 Annual International Symposium on Optical Science and Technology, Vol. 4479, San Jose, CA, pp. 12–23.

Milanese, R., Wechsler, H., Gill, S., Bost, J.M. and Pun, T. (1994) Integration of bottom-up and top-down cues for visual attention using non-linear relaxation. In: International Conference on Computer Vision and Pattern Recognition, Seattle, WA, pp. 781–785.

Mitchell, J.F., Stoner, G.R., Fallah, M. and Reynolds, J.H. (2003) Attentional selection of superimposed surfaces cannot be explained by modulation of the gain of color channels. Vision Res., 43: 1323–1328.

Moore, C.M., Yantis, S. and Vaughan, B. (1998) Object-based visual selection: evidence from perceptual completion. Psychol. Sci., 9: 104–110.

Motter, B.C. (1994) Neural correlates of attentive selection for color or luminance in extrastriate area V4. J. Neurosci., 14: 2178–2189.

Mozer, M.C. (1991) The Perception of Multiple Objects: A Connectionist Approach. MIT Press, Cambridge, MA.

Mozer, M.C. and Sitton, M. (1998) Computational modeling of spatial attention. In: Pashler H. (Ed.), Attention. Psychology Press, London, pp. 341–393.

Mutch, J. and Lowe D.G. (2006) Multiclass object recognition with sparse, localized features. In: IEEE International Conference on Computer Vision and Pattern Recognition, New York, NY, pp. 11–18.

Navalpakkam, V. and Itti, L. (2005) Modeling the influence of task on attention. Vision Res., 45: 205–231.

Navalpakkam, V. and Itti, L. (2007) Search goal tunes visual features optimally. Neuron, 53: 605–617.

O'Craven, K.M., Downing, P.E. and Kanwisher, N. (1999) fMRI evidence for objects as the units of attentional selection. Nature, 401: 584–587.

Oliva, A. and Torralba, A. (2006) Building the gist of a scene: the role of global image features in recognition. In: Progress in Brain Research: Visual Perception, Vol. 155. Elsevier, Amsterdam, pp. 23–36.

Oliva, A., Torralba, A., Castelhano, M. and Henderson, J. (2003) Top-down control of visual attention in object detection. In: International Conference on Image Processing, Barcelona, Spain.

Olshausen, B.A., Anderson, C.H. and Van Essen, D.C. (1993) A neurobiological model of visual attention and invariant pattern recognition based on dynamic routing of information. J. Neurosci., 13: 4700–4719.

Peters, R.J., Iyer, A., Itti, L. and Koch, C. (2005) Components of bottom-up gaze allocation in natural images. Vision Res., 45: 2397–2416.

Plato. (1968) Book vii in The Republic, p. 195. Basic Books, New York, NY.

Poggio, T. and Edelman, S. (1990) A network that learns to recognize three-dimensional objects. Nature, 343: 263–266.

Posner, M.I. (1980) Orienting of attention. Q. J. Exp. Psychol., 32: 3–25.

Raizada, R.D. and Grossberg, S. (2001) Context-sensitive bindings by the laminar circuits of V1 and V2: a unified model of perceptual grouping, attention, and orientation contrast. Vis. Cogn., 8: 431–466.

Rao, R.P.N. (1998) Visual attention during recognition. In: Advances in Neural Information Processing Systems, Vol. 10, pp. 80–86.

Rees, G., Frackowiak, R. and Frith, C. (1997) Two modulatory effects of attention that mediate object categorization in human cortex. Science, 275: 835–838.

Rensink, R.A. (2000a) The dynamic representation of scenes. Vis. Cogn., 7: 17–42.

Rensink, R.A. (2000b) Seeing, sensing, and scrutinizing. Vision Res., 40: 1469–1487.

Reynolds, J.H., Alborzian, S. and Stoner, G.R. (2003) Exogenously cued attention triggers competitive selection of surfaces. Vision Res., 43: 59–66.

Reynolds, J.H., Pasternak, T. and Desimone, R. (2000) Attention increases sensitivity of V4 neurons. Neuron, 26: 703–714.

Riesenhuber, M. and Poggio, T. (1999) Hierarchical models of object recognition in cortex. Nat. Neurosci., 2: 1019–1025.

Roelfsema, P.R., Lamme, V.A.F. and Spekreijse, H. (1998) Object-based attention in the primary visual cortex of the macaque monkey. Nature, 395: 376–381.

Rolls, E.T. and Deco, G. (2006) Attention in natural scenes: neurophysiological and computational bases. Neural Netw., 19: 1383–1394.

Rothenstein, A.L. and Tsotsos, J.K. (2007) Attention links sensing to recognition. Image Vis. Comput. (in press).

Rutishauser, U., Walther, D., Koch, C. and Perona, P. (2004) Is attention useful for object recognition? In: IEEE International Conference on Computer Vision and Pattern Recognition, Washington, DC, Vol. 2, pp., 37–44.

Rybak, I.A., Gusakova, V.I., Golovan, A.V., Podladchikova, L.N. and Shevtsova, N.A. (1998) A model of attention-guided visual perception and recognition. Vision Res., 38: 2387–2400.

Saenz, M., Buracas, G.T. and Boynton, G.M. (2002) Global effects of feature-based attention in human visual cortex. Nat. Neurosci., 5: 631–632.

Schill, K., Umkehrer, E., Beinlich, S., Krieger, G. and Zetzsche, C. (2001) Scene analysis with saccadic eye movements: top-down and bottom-up modeling. J. Electronic Imaging, 10: 152–160.

Serre, T., Kouh, M., Cadieu, C., Knoblich, U., Kreiman, G. and Poggio, T. (2005a) A theory of object recognition: computations and circuits in the feedforward path of the ventral stream in primate visual cortex, CBCL Paper #259/AI Memo #2005-036 Technical report, Massachusetts Institute of Technology.

Serre, T., Wolf, L. and Poggio, T. (2005b) Object recognition with features inspired by visual cortex. In: IEEE International Conference on Computer Vision and Pattern Recognition, San Diego, CA, Vol. 2, pp. 994–1000.

Serre, T., Kreiman, G., Kouh, M., Cadieu, C., Knoblich, U. and Poggio, T. (2007a) A quantitative theory of immediate visual recognition. In: Progress in Brain Research, pp. 33–56.

Serre, T., Wolf, L., Bileschi, S., Riesenhuber, M. and Poggio, T. (2007b) Object recognition with cortex-like mechanisms. IEEE Trans. Pattern Anal. Machine Intell., 29: 411–426.

Spitzer, H., Desimone, R. and Moran, J. (1988) Increased attention enhances both behavioral and neuronal performance. Science, 240: 338–340.

Tarr, M.J. and Bülthoff, H.H. (1995) Is human object recognition better described by geon structural descriptions or by multiple views? Comment on Biederman and Gerhardstein (1993)
J. Exp. Psychol. Hum. Percept. Perform., 21: 1494–1505.

Torralba, A., Murphy, K. and Freeman, W. (2004) Sharing features: efficient boosting procedures for multiclass object detection. In: IEEE International Conference on Computer Vision and Pattern Recognition, Washington, DC.

Treisman, A.M. and Gelade, G. (1980) A feature-integration theory of attention. Cogn. Psychol., 12: 97–136.

Treue, S. and Martinez Trujillo, J.C. (1999) Feature-based attention influences motion processing gain in macaque visual cortex. Nature, 399: 575–579.

Tsotsos, J.K., Culhane, S.M., Wai, W.Y.K., Lai, Y.H., Davis, N. and Nuflo, F. (1995) Modeling visual-attention via selective tuning. Artif. Intell., 78: 507–545.

Tsotsos, J.K., Liu, Y., Martinez-Trujillo, J.C., Pomplun, M., Simine, E. and Zhou, K. (2005) Attending to motion. Comput. Vis. Image Underst., 100: 3–40.

Ullman, S. (1979) The Interpretation of Visual Motion. MIT Press, Cambridge, MA.

Ullman, S. (2007) Object recognition and segmentation by a fragment-based hierarchy. Trends Cogn. Sci., 11: 58–64.

Ullman, S., Vidal-Naquet, M. and Sali, E. (2002) Visual features of intermediate complexity and their use in classification. Nat. Neurosci., 5: 682–687.

VanRullen, R. (2006) On second glance: still no high-level pop-out effect for faces. Vision Res., 46: 3017–3027.

Wallis, G. and Rolls, E.T. (1997) Invariant face and object recognition in the visual system. Prog. Neurobiol., 51: 167–194.

Walther, D.B. and Fei-Fei, L. (2007) Task-set switching with natural scenes: measuring the cost of deploying top-down attention. J. Vis., 7(11):9, 1–12, http://journalofvision.org/7/11/9/, doi: 10.1167/7.11.9.

Walther, D., Itti, L., Riesenhuber, M., Poggio, T. and Koch, C. (2002) Attentional selection for object recognition: a gentle way. In: Lecture Notes in Computer Science, Vol. 2525. Springer, Berlin, Germany, pp. 472–479.

Walther, D. and Koch, C. (2006) Modeling attention to salient proto-objects. Neural Netw., 19: 1395–1407.

Walther, D., Rutishauser, U., Koch, C. and Perona, P. (2005a) Selective visual attention enables learning and recognition of multiple objects in cluttered scenes. Comput. Vis. Image Underst., 100: 41–63.

Walther, D., Serre, T., Poggio, T. and Koch, C. (2005b) Modeling feature sharing between object detection and top-down attention [abstract]. J. Vis., 5: 1041a.

Wolfe, J.M. (1994) Guided Search 2.0: a revised model of visual search. Psychon. Bull. Rev., 1: 202–238.

Wolfe, J.M., Cave, K.R. and Franzel, S.L. (1989) Guided search: an alternative to the feature integration model for visual search. J. Exp. Psychol. Hum. Percept. Perform., 15: 419–433.

P. Cisek, T. Drew & J.F. Kalaska (Eds.)
Progress in Brain Research, Vol. 165
ISSN 0079-6123
Copyright © 2007 Elsevier B.V. All rights reserved

CHAPTER 6

Towards a unified theory of neocortex: laminar cortical circuits for vision and cognition

Stephen Grossberg*

Department of Cognitive and Neural Systems, Center for Adaptive Systems, and Center of Excellence for Learning in Education, Science, and Technology, Boston University, 677 Beacon Street, Boston, MA 02215, USA

Abstract: A key goal of computational neuroscience is to link brain mechanisms to behavioral functions. The present article describes recent progress towards explaining how laminar neocortical circuits give rise to biological intelligence. These circuits embody two new and revolutionary computational paradigms: Complementary Computing and Laminar Computing. Circuit properties include a novel synthesis of feedforward and feedback processing, of digital and analog processing, and of preattentive and attentive processing. This synthesis clarifies the appeal of Bayesian approaches but has a far greater predictive range that naturally extends to self-organizing processes. Examples from vision and cognition are summarized. A LAMINART architecture unifies properties of visual development, learning, perceptual grouping, attention, and 3D vision. A key modeling theme is that the mechanisms which enable development and learning to occur in a stable way imply properties of adult behavior. It is noted how higher-order attentional constraints can influence multiple cortical regions, and how spatial and object attention work together to learn view-invariant object categories. In particular, a form-fitting spatial attentional shroud can allow an emerging view-invariant object category to remain active while multiple view categories are associated with it during sequences of saccadic eye movements. Finally, the chapter summarizes recent work on the LIST PARSE model of cognitive information processing by the laminar circuits of prefrontal cortex. LIST PARSE models the short-term storage of event sequences in working memory, their unitization through learning into sequence, or list, chunks, and their read-out in planned sequential performance that is under volitional control. LIST PARSE provides a laminar embodiment of Item and Order working memories, also called Competitive Queuing models, that have been supported by both psychophysical and neurobiological data. These examples show how variations of a common laminar cortical design can embody properties of visual and cognitive intelligence that seem, at least on the surface, to be mechanistically unrelated.

Keywords: neocortex; laminar circuits; learning; grouping; attention; 3D vision; working memory; categorization; V1; V2; V4; prefrontal cortex

*Corresponding author. Tel.: +1 617-353-7858;
Fax: +1 617-353-7755; E-mail: steve@bu.edu

DOI: 10.1016/S0079-6123(06)65006-1

Introduction

Although there has been enormous experimental and theoretical progress on understanding brain or mind in the fields of neuroscience and psychology, establishing a mechanistic link between them has been very difficult, if only because these two levels of description often seem to be so different. Yet establishing a link between brain and mind is crucial in any mature theory of how a brain or mind works. Without such a link, the mechanisms of the brain have no functional significance, and the functions of behavior have no mechanistic explanation. Throughout the history of psychology and neuroscience, some researchers have tried to establish such a link by the use of metaphors or the application of classical concepts to the brain. These have included hydraulic systems, digital computers, holograms, control theory circuits, and Bayesian networks, to name a few. None of these approaches has managed to explicate the unique design principles and mechanisms that characterize biological intelligence. The present chapter summarizes aspects of a rapidly developing theory of neocortex that links explanations of behavioral functions to underlying biophysical, neurophysiological, and anatomical mechanisms. Progress has been particularly rapid towards understanding how the laminar circuits of visual cortex see (Grossberg et al., 1997; Grossberg, 1999a, 2003a; Grossberg and Raizada, 2000; Raizada and Grossberg, 2001, 2003; Grossberg and Howe, 2003; Grossberg and Seitz, 2003; Grossberg and Swaminathan, 2004; Yazdanbakhsh and Grossberg, 2004; Cao and Grossberg, 2005; Grossberg and Yazdanbakhsh, 2005; Grossberg and Hong, 2006).

This progress illustrates the introduction of qualitatively new computational paradigms, as might have been expected, given how long these problems have remained unsolved. These results overcome a conceptual impasse that is illustrated by the popular proposal that our brains possess independent modules, as in a digital computer. The brain's organization into distinct anatomical areas and processing streams supports the idea that brain processing is specialized, but that, in itself, does not imply that these streams contain independent modules. This hypothesis gained dominance despite the fact that much behavioral data argue against independent modules. For example, during visual perception, strong interactions are known to occur between perceptual qualities (Kanizsa, 1974; Egusa, 1983; Faubert and von Grunau, 1995; Smallman and McKee, 1995; Pessoa et al., 1996). In particular, form and motion can interact, as can brightness and depth, among other combinations of qualities.

Complementary Computing and Laminar Computing

At least two new computational paradigms have gradually been identified from the cumulative experiences of modeling many kinds of brain and behavior data over the past three decades: Complementary Computing and Laminar Computing (Grossberg, 1999a, 2000). *Complementary Computing* concerns the discovery that pairs of parallel cortical processing streams compute complementary properties in the brain. Each stream has complementary computational strengths and weaknesses, much as in physical principles like the Heisenberg Uncertainty Principle. Each cortical stream can also possess multiple processing stages. These stages realize a *hierarchical resolution of uncertainty*. "Uncertainty" here means that computing one set of properties at a given stage can suppress information about a complementary set of properties at that stage. The computational unit of brain processing that has behavioral significance is thus not a single processing stage, or any smaller entity such as the potential of a single cell, or spike or burst of spikes. Instead, hierarchical interactions within a stream and parallel interactions between streams resolve their complementary deficiencies to compute complete information about a particular type of biological intelligence. These interactions have been used to clarify many of the data that do not support the hypothesis of independent modules. To model how the brain controls behavior, one thus needs to know how these complementary streams are organized with respect to one another.

Understanding how the brain sees is one area where experimental and modeling work have advanced the furthest, and illustrate several types of

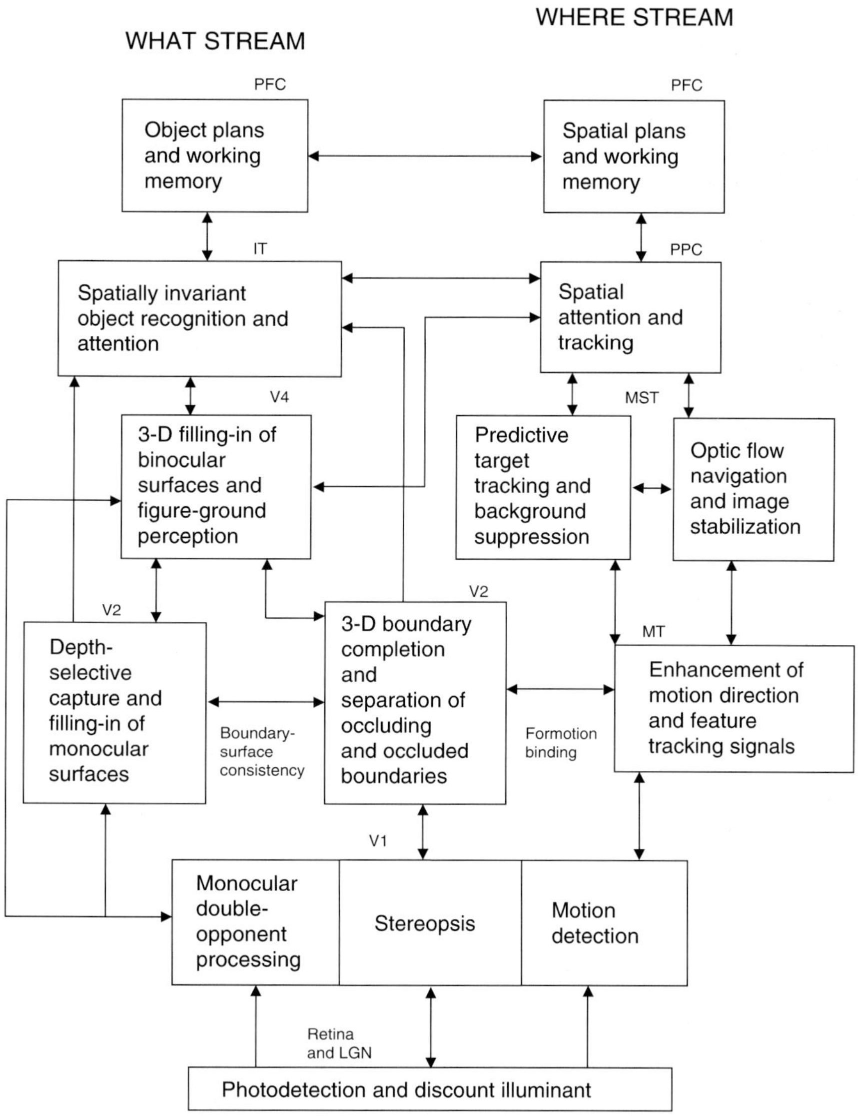

Fig. 1. Some visual processes and their anatomical substates that are being modeled as part of a unified vision system. LGN: Lateral Geniculate Nucleus; V1: striate visual cortex; V2, V4, MT, MST: prestriate visual cortex; IT: inferotemporal cortex; PPC: posterior parietal cortex; PFC: prefrontal cortex.

complementary interactions. Figure 1 provides a schematic macrocircuit of the types of processes that are being assembled into a unified theory of how the brain sees, including processes of vision, recognition, navigation, tracking, and visual cognition. In particular, matching and learning processes within the What and Where cortical streams have been proposed to be complementary: The What stream, through cortical areas V1-V2-V4-IT-PFC, learns to recognize *what* objects and events occur. The Where stream, through cortical areas V1-MT-MST-PPC-PFC, spatially localizes *where* they are, and acts upon them. Complementary processes also occur within each stream: What stream boundary grouping via the (V1 interblob)-(V2 pale stripe)-V4 stages, and surface formation via the (V1 blob)-(V2 thin stripe)-V4 stages, have complementary properties. Where stream target tracking via MT-(MST ventral) and navigation via MT-(MST dorsal) have complementary properties. Such complementary processes are predicted to arise from symmetry-breaking operations during cortical development.

Laminar Computing concerns the fact that cerebral cortex is organized into layered circuits (usually six main layers) which undergo characteristic bottom-up, top-down, and horizontal interactions, which have been classified into more than 50 divisions, or areas, of neocortex (Brodmann, 1909; Martin, 1989). The functional utility of such a laminar organization in the control of behavior has remained a mystery until recently. Understanding how different parts of the neocortex specialize the same underlying laminar circuit design in order to achieve all the highest forms of biological intelligence remains a long-term goal, although challenging data about both vision and cognitive information processing (Grossberg and Pearson, 2006; Pearson and Grossberg, 2005) have now been modeled as variations of this design.

Laminar computing by visual cortex: unifying adaptive filtering, grouping, and attention

A number of models have been proposed (Douglas et al., 1995; Stemmler et al., 1995; Li, 1998;

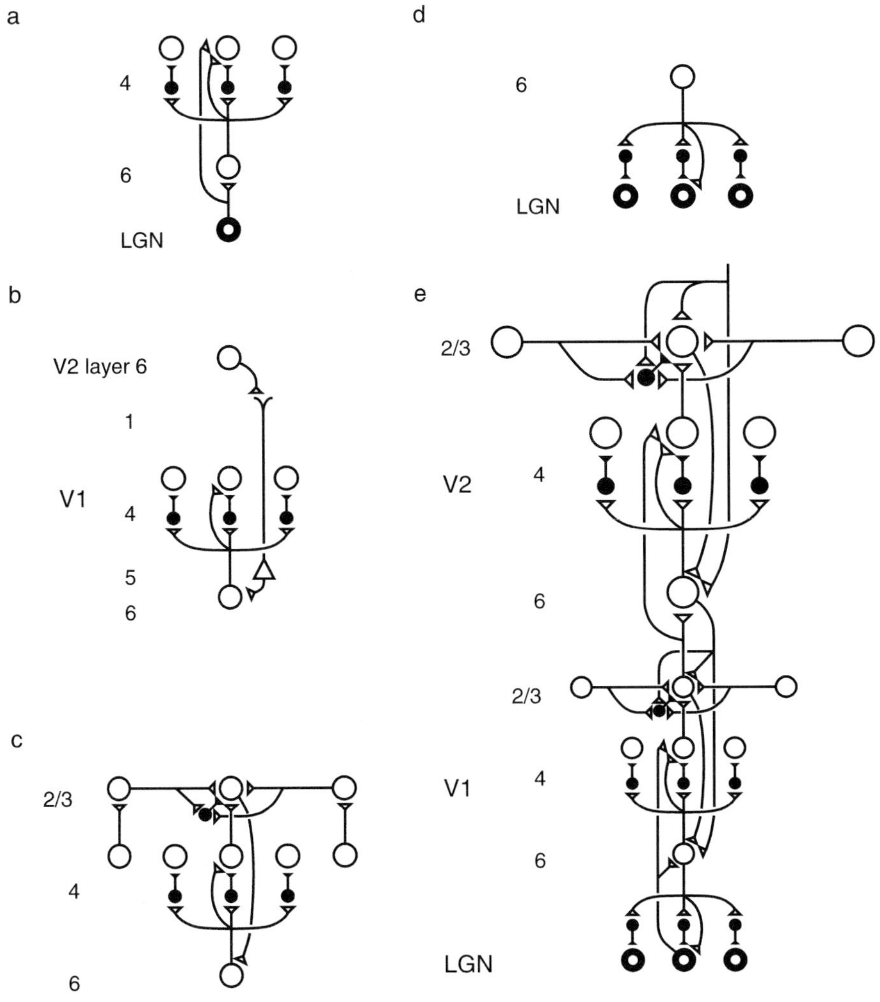

Somers et al., 1998; Yen and Finkel, 1998) to simulate aspects of visual cortical dynamics, but these models have not articulated why cortex has a laminar architecture. Our own group's breakthrough on this problem (Grossberg et al., 1997; Grossberg, 1999a) began with the suggestion that the laminar organization of visual cortex accomplishes at least three things: (1) the developmental and learning processes whereby the cortex shapes its circuits to match environmental constraints in a *stable* way through time; (2) the binding process whereby cortex groups distributed data into coherent object representations that remain sensitive to analog properties of the environment; and (3) the attentional process whereby cortex selectively processes important events.

These results further develop the proposal that even the earliest stages of visual cortex are not merely a bottom-up filtering device, as in the classical model of Hubel and Wiesel (1977). Instead, bottom-up filtering, horizontal grouping, and top-down attention are all joined together in laminar cortical circuits. Perceptual grouping, the process that binds spatially distributed and incomplete information into 3D object representations, starts at an early cortical stage; see Fig. 2c. These grouping interactions are often cited as the basis of "non-classical" receptive fields that are sensitive to the context in which individual features are found (von der Heydt et al., 1984; Peterhans and von der Heydt, 1989; Knierim and van Essen, 1992; Grosof et al., 1993; Kapadia et al., 1995; Sillito et al., 1995; Sheth et al., 1996; Bosking et al., 1997; Polat et al., 1998). Likewise, even early visual processing is modulated by system goals via top-down expectations and attention (Motter, 1993; Sillito et al., 1994; Roelfsema et al., 1998; Watanabe et al., 1998; Somers et al., 1999). The model proposes how mechanisms governing (1) in the infant lead to properties (2) and (3) in the adult, and properties (2) and (3) interact together intimately as a result.

The laminar model proposes that there is no strict separation of preattentive data-driven bottom-up filtering and grouping, from attentive task-directed top-down processes. The model shows how these processes may come together at a shared

Fig. 2. How known cortical connections join the layer 6→4 and layer 2/3 circuits to form an entire V1/V2 laminar model. Inhibitory interneurons are shown filled-in black. (a) The LGN provides bottom-up activation to layer 4 via two routes. First, it makes a strong connection directly into layer 4. Second, LGN axons send collaterals into layer 6, and thereby also activate layer 4 via the layer 6 → layer 4 on-center off-surround path. The combined effect of the bottom-up LGN pathways is to stimulate layer 4 via an on-center off-surround, which provides divisive contrast normalization (Grossberg, 1973, 1980; Heeger, 1992) of layer 4 cell responses. (b) Feedback carries attentional signals from higher cortex, via the modulatory layer 6 → layer 4 path, into layer 4 of V1. Because of the bend in this feedback pathway, I have called it "folded feedback." Corticocortical feedback axons tend preferentially to originate in layer 6 of the higher area and to terminate in layer 1 of the lower cortex (Salin and Bullier, 1995, p. 110), where they can excite the apical dendrites of layer 5 pyramidal cells whose axons send collaterals into layer 6. The triangle in the figure represents such a layer 5 pyramidal cell. Several other routes through which feedback can pass into V1 layer 6 exist (see Raizada and Grossberg (2001) for a review). Having arrived in layer 6, the feedback is then "folded" back up into the feedforward stream by passing through the 6 → 4 on-center off-surround path (Bullier et al., 1996). (c) Connecting the 6→4 on-center off-surround to the layer 2/3 grouping circuit: like-oriented layer 4 simple cells with opposite contrast polarities compete (not shown) before generating half-wave rectified outputs that converge onto layer 2/3 complex cells in the column above them. Just like attentional signals from higher cortex, as shown in (b), groupings that form within layer 2/3 also send activation into the folded feedback path, to enhance their own positions in layer 4 beneath them via the 6→4 on-center, and to suppress input to other groupings via the 6→4 off-surround. There exist direct layer 2/3→6 connections in macaque V1, as well as indirect routes via layer 5. (d) Top-down corticogeniculate feedback from V1 layer 6 to LGN also has an on-center off-surround anatomy, similar to the 6→4 path. The on-center feedback selectively enhances LGN cells that are consistent with the activation that they cause (Sillito et al., 1994), and the off-surround contributes to length-sensitive (endstopped) responses that facilitate grouping perpendicular to line ends. (e) The entire V1/V2 circuit: V2 repeats the laminar pattern of V1 circuitry, but at a larger spatial scale. In particular, the horizontal layer 2/3 connections have a longer range in V2, allowing above-threshold perceptual groupings between more widely spaced inducing stimuli to form (Amir et al., 1993). V1 layer 2/3 projects up to V2 layers 6 and 4, just as LGN projects to layers 6 and 4 of V1. Higher cortical areas send feedback into V2 which ultimately reaches layer 6, just as V2 feedback acts on layer 6 of V1 (Sandell and Schiller, 1982). Feedback paths from higher cortical areas straight into V1 (not shown) can complement and enhance feedback from V2 into V1. Top-down attention can also modulate layer 2/3 pyramidal cells directly by activating both the pyramidal cells and inhibitory interneurons in that layer. The inhibition tends to balance the excitation, leading to a modulatory effect. These top-down attentional pathways tend to synapse in layer 1, as shown in Fig. 2b. Their synapses on apical dendrites in layer 1 are not shown, for simplicity. (Adapted with permission from Raizada and Grossberg (2001).)

circuit, or interface, that is called the *preattentive–attentive interface*, which exists between layers 6 and 4 (Fig. 2a–c, e). Significantly, by indicating how mechanisms whereby the cortex can develop and learn in a stable way impose computational constraints that *define* key properties of adult visual information processing, the model begins to unify the fields of cortical development and adult perceptual learning and information processing. The model is called a LAMINART model (Fig. 2; Grossberg, 1999a; Raizada and Grossberg, 2003) because it clarifies how mechanisms of Adaptive Resonance Theory (ART), which have previously been predicted to stabilize cortical development and learning of bottom-up adaptive filters and top-down attentive expectations (Grossberg, 1980, 1999c; Carpenter and Grossberg, 1993) can be joined together in laminar circuits to processes of perceptual grouping through long-range horizontal interactions (Grossberg and Mingolla, 1985b).

A new way to compute: feedforward and feedback, speed and uncertainty, digital and analog

The LAMINART model proposes how laminar neocortex embodies a novel way to compute which exhibits at least three major new computational properties (Grossberg, 2003a). These new properties allow the fast, but stable, autonomous self-organization that is characteristic of cortical development and life-long learning in response to changing and uncertain environments. They go beyond the types of Bayesian cortical models that are so popular today, but also clarify the intuitive appeal of these models (Pilly and Grossberg, 2005; Grossberg and Pilly, 2007).

The first property concerns a new type of hybrid between *feedforward and feedback computing*. In particular, when an unambiguous scene is processed, the LAMINART model can quickly group the scene in a fast feedforward sweep of activation that passes directly through layer 4 to 2/3 and then on to layers 4 to 2/3 in subsequent cortical areas. This property clarifies how recognition can be fast in response to unambiguous scenes; e.g., Thorpe et al. (1996). If, however, there are multiple possible groupings, say in response to a complex textured scene, then competition among these possibilities due to inhibitory interactions in layers 4 and 2/3 can cause all cell activities to become smaller. This happens because the competitive circuits in the model are *self-normalizing*; that is, they tend to conserve the total activity of the circuit. This self-normalizing property emerges from on-center off-surround networks of cells that obey membrane, or *shunting*, equations. Such networks are capable of processing input contrasts over a large dynamic range without saturation (Grossberg, 1973, 1980; Heeger, 1992; Douglas et al., 1995).

In other words, these self-normalizing circuits carry out a type of real-time probability theory in which the amplitude and coherence of cell activity covaries with the certainty of the network's selection, or decision, about a grouping. Amplitude also covaries with processing speed. Low activation greatly slows down the feedforward processing in the circuit because it takes longer for cell activities to exceed output thresholds and to activate subsequent cells above threshold. In the model, network uncertainty is resolved through feedback: Weakly active layer 2/3 grouping cells feed back signals to layers 6-then-4-then-2/3 to close a cortical feedback loop that rapidly contrast enhances and amplifies a winning grouping. As the winner is selected, and weaker groupings are suppressed, its cells become more active, hence can again more rapidly exceed output thresholds and send the cortical decision to subsequent processing stages.

In summary, the LAMINART circuit behaves like a real-time probabilistic decision circuit that operates in a fast feedforward mode when there is little uncertainty, and automatically switches to a slower feedback mode when there is significant uncertainty. Feedback selects a winning decision that enables the circuit to speed up again. Activation amplitude and processing speed both increase with certainty. The large activation amplitude of a winning grouping is facilitated by the synchronization that occurs as the winning grouping is selected.

These concepts are illustrated within an emerging unified model of how cortical form and motion processes interact. This 3D FORMOTION model has quantitatively simulated and predicted the temporal dynamics of how the visual cortex responds to motion stimuli, including motion stimuli whose coherence is probabilistically defined (Chey et al., 1997; Grossberg et al., 2001; Berzhanskaya et al., 2007; Grossberg and Pilly, 2007). Grossberg and Pilly (2007) have, in particular, proposed a how Retina/LGN-V1-MT-MST-LIP-Basal Ganglia interactions can quantitatively explain and simulate data about probabilistic decision-making in LIP (Shadlen and Newsome, 2001; Roitman and Shadlen, 2002). These experiments have been presented as supportive of Bayesian processing.

The second property concerns a novel kind of hybrid computing that simultaneously realizes the *stability of digital computing and the sensitivity of analog computing*. This is true because the intracortical feedback loop between layers 2/3-6-4-2/3 that selects or confirms a winning grouping has the property of *analog coherence* (Grossberg et al., 1997; Grossberg, 1999a; Grossberg and Raizada, 2000); namely, this feedback loop can synchronously store a winning grouping without losing analog sensitivity to amplitude differences in the input pattern. The coherence of synchronous selection and storage provides the stability of digital computing. The sensitivity of analog computation can be traced to how excitatory and inhibitory interactions are balanced within layers 4 and 2/3, and to the shunting dynamics of the inhibitory interactions within these layers.

The third property concerns its ability to self-stabilize development and learning using the *intra*cortical feedback loop between layers 2/3-6-4-2/3 by selecting cells that fire together to wire together. As further discussed below, this intracortical decision circuit is predicted to help stabilize development in the infant and learning throughout life, as well as to select winning groupings in the adult (Grossberg, 1999a).

The critical role of the layer 6-to-4 decision circuits in the realization of all three properties clarifies that they are all different expressions of a shared circuit design.

Linking stable development to synchrony

The LAMINART model clarifies how excitatory and inhibitory connections in the cortex can develop in a stable way by achieving and maintaining a *balance* between excitation and inhibition (Grossberg and Williamson, 2001). Long-range excitatory horizontal connections between pyramidal cells in layer 2/3 of visual cortical areas play an important role in perceptual grouping (Hirsch and Gilbert, 1991; McGuire et al., 1991). The LAMINART model proposes how development enables the strength of long-range excitatory horizontal signals to become balanced against inhibitory signals that are mediated by short-range disynaptic inhibitory interneurons which target the same pyramidal cells (Fig. 2c). These balanced connections are proposed to realize properties of perceptual grouping in the adult. In a similar way, development enables the strength of excitatory connections from layer 6-to-4 to be balanced against those of inhibitory interneuronal connections (Wittmer et al., 1997); see Fig. 2a, c. Thus, the net excitatory effect of layer 6 on layer 4 is proposed to be modulatory. These approximately balanced excitatory and inhibitory connections exist within the on-center of a *modulatory* on-center, off-surround network from layer 6-to-4. This network plays at least three functional roles that are intimately linked: maintaining a contrast-normalized response to bottom-up inputs at layer 4 (Fig. 2a); forming perceptual groupings in layer 2/3 that maintain their sensitivity to analog properties of the world (Fig. 2c); and biasing groupings via top-down attention from higher cortical areas (Fig. 2b; also see Fig. 2d, e).

Balanced excitatory and inhibitory connections have been proposed by several models to explain the observed variability in the number and temporal distribution of spikes emitted by cortical neurons (Shadlen and Newsome, 1998; van Vreeswijk and Sompolinsky, 1998). The LAMINART model proposes that such variability may reflect mechanisms that are needed to ensure stable development and learning. If indeed "stability implies variability," how does the cortex convert these variable spikes, which are inefficient in driving responses from cortical neurons, into reliable responses to visual inputs? Within LAMINART circuits, such balanced excitatory and inhibitory connections respond to inputs by rapidly synchronizing their responses to input stimuli (Yazdanbakhsh and Grossberg, 2004; Grossberg and Versace, 2007; see also Grossberg and Somers, 1991;

Grossberg and Grunewald, 1997). In fact, the article that introduced ART predicted a role for synchronous cortical processing, including synchronous oscillations, which were there called "order-preserving limit cycles," as part of the process of establishing resonant states (Grossberg, 1976). Since the early experimental reports of Eckhorn et al. (1988) and Gray and Singer (1989), many neurophysiological experiments have reported synchronous cortical processing; e.g., Engel et al. (2001), Fries et al. (2001), and Sarnthein et al. (1998). The ART model further predicted a functional link between properties of stable development, adult perceptual learning, attention, and synchronous cortical processing, to which the LAMINART model adds perceptual grouping in laminar cortical circuits.

The Synchronous Matching ART (SMART) model of Grossberg and Versace (2005, 2006, 2007) further develops LAMINART to clarify how multiple levels of brain organization work together, ranging from individual spikes, through local field potentials and inter-areal synchronization, to cognitive learning dynamics. SMART proposes how higher-order specific and nonspecific thalamic nuclei are coordinated with multiple stages of cortical processing to control stable spike-timing-dependent plasticity (STDP). The model proposes how gamma oscillations can facilitate STDP learning, and how slower beta oscillations may be generated during reset events. It furthermore predicts that reset is mediated by the deeper layers of cortex. The model hereby predicts that "more gamma" can be expected through time in the superficial layers of cortex than the deeper layers.

Attention arises from top-down cooperative-competitive matching

Attention typically modulates an ongoing process. In order for the concept of attention to be scientifically useful, these processes need to be articulated and the way in which attention modulates them needs to be mechanistically explained. LAMINART, and ART before it, predicted that an intimate link exists between processes of attention, competition, and bottom-up/top-down matching. LAMINART predicts, in particular, that top-town signals from higher cortical areas, such as area V2, can attentionally prime, or modulate, layer 4 cells in area V1 by activating the on-center off-surround network from layer 6-to-4 (Fig. 2b, e). Because the excitatory and inhibitory signals in the on-center are balanced, attention can sensitize, or modulate, cells in the attentional on-center, without fully activating them, while also inhibiting cells in the off-surround.

The importance of the conclusion that top-down attention is often expressed through a top-down, modulatory on-center, off-surround network cannot be overstated. Because of this organization, top-down attention can typically provide only excitatory modulation to cells in the on-center, while it can strongly inhibit cells in the off-surround. As Hupé et al. (1997, p. 1031) have noted: "feedback connections from area V2 modulate but do not create center-surround interactions in V1 neurons." When the top-down on-center matches bottom-up signals, it can amplify and synchronize them, while strongly suppressing mismatched signals in the off-surround. This prediction was first made as part of ART in the 1970s (Grossberg, 1976, 1978, 1980, 1999a, 1999c). It has since received both of psychological and neurobiological empirical confirmation in the visual system (Downing, 1988; Sillito et al., 1994; Steinman et al., 1995; Bullier et al., 1996; Caputo and Guerra, 1998; Somers et al., 1999; Reynolds et al., 1999; Mounts, 2000; Smith et al., 2000; Vanduffel et al., 2000). Based on such data, this property has recently been restated, albeit without a precise anatomical realization, in terms of the concept of "biased competition" (Desimone, 1998; Kastner and Ungerleider, 2001), in which attention biases the competitive influences within the network. Figure 3 summarizes data of Reynolds et al. (1999) and a simulation of these data from Grossberg and Raizada (2000) that illustrate the on-center off-surround character of attention in macaque V2.

The preattentive–attentive interface and object-based attention

Top-down attention and preattentive perceptual grouping interact within the cortical layers to

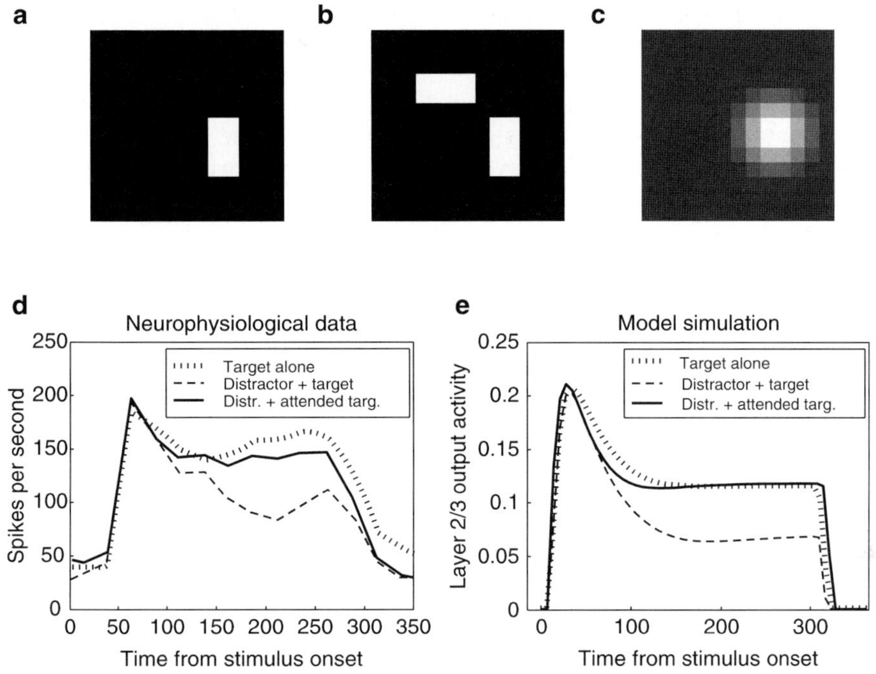

Fig. 3. The effect of attention on competition between visual stimuli. A target stimulus, presented on its own (a), elicits strong neural activity at the recorded cell. When a second, distracter stimulus is presented nearby (b), it competes against the target, and activity is reduced. Directing spatial attention to the location of the target stimulus (c), protects the target from this competition, and restores neural activity to the levels elicited by the target on its own. The stimuli shown here, based on those used in the neurophysiological experiments of Reynolds et al. (1999), were presented to the model neural network. Spatial attention (c), was implemented as a Gaussian of activity fed back into layer 6. (d) Neurophysiological data from macaque V2 that illustrate the recorded activity patterns described above: strong responses to an isolated target (dotted line), weaker responses when a competing distracter is placed nearby (dashed line) and restored levels of activity when the target is attended (solid line). (Adapted with permission from Reynolds et al., 1999, Fig. 5; see also Reynolds et al., 1995). (e) Model simulation of the Reynolds et al. data. The time-courses illustrated show the activity of a vertically oriented cell stimulated by the target bar. If only the horizontal distracter bar were presented on its own, this cell would respond very weakly. If both target and distracter were presented, but with the horizontal distracter attended, the cell would respond, but more weakly than the illustrated case where the distracter and target are presented together, with neither attended. (Adapted with permission from Grossberg and Raizada (2000).)

enable attention to focus on an entire object boundary, thereby enabling whole object boundaries to be selectively attended and recognized. This happens because the same layer 6-to-4 competition, or selection, circuit may be activated both by preattentive grouping cells in layer 2/3, and by top-down attentional pathways (Fig. 2b, c). Layer 4 cells can then, in turn, reactivate layer 2/3 cells (Fig. 2c). This layer 6-to-4 circuit "folds" the feedback from top-down attention or a layer 2/3 grouping back into the feedforward flow of bottom-up inputs to layer 4. It is thus said to embody a "folded feedback" process (Grossberg, 1999a). Thus, when ambiguous complex scenes are being processed, *intracortical* folded feedback enables stronger groupings in layer 2/3 to inhibit weaker groupings, whereas *intercortical* folded feedback enables higher-order attentive constraints to bias which groupings will be selected.

Figure 2e summarizes the hypothesis that top-down attentional signals to layer 1 may also directly modulate groupings via apical dendrites of both excitatory and inhibitory layer 2/3 cells in layer 1 (Rockland and Virga, 1989; Lund and Wu, 1997). By activating both excitatory and inhibitory cells in layer 2/3, the inhibitory cells may balance the excitatory cell activation, thereby enabling attention to directly modulate grouping cells in layer 2/3.

Because the cortex uses the same circuits to select groupings and to prime attention, attention can flow along perceptual groupings, as reported by Roelfsema et al. (1998). In particular, when attention causes an excitatory modulatory bias at some cells in layer 4, groupings that form in layer 2/3 can be enhanced by this modulation via their positive feedback loops from 2/3-to-6-to-4-to-2/3. The direct modulation of layer 2/3 by attention can also enhance these groupings. Figure 4 summarizes a LAMINART simulation of data from Roelfsema et al. (1998) of the spread of visual attention along an object boundary grouping. LAMINART has also been used to simulate the flow of attention along an illusory contour (Raizada and Grossberg, 2001), consistent with experimental data of Moore et al. (1998). The ability of attention to selectively light up entire object representations has an obviously important survival value in adults. It is thus of particular interest that the intracortical and intercortical feedback circuits that control this property have been shown in modeling studies to play a key role in stabilizing infant development and adult perceptual learning within multiple cortical areas, including cortical areas V1 and V2.

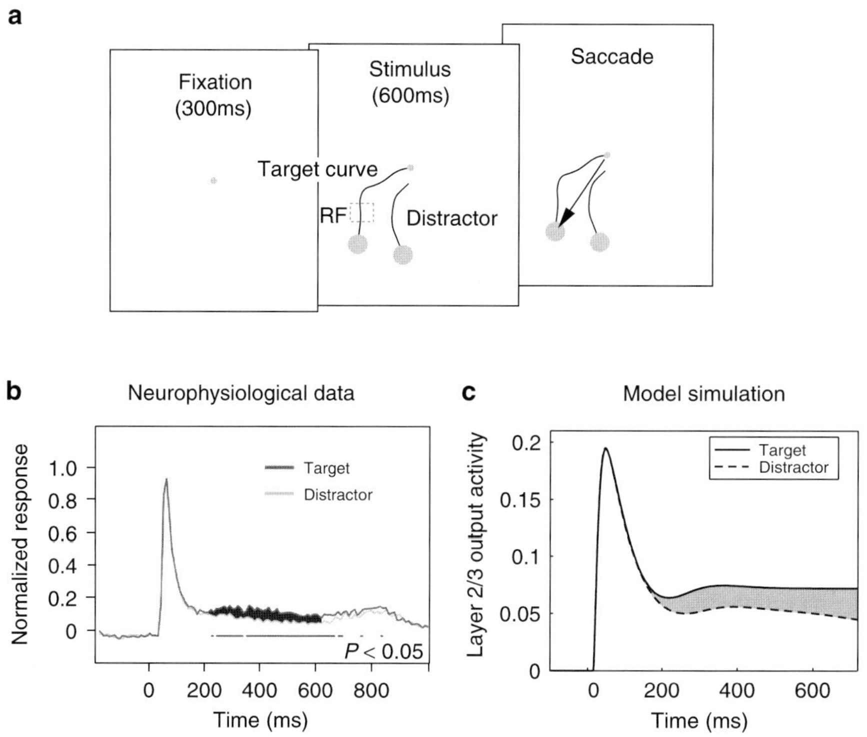

Fig. 4. Spread of visual attention along an object boundary grouping, from an experiment by Roelfsema et al. (1998). (a) The experimental paradigm. Macaque monkeys were trained to perform a mental curve-tracing task, during which physiological recordings were made in V1: A fixation spot was presented for 300 ms, followed by a target curve and a distracter curve presented simultaneously. The target was connected at one end to the fixation point. While maintaining fixation, the monkeys had to trace the target curve, then, after 600 ms, make a saccade to its endpoint. (b) Neurophysiological data showing attentional enhancement of the firing of a neuron when its receptive field (RF) lay on the target curve, as opposed to the distracter. Note that the enhancement occurs about 200 ms after the initial burst of activity. Further studies have indicated that the enhancement starts later in distal curve segments, far from the fixation point, than it does in proximal segments, closer to fixation (Pieter Roelfsema, personal communication). This suggests that attentional signals propagate along the length of the target curve. (Figures (a) and (b) adapted with permission from Roelfsema et al. (1998).) (c) Model simulation of the Roelfsema et al. data. (Adapted with permission from Grossberg and Raizada (2000).)

Stable development and learning through adaptive resonance

ART (Grossberg, 1980, 1995, 1999c; Pollen, 1999; Engel et al., 2001) is a cognitive and neural theory which addresses a general problem that faces all adaptive brain processes; namely, the *stability–plasticity dilemma*: how can brain circuits be plastic enough to be rapidly fine-tuned by new experiences, and yet simultaneously stable enough that they do not get catastrophically overwritten by the new stimuli with which they are continually bombarded?

The solution that ART proposes to this problem is to allow neural representations to be rapidly modified only by those incoming stimuli with which they form a sufficiently close match. If the match is close enough, then learning occurs. Because an approximate match is required, such learning fine-tunes the memories of existing representations, so that outliers cannot radically overwrite an already learned representation. ART proposes how a learning individual can flexibly vary the criterion, called *vigilance*, of how good a match is needed between bottom-up and top-down information in order for the presently active representation to be refined through learning. When coarse matches are allowed (low vigilance), the learned representations can represent general and abstract information. When only fine matches are allowed (high vigilance), the representations are more specific and concrete. If the active neural representation does not match with the incoming stimulus, then its neural activity is extinguished and hence unable to cause plastic changes. Suppression of an active representation enables mismatch-mediated memory search, or hypothesis-testing, to ensue whereby some other representation can become active through bottom-up signaling. This representation, in turn, reads out top-down signals that either gives rise to a match, thereby allowing learning, or a nonmatch, causing the search process to repeat until either a match is found or the incoming stimulus causes a totally new representation to be formed. For this to work, top-down expectations have large enough adaptive memory traces to enable an initial match to occur with a newly selected representation. It has been suggested that breakdowns in vigilance control can contribute to various disorders, including medial temporal amnesia (abnormally low vigilance; Carpenter and Grossberg, 1993) and autism (abnormally high vigilance; Grossberg and Seidman, 2006).

In both ART and its elaboration into LAMINART, attention is mediated by a top-down, modulatory on-center, off-surround network (e.g., Grossberg, 1980, 1982, 1999b), whose role is to select and enhance behaviorally relevant bottom-up sensory inputs (match), and suppress those that are irrelevant (nonmatch). Mutual excitation between the top-down feedback and the bottom-up signals that they match can amplify, synchronize, and maintain existing neural activity in a resonant state long enough for rapid synaptic changes to occur. Thus, attentionally relevant stimuli are learned, while irrelevant stimuli are suppressed and prevented from destabilizing existing representations. Hence the name *adaptive* resonance. Grossberg (1999c, 2003a) provides more detailed reviews.

The folded feedback layer 6-to-4 modulatory on-center, off-surround attentional pathway in the LAMINART model (Fig. 2b) satisfies the predicted properties of ART matching. The claim that bottom-up sensory activity is enhanced when matched by top-down signals is in accord with an extensive neurophysiological literature showing the facilitatory effect of attentional feedback (Sillito et al., 1994; Luck et al., 1997; Roelfsema et al., 1998), but not with models in which matches with top-down feedback cause suppression (Mumford, 1992; Rao and Ballard, 1999). The ART proposal raises two key questions: First, does top-down cortical feedback have the predicted top-down, modulatory on-center, off-surround structure in other neocortical structures, where again the stabilizing role of top-down feedback in learning would be required? Second, is there evidence that top-down feedback controls plasticity in the area to which it is directed?

Zhang et al. (1997) have shown that feedback from auditory cortex to the medial geniculate nucleus (MGN) and the inferior colliculus (IC) also has an on-center off-surround form, and Temereanca and Simons (2001) have produced evidence for a similar feedback architecture in the rodent barrel somatosensory system.

The link between attention and learning

Accumulating evidence also shows that top-down feedback helps to control cortical plasticity. Psychophysically, the role of attention in controlling adult plasticity and perceptual learning was demonstrated by Ahissar and Hochstein (1993). Gao and Suga (1998) reported physiological evidence that acoustic stimuli caused plastic changes in the IC of bats only when the IC received top-down feedback from auditory cortex. Plasticity is enhanced when the auditory stimuli were made behaviorally relevant, consistent with the ART proposal that top-down feedback allows attended, and thus relevant, stimuli to be learned, while suppressing unattended irrelevant ones. Cortical feedback also controls thalamic plasticity in the somatosensory system (Krupa et al., 1999; Parker and Dostrovsky, 1999). See Kaas (1999) for a review.

Models of intracortical feedback due to grouping, and of corticocortical and thalamocortical feedback due to attention, have shown that either type of feedback can rapidly synchronize the firing patterns of higher and lower cortical areas (Grossberg and Somers, 1991; Grossberg and Grunewald, 1997; Yazdanbakhsh and Grossberg, 2004; Grossberg and Versace, 2007). ART predicts that such synchronization phenomena underlie the type of resonances that can trigger cortical learning by enhancing the probability that "cells that fire together wire together." Engel et al. (2001) review data and related models that are consistent with the proposal that synchrony, attention, and learning are related.

View-invariant object category learning: coordinating object attention and surface-based spatial attention shrouds

The above summary has focused on object attention. It did not discuss spatial attention, how spatial and object attention work together, or how attention is hierarchically organized. The above summary also talks about category learning, but not the fact that view-invariant object categories may be learned from combinations of multiple object views. The present section sketches some results concerning these more global issues about brain organization.

One way in which attention may globally influence many brain regions is illustrated in Fig. 2e, which shows how attention can leap from higher cortical levels to multiple lower cortical areas via their layers 6. This anatomy proposes a solution to an otherwise challenging problem: How can attention prime so many cortical areas with higher-order constraints without inadvertently firing them all? Figure 2e shows that attention can leap between the layers 6 of different cortical areas without firing them all, because the layer 6-to-4 circuits that act intracortically are modulatory.

The above example illustrates how attention can act "vertically" between cortical regions. Many studies have analyzed how attention is spread "horizontally" across a given level of cortical processing, including how spatial attention may be simultaneously divided among several targets (Pylyshyn and Storm, 1988; Yantis, 1992), and how object and spatial attention may both influence visual perception (Posner, 1980; Duncan, 1984). The distinction between object and spatial attention reflects the organization of visual cortex into parallel What and Where processing streams (Fig. 1). Many cognitive neuroscience experiments have supported the hypotheses of Ungerleider and Mishkin (1982; see also Mishkin et al. (1983)) and of Goodale and Milner (1992) that inferotemporal cortex and its cortical projections learn to categorize and recognize what objects are in the world, whereas the parietal cortex and its cortical projections learn to determine where they are and how to deal with them by locating them in space, tracking them through time, and directing actions towards them. This design into parallel streams separates sensory and cognitive processing from spatial and motor processing.

The What stream strives to generate object representations that are independent of their spatial coordinates, whereas the Where stream generates representations of object location and action. The streams must thus interact to act upon recognized objects. Indeed, both object and spatial attention are needed to search a scene for visual targets and distracters using saccadic eye movements. Grossberg et al. (1994) illustrated how object and spatial attention may interact by quantitatively fitting a large human psychophysical database about visual search with a model, called the Spatial Object Search (SOS) model, that proposes how 3D boundary groupings and surface representations interact with object attention and spatial attention to find targets amid distracters. This analysis proposed that surface properties may engage spatial attention, as when search is restricted to all occurrences of a color on a prescribed depth plane (Egeth et al., 1984; Nakayama and Silverman, 1986; Wolfe and Friedman-Hill, 1992).

More recent modeling work has advanced the theoretical analysis of how spatial and object attention are coordinated by surface and boundary representations, by showing how they support the learning of view-invariant object categories (Fazl et al., 2005, 2006, 2007). This work advances the solution of the following problem: What is an object? How does the brain learn what an object is under both unsupervised and supervised learning conditions? How does the brain learn to bind multiple views of an object into a view-invariant representation of a complex object while scanning its various parts with active eye movements? How does the brain avoid the problem of erroneously classifying views of different objects as belonging to a single object, even before it has a concept of what the object is? How does the brain direct the eyes to explore an object's surface even before it has a concept of the object? The ARTSCAN model predicts how spatial and object attention work together to direct eye movements to explore object surfaces and to enable learning of view-invariant object categories from the multiple view categories that are thereby learned.

In particular, ARTSCAN predicts that spatial attention employs an *attentional shroud*, or form-fitting distribution of spatial attention, that is derived through feedback interactions with an object's surface representation. ARTSCAN modifies the original Tyler and Kontsevich (1995) concept of an attentional shroud in which the shroud was introduced as an alternative to the perception of simultaneous transparency, with evidence that only one plane is seen at a time within the perceptual moment. This concept focuses on object perception. ARTSCAN proposes that an attentional shroud also plays a fundamental role in regulating object learning.

Such a shroud is proposed to persist within the Where stream during active scanning of an object with attentional shifts and eye movements. This claim raises the basic question: How can the shroud persist during active scanning of an object, if the brain has not yet learned that there is an object there? ARTSCAN proposes how a *preattentively* formed surface representation leads to activation of a shroud, even before the brain can recognize the surface as representing a particular object. Such a shroud can be formed starting with either bottom-up or top-down signals. In the bottom-up route, a surface representation (e.g., in visual cortical area V4) directly activates a shroud, which conforms its shape to that of the surface, in a spatial attention cortical area (e.g., posterior parietal cortex). The shroud, in turn, can topographically prime the surface representation via top-down feedback. A surface-shroud resonance can hereby develop. In the top-down route, a volitionally controlled, local focus of spatial attention (an attentional spotlight) can send a top-down attentional signal to a surface representation. This spotlight of enhanced activation can then fill-in across the entire surface, being contained only by the surface boundary (Grossberg and Mingolla, 1985b). Surface filling-in generates a higher level of filled-in surface activation than did the bottom-up input to the surface alone. The filling-in of such a top-down attentional spotlight can hereby have an effect on the total filled-in surface activity that is similar to that caused by a higher bottom-up stimulus contrast (Reynolds and Desimone, 2003). The more highly active surface representation can reactivate the spatial attention region to define a surface form-fitting spatial locus of spatial attention; that is, a shroud. Again, the shroud is defined by a surface-shroud resonance.

Any surface in a scene can potentially sustain an attentional shroud, and surface representations dynamically compete for spatial attention. The winner of the competition at a given moment gains more activity and becomes the shroud.

As saccadic eye movements explore an object's surface, the surface-induced shroud modulates object learning in the What stream by maintaining activity of an emerging view-invariant category representation while multiple view-specific representations are linked to it through learning. Output from the shroud also helps to select the boundary and surface features to which eye movements will be directed, via a surface contour process that is predicted to play a key role in 3D figure-ground separation (Grossberg, 1994, 1997) and to be mediated via cortical area V3A (Nakamura and Colby, 2000a, b).

The model postulates that an active shroud weakens through time due to self-inhibitory inputs at selected target locations ("inhibition of return"; Grossberg, 1978; Koch and Ullman, 1985), combined with chemical transmitters that habituate, or are depressed, in an activity-dependent way (Grossberg, 1968; Francis et al., 1994; Abbott et al., 1997) and gate the signals that sustain the shroud. When an active shroud is weakened enough, it collapses and cannot any longer inhibit a tonically active reset signal. When a reset signal is disinhibited, it inhibits the active view-invariant object category in the What Stream, thereby preventing it from erroneously being linked to the view categories of subsequently foveated objects. Then a new shroud, corresponding to some other surface, is selected in the Where stream, as a new object category is activated in the What stream by the first view of the new object.

While a shroud remains active, the usual ART mechanisms direct object attention to ensure that new view categories and the emerging view-invariant object category are learned in a stable way through time. ARTSCAN hereby provides a new proposal for how surface-based spatial attention and object attention are coordinated to learn view-invariant object categories.

The ARTSCAN model learns with 98.1% accuracy on a letter database whose letters vary in size, position, and orientation. It does this while achieving a compression factor of 430 in the number of its category representations, compared to what would be required to learn the database without the view-invariant categories. The model also simulates reaction times (RTs) in human data about object-based attention: RTs are faster when responding to the noncued end of an attended object compared to a location outside the object, and slower engagement of attention to a new object occurs if attention has to first be disengaged from another object first (Brown and Denney, in press; Egly et al., 1994).

Learning without attention: the preattentive grouping is its own attentional prime

The fact that attentional feedback can influence cortical plasticity does not imply that unattended stimuli can never be learned. Indeed, abundant plasticity occurs during early development, before top-down attention has even come into being. Grossberg (1999a) noted that, were this not possible, an infinite regress could be created, since a lower cortical level like V1 might then not be able to stably develop unless it received attentional feedback from V2, but V2 itself could not develop unless it had received reliable bottom-up signals from V1. How does the cortex avoid this infinite regress so that, during development, plastic changes in cortex may be driven by stimuli that occur with high statistical regularity in the environment without causing massive instability, as modeled in the LAMINART simulations of Grossberg and Williamson (2001)? How does this process continue to fine-tune sensory representations in adulthood, even in cases where task-selective attention and awareness do not occur (Watanabe et al., 2001; Seitz and Watanabe, 2003)?

The LAMINART model clarifies how attention is used to help stabilize learning, while also allowing learning to slowly occur without task-selective attention and awareness. It also links these properties to properties of preattentive vision that are not obviously related to them. For example, how can preattentive groupings, such as illusory contours, form over positions that receive no bottom-up inputs? Although we take such percepts for granted, illusory contours seem to contradict the ART matching rule, which says that bottom-up inputs are needed to fire cells, while top-down feedback is modulatory. How, then, can cells that represent the illusory contour fire at positions that do not receive bottom-up inputs without destabilizing cortical development and learning? If the brain had not solved this problem, anyone could roam through the streets of a city and destabilize the brains of pedestrians by showing them images of Kanizsa squares! The absurdity of this possibility indicates how fundamental the issue at hand really is.

The LAMINART model proposes how the brain uses its laminar circuits to solve this problem using a *preattentive–attentive interface* in which both intercortical attentional feedback and *intra*cortical grouping feedback share the same selection circuit from layer 6-to-4: When a grouping starts to form in layer 2/3, it activates the intracortical feedback pathway from layer 2/3-to-6, which activates the modulatory on-center, off-surround network from layer 6-to-4. This intracortical feedback pathway helps to select which cells will remain active in a winning grouping. Attention uses this same network to stabilize cortical development and learning through intercortical interactions. In other words, the intracortical layer 6-to-4 selection circuit, which in the adult helps to choose winning groupings, is also predicted to help stabilize visually induced brain development by assuring that the ART matching rule holds at every position along a grouping. Because the matching rule holds, only the correct combinations of cells can "fire together and wire together," and hence stability is achieved. Intracortical feedback via layers 2/3-to-6-to-4-to-2/3 realizes this selection process even before intercortical attentional feedback can develop. I like to say that: "The preattentive grouping is its own attentional prime" (Grossberg, 1999a).

The LAMINART model hereby shows how, by joining together bottom-up (interlaminar) adaptive filtering, horizontal (intralaminar) grouping, top-down intracortical (but interlaminar) preattentive feedback, and top-down intercortical (and interlaminar) attentive feedback, some developmental and learning processes can be stabilized without top-down attention. This is realized by using intracortical feedback processes that activate the same stabilizing networks that top-down intercortical attentional processes use. Because of this intimate link between intracortical and intercortical feedback processes, attention can modulate the selection and activation level of preattentive grouping processes, as in the case of the Roelfsema et al. (1998) data.

Balanced excitatory and inhibitory circuits as a cortical design principle

The circuits that realize grouping and attentional processes compute balanced excitatory and inhibitory interactions. The excitatory/inhibitory balance within layer 2/3 circuits helps achieve perceptual grouping. The balance between excitatory and inhibitory interactions within the on-center of the network from layer 6-to-4 helps to do several things, among them render top-down attention modulatory. Figure 2 shows only these two types of balanced excitatory and inhibitory circuits. Other cortical interactions also balance excitation and inhibition, including the interactions that realize *monocular* simple cell receptive fields in layer 4 (data: Palmer and Davis, 1981; Pollen and Ronner, 1981; Liu et al., 1992; model: Olson and Grossberg, 1998). Balanced excitatory/inhibitory interactions within layer 3B also give rise to *binocular* simple cells that initiate stereopsis by matching monocular inputs from different eyes (Grossberg and Howe, 2003; Cao and Grossberg, 2005).

The balanced interactions within layer 2/3, and those from layer 6-to-4, as in Fig. 2, can explain data in which the excitatory/inhibitory balance is altered by sensory inputs. Figure 5 summarizes data of Polat et al. (1998) on contrast-dependent perceptual grouping in primary visual cortex, and a model simulation of Grossberg and Raizada (2000). The excitatory effects that enable colinear flankers to facilitate activation in response to a low-contrast target are mediated by layer 2/3 interactions, and the inhibitory effects that cause colinear flankers to depress activation in response to a high-contrast target are mediated by the layer 6-to-4 off-surround. These two types of effects propagate throughout the network via layer 4-to-2/3 and layer 2/3-to-6 interactions, among others. An important factor in the model simulation is that the inhibitory interactions are of shunting type (Grossberg, 1973, 1980; Heeger, 1992;

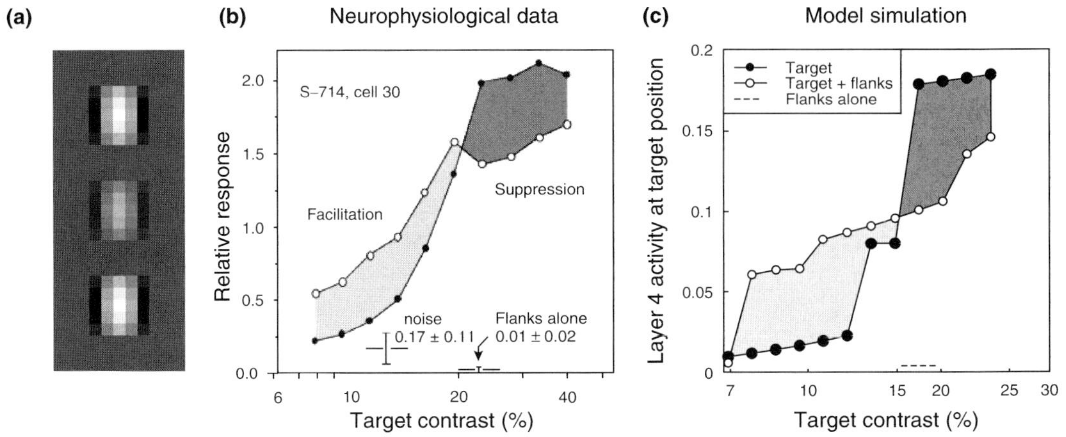

Fig. 5. Contrast-dependent perceptual grouping in primary visual cortex. (a) Illustrative visual stimuli. A variable-contrast oriented Gabor patch (middle stimulus pattern) stimulates the classical receptive field (CRF), with collinear flanking Gabors of fixed high-contrast outside of the CRF (the two end stimulus patterns). The stimulus shown here, based on those used (Polat et al., 1998), was presented to the model neural network. (b) Neural responses recorded from cat V1. The colinear flankers have a net facilitatory effect on weak targets which are close to the contrast-threshold of the cell, but they act to suppress responses to stronger, above-threshold targets. When the flankers are presented on their own, with no target present, the neural response stays at baseline levels. (Adapted with permission from Polat et al. (1998).) (c) Model simulation of the Polat et al. data. (Adapted with permission from Grossberg and Raizada (2000).)

Douglas et al., 1995) and thereby compute cell activations that are contrast-normalized.

How can perceptual grouping data be explained as a manifestation of excitatory/inhibitory balance? In cortical area V2 of monkeys, approximately colinear interactions from approximately co-oriented cells are capable of firing a cell that does not receive bottom-up inputs (von der Heydt et al., 1984; Peterhans and von der Heydt, 1989), as occurs when an illusory contour is perceived. The von der Heydt et al. (1984) experiment confirmed a prediction of Grossberg and colleagues (Cohen and Grossberg, 1984; Grossberg, 1984; Grossberg and Mingolla, 1985a, b) that perceptual grouping obeys a *bipole property* (Fig. 6); namely, such a cell can fire if it gets approximately colinear horizontal inputs from approximately co-oriented cells on both sides of its receptive field, even if it does not receive bottom-up input; or it can fire in response to bottom-up input alone, or to bottom-up input plus any combination of horizontal signals. The predicted bipole receptive field structure has been supported by later psychophysical experiments; e.g., Field et al. (1993) and Kellman and Shipley (1991), and anatomical experiments; e.g., Bosking et al. (1997). The LAMINART model (Grossberg et al., 1997; Grossberg, 1999a) extended this analysis by predicting how the bipole property may be realized by balanced excitatory/inhibitory interactions within layer 2/3, as summarized in Fig. 6. Without these balanced inhibitory interactions, the growth of horizontal connections during development could proliferate uncontrollably if inhibition is too weak, or could be suppressed entirely if inhibition is too strong (Grossberg and Williamson, 2001).

A synthesis of 3D vision, attention, and grouping

Our discussion so far has not considered how the brain sees the world in depth. Since the original LAMINART breakthrough in the mid-1990s, the model has been consistently extended into the 3D LAMINART model of 3D vision and figure-ground perception. This step was achieved by unifying two previous models: the LAMINART model, which had until that time focused on cortical development, learning, grouping, and attention, but did not consider binocular interactions and 3D vision; and the non-laminar FACADE model of 3D vision and figure-ground

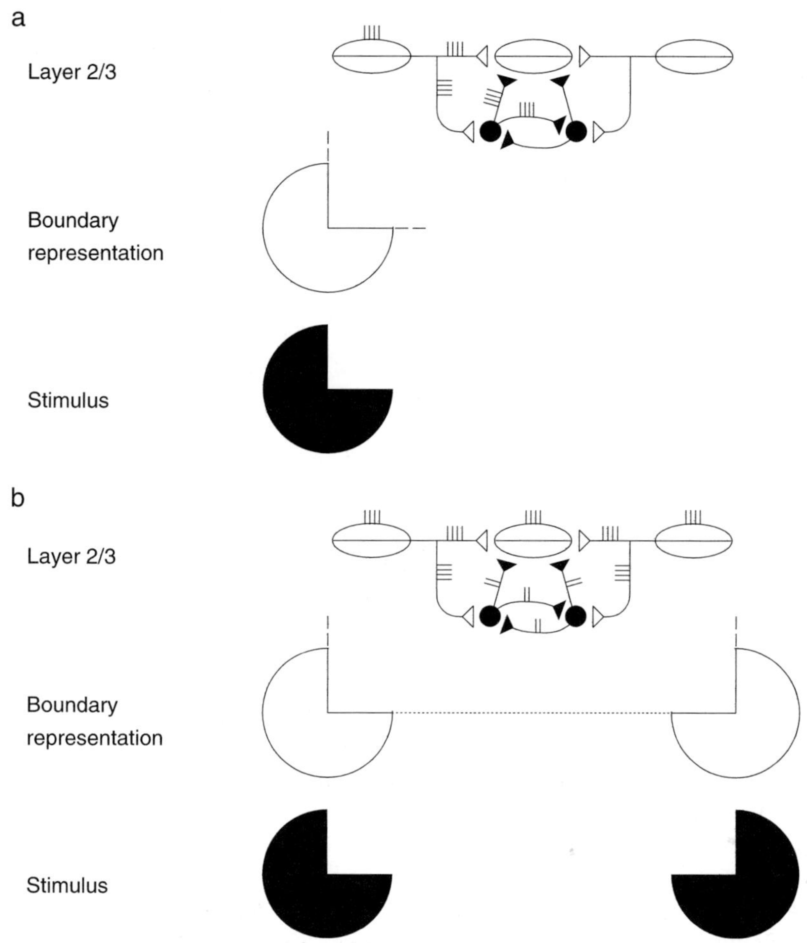

Fig. 6. Schematic of the boundary grouping circuit in layer 2/3. Pyramidal cells with colinear, co-oriented receptive fields (shown as ovals) excite each other via long-range horizontal axons (Bosking et al., 1997; Schmidt et al., 1997), which also give rise to short-range, disynaptic inhibition via pools of interneurons, shown filled-in black (McGuire et al., 1991). This balance of excitation and inhibition helps to implement the *bipole property*. (a) Illustration of how horizontal input coming in from just one side is insufficient to cause above-threshold excitation in a pyramidal cell (henceforth referred to as the target) whose receptive field does not itself receive any bottom-up input. The inducing stimulus (e.g., a Kanizsa 'pacman') excites the oriented receptive fields of layer 2/3 cells, which send out long-range horizontal excitation onto the target pyramidal. This excitation brings with it a commensurate amount of disynaptic inhibition. This balance of "one-against-one" prevents the target pyramidal cell from being excited above-threshold. The boundary representation of the solitary pacman inducer produces only weak, subthreshold colinear extensions (thin dashed lines). (b) When two colinearly aligned inducer stimuli are present, one on each side of the target pyramidal cell receptive field, a boundary grouping can form. Long-range excitatory inputs fall onto the cell from both sides, and summate. However, these inputs fall onto a shared pool of inhibitory interneurons, which, as well as inhibiting the target pyramidal, also inhibit each other (Tamas et al., 1998), thus normalizing the total amount of inhibition emanating from the interneuron pool, without any individual interneuron saturating. The combination of summating excitation and normalizing inhibition together create a case of "two-against-one," and the target pyramidal is excited above-threshold. This process occurs along the whole boundary grouping, which thereby becomes represented by a line of supra-threshold-activated layer 2/3 cells (thick dotted line). Boundary strength scales in a graded analog manner with the strength of the inducing signals. (Adapted with permission from Grossberg and Raizada (2000).)

perception (Grossberg, 1994, 1997; Grossberg and McLoughlin, 1997; McLoughlin and Grossberg, 1998; Kelly and Grossberg, 2000). The resulting unification was able to build upon LAMINART without having to discard any of its mechanisms, and to achieve a much broader explanatory and predictive range. Through this synthesis, the 3D LAMINART model has clarified how the laminar circuits of cortical areas V1, V2, and V4 are organized for purposes of stereopsis, 3D surface perception, and 3D figure-ground perception (Grossberg and Howe, 2003; Grossberg and Swaminathan, 2004; Cao and Grossberg, 2005; Fang and Grossberg, 2005; Grossberg and Yazdanbakhsh, 2005). As a result, the 3D LAMINART model predicts how cellular and network mechanisms of 3D vision and figure-ground perception are linked to mechanisms of development, learning, grouping, and attention. The following discussion merely hints at how this generalization builds seamlessly upon the already available LAMINART foundation. The original articles should be consulted for data support and model explanations and simulations of 3D vision data.

In the 3D LAMINART model, layer 4 no longer directly activates layer 2/3, as in Fig. 2c. Instead, layer 4 monocular simple cells first activate layer 3B binocular simple cells, which in turn activate layer 2/3A binocular complex cells, as shown in Fig. 7. The layer 2/3A cells can then interact via horizontal interactions, like those summarized in Figs. 2c, e, to enhance cell activations due to approximately co-oriented and colinear inputs. Second, binocular complex cells in layer 2/3A can represent different disparities, and thus different relative depths from an observer. Interactions between layer 2/3A cells that represent the same relative depth from the observer can be used to complete boundaries between object contours that lie at that depth.

Because binocular fusion begins already in layer 3B, the binocular boundaries that are formed in layers 3B and 2/3A may be positionally displaced, or shifted, relative to their monocular input signals from layers 6 and 4. Figure 2c illustrates that these layer 2/3 boundaries feed signals back to layer 6 in order to select the winning groupings that are formed in layer 2/3, but issues about binocular shifts did not need to be considered in data explanations of the original LAMINART model. Signals from the monocular layer 4 cells activate positionally shifted binocular cells in layer 3B, which in turn activate layer 2/3A binocular complex cells. This raises the question: How can the positionally displaced binocular boundaries in layer 2/3A of Fig. 6 contact the correct monocularly activated cells in layers 6 and 4, so that they can complete the feedback loop between layers 2/3-6-4-3B-2/3A that can select winning 3D groupings?

The 3D LAMINART model proposes that horizontal connections, which are known to occur in layers 5 and 6 (Callaway and Wiser, 1996), accomplish this. Feedback signals from layer 2/3A propagate vertically to layer 5, whose cells activate horizontal axons in this layer that contact the appropriate layer 6 cells. These layer 5-to-6 horizontal contacts are assumed to be selectively formed during development. Grossberg and Williamson (2001) and Grossberg and Seitz (2003) have simulated how layer 2/3 connections and layer 6-to-4 connections may be formed during development. The selective layer 5-to-6 contacts are proposed to form according to similar laws. In summary, *inward* horizontal layer 4-to-3B and 2/3A-to-2/3A connections are proposed to form binocular cells and their groupings, while *outward* layer 5-to-6 connections are proposed to close the feedback loops that help to select the correct 3D groupings.

Given how 3D groupings in layer 2/3A contact the correct layer 6 cells, the preattentive—attentive interface problem forces a proposal for how attention fits into the 3D circuit: namely, top-down attentional outputs from layer 6 of a higher cortical level like V2 activates the same layer 5 cells that contact monocular input sources in layer 6 via horizontal connections. Then the layer 6-to-4 modulatory on-center, off-surround network controls attentional priming and matching, just like in Fig. 2b. This proposal raises the question of how the top-down pathways from layer 6 of a higher cortical level know how to converge on the same layer 5 cells to which the layer 2/3 cells project at the lower cortical level? Since firing of the layer 2/3 cells activates the layer 5 cells as well as the layer 6 cells of the higher cortical level, this may occur due to associative learning.

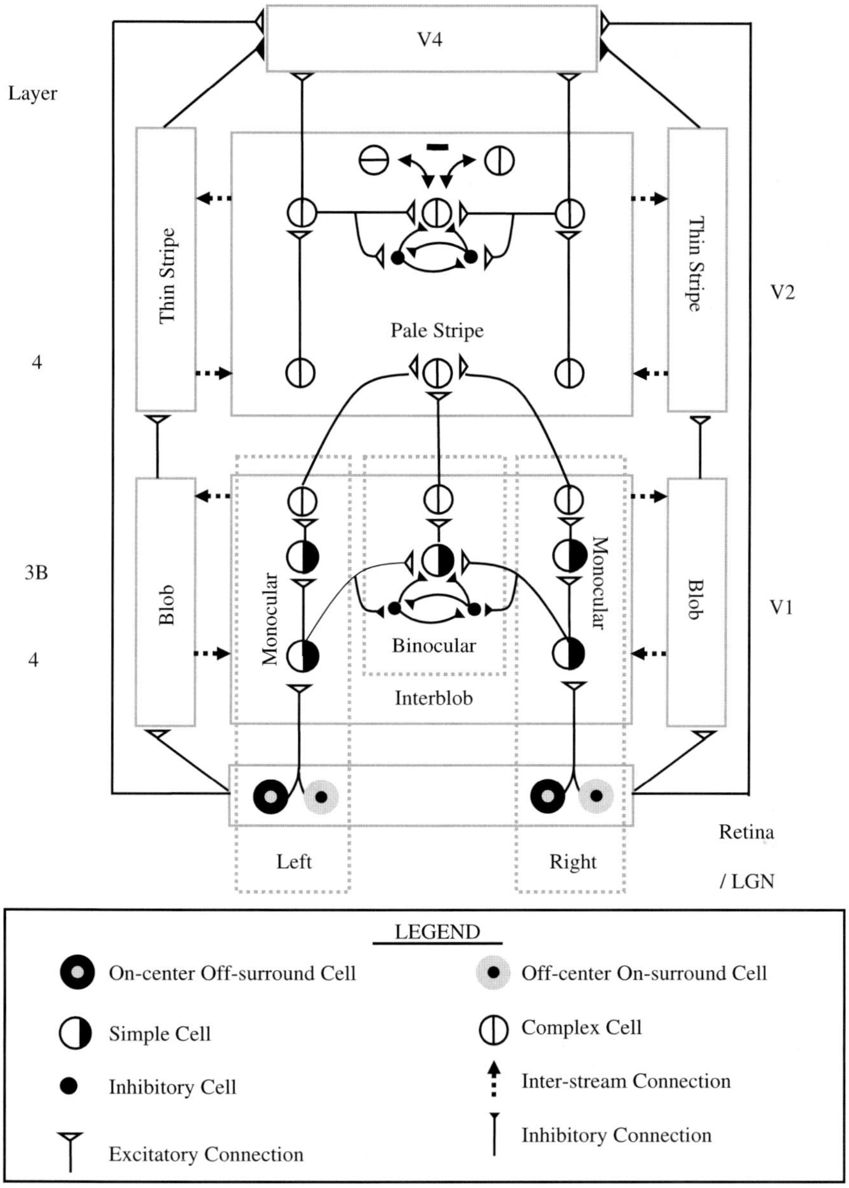

Fig. 7. Model circuit diagram. The full 3D LAMINART model consists of a boundary stream that includes V1 interblobs, V2 pale stripes, and part of V4, and computes 3D perceptual groupings; and a surface stream that includes V1 blobs, V2 thin stripes, and part of V4, and computes 3D surfaces that are infused with color and lightness in depth. These two streams both receive illuminant-discounted signals from Retina/LGN cells, and interact with each other to overcome their complementary deficiencies to create consistent 3D boundary and surface percepts in cortical area V4. Also, 3D boundary and surface representations formed in the pale stripes and thin stripes of cortical area V2, respectively, are amodally completed, and provide neural support for the object recognition process in inferotemporal cortex.

As noted above, Grossberg and Versace (2005, 2006, 2007) have proposed an elaboration of the LAMINART model, called the Synchronous Matching Adaptive Resonance Theory (SMART) model, in which such learning processes are studied down to the level of individual spikes and dendrites. This model extends laminar cortical modeling in a different direction by investigating how synchronization of neuronal spiking occurs within and across multiple brain regions, including how neocortical areas interact with higher-order specific and nonspecific thalamic nuclei, and how synchronization abets synaptic plasticity using STDP. The SMART extension of LAMINART also proposes a functional explanation for the differential expression of oscillation frequencies, notably gamma and beta, during match (gamma) or mismatch (beta) between bottom-up thalamic input and top-down cortical expectations, and of aggregate cell recordings such as current-source densities and local field potentials. The main fact for our present review is that a rational extension of LAMINART can bridge between all the processing levels that join individual spikes to cognitive information processing, and that SMART has begun to quantitatively simulate, and functionally rationalize, data on all these organizational levels.

Habituation, development, reset, and bistability

In addition to fast mechanisms of activation and slower mechanisms of learning, another intermediate time scale is needed to control cortical dynamics; notably, activity-dependent habituative mechanisms, as was noted above in the discussion of attentional shrouds. In particular, habituation of chemical transmitter gates has proved to be essential in studies of cortical development (Grunewald and Grossberg, 1998; Olson and Grossberg, 1998; Grossberg and Seitz, 2003); see Grossberg (2003b) for a review. The habituative mechanisms prevent the developmental process from "getting stuck" into activating, over and over, the cells that initially win the competition. Such perseveration would prevent multiple feature combinations from getting represented in a distributed fashion throughout the network. Habituative interactions help to solve this problem because habituation is activity-dependent: only those cells or connections habituate that are in active use. Thus, when habituation acts, it selectively weakens the competitive advantage of the initial winners, so that other cells can become activated to represent different input features.

Habituative mechanisms play an important role in adult vision by helping to *reset* previously active visual representations when the scenes or images that induced them change or disappear. Without such an active reset process, visual representations could easily persist for a long time due to the hysteresis that could otherwise occur in circuits with as many feedback loops as those in Figs. 2 and 7. In many examples of this reset process, offset of a previously active input leads to an *antagonistic rebound* of activation of previously inactive cells, and these newly activated cells help to inhibit the previously active cells, including grouping cells in layer 2/3. This reset process is not perfect, however, and there are large perceptual databases concerning residual effects of previously active representations. In fact, such a reset process has elsewhere been used to explain psychophysical data about visual aftereffects (Francis and Grossberg, 1996; Grunewald and Lankheet, 1996), visual persistence (Francis et al., 1994), and binocular rivalry (Grossberg, 1987; Arrington, 1993, 1995, 1996; Liang and Chow, 2002), among other data that are all proposed to be manifestations of the reset process. Ringach et al. (1999) have reported direct neurophysiological evidence for rebound phenomena using reverse correlation techniques to analyze orientational tuning in neurons of cortical area V1. Abbott et al. (1997) have provided direct experimental evidence in visual cortex of the habituative mechanisms that were predicted to cause the reset (Grossberg, 1968, 1969, 1980). Grossberg (1980, 1999b) also predicted that such reset processes play a role in driving the reset and memory search processes that help the adult brain to rapidly discover and learn new representations of the world, as part of ART.

The same habituative mechanisms that usually phasically reset active brain representations can also lead to persistent multistable percepts when two or more 3D interpretations of a 2D image are approximately equally salient, as in Necker cube percepts, and also during binocular rivalry. Grossberg and Swaminathan (2004) have used the same habituative and competitive mechanisms to simulate development of disparity-gradient cell receptive fields and how a 2D Necker cube image generates bi-stable 3D boundary and surface representations.

In summary, there is a predicted link, mediated by habituative transmitter mechanisms, between processes of cortical development in the infant and processes of perceptual and cognitive reset, learning, and bistability in the adult. This link is worthy of a lot more experimental study than it has received to date.

Towards a unified theory of laminar neocortex: from vision to cognition

The results above focus on vision, which is a spatial process, or more accurately, a SPATIO-temporal process. Can LAMINART principles be used to explain data about the temporal dynamics of cognitive information processing, which involves more spatio-TEMPORAL processes? In particular, how do the layered circuits of prefrontal and motor cortex carry out working memory storage, sequence learning, and voluntary, variable-rate performance of event sequences? A neural model called LIST PARSE (Pearson and Grossberg, 2005; Grossberg and Pearson, 2006) proposes an answer to this question that unifies the explanation of cognitive, neurophysiological, and anatomical data from humans and monkeys. It quantitatively simulates human cognitive data about immediate serial recall and free recall, and monkey neurophysiological data from the prefrontal cortex obtained during sequential sensory-motor imitation and planned performance. The human cognitive data include bowing of the serial position performance curves, error-type distributions, temporal limitations upon recall accuracy, and list length effects. LIST PARSE also qualitatively explains cognitive effects related to attention, temporal grouping, variable presentation rates, phonemic similarity, presentation of non-words, word frequency/item familiarity and list strength, distracters and modality effects.

The model builds upon earlier working memory models that predict why both spatial and nonspatial working memories share the same type of circuit design (Grossberg, 1978). These Item and Order working memories, also called Competitive Queuing models (Houghton, 1990), propose rules for the storage of event sequences in working memory as evolving spatial patterns of activation. LIST PARSE proposes how to embody an Item and Order cognitive working memory model into the laminar circuits of ventrolateral prefrontal cortex. Such Competitive Queuing models have gradually become the dominant model for how to temporarily store sequences of events in working memory.

Grossberg (1978) derived this class of models from an analysis of how to store sequences of speech or motor items in working memory in a manner that can be stably coded in long-term memory (e.g., word, language, and skill learning) without destabilizing previously learned list categories that are subcategories of the new ones being learned. For example, how do you learn a list category for the novel word MYSELF when you already know the words MY, SELF, and ELF? The main design principle is called the LTM Invariance Principle. An exciting consequence of the LTM Invariance Principle is that the following types of activity patterns naturally emerge across the items that are stored in working memory: (1) primacy gradients of activity across the stored items wherein the earliest items are stored with the greatest activity — a primacy gradient can control the correct order of recall; (2) recency gradients, which control a backwards order of recall; and (3) bowed gradients, which permit recall of items at the list ends before the items near the list middle, and with higher probability than the list middle. Even if a primacy gradient is stored for a short list, a bowed gradient will then always emerge for a sufficiently long list. As just noted, bowing means that the system is not able to reproduce the correct order from working memory, because items near the list end will be recalled before items in the middle. Thus, the inability to read-out the correct order of long lists from working memory can be traced to a constraint on the design of working memories that ensures stable learning of list categories, or chunks.

Any model of working memory needs to confront the question of how it evolved during natural selection. Happily, the LTM Invariance Principle can be realized by the same sort of shunting on-center off-surround network that is so frequently found in other parts of the brain, notably the visual cortex (Bradski et al., 1992; Grossberg, 1978, 1994). These on-center off-surround networks must be recurrent, or feedback, networks whose positive and negative feedback signals establish and store the spatial patterns of activity that represent the stored working memory. Specialization of how these recurrent networks sequentially rehearse their stored patterns and reset each rehearsed item is what sets them apart from other recurrent shunting on-center off-surround networks across the brain.

LIST PARSE is a LAMINART-style model that illustrates how variations on granular laminar cortical circuits can quantitatively simulate data about spatio-TEMPORAL cognitive processes as well as SPATIO-temporal visual processes. The family of LAMINART models now allows us to understand as variations of a shared cortical design brain processes that seem to be totally unrelated on the level of behavioral function. As just one example, LAMINART predicts that the volitional mechanism which allows humans to experience visual imagery and fantasy, is the same mechanism, suitably specialized, that regulates the storage of event sequences in working memory. The volitional gain control mechanism that is predicted to carry out this function may be realized by inhibition of inhibitory interneurons in layer 4 of both cortical areas. It remains to be seen how such LAMINART mechanisms are specialized within the laminar circuits of other cortical areas to realize a variety of intelligent behaviors.

Acknowledgments

Supported in part by the National Science Foundation (SBE-0354378) and the Office of Naval Research (ONR N00014-01-1-0624).

References

Abbott, L.G., Varela, J.A., Sen, K. and Nelson, S.B. (1997) Synaptic depression and cortical gain control. Science, 275: 220–224.

Ahissar, M. and Hochstein, S. (1993) Attentional control of early perceptual learning. Proc. Natl. Acad. Sci. U.S.A., 90: 5718–5722.

Amir, Y., Harel, M. and Malach, R. (1993) Cortical hierarchy reflected in the organization of intrinsic connections in Macaque monkey visual cortex. J. Comp. Neurol., 334: 19–46.

Arrington, K.F. (1993) Binocular rivalry model using multiple habituating nonlinear reciprocal connections. Neurosci. Abstr., 19: 1803.

Arrington, K.F. (1995) Neural model of rivalry between occlusion and disparity depth signals. Neurosci. Abstr., 21: 125.

Arrington, K.F. (1996) Stochastic properties of segmentation-rivalry alternations. Perception, 25(Supplement): 62.

Berzhanskaya, J., Grossberg, S. and Mingolla, E. (2007) Laminar cortical dynamics of visual form and motion interactions during coherent object motion perception. Spat. Vis., 20: 337–395.

Bosking, W.H., Zhang, Y., Schofield, B. and Fitzpatrick, D. (1997) Orientation selectivity and the arrangement of horizontal connections in the tree shrew striate cortex. J. Neurosci., 17: 2112–2127.

Bradski, G., Carpenter, G.A. and Grossberg, S. (1992) Working memory networks for learning temporal order, with application to 3-D visual object recognition. Neural Comput., 4: 270–286.

Brodmann, K. (1909) Vergleichende Lokalisationslehre der Grosshirnrinde in ihren Prinzipien dargestellt auf Grund des Zellenbaues. Barth, Leipzig.

Brown, J.M. and Denney, H.I. (in press) Shifting attention into and out of objects: evaluating the processes underlying the object advantage. Percept. Psychophys.

Bullier, J., Hupé, J.M., James, A. and Girard, P. (1996) Functional interactions between areas V1 and V2 in the monkey. J. Physiol. (Paris), 90: 217–220.

Callaway, E.M. and Wiser, A.K. (1996) Contributions of individual layer 2–5 spiny neurons to local circuits in macaque primary visual cortex. Vis. Neurosci., 13: 907–922.

Cao, Y. and Grossberg, S. (2005) A laminar cortical model of stereopsis and 3D surface perception: closure and da Vinci stereopsis. Spat. Vis., 18: 515–578.

Caputo, G. and Guerra, S. (1998) Attentional selection by distractor suppression. Vision Res., 38: 669–689.

Carpenter, G.A. and Grossberg, S. (1993) Normal and amnesic learning, recognition, and memory by a neural model of cortico-hippocampal interactions. Trends Neurosci., 16: 131–137.

Chey, J., Grossberg, S. and Mingolla, E. (1997) Neural dynamics of motion grouping: from aperture ambiguity to object speed and direction. J. Opt. Soc. Am. A, 14: 2570–2594.

Cohen, M.A. and Grossberg, S. (1984) Neural dynamics of brightness perception: Features, boundaries, diffusion, and resonance. Percept. Psychophys., 36: 428–456.

Desimone, R. (1998) Visual attention mediated by biased competition in extrastriate visual cortex. Philos. Trans. R. Soc. Lond., 353: 1245–1255.

Douglas, R.J., Koch, C., Mahowald, M., Martin, K.A.C. and Suarez, H.H. (1995) Recurrent excitation in neocortical circuits. Science, 269: 981–985.

Downing, C.J. (1988) Expectancy and visual-spatial attention: effects on perceptual quality. J. Exp. Psychol.: Hum. Percept. Perform., 14: 188–202.

Duncan, J. (1984) Selective attention and the organization of visual information. J. Exp. Psychol.: Gen., 113: 501–517.
Eckhorn, R., Bauer, R., Jordan, W., Brosch, M., Kruse, W., Munk, M. and Reitbock, H.J. (1988) Coherent oscillations: a mechanism of feature linking in the visual cortex? Biol. Cybern., 60: 121–130.
Egeth, H., Virzi, R.A. and Garbart, H. (1984) Searching for conjunctively defined targets. J. Exp. Psychol.: Hum. Percept. Perform., 10: 32–39.
Egusa, H. (1983) Effects of brightness, hue, and saturation on perceived depth between adjacent regions in the visual field. Perception, 12: 167–175.
Egly, R., Driver, J. and Rafal, R.D. (1994) Shifting visual attention between objects and locations: evidence from normal and parietal lesion subjects. J. Exp. Psychol. Gen., 123: 161–177.
Engel, A.K., Fries, P. and Singer, W. (2001) Dynamic predictions: oscillations and synchrony in top-down processing. Nat. Rev. Neurosci., 2: 704–716.
Fang, L. and Grossberg, S. (2005) A laminar cortical model of stereogram depth, lightness, and amodal completion. Soc. Neurosci. Abstr., 768.4.
Faubert, J. and von Grunau, M. (1995) The influence of two spatially distinct primers and attribute priming on motion induction. Vision Res., 35: 3119–3130.
Fazl, A., Grossberg, S. and Mingolla, E. (2005) Invariant object learning and recognition using active eye movements and attentional control. J. Vis., 5(8): 738a.
Fazl, A., Grossberg, S. and Mingolla, E. (2006) View-invariant object category learning: how spatial and object attention are coordinated using surface-based attentional shrouds. J. Vis., 6(6): 315a.
Fazl, A., Grossberg, S. and Mingolla, E. (2007) View-invariant object category learning, recognition, and search: How spatial and object attention are coordinated using surface-based attentional shrouds. Technical Report CAS/CNS-2007-011, Boston University.
Field, D.J., Hayes, A. and Hess, R.F. (1993) Contour integration by the human visual system: evidence for a local "association field." Vision Res., 33: 173–193.
Francis, G. and Grossberg, S. (1996) Cortical dynamics of boundary segmentation and reset: persistence, afterimages, and residual traces. Perception, 35: 543–567.
Francis, G., Grossberg, S. and Mingolla, E. (1994) Cortical dynamics of feature binding and reset: control of visual persistence. Vision Res., 34: 1089–1104.
Fries, P., Ryenolds, J.H., Rorie, A.E. and Desimone, R. (2001) Modulation of oscillatory neuronal synchronization by selective visual attention. Science, 291: 1560–1563.
Gao, E. and Suga, N. (1998) Experience-dependent corticofugal adjustment of midbrain frequency map in bat auditory system. Proc. Natl. Acad. Sci. U.S.A., 95: 12663–12670.
Goodale, M.A. and Milner, D. (1992) Separate visual pathways for perception and action. Trends Neurosci., 15: 10–25.
Gray, C.M. and Singer, W. (1989) Stimulus-specific neuronal oscillations in orientation columns of cat visual cortex. Proc. Natl. Acad. Sci. U.S.A., 86: 1698–1702.
Grosof, D.H., Shapley, R.M. and Hawken, M.J. (1993) Macaque V1 neurons can signal 'illusory' contours. Nature, 365: 550–552.
Grossberg, S. (1968) Some physiological and biochemical consequences of psychological postulates. Proc. Natl. Acad. Sci. U.S.A., 60: 758–765.
Grossberg, S. (1969) On the production and release of chemical transmitters and related topics in cellular control. J. Theor. Biol., 22: 325–364.
Grossberg, S. (1973) Contour enhancement, short term memory, and constancies in reverberating neural networks. Stud. Appl. Math., 52: 217–257. Reprinted in Grossberg, S. (1982) Studies of Mind and Brain. D. Reidel Publishing Company, Dordrecht, The Netherlands.
Grossberg, S. (1976) Adaptive pattern classification and universal recoding II: feedback, expectation, olfaction, and illusions. Biol. Cybern., 23: 187–202.
Grossberg, S. (1978) A theory of human memory: self-organization and performance of sensory-motor codes, maps, and plans. In: Rosen R. and Snell F. (Eds.), Progress in Theoretical Biology, Vol. 5. Wiley Press, New York, pp. 183–232.
Grossberg, S. (1980) How does a brain build a cognitive code? Psychol. Rev., 87: 1–51.
Grossberg, S. (1982) Studies of mind and brain. Kluwer, Amsterdam.
Grossberg, S. (1984) Outline of a theory of brightness, color, and form perception. In: Degreef E. and van Buggenhaut J. (Eds.), Trends in mathematical psychology. North-Holland, Amsterdam, pp. 59–86.
Grossberg, S. (1987) Cortical dynamics of three-dimensional form, color, and brightness perception: II. Binocular theory. Percept. Psychophys., 41: 117–158.
Grossberg, S. (1994) 3-D vision and figure-ground separation by visual cortex. Percept. Psychophys., 55: 48–120.
Grossberg, S. (1995) The attentive brain. Am. Sci., 83: 438–449.
Grossberg, S. (1997) Cortical dynamics of three-dimensional figure-ground perception of two-dimensional figures. Psychol. Rev., 104: 618–658.
Grossberg, S. (1999a) How does the cerebral cortex work? Learning, attention, and grouping by the laminar circuits of visual cortex. Spat. Vis., 12: 163–187.
Grossberg, S. (1999b) Pitch-based streaming in auditory perception. In: Griffith N. and Todd P. (Eds.), Musical Networks: Parallel Distributed Perception and Performance. MIT Press, Cambridge, MA, pp. 117–140.
Grossberg, S. (1999c) The link between brain learning, attention, and consciousness. Conscious. Cogn., 8: 1–44.
Grossberg, S. (2000) The complementary brain: unifying brain dynamics and modularity. Trends Cogn. Sci., 4: 233–246.
Grossberg, S. (2003a) How does the cerebral cortex work? Development, learning, attention, and 3D vision by laminar circuits of visual cortex. Behav. Cogn. Neurosci. Rev., 2: 47–76.
Grossberg, S. (2003b) Linking visual cortical development to visual perception. In: Hopkins B. and Johnson S. (Eds.), Neurobiology of Infant Vision. Ablex Press, pp. 211–271.

Grossberg, S. and Grunewald, A. (1997) Cortical synchronization and perceptual framing. J. Cogn. Neurosci., 9: 117–132.
Grossberg, S. and Hong, S. (2006) A neural model of surface perception: lightness, anchoring, and filling-in. Spat. Vis., 19: 263–321.
Grossberg, S. and Howe, P.D.L. (2003) A laminar cortical model of stereopsis and three-dimensional surface perception. Vision Res., 43: 801–829.
Grossberg, S. and McLoughlin, N. (1997) Cortical dynamics of three-dimensional surface perception: binocular and half-occluded scenic images. Neural Netw., 10: 1583–1605.
Grossberg, S. and Mingolla, E. (1985a) Neural dynamics of form perception: boundary completion, illusory figures, and neon color spreading. Psychol. Rev., 92: 173–211.
Grossberg, S. and Mingolla, E. (1985b) Neural dynamics of perceptual grouping: textures, boundaries and emergent segmentations. Percept. Psychophys., 38: 141–171.
Grossberg, S., Mingolla, E. and Ross, W.D. (1994) A neural theory of attentive visual search: interactions of boundary, surface, spatial, and object representations. Psychol. Rev., 101: 470–489.
Grossberg, S., Mingolla, E. and Ross, W.D. (1997) Visual brain and visual perception: How does the cortex do perceptual grouping? Trends Neurosci., 20: 106–111.
Grossberg, S., Mingolla, E. and Viswanathan, L. (2001) Neural dynamics of motion integration and segmentation within and across apertures. Vision Res., 41: 2521–2553.
Grossberg, S. and Pearson, L.R. (2006) Laminar cortical dynamics of cognitive and motor working memory, sequence learning, and performance: Toward a unified theory of how the cerebral cortex works. Technical Report CAS/CNS TR-2006-002, Boston University.
Grossberg, S. and Pilly, P.K. (2007) Temporal dynamics of decision-making during motion perception in the visual cortex. Technical Report BU CAS/CNS TR-2007-001, Boston University.
Grossberg, S. and Raizada, R.D.S. (2000) Contrast-sensitive perceptual grouping and object-based attention in the laminar circuits of primary visual cortex. Vision Res., 40: 1413–1432.
Grossberg, S. and Seidman, D. (2006) Neural dynamics of autistic behaviors: cognitive, emotional, and timing substrates. Psychol. Rev., 113: 483–525.
Grossberg, S. and Seitz, A. (2003) Laminar development of receptive fields, maps, and columns in visual cortex: the coordinating role of the subplate. Cereb. Cortex, 13: 852–863.
Grossberg, S. and Somers, D. (1991) Synchronized oscillations during cooperative feature linking in a cortical model of visual perception. Neural Netw., 4: 453–466.
Grossberg, S. and Swaminathan, G. (2004) A laminar cortical model for 3D perception of slanted and curved surfaces and of 2d images: development, attention, and bistability. Vision Res., 44: 1147–1187.
Grossberg, S. and Versace, M. (2005) Temporal binding and resonance in thalamocortical assemblies: Learning and cognitive information processing in a spiking neuron model. Soc. Neurosci. Abstr., 538.8.
Grossberg, S. and Versace, M. (2006) From spikes to interareal synchrony: how attentive matching and resonance control learning and information processing by laminar thalamocortical circuits. Soc. Neurosci. Abstr., 65.11/Z12.
Grossberg, S. and Versace, M. (2007) Spikes, synchrony, and attentive learning by laminar thalamocortical circuits. Submitted for publication.
Grossberg, S. and Williamson, J.R. (2001) A neural model of how horizontal and interlaminar connections of visual cortex develop into adult circuits that carry out perceptual groupings and learning. Cereb. Cortex, 11: 37–58.
Grossberg, S. and Yazdanbakhsh, A. (2005) Laminar cortical dynamics of 3D surface perception: stratification, transparency, and neon color spreading. Vision Res., 45: 1725–1743.
Grunewald, A. and Grossberg, S. (1998) Self-organization of binocular disparity tuning by reciprocal corticogeniculate interactions. J. Cogn. Neurosci., 10: 199–215.
Grunewald, A. and Lankheet, M.J. (1996) Orthogonal motion after-effect illusion predicted by a model of cortical motion processing. Nature, 384: 358–360.
Heeger, D.J. (1992) Normalization of cell responses in cat striate cortex. Vis. Neurosci., 9: 181–197.
von der Heydt, R., Peterhans, E. and Baumgartner, G. (1984) Illusory contours and cortical neuron responses. Science, 224: 1260–1262.
Hirsch, J.A. and Gilbert, C.D. (1991) Synpatic physiology of horizontal connections in the cat's visual cortex. J. Neurosci., 11: 1800–1809.
Houghton, G. (1990) The problem of serial order: a neural network model of sequence learning and recall. In: Dale R., Mellish C. and Zock M. (Eds.), Current Research in Natural Language Generation. Academic Press, London, pp. 287–319.
Hubel, D.H. and Wiesel, T.N. (1977) Functional architecture of macaque monkey visual cortex. Proc. Royal Soc. Lond. (Series B), 198: 1–59.
Hupé, J.M., James, A.C., Girard, D.C. and Bullier, J. (1997) Feedback connections from V2 modulate intrinsic connectivity within V1. Soc. Neurosci. Abstr., 406.15: 1031.
Kaas, J.H. (1999) Is most of neural plasticity in the thalamus cortical? Proc. Nat. Acad. Sci. U.S.A., 96: 7622–7623.
Kanizsa, G. (1974) Contours without gradients or cognitive contours. Ital. J. Psychol., 1: 93–113.
Kapadia, M.K., Ito, M., Gilbert, C.D. and Westheimer, G. (1995) Improvement in visual sensitivity by changes in local context: parallel studies in human observers and in V1 of alert monkeys. Neuron, 15: 843–856.
Kastner, S. and Ungerleider, L.G. (2001) The neural basis of biased competition in human visual cortex. Neuropsychologia, 39: 1263–1276.
Kellman, P.J. and Shipley, T.F. (1991) A theory of visual interpolation in object perception. Cogn. Psychol., 23: 141–221.
Kelly, F. and Grossberg, S. (2000) Neural dynamics of 3-D surface perception: figure-ground separation and lightness perception. Percept. Psychophys., 62: 1596–1618.
Knierim, J.J. and van Essen, D.C. (1992) Neuronal responses to static texture patterns in area V1 of the alert macaque monkey. J. Neurophysiol., 67: 961–980.

Koch, C. and Ullman, S. (1985) Shifts in selective visual attention: towards the underlying neural circuitry. Hum. Neurobiol., 4: 219–227.

Krupa, D.J., Ghazanfar, A.A. and Nicolelis, M.A. (1999) Immediate thalamic sensory plasticity depends on corticothalamic feedback. Proc. Natl. Acad. Sci. U.S.A., 96: 8200–8205.

Li, Z. (1998) A neural model of contour integration in the primary visual cortex. Neural Comput., 10: 903–940.

Liang, C.R. and Chow, C.C. (2002) A spiking neuron model for binocular rivalry. J. Comput. Neurosci., 12: 39–53.

Liu, Z., Gaska, J.P., Jacobson, L.D. and Pollen, D.A. (1992) Interneuronal interaction between members of quadrature phase and anti-phase pairs in the cat's visual cortex. Vision Res., 32: 1193–1198.

Luck, S.J., Chelazzi, L., Hillyard, S.A. and Desimone, R. (1997) Neural mechanisms of spatial selective attention in areas V1, V2, and V4 of macaque visual cortex. J. Neurophysiol., 77: 24–42.

Lund, J.S. and Wu, C.Q. (1997) Local circuit neurons of macaque monkey striate cortex: IV. Neurons of laminae 1-3A. J. Comp. Neurol., 384: 109–126.

Martin, J.H. (1989) Neuroanatomy: Text and Atlas. Appleton and Lange, Norwalk.

McGuire, B.A., Gilbert, C.D., Rivlin, P.K. and Wiesel, T.N. (1991) Targets of horizontal connections in macaque primary visual cortex. J. Comp. Neurol., 305: 370–392.

McLoughlin, N.P. and Grossberg, S. (1998) Cortical computation of stereo disparity. Vision Res., 38: 91–99.

Mishkin, M., Ungerleider, L.G. and Macko, K.A. (1983) Object vision and spatial vision: two cortical pathways. Trends Neurosci., 6: 414–417.

Moore, C.M., Yantis, S. and Vaughan, B. (1998) Object-based visual selection: evidence from perceptual completion. Psychol. Sci., 9: 104–110.

Motter, B.C. (1993) Focal attention produces spatially selective processing in visual cortical areas V1, V2 and V4 in the presence of competing stimuli. J. Neurophysiol., 70: 909–919.

Mounts, J.R.W. (2000) Evidence for suppressive mechanisms in attentional selection: feature singletons produce inhibitory surrounds. Percept. Psychophys., 62: 969–983.

Mumford, D. (1992) On the computational architecture of the neocortex. II. The role of corticocortical loops. Biol. Cybernet., 66: 241–251.

Nakamura, K. and Colby, C.L. (2000a) Visual, Saccade-related, and cognitive activation of single neurons in monkey extrastriate area V3A. J. Neurophysiol., 84: 677–692.

Nakamura, K. and Colby, C.L. (2000b) Updating of the visual representation in monkey striate and extrastriate cortex during saccades. Proc. Natl. Acad. Sci. U.S.A., 99: 4026–4031.

Nakayama, K. and Silverman, G.H. (1986) Serial and parallel processing of visual feature conjunctions. Nature, 320: 264–265.

Olson, S. and Grossberg, S. (1998) A neural network model for the development of simple and complex cell receptive fields within cortical maps of orientation and ocular dominance. Neural Netw., 11: 189–208.

Palmer, L.A. and Davis, T.L. (1981) Receptive field structure in cat striate cortex. J. Neurophysiol., 46: 260–276.

Parker, J.L. and Dostrovsky, J.O. (1999) Cortical involvement in the induction, but not expression, of thalamic plasticity. J. Neurosci., 19: 8623–8629.

Pearson, L.R. and Grossberg, S. (2005) Neural dynamics of motor sequencing in lateral prefrontal cortex. Soc. Neurosci. Abstr., 194.11.

Pessoa, L., Beck, J. and Mingolla, E. (1996) Perceived texture segregation in chromatic element-arrangement patterns: high intensity interference. Vision Res., 36: 1745–1760.

Peterhans, E. and von der Heydt, R. (1989) Mechanisms of contour perception in monkey visual cortex II. Contours bridging gaps. J. Neurosci., 9: 1749–1763.

Pilly, P. and Grossberg, S. (2005) Brain without Bayes: Temporal dynamics of decision-making in the laminar circuits of visual cortex. Soc. Neurosci. Abstr., 591.1.

Polat, U., Mizobe, K., Pettet, M.W., Kasamatsu, T. and Norcia, A.M. (1998) Collinear stimuli regulate visual responses depending on cell's contrast threshold. Nature, 391: 580–584.

Pollen, D.A. (1999) On the neural correlates of visual perception. Cereb. Cortex, 9: 4–19.

Pollen, D.A. and Ronner, S.F. (1981) Phase relationships between adjacent simple cells in the visual cortex. Science, 212: 1409–1411.

Posner, M.I. (1980) Orienting of attention. Q. J. Exp. Psychol., 32: 2–25.

Pylyshyn, Z.W. and Storm, R.W. (1988) Tracking multiple independent targets: evidence for a parallel tracking mechanism. Spat. Vis., 3: 179–197.

Raizada, R.D.S. and Grossberg, S. (2001) Context-sensitive bindings by the laminar circuits of V1 and V2: a unified model of perceptual grouping, attention, and orientation contrast. Vis. Cogn., 8: 341–466.

Raizada, R.D.S. and Grossberg, S. (2003) Towards a theory of the laminar architecture of cerebral cortex: computational clues from the visual system. Cereb. Cortex, 13: 100–113.

Rao, R.P.N. and Ballard, D.H. (1999) Predictive coding in the visual cortex: a functional interpretation of some extra-classical receptive field effects. Nat. Neurosci., 2: 79–87.

Reynolds, J., Chelazzi, L. and Desimone, R. (1999) Competitive mechanisms subserve attention in macaque areas V2 and V4. J. Neurosci., 19: 1736–1753.

Reynolds, J., Nicholas, J., Chelazzi, L. and Desimone, R. (1995) Spatial attention protects macaque V2 and V4 cells from the influence of non-attended stimuli. Soc. Neurosci. Abstr., 21.3: 1759.

Reynolds, J.H. and Desimone, R. (2003) Interacting roles of attention and visual salience in V4. Neuron, 37: 853–863.

Ringach, D.L., Hawken, M.J. and Shapley, R. (1999) Properties of macaque V1 neurons studied with natural image sequences. Invest. Ophthalmol. Vis. Sci., 40 Abstract 989.

Rockland, K.S. and Virga, A. (1989) Terminal arbors of individual 'feedback' axons projecting from area V2 to V1 in the macaque monkey: a study using immunohistochemistry of anterogradely transported phaseolus vulgaris-leucoagglutinin. J. Comp. Neurol., 285(1): 54–72.

Roelfsema, P.R., Lamme, V.A.F. and Spekreijse, H. (1998) Object-based attention in the primary visual cortex of the macaque monkey. Nature, 395: 376–381.

Roitman, J.D. and Shadlen, M.N. (2002) Response of neurons in the lateral intraparietal area during a combined visual discrimination reaction time task. J. Neurosci., 22: 9475–9489.

Salin, P. and Bullier, J. (1995) Corticocortical connections in the visual system: structure and function. Physiol. Rev., 75: 107–154.

Sandell, J.H. and Schiller, P.H. (1982) Effect of cooling area 18 on striate cortex cells in the squirrel monkey. J. Neurophysiol., 48: 38–48.

Sarnthein, J., Petsche, H., Rappelsberger, P., Shaw, G.L. and von Stein, A. (1998) Synchronization between prefrontal and posterior association cortex during human working memory. Proc. Natl. Acad. Sci. U.S.A., 95: 7092–7096.

Schmidt, K.E., Schlote, W., Bratzke, H., Rauen, T., Singer, W. and Galuske, R.A.W. (1997) Patterns of long range intrinsic connectivity in auditory and language areas of the human temporal cortex. Soc. Neurosci. Abstr., 415.13: 1058.

Seitz, A. and Watanabe, T. (2003) Is subliminal learning really passive? Nature, 422: 6927.

Shadlen, M.N. and Newsome, W.T. (1998) The variable discharge of cortical neurons: implications for connectivity, computation, and information coding. J. Neurosci., 18: 3870–3896.

Shadlen, M.N. and Newsome, W.T. (2001) Neural basis of a perceptual decision in the parietal cortex (area LIP) of the rhesus monkey. J. Neurophysiol., 86: 1916–1936.

Sheth, B.R., Sharma, J., Rao, S.C. and Sur, M. (1996) Orientation maps of subjective contours in visual cortex. Science, 274: 2110–2115.

Sillito, A.M., Grieve, K.L., Jones, H.E., Cudeiro, J. and Davis, J. (1995) Visual cortical mechanisms detecting focal orientation discontinuities. Nature, 378: 492–496.

Sillito, A.M., Jones, H.E., Gerstein, G.L. and West, D.C. (1994) Feature-linked synchronization of thalamic relay cell firing induced by feedback from the visual cortex. Nature, 369: 479–482.

Smallman, H.S. and McKee, S.P. (1995) A contrast ratio constraint on stereo matching. Proc. R. Soc. Lond. B, 260: 265–271.

Smith, A.T., Singh, K.D. and Greenlee, M.W. (2000) Attentional suppression of activity in the human visual cortex. Neuroreport, 11: 271–277.

Somers, D.C., Dale, A.M., Seiffert, A.E. and Tootell, R.B. (1999) Functional MRI reveals spatially specific attentional modulation in human primary visual cortex. Proc. Natl. Acad. Sci. U.S.A., 96: 1663–1668.

Somers, D.C., Todorov, E.V., Siapas, A.G., Toth, L.J., Kim, D. and Sur, M. (1998) A local circuit approach to understanding integration of long-range inputs in primary visual cortex. Cereb. Cortex, 8: 204–217.

Steinman, B.A., Steinman, S.B. and Lehmkuhle, S. (1995) Visual attention mechanisms show a canter-surround organization. Vision Res., 35: 1859–1869.

Stemmler, M., Usher, M. and Niebur, E. (1995) Lateral interactions in primary visual cortex: a model bridging physiology and psycho-physics. Science, 269: 1877–1880.

Tamas, G., Somogyi, P. and Buhl, E.H. (1998) Differentially interconnected networks of GABAergic interneurons in the visual cortex of the cat. J. Neurosci., 18: 4255–4270.

Temereanca, S. and Simons, D.J. (2001) Topographic specificity in the functional effects of corticofugal feedback in the whisker/barrel system. Soc. Neurosci. Abstr., 393.6.

Thorpe, S., Fize, D. and Marlot, C. (1996) Speed of processing in the human visual system. Nature, 381: 520–522.

Tyler, C.W. and Kontsevich, L.L. (1995) Mechanisms of stereoscopic processing: stereo attention and surface perception in depth reconstruction. Perception, 24(2): 127–153.

Ungerleider, L.G. and Mishkin, M. (1982) Two cortical visual systems: separation of appearance and location of objects. In: Ingle D.L., Goodale M.A. and Mansfield R.J.W. (Eds.), Analysis of Visual Behavior. MIT Press, Cambridge, MA, pp. 549–586.

Vanduffel, W., Tootell, R.B. and Orban, G.A. (2000) Attention-dependent suppression of metabolic activity in the early stages of the macaque visual system. Cereb. Cortex, 10: 109–126.

van Vreeswijk, C. and Sompolinsky, H. (1998) Chaotic balanced state in a model of cortical circuits. Neural Comput., 10: 1321–1371.

Watanabe, T., Nanez, J.E. and Sasaki, Y. (2001) Perceptual learning without perception. Nature, 413: 844–848.

Watanabe, T., Sasaki, Y., Nielsen, M., Takino, R. and Miyakawa, S. (1998) Attention-regulated activity in human primary visual cortex. J. Neurophysiol., 79: 2218–2221.

Wittmer, L.L., Dalva, M.B. and Katz, L.C. (1997) Reciprocal interactions between layer 4 and layer 6 cells in ferret visual cortex. Soc. Neurosci. Abstr., 651.5: 1668.

Wolfe, J.M. and Friedman-Hill, S.R. (1992) Part-whole relationships in visual search. Invest. Opthalmol. Vis. Sci., 33: 1355.

Yantis, S. (1992) Multielement visual tracking: attention and perceptual organization. Cogn. Psychol., 24: 295–340.

Yazdanbakhsh, A. and Grossberg, S. (2004) Fast synchronization of perceptual grouping in laminar visual cortical circuits. Neural Netw., 17: 707–718.

Yen, S.C. and Finkel, L.H. (1998) Extraction of perceptually salient contours by striate cortical networks. Vision Res., 38: 719–741.

Zhang, Y., Suga, N. and Yan, J. (1997) Corticofugal modulation of frequency processing in bat auditory system. Nature, 387: 900–903.

P. Cisek, T. Drew & J.F. Kalaska (Eds.)
Progress in Brain Research, Vol. 165
ISSN 0079-6123
Copyright © 2007 Elsevier B.V. All rights reserved

CHAPTER 7

Real-time neural coding of memory

Joe Z. Tsien*

Center for Systems Neurobiology, Departments of Pharmacology and Biomedical Engineering, Boston University, Boston, MA 02118, USA

Abstract: Recent identification of network-level functional coding units, termed neural cliques, in the hippocampus has allowed real-time patterns of memory traces to be mathematically described, intuitively visualized, and dynamically deciphered. Any given episodic event is represented and encoded by the activation of a set of neural clique assemblies that are organized in a categorical and hierarchical manner. This hierarchical feature-encoding pyramid is invariantly composed of the general feature-encoding clique at the bottom, sub-general feature-encoding cliques in the middle, and highly specific feature-encoding cliques at the top. This hierarchical and categorical organization of neural clique assemblies provides the network-level mechanism the capability of not only achieving vast storage capacity, but also generating commonalities from the individual behavioral episodes and converting them to the abstract concepts and generalized knowledge that are essential for intelligence and adaptive behaviors. Furthermore, activation patterns of the neural clique assemblies can be mathematically converted to strings of binary codes that would permit universal categorizations of the brain's internal representations across individuals and species. Such universal brain codes can also potentially facilitate the unprecedented brain–machine interface communications.

Keywords: neural code; declarative memory; episodic memory; semantic memory; hippocampus; ensemble recording; multiple discriminant analysis; hierarchical clustering method; neural clique; cell assembly; hierarchical and categorical organization; feature-encoding pyramid; binary code; concept; knowledge

Introduction: seeking the neural code

One fundamental goal of neuroscience is to understand the organizing principles and the neural network mechanisms by which the brain encodes and processes information in real time. Although valuable information can be obtained either by using EEG or field recording to map global brain responses or by recording the activity of one or few neurons at a time, neither approach provides direct means to investigate the network-encoding mechanisms underlying information processing. In EEG or field recording experiments, one can only study the summed neural responses across one or multiple areas. In single neuron studies, the recorded activity of single neurons typically needs to be averaged over many trials or even using different animals in order to overcome its firing variability and to identify its event-related responses and encoding properties. However, the brain is unlikely to accomplish its processing through many repetitions in order to seek out statistically meaningful results.

To explain how the brain might achieve its neural coding, Hebb postulated (1949) that information processing in the brain may involve the coordinated activity of large number of neurons,

*Corresponding author. Tel.: +1 617-414-2655;
Fax: +1 617-638-4329; E-mail: jtsien@bu.edu

DOI: 10.1016/S0079-6123(06)65007-3

or cell assemblies (Hebb, 1949). This notion, although rather vague, makes good sense both from the computational and cellular perspective (Wigstrom and Gustafsson, 1985; Bliss and Collingridge, 1993; Abbott and Sejnowski, 1999; Tsien, 2000a, b; Sanger, 2003; Shamir and Sompolinsky, 2004). However, little is known regarding the actual organizing principles and network architectures at the population level. Therefore, the major challenge to date has been to identify the actual patterns of activity of a large neuronal population during cognition, and then to extract the network-level organizing mechanisms that enable the brain to achieve its real-time encoding, processing, and execution of cognitive information. In other words, what are the neural network organizing principles that give rise to the neural code in the brain? Before we can address this question, for practical purposes let us first define what the neural code is.

The neural code is the set of rules and syntax that transform electrical impulses emitted by the brain cells into perceptions, memories, knowledge, decisions, and actions. Neuroscientists try to decipher the brain's neural codes by searching for reliable correlation between firing patterns of neurons and behavioral functions (Adrian, 1926; Gross et al., 1972; Fuster, 1973; Funahashi et al., 1989). As early as the 1920s, Edgar Adrian in his pioneering recording showed that the firing rate of a frog muscle's stretch receptor increases as a function of the load on the muscle (Adrian, 1926), suggesting that information is conveyed by the specific firing patterns of neurons. Two leading neural coding theories can be found in the literature: namely a "*rate code*" and a "*temporal code*" (Barlow, 1972; Softky, 1995; Eggermont, 1998; Van Rullen and Thorpe, 2001). In the *rate code*, all the information is conveyed in the changes in the firing of the cell. In the *temporal code*, information is also conveyed in the precise inter-spike intervals. However, due to a large amount of response variability at the single neuron level in the brain, even in response to identical stimuli (Bialek and Rieke, 1992; Lestienne, 2001), those two types of single neuron-based decoding schemes often produce significant errors in predictions about the stimulus identities or external information.

One good example is place cells in the hippocampus, which were originally discovered by John O'Keefe. These cells show "location-specific" firing when an animal navigates through familiar environments (O'Keefe and Dostrovsky, 1971). The discharge of place cells is shown to be extremely variable during individual passes through their place fields (Fenton and Muller, 1998). Moreover, identification of the place cells routinely requires additional data manipulations, such as excluding those recordings corresponding to periods in which the animals do not reach a certain running speed or simply stay in one location. The traditional way to deal with the response variability of single neurons is to average neuronal discharge over repeated trials. Although data averaging across trials permits the identification of tuning properties of individual neurons, unfortunately, this practice invariably loses crucial information regarding real-time encoding process in the brain.

Early efforts to overcome the poor prediction of using single neurons involved examining the firing pattern of several neurons at the same time. This required the researchers to either record multiple neurons simultaneously or to reconstruct ensembles of multiple neurons from serially, not simultaneously, recorded single neuron data. The reconstructed population approach has indeed been shown to improve the classification and prediction of datasets (Eskandar et al., 1992; Miller et al., 1993; Gochin et al., 1994; but see Hampson and Deadwyler, 1999). With technical developments over the past decades, simultaneous monitoring of the activity of many neurons has become more feasible (McNaughton et al., 1983; Schmidt, 1999; Harris et al., 2000; Buzsaki, 2004). Thus, researchers can rely more on the simultaneously recorded population data than the reconstructed population data. In the 1980s, Georgopoulos and his colleagues were among the first to apply a population-vector method to analyze ensemble firing patterns and show the improved population coding corresponding to arm movements (Georgopoulos et al., 1986). By calculating the mean firing rates for each neuron corresponding to arm movement, a set of population vectors can be obtained that correspond to specific angles of arm rotation. Subsequently, this population vector method has been successfully extended to other studies including place cells (Wilson and McNaughton, 1993). In most cases, the population vectors were typically constructed from all cells, including those cells that did not respond to stimuli. Although such practices did result in better classification, the underlying assumption is that information is represented by the activities of every cell in the population. This is known in the literature as the fully distributed population coding. Recognizing this potential pitfall, some researchers have come up with a compromise between the fully distributed population code and single neuron code by removing the non-responsive cells from the data analyses. The information obtained by this practice is called a sparsely distributed population coding. In other words, information is represented by a subset of cells in the population. Both types of the population coding schemes have been explored using computation simulation methods (Vinje and Gallant, 2000) and shown to be useful in term of improving prediction performances. However, the general rules and organizing principles underlying population encoding remain unknown.

Brief history of memory research

The ability to learn and to remember is one of the most fundamental features of the brain. Understanding how learning and memory work is important because what we learn and remember determine, to a great extent, what and who we are. Furthermore, the impact of learning and memory reaches far beyond the individual, and forms the very foundation for transmitting knowledge through generations, consequently, serving as the major force in driving cultural and social evolution.

From the semantic definition, learning is the acquisition of new information, whereas memory is the retention of acquired information. Driven by knowledge obtained by scientists before them, various disciples of neuroscience over the course of the past 100 years have learned much about various components of learning and memory, whether at the molecular level of synapses or at the systems level of brain circuitry. The concept of memory of mind has existed since the time of Aristotle. However, it is only during the past 50 years or so that scientists have begun to unravel some of the anatomical and cellular bases underlying such a complex mental process. Most neuroscientists regard the ideas and observations of Santiago Ramon Y. Cajal at the end of 19th century as the beginning of the cellular exploration of just how memory is retained in the brain. Upon his original observation of synaptic conjunction between neurons, he immediately entertained the idea that the modification of these conjunctions could form the anatomical basis responsible for the persistence of memory.

So where should we look for such changes? The answer to this seemingly simple question lies at the core effort of modern neuroscience. Scientists must first find out where memories reside in the brain. A breakthrough was made in mid-1950s by Wilder Penfield who had the opportunity to stimulate the cortical surface of over a thousand epileptic patients in the course of neurosurgery for removing epileptic tissue (Penfield and Jasper, 1954). He showed that electrical stimulation of specific limbic structures within the temporal lobe system, such as the hippocampus and amygdala, was capable of generating mental experiences that had a dream-like quality (Halgren et al., 1978). These fascinating reports were the first indications that the temporal lobe system may play a crucial role in representing memories and thoughts. In fact, we now know that memories are processed in many regions of the brain, far beyond the temporal lobe system (Fuster, 1994). The results of many studies suggest that memory is both distributed and localized. In other words, no single memory center exists in the brain, but rather memory is encoded along many pathways in the brain by a set of specific circuits.

Almost at the same period as Penfield's studies, Dr. Brenda Milner of the Montreal Neurological Institute examined a patient, known by his initials as H.M., who had undergone bilateral surgical removal of the temporal lobe (medial temporal cortex, amygdala, and two-thirds of the hippocampus). The surgery was apparently a success in terms of relieving his severe epilepsy, but left him with a devastating loss of his ability to form memories (Scoville and Milner, 1957). For example, although H.M. recognized his childhood pictures and remembered well his childhood events, he had trouble recalling major personal and social events that had taken place a couple of years before his operation. This inability to remember things that happened several years preceding the surgery is called retrograde amnesia.

More strikingly, in H.M.'s case, the surgery also produced severe anterograde amnesia – the inability to form new memories about events, places, and people he encountered after the operation. For example, H.M would not recognize Dr. Milner even though, following the surgery, she had been examining him very frequently for more than 40 years. His anterograde amnesia was so severe that H.M. could not even recognize current photos of himself despite the fact that he viewed himself in a mirror every morning. H.M. was the first human case in which specific amnesia could be linked to selective regions of the brain.

Since then, many patients have been identified as having selective lesions to the temporal lobe system, especially within the hippocampus. They exhibited similar amnesias similar to H.M.'s. For example, amnesic patient R.B., who had a specific lesion in the CA1 region of hippocampus, showed profound loss of the ability to form new memories of people, places, and events (Zola-Morgan et al., 1986). R.B. also lost memories regarding public and personal events that he had experienced 2 years before his CA1 lesion. Such clinical observations have established the view that the hippocampus system is critically involved in memory processes.

Based on the types of memory selectively affected in those amnesic patients, memory can be divided into two major classes: declarative memory and procedural memory. Declarative memory, also termed explicit memory, is memory of places, events, people, and facts and knowledge, and is dependent on the temporal lobe system, whereas procedural memory is memory of motor and procedural skills (such as playing the piano, riding a bike, etc.). In contrast to the unconscious recall of procedural memory, retrieval of declarative memories requires conscious recollection and this type of memory tends to form and be forgotten easily.

Further, declarative memory can be divided into two subclasses: namely episodic memory and semantic memory. Episodic memory refers to memory of episodic events that contain what, where and when information. This is the major type of memory encoded in our daily life. Semantic memory refers to memory of facts, and knowledge that are no longer ascribable to any particular occasion in life (without necessarily remembering where and when the person acquires it). Thus, semantic memory, created through either single or repeated experience, represents a more abstract generalization of experience that may give rise to abstract concepts and categorization. Lesions in the temporal lobe, such as the hippocampus, are known to greatly impair patients' ability to learn new facts, concepts, vocabulary, and knowledge about the world.

The critical role of the hippocampus in memory formation has led to intense investigation of the molecular and cellular processes occurring in the hippocampal circuitry. It is widely believed that neurons in the hippocampus detect coincident neural activity and then convert such coordinated activation into biochemical and synaptic changes that modify the strength of connection between those neurons. A series of sophisticated genetic experiments have firmly established that the NMDA receptor serves as the synaptic coincidence detector (Wigstrom and Gustafsson, 1985; Tsien, 2000a, b) and plays the central role in initiating synaptic plasticity and memory formation (Tsien et al., 1996a, b; Tang et al., 1999). Moreover, recent studies further suggest that reactivation of the NMDA receptor is also crucial for the systems-level consolidation and storage of the long-term memories in the brain (Shimizu et al., 2000; Wittenberg et al., 2002; Cui et al., 2004; Wang et al., 2006).

In search of memory traces

Changes in discharge frequency or latencies of neurons upon learning or electrical stimulation are well known. Some of the earliest experiments came from in vivo recordings in the hippocampus (Barlow, 1972; Berger et al., 1976), a region known to be crucial for memory (Squire, 1987; Cohen and Eichenbaum, 1993). For example, pioneering studies by Thompson and his colleagues showed that classical eye-blink conditioning induces increases in neuronal discharges in the hippocampus and cerebellum (Berger et al., 1976; McCormick and Thompson, 1984; Thompson, 2005). Such a learning paradigm is also reportedly associated with changes in response latency or membrane potential (Olds et al., 1972; Gabirel, 1976; Disterhoft et al., 1986).

Another major focus in the study of hippocampal functions is investigation of the place cells in the hippocampus (O'Keefe and Dostrovsky, 1971; Wilson and McNaughton, 1993; Eichenbaum et al., 1999; Best et al., 2001; Redish, 2001; Poucet et al., 2004). Experiments suggest that place cell activity is controlled by complex internal and external inputs and are modifiable by behaviors and long-term potentiation (McHugh et al., 1996; Dragoi et al., 2003; Jarosiewicz and Skaggs, 2004; Kentros et al., 2004; Lee et al., 2004; Moita et al., 2004; Yeshenko et al., 2004; Wills et al., 2005). A variety of models, such as the rate code, temporal code, population code, and reverberatory activity hypothesis, have been proposed to further test how the hippocampus might represent and process spatial information (Wilson and McNaughton, 1993; Huxter et al., 2003; Howard et al., 2005; Moser et al., 2005).

In addition, learning-related firing changes have also been found in the prefrontal cortex during working memory-related tests (Fuster, 1973; Funahashi et al., 1989). The persistent neural activity has been a subject for many computational modeling studies, including models of *recurrent excitation within cell assemblies*, *synfire chains*, and *single cell bistability* (for a review, see Durstewitz et al., 2000). Moreover, the head-direction cells in the limbic system which also exhibit persistent neural activity in relation to the animal's direct heading in space have been another focus for both experimental investigations and computational modeling of the underlying mechanisms (for a review, see Taub and Bassett, 2003). Thus, individual neurons in the memory systems are clearly responsive to external inputs and have many interesting firing properties. In light of those studies, a number of fundamental questions arise: what is a memory trace? Can the patterns of memory traces be visualized and decoded? Is there any organizing principle underlying the brain's ability to achieve real-time memory encoding and processing?

Visualizing network-level memory traces

Although tremendous progress has been made in terms of our understanding of where and how memory is formed, the question of what memory is has remained unknown. In other words, what are the network-level patterns and organizing principles underlying memory formation? Can those memory-encoding patterns be mathematically described and intuitively visualized?

To examine the real-time encoding mechanisms underlying the network-level representation of memories in the brain, one needs to develop the capability to monitor large number of neurons simultaneously in freely behaving animals, coupled with the clever design of robust memory paradigms and powerful mathematical analyses (Lin et al., 2006a).

Researchers have recently developed a large-scale ensemble recording technique in mice, capable of recording the activity of several hundreds of individual neurons simultaneously (Lin et al., 2006b). Moreover, since the brain is well-known to produce vivid and long-lasting memories about startling events such as devastating earthquakes, high-speed roller coaster rides, or attacks by a lion or a shark, the researchers have also designed similar versions of these startling episodes for mice, such as introducing laboratory-version earthquakes to mice by unexpectedly shaking their cage, or a sudden blast of air to the animal's back (mimicking an owl attack from sky), or a brief vertical freefall inside a plunging small elevator (Lin et al., 2005). Lin et al. simultaneously recorded 260, 210, 148, and 138 individual CA1 units in mice A, B, C, and D, respectively, while subjecting them to seven repetitions of each of the above-mentioned computer-controlled startling stimuli. These stimuli were observed to produce collective changes in firing rates and activity patterns within a subset of the recorded neuronal populations (Lin et al., 2005). Of the total of 756 single units that Lin et al. recorded (pooled from four animals), 13.5% exhibited transient increases, 31.7% showed prolonged increases, 1.9% had transient decreases, and 1.4% responded with a prolonged decrease in their firing frequency. Thus, the ratio of neurons showing increased vs. decreased firing is about 14:1. As an example, the spike rasters of 260 simultaneously recorded single units from mouse A show dynamic changes in the firing patterns of many CA1 neurons after the occurrence of single startling episodes of air-blow, drop and shake (Fig. 1A, epochs of 1 s prior to and 2 s after a drop is shown).

Robust memories induced by those startling events can be easily assessed through behavioral paradigms such as the place conditioning test. For example, using a two-compartment apparatus, researchers can deliver a sudden air-blow (as unconditioned stimulus) to the back of mice whenever they enter the conditioning compartment from the safe compartment. As shown in a 3 h retention test, the conditioned mice exhibited a clear tendency to avoid the conditioned (startled) compartment and spent significantly more time in the unconditioned safe compartment (Fig. 1B). Therefore, such behavioral paradigms enable the researchers to define clearly the categorical variables, consequently facilitating the search for the organizing principles underlying real-time memory encoding.

The existence of a variety of responsive individual neurons suggests that startling events may be represented through distinct activity patterns at the network level by an ensemble of individual neurons. However, it is evident that traditional approaches such as Peri-event histograms, cross-correlation methods, etc., are no longer suitable for dealing with the high dimensionality of the large datasets. Instead, it is necessary to apply statistical tools that are capable of integrating information from large number of units.

To provide an intuitive solution that would facilitate a search for the relevant network-encoding patterns that might be hidden among the activity of the hundreds of simultaneously recorded neurons, researchers used multiple discriminant analysis (MDA) to compute a highly informative low-dimensional subspace among the firing patterns of responsive neurons (Lin et al., 2005). MDA is a supervised dimensionality-reduction method that is well suited for identifying and integrating the classification-significant statistical features of neural population responses to distinct types of known stimuli. This method calculates a low-dimensional subspace that is maximally discriminating for the response matrix, and projects the individual startle responses onto this subspace. Projecting the neural population responses to given events onto single points in this subspace shows that repeated startle responses form clearly well-separated clusters (Fig. 2A), which are distinct from the cluster formed by the rest projections. In other words, CA1 ensemble activity patterns elicited by various startling stimuli can be mathematically described and conveniently visualized as specific startle clusters in a low-dimensional encoding subspace (here in three dimensions), achieving levels of startle discrimination not seen in individual CA1 neuron responses. In addition, air-blow and drop environmental context representations can be further separated using two additional classification steps, further defining where those startle events took place. Moreover, "leave-one-out" cross-validation method further indicates that the prediction accuracy is as high as 99% for mouse A in which 260 cells were recorded. In general, the prediction performance is positively correlated with the number of startle-responsive cells recorded.

To further confirm that various startle-triggered ensemble responses of individual CA1 neurons form distinct encoding patterns in low-dimensional subspaces, the researchers used an independent classification method, known as principal component analysis (PCA) to analyze the datasets (Lin et al., 2005). PCA is an unsupervised, linear dimensionality-reduction tool that is often used for identifying the structure that best represents the data in a least-square sense. Confirming the observations from MDA analysis, the encoding structure of the CA1 population activity is revealed again using this independent dimensionality-reduction method (Lin et al., 2005). The power of those mathematical tools can be further expanded by coupling them with a sliding-window technique (sliding through the recorded neural activity). As such, the researchers, for the first time, were able to directly visualize and dynamically monitor real-time network-level memory-encoding patterns (Lin et al., 2005) (Fig. 2).

The MDA-based sliding window-decoding method also detected the spontaneous reactivation of newly formed memory traces, represented by dynamical trajectories with similar geometric shapes but smaller amplitudes, occurring causally

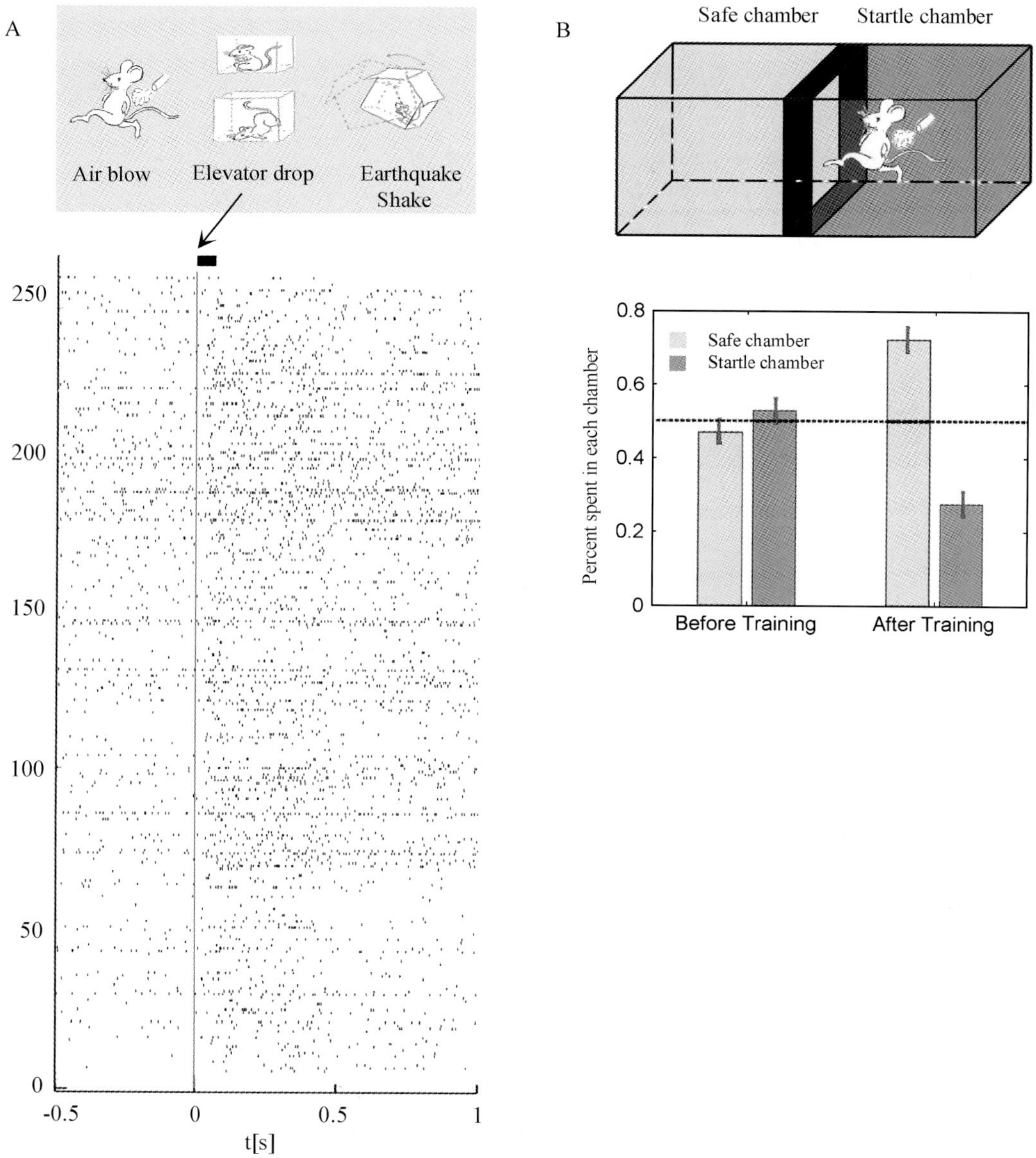

Fig. 1. Large-scale simultaneous monitoring of 260 CA1 cells in a mouse during various mnemonic startling episodes. (A) Three kinds of startling events (sudden air blow, elevator-drop, and earthquake-like shake) were used to produce episodic memories. A spike raster of 260 simultaneously recorded single units from mouse A during a period of 0.5 s prior to and 1 s after the occurrence of single startling episodes of elevator-drop is presented ($t = 0$ marked with vertical red line). X-axis: time-scale (seconds); Y-axis: the numbers of simultaneously recorded single units ($n = 260$). The startle stimulus duration is indicated as a bar next to the vertical red line above the spike raster. (B) The formation of robust memory about the startling events can be assessed by conditioned place conditioning paradigm. The mice spent equal amounts of time (~50%) between *the safe chamber* and *the startle chamber* during the pre-training session prior to startle conditioning. However, after startle conditioning the mice spent significantly more time in the unconditioned (safe) chamber as shown in the 3 h retention test (red bar, 130.1 ± 5.8 s out of the total 180 s, $p < 0.0005$; numbers of mice: $n = 14$).

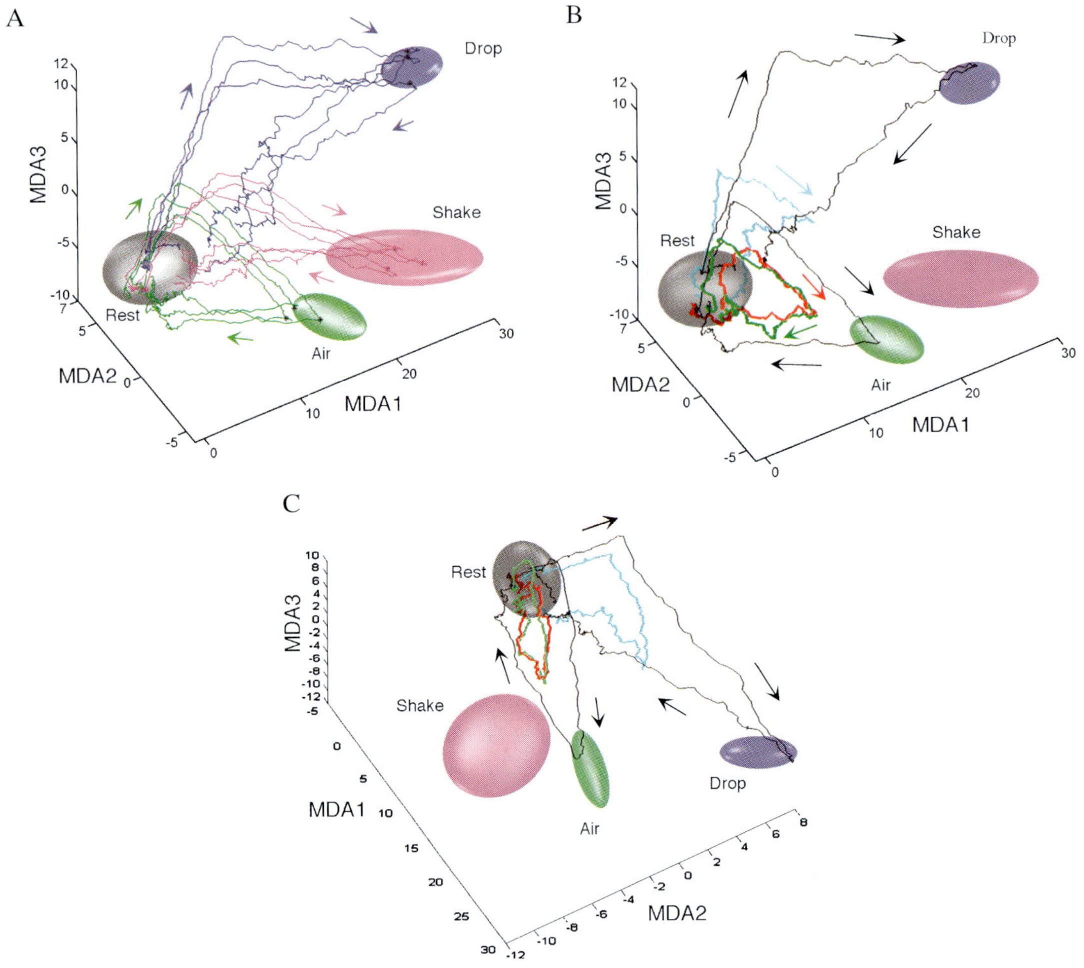

Fig. 2. Classification, visualization, and dynamical decoding of CA1 ensemble representations of startle episodes by *multiple discriminant analysis* (MDA) method. (A) Ensemble patterns during awake rest (gray ellipsoid), air-blow (yellow ellipsoid), drop (blue ellipsoid) and earthquake (magenta ellipsoid) epochs are shown in a three-dimensional sub-encoding space obtained using MDA for a mouse in which 260 CA1 cells were simultaneously recorded; MDA1–3 denote the discriminant axes. Three representative dynamical trajectories of network patterns, revealed by the sliding-window/MDA method, during the encoding of each type of startling events are shown. (B) Dynamical monitoring of post-learning spontaneous reactivations of network traces during and after the actual startling events. 3-D subspace trajectories of the ensemble encoding patterns during drop and air-blow episode in the same mouse are shown. The initial response to an actual air-blow or drop event (black lines) is followed by spontaneous reactivations (red and green lines for two air-blow reactivations, and blue line for drop pattern reactivation), characterized by coplanar, geometrically similar lower amplitude trajectories (directionality indicated by arrows). (C) The same trajectories of reactivation traces from a different orientation after a 3-D rotation show that the trajectories are highly specific toward its own startle clusters. These post-learning dynamical trajectories are typically smaller in amplitude and take place without any time compression, and the numbers of reactivations within the initial several minutes seem to be in the range of one to five, with random intervals.

at intervals ranging from several seconds to minutes after the actual event (Fig. 2). The finding of these reactivations suggests that memory formation is a highly dynamic process, and that this reactivation might play a crucial role in the immediate post-learning fixation of newly formed memory traces (Figs. 2B and C). Previous studies, based on the comparison of firing covariance value of place cells with overlapping fields between the running sessions and the post-running sleep period, imply that place cells participate in reactivations during sleep (Wilson and McNaughton, 1994). The detection of awake-state reactivations of network-level memory-encoding patterns immediately following the startling events further illustrates the unprecedented sensitivity of this new decoding method.

The spontaneous reactivations of the neural patterns can explain what all of us often mentally experience after undergoing such dramatic events: such as coming down from a ride of Tower of Terror, one can not help thinking and talking to one's friends about how scary it was. We believe that these spontaneous reactivations might play a crucial role in the post-learning fixation of newly formed memory traces into long-lasting memories (Wittenberg et al., 2002; Wang et al., 2006). Thus, the combined applications of large-scale ensemble recording and new decoding algorithms open a door to direct visualization and precise measurement of memory traces and their dynamic temporal evolution in the brain.

Identification of neural cliques as real-time memory-coding units

To further identify the internal structures underlying real-time memory encoding, we have employed the agglomerative hierarchical clustering and sequential MDA methods (Lin et al., 2005). These analyses reveal that the encoding power at the population level is actually derived from a set of network-level functional coding units, termed *neural cliques* – a group of neurons with the similar response properties and selectivities, in the CA1 cell population (Fig. 3). For example, *the general startle neural clique* consists of individual cells capable of responding to all types of startling stimuli including the elevator-drop, earthquake, and air-blow, whereas the *sub-general startle cliques* are neural groups that respond to a combination of only two types of startling events. In addition, there are neuron groups that exhibit high specificity toward one specific type of startling events, such as elevator-drop (*the drop-specific neural clique*), earthquake (*earthquake-specific neural clique*), or sudden air-blow events (*air-puff specific neural clique*).

One can mathematically evaluate the contribution of these neural cliques to the CA1 representations by repeating the MDA analysis while sequentially adding clique members to an initial set of non-responsive neurons. For example, a random selection of 40 non-responsive cells as an initial set provides no discriminating power, yielding only overlapping representations (Lin et al., 2005). In contrast, inclusion of the 10 most responsive cells from the "*general startle clique*" leads to good separation between the rest cluster and the startle clusters, but not among startle clusters. The selective discrimination of "drop" startles is obtained by the addition of as few as 10 of the top neurons from the "*drop clique*." Similarly, the inclusion of 10 *air-blow clique* and 10 *shake clique* top neurons subsequently leads to full discrimination between all startle types. Thus, these neural cliques indeed constitute the basic functional coding units for encoding the identity of different startling episodes.

One crucial feature of neural cliques is that the individual neurons belonging to a given clique exhibit "collective co-spiking" temporal dynamics (Fig. 4). This collective co-spiking dynamics among neural clique members enable the memory-coding units to achieve real-time network-level encoding robustness by overcoming the response variability of individual neurons (Fig. 4). Moreover, based on their temporal dynamics, neurons within each clique can be further subgrouped into the four major subtypes: namely, (1) *transient increase*, (2) *prolonged increase*, (3) *transient decrease*, and (4) *prolonged decrease*. The existence of four types of neurons can greatly enhance the real-time encoding robustness as well as provide potential means for modifying clique membership via synaptic plasticity. Finally, neural cliques, as network-level functional coding units, should also be less vulnerable to the death of one or a few neurons, and therefore exhibit graceful degradation should such conditions arise during the aging process or disease states.

Hierarchical and categorical organization of memory-encoding neural clique assemblies

Through examining the overall organization of neural clique assembly involved in startle memory encoding, it is clear that the internal CA1

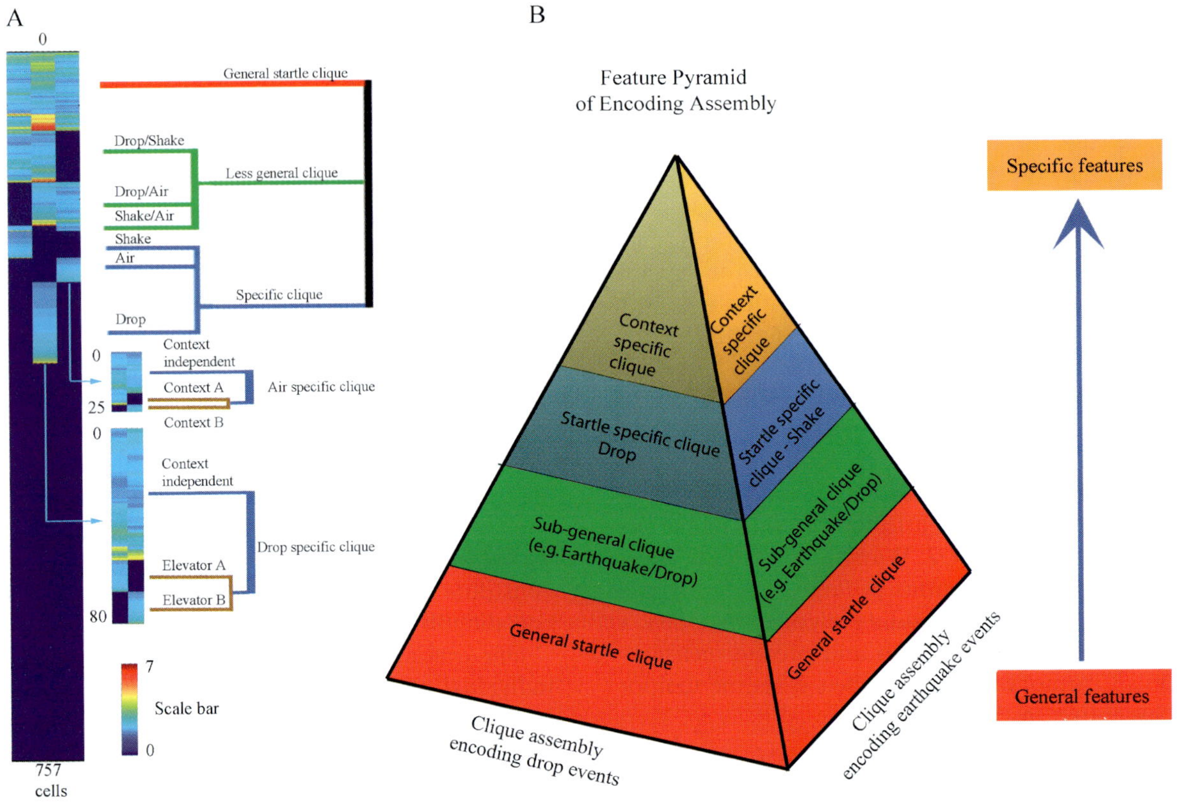

Fig. 3. Categorical and hierarchical organization within memory-encoding neural clique assemblies. Memory-coding units are organized in a categorical and hierarchical manner. The hierarchical clustering analysis of responses of a total of 757 CA1 neurons from four mice to the three different types of startling episodes reveals the existence of seven major neural cliques. (Panel A): General startle clique; sub-general startle cliques which has three combination – drop-shake clique, air blow-drop cliques, shake-air blow clique; three startle type-specific cliques which are drop-specific clique, shake-specific clique, and air blow-specific clique. Furthermore, within startle-specific neural cliques, neurons can be further divided into startle context-specific clique (air-blow in context A-specific clique, air-blow in context B-specific clique, drop in elevator A-specific clique, and drop in elevator B-specific clique). Non-responsive units are grouped in the bottom half. The color scale bar indicates the normalized response magnitude (1–7). It is clearly evident that those memory-coding units are organized in a hierarchical and categorical fashion (Panel B), and any given startling episode is encoded by a combinatorial assembly of a series of neural cliques, invariantly consisting of the general startle clique, sub-general startle clique, startle identity-specific clique, and context-specific startle clique. In this feature pyramid of the encoding clique assembly, the neural clique representing the most general, abstract features (to all categories) is at the bottom and it forms a common building block for all types of startle event encoding. The next layer of the pyramid is made by neural cliques responding to less general features (covering multiple, but not all, categories); those sub-general cliques are present in a subset of the neural clique assemblies. As moving up along this encoding feature pyramid, neural cliques become more and more specific. The neural clique at the top of the pyramid encodes the most specific and highly discriminate features, thereby defining a particular event or experience. Please note that the number of neurons for each clique does not necessarily correspond to its position in the feature pyramid. In other words, the neural clique encoding the general features does not necessarily have more neurons than the neural cliques encoding more specific features.

representations of any given startle episode involve a set of neural clique assemblies that are invariantly organized in a categorical hierarchy manner. This neural architecture, termed *feature-encoding pyramid* (Fig. 3), starts with the neural clique representing the most general and common features (to all categories) at the bottom layer, followed by neural cliques responding to less general features (covering multiple, but not all, common categories), and then moves gradually up towards more and more specific and discriminating features (responding to a specific category), and eventually

A

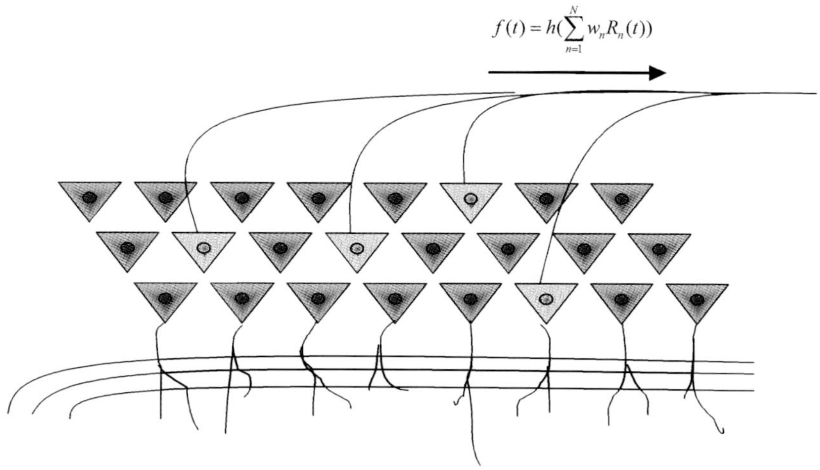

B

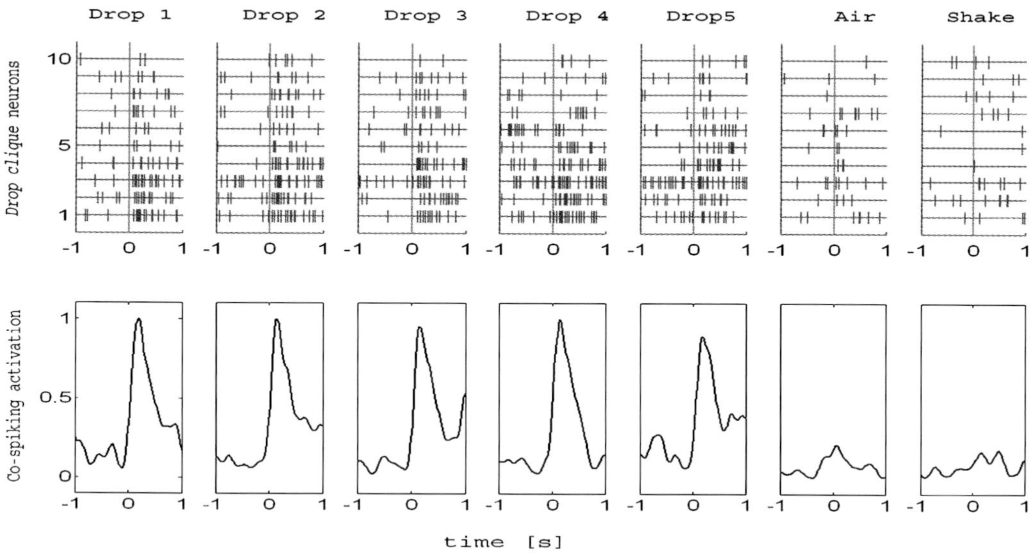

Fig. 4. Real-time encoding robustness of memory-coding units is achieved through "collective co-spiking" of its individual members within each neural clique. (A) Illustration of a *neural clique* in the CA1 network. The activation function of the clique to drive the downstream clique can be mathematically described (equation is listed on top) as a threshold function of the integrated inputs from the upstream clique members, where R is input, w is the weighting factor, h is the threshold function (e.g., sigmoid), and t is time. (B) Spike rasters and weighted responses of the top 10 neurons within the *drop-specific neural clique* (listed in Y-axis) during five elevator-drop events (1 s before and after the startle, X-axis) are shown as an example. Although responses of the individual member (neuron) are quite variable from trial to trial, the consistency and specificity of the collective co-spiking of the clique responses is evident from each drop episode (five episodes are listed). The *drop-specific neural clique* exhibited no significant responses to air-blow or shake episodes (last two insets on the right). Robust co-spiking of membership neurons in the cliques is also preserved at the finer time-scale (20–30 ms).

with the most discriminating feature clique (corresponding to context specificity) on the top of the feature-encoding pyramid.

According to this hierarchical structure of network-level memory encoding (Fig. 3), the *general startle neural clique* represents the neurons engaged in the extraction of the common features among various episodes (e.g., encoding abstract and generalized knowledge that "such events are scary and dangerous," by integrating neural inputs from the amygdala). The *sub-general neural cliques* are involved in identifying sub-common features across a subset of startling episodes (e.g., perhaps, the *earthquake and drop-specific clique* for encoding the semantic memory of the fact that "those events involve shaking and motion disturbances," by integrating inputs from the vestibular system); whereas the *startle identity-specific cliques* encode discriminative information about startle types (defining "what type" of event has happened); and the *startle context-specific cliques* provide even more specific feature, such as contextual information about "where" a particular startling event has happened.

This invariant *feature-encoding pyramid* of neural clique assemblies reveals four basic principles for the organization of memory encoding in the brain (Fig. 3).

First, the neural networks in the memory systems employ a categorical and hierarchical architecture in organizing memory-coding units.

Second, the internal representations of external events in the brain through such a feature-encoding pyramid is achieved not by recording exact details of the external event, but rather by recreating its own selective pictures based on the importance for survival and adaptation.

Third, the "*feature-encoding pyramid*" structure provides a network mechanism, through a combinatorial and self-organizing process, for creating seemingly unlimited numbers of unique internal patterns, capable of dealing with potentially infinite numbers of behavioral episodes that an animal or human may encounter during its life.

Fourth, in addition to its vast memory storage capacity, this neural clique-based hierarchical extraction and parallel binding processes also enable the brain to achieve abstraction and generalization, cognitive functions essential for dealing with the complex, ever-changing situations. Recent identification of hippocampal cells encoding abstract concept of nest or bed provides further experimental validation for the existence of neurons in the memory network for extracting abstract knowledge from episodic experiences (Lin et al., 2007).

The finding that the memory-encoding neural clique assembly appears to invariantly contain the coding units for processing the abstract and generalized information (Lin et al., 2006a) is interesting. It fits well with the anatomical evidence that virtually all of the sensory input that the hippocampus receives arises from higher-order, multimodal cortical regions and the hippocampus has a high degree of sub-regional divergence and convergence at each loop. This unique anatomical layout supports the notion that whatever processing is achieved by the hippocampus in the service of long-term memory formation should have already engaged with fairly abstract, generalized representations of events, people, facts, and knowledge.

The observed "*feature-encoding pyramid*" structure of the neural clique assembly is likely to represent a general mechanism for memory encoding across different animal species. For example, single-unit recordings in human hippocampus show that some hippocampal cells fire in response to faces, or more selectively, to a certain type of human facial emotions; other seems to exhibit highly selective firing to one specific person (e.g., "actress Halle Berry cell" which fires selectively to her photo portraits, Cat-woman character, and even a string of her name (Fried et al., 1997; Quiroga et al., 2005)). Although those cells were not recorded simultaneously, the findings are nonetheless consistent with the general-to-specific *feature-pyramid structure*. In addition, it is also reported that while some place cells in the rat hippocampus exhibit location-specific firing regardless of whether the animals engage in a random forage or goal-oriented food retrieval (or make a left- or right-turn in a T-maze), others seem to fire selectively at their place fields only in association with a particular kind of experience (Markus et al., 1995; Wood et al., 2000). Thus, those studies also seem to support the existence of a hierarchical structure involved in space coding. Therefore, the hierarchical organization of the neural clique assembly, revealed through large-scale recording of startling episodes and mathematical analyses, may represent a general feature for memory encoding in the brain. In addition, it further suggests that episodic memory is intimately linked with and simultaneously converted to semantic memory and generalized knowledge.

This form of hierarchical extraction and parallel binding along CNS pathways into memory and other higher cognition systems is fundamentally different from the strategies used in current computers, camcorders, or intelligent machines. These unique design principles allow the brain to extract the commonalities through one or multiple exposures and to generate more abstract knowledge and generalized experiences. Such generalization and abstract representation of behavioral experiences has enabled humans and other animals to avoid the burden of remembering and storing each mnemonic detail. More importantly, by extracting the essential elements and abstract knowledge, animals can apply past experiences to future encounters that share the same essential features but may vary greatly in physical detail. These higher cognitive functions are obviously crucial for survival and reproduction of animal species.

Universal activation codes for the brain's real-time neural representations across individuals and species

With the identification of the neural clique as a basic coding unit and the *feature-encoding pyramid* within the clique assemblies, we can further convert (through matrix inversion) those distinct ensemble representations observed in a low-dimensional encoding-subspace into a string of binary activation codes with 1s and 0s (Fig. 5). This binary assignment, 1 for the active state and 0 for the inactive state of neural cliques, is based on the idea that the activity state of a neural clique can be monitored by downstream neurons using a biologically plausible binary activation function (McCulloch and Pitts, 1990). This mathematical conversion of the activation patterns of the neural clique assembly into a binary code of 1s and 0s creates a simple and convenient way for universally comparing and categorizing network-level representations from brain to brain (Lin et al., 2006a).

This type of the universal binary code can provide a potentially unifying framework for the study of cognition even across animal species. For example, should a mouse, dog, and human all experience a sudden free-fall in a plunging elevator, the activation patterns of the *general startle neural clique*, *drop-specific clique*, *air-puff clique*, and *earthquake clique* in their brains would produce the identical real-time binary activation code (1, 1, 0, 0), according to the above permutation and arrangement of the coding unit assembly. Yet, since the mouse, dog, and human may perceive other subtle information differently during the incident, the subsequent digits may differ. For example, the dog may sense a trace amount of smell of burning wires, whereas the human may see erratic flicking of elevator buttons, and the mouse may have a flying candy wrap hit its face. As such, the binary activation codes would permit the universal measurement and categorization of neural representations between those three species, with the initial four digits defining the common experience of free falling, and the subsequent digits corresponding to different subtle details.

The proposed binary codes, derived from the activation patterns of the neural clique assembly, offer a concise way to universally categorize the neural representations of cognition in various brains. In the meantime, it is important for us to point out the fundamental differences between the neural clique pattern-based brain codes and the nucleotide-based genetic codes (Lin et al., 2006a). Specifically, the neural clique-based brain codes have at least four distinct properties:

1. *Un-inheritable*: Genetic codes are directly transferred through reproduction, whereas brain codes, by and large, are not inheritable and can only be acquired through experiences (perhaps with the exception of those neural codes controlling primitive functions such as breathing, heartbeat, and the knee-jerk reflex, etc.; those may have been genetically programmed).
2. *Self-organizing*: Genetic codes act like predetermined scaffolds, providing blueprints

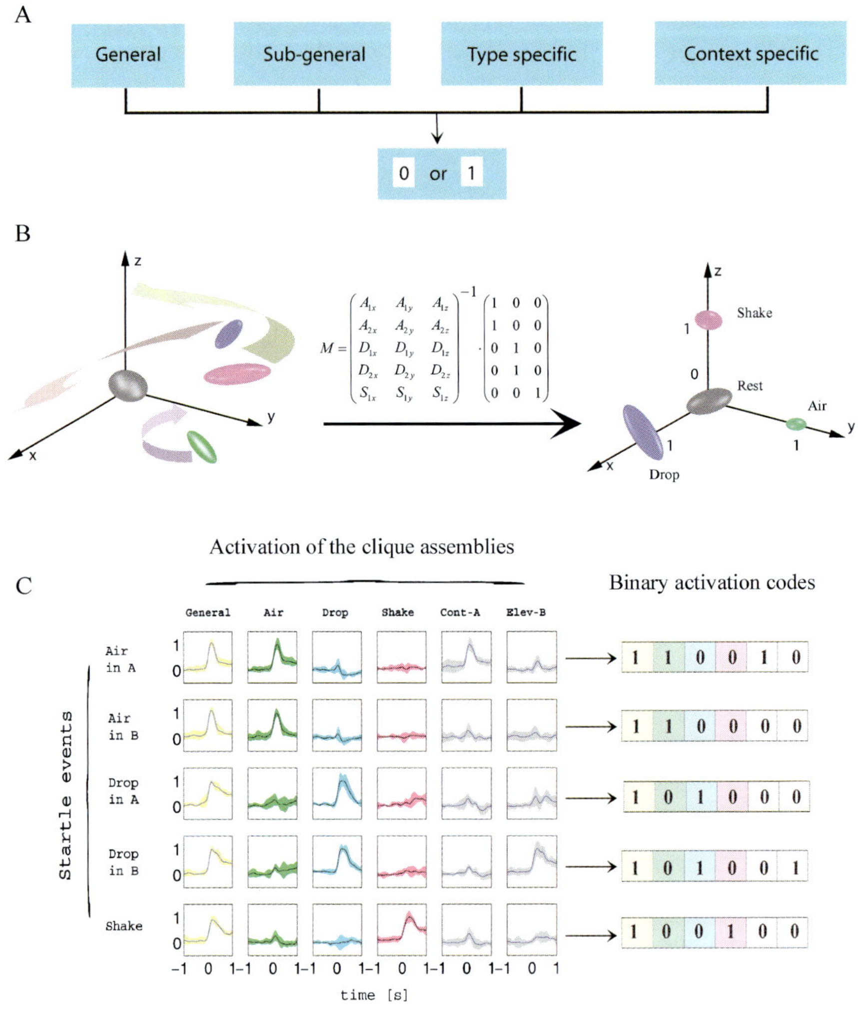

Fig. 5. Conversion of activation patterns of neural clique assemblies into a binary code. (A) Conversion of the activation state of a neural clique assembly that encodes one type of startling events into binary digits 1 or 0. (B) Mathematical transformation of MDA pattern into a startle type-specific binary encoding system. While the MDA method provides an efficient separation of the startle episodes, each of the discriminant axes at those MDA encoding subspaces (on the left) is no longer corresponding to functional meaning. Thus, we used matrix inversion to translate the ensemble patterns into a startle-specific encoding coordinate system (on the right). This is achieved by assigning new positions for the cluster centers so that they are linearly mapped into a "clique-space," where each axis directly corresponds to a particular clique, thus projecting specific activation patterns to 1 and the absence of activation to 0 (top panel). This mathematical operation allows us to map the encoding subspace into one where the startle representations can directly correspond to neural clique activity patterns and subsequently, to translate the collective activity patterns of neural clique assembly into unique and efficient network-level binary activation codes with a string of binary digits (1s and 0s). (C) Conversion of activation patterns of multiple neural clique assemblies into real-time binary codes. Responses of neural cliques are illustrated in different colors. The activation function of a given clique at each network level can be mathematically described. Rows correspond to the different startling episodes, while columns indicate the different neural cliques (general startle, air-blow, drop, shake, air-blow context-A and drop context-B). The binary activation patterns corresponding to each event can be mathematically converted to a set of binary codes (on the right column, following the defined sequence of the cliques). As such the clique activation codes are: 110010 for air-blow in context A; 110000 for air-blow in context B; 101000 for drop in elevator A; 101001 for drop in elevator B; and 100100 for shake. This binary code can allow us to accurately predict the behavioral experiences by just sliding through the recorded neural population activity and calculating the hit ratio of matching those binary codes with the occurrences of each startling event.

for the development and basic functionality of the organism, whereas brain codes are dynamical and self-organizing, arising out of internal structures and connectivity of neural networks upon behavioral experiences.

3. *Variable sizes*: The numbers of genes are exactly fixed for each individual and species, whereas the number of brain codes is highly variable in each brain, and in theory, it is only limited by the network capacity (which is determined by the convergence and divergence in connectivity), as well as the amount of behavioral experiences that an individual encounters.
4. *Modifiable*: Unless mutated, the genetic code remains static, whereas the membership of individual neurons within a given neural clique is modifiable by experience-dependent synaptic plasticity or disease states.

Thus, the above features of brain codes are set apart in a fundamental way from genetic codes.

The identification of neural cliques as memory-coding units in the hippocampus prompts us to entertain that the concept of neural cliques as basic, self-organizing processing units may be applicable to many, if not all, neural networks in the brain (Lin et al., 2006a). Under this neural clique code model (Fig. 6), the functionality implemented by neural cliques in a given network depends on the specializations of the corresponding regions. Neural cliques in the primary sensory regions (perhaps organized inside cortical columns) encode piecemeal information by decomposing external events into

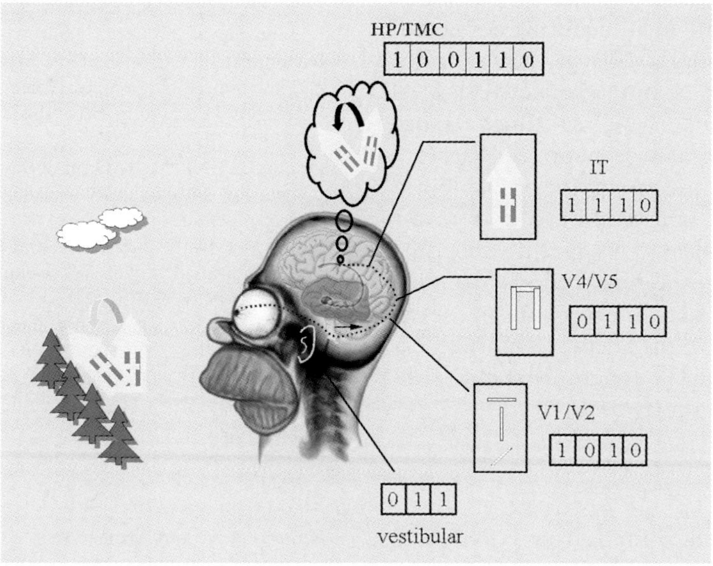

Fig. 6. *Neural clique code*-based real-time information processes in the brain. Through a series of hierarchical-extraction and parallel-binding processes, the brain achieves coherent internal encoding and processing of the external events. For example, when a person experiences a sudden earthquake, neural cliques in his primary visual cortex encode the decomposed features about edge orientation, movement, and eventually shapes of visual objects, whereas the neural cliques in the vestibular nuclei detect sudden motion disturbances. As information is processed along its pathways into deeper cortex such as the inferior temporal cortex (IT), neural cliques begin to exhibit complex encoding features such as houses. By the time it reaches high association cortices such as the hippocampus (HP) and temporal medial cortex (TMC), the neural clique assembly encodes earthquake experience and its location, with a selective set of "what and where" information. At this level, abstract and generalized information such as semantic memories of "*the earthquake is dangerous and scary*" have emerged. As information is further processed into other cortical regions involving decision making and motor planning, a series of phased firing among various neural clique assemblies lead to adaptive behaviors such as screaming and running away from the house, or hiding under a dining table. As illustrated, the activation patterns of neural clique assemblies in each brain region can be also converted into a binary code for universally comparing and categorizing network-level representations from brain to brain. Such universal brain codes can also allow more seamless brain–machine interface communications.

various basic features (e.g., the primary visual cortex for detecting edge orientation, color, or size of visual objects, whereas the vestibular nuclei for detecting motion, etc.) (Fig. 6). As information is further processed along its pathways into deeper regions, neural cliques (although no longer organized in their anatomically distinguishable maps or columns) start to encode more complex features (e.g., shapes and complex objects such as houses and faces in the inferior temporal cortex). By the time it reaches high association cortices such as the hippocampus, neural cliques have already contained both specific and generalized mnemonic information about events, places, and people with a significant amount of abstraction and generalization (Fig. 6). Eventually, the brain areas involved in decision making, executive function, and motor planning may start coherent and phased firings among various neural cliques, thereby generating behaviors.

In summary, recent identification of neural cliques as the basic coding units in the brain has provided crucial insights into the network-level organizing principles underlying real-time memory encoding (Tsien, 2007). Those neural cliques are self-organized through a combinatorial fashion to form a memory-encoding assembly with an invariant hierarchical structure. This feature-encoding hierarchical structure of the neural clique assembly immediately suggests a network mechanism for the brain to achieve both large memory storage capacity and higher cognitive functions such as abstraction and generalization.

Acknowledgments

This work was supported by funds from NIMH and NIA, Burroughs Welcome Fund, ECNU Award, W.M. Keck Foundations, special funds for Major State Basic Research of China (NO2003CB716600), the key project of Chinese Ministry of Education (NO104084), and grants from Shanghai Commissions on Science and Technology, and Education.

References

Abbott, L.E. and Sejnowski, T.J. (1999) Neural Codes and Distributed Representations. The MIT Press, Cambridge, MA.

Adrian, E.G. (1926) The impulses produced by sensory nerve endings: Part 1. J. Physiol., 61: 49–72.

Barlow, H. (1972) Single units and sensation: a doctrine for perceptual psychology? Perception, 1: 371–394.

Berger, T.W., Alger, B. and Thompson, R.F. (1976) Neuronal substrate of classical conditioning in the hippocampus. Science, 192: 483–485.

Best, P.J., White, A.M. and Minai, A. (2001) Spatial processing in the brain: the activity of hippocampal place cells. Annu. Rev. Neurosci., 24: 459–486.

Bialek, W. and Rieke, F. (1992) Reliability and information transmission in spiking neuron. Trends Neurosci., 15: 428–433.

Bliss, T.V. and Collingridge, G.L. (1993) A synaptic model of memory: long-term potentiation in the hippocampus. Nature, 361: 31–39.

Buzsaki, G. (2004) Large-scale recording of neuronal ensembles. Nat. Neurosci., 7: 446–451.

Cohen, N.J. and Eichenbaum, H. (1993) Memory, Amnesia, and the Hippocampal System. The MIT Press, Cambridge, MA.

Cui, Z.Z., Wang, H., Tan, Y., Zaia, K.A., Zhang, S. and Tsien, J.Z. (2004) Inducible and reversible NR1 knockout reveals crucial role of the NMDA receptor in preserving remote memories in the brain. Neuron, 41: 781–793.

Disterhoft, J.F., Coutler, D.A. and Alkon, D.L. (1986) Conditioning-specific membrane changes of rabbit hippocampal neurons measured in vitro. Proc. Natl. Acad. Sci. U.S.A., 83: 2733–2737.

Dragoi, G., Harris, K.D. and Buzsaki, G. (2003) Place representation within hippocampal network is modified by long-term potentiation. Neuron, 39: 843–853.

Durstewitz, D., Seamans, K.J. and Sejnowki, T.J. (2000) Neurocomputational models of working memory. Nat. Neurosci., Suppl. 3: 1184–1191.

Eggermont, J.J. (1998) Is there a neural code? Neurosci. Biobehav. Rev., 22: 355–370.

Eichenbaum, H., Dudchenko, P., Wood, E. Shapiro, M. and Tanila, H. (1999) The hippocampus, memory, and place cells: is it spatial memory or a memory space? Neuron, 23: 209–226.

Eskandar, E.N., Richmond, B.J. and Optican, L.M. (1992) Role of inferior temporal neurons in visual memory. J. Neurophysiol., 68: 1277–1296.

Fenton, A.A. and Muller, R.U. (1998) Place cell discharge is extremely variable during individual passes of the rat through the firing field. Proc. Natl. Acad. Sci. U.S.A., 95: 3182–3187.

Fried, I., MacDonald, K.A. and Wilson, C.L. (1997) Single neuron activity in human hippocampus and amygdala during recognition of faces and objects. Neuron, 18: 753–765.

Funahashi, S., Bruce, C.J. and Goldman-Rakic, P.S. (1989) Mnemonic coding of visual space in the monkey's dorsolateral prefrontal cortex. J. Neurophysiol., 61: 331–349.

Fuster, J.M. (1973) Unit activity in prefrontal cortex during delayed-response performance: neuronal correlates of transient memory. J. Neurophysiol., 36: 61–78.

Fuster, J.M. (1994) Memory in the Cerebral Cortex: An Empirical Approach to Neural Networks in the Human and Nonhuman Primate. The MIT Press, Cambridge, MA.

Gabirel, M. (1976) Short-latency discriminative unit responses: engram or bias? Physiol. Psychol., 4: 275–280.

Georgopoulos, A.P., Schwartz, A.B. and Kettner, R.E. (1986) Neuronal population coding of movement direction. Science, 233: 1416–1419.

Gochin, P.M., Colombo, M., Dorfman, G.A., Gerstein, G.L. and Gross, C.G. (1994) Neural ensemble coding in inferior temporal cortex. J. Neurophysiol., 71: 2325–2335.

Gross, C.G., Rocha-Miranda, C.E. and Bender, D.B. (1972) Visual properties of neurons in inferotemporal cortex of the macaque. J. Neurophysiol., 35: 96–111.

Halgren, E., Walter, R.D., Cherlow, A.G. and Crandall, P.H. (1978) Mental phenomena evoked by electrical stimulation of the human hippocampal formation and amygdala. Brain, 101: 83–117.

Hampson, E.R. and Deadwyler, S.A. (1999) Pitfalls and problems in the analysis of neuronal ensemble recordings during behavioral tasks. In: Nicolels M. (Ed.), Methods for Neural Ensemble Recordings. CRC Press, New York, pp. 229–248.

Harris, K.D., Henze, D.A., Csiscvari, J., Hirase, H. and Buzsaki, G. (2000) Accuray of tetrode spike separation as determined by simultaneous intracellular and extracellular measurement. J. Neurophysiol., 84: 401–414.

Hebb, D.O. (1949) The Organization of Behavior. Wiley, New York.

Howard, M.W., Fotedar, M.S., Datey, A.V. and Hasselmo, M.E. (2005) The temporal context model in spatial navigation and relational learning: toward a common explanation of medial temporal lobe function across domains. Psychol. Rev., 112: 75–116.

Huxter, J., Burgess, N. and O'Keefe, J. (2003) Independent rate and temporal coding in hippocampal pyramidal cells. Nature, 425: 828–832.

Jarosiewicz, B. and Skaggs, W.F. (2004) Hippocampal place cells are not controlled by visual input during the small irregular activity state in the rat. J. Neurosci., 24: 5070–5077.

O'Keefe, J. and Dostrovsky, J. (1971) The hippocampus as a spatial map. Preliminary evidence from unit activity in the freely-moving rat. Brain Res., 34: 171–175.

Kentros, C.G., Agnihotri, N.T., Streater, S., Hawkins, R.D. and Kandel, E.R. (2004) Increased attention to spatial context increases both place field stability and spatial memory. Neuron, 42: 283–295.

Lee, I., Rao, G. and Knierim, J.J. (2004) A double dissociation between hippocampal subfields: differential time course of CA3 and CA1 place cells for processing changed environments. Neuron, 42: 803–815.

Lestienne, R. (2001) Spike timing, synchronization and information processing on the sensory side of the central nervous system. Prog. Neurobiol., 65: 545–591.

Lin, L., Chen, G., Kuang, H., Wang, D. and Tsien, J.Z. (2007) Neural encoding of the concept of nest in the mouse brain. Proc. Natl. Acad. Sci. U.S.A., 104: 6066–6071.

Lin, L., Chen, G., Xie, K., Zaia, K., Zhang, S. and Tsien, J.Z. (2006a) Large-scale neural ensemble recording in the brains of freely behaving mice. J. Neurosci. Methods, 155: 28–38.

Lin, L., Osan, R., Shoham, S., Jin, W., Zuo, W. and Tsien, J.Z. (2005) Identification of network-level coding units for real-time representation of episodic experiences in the hippocampus. Proc. Natl. Acad. Sci. U.S.A., 102: 6125–6130.

Lin, L., Osan, R. and Tsien, J.Z. (2006b) Organizing principles of real-time memory encoding: neural clique assemblies and universal neural codes. Trends Neurosci., 29: 48–56.

Markus, E.J., Qin, Y.L., Leonard, B., Skaggs, W.E. and McNaughton, B.L. (1995) Interactions between location and task affect the spatial and directional firing of the hippocampal neurons. J. Neurosci., 15: 7079–7794.

McCormick, D.A. and Thompson, R.F. (1984) Neuronal responses of the rabbit cerebellum during acquisition and performance of a classically conditioned nictitating membrane-eyelid response. J. Neurosci., 4: 2811–2822.

McCulloch, W.S. and Pitts, W. (1990) A logical calculus of the ideas immanent in nervous activity. 1943. Bull. Math. Biol., 52: 99–115 Discussion 173–197.

McHugh, T., Blum, K.I., Tsien, J.Z., Tonegawa, S. and Wilson, M.A. (1996) Impaired hippocampal representation of space in CA1-specific NMDAR1 knockout mice. Cell, 87: 1339–1349.

McNaughton, B.L., O'Keefe, J. and Barnes, C.A. (1983) The stereotrode: a new technique for simultaneous isolation of several single units in the central nervous system from multiple unit records. J. Neurosci. Methods, 8: 391–397.

Miller, E.K., Li, L. and Desimone, R. (1993) Activity of neurons in anterior inferior temporal cortex during a short-term memory task. J. Neurosci., 13: 1460–1468.

Moita, M.A., Rosis, S., Zhou, Y., LeDoux, J.E. and Blair, H.T. (2004) Putting fear in its place: remapping of hippocampal place cells during fear conditioning. J. Neurosci., 24: 7015–7023.

Moser, E.I., Moser, M.B., Lipa, P., Newton, M., Houston, F.P., Barnes, C.A. and McNaughton, B.L. (2005) A test of the reverberatory activity hypothesis for hippocampal place cells. Neuroscience, 130: 519–526.

Olds, J., Disterhoft, J.F., Segal, M., Komblith, C.L. and Hirsh, R. (1972) Learning centers of rat brain mapped by measuring latencies of conditioned unit responses. J. Neurophysiol., 35: 202–219.

Penfield, W.W. and Jasper, H. (1954) Epilepsy and the Functional Anatomy of the Human Brain. Brown, Boston, MA.

Poucet, B., Lenck-Santini, P.P., Hok, V., Save, E., Banquet, J.P., Gaussier, P. and Muller, R.U. (2004) Spatial navigation and hippocampal place cell firing: the problem of goal encoding. Rev. Neurosci., 15: 89–107.

Quiroga, R.Q., Reddy, L., Jreiman, G., Koch, C. and Fried, I. (2005) Invariant visual representation by single neurons in the human brain. Nature, 435: 1102–1107.

Redish, A.D. (2001) The hippocampal debate: are we asking the right questions? Behav. Brain Res., 127: 81–98.

Sanger, T.D. (2003) Neural population codes. Curr. Opin. Neurobiol., 13: 238–249.

Schmidt, E.M. (1999) Electrodes for many single neuron recordings. In: Nicolels M. (Ed.), Methods for Neural Ensemble Recordings. CRC Press, New York, pp. 1–23.

Scoville, W.B. and Milner, B. (1957) Loss of recent memory after bilateral hippocampal lesions. J. Neurol. Neurosurg. Psychiatry, 20: 11–21.
Shamir, M. and Sompolinsky, H. (2004) Nonlinear population codes. Neural Comput., 16: 1105–1136.
Shimizu, E., Tang, Y., Rampon, C. and Tsien, J.Z. (2000) NMDA receptor-dependent synaptic reinforcement as a crucial process for memory consolidation. Science, 290: 1170–1174.
Softky, W.R. (1995) Simple codes versus efficient codes. Curr. Opin. Neurobiol., 5: 239–247.
Squire, L. (1987) Memory and Brain. Oxford University Press, New York.
Tang, Y.P., Shimizu, E., Dube, G.r., Rampon, C., Kerchner, G.A., Zhuo, M., Liu, G. and Tsien, J.Z. (1999) Genetic enhancement of learning and memory in mice. Nature, 401: 63–69.
Taub, J.S. and Bassett, J.P. (2003) Persistent neural activity in head direction cells. Cereb. Cortex, 13: 1162–1172.
Thompson, R.F. (2005) In search of memory traces. Annu. Rev. Psychol., 56: 1–23.
Tsien, J.Z. (2000a) Linking Hebb's coincidence-detection to memory formation. Curr. Opin. Neurobiol., 10: 266–273.
Tsien, J.Z. (2000b) Building a brainier mouse. Sci. Am., 282: 62–68.
Tsien, J.Z. (2007) The memory code. Scientific American, July Issue, 52–59.
Tsien, J.Z., Chen, D., Gerber, C., Mercer, D., Anderson, D., Kandel, E.R. and Tonegawa, S. (1996a) Subregion- and cell type-restricted gene knockout in mouse brain. Cell, 87: 1317–1326.
Tsien, J.Z., Huerta, P.T. and Tonegawa, S. (1996b) The essential role of hippocampal CA1 NMDA receptor-dependent synaptic plasticity in spatial memory. Cell, 87: 1327–1338.
Van Rullen, R. and Thorpe, S.J. (2001) Rate-coding versus temporal order coding: what the retinal ganglion cells tell the visual cortex. Neural Comput., 13: 1255–1283.
Vinje, W.E. and Gallant, J.L. (2000) Sparse coding and decorrelation in primary visual cortex during natural vision. Science, 287: 1273–1276.
Wang, H., Hu, Y. and Tsien, J.Z. (2006) Molecular and systems mechanisms of memory consolidation and storage. Prog. Neurobiol., 79: 123–135.
Wigstrom, H. and Gustafsson, B. (1985) On long-lasting potentiation in the hippocampus: a proposed mechanism for its dependence on coincident pre- and postsynaptic activity. Acta Physiol. Scand., 123: 519–522.
Wills, T.J., Lever, C., Caucci, F., Burgess, N. and O'Keefe, J. (2005) Attractor dynamics in the hippocampal representation of the local environment. Science, 308: 873–876.
Wilson, M.A. and McNaughton, B.L. (1993) Dynamics of the hippocampal ensemble code for space. Science, 261: 1055–1059.
Wilson, M.A. and McNaughton, B.L. (1994) Reactivation of hippocampal ensemble memories during sleep. Science, 265: 676–679.
Wittenberg, G.M., Sullivan, M.R. and Tsien, J.Z. (2002) An emerging molecular and cellular framework for memory processing by the hippocampus. Trends Neurosci., 25: 501–505.
Wood, E., Dudchenko, P.A., Robitsek, R.J. and Eichenbaum, H. (2000) Hippocampal neurons encode information about different types of memory episodes occurring in the same location. Neuron, 27: 623–633.
Yeshenko, O., Guazzelli, A. and Mizumori, S.J. (2004) Context-dependent reorganization of spatial and movement representations by simultaneously recorded hippocampal and striatal neurons during performance of allocentric and egocentric tasks. Behav. Neurosci., 118: 751–769.
Zola-Morgan, S., Squire, L.R. and Amaral, D. (1986) Human amnesia and the medial temporal region: enduring memory impairment following a bilateral lesion limited to the CA1 field of the hippocampus. J. Neurosci., 6: 2950–2967.

P. Cisek, T. Drew & J.F. Kalaska (Eds.)
Progress in Brain Research, Vol. 165
ISSN 0079-6123
Copyright © 2007 Elsevier B.V. All rights reserved

CHAPTER 8

Beyond timing in the auditory brainstem: intensity coding in the avian cochlear nucleus angularis

Katrina M. MacLeod* and Catherine E. Carr

Department of Biology, University of Maryland, College Park, MD 20742, USA

Abstract: Many of the computational principles for sound localization have emerged from the study of avian brains, especially for the construction of codes for interaural timing differences. Our understanding of the neural codes for interaural level differences, and other intensity-related, non-localization sound processing, has lagged behind. In birds, cochlear nucleus angularis (NA) is an obligatory relay for intensity processing. We present our current knowledge of the cell types found in NA, their responses to auditory stimuli, and their likely coding roles. On a cellular level, our recent experimental and modeling studies have shown that short-term synaptic plasticity in NA is a major player in the division of intensity and timing information into parallel pathways. NA projects to at least four brain stem and midbrain targets, suggesting diverse involvement in a range of different sound processing circuits. Further studies comparing processing in NA and analogous neurons in the mammalian cochlear nucleus will highlight which features are conserved and perhaps may be computationally advantageous, and which are species- or clade-specific details demonstrating either disparate environmental requirements or different solutions to similar problems.

Keywords: sound localization; interaural intensity difference; short-term synaptic plasticity; auditory nerve; facilitation; depression; cochlear nucleus

Introduction

Individual acoustic stimulus waveforms in the environment sum together to form a complex composite waveform that arrives at the ear as a single, time-varying pressure amplitude wave. The fundamental problem of hearing lies in how the brain decomposes that waveform into information useful for performing auditory tasks necessary for the survival of the animal. These tasks include sound localization and vocal communication. Many of the computational principles for sound localization have emerged from the study of avian brains, especially that of the barn owl, a specialized auditory predator (Konishi et al., 1988). Additional data from less specialized birds, such as the chicken, and from mammals have revealed a suite of cellular and synaptic specializations in common for the temporal coding of sound, necessary for the detection of one important cue for localization, interaural time difference (ITD). These specializations include fast glutamatergic neurotransmission, endbulb synaptic morphology, low-threshold voltage-gated potassium conductances, bipolar dendritic structures, and axonal delay lines (Carr, 1993; Oertel, 1999; Trussell, 1999). Such commonalities arising in widely disparate animal clades suggest that there is a computational advantage to that form, whether it is dendritic structure,

*Corresponding author. Tel.: +1 301 405 7174;
Fax: +1 301 314 9358; E-mail: macleod@umd.edu

DOI: 10.1016/S0079-6123(06)65008-5

expression of a suite of ion channels, or the temporal patterning of activity. Brains in both birds and mammals experience similar constraints in detecting sound, and because hearing of airborne sound arose separately in these two groups, similarities in structure, function, and coding between them suggest common coding principles at work, and common solutions arrived at through parallel evolution (Carr and Soares, 2002).

Similarly comprehensive computational solutions to the questions of coding non-ITD aspects of sound have lagged behind. Early work in the barn owl recognized a division of labor between coding interaural timing differences and interaural intensity differences (also known as interaural level differences, or ILDs), beginning with the two divisions of the avian cochlear nucleus (CN): nucleus magnocellularis (NM) as the origin of the "timing pathway" and nucleus angularis (NA) as the origin of the "intensity pathway" (Fig. 1) (Sullivan and Konishi, 1984; Takahashi et al., 1984; Carr and Friedman, 1999; Konishi, 2003). Both cochlear nuclei are monaural and project to binaural targets that compare inputs from the two ears (Manley et al., 1988; Takahashi and Konishi, 1988). NM is a homogeneous nucleus whose neurons are similar to the spherical bushy cells of the mammalian CN. NA is a heterogeneous nucleus whose neurons have many similarities to the multipolar neurons in the mammalian cochlear nuclei (Oertel, 1999; Soares and Carr, 2001; Carr and Soares, 2002; Soares et al., 2002; Köppl and Carr,

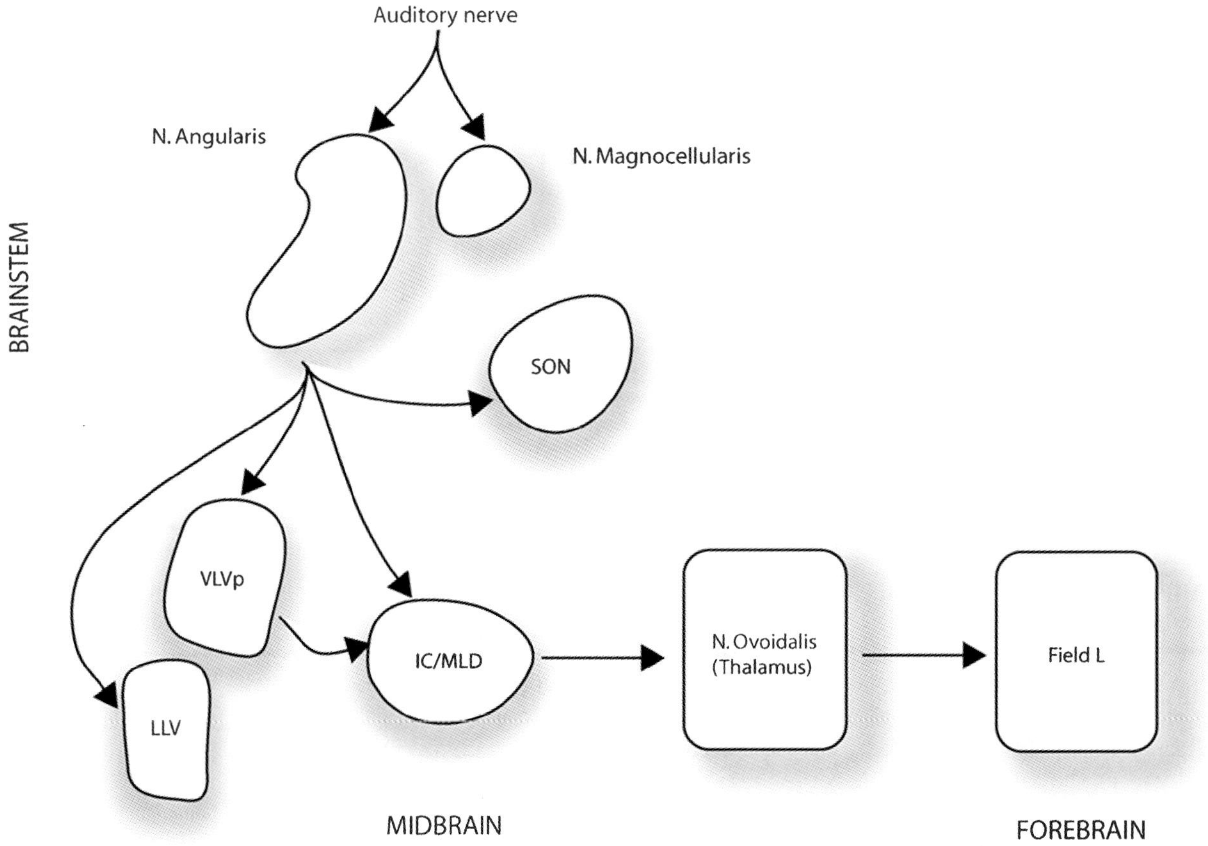

Fig. 1. Projections of the cochlear nucleus angularis and ascending pathways to the forebrain. NA and NM receive input from the auditory nerve. NA projects to SON, VLVp, LLV, and IC (specifically, IC, central nucleus, lateral shell). VLVp in turn projects to IC, which sends ascending information via the thalamus to the forebrain. The ascending inputs to SON are bilateral, but are heavier to the ipsilateral SON; the ascending inputs to VLVp, LLV, and IC are heavily contralateral.

2003). This heterogeneity, coupled with the remarkable specialization of one pathway for timing cues beginning with NM, suggests that NA is largely responsible for encoding non-localization aspects of sound in addition to its role in the ILD pathway. A major challenge lies in determining how each component cell type contributes to different aspects of sound recognition.

In this paper, we briefly review the cell types found in NA, their responses to auditory stimuli, and their likely coding roles, and compare them to those of the mammalian CN. Our recent experimental and modeling studies have shown that short-term synaptic plasticity in NA is a major player in the division of intensity and timing information into parallel pathways. Further computational studies of auditory coding in the cochlear nuclei should reveal common principles of auditory coding above and beyond the well-known algorithms for encoding interaural time differences. A better understanding of sound coding at the brainstem level will contribute to our understanding of complex auditory functions, such as birdsong recognition and learning.

Morphological and physiological characteristics of the neurons of the cochlear nuclei

In both birds and mammals, the auditory nerve forms endbulb synapses on one cell type and bouton-like terminals on all other CN targets (Ryugo and Parks, 2003). In mammals, endbulb of Held terminals are formed on the bushy cells of the ventral CN. In birds, the auditory nerve forms endbulb terminals on the cells of the NM and bouton-like terminals on cell types of the cochlear NA. These different terminal types are the origin of the differentiation between the ascending neural pathway that encodes timing information, and the parallel pathway for encoding sound level in NA (Takahashi et al., 1984). In addition to two types of auditory nerve terminals, the morphological and physiological characteristics of their target neurons contribute to the input–output functions of the CN, and the encoding of different components of the auditory stimulus.

Both brain slice and in vivo studies in the barn owl, redwing blackbird, and chicken show that NA is physiologically and morphologically much more heterogeneous than NM, with five major response types (Fig. 2) (Sachs and Sinnott, 1978; Sullivan, 1985; Warchol and Dallos, 1990; Köppl and Carr, 2003). The most common response pattern in NA is a primary-like post-stimulus time histogram with a transient-sustained rate response similar to that of auditory nerve fibers (Fig. 2A). NA contains onset units with characteristically low discharge rates (Fig. 2B), and a complex response type with a pronounced inhibitory component, similar to the mammalian type IV found in the mammalian dorsal cochlear nucleus (DCN) (Fig. 2C) (Sachs and Sinnott, 1978; Köppl and Carr, 2003). There are also two types of "chopper" responses, neurons that show regular firing unrelated to the auditory stimulus phase (Fig. 2D, E). Despite the striking similarities in auditory response types between birds and mammals, however, parallel morphological and brain slice studies show that the cell types in bird and mammal cochlear nuclei are not completely analogous (Soares and Carr, 2001; Soares et al., 2002).

There appears to be parallel evolution of neurons specialized for encoding different, behaviorally relevant features of the auditory stimulus (Köppl and Carr, 2003). The cells and circuits in birds and mammals appear to have independently evolved similar algorithms for encoding the salient features of the stimulus. This is similar to the appearance of neurons selective for orientation, direction of movement, and binocular disparity of straight-line contours in the barn owl visual Wulst and primate and cat visual cortex (Pettigrew and Konishi, 1976). These examples suggest that natural selection leads to the emergence of similar neural codes in systems with similar constraints.

Synaptic mechanisms in NA mediate parallel processing of intensity and timing information

Major differences exist in the synaptic physiological properties of the primary glutamatergic inputs coming from the auditory nerve to NM and NA

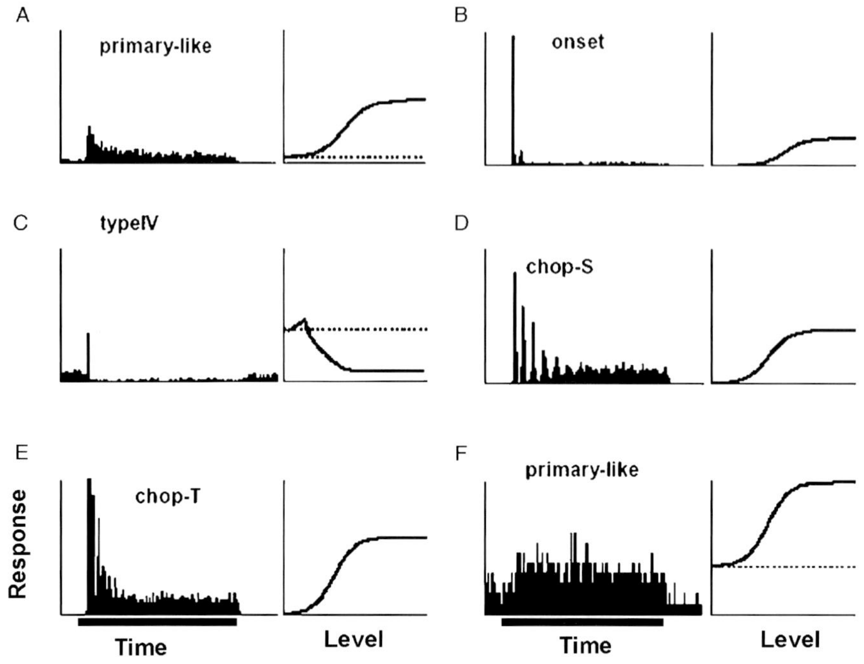

Fig. 2. Auditory response properties of NA neurons in vivo. Post-stimulus time histograms (PSTHs) (left panels) and rate intensity curves (right panels) illustrate the different response types to pure-tone stimulation in NA (A–E) and NM (F). Stimuli were 50 ms tones (horizontal bar) at characteristic frequency, 20–35 dB above threshold. The dashed lines in the rate intensity curves indicate the spontaneous rate, where applicable. All PSTHs and rate curves are scaled identically. (A) A primary-like unit in NA shows a similar phasic-tonic shape characteristic of auditory nerve units and NM units (F). (B) An onset unit shows a very sharp onset peak but low tonic activity. (C) Type IV unit displays inhibition below a relatively high spontaneous rate and a non-monotonic rate-intensity curve. Similar to Type IV units as described in the mammalian dorsal cochlear nucleus. (D, E) Two types of chopping responses. Chop-S: sustained chopper; Chop-T: transient chopper. These cell types show poor phase locking. (F) All units in NM show a primary-like response. Adapted with permission from Köppl and Carr (2003) and Grothe et al. (2005).

that underlie the emergence of parallel pathways that encode timing and sound level information. These differences fall into two categories: integrative properties and dynamic regulation of synaptic strength. Many of the synaptic properties found at the auditory nerve inputs to NA neurons lead to spatial and temporal integration. The short-term synaptic plasticity properties of these inputs also result in maintenance of the synaptic strength across a wide range of input rates, which supports transmission of the rate coded information about sound intensity.

Large differences between the two nuclei in the basic synaptic currents contribute to the differential representation of timing in NM and intensity in NA. Stimulation of the auditory nerve inputs to NA neurons results in many small- to moderate-amplitude, graded excitatory postsynaptic currents (EPSCs), in contrast to the all-or-none, very large synaptic currents evoked in NM neurons (Hackett et al., 1982; Zhang and Trussell, 1994; MacLeod and Carr, 2005). Each NA neuron receives input from multiple nerve fibers, and each auditory nerve fiber probably makes multiple contacts onto the multipolar dendrites of the postsynaptic NA neuron (MacLeod and Carr, 2005). Auditory nerve inputs are mediated entirely by alpha-amino-3-hydroxy-5-methyl-4-isoxazolepropionic acid (AMPA)- and *N*-methyl-D-aspartate (NMDA)-type glutamate receptors. The AMPA-receptor mediated currents in NA neurons have very fast decay kinetics similar to those in NM; however, the excitatory postsynaptic potentials (EPSPs) are slowed due to a slower membrane time constant, resulting in temporal integration. A large NMDA-receptor mediated component will further promote integration in NA neurons.

Information transmission in neural networks is partially determined by the rapid, activity-dependent alterations of synaptic strength in response to patterns of presynaptic action potentials, known as short-term synaptic plasticity. This form of plasticity is brief (milliseconds to seconds), occurs on the time scale of perceptual processes, and is ubiquitous throughout the nervous system (Zucker and Regehr, 2002). Short-term synaptic plasticity has been proposed to contribute to neural coding by acting as a filter of the presynaptic spike train, determining what information is passed via the synapse (O'Donovan and Rinzel, 1997; Markram et al., 1998; Abbott and Regehr, 2004). Recent work in the sound localization circuits shows that short-term synaptic plasticity may play a computational role in auditory processing. Auditory nerve fiber synapses in NA displayed a radically different short-term plasticity profile compared to the auditory nerve endbulbs in NM, or the bouton synapses in nucleus laminaris (NL) (Fig. 3A) (MacLeod et al., 2007). The short-term plasticity expressed at auditory nerve synapses in NA showed weaker net depression and little variation in the steady state EPSC amplitude with input rate (Fig. 3B, left panel). Typical responses included mixed facilitation and depression; a balance of these mechanisms results in the maintenance of EPSC amplitude across the train and across a range of high frequencies. In contrast, the short-term plasticity expressed by auditory nerve synapses in NM is characterized by depression, which monotonically deepens with increasing stimulation rates (Zhang and Trussell, 1994). Similar depression profiles are found at mammalian endbulb synapses (Wang and Kaczmarek, 1998; Oleskevich and Walmsley, 2002; Schneggenburger et al., 2002; von Gersdorff and Borst, 2002; Wong et al., 2003) and at the bouton-like synapses between NM afferents and NL neurons (Kuba et al., 2002; Cook et al., 2003), suggesting that depression is characteristic of the timing circuits.

In all these cases, with the exception of NA, the steady state levels of depression have a monotonic relationship with the input firing rate, in that an increase in the firing rate is matched by a nearly inversely proportional decrease in the steady state EPSC amplitude (Fig. 3B, left panel). As a result, the current drive per unit time experienced by the postsynaptic neuron, defined as the product of the steady state amplitude and the input rate, has a nearly flat relationship with input rate, meaning that increases in firing rate produces no further increase in postsynaptic drive (Fig. 3B, right panel). This relationship has been described in neocortical pyramidal synapses as a synapse-specific gain control (Abbott et al., 1997; Tsodyks and Markram, 1997). Auditory nerve synapses in NA, however, have a non-monotonic, or relatively flat, profile of steady state EPSC amplitudes with input rate (Fig. 3B, left panel), and therefore produce a linear increase in synaptic drive with input rate (Fig. 3B, right panel).

How does the short-term synaptic plasticity relationships affect intensity coding in the brainstem? Information about sound intensity is encoded in the auditory nerve firing rate (Salvi et al., 1992; Saunders et al., 2002). Because increases in the firing rate of the auditory nerve inputs will cause a proportional increase in drive to the postsynaptic NA neuron, the rate information contained in the auditory nerve fibers will be transmitted linearly. Therefore, the short-term plasticity expressed will result in the transmission of sound intensity at a typical NA synapse.

Auditory nerve firing rates are dynamic when stimulated with natural tone stimuli (the PSTHs have a phasic-tonic shape, dependent on sound intensity; Fig. 2). Furthermore, as shown in the trace in Fig. 3A, synapses in NA often show transient facilitation, and are therefore also dynamic over the course of the stimulus. Simulations using more naturalistic stimuli suggest that the effects described above for steady state are not limited to non-physiological constant-frequency stimuli (MacLeod et al., 2007). These simulations suggest that the combination of facilitation and depression shown in the synaptic plasticity in NA maintains ongoing intensity information nearly as well as the auditory nerve inputs. Synapses that expressed simple depression lost all intensity information during the tonic component of the primary-like input, although some intensity information is

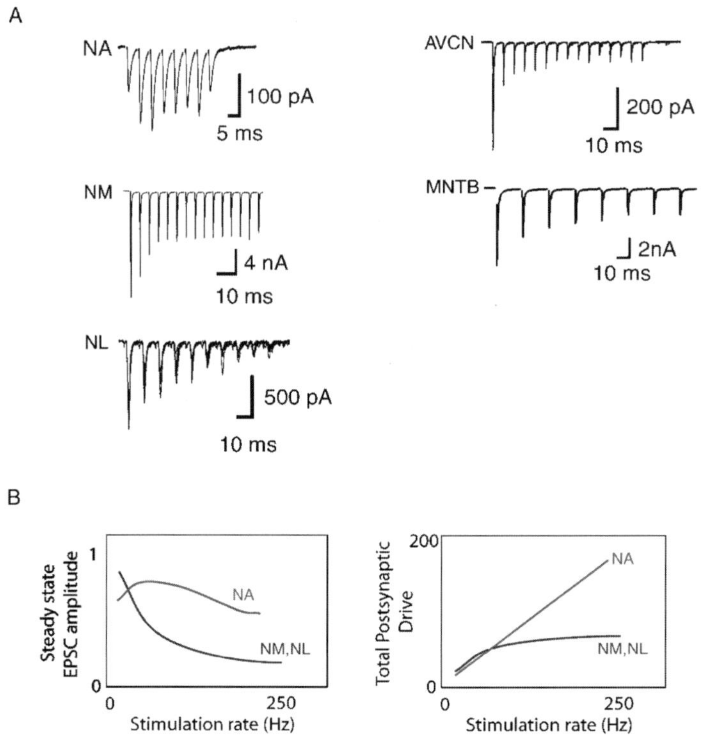

Fig. 3. Short-term synaptic plasticity in the auditory brainstem. (A) Traces of EPSC responses to trains of input stimulation at avian brainstem nuclei (NA, NM, NL), and mammalian brainstem nuclei (AVCN bushy cells, MNTB). All three endbulb synapses (NM, bushy, MNTB) and the synapses from NM to NL show strong depression, while the inputs to NA neurons can show a mixture of facilitation and depression, sometimes including transient net facilitation. (B) Cartoon of two types of input firing rate dependence of the plasticity. Steady state EPSC amplitude versus stimulus train input rate. The depressing synapses display monotonic profile, in which the steady state amplitude declines progressively with input rate ("NM, NL"). Synapses in NA on average show a non-monotonic, or flat, relationship with input rate ("NA"). (C) Total postsynaptic drive at steady state has a linear relationship with input rate for synapses in NA, but the relationship for depressing synapses is nonlinear and saturating as higher input rates. Total drive is calculated as the product of the normalized steady state amplitude and input rate. Traces in A adapted with permission from: NM, Brenowitz and Trussell (2001); NL, Kuba et al. (2002); AVCN bushy cell, Oleskevich and Walmsley (2002); MNTB, Wong et al. (2003); NA, MacLeod et al. (2007).

contained in the phasic component within <10 ms of the tone onset. This suggests that synaptic depression could signal changes in intensity or onsets. However, processing of ongoing intensity information, or comparisons of intensity information across different channels, requires the balanced synapses.

Intensity information is important not only for binaural intensity comparisons (ILD), but for other tasks that require knowledge of sound intensity. Any analysis of spectral cues, for example, must require a comparison of the intensity across frequency channels (Takahashi et al., 2003). Similarly, amplitude modulated signals are dynamic changes in intensity and a major component of song and speech. Despite perceptual adaptation to the level of environmental noise, overall loudness modifies vocal output. This is known as the Lombard effect, or the tendency to increase one's vocal intensity in noise.

What is the role of short-term depression in the timing circuits? The significance of depression at endbulb synapses is unclear, because even strongly depressed EPSPs are supra-threshold and therefore depression should have little impact on the postsynaptic response. However, short-term depression may contribute to ITD coding at the smaller synapses in the third-order brain stem NL.

NL receives input from both the ipsilateral and contralateral NM axons (Fig. 4C), and is the first brain area to display sensitivity to ITD (Moiseff and Konishi, 1983; Carr and Konishi, 1990). The type of gain control described above for short-term depression could improve ITD tuning by scaling the incoming synaptic amplitudes to keep the inputs within a range suitable for coincidence detection (Kuba et al., 2002; Cook et al., 2003). If one ear were driven more strongly than the other, due to interaural intensity differences, synaptic depression would act to reduce the more strongly activated inputs relatively more, equalizing the input amplitudes, and favoring binaural coincidence detection. Thus, synaptic depression acts as an adaptive mechanism that maintains interaural timing information irrespective of interaural intensity differences, and may account for the insensitivity of ITD tuning in NL to changes in sound intensity (Pena et al., 1996; Viete et al., 1997).

These data suggest that the short-term plasticity expressed at a synapse is related to its functional role, rather than to the size of the synapse. The effect on coincidence detection described above depends heavily on NL synapses showing increased depression with input rate (with a nearly inversely proportional relationship). Thus the

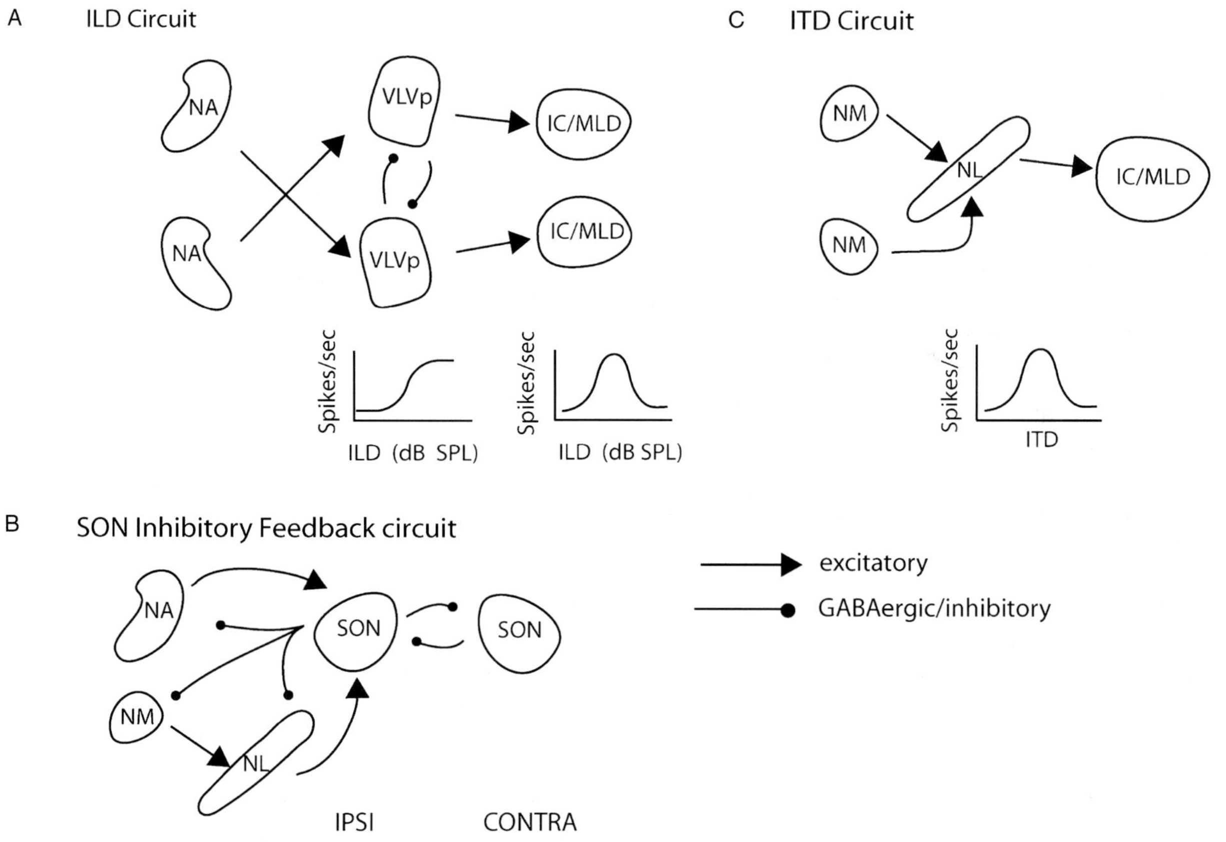

Fig. 4. (A) The ILD circuit. NA projects contralaterally to the VLVp, which in turn projects to the contralateral VLVp with an inhibitory connection. Neurons in the VLVp are sensitive to ILD, with a sigmoidally shaped response curve that is most excited by contralateral stimuli. VLVp sends input to the IC, where there are neurons sensitive to ILD (contingent on ITD). (B) "Gain control" circuit of the SON. NA and NL project ipsilaterally to the SON, which makes ipsilateral inhibitory feedback projects to NM, NL, and NA. SON also makes presumed inhibitory (GABAergic) connections onto the contralateral SON. (C) The ITD circuit. Nucleus laminaris receives input from both ispilateral and contralateral NM, and projects to the IC (central nucleus, core region). Arrows represent excitatory connections; circles represent GABAergic, presumed inhibitory connections.

relationship to input rate is the important difference between the NL and NA synaptic plasticity, rather than the overall degree of depression at any given input rate, although synapses in NA showed significantly less net depression. Taken together, these data suggest that the short-term plasticity expressed at the synapses in NL and NA may be computationally advantageous and directly related to function.

Functional roles for NA in coding sound intensity: a multi-faceted nucleus

The idea that NA may play multiple roles in sound processing is reinforced by the fact that NA projects to multiple nuclei in the ascending pathway. These projection targets include the superior olivary nucleus (SON), two lemniscal nuclei, and a direct projection to the inferior colliculus (IC) (Fig. 1). Three of these targets are associated with different roles; the superior olive mediates descending control of gain (Pena et al., 1996; Monsivais et al., 2000; Takahashi and Konishi, 2002; (for review, see Hyson, 2005)), the lemniscal nuclei encode sensitivity to ILDs (Manley et al., 1988; Adolphs, 1993; Mogdans and Knudsen, 1994; Takahashi et al., 1995), and the IC mediates the emergence of responses to biologically relevant stimuli (for review, see Konishi, 2003). Further processing of sound information follows an ascending pathway via the avian thalamic nucleus ovoidalis to the forebrain region Field L (Fig. 1).

NA's most established role is as the origin of the ILD pathway. Takahashi et al. (1984) showed that injection of lidocaine into NA altered ILD sensitivity in space-specific neurons in the IC. Processing of ILDs has been extensively examined in the barn owl, where the vertical asymmetry in ear directionality makes ILD a cue for sound source elevation (Knudsen and Konishi, 1980; Keller et al., 1998). The current hypothesis proposes that level is encoded by neurons in NA (Köppl and Carr, 2003) and binaural level difference sensitivity then emerges in one of the lemniscal targets of NA, the nucleus ventralis lemnisci lateralis, pars posterior, or VLVp (also referred to as the dorsal nucleus of the lateral lemniscus, posterior portion, or LLDp; Fig. 4A). VLVp neurons exhibit discharge rates that are sigmoidal functions of ILD (Manley et al., 1988; Adolphs, 1993; Takahashi et al., 1995). The neural responses in VLVp are similar to those in the mammalian lateral superior olive (LSO) (Tsuchitani, 1977; Takahashi and Keller, 1992). VLVp projects bilaterally to the IC, specifically to the lateral shell of the central nucleus of the IC, endowing the neurons there with sensitivity to ILD (Adolphs, 1993).

NA also projects bilaterally to the SON, with its heaviest projection to the ipsilateral SON (Fig. 4B) (Conlee and Parks, 1986; Takahashi and Konishi, 1988). Auditory responses in SON are biased toward ipsilateral excitation (Moiseff and Konishi, 1983). One class of SON neurons is GABAergic, and appears to be the major source of inhibitory feedback to NL, NA, and also NM (Lachica et al., 1994; Yang et al., 1999; Burger et al., 2005). A separate population of SON neurons project contralaterally to the opposite SON (Burger et al., 2005). The inhibitory feedback from SON onto ipsilateral NM could help to isolate the ITD response from differences in interaural level by reducing the monaural input strength to NL and favoring coincident inputs (Fujita and Konishi, 1991; Pena et al., 1996; Viete et al., 1997; Burger et al., 2005). Alternative, or additional, functions for a GABAergic feedback to the cochlear nuclei include enhancement of the phase-locking fidelity of NM neurons (by a reduction in membrane time constant indirectly through activation of low threshold K^+ channels) (Monsivais et al., 2000), and as a potential source of the inhibitory component of the type IV auditory responses found in NA (Köppl and Carr, 2003).

Which NA neurons encode the level information for which auditory tasks? Tracing studies showed that all four morphological types of NA neurons project to both the VLVp and IC (Soares and Carr, 2001). At this time it is unknown whether the different cell types in NA form parallel pathways, whether there are differences in their projection fields, and what role is played by the direct projection from NA to IC. Further studies relating the in vivo response classes to in vitro and morphological types through intracellular recordings will be needed to clarify this question.

Physiological responses in NA show that primary like, onset and chopper responses did not differ in dynamic range from the auditory nerve (Köppl and Carr, 2003), and thus any or all of these response types may encode changes in intensity. How intensity is represented remains an important issue in auditory neurobiology.

Conclusions

The study of auditory coding in birds provides an excellent opportunity to combine behavioral, systems-level and cellular analyses of hearing. The established behavioral paradigms in birds, particularly sound localization in the barn owl, and birdsong recognition in zebra finches and other songbirds, provide a rich context and framework for understanding the physiological, biophysical, and synaptic properties of the auditory brainstem neurons. An understanding of the cochlear NA, along with the organization of its outputs, is critically important to forming a more comprehensive theory of sound processing. This will necessarily expand and blur the simplistic dualistic construct of "timing" and "intensity" pathways, localization and non-localization tasks. Comparative studies between avian and mammalian CN will furthermore highlight which features are conserved and perhaps may be computationally advantageous, and which are species- or clade-specific details demonstrating either disparate environmental requirements or, alternatively, different solutions to similar problems.

Abbreviations

AMPA	alpha-amino-3-hydroxy-5-methyl-4-isoxazolepropionic acid
AVCN	anterior ventral cochlear nucleus
CN	cochlear nucleus
DCN	dorsal cochlear nucleus
EPSC	excitatory postsynaptic current
EPSP	excitatory postsynaptic potential
GABA	gamma-aminobutyric acid
IC	inferior colliculus (also referred to as MLD)
ILD	interaural level difference
IPSP	inhibitory postsynaptic potential
ITD	interaural time difference
LLV	nucleus lemnisci lateralis, pars ventralis
LSO	lateral superior olive
MLD	nucleus mesencephalicus lateralis, pars ventralis (also referred to as IC)
MNTB	medial nucleus of the trapezoid body
NA	nucleus angularis
NL	nucleus laminaris
NM	nucleus magnocellularis
NMDA	*N*-methyl-D-aspartate
SON	superior olivary nucleus
VLVp	nucleus ventralis lemnisci lateralis, pars posterior

Acknowledgment

The authors acknowledge the support of the National Institutes of Health (grants R01-DC000436 and R03-007972) and the Center for the Comparative and Evolutionary Biology Hearing (NIH grant DC04664).

References

Abbott, L.F. and Regehr, W.G. (2004) Synaptic computation. Nature, 431: 796–803.

Abbott, L.F., Varela, J.A., Sen, K. and Nelson, S.B. (1997) Synaptic depression and cortical gain control. Science, 275: 220–224.

Adolphs, R. (1993) Bilateral inhibition generates neuronal response tuned to interaural level differences in the auditory brainstem of the barn owl. J. Neurosci., 13: 3647–3668.

Brenowitz, S. and Trussell, L.O. (2001) Maturation of synaptic transmission at end-bulb synapses of the cochlear nucleus. J. Neurosci., 21: 9487–9498.

Burger, R.M., Cramer, K.S., Pfeiffer, J.D. and Rubel, E.W. (2005) Avian superior olivary nucleus provides divergent inhibitory input to parallel auditory pathways. J. Comp. Neurol., 481: 6–18.

Carr, C.E. (1993) Processing of temporal information in the brain. Annu. Rev. Neurosci., 16: 223–243.

Carr, C.E. and Friedman, M.A. (1999) Evolution of time coding systems. Neural Comput., 11: 1–20.

Carr, C.E. and Konishi, M. (1990) A circuit for detection of interaural time differences in the brain stem of the barn owl. J. Neurosci., 10: 3227–3246.

Carr, C.E. and Soares, D. (2002) Evolutionary convergence and shared computational principles in the auditory system. Brain Behav. Evol., 59: 294–311.

Conlee, J.W. and Parks, T.N. (1986) Origin of ascending auditory projections to the nucleus mesencephalicus lateralis pars dorsalis in the chicken. Brain Res., 367: 96–113.

Cook, D.L., Schwindt, P.C., Grande, L.A. and Spain, W.J. (2003) Synaptic depression in the localization of sound. Nature, 421: 66–70.

O'Donovan, M.J. and Rinzel, J. (1997) Synaptic depression: a dynamic regulator of synaptic communication with varied functional roles. Trends Neurosci., 20: 431–433.

Fujita, I. and Konishi, M. (1991) The role of GABAergic inhibition in processing of interaural time difference in the owl's auditory system. J. Neurosci., 11: 722–739.

von Gersdorff, H. and Borst, J.G. (2002) Short-term plasticity at the calyx of held. Nat. Rev. Neurosci., 3: 53–64.

Grothe, B., Carr, C.E., Casseday, J., Fritzsch, B. and Köppl, C. (2005) The evolution of central pathways and their neural processing patterns. In: Manley G., Popper A. and Fay R. (Eds.), Evolution of the vertebrate auditory system. Springer, New York, pp. 289–359.

Hackett, J.T., Jackson, H. and Rubel, E.W. (1982) Synaptic excitation of the second and third order auditory neurons in the avian brain stem. Neuroscience, 7: 1455–1469.

Hyson, R.L. (2005) The analysis of interaural time differences in the chick brain stem. Physiol. Behav., 86: 297–305.

Keller, C.H., Hartung, K. and Takahashi, T.T. (1998) Head-related transfer functions of the barn owl: measurement and neural responses. Hear. Res., 118: 13–34.

Knudsen, E.I. and Konishi, M. (1980) Monaural occlusion shifts receptive-field locations of auditory midbrain units in the owl. J. Neurophysiol., 44: 687–695.

Konishi, M. (2003) Coding of auditory space. Annu. Rev. Neurosci., 26: 31–55.

Konishi, M., Takahashi, T., Wagner, H., Sullivan, W.E. and Carr, C.E. (1988) Neurophysiological and anatomical substrates of sound localization in the owl. In: Edelman G.M., Gan W.E. and Cowan W.M. (Eds.), Auditory Function: Neurobiological Bases of Hearing. Wiley, New York, pp. 721–745.

Köppl, C. and Carr, C.E. (2003) Computational diversity in the cochlear nucleus angularis of the barn owl. J. Neurophysiol., 89: 2313–2329.

Kuba, H., Koyano, K. and Ohmori, H. (2002) Synaptic depression improves coincidence detection in the nucleus laminaris in brainstem slices of the chick embryo. Eur. J. Neurosci., 15: 984–990.

Lachica, E.A., Rubsamen, R. and Rubel, E.W. (1994) GABAergic terminals in nucleus magnocellularis and laminaris originate from the superior olivary nucleus. J. Comp. Neurol., 348: 403–418.

MacLeod, K.M. and Carr, C.E. (2005) Synaptic physiology in the cochlear nucleus angularis of the chick. J. Neurophysiol., 93: 2520–2529.

MacLeod, K.M., Horiuchi, T.K. and Carr, C.E. (2007) A role for short-term synaptic facilitation and depression in the processing of intensity information in the auditory brainstem. J. Neurophysiol., 97: 2863–2874.

Manley, G.A., Koppl, C. and Konishi, M. (1988) A neural map of interaural intensity differences in the brain stem of the barn owl. J. Neurosci., 8: 2665–2676.

Markram, H., Gupta, A., Uziel, A., Wang, Y. and Tsodyks, M. (1998) Information processing with frequency-dependent synaptic connections. Neurobiol. Learn. Mem., 70: 101–112.

Mogdans, J. and Knudsen, E.I. (1994) Representation of interaural level difference in the VLVp, the first site of binaural comparison in the barn owl's auditory system. Hear. Res., 74: 148–164.

Moiseff, A. and Konishi, M. (1983) Binaural characteristics of units in the owl's brainstem auditory pathway: precursors of restricted spatial receptive fields. J. Neurosci., 3: 2553–2562.

Monsivais, P., Yang, L. and Rubel, E.W. (2000) GABAergic inhibition in nucleus magnocellularis: implications for phase locking in the avian auditory brainstem. J. Neurosci., 20: 2954–2963.

Oertel, D. (1999) The role of timing in the brain stem auditory nuclei of vertebrates. Annu. Rev. Physiol., 61: 497–519.

Oleskevich, S. and Walmsley, B. (2002) Synaptic transmission in the auditory brainstem of normal and congenitally deaf mice. J. Physiol., 540: 447–455.

Pena, J.L., Viete, S., Albeck, Y. and Konishi, M. (1996) Tolerance to sound intensity of binaural coincidence detection in the nucleus laminaris of the owl. J. Neurosci., 16: 7046–7054.

Pettigrew, J.D. and Konishi, M. (1976) Neurons selective for orientation and binocular disparity in the visual Wulst of the barn owl (*Tyto alba*). Science, 193: 675–678.

Ryugo, D.K. and Parks, T.N. (2003) Primary innervation of the avian and mammalian cochlear nucleus. Brain Res. Bull., 60: 435–456.

Sachs, M.B. and Sinnott, J.M. (1978) Responses to tones of single cells in nucleus magnocellularis and nucleus angularis of the redwing blackbird (*Agelaius phoeniceus*). J. Comp. Physiol. A, 126: 347–361.

Salvi, R.J., Saunders, S.S., Powers, N.L. and Boettcher, F.A. (1992) Discharge patterns of cochlear ganglion neurons in the chicken. J. Comp. Physiol. A, 170: 227–241.

Saunders, J., Ventetuolo, C., Plontke, S. and Weiss, B. (2002) Coding of sound intensity in the chick cochlear nerve. J. Neurophysiol., 88: 2887–2898.

Schneggenburger, R., Sakaba, T. and Neher, E. (2002) Vesicle pools and short-term synaptic depression: lessons from a large synapse. Trends Neurosci., 25: 206–212.

Soares, D. and Carr, C.E. (2001) The cytoarchitecture of the nucleus angularis of the barn owl (*Tyto alba*). J. Comp. Neurol., 429: 192–205.

Soares, D., Chitwood, R.A., Hyson, R.L. and Carr, C.E. (2002) Intrinsic neuronal properties of the chick nucleus angularis. J. Neurophysiol., 88: 152–162.

Sullivan, W.E. (1985) Classification of response patterns in cochlear nucleus in the barn owl: correlation with functional response properties. J. Neurophysiol., 53: 201–216.

Sullivan, W.E. and Konishi, M. (1984) Segregation of stimulus phase and intensity coding in the cochlear nucleus of the barn owl. J. Neurosci., 4: 1787–1799.

Takahashi, T., Moiseff, A. and Konishi, M. (1984) Time and intensity cues are processed independently in the auditory system of the owl. J. Neurosci., 4: 1781–1786.

Takahashi, T.T., Bala, A.D., Spitzer, M.W., Euston, D.R., Spezio, M.L. and Keller, C.H. (2003) The synthesis and use of the owl's auditory space map. Biol. Cybern., 89: 378–387.

Takahashi, T.T., Barberini, C.L. and Keller, C.H. (1995) An anatomical substrate for the inhibitory gradient in the VLVp of the owl. J. Comp. Neurol., 358: 294–304.

Takahashi, T.T. and Keller, C.H. (1992) Commissural connections mediate inhibition for the computation of interaural level difference in the barn owl. J. Comp. Physiol. A, 170: 161–169.

Takahashi, T.T. and Konishi, M. (1988) Projections of nucleus angularis and nucleus laminaris to the lateral lemniscal nuclear complex of the barn owl. J. Comp. Neurol., 274: 212–238.

Takahashi, Y. and Konishi, M. (2002) Manipulation of inhibition in the owl's nucleus laminaris and its effects on optic tectum neurons. Neuroscience, 111: 373–378.

Trussell, L.O. (1999) Synaptic mechanisms for coding timing in auditory neurons. Annu. Rev. Physiol., 61: 477–496.

Tsodyks, M.V. and Markram, H. (1997) The neural code between neocortical pyramidal neurons depends on neurotransmitter release probability. Proc. Natl. Acad. Sci. U.S.A., 94: 719–723.

Tsuchitani, C. (1977) Functional organization of lateral cell groups of cat superior olivary complex. J. Neurophysiol., 40: 296–318.

Viete, S., Pena, J.L. and Konishi, M. (1997) Effects of interaural intensity difference on the processing of interaural time difference in the owl's nucleus laminaris. J. Neurosci., 17: 1815–1824.

Wang, L.-Y. and Kaczmarek, L.K. (1998) High-frequency firing helps replenish the readily releasable pool of synaptic vesicles. Nature, 394: 384–388.

Warchol, M.E. and Dallos, P. (1990) Neural coding in the chick cochlear nucleus. J. Comp. Physiol. A, 166: 721–734.

Wong, A.Y., Graham, B.P., Billups, B. and Forsythe, I.D. (2003) Distinguishing between presynaptic and postsynaptic mechanisms of short-term depression during action potential trains. J. Neurosci., 23: 4868–4877.

Yang, L., Monsivais, P. and Rubel, E.W. (1999) The superior olivary nucleus and its influence on nucleus laminaris: a source of inhibitory feedback for coincidence detection in the avian auditory brainstem. J. Neurosci., 19: 2313–2325.

Zhang, S. and Trussell, L.O. (1994) Voltage clamp analysis of excitatory synaptic transmission in the avian nucleus magnocellularis. J. Physiol., 480: 123–136.

Zucker, R.S. and Regehr, W.G. (2002) Short-term synaptic plasticity. Annu. Rev. Physiol., 64: 355–405.

P. Cisek, T. Drew & J.F. Kalaska (Eds.)
Progress in Brain Research, Vol. 165
ISSN 0079-6123
Copyright © 2007 Elsevier B.V. All rights reserved

CHAPTER 9

Neural strategies for optimal processing of sensory signals

Leonard Maler*

Department of Cell and Molecular Medicine and Center for Neural Dynamics, University of Ottawa, 451 Smyth Rd, Ottawa, ON K1H 8M5, Canada

Abstract: The electrosensory system is used for both spatial navigation tasks and communication. An electric organ generates a sinusoidal electric field and cutaneous electroreceptors respond to this field. Objects such as prey or rocks cause a local low-frequency modulation of the electric field; this cue is used by electric fish for navigation and prey capture. The interference of the electric fields of conspecifics produces beats, often with high frequencies, that are also sensed by the electroreceptors; furthermore, these electric fish can transiently modulate their electric discharge as a communication signal. Thus these fish must therefore detect a variety of low-intensity signals that differ greatly in their spatial extent, frequency, and duration. Behavioral studies suggest that they are highly adapted to these tasks. Experimental and theoretical analyses of the neural circuitry for the electrosense has demonstrated many commonalities with the more common senses, e.g., topographic mapping and receptive fields with On or Off centers and surround inhibition. The integration of computational and experimental analyses has demonstrated novel mechanisms that appear to optimize weak signal detection in the electrosense including: noise shaping by correlations within single spike trains, induction of oscillations by delayed feedback inhibition, the requirement for maps with differing receptive field sizes tuned for different stimulus parameters, and the role of non-plastic feedback for adaptive cancellation of redundant signals. It is likely that these mechanisms will also be operative in other sensory systems.

Keywords: sensory coding; noise shaping; fisher information; synchrony; topographic maps

Introduction

Sensory systems have evolved to enhance species fitness, i.e., to efficiently detect food, avoid predators, and guide navigation, as well as to communicate with potential mates or rivals. The exquisite sensitivity of sensory systems such as bat echolocation or insect olfaction suggests that they have reached optimal performance levels. In many cases it is difficult to evaluate sensory processing since even a description of sensory signals may be problematic (e.g., primate audition or vision); furthermore, the immense complexity of most vertebrate nervous systems makes it very difficult to derive general principles for optimal and biologically realistic neural processing of naturalistic signals. Below I briefly review neural processing in the electrosensory system. Natural electrosensory signals are well characterized and can be readily mimicked in the laboratory; further, the relatively simple laminar structure of the first order

*Corresponding author. Tel.: +1 613 562 5800 8189; Fax: +1 613 562 5434; E-mail: lmaler@uottawa.ca

DOI: 10.1016/S0079-6123(06)65009-7

hindbrain processing center simplifies both in vivo and in vitro studies of its neural circuitry. Despite its apparently unique nature, the electrosense must solve many of the same problems as other senses (i.e., detecting weak signals in the presence of noise). As discussed below, our studies to date suggest that this sense has evolved solutions similar to that of other senses, and that we can expose principles for optimal sensory processing that may have general applicability.

Electroreception is based on receptors whose structure resembles that of lateral line or auditory hair cells prompting the speculation that there is an evolutionary link between these senses. Remarkably, the processing of some types of electric signals (communication) resembles that reported for neurons in the auditory pathways. As described below electroreceptors are distributed over the surface of the body and other types of electric signals (e.g., prey) can excite small local groups of electroreceptors. It is also remarkable that the processing of these signals strongly resembles the neural operations described at lower levels of the visual system. This review therefore focuses on comparisons of the computations performed by the electrosensory and auditory and visual systems.

Electroreception is an ancient sense of aquatic vertebrates (Ronan, 1986). In sharks (elasmobranches), for example, specialized cutaneous electroreceptors detect the weak low-frequency electric fields produced by prey (Wilkens and Hofman, 2005). Electroreceptors sensitive to exogenous electric fields are named ampullary type receptors and this form of electroreception is generally referred to as passive electroreception (Bodznick and Montgomery, 2005). Passive electroreception was lost at the beginning of the teleost lineage. Two modern teleost families (catfish and notopterids) have, however, independently evolved ampullary receptors functionally similar to those of elasmobranches. The passive electric sense has further independently evolved, in both families, into an active electric sense. Thus, both the South American family of gymnotiform fish and the African family of mormyriform fish, have an electric organ — a modified muscle that, upon stimulation by motorneuron axons, emits an electric organ discharge (EOD) instead of contracting (Alves-Gomez, 1999). This electromotor system has evolved in tandem with the evolution of a subset of ampullary receptors into electroreceptors tuned to the EOD waveform (Kawasaki, 2005). These electroreceptors are named tuberous organs and, since they respond to the electric field produced by the fish's own discharge, are said to implement active electroreception (both families of electric fish still retain their ampullary receptors). The details of control of EOD production as well as the coding properties of electroreceptors and the central representation of electrosensory input has been worked out in some detail for a small number of gymnotiform and mormyriform species; detailed reviews of this material can be found in three recent review volumes (Bullock and Heiligenberg, 1986; Turner et al., 1999; Bullock and Hopkins, 2005). This review will concentrate on gymnotiform fish.

Gymnotiform fish are abundant in central and South America with many very different species (Albert and Crampton, 2005). A useful functional distinction is between pulse and wave species: a pulse species emits brief EOD pulses separated by long and often variable inter-pulse intervals; in contrast, a wave species emits a sinusoidal EOD. This review concentrates on the sensory adaptations of wave species, primarily that of *Apteronotus leptorhynchus*. This species has a continuous quasi-sinusoidal EOD that ranges from ~650 to 1000 Hz; the frequency of an individual fish's EOD is remarkably constant. Mature males have a higher EOD frequency (> 800 Hz) than mature females (< 800 Hz). The EOD is detected by a large number of tuberous electroreceptors that are tuned to the fish's own EOD frequency. Most of these receptors respond in a phase-locked but probabilistic manner to the EOD with probabilities ranging from 0.1 to 0.5; for this reason they are designated P-units. A second T-type tuberous electroreceptor discharges on each EOD cycle and signals EOD amplitude by the phase of its spikes (with respect to the EOD); since T-type receptors are rare in *A. leptorhynchus*, they will not be further considered in this review. There are ~15,000 P-units distributed over the body of the fish with the highest density over the head

(Carr et al., 1982). P-units are continuously stimulated by the EOD and their baseline discharge ranges from ~100 to 500 Hz (Nelson et al., 1997). The EOD is a carrier waveform and EOD amplitude modulations (AMs) are the signals that drive changes in P-unit discharge. The active electrosense is used for two very different functions: electrolocation and electrocommunication. These activities result in very different spatiotemporal patterns of AMs and a major challenge for the electrosense, as for other senses, is to discriminate communication signals from animate (prey, plants, predators) and inanimate (rocks) objects.

Below I discuss the nature of electrolocation and communication signals and the dynamic properties of the electroreceptors that optimize processing of both classes of signal. I then review the organization of neural circuitry that decodes the spatiotemporal patterns of electroreceptor activity with an emphasis on general principles of neural dynamics and computation.

Electrosensory signals

Here I briefly describe the nature of electrosensory signals with emphasis on the computational constraints they impose on their target neural circuitry.

Electrolocation

A. leptorhynchus feeds on small crustaceans and will make stereotypic scanning movements while foraging for their prey (Nelson and MacIver, 1999; MacIver et al., 2001). Crustaceans have a conductivity greater than that of the ambient water and the presence of such prey in proximity to the fish's skin (<3 cm) will therefore cause a local AM increase; the frequency of this AM depends on the fish's scanning speed and is typically <20 Hz. Because of the physics of the fish's electric field, a prey object will produce a blurred electric "image" on the fish's skin below it (Fig. 1A). The intensity and spread of the image depend in a complex way on the size and conductivity of the object as well as its distance from the fish, but accurate models are now available to compute the electric image (Chen et al., 2005; Nelson, 2005; Babineau et al., 2006). Gymnotiform fish are characterized by a long ribbon fin at their ventral surface. Undulations of this fin are used to propel the fish forward or backward with equal ease (Lannoo and Lannoo, 1992). *A. leptorhynchus* scans its environment in a highly stereotypical manner: it first swims forward at low speeds (<10 cm/s); when it encounters a prey item, the fish will rapidly decelerate and then swim backward so as to capture the prey (Nelson and MacIver, 1999). During the latter phase of prey capture, the fish typically orients its body so that the prey traverses a longitudinal trajectory along the fish's dorsal body surface — a region with a high P-unit receptor density (Carr et al., 1982; MacIver et al., 2001). Thus, as the fish scans the environment, it will sense the slow local prey-induced AM (increase) by the increased discharge of the P-units underneath the prey. Scanning past rocks (non-conductors) will produce slow local decreases in EOD amplitude that decrease the discharge of P-units. The resulting activation of a spatial sequence of P-units is processed by the electrosensory circuitry and controls the motor system (via the optic tectum) in a manner similar to that of visually guided movements (see below).

Electrocommunication

When two fish are in proximity their electric fields will summate to produce an AM (beat) whose frequency equals the difference of EOD frequencies (Heiligenberg, 1991); this AM will activate an appreciable fraction of the P-unit population and is therefore described as a global signal (Fig. 1B). Thus male–female interactions will typically produce global beats with frequencies >50 Hz (up to ~250 Hz) while same-sex interactions will produce beats with frequencies <50 Hz. There is evidence from related wave-type electric fish that beat frequency is in fact used to discriminate the sex of interacting conspecifics and can therefore be considered as a simple communication signal (Kramer, 1999). Electric fish typically also modulate their EOD frequency as a communication signal; a large variety of such electrocommunication signals have been described in *A. leptorhynchus*, the most commonly observed being chirps — these are transient (typically <25 ms) increases in EOD frequency (Zakon et al., 2002; Zupanc, 2002). Chirps are commonly emitted by males and come in two flavors: small chirps (EOD frequency increases of ~100 Hz) and large chirps (EOD frequency increases >300 Hz). Small chirps are usually elicited by beat with frequencies <30 Hz (other males) and are assumed to be an aggressive

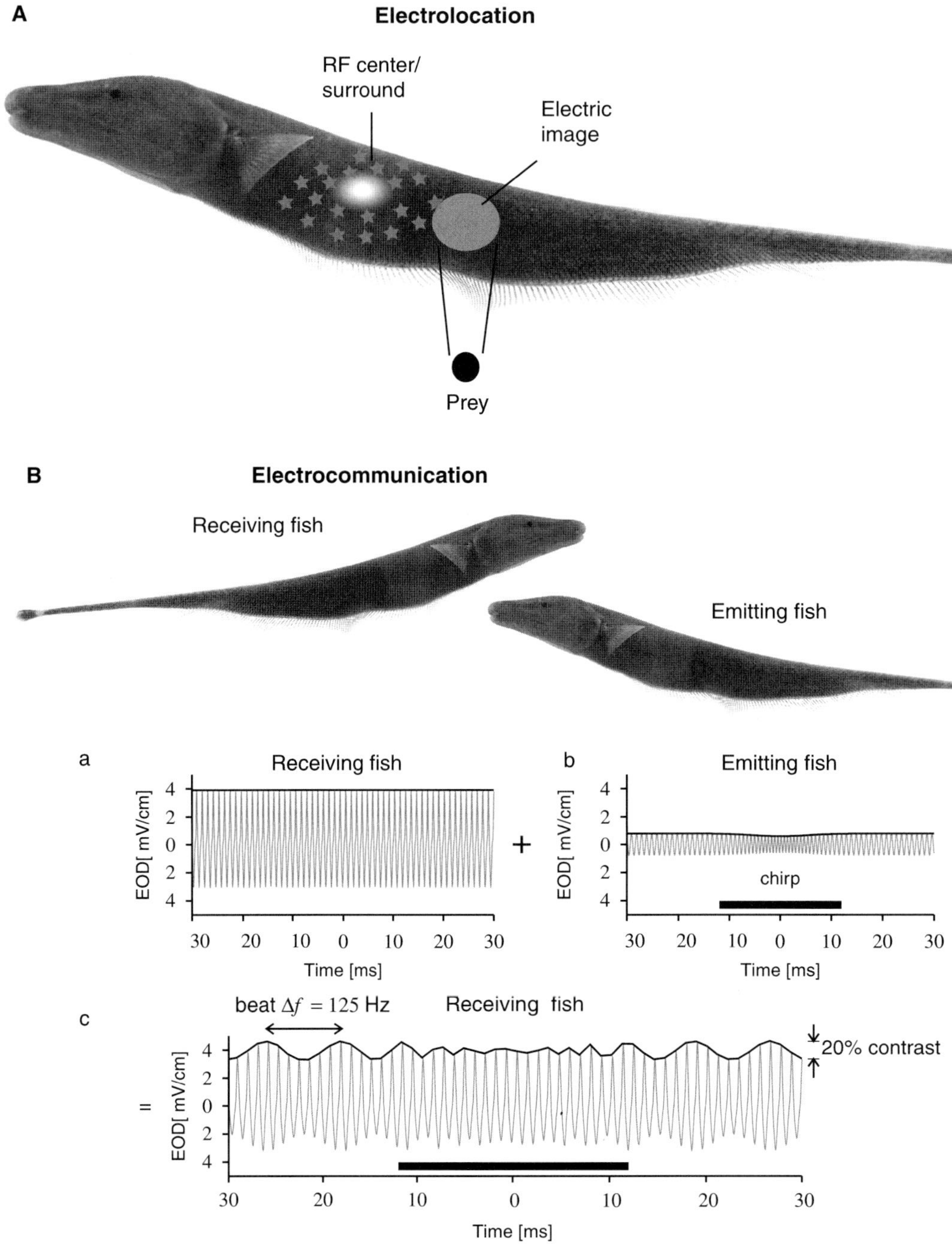

signal; large chirps are elicited by beat with frequencies > 50 Hz and are assumed to be associated with courtship (Bastian et al., 2001; Engler and Zupanc, 2001; Triefenbach and Zakon, 2003).

Recent studies have shown that electric fish often forage in small groups (Tan et al., 2005). The interacting EODs of multiple fish can produce complex AMs known as envelope signals (Middleton et al., 2006) similar to those seen in both auditory (Joris et al., 2004) and visual (Mareschal and Baker, 1998) systems. Decoding such signals can inform the fish of the identity of the foraging group members (Middleton et al., 2006).

Synopsis

Electric signals associated with moving prey involve activation of small localized patch of electroreceptors — this activity "bump" moves across the skin; since the fish's movements are relatively slow, only low frequencies are generated by prey. Detecting these inputs is therefore analogous to the problem faced by the visual system in detecting small, relatively slowly moving objects and directing the motor system towards them. In contrast, electrocommunication signals are always spatially diffuse (global) and often of high frequency. Detecting such signals involves detection of synchronous activation of many receptors — a problem similar to that faced by the auditory system in the processing of acoustic communications signals.

It is possible that electrocommunication signals may have evolved to be maximally different from electrolocation signals. A major problem for electrosensory research is how such spatially and temporally different signals can both be simultaneously processed in an optimal manner.

Electroreceptors and optimal encoding of natural signals

A key question is how P-units and their central targets can detect a wide AM frequency range (1–250 Hz) as well as transient signals (chirps) superimposed on an ongoing beat.

The initial studies of P-units emphasized their high pass gain characteristics thus creating an apparent paradox: the low-frequency AMs resulting from scanning prey would not be well transmitted. A recent study has resolved this issue by using stimulus–response coherence (correlation of stimulus and response in the frequency domain) to quantify the frequency dependence of P-unit response; the coherence measure shows that P-units respond well to a wide frequency range (a few to > 100 Hz) that includes both prey and

Fig. 1. (A) Objects cause local distortions in the electric field generated by the fish's electric organ. In the case of prey (conductivity greater than that of the ambient water) the local amplitude of the electric field is increased and forms an electric image of the prey. The electroreceptors under the prey respond to this electric image by increasing their rate of discharge. These electroreceptors then project to the ELL where they contact both pyramidal cells and local inhibitory interneurons. The direct synaptic contacts of electroreceptors onto (E type) pyramidal cells are glutamatergic (mainly AMPA receptors) and therefore form an excitatory receptive field center (RF). The di-synaptic inhibition results in an inhibitory surround (stars). Thus a subset of ELL pyramidal cells have a classic On center-Off surround RF organization; other pyramidal cells (I type) display an Off center-On surround type organization but the more complex circuitry responsible is not discussed in this review or illustrated here. The size of the RF center varies across the ELL maps: it is largest in the lateral map and smallest in the centromedial map. (B) The EOD of these fish is usually of constant amplitude (black line in (a)) and frequency. When two fish are in proximity their EODs will interfere to create a beat whose frequency equals the difference of the EOD frequencies. In this species males have higher EOD frequencies than females. Here I illustrate the case when the female ((a) Receiving fish) has an EOD frequency of 800 Hz while the male ((b) Emitting fish) has an EOD frequency of 925 Hz. The two EODs summate resulting in a beat frequency of 125 Hz (c). Since the male is several centimeters from the female, her electroreceptors respond to the 125 Hz amplitude modulation of her baseline EOD; the amplitude of this modulation depends on the distance between the fish; in this case it is 20% of the female's EOD amplitude. This strong signal causes the female's electroreceptors to synchronize without much change in their mean rate of discharge. The male modulates his EOD amplitude (~10% decrease) and frequency (~400 Hz increase) to produce a brief (< 20 ms) large chirp (black line in (b)). During the chirp, the female ((c) receiving fish) senses a disruption of the ongoing 125 Hz beat frequency. In this case (a large chirp) the female's electroreceptors respond by desynchronizing, again without any change in their mean discharge rate (Benda et al., 2006). In the electrosensory system transient signals (chirps) are encoded as changes in afferent fiber synchrony. (Part B courtesy of Jan Benda, modified from Fig. 1 of Benda et al., 2006.)

male–female beat signals (Chacron et al., 2005a). The reason for the discrepancy between gain and coherence measures lies in the fact that the coherence measure takes into account the intrinsic frequency-dependent noise of the P-units (Chacron et al., 2005a). Although P-unit discharge appears random, their interspike intervals (ISI) have a strong negative serial correlation (Ratnam and Nelson, 2000) and they show rapid adaptation to sustained input (Xu et al., 1996) — the P-unit spike train is therefore highly nonrenewal. As a consequence of the "memory" in the P-unit spike train, P-unit noise is reduced at low frequencies optimizing the detection of prey (Ratnam and Nelson, 2000; Chacron et al., 2001). In vivo spike trains recorded from many types of neurons have non-renewal statistics (ISI correlations) similar to those of P-units. Theoretical analysis based on P-unit biology has clarified the theoretical basis by which ISI correlations can increase information transfer (Chacron et al., 2004) and this analysis may be highly relevant to the nonrenewal spike trains of cortical and other neurons.

The rapid adaptation of P-units also permits them to respond in an enhanced manner to small chirps occurring within a low-frequency beat (Benda et al., 2005); this optimizes the ability of male fish to engage in agonistic interactions. P-units respond to high-frequency beats (male–female interactions) by synchronizing; remarkably they then respond to the large (putative courtship) chirps by desynchronizing (Benda et al., 2006). How a P-unit synchronization–desynchronization code for communication signals is read out by target neurons is an exciting direction of current studies.

Synopsis

P-unit adaptation dynamics effectively permit the fish to detect a wide range of continuous AM frequencies as well as transient modulations (chirps) superimposed on a background slow modulation. These dynamics are revealed in the ISI statistics and in the frequency-dependent population synchronization. Similar principles are likely operative in auditory coding of complex sounds.

Neural processing of electrosensory signal's overall anatomical organization (Fig. 2)

This section briefly outlines the anatomy and physiology of the central circuitry that decodes the P-unit input (Fig. 2).

Electroreceptors are innervated by the peripheral nerves emanating from cells within the anterior lateral line nerve ganglion (adjacent to the vestibular/cochlear ganglion). These ganglion cells project in a topographic manner solely to an ipsilateral hypertrophied dorsal medullary structure — the electrosensory lateral line lobe (ELL). The ELL is both laminated and segmented as discussed below. The ELL projects to only two contralateral (mainly) brain regions: the nucleus praeminentialis (nP) and the torus semicircularis (TS). Both pathways are topographically organized (see Carr and Maler, 1986; Bell and Maler, 2005, for reviews of the electrosensory circuitry described below).

The nP is involved strictly in the feedback regulation of electrosensory input. There are three feedback pathways. Inhibitory and excitatory pathways project from nP directly back to the ELL (direct feedback) while an excitatory pathway projects to a mass of cerebellar granule cells overlying the ELL (EGp); these granule cells then project back to the ELL — this latter pathway is therefore referred to as the indirect feedback pathway.

The TS is a complex hypertrophied midbrain region analogous to the inferior colliculus of other vertebrates. The output of the TS clearly reflects the dual use of the electrosense in that its projections are segregated into those responsible for electrolocation versus electrocommunication. One TS output is a topographically organized projection (ipsilateral) to the optic tectum in topographic register with the tectal retinal input. It is thus via the tectum that electroreceptor input can direct the fish toward prey objects or around obstacles; this organization is evidently similar to that of the auditory control of spatially directed movement via the tectum. Secondly, the TS has a diffuse (non-topographic) projection to the nucleus electrosensorius (nE). The nE contains neurons responsive to a variety of electrocommunication signals and in turn projects to the prepacemaker

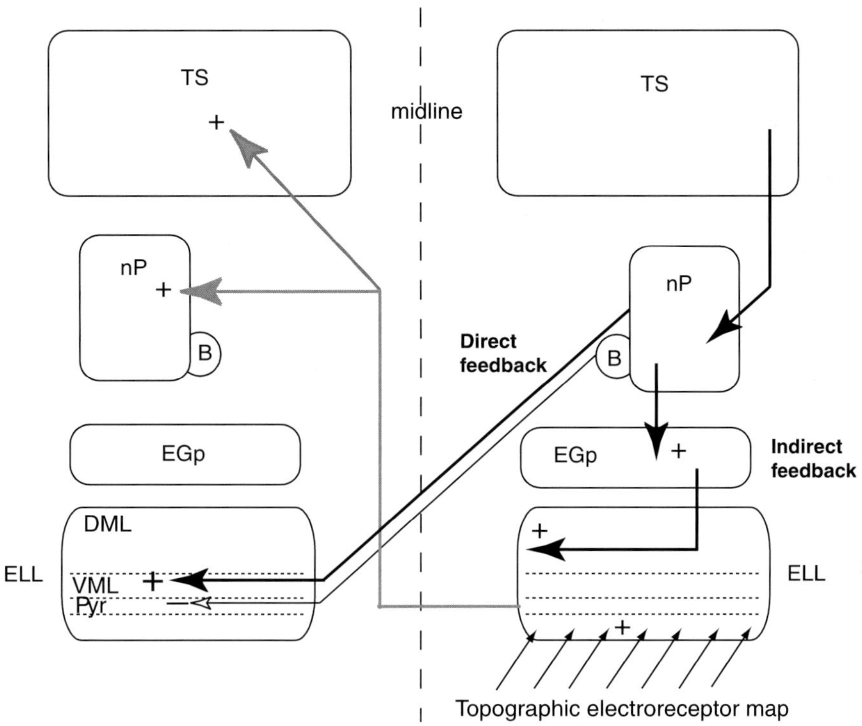

Fig. 2. Anatomical organization of feedforward (left side, gray) and feedback (right side, black) electrosensory pathways. Electroreceptor afferents terminate in the ELL (of all three maps) to form a topographic representation of the body surface. ELL pyramidal cells project topographically to two contralateral (predominantly) brain structures: the nucleus praeminentialis (nP) and the torus semicircularis (TS). The nP is involved strictly in feedback regulation of the ELL itself; the TS (similar to the inferior colliculus of mammals) is a complex hypertrophied structure involved both in the tracking of moving electrosensory stimuli (e.g., prey) and in detecting electric communication signals (ascending projections of the TS are not illustrated here). The TS projects back to nP; neither the anatomical organization nor the physiological properties of this feedback pathway have been investigated. Two types of nP neuron project directly back to the ELL. Bipolar cells (B) project in a diffuse manner to the ELL pyramidal cell layer (Pyr) where they make inhibitory (GABAergic) synaptic connections (delayed global inhibitory feedback). Numerous nP stellate cells (not specifically illustrated) project in a topographic manner to the ventral molecular layer (VML) of the ELL where they make excitatory synaptic contacts on the proximal apical dendrites of pyramidal cells (positive local feedback) as well as onto local inhibitory interneurons. Several populations of nP cells (including multipolar cells) project back in a fairly diffuse manner onto a mass of cerebellar granule cells (EGp) overlying the ELL. Parallel fiber emanating from these cerebellar granule cells project to the ELL where they form the dorsal molecular layer (DML). These parallel fibers run transversely across the entire ELL and form excitatory synapses on pyramidal cell distal apical dendrites (positive global feedback) as well as onto local inhibitory interneurons.

nucleus (and hypothalamic regions) responsible for evoking the EOD modulations used in electrocommunication. Lastly, the TS also projects back massively to nP and can thus regulate feedback control of ELL itself; the function of this latter projection has not yet been experimentally analyzed. Detailed physiological studies have begun to reveal the function of the TS (Heiligenberg, 1991; Rose and Fortune, 1999a, b; Fortune and Rose, 2000, 2001, 2003; Ramcharitar et al., 2006) but this work is beyond the scope of this review.

Structure of the ELL (Fig. 3)

The laminar structure of the ELL (Fig. 3) is organized so as to separate electroreceptor from intrinsic and feedback synaptic input to the main populations of output neurons — pyramidal cells. Electroreceptor afferents and their terminal boutons form the deepest layer of the ELL. These afferents terminate on both the pyramidal (projection) cells and interneurons. The majority of interneurons are found in a granular lamina

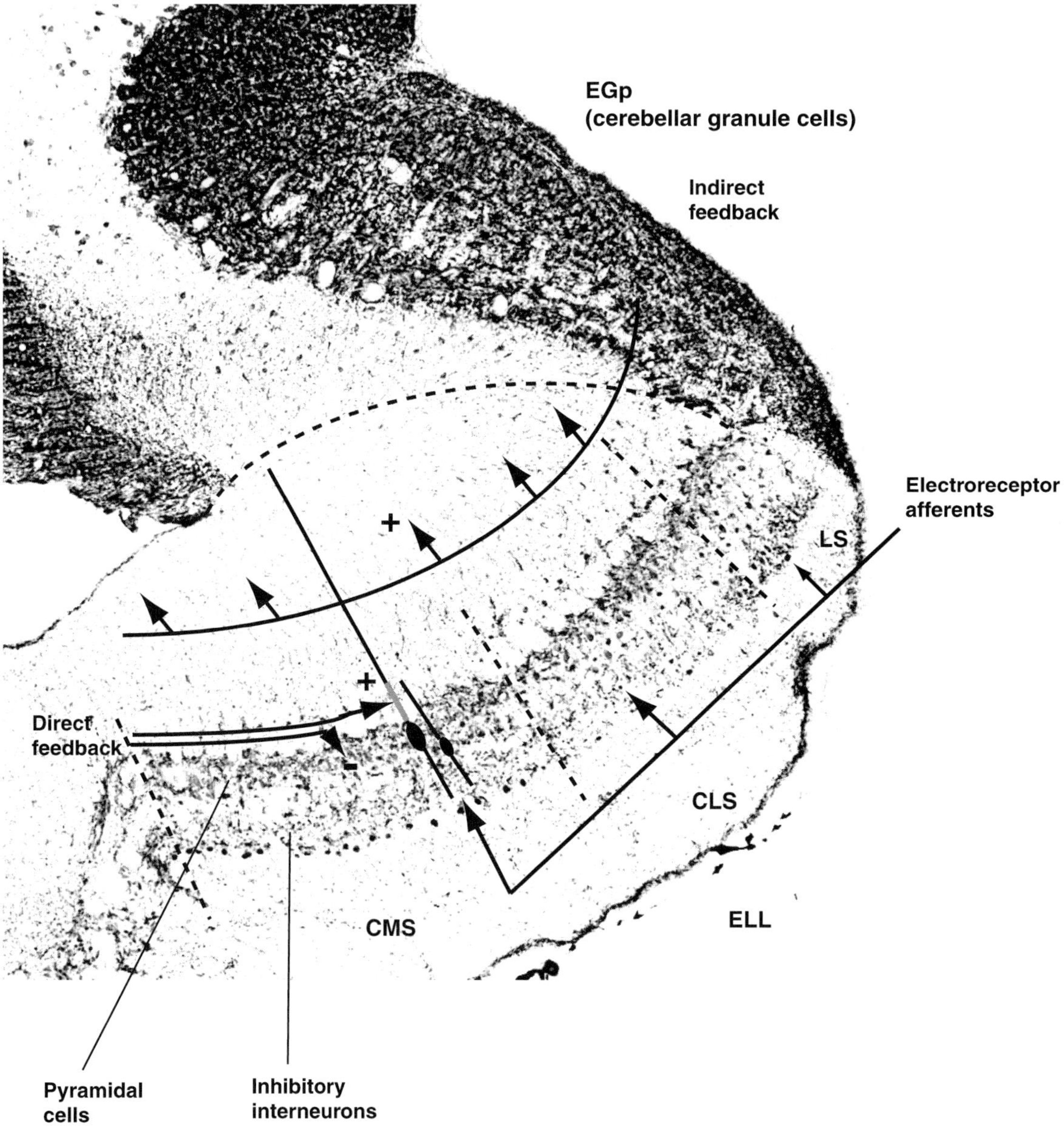

Fig. 3. The ELL is subdivided into three tuberous segments (CMS, CLS, LS); one class of electroreceptor afferent (P-units) enters the ELL and trifurcates so as to form a topographic map within each segment (a second class of electroreceptor afferent, ampullary receptors, terminates topographically in a medial segment; this is not illustrated). The laminar organization of the three maps is identical. Electroreceptor afferents terminate on the basal dendrites of pyramidal cells (receptive field center) as well as on inhibitory interneurons (surround inhibition and temporal sharpening). There are several classes of pyramidal cells (a superficial and a deep basilar cell are illustrated) organized in a columnar arrangement; the members of a column receive the same electroreceptor input. Apical dendrites of pyramidal cells extend into a molecular layer. The direct feedback pathway contains a diffuse inhibitory component terminating on pyramidal cell somata and a topographic excitatory component terminating on the proximal apical dendrites of pyramidal cells (ventral molecular layer). The proximal apical dendrites are endowed with Na^+ channels that support backpropagating dendritic action potentials (gray region); the dendritic currents give rise to somatic depolarizing after-potentials (DAPs) responsible for spike bursts. An indirect feedback pathway from cerebellar granule cells terminates in the dorsal molecular layer on the distal dendrites of pyramidal cells (molecular layer interneurons are not illustrated).

dorsal to the electroreceptor afferent terminals while the pyramidal cell somata are found mainly in their own lamina dorsal to that of the interneurons. Pyramidal cell apical dendrites ramify dorsally within a large molecular layer. These dendrites receive input from the direct and indirect feedback pathways (see below).

The ELL is composed of four segments: medial (MS), centromedial (CMS), centrolateral (CLS), and lateral (LS). The MS receives input from ampullary receptors while the CMS, CLS, and LS receive input from tuberous receptors (mainly P-units). Each P-unit trifurcates as it enters the ELL so that all three maps (CMS, CLS, LS) receive identical input. The electrosensory afferent input to all maps runs in precisely organized fiber bundles that terminate in a topographic manner (Lannoo et al., 1989). Thus, the electrosensory system (both ampullary and tuberous) contains topographic maps of the electrosensory body surface. Multiple topographic maps that preserve the topological organization of a receptor surface but process this input in different ways are a ubiquitous feature of sensory systems: auditory (tonotopy — maps in dorsal and ventral cochlear nuclei), somatosensory (somatotopy — multiple representations of various receptors for the same skin region), and visual (retinotopy — multiple classes of retinal ganglion cells). The size of the three P-unit maps is different with CMS > CLS > CMS. The effect of this variation on receptive field (RF) size and signal estimation will be discussed below.

The projection of ELL to nP preserves both the topography and segregation of the maps while, in the TS, the maps all converge onto a common topographic representation of the electroreceptive body surface (Carr and Maler, 1986; Bell and Maler, 2005).

Synopsis

The electroreceptive periphery is centrally represented in multiple maps that process the same input in different ways. This arrangement is similar to that of other sensory systems; its utility will be discussed below.

Higher order electrosensory circuitry segregates electrolocation (tectum) and electrocommunication (n. electrosensorius, hypothalamus) circuitry much in the same way as auditory input is eventually directed to auditory cortex and amygdala.

Physiology of the ELL

Organization of receptive fields (RFs)

Here I describe how ELL circuitry creates RFs appropriate for the detection of conductive and non-conductive objects as well as a wide range of AMs.

There are two major subtypes of ELL pyramidal cells: basilar (E type) and non-basilar (I type) pyramidal cells (Maler, 1979; Saunders and Bastian, 1984). The basilar pyramidal cells have a single thick basal trunk that ramifies as a small dendritic bush in direct receipt of P-unit input. P-units make synaptic contacts on these basal bushes that utilize mainly AMPA receptors evoking fast EPSPs (Berman and Maler, 1998a). These cells respond with excitation to stimulation of P-units within a circumscribed patch of skin (Bastian et al., 2002) and therefore have classic RF centers (Fig. 1A) — they are therefore also described as E type cells (excited by stimulation of their RF). Non-basilar pyramidal cells have no basal dendrite and receive P-unit input indirectly via inhibitory interneurons (mostly GABAergic and located in the granular cell layer, see below) that make synaptic contact with pyramidal cell somata (Berman and Maler, 1998a). These pyramidal cells also have well delimited RF centers but, because they are driven by inhibitory interneurons, respond to increased P-unit activity with reduced discharge (Bastian et al., 2002) — they are therefore also described as I type cells (inhibited by stimulation of their RF). Because of this sign inversion, I cells respond with increased discharge when P-unit activity within their RF center is decreased — this will occur whenever a nonconductor is present over their RF centers. Thus, E cells respond to prey (more conductive than the ambient water) while I cells respond to rocks (less conductive). The E and I type cells are found adjacent to one another and both classes tile the electroreceptive surface. The functional analogy of E and I type pyramidal cells to the On and Off center retinal ganglion cells is striking, although the biophysical mechanisms that generate the E and I center responses are different (Maler et al., 1981).

Within the E versus I cell classes there also exists a more subtle subdivision based on the position of the pyramidal cell's somata; thus, for both E and I type cells, there exist superficial- (SP), intermediate- (IP), and deep pyramidal (DP) cells (Bastian and Courtright, 1991). SP cells are found most dorsally in the pyramidal cell layer just underneath a major myelinated feedback fiber bundle, the stratum fibrosum (see below), while the DP cells are found within the granular layer; the IPs have an intermediate location. This classification is also consistent with many other morphological, physiological, and biochemical properties as discussed below. It is also remarkable that these different types of pyramidal cells have a columnar organization: each column receives input from the same patch of skin and contains E and I versions of SP, IP, and DP cells; thus each column represents a processing unit that performs several operations (see below) on the same input (unpublished observations).

SPs have a prominent inhibitory surround and a phasic response to electrosensory input while DPs have no surround inhibition and a sustained response to the same input (Bastian and Courtright, 1991; Bastian et al., 2002); both properties are due in part to local ELL inhibitory circuits that terminate on the somata of pyramidal cells. It has been hypothesized that Type 1 of interneurons are responsible for surround inhibition while Type 2 interneurons are responsible for the phasic response (Maler, 1979; Maler et al., 1981; Shumway and Maler, 1989; Maler and Mugnaini, 1994; Berman and Maler, 1998a; Bastian et al., 2002).

The classic view of surround inhibition is that it will sharpen spatial localization and this would suggest that the SPs would be optimized for the precise localization of prey (E cells) and rocks (I cells). However a recent theoretical analysis has indicated that, in cortex, surround inhibition will generate long-range spatial correlations of pyramidal cell discharge that degrade signal (orientation) estimation (Series et al., 2004). From this perspective the DPs would provide a better estimate of an objects position since they would presumably not have such long-range spatial correlations. This is however contradicted by the fact that SPs have a stronger response to the local low-frequency input produced by objects (Chacron et al., 2005b). An interesting direction for theoretical analyses is the hypothesis that better estimates of location can be achieved by computations combining the output of neurons both with and without strong inhibitory surrounds.

Type 2 interneurons are GABAergic and their synapses utilize GABA-A receptors (Maler and Mugnaini, 1994; Berman and Maler, 1998a). The fast IPSPs generated by these cells truncate electroreceptor evoked EPSPs and an early study suggested that this might be responsible for the phasic response of pyramidal cells (Shumway and Maler, 1989). Consistent with this theory SPs have a far more phasic response than DPs (Bastian and Courtright, 1991). The presence of both tonic and phasic responses presumably confers some advantage to the estimate of prey (and rock) location, but this issue has not been explored theoretically.

Synopsis

The ELL contains E and I type pyramidal cells with On center-Off surround and Off center-On surround organization; the RFs of these cells tile the body surface. In addition, subsets of pyramidal cells vary with respect to both the strength of their RF surrounds and their frequency tuning. This functional organization is quite similar to that of retina, which suggests that it is an optimal strategy evolved by neural networks designed for spatial localization. A subset of pyramidal cells lack surround inhibition; it is possible that downstream networks can use the output of these cells to ameliorate the negative consequences of surround-induced correlations on population coding.

Multiple maps: variations in RF tuning width and the estimation of prey location

Here I present data suggesting that there is no single optimal RF size; rather, RF tuning width variation might optimize the estimation of different stimulus features.

The different size of the three ELL segments receiving P-unit input suggests that the RF centers of their pyramidal cells might differ and this has been confirmed both physiologically (Shumway, 1989a) and anatomically (Shumway, 1989b and unpublished observations). Several theoretical studies have addressed the issue of the optimal size of RFs. For estimates of a location parameter(s) the optimal RF size depends on the dimension of the parameter space (Abbott and Dayan, 1999; Zhang and Sejnowski, 1999). For a one-dimensional parameter (e.g., orientation of bars, sound frequency) a small RF is optimal. For a two-dimensional parameter (e.g., somatotopic or retinotopic localization) the RF size is irrelevant, while for parameter dimensions >2, large RFs are best. Thus, for localizing prey (or other objects) in the dorso-ventral and rostro-caudal body axes (two-dimensional: X, Y axes), it would appear that there is no need for three ELL maps. These fish are, however, also capable of estimating the prey's distance from the body (Z axis) (MacIver et al., 2001). The intensity of the electric image decreases with distance from body and might serve to estimate the Z value; this cue is, however, ambiguous since a large distant object can produce the same intensity as a smaller closer one. The spread (blur) of an electric image also increases with Z value. Thus, to estimate the distance of prey, the fish must estimate both the intensity of the electric image and its two-dimensional spread. We have used Fisher information to demonstrate that larger RFs are best suited for intensity estimate while smaller RFs give the best estimates of electric image spread (Lewis and Maler, 2001). At least two ELL maps might therefore be required for estimating the distance (via stimulus intensity and blur) of prey from the fish's body and for accurately navigating among rocks. Remarkably blur is also a cue for visual depth perception (Lewis and Maler, 2002a), suggesting that the wide range of RF sizes (spatial frequencies) in visual cortex might have a similar computational role to that of the three ELL maps.

In addition to their different putative roles in localizing prey, the three ELL tuberous maps also have different roles in electrocommunication. The frequency tuning of pyramidal cells decreases from LS to CMS (Shumway, 1989a) suggesting that there may be a trade-off between spatial and temporal frequency resolution. A number of cellular mechanisms probably also contribute to this difference. For example, a potassium channel (Kv3.1) hypothesized to contribute to high-frequency tuning in the auditory system has been found to be highly enriched in the LS (high frequency) compared to the CMS (low frequency) (Deng et al., 2005). The LS has also been implicated in the detection of transient chirp signals (Metzner and Heiligenberg, 1992; Metzner and Juranek, 1997) but computational analyses of the mechanisms involved are not available.

There are now many sophisticated mathematical analyses (using either Fisher or Shannon information) of the effect of RF size and correlations on population estimates of sensory parameters (Abbott and Dayan, 1999; Eurich and Wilke, 2000; Sompolinsky et al., 2001; Wilke and Eurich, 2002; Averbeck and Lee, 2004, 2006; Shamir and Sompolinsky, 2004). From the viewpoint of a sensory physiologist or neuroethologist these analyses appear somewhat naïve in that they typically assume that sensory neurons are tuned to only a single parameter; orientation tuning (one-dimensional) in pyramidal cells of primary visual cortex (V1) has been a favorite parameter for many of these studies. It is well known, however, that cells in V1 are tuned to many other parameters than orientation including spatial frequency. In fact all possible combinations of orientation and spatial frequency are represented in V1 (Issa et al., 2000), i.e., multiple maps responsive to various combinations of at least these two parameters are present in V1 and used for their estimation. It is likely that computational studies of population coding based on a deeper appreciation of the multiple parameters available in natural signals might prove advantageous.

Synopsis

Multiple representations (maps) of the same sensory input are ubiquitous in sensory systems and these maps typically contain neurons with very different tuning curve widths (RF sizes). There have been extensive information-theoretical analyses of the effect of RF size on coding properties but these analyses have assumed that there is some optimal size. Our studies suggest that, when complex stimuli must be estimated (e.g., in the visual system: orientation, spatial and temporal frequency, and contrast) it is advantageous to have multiple RF tuning widths. The important theoretical issue is not which tuning width is optimal but rather how the output of multiple maps can be integrated for optimal estimation/detection of complex natural signals.

Spike bursting in ELL pyramidal cells

Here I describe pyramidal cell burst dynamics and its computational consequences.

Mathieson and Maler first developed an in vitro (slice) preparation of the ELL (Mathieson and Maler, 1988). This initial study suggested that pyramidal cell somata expressed, in addition to the fast Na^+ and K^+ currents responsible for spiking, additional currents important for their subthreshold activity: a persistent Na^+ current, an I_A-like current that retarded the delayed onset to discharge and an apamin-sensitive current that appeared to contribute to spike frequency adaptation. Subsequently ELL pyramidal cells were shown to support a strong back-propagating spike dependent on Na^+ channels on their proximal apical dendrites; furthermore, the reflection of dendritic spike currents back to the soma resulted in a depolarizing after-potential (DAP) that could lead to spike bursting (Turner et al., 1994). A series of experimental (Lemon and Turner, 2000; Rashid et al., 2001; Doiron et al., 2003b; Noonan et al., 2003) and modeling (Doiron et al., 2001) studies elucidated the biophysical basis of bursting in these cells and the contribution of various somatic and dendritic Na^+ and K^+ channels; one key conclusion of these studies is that low-intensity input initially causes regular firing and that sustained or strong input can switch these pyramidal cells into a burst mode. A combined experimental and modeling analysis of pyramidal cell bursting revealed that a slow cumulative inactivation of dendritic Na^+ channels was essential for spike bursting (Fernandez et al., 2005); this inactivation causes a delay in the onset of the DAP that moves it past the refractory period due to Kv3 and delayed rectifier K^+ channels. Theoretical studies (Doiron et al., 2003a; Laing et al., 2003) treated this as a dynamical system with fast–slow dynamics and demonstrated that bursting was caused by a bifurcation (saddle node of a limit cycle) from a limit cycle (regular discharge) to chaotic burst discharge. There were two major predictions of this theory: that the onset of bursting should be influenced by the now unstable limit cycle (ghost of the attractor — for this reason these dynamics are now referred to as "ghostbursting") and that there should be a sharp transition (bifurcation) from regular to burst discharge as the input current is gradually increased. Remarkably, both predictions were verified (Doiron et al., 2002; Laing et al., 2003). Ghostbursting appears to be different from other kinds of experimentally or theoretically demonstrated neuronal bursting (Izhikevich, 2000); since DAPs are common in cortical pyramidal cells, it is possible that ghostbursting is a more broadly distributed mechanism that has simply not been looked for in other neurons.

A seminal study by Gabbiani (Gabbiani et al., 1996) had proposed that ELL pyramidal cell bursts acted as feature detectors for increases (E cells) or decreases (I cells) of EOD amplitude; this study hypothesized that the bursts seen in vivo were in fact caused by the DAPs revealed in vitro. Oswald et al. (2004) then used broadband Gaussian noise (0–60 Hz) delivered either as a sensory input (in vivo) or via intracellular current injection (in vitro) to more carefully analyze the role of isolated spikes and spike bursts; previous studies had demonstrated that the pyramidal cell's membrane potential can closely follow this type of sensory input (nature or neuron) thus justifying the substitution of current injection for sensory input (Chacron et al., 2003; Bastian et al., 2004; Middleton et al., 2006). This study directly revealed that DAP-induced bursts were highly coherent with the low frequencies in the signal while isolated spikes could also code for high frequencies. Further the bursts did act as feature detectors while the single spikes were effectively estimating the signal over its entire frequency range. Pyramidal cells can therefore simultaneously transmit both low (e.g., prey, bursts) and high (communication, isolated spikes) frequency signals. Several recent studies have suggested that relay neurons in the mammalian lateral geniculate nucleus (LGN) also code for low-frequency input with spike bursts (Lesica and Stanley, 2004; Grubb and Thompson, 2005; Bezdudnaya et al., 2006). LGN burst dynamics are due to the interplay of a somatic low threshold Ca^{2+} current and a hyperpolarization activated K^+ current (Llinas and Steriade, 2006), and therefore entirely different from pyramidal cell ghostbursting. The question therefore arises whether the specific form of burst dynamics are in any way critical to their role in the detection of low-frequency signals. Krahe and Gabbiani (2004) proposed that the essential role of dendritic currents in ghostbursting might permit its regulation by feedback targeting the apical dendrites of pyramidal cells; experimental evidence in favor of this idea is presented below.

Synopsis

Spike bursting in ELL pyramidal cells is caused by somatic–dendritic interactions; these ghostburst dynamics allow these cells to encode, in parallel, low (bursts) and high (single spikes) frequency signals. This division of labor might also apply to the visual system, although implemented by different burst dynamics.

Feedback to ELL pyramidal cells

Here I review the multiple feedback pathways to ELL that control sensory acquisition.

As reviewed above, there are three feedback pathways to ELL, all three arising from distinct cell types in one nucleus — nP. (a) The direct inhibitory pathway (bipolar cells), (b) the direct excitatory pathway (stellate cells), and (c) the indirect excitatory pathway (multiple cell types including multipolar cells). Below I discuss the dynamics and computations associated with these different feedback sources.

Diffuse inhibitory feedback

Here I describe a direct inhibitory pathway that induces gamma oscillations and speculate on the potential role of the oscillation in optimizing prey detection in the presence of conspecifics.

Bipolar cells (nP) receive ascending input from ELL (non-topographic, representing the entire body); these are GABAergic and terminate diffusely in the pyramidal cell layer (Maler and Mugnaini, 1994). In vitro studies have shown that they act via both GABA-A and GABA-B receptors to reduce pyramidal cell discharge (Berman and Maler, 1998b); this pathway therefore implements delayed feedback inhibition. In vivo global stimulation with broadband Gaussian noise induces a stochastic oscillatory discharge in the gamma range (~30 Hz) in pyramidal cells; the same stimulus delivered to the RF of the cell (local stimulation) failed to induce an oscillation (Doiron et al., 2003a). Blockade of the bipolar cell feedback fibers prevents the induction of oscillatory activity; modeling and theoretical analyses have demonstrated that key requirements for this effect are the transmission delays of the pathway and correlated negative feedback from a large fraction of the electroreceptors (Doiron et al., 2003a, 2004; Lindner et al., 2005). The function of this signal-induced gamma oscillation is not known with certainty. In other sensory systems gamma oscillations are known to enhance sensory processing (Fries et al., 2002; Cardin et al., 2005; Ishikane et al., 2005; Palva et al., 2005; Taylor et al., 2005; Bauer et al., 2006; Womelsdorf et al., 2006), although the mechanism responsible for this effect is not known. Recent studies have demonstrated that oscillations in this range can occur naturally when conspecifics of a related electric fish species interact and that these oscillations can enhance the directionally selective responses of TS cells to moving objects (Ramcharitar et al., 2006); the authors present plausible evidence that this enhancement is due to synaptic depression and link it to a similar mechanism proposed for directional selectivity in visual cortex (Chance et al., 1998). The mechanism underlying the enhanced response was shown to involve synaptic depression in the TS. Wave type electric fish typically forage in small groups (Tan et al., 2005) and this induced oscillation may represent a mechanism to enhance the detection of prey under these conditions.

Synopsis

Delayed inhibitory feedback generates a gamma oscillation in ELL pyramidal cells. By itself this is an expected finding since it is already known that cortical inhibitory loops are able to induce gamma oscillations. However the function of gamma oscillations in cortex is still a matter of debate. In the case of the electrosensory system, evidence is accumulating that these oscillations specifically enhance the detection of prey objects and that they do so by interacting with short-term synaptic plasticity dynamics of the pyramidal cell synapses on midbrain (TS) neurons. The generation of directional responses via synaptic depression has previously been proposed for visual cortex; more detailed comparisons of these very different neural circuits would therefore be of great theoretical interest.

Direct topographic excitatory feedback

Here I describe a reciprocal, excitatory topographic feedback system. I present evidence supporting the hypothesis that this feedback implements a "searchlight" that enable the electric fish to enhance the detection of prey during its scanning movements.

The ELL projection to nP is topographic. The stellate cells are glutamatergic and project to the proximal apical dendrites of ELL pyramidal cells (Sas and Maler, 1983; Berman et al., 1997); this pathway is also characterized by the presence of presynaptic CaMKIIα (Maler and Hincke, 1999), a kinase known to be important for both pre- and post-synaptic plasticity in many types of neurons (Lisman et al., 2002; Ninan and Arancio, 2004). The ELL-stellate cell connections are segment specific, topographic, and reciprocal and the RFs of stellate cells are slightly larger than those of ELL pyramidal cells (Maler et al., 1982; Bratton and Bastian, 1990). The synaptic connections from this pathway are mainly onto SP and IP cells and utilize AMPA and NMDA receptors (Berman et al., 1997); this connectivity therefore implements a form of local positive feedback circuit. In addition these feedback fibers also provide a strong input to a local interneuron (VML cell) that in turn projects to the proximal apical dendrites of pyramidal cells; the VML cell is GABAergic and its activation produces GABA-A mediated IPSPs (Maler and Mugnaini, 1994; Berman and Maler, 1998c). The excitatory direct feedback pathway therefore produces direct excitation but is kept from instability by di-synaptic inhibition.

Stellate cells (nP) have fairly restricted RFs and respond strongly, though transiently, to moving objects (low-frequency AMs) and are inhibited by high frequency communication signal (Bratton and Bastian, 1990). The anatomy and physiology of this system have led to the hypothesis that the direct excitatory feedback pathway implements a positive feedback searchlight mechanism that enhances the fish's ability to detect prey or other objects while scanning its environment (Berman and Maler, 1999).

Both in vivo and in vitro studies have demonstrated that the direct excitatory feedback strongly potentiates when stimulated at rates that mimic their discharge to natural stimuli (Wang and Maler, 1997; Bastian, 1998b). In vitro studies have further shown that there are multiple time constants of short-term presynaptic plasticity (both potentiation and depression) including a fairly slow potentiation that requires CaMKIIα (Wang and Maler, 1998; Oswald et al., 2002). A hypothesis suggested by models of these processes is that the "searchlight" may itself become enhanced during the fish's scanning movements as it hunts for prey.

Both the direct feedback pathway and its associated di-synaptic inhibition (VML cell) terminate on the proximal apical dendritic region of pyramidal cells — the site that produces back-propagating spikes that result in DAPs and can therefore initiate bursting. Preliminary studies have indicated that strong activation of the excitatory component of the direct feedback pathway in vitro (mimicking natural in vivo activity caused by movement of objects across the RFs of stellate cells) can induce ghostbursting, presumably due to the summation of the feedback EPSP and the DAP (Turner et al., 2002). This result is intuitively appealing since it implies that the putative "searchlight" will increase the spike bursts that signal the presence of prey and other objects.

In contrast, application of muscimol (a GABA-A receptor agonist — this mimics the effect of stimulation of the VML cell) decreases the amplitude of the DAP resulting in a divisive inhibition of pyramidal cell spiking (Mehaffey et al., 2005). Both spike bursts and single spikes are inhibited. The combined effect of the excitatory and inhibitory components of the direct feedback pathway has not been modeled, but it possibly acts to enhance burst production for the brief time period that an object takes to pass through the RF of a pyramidal cell.

Synopsis

The direct feedback pathway appears to implement a searchlight that optimizes the detection of prey under natural conditions, i.e., while the fish scans its environment with stereotypical movements. There are numerous mechanisms designed to enhance the response to prey: these include the topography of the feedback, the synaptic plasticity dynamics, and the interaction of the synaptic feedback with the intrinsic bursting dynamics of pyramidal cells. This last point is perhaps of the most general interest since the regulation of intrinsic burst dynamics by synaptic input is likely to be widespread in cortico-thalamic feedback pathways.

Indirect diffuse excitatory feedback

Here I describe a diffuse excitatory feedback system that implements the cancellation of redundant sensory input. The cellular mechanisms that underlie this computation have been described and network models incorporating these dynamics can account for our experimental results.

The indirect feedback emanates from several types of nP cells (Sas and Maler, 1983) only one of which, the multipolar cell, has been physiologically characterized (Bastian and Bratton, 1990). These nP cells project to the granule cells of the caudal cerebellum (eminentia granularis posterior, EGp) with only a crude rostro-caudal topography (Sas and Maler, 1987). EGp granule cells form typical cerebellar parallel fibers (Maler, 1979; Maler et al., 1981). After their T-junction division one branch of the parallel fiber stays within the cerebellum while the other enters the dorsal molecular layer (DML) of the ELL and terminates on distal apical dendrites of pyramidal cells. Within the DML the parallel fibers terminate on dendrites of all ELL maps. The indirect feedback pathway thus appears to have little regard for the topographic organization of the ELL and is therefore said to implement a more "global" feedback.

In vitro studies have demonstrated that the parallel fiber feedback input to ELL molecular layer utilizes both AMPA and NMDA receptors boosted by somatic persistent Na^+ currents (Berman et al., 2001) and also causes GABA-A receptor mediated inhibition via molecular layer interneurons (Berman and Maler, 1998c). The dynamics of this projection are complex and involve both facilitation and depression with several time constants (Lewis and Maler, 2002b, 2004). The net result appears to be frequency tuning that optimizes feedback strength at particular stimulation frequencies; the role of such tuning in vivo is not known.

Bastian has shown that tail bending can produce global low-frequency AMs of the fish's EOD and that these AMs are detected by the P-units. In contrast, ELL pyramidal cells do not respond to such global low-frequency AMs demonstrating the presence of a cancellation mechanism that removes redundant sensory input (Bastian, 1995). Further studies then demonstrated that this cancellation was due to the feedback pathways via the formation of a "negative image" of the global electrosensory input (Bastian, 1996a, b, 1998b) and this process was in part mediated by postsynaptic Ca^{2+} (Bastian, 1998a); this work is reviewed in detail by Bastian (1999). Basically, the low-frequency input to the RF center of a pyramidal cell is cancelled by a phase matched negative image from a non-classic RF of the cell generated primarily by the indirect feedback pathway (Chacron et al., 2003).

The cancellation process is dynamic and a subset of ELL pyramidal cells can adaptively learn to cancel new global input: the SP cells are the most plastic while the DP cells are not at all plastic (Bastian et al., 2004). This is consistent with the far more extensive apical dendritic tree of SP versus DP cells as well as with the presence of intracellular signaling proteins in SP/IP but not DP cells (Zupanc et al., 1992; Berman et al., 1995). Models of the indirect feedback circuitry predicted that the input to cerebellar granule cells should be predominantly from the nonplastic DP cells since, as the cancellation proceeded, the output of the SP cells would be expected to diminish. Remarkably this prediction was confirmed: the DP cells (nonplastic) are the major source of input to nP and therefore to the indirect feedback pathway (Bastian et al., 2004). Studies of mammalian cortex have revealed that a key second messenger pathway (CaMKIIα) is expressed in only a subset of pyramidal cells (Jones et al., 1994). Thus it is possible that this principle — non-plastic neurons providing modifiable input to plastic neurons — might be more generally applicable.

Overall it appears that the indirect feedback pathway allows large areas of the electroreceptive surface (non-classic RF) to influence processing of input to the RF center. One purpose of this pathway is to adaptively cancel redundant signals caused by spatially global input due to tail bending or the low-frequency beats of conspecifics; this cancellation increases the salience of the spatially localized signals due to prey. Non-classic RFs have often been reported in other sensory systems and it has been suggested that this input can enhance information transfer in these systems as well (Vinje and Gallant, 2002). The cortico-cortical connections believed to mediate the non-classic RF are likely highly plastic and it will be important to investigate whether similar principles of adaptive cancellation or enhancement are operative in cortex as well.

Synopsis

The indirect feedback input to ELL serves to adaptively cancel predictable signals. The circuitry that does this involves nonplastic DP cells providing feedback to plastic SP cells. It is possible that a similar principle applies in neocortex as well.

Summary

Although the electrosense might appear to be an exotic sense, it has become clear that its organization is functionally very similar to that of the more familiar senses of vision, touch, and hearing. Some common principles include a topographic organization of sensory receptor input, the presence of separate "On" and "Off" type pathways that segregate both kinds of contrast available in the sensory input, center-surround organization of RFs, multiple maps with different RF tuning widths, as well as the major role that feedback plays in shaping early sensory processing.

Some general new principles of sensory processing have also emerged from our early explorations of the electrosense. These include: (a) the importance of the statistical structure of spike trains for noise shaping so as to maximize information transfer of a wide range of signals; (b) the presence of maps that combine different RF sizes with different biophysical properties to optimize estimation of the multiple parameters present in sensory input; (c) the columnar organization of projection neurons with very different, cellular and circuit properties — thus not all neurons need have surround inhibition and computations combining those with and without this property may be far more powerful than currently envisioned; (d) the presence of both plastic and non-plastic neurons within a column — again the combination of both types appears to allow for richer computations than possible with one type alone; (e) the multiple important roles of feedback that ranges from induction of oscillatory activity and the enhancement of localized input to the adaptive cancellation of redundant input. The ELL is, in comparison to the cortex of mammals, a relatively simple structure; it is therefore likely that all these principles will also be manifested in cortex as well.

Acknowledgments

I would like to thank Joe Bastian, Jan Benda, Maurice Chacron, Brent Doiron, Rob Dunn, John Lewis, Andre Longtin, Jason Middleton,

Anne-Marie Oswald, and Ray Turner for many discussions of electrosensory processing. The research in my laboratory was supported by grants from the Canadian Institutes of Health Research.

References

Abbott, L.F. and Dayan, P. (1999) The effect of correlated variability on the accuracy of a population code. Neural Comput., 11: 91–101.

Albert, J.S. and Crampton, W.G.R. (2005) Diversity and phylogeny of neotropical electric fishes. In: Bullock T.H. and Hopkins C.D. (Eds.), Electroreception. Springer, New York.

Alves-Gomez, J.A. (1999) Systematic biology of gymnotiform and mormyriform electric fishes: phylogenetic relationships, molecular clocks and rates of evolution in the mitochondrial rRNA genes. J. Exp. Biol., 202: 1167–1183.

Averbeck, B.B. and Lee, D. (2004) Coding and transmission of information by neural ensembles. Trends Neurosci., 27: 225–230.

Averbeck, B.B. and Lee, D. (2006) Effects of noise correlations on information encoding and decoding. J. Neurophysiol., 95: 3633–3644.

Babineau, D., Longtin, A. and Lewis, J.E. (2006) Modeling the electric field of weakly electric fish. J. Exp. Biol., 209: 3636–3651.

Bastian, J. (1995) Pyramidal-cell plasticity in weakly electric fish: a mechanism for attenuating responses to reafferent electrosensory inputs. J. Comp. Physiol. A Sens. Neural Behav. Physiol., 176: 63–73.

Bastian, J. (1996a) Plasticity in an electrosensory system I: general features of a dynamic sensory filter. J. Neurophysiol., 76: 2483–2496.

Bastian, J. (1996b) Plasticity in an electrosensory system II: postsynaptic events associated with a dynamic sensory filter. J. Neurophysiol., 76: 2497–2507.

Bastian, J. (1998a) Modulation of calcium-dependent postsynaptic depression contributes to an adaptive sensory filter. J. Neurophysiol., 80: 3352–3355.

Bastian, J. (1998b) Plasticity in an electrosensory system III: contrasting properties of spatially segregated dendritic inputs. J. Neurophysiol., 79: 1839–1857.

Bastian, J. (1999) Plasticity of feedback inputs in the apteronotid electrosensory system. J. Exp. Biol., 202: 1327–1337.

Bastian, J. and Bratton, B. (1990) Descending control of electroreception I: properties of nucleus praeeminentialis neurons projecting indirectly to the electrosensory lateral line lobe. J. Neurosci., 10: 1226–1240.

Bastian, J., Chacron, M.J. and Maler, L. (2002) Receptive field organization determines pyramidal cell stimulus-encoding capability and spatial stimulus selectivity. J. Neurosci., 22: 4577–4590.

Bastian, J., Chacron, M.J. and Maler, L. (2004) Plastic and nonplastic pyramidal cells perform unique roles in a network capable of adaptive redundancy reduction. Neuron, 41: 767–779.

Bastian, J. and Courtright, J. (1991) Morphological correlates of pyramidal cell adaptation rate in the electrosensory lateral line lobe of weakly electric fish. J. Comp. Physiol. A Sens. Neural Behav. Physiol., 168: 393–407.

Bastian, J., Schneiderjen, S. and Nguyenkim, J. (2001) Arginine vasotocin modulates a sexually dimorphic communication behavior in the weakly electric fish, *Apteronotus leptorhynchus*. J. Exp. Biol., 204: 1909–1923.

Bauer, M., Ostenveld, R., Peeters, M. and Fries, P. (2006) Tactile spatial attention enhances gamma-band activity in somatosensory cortex and reduces low-frequency activity in parieto-occipital areas. J. Neurosci., 26: 490–501.

Bell, C. and Maler, L. (2005) Central neuroanatomy of electrosensory systems in fish. In: Bullock T.H. and Hopkins C. (Eds.), Electroreception. Springer, New York.

Benda, J., Longtin, A. and Maler, L. (2005) Spike-frequency adaptation separates transient communication signals from background oscillations. J. Neurosci., 25: 2312–2321.

Benda, J., Longtin, A. and Maler, L. (2006) A synchronization-desynchronization code for natural communication signals. Neuron, 52: 347–358.

Berman, N., Dunn, R.J. and Maler, L. (2001) Function of NMDA receptors and persistent sodium channels in a feedback pathway of the electrosensory system. J. Neurophysiol., 86: 1612–1621.

Berman, N.J., Hincke, M.T. and Maler, L. (1995) Inositol 1,4,5-trisphosphate receptor localization in the brain of a weakly electric fish (*Apteronotus leptorhynchus*) with emphasis on the electrosensory system. J. Comp. Neurol., 361: 512–524.

Berman, N.J. and Maler, L. (1998a) Inhibition evoked from primary afferents in the electrosensory lateral line lobe of the weakly electric fish (*Apteronotus leptorhynchus*). J. Neurophysiol., 80: 3173–3196.

Berman, N.J. and Maler, L. (1998b) Interaction of GABA B-mediated inhibition with voltage-gated currents of pyramidal cells: computational mechanism of a sensory searchlight. J. Neurophysiol., 80: 3197–3213.

Berman, N.J. and Maler, L. (1998c) Distal versus proximal inhibitory shaping of feedback excitation in the electrosensory lateral line lobe: implications for sensory filtering. J. Neurophysiol., 80: 3214–3232.

Berman, N.J. and Maler, L. (1999) Neural architecture of the electrosensory lateral line lobe: adaptations for coincidence detection, a sensory searchlight and frequency-dependent adaptive filtering. J. Exp. Biol., 202: 1243–1253.

Berman, N.J., Plant, J., Turner, R. and Maler, L. (1997) Excitatory amino acid transmission at a feedback pathway in the electrosensory system. J. Neurophysiol., 78: 1869–1881.

Bezdudnaya, T., Cano, M., Bereshpolova, Y., Stoelzel, C.R., Alonso, J.M. and Swadlow, H.A. (2006) Thalamic burst mode and inattention in the awake LGNd. Neuron, 49: 421–432.

Bodznick, D. and Montgomery, J.C. (2005) The physiology of low frequency electrosensory systems. In: Bullock T.H.

and Hopkins C.D. (Eds.), Electroreception. Springer, New York.

Bratton, B. and Bastian, J. (1990) Descending control of electroreception II: properties of nucleus praeeminentialis neurons projecting directly to the electrosensory lateral line lobe. J. Neurosci., 10: 1241–1253.

Bullock, T.H. and Heiligenberg, W. (1986) Electroreception. Wiley, New York.

Bullock, T.H. and Hopkins, C.D. (Eds.). (2005) Electroreception. Springer, New York.

Cardin, J.A., Palmer, L.A. and Contreras, D. (2005) Stimulus-dependent gamma (30–50 Hz) oscillations in simple and complex fast rhythmic bursting cells in primary visual cortex. J. Neurosci., 25: 5339–5350.

Carr, C.E. and Maler, L. (1986) Electroreception in gymnotiform fish: central anatomy and physiology. In: Bullock T.H. and Heiligenberg W. (Eds.), Electroreception. Wiley, New York.

Carr, C.E., Maler, L. and Sas, E. (1982) Peripheral organization and central projections of the electrosensory organs in gymnotiform fish. J. Comp. Neurol., 211: 139–153.

Chacron, M.J., Doiron, B., Maler, L., Longtin, A. and Bastian, J. (2003) Non-classical receptive field mediates switch in a sensory neuron's frequency tuning. Nature, 423: 77–81.

Chacron, M.J., Lindner, B. and Longtin, A. (2004) Noise shaping by interval correlations increases information transfer. Phys. Rev. Lett., 92: 080601.

Chacron, M.J., Longtin, A. and Maler, L. (2001) Negative interspike interval correlations increase the neuronal capacity for encoding time-dependent stimuli. J. Neurosci., 21: 5328–5343.

Chacron, M.J., Maler, L. and Bastian, J. (2005a) Electroreceptor neuron dynamics shape information transmission. Nat. Neurosci., 8: 673–678.

Chacron, M.J., Maler, L. and Bastian, J. (2005b) Feedback and feedforward control of frequency tuning to naturalistic stimuli. J. Neurosci., 25: 5521–5532.

Chance, F.S., Nelson, S.B. and Abbott, L.F. (1998) Synaptic depression and the temporal response characteristics of V1 cells. J. Neurosci., 18: 4785–4799.

Chen, L., House, J.L., Krahe, R. and Nelson, M.E. (2005) Modeling signal and background components of electrosensory scenes. J. Comp. Physiol. A Neuroethol. Sens. Neural Behav. Physiol., 191: 331–345.

Deng, Q., Rashid, A.J., Fernandez, F.R., Turner, R.W., Maler, L. and Dunn, R.J. (2005) A C-terminal domain directs Kv3.3 channels to dendrites. J. Neurosci., 25: 11531–11541.

Doiron, B., Chacron, M.J., Maler, L., Longtin, A. and Bastian, J. (2003a) Inhibitory feedback required for network oscillatory responses to communication but not prey stimuli. Nature, 421: 539–543.

Doiron, B., Laing, C., Longtin, A. and Maler, L. (2002) Ghostbursting: a novel neuronal burst mechanism. J. Comp. Neurosci., 14: 5–25.

Doiron, B., Lindner, B., Longtin, A., Maler, L. and Bastian, J. (2004) Oscillatory activity in electrosensory neurons increases with the spatial correlation of the stochastic input stimulus. Phys. Rev. Lett., 93: 048101.

Doiron, B., Longtin, A., Turner, R.W. and Maler, L. (2001) Model of gamma frequency burst discharge generated by conditional backpropagation. J. Neurophysiol., 86: 1523–1545.

Doiron, B., Noonan, L., Lemon, N. and Turner, R.W. (2003b) Persistent Na(+) current modifies burst discharge by regulating conditional backpropagation of dendritic spikes. J. Neurophysiol., 89: 324–337.

Engler, G. and Zupanc, G.K.H. (2001) Differential production of chirping behavior evoked by electrical stimulation of the weakly electric fish, *Apteronotus leptorhynchus*. J. Comp. Physiol. A, 187: 747–756.

Eurich, C.W. and Wilke, S.D. (2000) Multidimensional coding strategy of spiking neurons. Neural Comput., 12: 1519–1529.

Fernandez, F.R., Mehaffey, W.H. and Turner, R.W. (2005) Dendritic Na^+ current inactivation can increase cell excitability by delaying a somatic depolarizing afterpotential. J. Neurophysiol., 94: 3836–3848.

Fortune, E.S. and Rose, G.J. (2000) Short-term synaptic plasticity contributes to the temporal filtering of electrosensory information. J. Neurosci., 20: 7122–7130.

Fortune, E.S. and Rose, G.J. (2001) Short-term synaptic plasticity as a temporal filter. Trends Neurosci., 24: 381–385.

Fortune, E.S. and Rose, G.J. (2003) Voltage-gated Na^+ channels enhance the temporal filtering properties of electrosensory neurons in the torus. J. Neurophysiol., 90: 924–929.

Fries, P., Schroeder, J.H., Roelfsema, P.R., Singer, W. and Engel, A.K. (2002) Oscillatory neuronal synchronization in primary visual cortex as a correlate of stimulus selection. J. Neurosci., 22: 3739–3754.

Gabbiani, F., Metzner, W., Wessel, R. and Koch, C. (1996) From stimulus encoding to feature extraction in weakly electric fish. Nature, 384: 564–567.

Grubb, M.S. and Thompson, I.D. (2005) Visual response properties of burst and tonic firing in the mouse dorsal lateral geniculate nucleus. J. Neurophysiol., 93: 3224–3247.

Heiligenberg, W. (1991) Neural Nets in Electric Fish. MIT Press, Cambridge, MA.

Ishikane, H., Gangi, M., Honda, S. and Tachibana, M. (2005) Synchronized retinal oscillations encode essential information for escape behavior in frogs. Nat. Neurosci., 8: 1087–1095.

Issa, N.P., Trepel, C. and Stryker, M.P. (2000) Spatial frequency maps in cat visual cortex. J. Neurosci., 20: 8504–8514.

Izhikevich, E.M. (2000) Neural excitability, spiking, and bursting. Int. J. Bifurcat. Chaos, 10: 1171–1266.

Jones, E.G., Huntley, G.W. and Benson, D.L. (1994) Alpha calcium/calmodulin-dependent protein kinase II selectively expressed in a subpopulation of excitatory neurons in monkey sensory-motor cortex: comparison with GAD-67 expression. J. Neurosci., 14: 611–629.

Joris, P.X., Schreiner, C.E. and Rees, A. (2004) Neural processing of amplitude-modulated sounds. Physiol. Rev., 84: 541–577.

Kawasaki, M. (2005) Physiology of tuberous electrosensory systems. In: Bullock T.H. and Hopkins C.D. (Eds.), Electroreception. Springer, New York.

Krahe, R. and Gabbiani, F. (2004) Burst firing in sensory systems. Nat. Rev. Neurosci., 5: 13–23.

Kramer, B. (1999) Waveform discrimination, phase sensitivity and jamming avoidance in a wave-type electric fish. J. Exp. Biol., 202: 1387–1398.

Laing, C.R., Doiron, B., Longtin, A., Noonan, L., Turner, R.W. and Maler, L. (2003) Type I burst excitability. J. Comput. Neurosci., 14: 329–342.

Lannoo, M.J. and Lannoo, S.J. (1992) Why do electric fish swim backwards? An hypothesis based on gymnotiform behavior, interpreted through sensory constraints. Environ. Biol. Fishes, 36: 157–165.

Lannoo, M.J., Maler, L. and Tinner, B. (1989) Ganglion cell arrangement and axonal trajectories in the anterior lateral line nerve of the weakly electric fish *Apteronotus leptorhynchus* (*Gymnotiformes*). J. Comp. Neurol., 280: 331–342.

Lemon, N. and Turner, R.W. (2000) Conditional spike backpropagation generates burst discharge in a sensory neuron. J. Neurophysiol., 84: 1519–1530.

Lesica, N.A. and Stanley, G.B. (2004) Encoding of natural scene movies by tonic and burst spikes in the lateral geniculate nucleus. J. Neurosci., 24: 10731–10740.

Lewis, J.E. and Maler, L. (2001) Neuronal population codes and the perception of distance in weakly electric fish. J. Neurosci., 21: 2842–2850.

Lewis, J.E. and Maler, L. (2002a) Blurring of the senses: common cues for distance perception in diverse sensory systems. Neuroscience, 114: 19–22.

Lewis, J.E. and Maler, L. (2002b) Dynamics of electrosensory feedback: short-term plasticity and inhibition in a parallel fiber pathway. J. Neurophysiol., 88: 1695–1706.

Lewis, J.E. and Maler, L. (2004) Synaptic dynamics on different time scales in a parallel fiber feedback pathway of the weakly electric fish. J. Neurophysiol., 91: 1064–1070.

Lindner, B., Doiron, B. and Longtin, A. (2005) Theory of oscillatory firing induced by spatially correlated noise and delayed inhibitory feedback. Phys. Rev. E Stat. Nonlin. Soft Matter Phys., 72: 061919.

Lisman, J., Schulman, H. and Cline, H. (2002) The molecular basis of CaMKII function in synaptic and behavioral memory. Nat. Neurosci. Rev., 3: 175–190.

Llinas, R.R. and Steriade, M. (2006) Bursting of thalamic neurons and states of vigilance. J. Neurophysiol., 95: 3297–3308.

MacIver, M.A., Sharabash, N.M. and Nelson, M.E. (2001) Prey capture behavior in gymnotid electric fish: motion analysis and effects of water conductivity. J. Exp. Biol., 204: 543–557.

Maler, L. (1979) The posterior lateral line lobe of certain gymnotiform fish: quantitative light microscopy. J. Comp. Neurol., 183: 323–363.

Maler, L. and Hincke, M. (1999) The distribution of calcium/calmodulin-dependent kinase 2 in the brain of *Apteronotus leptorhynchus*. J. Comp. Neurol., 408: 177–203.

Maler, L. and Mugnaini, E. (1994) Correlating gamma-aminobutyric acidergic circuits and sensory function in the electrosensory lateral line lobe of a gymnotiform fish. J. Comp. Neurol., 345: 224–252.

Maler, L., Sas, E., Carr, C. and Matsubara, J. (1982) Efferent projections of the posterior lateral line lobe in a gymnotiform fish. J. Comp. Neurol., 21: 154–164.

Maler, L., Sas, E.K. and Rogers, J. (1981) The cytology of the posterior lateral line lobe of high frequency weakly electric fish (*Gymnotoidei*): dendritic differentiation and synaptic specificity in a simple cortex. J. Comp. Neurol., 195: 87–139.

Mareschal, I. and Baker Jr., C.L. (1998) A cortical locus for the processing of contrast-defined contours. Nat. Neurosci., 1: 150–154.

Mathieson, W.B. and Maler, L. (1988) Morphological and electrophysiological properties of a novel in vitro preparation: the electrosensory lateral line lobe brain slice. J. Comp. Physiol. A, 163: 489–506.

Mehaffey, W.H., Doiron, B., Maler, L. and Turner, R.W. (2005) Deterministic multiplicative gain control with active dendrites. J. Neurosci., 25: 9968–9977.

Metzner, W. and Heiligenberg, W. (1992) The coding of signals in the gymnotiform fish *Eigenmannia*: from electroreceptors to neurons in the torus semicircularis of the midbrain. J. Comp. Physiol. A Sens. Neural Behav. Physiol., 169: 135–150.

Metzner, W. and Juranek, J. (1997) A sensory brain map for each behavior? Proc. Natl. Acad. Sci. U.S.A., 94: 14798–14803.

Middleton, J.W., Longtin, A., Benda, J. and Maler, L. (2006) The cellular basis for parallel neural transmission of a high-frequency stimulus and its low-frequency envelope. Proc. Natl. Acad. Sci. U.S.A., 103: 14596–14601.

Nelson, M.E. (2005) Target detection, image analysis, and modeling. In: Bullock T.H. and Hopkins C.D. (Eds.), Electroreception. Springer, New York.

Nelson, M.E. and MacIver, M.A. (1999) Prey capture in the weakly electric fish *Apteronotus leptorhynchus*: sensory acquisition strategies and electrosensory consequences. J. Exp. Biol., 202: 1195–1203.

Nelson, M.E., Xu, Z. and Payne, J.R. (1997) Characterization and modeling of P-type electrosensory afferent responses to amplitude modulations in a wave-type electric fish. J. Comp. Physiol. A Sens. Neural Behav. Physiol., 181: 532–544.

Ninan, I. and Arancio, O. (2004) Presynaptic CaMKII is necessary for synaptic plasticity in cultured hippocampal neurons. Neuron, 42: 129–141.

Noonan, L., Doiron, B., Laing, C., Longtin, A. and Turner, R.W. (2003) A dynamic dendritic refractory period regulates burst discharge in the electrosensory lobe of weakly electric fish. J. Neurosci., 23: 1524–1534.

Oswald, A.M., Chacron, M.J., Doiron, B., Bastian, J. and Maler, L. (2004) Parallel processing of sensory input by bursts and isolated spikes. J. Neurosci., 24: 4351–4362.

Oswald, A.M., Lewis, J.E. and Maler, L. (2002) Dynamically interacting processes underlie synaptic plasticity in a feedback pathway. J. Neurophysiol., 87: 2450–2463.

Palva, J.M., Palva, S. and Kaila, K. (2005) Phase synchrony among neuronal oscillations in the human cortex. J. Neurosci., 25: 3962–3972.
Ramcharitar, J.U., Tan, E.W. and Fortune, E.S. (2006) Global electrosensory oscillations enhance directional responses of midbrain neurons in *Eigenmannia*. J. Neurophysiol., 96: 2319–2326.
Rashid, A.J., Morales, E., Turner, R.W. and Dunn, R.J. (2001) The contribution of dendritic Kv3 K^+ channels to burst threshold in a sensory neuron. J. Neurosci., 21: 125–135.
Ratnam, R. and Nelson, M.E. (2000) Non-renewal statistics of electrosensory afferent spike trains: implications for the detection of weak sensory signals. J. Neurosci., 20: 6672–6683.
Ronan, M. (1986) Electroreception in cyclostomes. In: Bullock T.H. and Heiligenberg W. (Eds.), Electroreception. Wiley, New York.
Rose, G.J. and Fortune, E.S. (1999a) Frequency-dependent PSP depression contributes to low-pass temporal filtering in *Eigenmannia*. J. Neurosci., 19: 7629–7639.
Rose, G.J. and Fortune, E.S. (1999b) Mechanisms for generating temporal filters in the electrosensory system. J. Exp. Biol., 202: 1281–1289.
Sas, E. and Maler, L. (1983) The nucleus praeeminentialis: a golgi study of a feedback center in the electrosensory system of gymnotid fish. J. Comp. Neurol., 221: 127–144.
Sas, E. and Maler, L. (1987) The organization of afferent input to the caudal lobe of the cerebellum of the gymnotid fish *Apteronotus leptorhynchus*. Anat. Embryol., 177: 55–79.
Saunders, J. and Bastian, J. (1984) The physiology and morphology of two classes of electrosensory neurons in the weakly electric fish *Apteronotus leptorhynchus*. J. Comp. Physiol. A, 154: 199–209.
Series, P., Latham, P.E. and Pouget, A. (2004) Tuning curve sharpening for orientation selectivity: coding efficiency and the impact of correlations. Nat. Neurosci., 7: 1129–1135.
Shamir, M. and Sompolinsky, H. (2004) Nonlinear population codes. Neural Comput., 16: 1105–1136.
Shumway, C. (1989a) Multiple electrosensory maps in the medulla of weakly electric gymnotiform fish I: physiological differences. J. Neurosci., 9: 4388–4399.
Shumway, C. (1989b) Multiple electrosensory maps in the medulla of weakly electric gymnotiform fish II: anatomical differences. J. Neurosci., 9: 4400–4415.
Shumway, C.A. and Maler, L. (1989) GABAergic inhibition shapes temporal and spatial response properties of pyramidal cells in the electrosensory lateral line lobe of gymnotiform fish. J. Comp. Physiol. A, 164: 391–407.
Sompolinsky, H., Yoon, H., Kang, K. and Shamir, M. (2001) Population coding in neuronal systems with correlated noise. Phys. Rev. E, 64: 051904-1-051904-10.
Tan, E.W., Nizar, J.M., Carrera, G.E. and Fortune, E.S. (2005) Electrosensory interference in naturally occurring aggregates of a species of weakly electric fish, *Eigenmannia virescens*. Behav. Brain Res., 164: 83–92.
Taylor, K., Mandon, S., Freiwald, W.A. and Kreiter, A.K. (2005) Coherent oscillatory activity in monkey area v4 predicts successful allocation of attention. Cereb. Cortex, 15: 1424–1437.
Triefenbach, F. and Zakon, H. (2003) Effects of sex, sensitivity and status on cue recognition in the weakly electric fish *Apteronotus leptorhynchus*. Anim. Behav., 65: 19–28.
Turner, R.W., Lemon, N., Doiron, B., Rashid, A.J., Morales, E., Longtin, A., Maler, L. and Dunn, R.J. (2002) Oscillatory burst discharge generated through conditional backpropagation of dendritic spikes. J. Physiol. Paris, 96: 517–530.
Turner, R.W., Maler, L. and Burrows, M. (1999) Electroreception and electrocommunication. J. Exp. Biol., 202.
Turner, R.W., Maler, L., Deerinck, T., Levinson, S.R. and Ellisman, M.H. (1994) TTX-sensitive dendritic sodium channels underlie oscillatory discharge in a vertebrate sensory neuron. J. Neurosci., 14: 6453–6471.
Vinje, W.E. and Gallant, J.L. (2002) Natural stimulation of the nonclassical receptive field increases information transmission efficiency in v1. J. Neurosci., 22: 2904–2915.
Wang, D. and Maler, L. (1997) In vitro plasticity of the direct feedback pathway in the electrosensory system of *Apteronotus leptorhynchus*. J. Neurophysiol., 78: 1882–1889.
Wang, D. and Maler, L. (1998) Differential role of Ca^{2+}/calmodulin-dependent kinases in posttetanic potentiation at input selective glutamatergic pathways. Proc. Natl. Acad. Sci. U.S.A., 95: 7133–7138.
Wilke, S.D. and Eurich, C.W. (2002) Representational accuracy of stochastic neural populations. Neural Comput., 14: 155–189.
Wilkens, L.A. and Hofman, M.H. (2005) Behavior of animals with passive, low frequency electrosensory systems. In: Bullock T.H. and Hopkins C.D. (Eds.), Electroreception. Springer, New York.
Womelsdorf, T., Fries, P., Mitra, P.P. and Desimone, R. (2006) Gamma-band synchronization in visual cortex predicts speed of change detection. Nature, 439: 733–736.
Xu, Z., Payne, J.R. and Nelson, M.E. (1996) Logarithmic time course of sensory adaptation in electrosensory afferent nerve fibers in a weakly electric fish. J. Neurophysiol., 76: 2020–2032.
Zakon, H., Oestreich, J., Tallarovic, S. and Triefenbach, F. (2002) EOD modulations of brown ghost electric fish: JARs, chirps, rises, and dips. J. Physiol. Paris, 96: 451–458.
Zhang, K. and Sejnowski, T.J. (1999) Neuronal tuning: to sharpen or broaden. Neural Comput., 11: 75–84.
Zupanc, G.K. (2002) From oscillators to modulators: behavioral and neural control of modulations of the electric organ discharge in the gymnotiform fish, *Apteronotus leptorhynchus*. J. Physiol. Paris, 96: 459–472.
Zupanc, G.K.H., Airey, J.A., Maler, L., Sutko, J.L. and Ellisman, M.H. (1992) Immunohistochemical localization of ryanodine binding protein in the central nervous system of gymnotiform fish. J. Comp. Neurol., 325: 135–151.

P. Cisek, T. Drew & J.F. Kalaska (Eds.)
Progress in Brain Research, Vol. 165
ISSN 0079-6123
Copyright © 2007 Elsevier B.V. All rights reserved

CHAPTER 10

Coordinate transformations and sensory integration in the detection of spatial orientation and self-motion: from models to experiments

Andrea M. Green[1,*] and Dora E. Angelaki[2]

[1]*Département de Physiologie, Université de Montréal, 2960 Chemin de la Tour, Rm 2140, Montreal, QC, H3T 1J4, USA*
[2]*Department of Anatomy and Neurobiology, Box 8108, Washington University School of Medicine, 660 South Euclid Avenue, St. Louis, MO 63110, USA*

Abstract: An accurate internal representation of our current motion and orientation in space is critical to navigate in the world and execute appropriate action. The force of gravity provides an allocentric frame of reference that defines one's motion relative to inertial (i.e., world-centered) space. However, movement in this environment also introduces particular motion detection problems as our internal linear accelerometers, the otolith organs, respond identically to either translational motion or changes in head orientation relative to gravity. According to physical principles, there exists an ideal solution to the problem of distinguishing between the two as long as the brain also has access to accurate internal estimates of angular velocity. Here, we illustrate how a nonlinear integrative neural network that receives sensory signals from the vestibular organs could be used to implement the required computations for inertial motion detection. The model predicts several distinct cell populations that are comparable with experimentally identified cell types and accounts for a number of previously unexplained characteristics of their responses. A key model prediction is the existence of cell populations that transform head-referenced rotational signals from the semicircular canals into spatially referenced estimates of head reorientation relative to gravity. This chapter provides an overview of how addressing the problem of inertial motion estimation from a computational standpoint has contributed to identifying the actual neuronal populations responsible for solving the tilt-translation ambiguity and has facilitated the interpretation of neural response properties.

Keywords: vestibular; coordinate transformation; eye movement; reference frame; integrator; sensorimotor; sensory ambiguity; spatial motion

Introduction

To orient and navigate in the world the brain must process available sensory cues to construct an internal representation of inertial motion, in which the force of gravity provides a fixed reference. The construction of such a representation poses a computational challenge for two reasons. First, sensors typically encode information in egocentric (i.e., body-referenced) coordinate frames, rather than a common, allocentric (inertial) reference frame. As a result, implicit or explicit reference frame transformations are necessary for accurate

*Corresponding author. Tel.: +1 514 343 6111 Ext. 3301; Fax: +1 514 343 2111; E-mail: andrea.green@umontreal.ca

DOI: 10.1016/S0079-6123(06)65010-3

spatial motion estimation (Andersen et al., 1993). Second, an individual sensor often encodes an ambiguous representation of a physical stimulus. To resolve these ambiguities the brain typically must rely on computations that involve the integration of information from multiple sensory sources or multiple processing channels.

A well-known example of such sensory ambiguities arises in visual motion detection where the limited extent of sensory receptive fields gives rise to uncertainty in detecting the motion direction of an untextured contour moving within a small aperture (i.e., "the aperture problem," Movshon et al., 1985; Shimojo et al., 1989; Pack and Born, 2001). Similarly, in sound localization, a "phase ambiguity" arises in neurons with sharp frequency tuning that respond equally well to sounds with a particular interaural difference and their phase equivalents (Mazer, 1998). Another example of sensory ambiguity in spatial processing exists in the vestibular system. As in any man-made inertial guidance system, information about linear and angular accelerations is provided by distinct sets of sensors. The otolith organs function as *linear* accelerometers (Fernández and Goldberg, 1976a, b; Angelaki and Dickman, 2000), whereas the semicircular canals act as integrating *angular* accelerometers that provide the brain with an estimate of rotational head velocity (Fernández and Goldberg, 1971). Everyday activities typically give rise to both rotations and translations of the head, implying the integration of signals from both sets of sensors. However, the specific manner in which these signals combine to estimate spatial orientation and self-motion is far from trivial because of both the reference frame and sensory ambiguity problems outlined above.

In particular, the "reference frame problem" arises because the vestibular sensors are physically fixed in the head. The three sets of roughly orthogonal semicircular canals thus provide the brain with three-dimensional (3D) estimates of angular velocity in a head-centered reference frame. However, they provide no information about how the head moves relative to the outside world. Rotation about an axis aligned with the body (yaw-axis), for example, stimulates the horizontal semicircular canals in an identical fashion regardless of body orientation relative to gravity. The situation in the case of the otolith organs is even more complicated because not only are they fixed in the head but, as is true of any linear accelerometer, they respond equivalently to inertial (i.e., translational) and gravitational accelerations (Einstein's equivalence principle; Einstein, 1908). Thus, otolith afferents provide inherently ambiguous sensory information, as the encoded acceleration could have been generated during either actual translation or a head reorientation relative to gravity (Fernández and Goldberg, 1976a, b). For example, a head displacement to the right activates otolith afferents in an equivalent fashion to a leftward roll tilt from an upright to a left-ear-down orientation. Inertial motion detection in a gravitational environment thus requires a resolution of the "tilt-translation ambiguity" that exists in the interpretation of sensory otolith signals. Furthermore, as will be outlined in more detail in subsequent sections, the reference frame problem is intimately related to resolution of the otolith tilt-translation ambiguity such that the two problems are interdependent (e.g., see Green and Angelaki, 2004; Green et al., 2005; Yakusheva et al., 2007).

Under impoverished sensory conditions perceptual illusions do occur. For example, under conditions where only vestibular sensory cues are available, low-frequency translational stimuli (e.g., <0.1 Hz) are often misperceived as tilts in both ocular and perceptual responses (Seidman et al., 1998; Paige and Seidman, 1999) giving rise to the oculogravic and somatogravic illusions that are well known to aircraft pilots and astronauts (Graybiel et al., 1979; Clement et al., 2001). Nevertheless, under most circumstances the tilt-translation ambiguity is solved as evident from the fact that behavioral responses to tilts and translation are different. In the oculomotor system, for example, lateral translation elicits horizontal eye movements (Paige and Tomko, 1991; Schwarz and Miles, 1991; Telford et al., 1997; Angelaki, 1998) whereas roll tilt generates mainly ocular torsion (Crawford and Vilis, 1991; Seidman et al., 1995). Similarly, under conditions where the otoliths and canals provide accurate estimates of linear acceleration and angular velocity, respectively, tilts and translations are appropriately distinguished in perceptual responses (Merfeld et al., 2005a, b).

Conclusions from behavioral studies

Like other sensory ambiguities, the solution to the tilt-translation problem relies on neural computations involving the integration of multiple sensory cues. Insight as to how the problem is solved has been provided by several behavioral investigations that demonstrated a key role for sensory signals from the semicircular canals (Merfeld and Young, 1995; Angelaki et al., 1999; Merfeld et al., 1999, 2001; Zupan et al., 2000; Green and Angelaki, 2003; for a review see Green et al., 2005). Figure 1 illustrates an example from one such study that employed combinations of transient translation and tilt stimuli that were chosen to explicitly dissociate otolith and canal sensory contributions to horizontal eye movement responses (Green and Angelaki, 2003). These stimuli, all provided in complete darkness with the head fixed rigidly relative to the body (to avoid neck proprioceptive contributions), included lateral translation (*Translation only*), roll tilt (*Roll tilt only*) and simultaneous combinations of the two motions (*Translation + Roll tilt* and *Translation−Roll tilt* motions; Fig. 1, top). Importantly, the translation stimulus was adjusted such that the interaural (y-axis) net acceleration profile during the first 600 ms of motion closely matched that induced by the head reorientation relative to gravity (compare net acceleration traces in columns 1 and 2 of

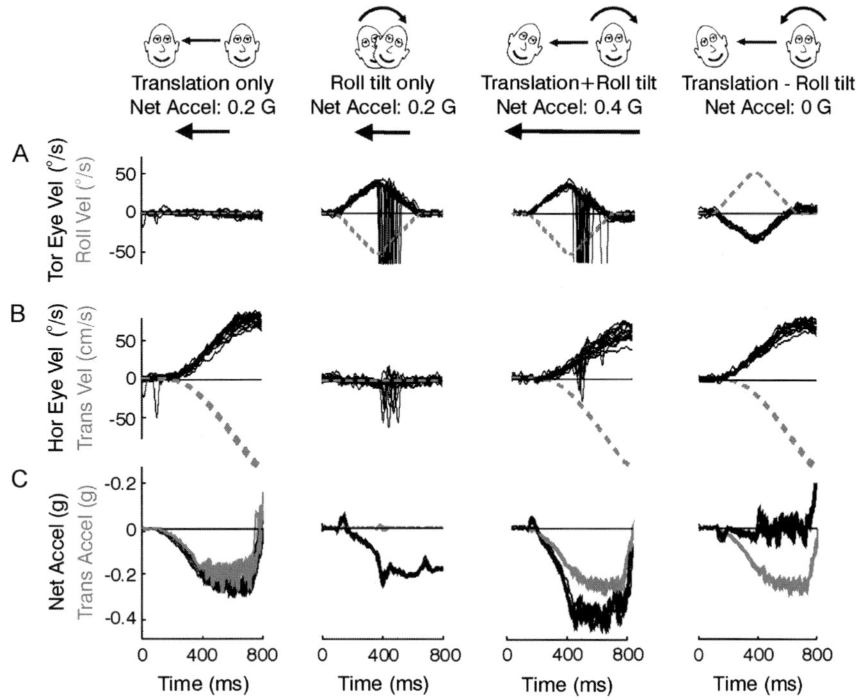

Fig. 1. Eye movement responses of a rhesus macaque monkey to combinations of transient roll tilt and lateral translation motions. (A) Torsional and (B) horizontal eye velocities for multiple trials of *Translation only*, *Roll tilt only*, *Translation + Roll tilt,* and *Translation−Roll tilt* motions. Dashed gray traces in (A) and (B) represent head roll velocity and the translational velocity of the linear sled on which the animal was mounted, respectively. (C) The translational acceleration of the sled (gray traces) is superimposed on the net interaural acceleration measured by a linear accelerometer mounted on the animal's head (black traces). Note that torsional eye movements represent the compensatory response to roll rotation (rotational vestibulo-ocular reflex; RVOR) whereas horizontal eye movements represent the reflexive response to lateral translation (translational vestibulo-ocular reflex; TVOR). Large deviations in eye velocity represent fast phases (saccades). Adapted with permission from Green and Angelaki (2003). Copyright 2003 by the Society for Neuroscience.

Fig. 1C). Thus, the net interaural acceleration stimulus to the otoliths was similar during either roll tilt or translation. Notice that whereas *Translation only* motion elicited large reflexive horizontal eye movements appropriate to compensate for the lateral motion (i.e., the translational vestibulo-ocular reflex, TVOR; Fig. 1B, column 1) such large horizontal eye movements were not elicited during *Roll tilt* (Fig. 1B, column 2). Instead torsional eye movements were evoked to compensate for the head rotation (i.e., rotational vestibulo-ocular reflex, RVOR; Fig. 1A, column 2). Thus, despite a similar interaural stimulus to the otoliths in both cases, the brain was able to resolve the otolith sensory ambiguity and detect the appropriate motion.

During the combined stimuli, the translational acceleration combined with the gravitational acceleration elicited by the roll tilt in either an additive or subtractive fashion, resulting in either a doubled (*Translation + Roll tilt*) or close to zero (*Translation−Roll tilt*) net interaural acceleration stimulus to the otoliths. The *Translation−Roll tilt* stimulus is of particular relevance since in this case the body translated in space but there was no net interaural acceleration stimulus (Fig. 1C, column 4). Thus, if otolith cues alone contributed to driving the TVOR there should be no compensatory response to the translational motion. Despite the absent otolith stimulus, horizontal eye velocity responses similar to those during the pure translation were elicited, providing evidence for an extra-otolith contribution to translational motion estimation (Fig. 1B, compare columns 1 and 4). In the absence of other nonvestibular sensory cues, this extra-otolith contribution must come from the semicircular canals. This supposition has been explicitly confirmed by the observation that such horizontal eye movements during *Translation−Roll tilt* motion are absent in canal-plugged animals (Angelaki et al., 1999). Furthermore, quantitative analyses of the horizontal eye movement profiles illustrated in Fig. 1 showed that the extra-otolith driven TVOR was best correlated with angular head *position*, suggesting that the sensorimotor processing involved a temporal integration of angular velocity signals from the semicircular canals (Green and Angelaki, 2003).

Collectively, such behavioral investigations have confirmed that the semicircular canals play a critical role in inertial motion estimation and that the processing of these signals involves a temporal integration. Furthermore, detailed computational modeling studies have illustrated the ability to interpret various complex aspects of 3D eye movements in the context of the tilt-translation discrimination problem (Merfeld, 1995; Glasauer and Merfeld, 1997; Merfeld and Zupan, 2002; Zupan et al., 2002; Zupan and Merfeld, 2005). However, until recently, progress in understanding how the required computations are implemented neurophysiologically has been limited. In particular, although the response properties of motion-sensitive neurons in the vestibular nuclei (Baker et al., 1984a, b; Schor et al., 1984, 1985; Kasper et al., 1988; Wilson et al., 1990, 1996; Angelaki and Dickman, 2000; Brettler and Baker, 2001; Dickman and Angelaki, 2002; Chen-Huang and Peterson, 2006; Yakushin et al., 2006; Zhou et al., 2006) and cerebellum (Siebold et al., 1997, 1999, 2001; Zhou et al., 2001; Shaikh et al., 2005a, b) have been previously characterized during rotation and/or translation, the interpretation of what exactly these cells encode has suffered from a lack of understanding of exactly what types of response properties experimentalists should be looking for and what types of paradigms should be used to isolate these properties.

Recently, the theoretical implications of solving the inertial motion estimation problem have been reconsidered from a more physiologically relevant perspective through the development of a computational model that emphasizes the predicted response properties of the neuronal elements that compute the solution (Green and Angelaki, 2004). The goal in this chapter is to provide an overview of how the development of such a computational model has contributed to identifying the experimental approaches required to isolate the neuronal populations responsible for solving the tilt-translation ambiguity and has facilitated interpretation of neural response properties.

Theory

Otolith afferents provide the brain with *net* acceleration, $\boldsymbol{\alpha}$, which is the vectorial combination of translational, **t**, and gravitational, **g**, components (Fernández and Goldberg, 1976a, b):

$$\boldsymbol{\alpha} = \mathbf{t} - \mathbf{g} \tag{1}$$

According to physical principles, there exists an ideal solution to the problem of distinguishing between translational motion and reorientations relative to gravity, as long as the brain also has access to an accurate internal estimate of angular velocity. Specifically, the solution relies on the idea that angular rotation, $\boldsymbol{\omega}$, can be used to compute a dynamic estimate of gravity, **g**, which when combined with linear acceleration, $\boldsymbol{\alpha}$, yields translational acceleration, **t** [Eq. (1)]. In particular, when the head reorients relative to gravity an independent dynamic estimate of **g** can be obtained using information about angular head velocity (derived from vestibular, visual and/or proprioceptive sensory cues) to keep track of changes in orientation of the gravity vector relative to the head. Specifically, the rate of change of the gravity vector in *head-centered* coordinates can be described by the following vector differential equation (e.g., Goldstein, 1980):

$$\frac{d\mathbf{g}}{dt} = -\boldsymbol{\omega} \times \mathbf{g} \tag{2}$$

where **g** and $\boldsymbol{\omega}$ are vector representations of gravity and angular velocity, respectively, and $\times$ denotes a vector cross-product. Solving the vector differential Eq. (1) yields a dynamic estimate of gravitational acceleration, $\mathbf{g} = -\int \boldsymbol{\omega} \times \mathbf{g}\, dt$, (assuming known initial conditions for **g** that can be derived from static otolith cues; e.g., Green and Angelaki, 2004; Green et al., 2005). The translational acceleration component, **t**, can subsequently be obtained by substituting this gravity estimate into Eq. (1):

$$\mathbf{t} = \boldsymbol{\alpha} - \int \boldsymbol{\omega} \times \mathbf{g}\, dt \tag{3}$$

To gain insight into the implications of computing a neural solution to Eq. (3), the vector cross product can be expanded to yield the individual vector components $\mathbf{t} = (t_x, t_y, t_z)$ in 3D:

$$t_x\mathbf{i} = \alpha_x\mathbf{i} - \int [(\omega_y g_z - \omega_z g_y)\mathbf{i}\, dt$$
$$t_y\mathbf{j} = \alpha_y\mathbf{j} - \int [(\omega_z g_x - \omega_x g_z)\mathbf{j}\, dt$$
$$t_z\mathbf{k} = \alpha_z\mathbf{k} - \int [(\omega_x g_y - \omega_y g_x)\mathbf{k}\, dt \tag{4}$$

where **i**, **j** and **k** are unit vectors along the x (naso-occipital, NO), y (interaural, IA) and z (dorsoventral, DV) head axes that define a right-handed coordinate system in a head-centered reference frame with forward, leftward and upward directions considered positive. Accordingly, angular head deviations about the x, y, and z axes (roll, pitch and yaw rotations, respectively) are positive for clockwise, downward and leftward rotations (from the subjective viewpoint; Fig. 2, top).

Equations (3) and (4) are perhaps better appreciated by their schematic representation in Fig. 2, which emphasizes several key points. First, the neural elements involved in resolving the tilt-translation ambiguity should exhibit a convergence of sensory information about both linear acceleration (α_x, α_y, α_z; e.g., otolith-derived signals) and angular velocity (ω_x, ω_y, ω_z; e.g., semicircular canal-derived signals). Second, the required computations involve a temporal integration of processed angular velocity information suggesting that the neural networks involved in performing the computations may act as distributed neural integrators. Third, calculation of the instantaneous translational acceleration along any given axis requires estimates of the components of gravitational acceleration along the two other axes, illustrating the 3D nature of the problem (i.e., coupled integrative networks in Fig. 2). Finally, because the computations involve multiplicative interactions between head-referenced internal estimates of angular velocity and gravitational acceleration, a nonlinear processing of sensory signals is required to implement the solution.

A number of theoretical studies have proposed models for tilt-translation discrimination that

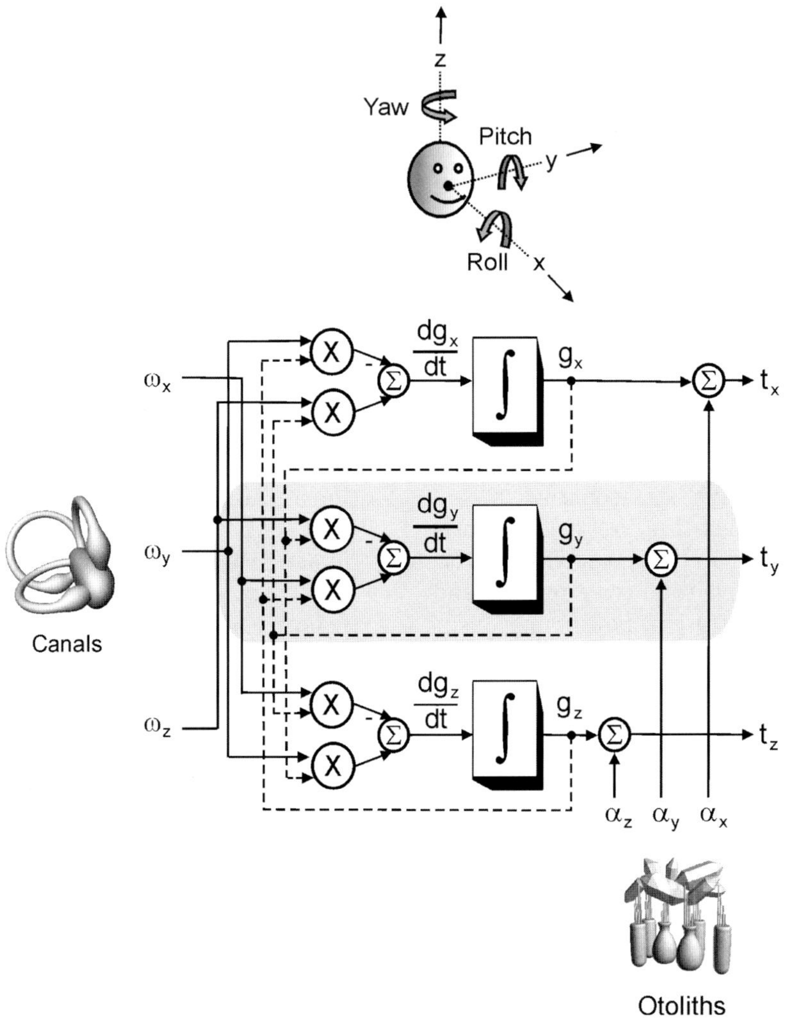

Fig. 2. Schematic illustration of the computations necessary to calculate translational acceleration, **t** (see Eqs. (3) and (4)). The computations involve using angular velocity estimates **ω** (e.g., provided by the semicircular canals; left) to compute the gravitational component of acceleration, **g**, in a head-fixed reference frame as the head reorients relative to gravity (i.e., $\mathbf{g} = -\int \boldsymbol{\omega} \times \mathbf{g}$). This estimate of **g** is then combined with the net gravito-inertial acceleration signal, **α** (provided by the otoliths; bottom) to calculate translational acceleration, **t**. Notice that calculation of the gravitational acceleration along any particular axis depends on the components of gravitational acceleration along the two other axes, illustrating the requirement for coupled integrative networks in 3D. The shaded area indicates the computations involved in estimating the translational acceleration component along the interaural axis, t_y, that are approximated in the model implementation of Fig. 3. The top inset illustrates the right-handed, head-centered coordinate frame assumed throughout the text. Thick arrows represent positive rotation directions while thin arrows represent positive translation directions. (Adapted from Green and Angelaki, 2004, used with permission.)

incorporate the idea that the brain uses sensory estimates of angular rotation and linear acceleration to create an internal representation or "internal model" of the solution to Eq. (3) (Merfeld, 1995; Glasauer and Merfeld, 1997; Angelaki et al., 1999; Mergner and Glasauer, 1999; Bos and Bles, 2002; Merfeld and Zupan, 2002; Zupan et al., 2002; Green and Angelaki, 2004; Zupan and Merfeld 2005; Holly et al., 2006). However, the majority of these models are abstract and do not easily provide an intuition about how individual neurons might be implementing these vector computations. To gain further insight into what should be expected from neural responses, it is helpful to simplify the problem and its solution. In particular, let's consider the otolith sensory ambiguity for tilts and translations along the interaural axis (i.e., y-axis associated with unit vector **j**; shaded region of Fig. 2) and the computation of

$$t_y = \alpha_y + g_y = \alpha_y - \int [(\omega_z g_x - \omega_x g_z) \mathrm{d}t$$
$$= \alpha_y - \int \omega_z(t_x - \alpha_x) - \omega_x(t_z - \alpha_z) \mathrm{d}t \quad (5)$$

Although, in general, calculation of translational acceleration along the interaural axis (y-axis) requires estimates of the gravitational acceleration components along the two other axes (i.e., g_x and g_z), the solution can be simplified further if we decouple computation of t_y from its dependency on g_x and g_z by limiting consideration to conditions where x- and z-axis translations are small (i.e., $t_x = t_z \approx 0$). Under these conditions (where $g_x \approx -\alpha_x$ and $g_z \approx -\alpha_z$) t_y can be approximated as:

$$t_y = \alpha_y + g_y \approx \alpha_y + \int [\omega_z \alpha_x - \omega_x \alpha_z] \mathrm{d}t \quad (6)$$

According to Eq. (6), the translational acceleration component along the interaural axis can be computed by combining the net otolith interaural acceleration signal with temporally integrated canal estimates of yaw and roll head velocities that have been premultiplied by the net instantaneous accelerations sensed by otolith afferents along the naso-occipital and dorsoventral axes, respectively. For the restricted conditions we consider here, the computation in Eq. (6) thus depends strictly on available sensory signals. In the following section we describe a model (Fig. 3) that implements this simplified equation, with the goal of illustrating several fundamental predictions with respect to the expected properties of neural populations that participate in inertial motion detection. Note that the simplified model that will be presented can easily be extended fully to 3D by appropriately interconnecting the network illustrated in Fig. 3 for extracting translational motion along the interaural (y) axis with similar networks for the naso-occipital (x) and dorsoventral (z) axes.

Model description

Any model implementing Eq. (6) must include: (1) a central neural integration of canal signals and (2) a head-orientation-dependent coupling between canal and otolith-derived sensory information. Thus, while the model structure illustrated in Fig. 3 represents only one of many possible integrative networks that could perform the required computations, the conclusions reached by examining its predictions have general applicability for interpreting neural response properties.

Inputs to the model of Fig. 3 include yaw and roll head velocities, ω_z and ω_x, sensed mainly by the horizontal and vertical semicircular canals, respectively, and the interaural acceleration, α_y, sensed by the otoliths (mainly the utricles, the otolith organs which sense linear accelerations in the horizontal head plane). The model incorporates 5 cell types (indicated by circles; V1–V5) that we might predict to be found in a network that computes a solution to the inertial motion detection problem. These include cell populations that encode head rotation (e.g., V1 and V2), cells that extract information about either head translation (e.g., V3) or tilt (e.g., V5) and cells that encode an intermediate combination of these signals (e.g., V4). Each circle represents the mean firing rate of a neural population and is treated mathematically in the model as a simple summing junction.

Boxes in the schematic represent dynamic elements or filters that can be represented in the Laplace domain (where s denotes the complex Laplace variable). These include first-order dynamic approximations of the semicircular canals, $C(s) = T_c s/(T_c s + 1)$, (Fernández and Goldberg, 1971) and the otolith organs, $O(s) = 1/(T_o s + 1)$, (Fernández and Goldberg, 1976b) as well as the neural filter, $C_{\mathrm{LP}}(s)$, which represents a low-pass internal model of the semicircular canals $[C_{\mathrm{LP}}(s) = 1/(T_c s + 1)]$. The parameters associated with each pathway represent the strength or weight of the projection. A key feature of the model is that the neural populations are interconnected

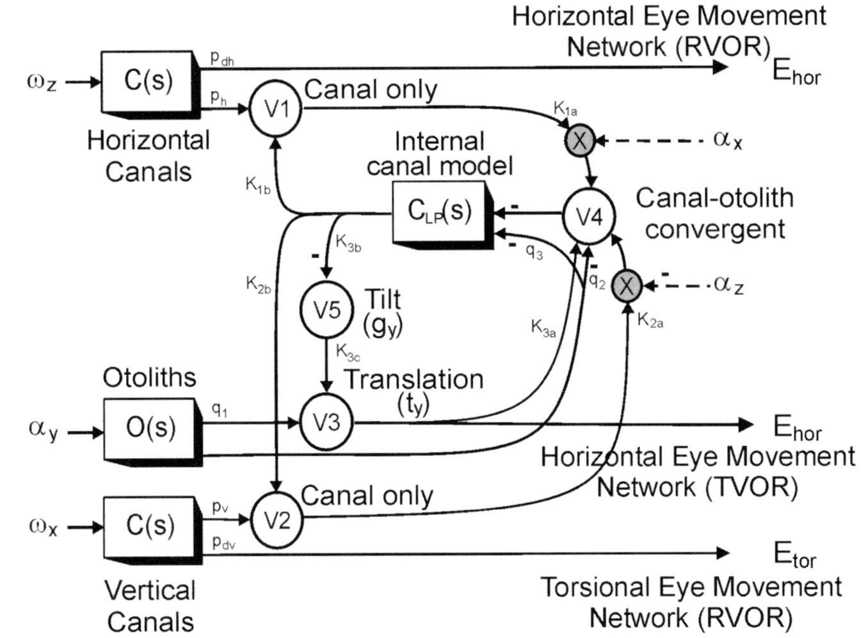

Fig. 3. Proposed model for estimating translational acceleration along the interaural (y) axis. (A) Circles labeled V1 through V5 represent different populations of motion-sensitive cells presumed to be located in the vestibular nuclei and cerebellum. Inputs to the model include yaw and roll velocities, ω_z and ω_x, sensed by the horizontal and vertical canals [$C(s) = T_c s/(T_c s{+}1)$], respectively, and interaural acceleration, α_y, sensed by the otolith organs [$O(s) = 1/(T_o s+1)$]. The box labeled $C_{LP}(s)$ represents a low-pass internal model of the semicircular canals [$C_{LP}(s) = 1/(T_c s+1)$]. X's indicate a multiplicative modulation in the strengths of the projections from cells V1 and V2 onto cell V4 by naso-occipital and dorsoventral accelerations, α_x and α_z, respectively. Parameters associated with different pathways represent the strength or weight of the projection. Inputs ω_z and ω_x are in units of deg/s, α_y is in units of cm/s^2 and α_z and α_x are in units of G (G = 9.81 cm/s^2). To simulate the compensatory eye movements (RVOR: rotational vestibulo-ocular reflex; TVOR: translational vestibulo-ocular reflex) that would be generated for different motion stimuli, the outputs of the network were conveyed to simple implementations of the premotor networks for generating horizontal and torsional eye movements (E_{hor} and E_{tor}) that have been previously described in detail (Green and Galiana 1998; Angelaki et al., 2001; Green and Angelaki, 2004). Model parameters are: $p_{dh} = 1$, $p_h = 2.568$, $p_{dv} = 0.56$, $p_v = 2.568$, $q_1 = 0.25$, $q_2 = 0.22$, $q_3 = 0.1975$, $K_{1a} = 1$, $K_{2a} = 1$, $K_{1b} = 0.1$, $K_{2b} = 0.1$, $K_3 = 0.061$, $K_{3b} = 1.6$, $K_{3c} = 6.25$. (Adapted from Green and Angelaki, 2004, used with permission.)

in dominantly positive feedback loops with the low-pass filter, $C_{LP}(s)$, to form a distributed neural integrator as required to compute a solution to Eq. (6). The required head-orientation-dependent sensory coupling [e.g., see Eq. (6)] is implemented in cell population V4 by modulating semicircular canal-derived signals as a function of the two orthogonal accelerations, α_x and α_z, also presumed to be provided by the otoliths (i.e., multiplicative interactions denoted by X's on cell V4).

The main model output is an estimate of translational acceleration, encoded by cell population V3, that can subsequently be used for both motor and perceptual purposes. To illustrate that the model can reproduce appropriate behavioral responses, we assume that this signal is conveyed to premotor eye movement networks that generate the TVOR. Similar projections from canal signals are presumed to drive the RVORs (see Green and Angelaki, 2004, for details).

Focusing our discussion on mid-high frequencies (>0.1 Hz) where the canals provide an accurate estimate of angular head velocity, the basic functioning of the model can be briefly summarized by considering the expected responses of neurons V3 and V5 for the chosen parameter set (see Fig. 3 caption). In particular, because the combination of otolith- and canal-derived sensory signals that converge on cell V4 appears temporally integrated at the output of the neural filter,

$C_{LP}(s)$, the response of cell V5 can be approximated as (see Green and Angelaki, 2004, for details):

$$V5 \approx G_{V5} \int (\omega_z \alpha_x - \omega_x \alpha_z)\,dt \qquad (7)$$

where G_{V5} is a static gain term that is a function of model parameters. Comparison of Eqs. (7) and (6) illustrates that cell V5 encodes a scaled estimate of gravitational acceleration, g_y. The network solution to Eq. (6) arises from cell V3, which performs the addition implied by Eq. (1) to extract the translational acceleration, t_y:

$$\begin{aligned} V3 &\approx q_1\alpha_y + K_{3c}V5 \\ &\approx q_1\left[\alpha_y + \frac{K_{3c}G_{V5}}{q_1}\int(\omega_z\alpha_x - \omega_x\alpha_z)\,dt\right] \\ &\approx q_1(\alpha_y + g_y) \approx q_1 t_y\left(\text{for } \frac{K_{3c}G_{V5}}{q_1} \approx 1\right) \qquad (8) \end{aligned}$$

Note that Eqs. (7) and (8) represent high-frequency, small angle approximations to the more general dynamic expressions for cells V5 and V3. Further details of the analytic descriptions of cell response dynamics and the criteria for choosing model parameters can be found in Green and Angelaki (2004). Next we describe simulations of the model output (e.g., eye movement and predicted responses of model neurons).

Model predictions

Simulated behavioral responses to tilt-translation combinations

To evaluate the ability of the model to resolve the tilt-translation ambiguity and reproduce responses consistent with experimental observations we begin by considering the eye velocity responses (Fig. 4) that would be predicted for the lateral translation (*Translation only*), roll tilt (*Roll tilt only*), and combined lateral translation and roll tilt motion stimuli (*Roll tilt + Translation* motion and *Roll tilt−Translation* motion) illustrated in Fig. 1. However, in keeping with the fact that neural response properties have been considered most frequently for sinusoidal motion stimulation we have used sinusoidal rather than transient stimuli in our simulations here. In the simulations, the interaural acceleration (α_y) stimulus to the otoliths during each of the sinusoidal *Translation* and *Roll tilt* motions was set to a peak of 0.2 g at 0.5 Hz (Fig. 4, bottom row).

The simulated eye movement responses illustrated in Fig. 4 are compatible with the experimentally observed results in Fig. 1 (also see Angelaki et al., 1999). Specifically, compensatory horizontal eye velocity responses (TVOR) of similar amplitude are predicted during all movements that include a translational motion component, while torsional eye velocity responses (RVOR) are elicited whenever the movement includes a roll component. Notice, in particular, that an appropriate TVOR is elicited for the *Roll tilt−Translation* motion when the interaural acceleration stimulus to otoliths is absent and the compensatory response must be driven by extra-otolith signals of semicircular canal origin. In contrast, no horizontal TVOR is predicted during pure roll tilt despite an interaural acceleration stimulus to the otoliths that is identical to that during pure translation (Fig. 4, compare columns 1 and 2). Thus, the model is able to resolve the otolith sensory ambiguity to distinguish between translational motion and head reorientations relative to gravity.

Predicted neural responses

Experimental background

Given a model that predicts appropriate behavioral responses the question is then what types of response properties are predicted for the neural populations that perform the required computations. To better appreciate the relevance of examining these predictions, it is worth first reviewing the types of problems encountered in interpreting experimental observations. At the time the model was developed, most neurophysiological investigations had examined vestibular and cerebellar neural response properties for a limited set of experimental protocols that included purely rotational and translational motions in an upright orientation. Such investigations typically revealed a broad array of signaling properties that were

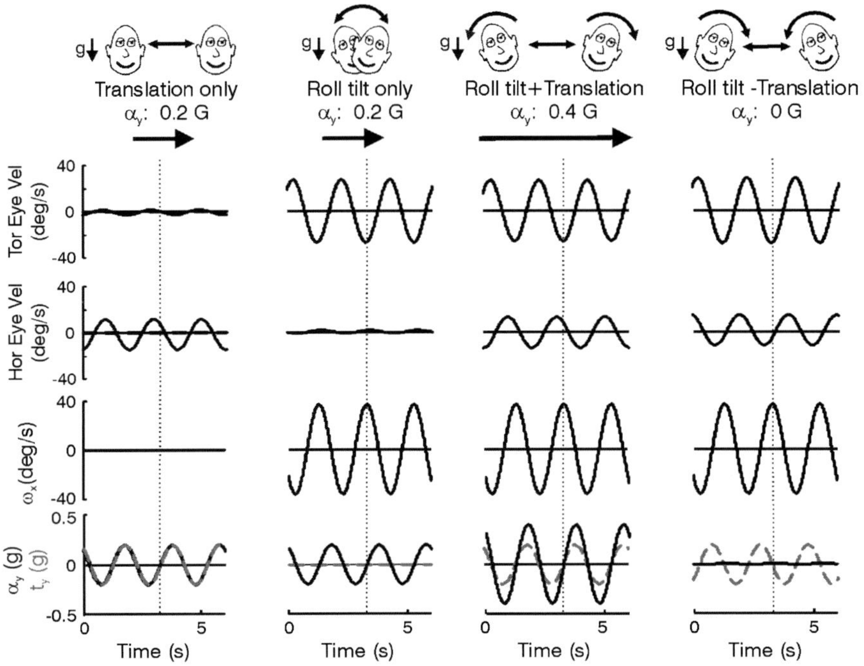

Fig. 4. Simulated behavioral responses. Horizontal and torsional eye velocity responses to *Translation only*, *Roll tilt only*, *Roll tilt + Translation* and *Roll tilt−Translation* motion stimuli during sinusoidal motion at 0.5 Hz. The bottom two traces illustrate the stimuli: roll angular velocity (ω_x, solid black); net and translational components of the interaural acceleration (α_y, solid black; t_y, dashed gray lines). Model simulations were performed using a fixed-step Runge–Kutta numerical integration routine (ode4 in SIMULINK) with time steps fixed at 0.01 s. (Adapted from Green and Angelaki, 2004, used with permission.)

often difficult to interpret. For example, some neurons exhibited rather simple properties, apparently responding only to rotational stimuli ("canal-only" neurons; e.g., Dickman and Angelaki, 2002). However, the majority of neurons that responded to translational stimuli were more complex. Most cells did not respond exclusively to tilts or translations but to both motions. However, because they typically responded differently to tilts versus translations they did not simply encode the otolith afferent signal either, suggesting a convergence of sensory information from both the otoliths and the canals ("canal-otolith" convergent cells; Dickman and Angelaki, 2002; Shaikh et al., 2005a).

To understand how the responses of these cells arose, attempts were made to parse out how each sensory signal contributed to the responses. Any cell that responded to translational motion clearly included a component of otolith origin. However, a problem with isolating the canal contribution to cell responses is that any rotation that reorients the head relative to gravity simultaneously stimulates both the otoliths and semicircular canals, confounding the interpretation of which sensory signals contributed to the response. One potential way to isolate the canal component is to examine neural responses during earth-vertical-axis rotations with the head in different orientations. As illustrated schematically in Fig. 6A, at each different static head orientation relative to gravity, rotation about an earth-vertical axis stimulates a different combination of the semicircular canals (which are fixed in the head). For example, earth-vertical-axis rotation with the head upright represents a yaw (z-axis) rotation that stimulates mainly the horizontal semicircular canals. However, when the head is pitched backward by 90° (supine orientation) rotation about this same earth-referenced vertical axis now represents a roll (x-axis) rotation in head coordinates and it is mainly a combination of the vertical semicircular canals that are stimulated. The advantage of considering earth-vertical-axis rotations is that they do not reorient the head relative to gravity and thus there is no concurrent dynamic stimulation of the otoliths (i.e., $\alpha_y = 0\,\mathrm{G}$). It was therefore expected that such a protocol could be used to isolate the semicircular canal contributions to neural activities (e.g., Siebold et al., 2001; Dickman and Angelaki, 2002).

When neural responses to earth-vertical-axis rotations were examined experimentally, however, the results generally failed to provide a clear explanation for observed cell activities. Cell responses often did not reflect a simple summation of the signals presumed to be of otolith and semicircular canal origin (Dickman and Angelaki, 2002). For example, in an extreme case, a cell could fail to respond to earth-vertical-axis rotations, suggesting no contribution from the semicircular canals (i.e., an "otolith-only" neuron), yet still exhibit completely different responses for tilts versus translations, suggesting that the cell did not simply encode otolith signals either. The goal of the following sections will be to illustrate how the proposed model can help explain such observations.

Simulated neural responses

In this section, we will first consider the predicted responses of the model neurons, V1–V5, for typical experimentally employed stimulus paradigms (e.g., Fig. 5, columns 1–3; Fig. 6). These will then be compared with the responses predicted for additional more novel stimulus conditions including the *Tilt–Translation* protocol (Fig. 5, column 4) and rotations in a different head orientation (Fig. 5, columns 5 and 6). The goal here will be to illustrate why traditional interpretations of neural response properties have led to incorrect conclusions regarding the signals encoded by central neurons and why considering predicted neural responses from a computational modeling approach has helped to explain what motion-sensitive cells in the vestibular nuclei and cerebellum do in fact encode.

As illustrated in Fig. 5, model cell populations V1 and V2 ("canal-only" neurons) reflect a simple encoding of rotation in head coordinates. Cell V1 responds to yaw rotation whereas cell V2 responds to roll rotation, regardless of head orientation. Neither cell type modulates during a pure translational stimulus. Cells V3, V4 and V5 ("translation," "canal-otolith convergent" and "tilt" neurons) reflect a somewhat more complicated processing of sensory information. Cell V3 responds to translation (Fig. 5, column 1), but not to tilt (Fig. 5, column 2) thus encoding, t_y, as predicted analytically (see Model Description section), whereas cell V5 responds to tilt but not to translation. Cell V4, on the other hand, appears to encode an intermediate signal, modulating for both tilts and translations but with different response amplitudes. None of these model cell types encodes otolith afferent-like information, suggesting that semicircular canal signals must make a contribution to their responses.

To examine these canal contributions in more detail we can consider the responses of each neuron type during earth-vertical-axis rotations. Fig. 6B plots predicted cell sensitivities to rotation (i.e., response gain: ratio of cell response to stimulus amplitude) as a function of static pitch orientation relative to the earth-vertical rotation axis. As expected, cells V1 and V2 exhibit responses that are consistent with simple canal afferent-like behavior. Notice that cell V1 responds maximally to earth-vertical-axis rotations in the upright orientation (pitch angle $= 0°$) when the rotation represents a z-axis or "yaw" rotation in head coordinates. The neuron's sensitivity to rotation drops off with the cosine of pitch angle, consistent with a cell dominantly sensitive to horizontal canal input. Cell V2 instead exhibits no modulation during earth-vertical-axis rotation in the upright orientation but maximal responses in prone and supine positions (pitch angles of $\pm 90°$) when the rotation represents an x-axis or "roll" rotation in head coordinates. The neuron's response is thus consistent with the cell receiving dominantly vertical canal signals.

The situation is quite different for cells V3, V4 and V5. Their response properties during upright tilts and translations (Fig. 5) suggest the

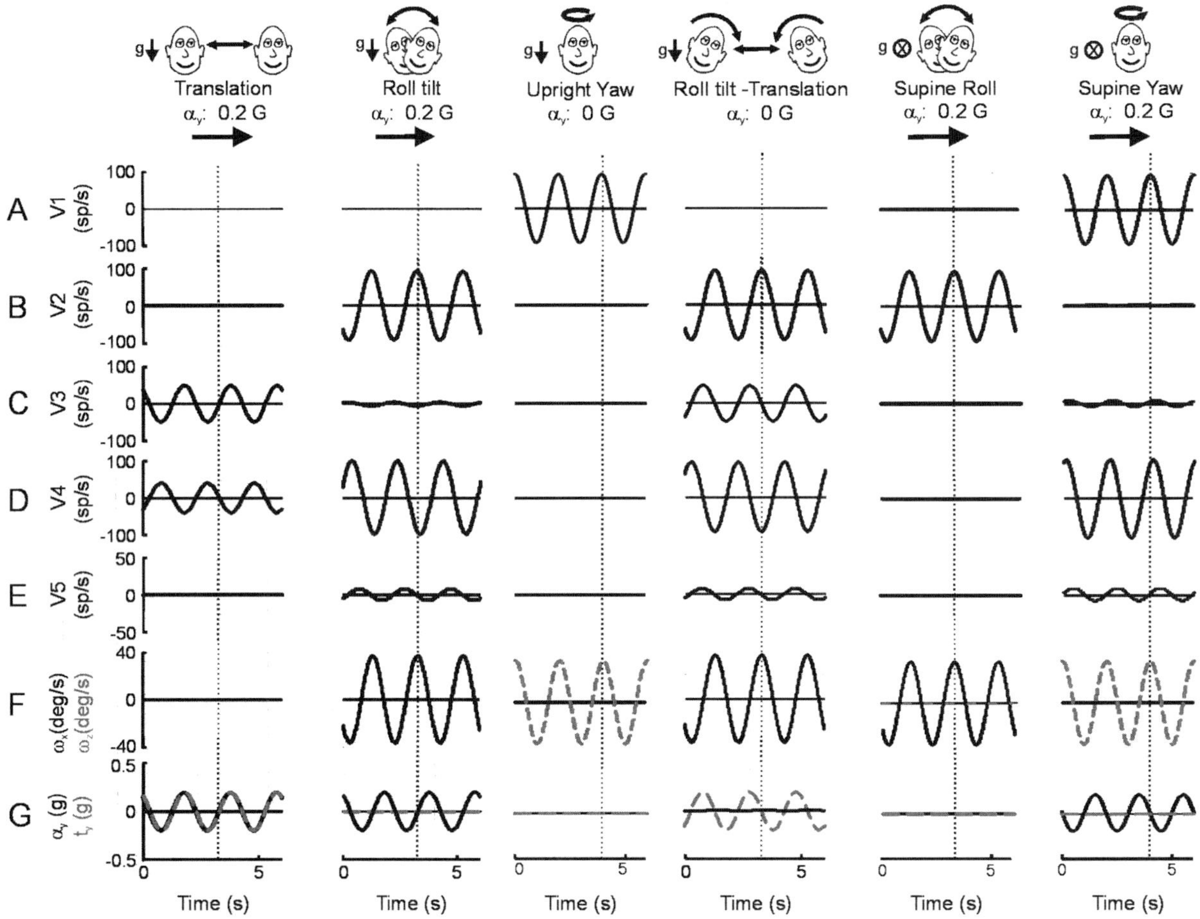

Fig. 5. Simulations of model cell responses during traditional stimulus paradigms including lateral translation, roll tilt and yaw rotation in an upright orientation (columns 1–3) as well as during more novel paradigms including the *Roll tilt–Translation* stimulus (column 4) and roll and yaw rotations in supine orientation (columns 5–6; ⊗ indicates that the gravity vector points into the page). (A)–(E) Firing rate modulation of cells V1 through V5. (F) Angular roll and yaw velocities (ω_x, solid black; ω_z, dashed gray lines). (G) Net and translational interaural accelerations (α_y, solid black; t_y, dashed gray lines). (Adapted from Green and Angelaki, 2004, used with permission.)

contribution of semicircular canal signals to their activities. Surprisingly, however, when their predicted responses during earth-vertical-axis rotations are examined, all three-cell types fail to demonstrate any modulation (Fig. 6B). If such observations were made in a traditional experimental setting it would likely have been concluded that in fact semicircular canal signals do not contribute to the activities of cells V3, V4 and V5. But how then did their response properties arise?

Examining the model neural populations under additional conditions lends insight as to the combination of signals they encode. Although earth-vertical-axis rotations do not reveal the contribution of signals originating from the semicircular canals to V3, V4 and V5, notice that all three cell populations modulate during *Roll tilt–Translation* motion (Fig. 5, column 4). Recall that during this unique motion combination the translational and gravitational acceleration components along the interaural axis cancel each other out such that there is no net linear acceleration stimulus along this axis to the otoliths. Any modulation in neural firing rates must therefore be

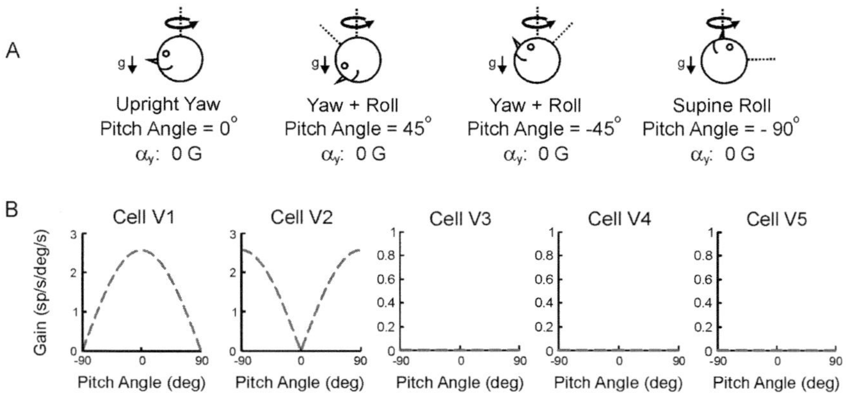

Fig. 6. Predicted neural response sensitivities (gains) to rotations about an earth-vertical-axis. (A) Schematic representation of the stimuli. As the head assumes different static orientations, different combinations of the horizontal and vertical semicircular canals are stimulated. When upright, rotation about the earth-vertical axis produces a yaw (z-axis) rotation in head coordinates. When supine, rotation about an earth-vertical axis now produces a roll (x-axis) rotation in head coordinates. At intermediate static pitch angles, earth-vertical-axis rotation results in a rotation axis in between the z- and x-axes in head coordinates (i.e., having both z- and x-axis components; yaw + roll stimulation). (B) Predicted gains (ratio of cell response amplitude to stimulus amplitude) for each cell type during rotation about an earth-vertical axis (gray dashed traces) plotted as a function of head orientation in pitch. Response gains are predicted to be zero across all head orientations for cells V3, V4 and V5 (dashed lines along abscissa). (Adapted from Green and Angelaki, 2004, used with permission.)

attributed to semicircular canal-derived signals. The *Roll tilt—Translation* protocol thus appears to unmask a "hidden" semicircular canal contribution to the responses of these cells that is not observed during earth-vertical-axis rotations.

Investigation of the predicted cell responses during simulated rotations in another head orientation yields further insight as to why this is the case. When neural responses are reexamined during simulated rotations in the supine body orientation (Fig. 5, columns 5 and 6), it becomes clear that some cell types exhibit responses to rotation that depend on current head orientation relative to gravity. Notice that cells V2 and V5 exhibit similar responses in the upright orientation; both cells respond exclusively to roll rotations. However, in the supine body orientation a clear difference in their response properties appears. Whereas cell V2 continues to encode roll rotations, cell V5 now fails to modulate during roll but instead responds during yaw rotation. Similar head-orientation-dependent behavior applies in the case of cell V4. Both the V4 and V5 cell populations thus appear to only encode rotations that reorient the head relative to gravity. Note, in particular, that cell V5 responds exclusively to head tilts, encoding an estimate of the gravitational acceleration component, g_y, as predicted analytically (see Model Description section). Thus, semicircular canal signals do in fact contribute to the responses of cells V3, V4, and V5. However, it appears that they have been used to construct an estimate of the component of rotation that reorients the head relative to gravity, a spatially referenced rotation component aligned with the earth-horizontal axis that is not observed during earth-vertical-axis rotations.

Reference frame transformation of semicircular canal signals

To explain in more detail why cells V3, V4 and V5 fail to respond to earth-vertical-axis rotations, yet exhibit clear evidence for a semicircular canal contribution to their activities during the *Roll tilt—Translation* protocol, we may return to the theoretical computations required to extract translational motion information. According to Eq. (3), translational motion estimation requires computation of the term

$$-\boldsymbol{\omega} \times \mathbf{g} = -[(\omega_y g_z - \omega_z g_y)\mathbf{i}, (\omega_z g_x - \omega_x g_z)\mathbf{j}, (\omega_x g_y - \omega_y g_x)\mathbf{k}] \quad (9)$$

where $\boldsymbol{\omega}$ and $\mathbf{g}$ are defined in head coordinates. An examination of the general expression for $-\boldsymbol{\omega} \times \mathbf{g}$ given by Eq. (9) reveals that for small reorientations of the head relative to gravity, the *amplitude* of $-\boldsymbol{\omega} \times \mathbf{g}$ approximates the amplitude of the *earth-horizontal* component of angular velocity. The *direction* of $-\boldsymbol{\omega} \times \mathbf{g}$ is orthogonal to both $\boldsymbol{\omega}$ and $\mathbf{g}$ and aligned with the dominant head axis along which a change in gravitational acceleration is introduced by the rotation. Since the computation of $-\boldsymbol{\omega} \times \mathbf{g}$ is approximated in the model cell population V4 (i.e., $-(\omega_z g_x - \omega_x g_z) \approx (\omega_z \alpha_x - \omega_x \alpha_z)$ for t_x, $t_z \approx 0$) these cells thus extract an estimate of only the earth-horizontal component of angular velocity.

To see this more clearly we can consider what happens during rotation about an earth-horizontal axis. For example, for small amplitude roll (x-axis) rotations from upright such that $\omega = \omega_x \mathbf{i}$, $\mathbf{g} \approx g_z \mathbf{k} \approx -1\,\mathrm{G}$ ($\mathrm{G} = 9.81\,\mathrm{m/s^2}$) and g_x, $g_y \approx 0\,\mathrm{G}$, Eq. (9) would yield $-\boldsymbol{\omega} \times \mathbf{g} \approx -\omega_x \mathbf{j}$. Cell V4 then faithfully encodes the angular velocity signal of canal origin that can be observed in the absence of a concurrent otolith signal during *Roll tilt–Translation* motion. However, if the same roll rotations are performed when supine (i.e., tilted 90° backwards) the rotation is now about an earth-vertical axis. In this case, when $\boldsymbol{\omega} = \omega_x \mathbf{i}$, $\mathbf{g} \approx g_x \mathbf{i} \approx -1\,\mathrm{G}$ and g_y, $g_z \approx 0\,\mathrm{G}$, then $-\boldsymbol{\omega} \times \mathbf{g} = 0$. Cell V4 fails to respond because the rotation did not reorient the head relative to gravity (i.e., no earth-horizontal component). When supine, it is instead an earth-horizontal-axis rotation about the head z-axis (i.e., supine yaw rotation) that stimulates a response in cell V4. More generally, it can be verified that for an arbitrary head orientation relative to gravity the computation $-\boldsymbol{\omega} \times \mathbf{g}$ extracts an estimate of only the *earth-horizontal* component of angular velocity (but rotated in terms of direction by 90° in the earth-horizontal plane).

The explanation for the failure to observe responses to earth-vertical-axis rotations in cell populations V3, V4 and V5 is now clear. Cell population V4 approximates the $-\boldsymbol{\omega} \times \mathbf{g}$ computation to construct a spatially referenced signal indicating the velocity with which the head reorients relative to gravity. Temporal integration of this signal by the model network yields a dynamic estimate of the gravitational acceleration, g_y, on cell V5. This signal is subsequently combined with the otolith-derived estimate of net acceleration on cell V3 to estimate translational acceleration, t_y.

All three cell types thus do carry signals of semicircular canal origin (the signal observed during *Tilt–Translation* motion). However, the assumption that the semicircular canal contribution could be observed during earth-vertical-axis rotations was incorrect. Specifically, although it has traditionally been assumed that central cells encode signals of semicircular canal origin in head-referenced coordinates, the computations for inertial motion estimation imply that this cannot be true of the cells that are used to resolve the tilt-translation ambiguity. Canal sensory information signaling rotation in a head-fixed reference frame must be transformed into a spatially referenced estimate of only the earth-horizontal angular velocity component. Thus, these canal-derived signals cannot normally be observed in the absence of concurrent otolith stimulation unless unique protocols like the *Tilt–Translation* protocol are employed.

More realistic neural response predictions

So far, we have considered the predictions of one particular model structure and have examined the predicted responses of several average neural populations that exhibit quite distinct properties. These include neurons that respond exclusively to translation or tilt or to some combination of the two motions. Most importantly, we have illustrated that the classical assumption that cells always encode canal-derived rotational signals in head-referenced coordinates is incorrect. If canal and otolith signals are combined as required to estimate translational motion then at least some cell populations must extract only the earth-horizontal component of rotation, a spatially referenced signal that indicates when the head reorients relative to gravity. This observation emphasizes the requirement for reinvestigating cell responses using new experimental paradigms.

Nonetheless, even when such paradigms are employed, the majority of individual cells are unlikely to exhibit the idealized properties of the model neurons. This is because each model neuron is intended to represent the average activity of a large population of cells. However, each individual cell may exhibit properties that differ somewhat from the idealized behavior of the population average. Here, we will briefly consider the impact of relaxing particular parametric assumptions in the model with the goal of gaining insight as to what the properties of *individual* neurons within our average populations might be. The range of individual neuron properties predicted by the model will then be compared with those of actual cells that have recently been recorded experimentally.

Variable responses to earth-vertical-axis rotations

Model cell populations V3, V4 and V5 all failed to respond to earth-vertical-axis rotations because they encoded only the earth-horizontal component of rotation. However, the majority of neurons recorded in the brainstem vestibular nuclei and fastigial nuclei of the cerebellum do indeed exhibit at least some response to earth-vertical-axis rotations (Siebold et al., 2001; Dickman and Angelaki, 2002; Shaikh et al., 2005a). Does this mean that these brain areas are not involved in the spatial transformation of canal signals required to estimate inertial motion? As we will illustrate here, neurons in these areas could still be involved in the required transformations since a *neural population* can perform the required computations even when the solution does not appear in individual neurons.

In the model, cell population V4 computes an approximation of $-\boldsymbol{\omega} \times \mathbf{g}$ to extract only the earth-horizontal rotation component. To make this computation V4 must receive horizontal and vertical canal-derived signals (i.e., estimates of ω_z and ω_x) in equal proportions. The projections onto cell V4 from cells V1 and V2 that convey these signals were thus assumed to be of equal strength (i.e., $K_{1a} = K_{2a}$ in Fig. 3). However, this need not be the case for individual neurons within the V4 population. Specifically, as illustrated in Fig. 7A, an individual neuron (labeled as $V4_i$ in Fig. 7A) may receive horizontal and vertical canal-related inputs that are of unequal strengths (i.e., $K_{1ai} \neq K_{2ai}$). In addition, some proportion of the canal inputs to an individual neuron may not be processed by the nonlinear multiplicative interactions required to estimate the earth-horizontal rotation component (K_{1oi}, K_{2oi} not multiplied by signals α_x and $-\alpha_z$). The overall computation will be the same as long as these contributions balance out appropriately across the neural population (i.e., $\sum_{i=1}^{N} K_{1oi} = 0$, $\sum_{i=1}^{N} K_{2oi} = 0$ and $\sum_{i=1}^{N} K_{1ai} = \sum_{i=1}^{N} K_{2ai}$ for a population of N neurons).

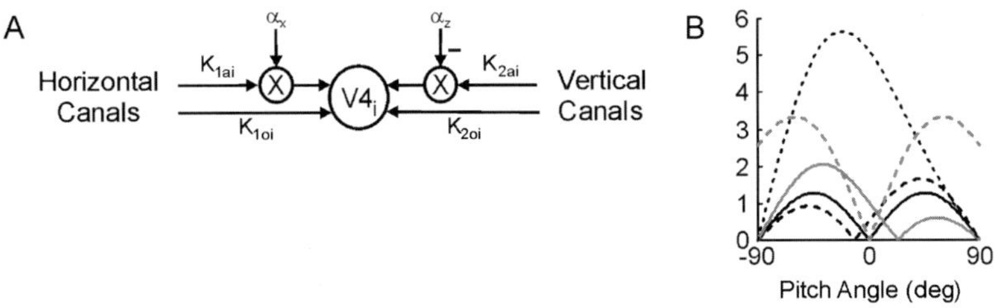

Fig. 7. Predicted earth-vertical-axis rotation responses for cell population V4, allowing for variability in individual cell characteristics. (A) Schematic of an individual cell within the V4 population ($V4_i$) that could potentially receive a combination of both head-orientation-dependent and head-orientation-independent horizontal and vertical canal inputs (weights K_{1ai}, K_{2ai} and K_{1oi}, K_{2oi}, respectively). Head-orientation-dependent inputs are shown to be multiplied by naso-occiptal and dorsoventral accelerations, α_x and α_z, as in Fig. 3. (B) Potential variations in individual V4 cell responses to earth-vertical-axis rotations under conditions where K_{1oi}, $K_{2oi} \neq 0$ and/or $K_{1ai} \neq K_{2ai}$ for different combinations of head-orientation-independent and dependent projection strengths. Solid black: $K_{1ai} = 2$, $K_{2ai} = 1$, $K_{1oi} = K_{2oi} = 0$; solid gray: $K_{1ai} = 2$, $K_{2ai} = 1$, $K_{1oi} = 0.4$, $K_{2oi} = 0$; dashed black: $K_{1ai} = 1$, $K_{2ai} = 2$, $K_{1oi} = 0.2$, $K_{2oi} = 0$; dashed gray: $K_{1ai} = 1$, $K_{2ai} = 2$, $K_{1oi} = 0$, $K_{2oi} = -1$; dotted black: $K_{1ai} = 2$, $K_{2ai} = 1$, $K_{1oi} = 2$, $K_{2oi} = 0$. (Adapted from Green and Angelaki, 2004, used with permission.)

As illustrated in Fig. 7B, such "unbalanced" canal contributions (i.e., $K_{1ai} \neq K_{2ai}$, K_{1oi}, $K_{2oi} \neq 0$) result in a broad range of potential response properties for individual neurons during earth-vertical-axis rotations. Notice, in particular, that many neurons do not reflect the simple cosine-tuned responses illustrated for cells V1 and V2 in Fig. 6B. The more complex patterns of response gains as a function of head orientation provide evidence for an underlying nonlinear processing of canal signals (i.e., multiplicative interactions) required to compute the required coordinate transformations (i.e., see Green and Angelaki, 2004, for details).

The observed distribution of response properties arises because each individual neuron might encode neither a completely spatially referenced, earth-horizontal velocity signal, nor the head-referenced angular velocity signal provided by the canal sensors, but rather some intermediate representation of rotation. This implies that the neurons actually involved in constructing a spatial rotation estimate may exhibit a broad range of response properties with the earth-horizontal rotation component only apparent at the population level. What is important to appreciate is that earth-vertical-axis rotations will reveal only a portion of canal inputs to a cell. Any component that has already been transformed into a spatially referenced estimate of head reorientation relative to gravity will remain "hidden." These observations thus emphasize that both an examination of response properties in different head orientations and the use of paradigms such as *Tilt—Translation* are necessary to characterize the sensory contributions to neural activities and reveal any associated transformations of these signals that may have taken place.

Distributed dynamic response properties

Finally, an additional puzzling but consistent experimental observation is that central motion-sensitive neurons exhibit a broad distribution of response dynamics during translation (Angelaki and Dickman, 2000; Zhou et al., 2001, 2006; Dickman and Angelaki, 2002; Musallam and Tomlinson, 2002). Whereas a few central neurons respond to translations like otolith afferents and modulate in phase with linear acceleration, the majority of cells exhibit a broad range of response dynamics with many neurons appearing to more closely encode translational velocity. The model can help explain this observation.

A key feature of the proposed model is that it acts as a distributed neural integrator. An important consequence of coupling otolith signals to such an integrative network is that relatively small parameter changes can yield a wide range of response properties. For example, with the parameter set chosen for the model simulations of Figs. 4 and 5, cells V3 and V4 were predicted to modulate in phase with linear acceleration during translation. However, random variation of a single model parameter, weight q_3, about its nominal value (see Fig. 8 legend) is sufficient to predict a broad range of response dynamics during translation. This is illustrated in Fig. 8A which plots the distribution of response gains and phases that could be observed on individual V3 cells during sinusoidal translation at 0.5 Hz. A similar variability in response dynamics is predicted for cells V4 and V5 (see Green and Angelaki, 2004).

Of course, if the V3 population is to construct an estimate of translational motion then processed canal signals must also reflect a broad distribution of response dynamics. This is the case because, according to Eq. (3), to construct an estimate of translation, $\mathbf{t}$, an otolith estimate of net linear acceleration, $\boldsymbol{\alpha}$, must be combined with a canal-derived estimate of gravity, $-\int \boldsymbol{\omega} \times \mathbf{g}$. Any dynamic variability in the otolith signal (in terms of gain and phase) must therefore be matched by similar variability in the canal-derived estimate of $-\int \boldsymbol{\omega} \times \mathbf{g}$ if the population of neurons is to extract information about translation.

One way to achieve this in the model is to presume that, in addition to the canal-derived "tilt" signals (i.e., estimates of $-\int \boldsymbol{\omega} \times \mathbf{g}$) conveyed from cell V5, V3 cells also receive additional projections directly from the horizontal and vertical semicircular canals. Given this assumption, V3 neurons will now exhibit a distribution of response gains and phases associated with canal-derived estimates of gravitational acceleration (Fig. 8B) that is

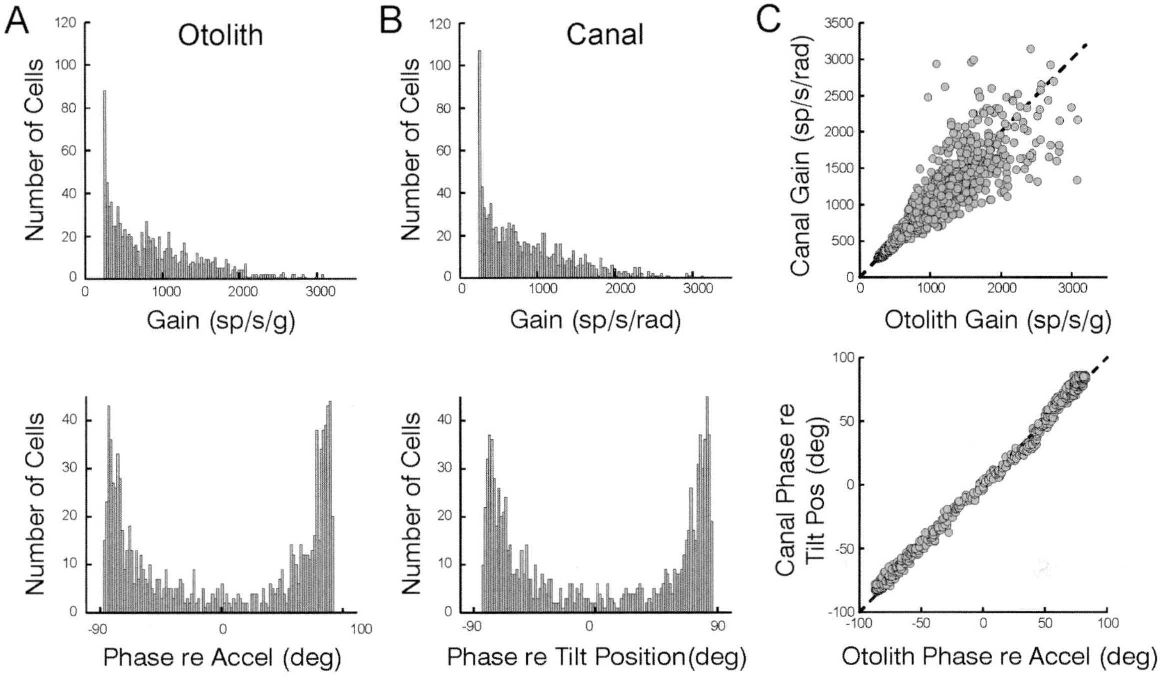

Fig. 8. Predicted variability in response dynamics of cells within population V3. (A) Histograms illustrating the distributions of 0.5 Hz response gains and phases associated with the otolith component of individual V3 cell responses (e.g., during pure translational motion) that are predicted when weight q_3 was varied about its nominal mean value of 0.1975 according to a normal distribution with a standard deviation equal to 10 times this parameter (i.e., $\approx$68% of the randomly chosen q_3 values fell within $\pm$10 times the mean value of 0.1975). Each histogram displays a total of 1000 gains or phases, associated with different values of q_3 that are grouped into 100 bins. Gains and phases are expressed relative to linear acceleration, $\boldsymbol{\alpha}$. (B) Distribution of the gains and phases associated with the canal-derived component of V3 cell responses that approximates $-\int \boldsymbol{\omega} \times \mathbf{g}$ (e.g., as can be observed during *Tilt–Translation* motion). To create this distribution, V3 was presumed to receive additional direct projections from the semicircular canals (i.e., additional direct projections from the outputs of boxes *C*(*s*) in Fig. 3). These were presumed to vary in strength about a mean value of 0 according to a normal distribution with a standard deviation equal to 14. Gains and phases are expressed relative to head tilt (i.e., relative to an estimate of $-\int \boldsymbol{\omega} \times \mathbf{g}$). Note that the histograms only illustrate the response dynamics associated with all contributing canal inputs. However, to simulate the appropriate contributions from each endorgan in different head orientations, the additional direct inputs from the vertical and horizontal canals onto cell V3 must be multiplied by $-\alpha_z$ and α_x, respectively (i.e., similar to the inputs to cell V4). (C) Gains and phases associated with the canal-derived estimate of $-\int \boldsymbol{\omega} \times \mathbf{g}$ on individual V3 cells plotted versus those associated with the otolith-derived estimate of $\boldsymbol{\alpha}$ when sensory contributions with similar dynamics were partially "matched" to simulate how individual cells might "learn" to approximate translational motion at the single cell level. Here the matching was accomplished artificially by arranging otolith [from the distribution in (A)] and canal [from the distribution in (B)] contributions in order of phase using a noisy sorting algorithm and assigning these partially matched contributions to individual cells. Notice that the matching of phase results in a much more "noisy" matching of response gains. However, both gains and phases are distributed about the unity-slope lines illustrating that otolith and canal signal contributions to estimating $\boldsymbol{\alpha}$ and $-\int \boldsymbol{\omega} \times \mathbf{g}$ appear in the appropriate proportions across the population to extract an estimate of translational motion.

similar to the distribution of otolith-derived estimates of the net acceleration (Fig. 8A). Importantly, the similarity in distributions guarantees that the *population* of neurons can be used to extract translational motion information. However, we can take this one step further. If translational motion estimates are eventually to be extracted at the level of individual neurons one might expect that at some stage in processing we should observe an at least partial "matching" of otolith- and canal-derived signal dynamics on single cells. This is illustrated in Fig. 8C. Otolith- and canal-derived contributions to V3 chosen from the distributions in Fig. 8A, B were ordered in terms of phase using a noisy sorting process (see Fig. 8 legend for details). The partially matched phases and associated gains were then assigned to individual neurons. As a result of this ordering, when the canal-derived phases for individual neurons are plotted versus the otolith phases, they appear similarly matched on individual neurons (Fig 8C, bottom). When the associated canal-derived gain contributions are plotted versus the otolith component gains they are also well matched although the distribution is much broader (Fig. 8C, top). Because the canal- and otolith-derived signal gains are not perfectly matched on most individual neurons, most cells do not explicitly encode translational motion. Notice, however, that both the gains and phases are nonetheless distributed about the unity slope lines. Thus, it is clear that otolith and canal-related signal contributions appear in the appropriate proportions *across the neural population* to extract an estimate of translational motion.

These observations have several important implications. First, they imply that in the real physiological system we might expect to observe a broad distribution of sensory response dynamics. Second, as a result of this broad distribution, the extracted information about translational motion need not explicitly be a translational acceleration signal [i.e., as implied by Eq. (3)]. Instead, translational motion estimates could range dynamically from being more acceleration- to more velocity-like. Finally, most individual neurons may not exclusively encode either head translation or tilt since a dynamic matching of otolith and canal signals need not be perfect at the individual neuron level. However, there should nonetheless exist cell populations in which otolith- and canal-derived signals combine in the appropriate proportions (in terms of both gain and phase) to extract translational motion information across the population. In the following section we will consider how these predictions correspond with actual experimental observations.

Comparison with experimentally observed neural responses

The neural basis for resolving the tilt-translation ambiguity has recently been investigated by explicitly taking into account the theoretical requirements for inertial motion estimation and by employing stimulus paradigms appropriate to reveal the required neural computations (Angelaki et al., 2004; Green et al., 2005; Shaikh et al., 2005b; Yakusheva et al., 2007). These studies have in fact revealed neural response properties compatible with the predictions of the model network. Figure 9A illustrates a neuron that was recorded in the vestibular nucleus of a rhesus macaque monkey during the four *Translation*, *Tilt*, *Tilt + Translation* and *Tilt − Translation* stimulus protocols. This neuron responded similarly during all motions that included a translational component but its response was highly attenuated during head tilt. Thus, unlike primary otolith afferents, which respond identically to tilts and translations, this neuron appeared to extract information about translational motion, as predicted for cell V3 in the model. Notice, in particular, that like cell V3, this vestibular neuron demonstrated a robust response during *Tilt − Translation* motion, revealing a semicircular canal contribution to constructing an estimate of translational motion. Such robust responses to *Tilt − Translation* motion were generally observed in most neurons recorded in the brainstem vestibular and deep cerebellar (fastigial)

Fig. 9. Experimental results. (A) Translation-encoding neuron recorded in the vestibular nucleus of a rhesus macaque monkey during the four tilt-translation motion paradigms in darkness. Data adapted from Angelaki et al. (2004). (B) Relationship between the k_1 and k_2 regression coefficients associated with modeling cell responses as a combination of terms related to the net acceleration signal encoded by otolith afferents (associated with k_1) and a gravitational acceleration estimate constructed centrally using semicircular canal-derived signals (associated with k_2). The plots show the relationship between both the amplitude and phase aspects of the coefficients for individual cells in the vestibular (open circles) and fastigial (gray shaded squares) nuclei. Black (vestibular) and gray (fastigial) solid lines illustrate linear regressions for amplitude and phase. Data adapted from Shaikh et al. (2005b) with permission from Elsevier. (C) Example responses of a cell recorded in the posterior cerebellar vermis (nodulus) during *Tilt − Translation* motion and earth-vertical-axis rotations.

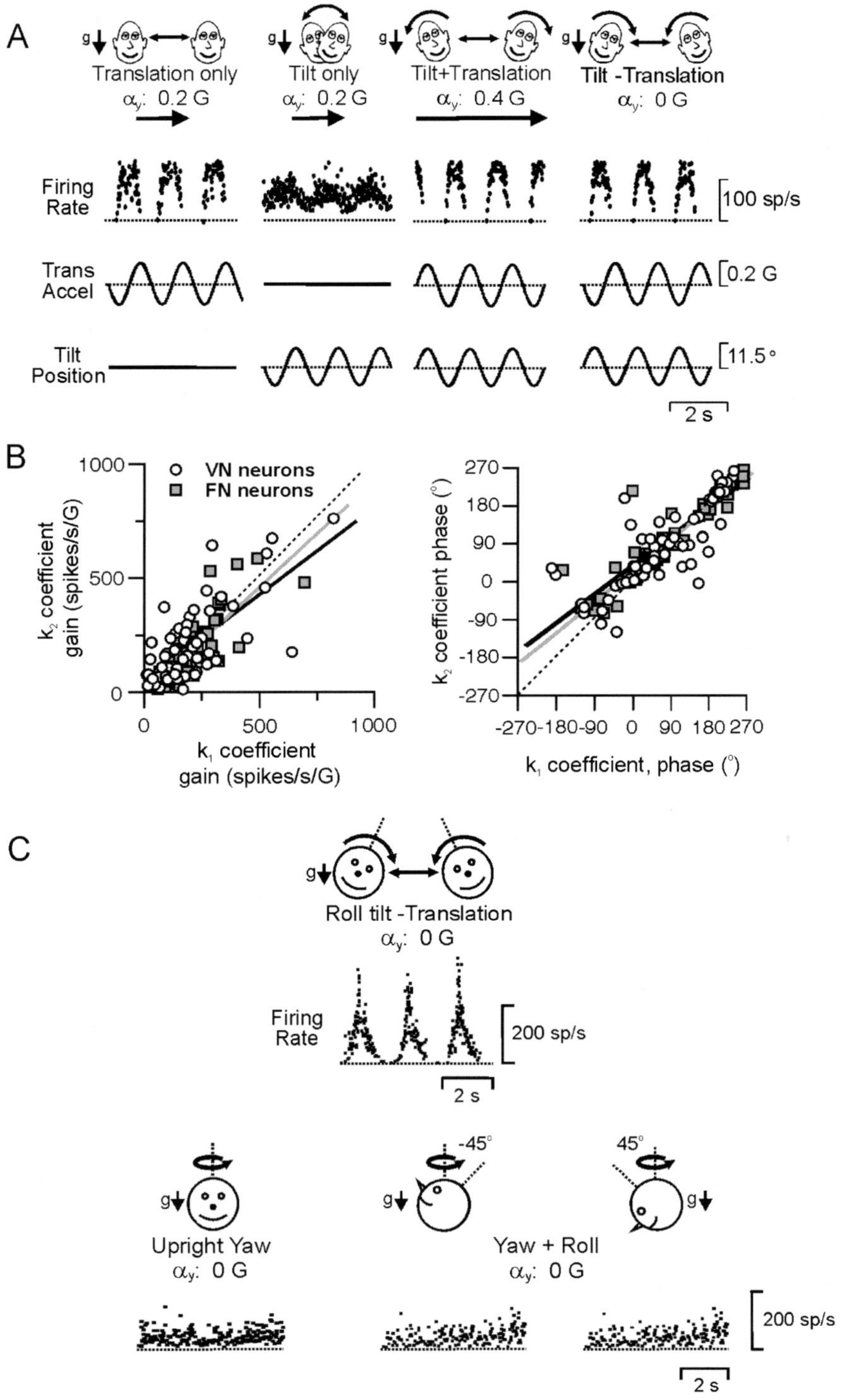
A
Translation only
αy: 0.2 G
Tilt only
αy: 0.2 G
Tilt+Translation
αy: 0.4 G
Tilt -Translation
αy: 0 G
Firing Rate
100 sp/s
Trans Accel
0.2 G
Tilt Position
11.5°
2 s
B
VN neurons
FN neurons
k2 coefficient gain (spikes/s/G)
k1 coefficient gain (spikes/s/G)
k2 coefficient phase (°)
k1 coefficient, phase (°)
C
Roll tilt -Translation
αy: 0 G
Firing Rate
200 sp/s
2 s
-45°
45°
Upright Yaw
αy: 0 G
Yaw + Roll
αy: 0 G
200 sp/s
2 s

nuclei (Angelaki et al., 2004; Shaikh et al., 2005b). The fact that these indeed reflected signals of semicircular canal origin was explicitly confirmed by demonstrating that responses to *Tilt–Translation* motion disappeared after the semicircular canals were inactivated by plugging (Shaikh et al., 2005b).

In contrast to the translation-encoding cell of Fig. 9A, however, most neurons recorded in the vestibular and fastigial nuclei responded to both tilts and translations with different response amplitudes and a range of dynamic properties (Angelaki et al., 2004; Green et al., 2005; Shaikh et al., 2005b). Note that this observation is compatible with the predictions of the model, suggesting the likelihood of observing a broad range of response dynamics in a distributed integrative network for computing inertial motion (see Distributed Dynamic Response Properties section). The key question to be addressed was thus whether, despite this broad range of responses, evidence could be provided that a translational motion estimate was nonetheless constructed across *the neural populations* (e.g., see Fig. 8).

To explicitly investigate this hypothesis a first step was to examine whether individual cells indeed encoded the necessary signals to estimate translational motion. Specifically, individual neurons were examined to see if they could be modeled by a combination of the terms of Eq. (3). For the small amplitude lateral (y-axis) translations and roll (x-axis) rotations from upright employed in the study, Eq. (3) simplifies to the scalar expression

$$t_y = \alpha_y + g_y \approx \alpha_y + \int \omega_x g_z \qquad (10)$$

Neural responses were thus investigated to see if they could be explained by the equation

$$\mathrm{FR} = k_1 \alpha_y + k_2 \int \omega_x g_z \qquad (11)$$

where FR denotes the neural firing rate. k_1 and k_2 represent the regression coefficients associated with response terms related to a net acceleration component, α_y, encoded by otolith afferents, and a gravitational acceleration component ($g_y = \int \omega_x g_z$) computed by temporal integration of semicircular canal signals. Note that, despite the simplified expression in Eq. (11), the actual equation used was somewhat more complex in that it took into account the observed variability in neural response dynamics predicted by the model of Fig. 3. This was accomplished by allowing regression coefficients k_1 and k_2 to vary in terms of both amplitude and phase (see Shaikh et al., 2005b; Green et al., 2005, for details). Comparison of Eqs. (10) and (11) shows that when $k_1 = k_2$ (in terms of both amplitude and phase) the neuron extracts an estimate of translational motion.

When the ability of Eq. (11) to predict vestibular and fastigial neural responses was examined across all four tilt-translation stimulus paradigms the results were striking. Not only did this equation provide an excellent fit to almost all neurons (see Angelaki et al., 2004; Shaikh et al., 2005b, for details) but there was a clear relationship observed between coefficients k_1 and k_2. As illustrated in Fig. 9B, the gain aspects of these coefficients were not typically matched at the level of individual neurons (i.e., $k_1 \neq k_2$) accounting for the fact that most individual neurons did not simply encode translational motion. However, the coefficients were linearly correlated with regression slopes for both vestibular and fastigial neurons that were not significantly different from unity. Even more striking was the observation that despite a broad distribution of coefficient phases, these too tended to be matched even at the level of individual neurons. Notice that these observations are identical to the predictions of Fig. 8C. They illustrate that although individual neurons do not explicitly encode translation, they nonetheless carry the appropriate combination of net linear acceleration, α_y, and gravitational acceleration signals, $g_y = \int \omega_x g_z$. These signal components appear dynamically matched in terms of phase even at the level of individual neurons (e.g., compare Fig. 8C, bottom and Fig. 9B, right) and appear in equal proportions in terms of gain (e.g., compare Fig. 8C, top and Fig. 9B, left) to extract information about translational motion at the population level. Subsequent experimental extensions to this work to investigate neural responses in 2D (i.e., for tilts and translations in multiple horizontal-plane directions) revealed that canal and otolith contributions to cell activities were matched not only temporally but also spatially in exactly the fashion expected if these neurons construct a neural representation of the computations for estimating translational motion (Green et al., 2005).

The studies described above have illustrated that vestibular and fastigial neurons carry the appropriate signals to construct a distributed translational motion estimate in 2D during tilts and translations from an upright orientation. However, some of the most exciting new results come from recent work demonstrating evidence for the reference frame transformation of canal signals required to estimate inertial motion in 3D (Yakusheva et al., 2006). The posterior vermis of the cerebellum (nodulus and ventral uvula) has long been implicated as part of an integrative network that among other things participates in the spatial transformation of rotation signals during the vestibulo-ocular reflex (Angelaki and Hess, 1994, 1995; Wearne et al., 1997, 1998; Cohen et al., 2002). It has recently been suggested that this same network is involved in resolving the tilt-translation ambiguity (Green and Angelaki, 2003).

To investigate what role this cerebellar cortical region might play in inertial motion estimation, nodulus/ventral uvula neuron responses were examined using the same tilt-translation combinations used to investigate vestibular and fastigial neuron responses. Remarkably, in contrast to the more distributed response properties observed in vestibular and fastigial neurons (e.g., as reflected in the plots of Fig. 9B), individual Purkinje cells in the posterior vermis all appeared to encode a faithful estimate of translation motion (e.g., like the atypical vestibular cell in Fig. 9A and model cell V3). Furthermore, these neurons showed clear evidence for the reference frame transformation of canal signals predicted by the model. This is illustrated for an example nodulus neuron in Fig. 9C. The cell exhibited a robust response to *Tilt–Translation* motion, providing evidence for a semicircular canal contribution to its response. However, during earth-vertical-axis rotations with the head in different orientations the cell failed to respond (e.g., like model cells V3, V4 and V5; Fig. 6B) and the same was generally true for all recorded neurons. Thus, compatible with model predictions, these posterior vermis cells appeared to exclusively encode an estimate of the earth-horizontal rotation component of sensory canal signals, the spatially referenced signal required to estimate head reorientation relative to gravity.

Discussion

Estimation of spatial orientation and inertial motion is essential for the planning and coordination of appropriate action. Neurophysiologists have long attempted to understand how the brain constructs such estimates by examining the activities of neurons in areas known to be involved in processing motion information (e.g., vestibular nuclei and cerebellum) for varieties of different motion stimuli. However, the interpretation of neural activities has been complicated by the fact that individual sensors provide inherently ambiguous spatial motion information. Not only do sensory signals provide motion information in body-centered coordinates rather than in a common spatial reference frame, but the detection of motion in a gravitational field poses additional problems because inertial and gravitational linear accelerations are sensed equivalently by the otoliths. To accurately distinguish between the two requires neural computations involving the integration of information from multiple sensory sources. Here, we have illustrated why considering the problem from a theoretical perspective that takes into account the required computations for inertial motion estimation has been essential to understanding the neural correlates for its solution.

In particular, we have illustrated that the computations for discriminating tilt and translation can be performed by a distributed integrative network that combines canal and otolith sensory signals in a nonlinear fashion to compute an internal estimate of head orientation relative to gravity. This estimate, in conjunction with the net linear acceleration signals provided by the otoliths, is used to extract the translational acceleration component of motion. By examining the predicted responses of neural populations in this model we have illustrated why relying on classically employed stimuli that embed traditional assumptions with respect to what central neurons encode has led to inappropriate conclusions with respect to observed response patterns. Perhaps the most important model prediction is that populations of the neurons involved in inertial motion estimation should exhibit responses to rotational stimuli from the canals that depend on head orientation. In contrast to the head-referenced estimates of rotation provided by the semicircular canals, such cells should instead encode mainly the earth-horizontal component of rotation. This spatially referenced signal that indicates when the head reorients relative to gravity is necessary to estimate the gravitational component of otolith-sensed net acceleration.

Recent investigations that have used stimulus paradigms and analysis approaches which take into account the required computations for inertial motion estimation have now revealed populations of neurons in the vestibular and deep cerebellar (fastigial) nuclei that appear to encode a distributed solution to the inertial motion detection problem (Angelaki et al., 2004; Green et al., 2005; Shaikh et al., 2005b). Most individual neurons in these areas don't explicitly encode translation, but information about translational motion can be extracted at the population level. In contrast, new evidence suggests that cerebellar cortical neurons in the posterior vermis reflect the solution to the problem. Individual neurons in the posterior vermis not only explicitly encode translational motion but demonstrate clear evidence that semicircular canal-derived estimates of rotation in head coordinates have been transformed into spatially referenced estimates of the earth-horizontal rotation component (Yakusheva et al., 2007). Nonetheless, although significant progress has been made in identifying populations of neurons in different brain areas that are involved in solving the inertial motion detection problem, much work remains to address exactly how each of these populations actually contributes to the required computations.

A key question surrounds where in fact the required reference frame transformations actually take place. Purkinje cells in the posterior cerebellar vermis appear to represent a homogenous population that encodes translational motion and show evidence for canal-derived rotational signals that have been transformed into a spatial reference frame. Thus, these cells seem to reflect the output of the required computations, as reflected in the "translation-encoding" model cell population V3. However, one might expect the individual neurons involved in *effecting* the required reference frame transformations to exhibit more variable response properties that reflect neither completely spatially referenced, nor head-referenced rotation estimates but rather some intermediate representation (e.g., see predictions in Fig. 7B for individual V4 cells). One possibility, therefore, is that many of the required computations for inertial motion estimation are performed locally by networks of neurons within the cerebellar cortex, with the solution apparent in the cerebellar output on Purkinje cells. However, the fact that vestibular and deep cerebellar (fastigial) neurons also reflect a population level solution to the inertial motion estimation problem suggests that the problem is likely to be solved in a more distributed fashion involving several different brain areas.

This notion is consistent with known anatomy demonstrating that the cerebellar vermis has direct reciprocal connections with the vestibular nuclei and also projects to the fastigial nuclei (see Barmack, 2003, for a review). It is also compatible with the general structure of the proposed model in which the translation-encoding V3 population is presumed reciprocally connected in feedback loops with other cell types that implement the required nonlinear sensory interactions (e.g., cell population V4). The observed distributed response properties of vestibular and fastigial cells that reflect complex canal-otolith interactions in 3D (e.g., Siebold et al., 1997, 1999, 2001; Dickman and Angelaki, 2002; Shaikh et al., 2005a) supports the possibility that many of these neurons may be best represented by model cell population V4. If true, then it could be the vestibular and fastigial populations that are most important in executing the required transformation of canal signals into a spatial reference frame. However, the critical tests of this hypothesis, involving detailed examinations of the responses of these cells over a broad range of head orientations, remain to be performed. Further insight may also be provided by examining how inactivation of the posterior cerebellar vermis impacts on the ability to discriminate tilts and translations.

The extent of the transformations that take place also remains in question. Specifically, it remains to be shown whether the populations of neurons that have been examined so far are more generally involved in constructing estimates of spatial motion in 3D or whether they are explicitly involved in solving only an aspect of this problem, that of resolving the tilt-translation ambiguity. So far, experiments have provided evidence that posterior vermis cells encode only the *earth-horizontal* component of rotation, compatible with the requirements for resolving the tilt-translation ambiguity. On the other hand, behavioral recordings of eye movement responses have implicated the posterior cerebellar vermis as an essential part of an integrative neural network thought to play a role in estimating the *earth-vertical* component of rotation (Angelaki and Hess, 1994, 1995; Wearne et al., 1997, 1999; Cohen et al., 2002). Such observations might imply a more general role for the brainstem-cerebellar circuitry in transforming rotational information into a spatial reference frame (i.e., a full coordinate transformation) from which both earth-horizontal and earth-vertical rotation estimates could be extracted for the purposes of estimating spatial motion.

Alternatively, the role of the neural populations investigated so far may be more limited to the specific problem of resolving the otolith sensory ambiguity with spatial motion estimates arising as an indirect consequence of the required computations (Merfeld, 1995; Merfeld and Zupan, 2002; Zupan et al., 2002; Green and Angelaki, 2003, 2004; Zupan and Merfeld, 2005). Specifically, as suggested in the model, the neural populations studied thus far could simply be involved in computing the vector cross-product of Eq. (3) without implementing a full coordinate transformation. This latter possibility is in fact more computationally accurate since the output of the vector cross-product computation can only be approximated by an earth-horizontal rotation estimate for small head reorientations relative to gravity. Future experimental work will thus be required to investigate what computations are actually performed, how/where they are implemented and the extent to which the computations made by subcortical neural populations contribute more generally to the problems of spatial motion perception and motor planning at the cortical level.

Finally, while we have focused here exclusively on vestibular contributions to spatial motion estimation, other sensory cues (e.g., visual and somatosensory), as well as efference copies of the motor commands for active movements, are also likely to contribute to the computations for detecting inertial motion and spatial orientation. For example, it is well known that visual cues can significantly influence our percept of head orientation relative to gravity (Dichgans et al., 1972; Howard, 1986; Howard and Hu, 2001). Such extra-vestibular contributions will be particularly important in constructing rotational motion estimates at low frequencies (<0.1 Hz) where the semicircular canals provide inaccurate angular velocity information (due to their high-pass dynamic characteristics). Recently, it has been illustrated that this is indeed the case for visual rotational cues that contribute to the computation of central estimates of inertial motion in a similar fashion to canal signals (Zupan and Merfeld, 2003). Whether this contribution is also mediated in a similar fashion at the neuronal level remains to be investigated.

In summary, the important question of how and where multisensory cues are integrated centrally to construct internal representations of spatial self-motion and orientation has only just begun and represents a challenging and exciting area of investigation. Here, we have focused on the computational implications of one aspect of spatial motion estimation in the context of vestibular sensory signals within a brainstem-cerebellar network. However, spatial motion estimates that rely on different combinations of sensory signals to different extents must be constructed in multiple subcortical and cortical brain areas to subserve different motor and perceptual functions (e.g., Andersen et al., 1993; Indovina et al., 2005). Such estimates are likely to rely on the integration of multiple sensory cues and combined theoretical and experimental approaches will be essential to interpreting the neural computations involved.

References

Andersen, R.A., Snyder, L.H., Li, C.S. and Stricanne, B. (1993) Coordinate transformations in the representation of spatial information. Curr. Opin. Neurobiol., 3: 171–176.

Angelaki, D.E. (1998) Three-dimensional organization of otolith-ocular reflexes in rhesus monkeys. III. Responses to translation. J. Neurophysiol., 80: 680–695.

Angelaki, D.E. and Dickman, J.D. (2000) Spatiotemporal processing of linear acceleration: primary afferent and central vestibular neuron responses. J. Neurophysiol., 84: 2113–2132.

Angelaki, D.E., Green, A.M. and Dickman, J.D. (2001) Differential sensorimotor processing of vestibulo-ocular signals during rotation and translation. J. Neurosci., 21: 3968–3985.

Angelaki, D.E. and Hess, B.J. (1994) Inertial representation of angular motion in the vestibular system of rhesus monkeys. I. Vestibuloocular reflex. J. Neurophysiol., 71: 1222–1249.

Angelaki, D.E. and Hess, B.J. (1995) Inertial representation of angular motion in the vestibular system of rhesus monkeys. II. Otolith-controlled transformation that depends on an intact cerebellar nodulus. J. Neurophysiol., 73: 1729–1751.

Angelaki, D.E., McHenry, M.Q., Dickman, J.D., Newlands, S.D. and Hess, B.J.M. (1999) Computation of inertial motion: neural strategies to resolve ambiguous otolith information. J. Neurosci., 19: 316–327.

Angelaki, D.E., Shaikh, A.G., Green, A.M. and Dickman, J.D. (2004) Neurons compute internal models of the physical laws of motion. Nature, 430: 560–564.

Baker, J., Goldberg, J., Hermann, G. and Peterson, B. (1984a) Optimal response planes and canal convergence in secondary neurons in vestibular nuclei of alert cats. Brain Res., 294: 133–137.

Baker, J., Goldberg, J., Hermann, G. and Peterson, B. (1984b) Spatial and temporal response properties of secondary neurons that receive convergent input in vestibular nuclei of alert cats. Brain Res., 294: 138–143.

Barmack, N.H. (2003) Central vestibular system: vestibular nuclei and posterior cerebellum. Brain Res. Bull., 60: 511–541.

Bos, J.E. and Bles, W. (2002) Theoretical considerations on canal-otolith interaction and an observer model. Biol. Cybern., 86: 191–207.

Brettler, S.C. and Baker, J.F. (2001) Directional sensitivity of anterior, posterior, and horizontal canal vestibulo-ocular neurons in the cat. Exp. Brain Res., 140: 432–442.

Cohen, B., John, P., Yakushin, S.B., Buettner-Ennever, J. and Raphan, T. (2002) The nodulus and uvula: source of cerebellar control of spatial orientation of the angular vestibulo-ocular reflex. Ann. N.Y. Acad. Sci., 978: 28–45.

Chen-Huang, C. and Peterson, B.W. (2006) Three dimensional spatial-temporal convergence of otolith related signals in vestibular only neurons in squirrel monkeys. Exp. Brain Res., 168: 410–426.

Clement, G., Moore, S.T., Raphan, T. and Cohen, B. (2001) Perception of tilt (somatogravic illusion) in response to sustained linear acceleration during space flight. Exp. Brain Res., 138: 410–418.

Crawford, J.D. and Vilis, T. (1991) Axes of eye rotation and Listing's law during rotations of the head. J. Neurophysiol., 65: 407–423.

Dichgans, J., Held, R., Young, L.R. and Brandt, T. (1972) Moving visual scenes influence the apparent direction of gravity. Science, 178: 1217–1219.

Dickman, J.D. and Angelaki, D.E. (2002) Vestibular convergence patterns in vestibular nuclei neurons of alert primates. J. Neurophysiol., 88: 3518–3533.

Einstein, A. (1908) Uber das Relativitätsprinzip und die aus demselben gezogenen Folgerungen. Jahrb. Radioakt., 4: 411–462.

Fernández, C. and Goldberg, J.M. (1971) Physiology of peripheral neurons innervating semicircular canals of the squirrel monkey. II. Response to sinusoidal stimulation and dynamics of peripheral vestibular system. J. Neurophysiol., 34: 661–675.

Fernández, C. and Goldberg, J.M. (1976a) Physiology of peripheral neurons innervating otolith organs of the squirrel monkey. I. Response to static tilts and to long-duration centrifugal force. J. Neurophysiol., 39: 970–984.

Fernández, C. and Goldberg, J.M. (1976b) Physiology of peripheral neurons innervating otolith organs of the squirrel monkey. III. Response dynamics. J. Neurophysiol., 39: 996–1008.

Glasauer, S. and Merfeld, D.M. (1997) Modelling three-dimensional vestibular responses during complex motion stimulation. In: Fetter M., Haslwanter T., Misslisch H. and Tweed D. (Eds.), Three-Dimensional Kinematics of Eye, Head and Limb Movements. Harwood Academic, Amsterdam, pp. 387–398.

Goldstein, H. (1980) Classical Mechanics. Addison Wesley, Reading, MA.

Graybiel, A., Johnson, W.H., Money, K.E., Malcolm, R.E. and Jennings, G.L. (1979) Oculogravic illusion in response to straight-ahead acceleration of CF-104 aircraft. Aviat. Space Environ. Med., 50: 382–386.

Green, A.M. and Angelaki, D.E. (2003) Resolution of sensory ambiguities for gaze stabilization requires a second neural integrator. J. Neurosci., 23: 9265–9275.

Green, A.M. and Angelaki, D.E. (2004) An integrative neural network for detecting inertial motion and head orientation. J. Neurophysiol., 92: 905–925.

Green, A.M. and Galiana, H.L. (1998) Hypothesis for shared central processing of canal and otolith signals. J. Neurophysiol., 80: 2222–2228.

Green, A.M., Shaikh, A.G. and Angelaki, D.E. (2005) Sensory vestibular contributions to constructing internal models of self-motion. J. Neural Eng., 2: S164–S179.

Holly, J.E., Pierce, S.E. and McCollum, G. (2006) Head tilt-translation combinations distinguished at the level of neurons. Biol. Cybern., 95: 311–326.

Howard, I.P. (1986) The perception of posture, self-motion and the visual vertical. In: Boff K.R., Kaufman L. and Thomas J.P. (Eds.), Handbook of Perception and Human Performance. Wiley, New York, pp. 1–50 Chapter 18.

Howard, I.P. and Hu, G. (2001) Visually induced reorientation illusions. Perception, 30: 583–600.

Indovina, I., Maffei, V., Bosco, G., Zago, M., Macaluso, E. and Lacquaniti, F. (2005) Representation of visual gravitational motion in the human vestibular cortex. Science, 308: 416–419.

Kasper, J., Schor, R.H. and Wilson, V.J. (1988) Response of vestibular neurons to head rotations in vertical planes. I. Response to vestibular stimulation. J. Neurophysiol., 60: 1753–1764.

Mazer, J.A. (1998) How the owl resolves auditory coding ambiguity. Proc. Natl. Acad. Sci., 95: 10932–10937.

Merfeld, D.M. (1995) Modeling the vestibulo-ocular reflex of the squirrel monkey during eccentric rotation and roll tilt. Exp. Brain Res., 106: 123–134.

Merfeld, D.M., Park, S., Gianna-Poulin, C., Black, F.O. and Wood, S. (2005a) Vestibular perception and action employ qualitatively different mechanisms. I. Frequency response of VOR and perceptual responses during translation and tilt. J. Neurophysiol., 94: 186–198.

Merfeld, D.M., Park, S., Gianna-Poulin, C., Black, F.O. and Wood, S. (2005b) Vestibular perception and action employ qualitatively different mechanisms. II. VOR and perceptual responses during combined Tilt & Translation. J. Neurophysiol., 94: 199–205.

Merfeld, D.M. and Young, L.R. (1995) The vestibulo-ocular reflex of the squirrel monkey during eccentric rotation and roll tilt. Exp. Brain Res., 106: 111–122.

Merfeld, D.M. and Zupan, L.H. (2002) Neural processing of gravitoinertial cues in humans. III. Modeling tilt and translation responses. J. Neurophysiol., 87: 819–833.

Merfeld, D.M., Zupan, L.H. and Gifford, C.A. (2001) Neural processing of gravitoinertial cues in humans. II. Influence of the semicircular canals during eccentric rotation. J. Neurophysiol., 85: 1648–1660.

Merfeld, D.M., Zupan, L.H. and Peterka, R.J. (1999) Humans use internal models to estimate gravity and linear acceleration. Nature, 398: 615–618.

Mergner, T. and Glasauer, S. (1999) A simple model of vestibular canal-otolith signal fusion. Ann. N.Y. Acad. Sci., 871: 430–434.

Movshon, J.A., Adelson, E.H., Gizzi, M.S. and Newsome, W.T. (1985) The analysis of moving visual patterns. In: Chagas C., Gattass R. and Gross C. (Eds.), Study Group on Pattern Recognition Mechanisms. Pontifica Academia Scientiarum, Vatican City, pp. 117–151.

Musallam, S. and Tomlinson, R.D. (2002) Asymmetric integration recorded from vestbular-only cells in response to position transients. J. Neurophysiol., 88: 2104–2113.

Pack, C.C. and Born, R.T. (2001) Temporal dynamics of a neural solution to the aperature problem in visual area MT of macaque brain. Nature, 409: 1040–1042.

Paige, G.D. and Seidman, S.H. (1999) Characteristics of the VOR in response to linear acceleration. Ann. N.Y. Acad. Sci., 871: 123–135.

Paige, G.D. and Tomko, D.L. (1991) Eye movement responses to linear head motion in the squirrel monkey. I. Basic characteristics. J. Neurophysiol., 65: 1170–1182.

Schor, R.H., Miller, A.D., Timerick, S.B.J. and Tomko, D.L. (1985) Responses to head tilt in cat central vestibular neurons. II. Frequency dependence of neural response vectors. J. Neurophysiol., 53: 1444–1452.

Schor, R.H., Miller, A.D. and Tomko, D.L. (1984) Responses to head tilt in cat central vestibular neurons. I. Direction of maximum sensitivity. J. Neurophysiol., 51: 136–146.

Schwarz, U. and Miles, F.A. (1991) Ocular responses to translation and their dependence on viewing distance. I. Motion of the observer. J. Neurophysiol., 66: 851–864.

Seidman, S.H., Telford, L. and Paige, G.D. (1995) Vertical, horizontal and torsional eye movement responses to head roll in the squirrel monkey. Exp. Brain Res., 104: 218–226.

Seidman, S.H., Telford, L. and Paige, G.D. (1998) Tilt perception during dynamic linear acceleration. Exp. Brain Res., 119: 307–314.

Shaikh, A.G., Ghasia, F.F., Dickman, J.D. and Angelaki, D.E. (2005a) Properties of cerebellar fastigial neurons during translation, rotation, and eye movements. J. Neurophysiol., 93: 853–863.

Shaikh, A.G., Green, A.M., Ghasia, F.F., Newlands, S.D., Dickman, J.D. and Angelaki, D.E. (2005b) Sensory convergence solves a motion ambiguity problem. Curr. Biol., 15: 1657–1662.

Shimojo, S., Silverman, G.H. and Nakayama, K. (1989) Occlusion and the solution to the aperature problem for motion. Vision Res., 29: 619–626.

Siebold, C., Anagnostou, E., Glasauer, S., Glonti, L., Kleine, J.F., Tchelidze, T. and Büttner, U. (2001) Canal-otolith interaction in the fastigial nucleus of the alert monkey. Exp. Brain Res., 136: 169–178.

Siebold, C., Glonti, L., Glasauer, S. and Büttner, U. (1997) Rostral fastigial nucleus activity in the alert monkey during three dimensional passive head movements. J. Neurophysiol., 77: 1432–1446.

Siebold, C., Kleine, J.F., Glonti, L., Tchelidze, T. and Büttner, U. (1999) Fastigial nucleus activity during different frequencies and orientations of vertical vestibular stimulation in the monkey. J. Neurophysiol., 82: 34–41.

Telford, L., Seidman, S.H. and Paige, G.D. (1997) Dynamics of squirrel monkey linear vestibuloocular reflex and interactions with fixation distance. J. Neurophysiol., 78: 1775–1790.

Wearne, S., Raphan, T. and Cohen, B. (1998) Control of spatial orientation of the angular vestibuloocular reflex by the nodulus and uvula. J. Neurophysiol., 79: 2690–2715.

Wearne, S., Raphan, T., Waespe, W. and Cohen, B. (1997) Control of three-dimensional dynamic characteristics of the angular vestibulo-ocular reflex by the nodulus and uvula. Prog. Brain Res., 114: 321–334.

Wilson, V.J., Ikegami, H., Schor, R.H. and Thomson, D.B. (1996) Tilt responses of neurons in the caudal descending nucleus of the decerebrate cat: influence of the caudal cerebellar vermis and of neck receptors. J. Neurophysiol., 75: 1242–1249.

Wilson, V.J., Yamagata, Y., Yates, B.J., Schor, R.H. and Nonaka, S. (1990) Response of vestibular neurons to head rotations in vertical planes. III. Response of vestibulocollic neurons to vestibular and neck stimulation. J. Neurophysiol., 64: 1695–1703.

Yakusheva, T.A., Shaikh, A.G., Green, A.M., Blazquez, P.M., Dickman, J.D. and Angelaki, D.E. (2007) Purkinje cells in the posterior vermis encode motion in an inertial reference frame. Neuron, 54: 973–985.

Yakushin, S.B., Raphan, T. and Cohen, B. (2006) Spatial properties of central vestibular neurons. J. Neurophysiol., 95: 464–478.

Zhou, W., Tang, B.F. and King, W.M. (2001) Responses of rostral fastigial neurons to linear acceleration in an alert monkey. Exp. Brain Res., 139: 111–115.

Zhou, W., Tang, B.F., Newlands, S.D. and King, W.M. (2006) Responses of monkey vestibular-only neurons to translation and angular rotation. J. Neurophysiol., 96: 2915–2930.

Zupan, L.H. and Merfeld, D.M. (2003) Neural processing of gravito-inertial cues in humans. IV. Influence of visual rotational cues during roll optokinetic stimuli. J. Neurophysiol., 89: 390–400.

Zupan, L.H. and Merfeld, D.M. (2005) An internal model of head kinematics predicts the influence of head orientation on reflexive eye movements. J. Neural Eng., 2: S180–S197.

Zupan, L.H., Merfeld, D.M. and Darlot, C. (2002) Using sensory weighting to model the influence of canal, otolith and visual cues on spatial orientation and eye movements. Biol. Cybern., 86: 209–230.

Zupan, L.H., Peterka, R.J. and Merfeld, D.M. (2000) Neural processing of gravito-inertial cues in humans. I. Influence of the semicircular canals following post-rotatory tilt. J. Neurophysiol., 84: 2001–2015.

P. Cisek, T. Drew & J.F. Kalaska (Eds.)
Progress in Brain Research, Vol. 165
ISSN 0079-6123
Copyright © 2007 Elsevier B.V. All rights reserved

CHAPTER 11

Sensorimotor optimization in higher dimensions

Douglas Tweed*

Departments of Physiology and Medicine, University of Toronto, Centre for Vision Research, York University, Toronto, ON, Canada

Abstract: Most studies of neural control have looked at constrained tasks, with only a few degrees of freedom, but real sensorimotor systems are high dimensional — e.g. gaze-control systems that coordinate the head and two eyes have to work with 12 degrees of freedom in all. These extra degrees of freedom matter, because they bring with them new issues and questions, which make it hard to translate low-dimensional findings into theories of real neural control. Here I show that it is possible to predict high-dimensional behavior if we apply the optimization principles introduced by 19th-century neuroscientists like Helmholtz, Listing, and Wundt. Using three examples — the vestibulo-ocular reflex, saccadic eye movements, and depth vision — I show how simple optimization theories can predict complex, unexpected behaviors and reveal fundamental features of sensorimotor control, e.g. that neural circuits perform non-commutative algebra; that in rapid gaze shifts the eye controllers deliver commands with three degrees of freedom, not two; and that the eyes roll about their lines of sight in a way that may simplify stereopsis.

Keywords: sensorimotor; optimization; degrees of freedom; control; oculomotor; vision; vestibulo-ocular reflex (VOR); saccades; stereopsis; computational; behavioral

In sensorimotor control, as in science fiction, strange things happen in higher dimensions. For simplicity, most studies of neural control have focused on low-dimensional tasks, meaning ones with few degrees of freedom, such as purely horizontal movements of an eye or flexions of a single joint. But real sensorimotor systems are high dimensional. An arm, for instance, has 7 degrees of freedom — 3 for the shoulder, 2 for the elbow, and 2 for the wrist. A single eye rotates with 3 degrees of freedom — horizontal, vertical, and torsional. The head moves with 6 degrees of freedom, and so gaze-control systems that coordinate the head and two eyes have to work with 12 dimensions in all. In these cases and others, the hope has been that if we can first manage to understand the system in a simple, constrained setting, we can then extrapolate to higher dimensions. But the extrapolation has often proved difficult. The key problem is that fundamentally new issues arise in higher dimensions, making it hard to generalize from low-dimensional findings. Here I will give examples of new concepts emerging in this way, but I will also argue that it is possible to predict high-dimensional behavior if we extrapolate in the right way, based on optimization.

Optimization theories of the brain go back to Helmholtz, Listing, Wundt, and other oculomotor pioneers of the 19th century (Helmholtz, 1867). To analyze a neural system by this approach, you first figure out what it is trying to do and state your guess precisely, in the form of a cost function. In an eye or arm movement, for instance, the cost

*Corresponding author. Tel.: +1-416-978-2603;
Fax: +1-416-978-4373; E-mail: douglas.tweed@utoronto.ca

DOI: 10.1016/S0079-6123(06)65011-5

might be the time to reach the target. To test the theory, you devise a controller that minimizes that cost function, and compare it to real human or animal behavior. Here I will show how simple optimization theories have predicted complex, high-dimensional behaviors that might otherwise have been inscrutable, or might never have been found. For instance, they predicted that certain random-dot stereograms are perceived as three-dimensional (3-D) only when they are viewed looking up, not down; and that in some gaze shifts the eyeballs twirl about their lines of sight at up to 200°/s for a fraction of a second and then unwind again. More importantly, these theories have revealed fundamental features of sensorimotor control, e.g. that in depth vision the brain searches for matching images in the two eyes over fixed rather than mobile patches of retina and that the two eyes are coordinated so as to shrink these patches; that eye control during gaze shifts is 3-D even though the line of sight has just 2 degrees of freedom; and that there is noncommutative computation in the sensorimotor circuitry of the brain.

Noncommutativity in the brain

One example of a concept that emerges in higher dimensions is noncommutativity. A process is said to be noncommutative if order matters when things combine; if order makes *no* difference, the process is commutative (Hamilton, 1853). For instance ordinary multiplication of numbers is commutative because the order of factors is irrelevant, e.g. $5 \times 7 = 7 \times 5$. 1-D rotations are also commutative — turning first 10° right and then 20° right yields the same outcome as turning first 20° and then 10°. But 3-D rotations are noncommutative: the same two rotations applied in different orders can yield different overall rotations (Westheimer, 1957; Tweed and Vilis, 1987; Carpenter, 1988; McCarthy, 1990). This point is illustrated in Fig. 1, where the chess knight, starting from the same orientation, undergoes the same two rotations, 90° right and 90° down, in different orders and winds up in different final positions (in this figure the motions are defined in a knight-fixed frame, but rotations defined in a space-fixed frame do not commute either).

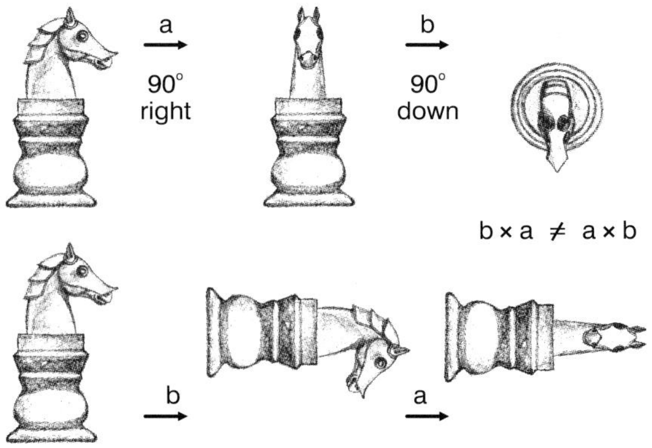

Fig. 1. Three-dimensional rotations do not commute. (Adapted with permission from Tweed (2003).)

Why does noncommutativity of rotations matter for the brain? Because many brain processes have to deal with rotations, e.g. processes such as spatial perception, navigation, and the control of rotary joints. If they are to do their jobs even half decently, these systems have to represent and compute rotations, and for this they need noncommutative algebra (Westheimer, 1957; Tweed and Vilis, 1987, 1990; Crawford and Vilis, 1991; Minken et al., 1993; Hestenes, 1994a, b; Tweed et al., 1994; Tweed 1997a; Henriques et al., 1998; Smith and Crawford 1998), though for a long time this idea was controversial (van Opstal et al., 1991; Tweed et al., 1994, 1999; Straumann et al., 1995; Raphan, 1997, 1998; Quaia and Optican, 1998; Smith and Crawford, 1998; Schnabolk and Raphan, 1994).

Of all the neural systems that deal with rotations, maybe the simplest is the vestibulo-ocular reflex, or VOR. This reflex acts like a Steadicam for the eyeballs, stabilizing the retinal images when the head moves. Sense organs in the inner ear measure head velocity and send commands to the eye muscles, moving the eyes in the opposite direction when the head turns, so as to prevent the eyeballs rotating relative to space (Carpenter, 1988).

That the VOR needs noncommutative computation is illustrated in Fig. 2. Here a subject sits in a rotary capsule and looks out through a viewing screen at a space-fixed target, the black disk. In

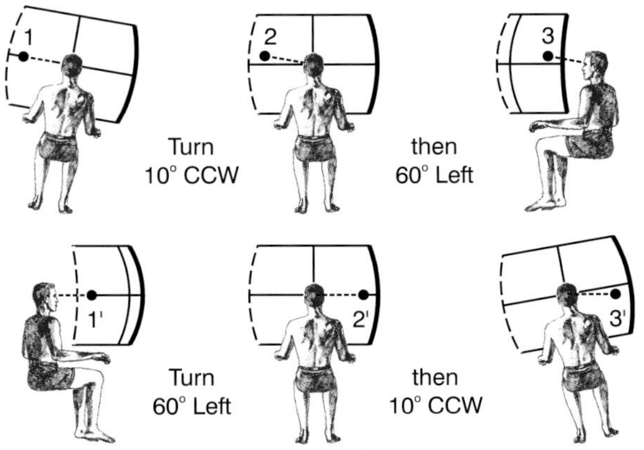

Fig. 2. An optimal VOR must be noncommutative. In both the upper and lower series, the subject sits in a rotary capsule viewing a space-fixed object (the black disk), and in both series the target's initial location relative to the subject is the same: 30° directly to the left. Then the lights go out, and the subject tries to keep looking at the unseen disk while undergoing two rotations. In the two series the rotations are identical but are applied in opposite orders. Because of noncommutativity, the target's final locations relative to the subject are different. (Adapted with permission from Tweed (2003).)

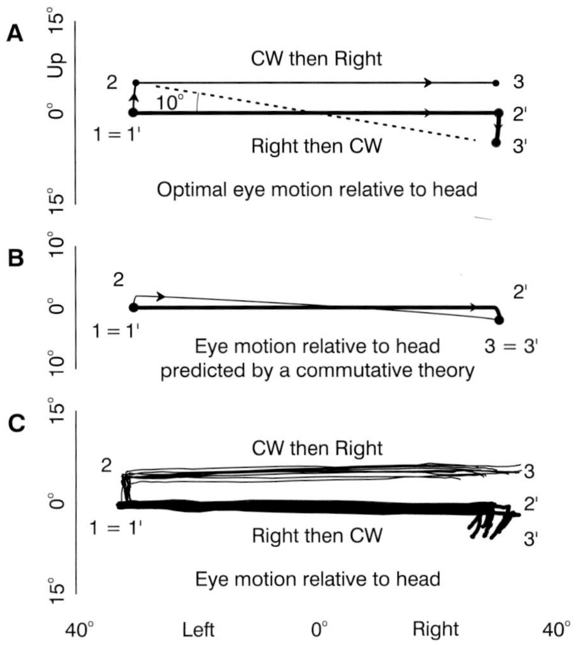

Fig. 3. The real VOR is noncommutative. (A) Motion of the eye in the head, during the task from Fig. 2, as predicted by a theory of the VOR where the retinal image is perfectly stabilized. The system is noncommutative, yielding different final eye positions depending on the order of head rotations. (B) Performance, on the same task, of a VOR model in which all neural processing is commutative. The final eye positions do not depend on the sequence of body rotations. (C) A real human subject shows noncommutativity: final eye Positions 3 and 3′ differ by about 10°, as predicted by the optimization theory. (Adapted with permission from Tweed et al. (1999).)

both the upper and lower sequences, the subject starts out in the same position relative to the target: looking at it 30° directly to his left. In the upper series the subject turns first 10° counterclockwise (CCW) and then 60° left, so to keep his eyes on the target; he has to end up looking 30° right and 5° up. In the lower series, the subject undergoes the same two rotations in reverse order and winds up looking right and *down*. In other words the VOR must compute different final eye-position commands when the subject goes through the same rotations in different orders (Tweed et al., 1999).

The motion of the eye in the head is plotted in Fig. 3A. If the subject turns first CCW then left, the eyes counter-rotate first clockwise (CW) then right, winding up at Position 3. When the subject rotates in the reverse order the eyes turn right and then CW, winding up at Position 3′. These trajectories are simulations of a theory of the 3-D VOR (Tweed, 1997b; Tweed et al., 1999) that was extrapolated from earlier, 1-D theories where the eye moved purely horizontally. There are many ways to extrapolate from low dimensions to high, but here the extrapolation preserved the optimization principle that the VOR acts to minimize retinal-image slip. So noncommutativity emerged as a necessary feature.

This optimal behavior was not predicted by previous 3-D models of the VOR, because they were extrapolated from 1-D theories in a different way, by preserving the 1-D principle that eye-position commands are integrals of eye-velocity commands. But integration is commutative in the sense that the final value of an integral does not depend on the temporal order of its inputs, and therefore models based on this principle neglect noncommutativity and are incompatible with optimal image stabilization in 3-D. For instance, Fig. 3B shows one such commutative model (Raphan, 1997): regardless of the order of rotations, it brings the eye to the same final orientation relative to the head, and so relative to space the eye is incorrectly positioned, off the target.

On this same task, real human subjects closely matched the optimization theory, adopting different final eye positions that depended on the order of rotations. For the subject shown in Fig. 3C, the difference between Positions 3 and 3′ (averaged over several trials) was 9.0° vertically, as compared to the optimal value of 10°. Averaged across all five subjects, the difference was 10.3° (range 7.4–12.6), and it was significant for each individual subject.

These findings established that there is non-commutativity in the VOR: the reflex correctly computes different final eye-position commands when put through identical rotations in different orders. And the broader point is that a simple optimization theory, based on minimizing retinal slip, predicted a fundamental feature of eye control that was absent in 1-D and was missed by other approaches. This theory (for details, see Tweed et al., 1994; Tweed, 1997a, b) has predicted many features of ocular control (e.g. Tweed, 1997b; Tweed et al., 1999; Misslisch and Tweed, 2000) and continues to find experimental support; e.g. Klier et al. (2006) recently showed that stimulating the abducens nerve rotates the eyeball around an axis that tilts as a function of eye position, in the pattern predicted by this theory.

Optimizing gaze control in three dimensions

In this section we focus on another high-dimensional concept, kinematic redundancy. We say a system is kinematically redundant if it has more degrees of freedom that it needs for some job. For example an arm has 7 degrees of freedom, but it needs only 6 to place the hand in any possible position (within a reasonable range near the shoulder joint). And in 3-D, an eye also has kinematic redundancy: it rotates with 3 degrees of freedom, but the line of sight has just 2, so there are infinitely many different eye positions that are all compatible with any one gaze direction (Fig. 4A shows three possible eye positions for straight-ahead gaze).

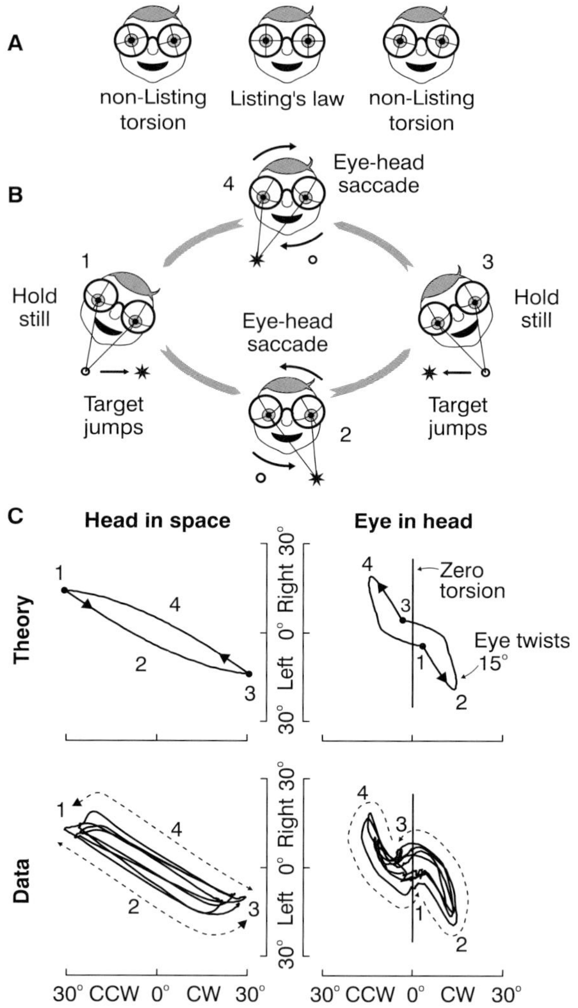

Fig. 4. (A) Listing's law. The three heads show three of the infinitely many different possible eye orientations for the same gaze direction, but Listing' law says the eye always chooses the orientation where torsion is zero. (B) A twisting-saccade task shows that a time-optimal saccadic system must control ocular torsion independently of horizontal and vertical eye position (see text). (C) Experimental data confirm that the eye controller for saccades has independent control of torsion. In the top row, a time-optimal controller performs the twisting-saccade task (number labels 1–4 correspond to Stages 1–4 in the illustration above). The controller does not move the eye directly (in the head frame) from its initial Position 1 to its final Position 3, but instead drives the eye out to 15° of torsion, which is the maximal allowable in this simulation, and then unwinds to its destination. This way, the eye stays near zero torsion between movements, and during gaze shifts it takes the fastest possible path to its final position in space. In the bottom row, a real human subject shows the same pattern. (Adapted with permission from Tweed et al. (1998).)

From this infinity of potential positions, the brain consistently chooses just one — the orientation in which the eye's torsional angle is zero, as shown by the central face in Fig. 4A. This zero-torsion rule is known as Listing's law, and it holds to within a degree or two during fixation and in the rapid gaze shifts known as saccades, as long as the head stays still (Helmholtz, 1867; Tweed and Vilis, 1990; Minken et al., 1993; Straumann et al., 1995).

Listing's law has been taken to mean that the eye controller for gaze shifts is 2-D, generating only horizontal and vertical commands. No torsional commands are needed, in this view, because torsion just stays at zero. But from an optimization viewpoint, there is reason to suspect that torsion is under separate, active neural control, and that the torsional command might be revealed by looking at saccades involving eye and head. More precisely, we need torsional control if the saccadic system is even roughly time-optimal, bringing the eye to its target position quickly (Tweed et al., 1998).

The crucial thought experiment is shown in Fig. 4B. The subject starts out in Position 1, with the head tilted 30° left-ear-down, looking at a target light which is 20° straight down relative to the head, and 1 m away. Then the target jumps sideways and the subject makes a twisting eye-head gaze shift to refixate it, passing through Position 2 in mid-saccade and ending up in Position 3. The interesting stage is 2: the eye is quicker than the head — it reorients more swiftly when an interesting object appears in the visual periphery (Roucoux et al., 1980; Laurutis and Robinson, 1986; Guitton and Volle, 1987; Tweed et al., 1995) — so a time-optimal controller would exploit that speed, flicking the eye to the target and locking on while the head catches up. The eye should move quickly to its final 3-D orientation in space, turning not just horizontally and vertically but also torsionally, so that midway through the head movement, the eye is twisted in its socket in the CW direction (from the subject's viewpoint), as shown in Stage 2. It should then hold still in space as the head completes its motion. If the target then jumps back to its original location, we should expect a similar return trajectory, this time with a strong CCW twist in mid-saccade, as in Stage 4. This is the time-optimal strategy, and clearly it requires a torsional controller that can twist the eye rapidly in its orbit.

The top row of Fig. 4C shows a simulated time-optimal controller (Tweed, 1997a; Tweed et al., 1998) performing this task. Again, the interesting thing is the eye's path relative to the head: it does not simply jump from Position 1 to 3, but takes a wide horizontal and torsional detour through Position 2, twisting through about 15° and then unwinding back to near-zero torsion (and similarly on the return trip through Position 4).

Faced with this same task, real human subjects behave like the time-optimal model. The subject in Fig. 4C showed the predicted torsional loops, ranging from 17° CW to 15° CCW. Across all four subjects, the torsional range averaged 29°, which far exceeds the 2–4° seen during head-fixed gaze shifts (Helmholtz, 1867; Tweed and Vilis, 1990; Minken et al., 1993; Straumann et al., 1995). And these huge torsional excursions really were visually evoked gaze shifts, not vestibular reactions to head motion, because they usually began 20–60 ms before the head started moving (Tweed et al., 1998). The eye spun about its line of sight at up to 200°/s for 80 ms and then unwound to near-zero torsion over 200 ms (Tweed et al., 1998), so that Listing's law was in force at the end of the movement.

Obeying Listing's law brings advantages: it likely requires less muscle force to hold torsion near zero; and the eye, at the center of its torsional range, is optimally placed for the next gaze shift, which may go either CW or CCW (Hepp, 1990; Tweed, 1997c). So why does the eye break Listing's law during the gaze shift? As shown in Fig. 4B, the eye twists to anticipate the impending torsional motion of the head. This way, it reaches its final position in space while the head is still in mid-movement. From then on, the eye holds a stable orientation in space, so the visual world remains stationary on the retina, blur is reduced, and visual analysis is simplified in other ways as well (Tweed et al., 1998). So there is more to torsional eye movement than simply holding at zero, and this study shows how an optimization model led to the discovery of an independent torsion-control system that helps drive saccades and underlies Listing's law.

The motor side of depth vision

For our final example we turn to stereopsis, where the visual system computes the 3-D locations of objects based on their images in the two eyes. The first step is to identify corresponding image features on the two retinas (Julesz, 1960). Figure 5 illustrates the problem: the eyes view a cloud of 21 dots, which cast 21 identical images on each retina. How does the brain know which dot on the right retina corresponds to which one on the left? We know the brain can find these matches, even when the images are thousands of identical dots, as in random-dot stereograms.

How does it manage? Geometry may help: as shown in Fig. 5, the optics of the situation restrict matching images to what are called epipolar lines (Ogle, 1950; Rogers and Bradshaw, 1996). So if it could locate these epipolar lines, the brain could simplify its quest for matching images: it would not have to search the entire retina for a match, but could carry out a 1-D search along the epipolar line, like looking for lost hikers along a single trail rather than combing the whole forest.

Most theories of stereopsis have proposed that the brain searches along epipolar lines. But these theories were worked out assuming stationary eyes. When we consider that the eyes move, the theories hits a snag: the epipolar lines migrate on the retinas (Garding et al., 1995; Stevenson and Schor, 1997; Tweed, 1997c). As shown in Fig. 5, the same point on one retina corresponds to different epipolar lines on the other retina, depending on the configuration of the eyes (in the figure, the eyes rotate about their own lines of sight, but other sorts of rotations also shift the epipolar lines). Again this is a problem of dimensionality: earlier theories neglected all three dimensions of eye rotation (or all six, counting both eyes), and new issues arise when we consider these extra degrees of freedom.

In light of this complication, there are two ways the brain might find matching images in mobile eyes (Schreiber et al., 2001). The options are illustrated in Fig. 6. Given an image falling on some locus in one retina, the brain could use eye-position information to locate its epipolar line on the other retina. The other option is to forget about finding epipolar lines and instead search a 2-D patch of retina large enough to encompass all possible locations of the epipolar line in any likely eye configuration. This way, the stereoptic system would not have to monitor eye position, but it would lose the advantage of a 1-D search. So the question is: Does the brain search for matches along epipolar lines, or over retina-fixed 2-D zones?

We can answer this question using rotated stereograms, as shown in Fig. 7. We construct a random-dot stereogram in the usual way and then rotate the disks. If the disk viewed by the right eye is turned CCW, and the other CW, as in Fig. 7, the stereogram is incyclorotated. If the rotations are reversed, it is excyclorotated. Why are these stereograms useful? We know that when people converge their eyes and look up — when they look at something close to their forehead — they excycloverge, rotating the upper poles of both eyeballs outward (Allen, 1954; Mok et al., 1992; Van Rijn and Van den Berg, 1993; Minken and Van Gisbergen, 1994; Tweed, 1997c; Kapoula et al., 1999; Steffen et al., 2000; Schreiber et al., 2001); and when they converge and look down, they incycloverge. So the prediction is this: if our stereo search zones are retina-fixed, we should be better able to see excyclorotated stereograms on upgaze, incyclorotated on downgaze.

For example, suppose you view a stereogram that is excyclorotated by 5°. When your eyes are also excycloverged 5°, just like the stereogram disks, then the optical correspondence should be normal, just as if you were viewing a normal, nonrotated stereogram with zero cyclovergence, so the image should be easy to see. But when your eyes are cycloverged 0° then corresponding dots in the two excyclorotated disks will project onto odd locations on your two retinas, making the stereoimage hard to see. And again, this is the prediction *if* stereo search zones are retina-fixed; if instead the search zones move with the epipolar lines then eye position should not affect visibility.

Figure 8A confirms that the search zones move with the retinas, not with the epipolar lines. It plots the probability of stereoptic vision versus cyclorotation of the stereogram at three eye elevations for a typical subject. For instance when a

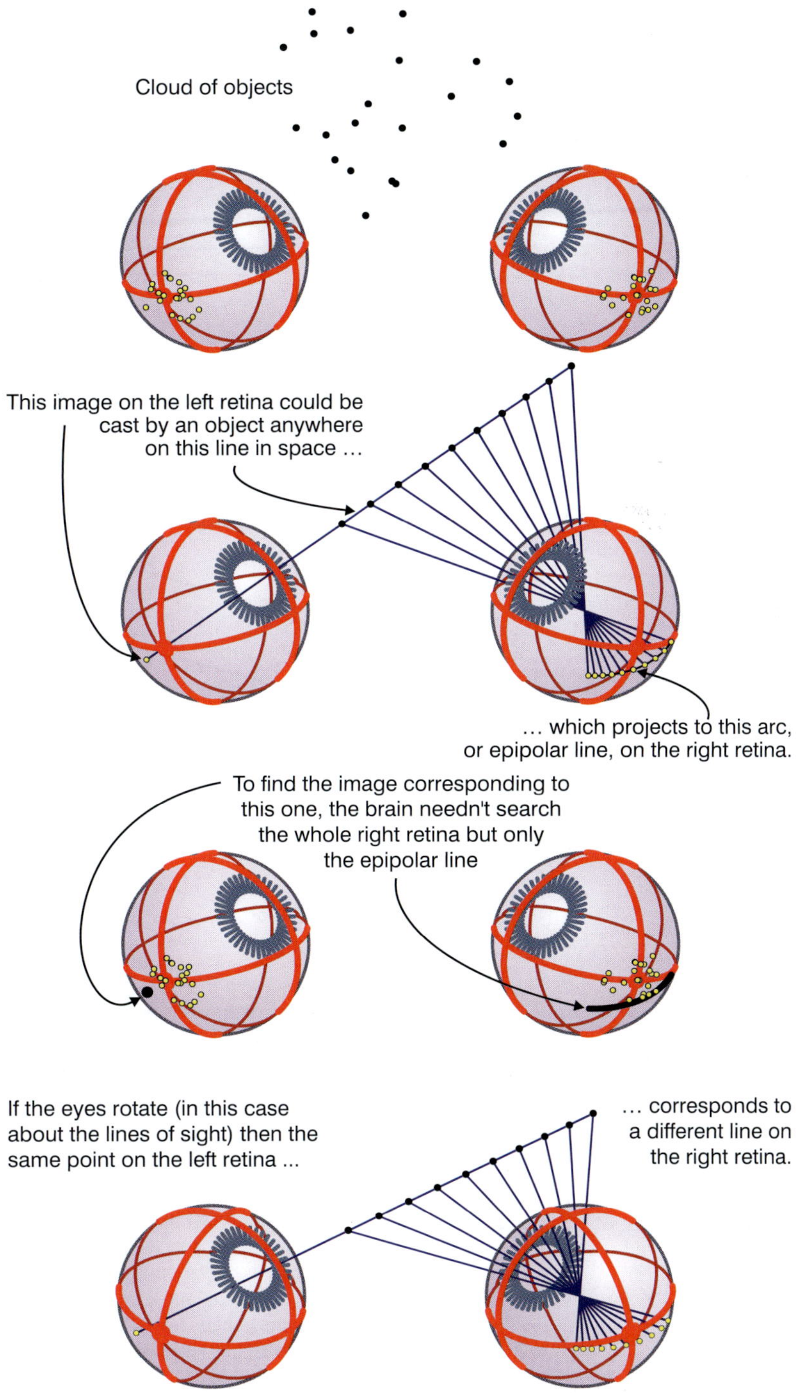

Fig. 5. Any animal with stereopsis must solve the stereo-matching problem, deducing which images on the right retina correspond to which ones on the left. The task can be simplified using epipolar lines, but when the eyes move, the epipolar lines migrate on the retinas.

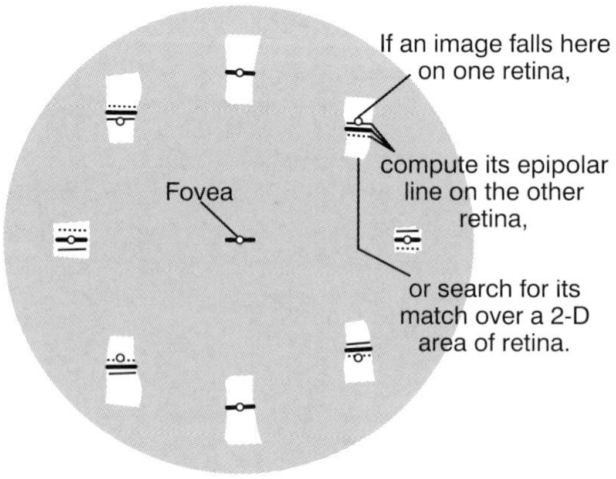

Fig. 6. There are two ways the visual system could look for matching images in mobile eyes. The nine small circles are nine images projected onto the right retina, one foveal and the others 15° eccentric (the large gray disk is the region within 22.5° of the fovea). Corresponding images on the left retina must lie somewhere on the line segments, which are pieces of epipolar lines, but the lines are in different places depending on the positions of the eyes. White patches cover the ranges of motion of the epipolar segments when the eyes move over a realistic range. (Adapted with permission from Schreiber et al. (2001).)

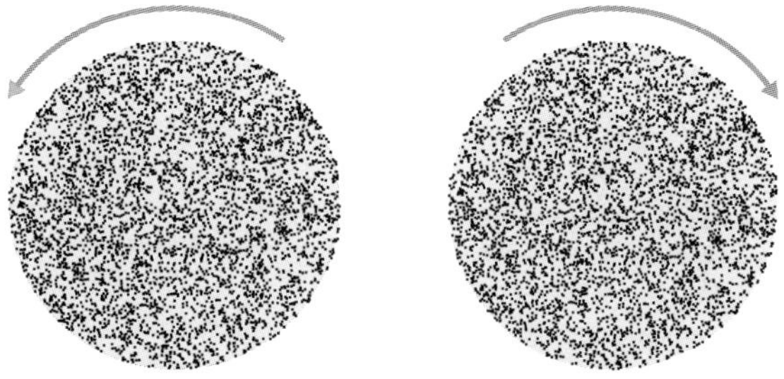

Fig. 7. Top: cyclorotated stereograms are visible only in certain eye positions. Cross-fuse the disks from 30 cm away and depress your gaze as far as possible, holding the paper orthogonal to the plane of your sight lines. You should see a depth image (a triangle) in this position, but not when you do the same on upward gaze. If the image never disappears, your search zones are too large for this stereogram; try the examples in Schreiber et al. (2001).

stereogram is incyclorotated by 6° then it is perceived with probability 1 when the eyes are directed 30° down (dotted line) and with probability 0 when the eyes are 30° up (thin gray line) — i.e. this stereogram is visible on downgaze but not on upgaze.

Figure 8B plots perception thresholds — the angles of cyclorotation at which stereograms were

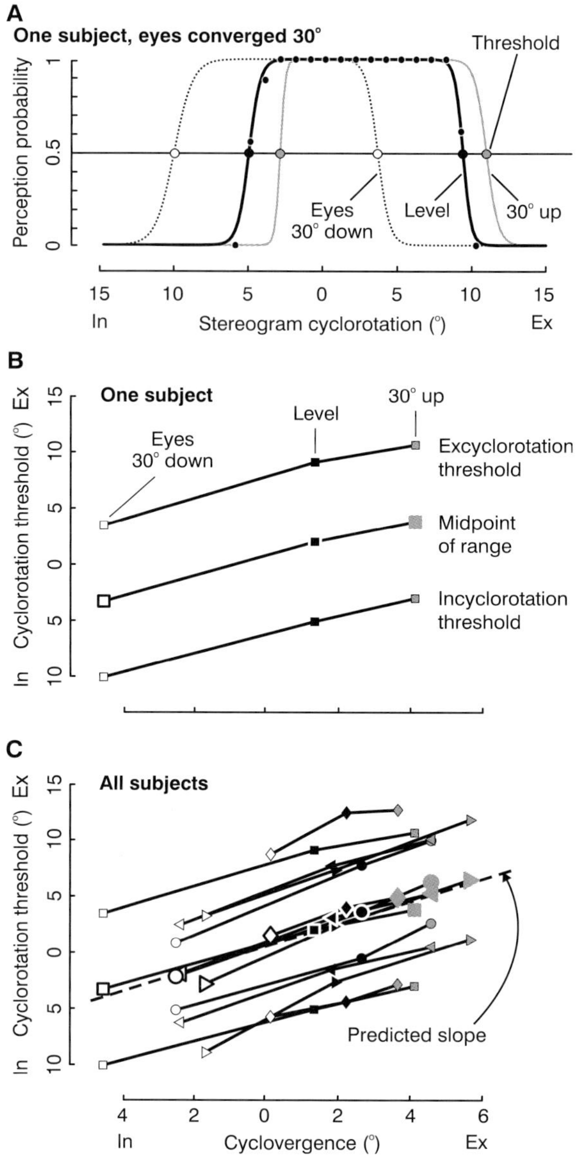

Fig. 8. Stereopsis depends on gaze elevation. (A) For this typical subject, the small black dots show the probability of seeing the stereoimage as a function of the cyclorotation of the stereogram when the elevation of the eyes is 0°. Curves plot performance at three gaze elevations. Larger circles are perceptual thresholds — cyclorotation angles at which stereoimages were perceived with probability 0.5. (B) Stereoptic thresholds depend on gaze elevation and cyclovergence. (C) For all five subjects, these thresholds varied significantly with gaze elevation, shifting toward incyclorotation on downgaze (leftmost symbol in each string of three) and toward excyclorotation on upgaze (rightmost symbol). The average slope is 1.06, very close to the slope of 1 predicted if stereo search zones are retina-fixed. (Adapted with permission from Schreiber et al. (2001).)

perceived 50% of the time — versus ocular cyclovergence, for the same subject as in Fig. 8A, for the same three eye elevations. And as in Fig. 8A, white symbols indicate data collected on downgaze, gray means upgaze, and black means level. So for instance, of the nine plotted points in this panel, the small white one at the lower left corner means that this subject, when looking 30° down (and converging 30°) had about 4.5° of incyclovergence, and an incyclorotation threshold of 10°; i.e. the subject perceived stereoimages with probability 0.5 when the stereogram was incyclorotated by that amount. Similarly, the leftmost point on the upper line of the plot means this subject's excyclorotation threshold under these conditions was about 4°. The large dot halfway between the in- and excyclorotation thresholds is the average of the two thresholds.

Figure 8C shows thresholds and midpoints for all five subjects. If stereo search zones were perfectly fixed on the retinas then the line of midpoints would have a slope of 1 (because the cyclorotation thresholds would rotate exactly as far as the eyes), as indicated by the dashed line. The actual slope, averaged over all subjects, was 1.06, and not significantly different from 1. So the data indicate that stereo search zones are retina-fixed.

This finding suggests that eye control plays a central part in stereopsis. An optimized controller could coordinate the eyes so as to minimize the motion of the epipolar lines, allowing stereopsis to get by with the smallest possible search zones. The normal pattern of eye control when viewing distant objects is Listing's law (Helmholtz, 1867; Carpenter, 1988), but on near gaze the law is broken (Allen, 1954; Mok et al., 1992; Van Rijn and Van den Berg, 1993; Minken and Van Gisbergen, 1994; Tweed, 1997c; Kapoula et al., 1999; Steffen et al., 2000; Schreiber et al., 2001) and it can be shown that the deviations from Listing's law shrink the required search zones (Schreiber et al., 2001). The zones are not precisely minimized — the eye's deviations from Listing's law are not large enough for that — but the reason may be that the controller is balancing the benefits of small zones against the advantages of Listing's law (Carpenter, 1988; Hepp, 1990; Tweed, 1997c).

Conclusion

Most studies of neural control have focused on low-dimensional tasks, with few degrees of freedom, but real sensorimotor systems are high dimensional. I have argued that new issues arise in higher dimensions, but I have also shown, with three examples, that it is nevertheless possible to extrapolate usefully from low-dimensional findings if we do it based on optimization principles. Each of these three examples suggests further questions and generalizations. I have shown that there is noncommutative computation in the circuitry of the VOR, and by similar reasoning, one would expect noncommutativity also in many other brain systems that deal with rotations, such as those for head and limb control, auditory and visual localization, space constancy, and mental rotation of objects (Hestenes, 1994b; Tweed, 1997a; Henriques et al., 1998). Optimization ideas clarified the implementation of Listing's and Donders' laws, and there are doubtless, waiting to be discovered, many higher-dimensional analogs of these laws, constraining the motions of the eyes, head, and limbs in complex tasks. An optimization model clarified the relation between stereopsis and eye control in six dimensions, and this model, too, leads to further predictions, for instance that the layout of stereo search zones on the retinas should resemble the optimal pattern in Fig. 6 (Schreiber et al., 2001). And optimization methods have been applied with great success to many other sensorimotor problems besides my specific examples. From the pioneering work of Helmholtz to the present day, probably no other approach has been so successful at illuminating the complex control systems of the brain.

Acknowledgments

I thank my co-authors on the original studies, J. D. Crawford, M. Fetter, T. Haslwanter, V. Happe, and K. Schreiber. For comments and technical help I thank M. Abdelghani, K. Beykirch, D. Broussard, L. Chinta Venkataswararao, J. Dichgans, S. Ferber, K. Fortney, D. Goche Montes, P. Hallett, C. Hawkins, D. Henriques, I. Howard,

E. Klier, H. Misslisch, P. Nguyen, M. Niemeier, J. Sharpe, T. Vilis, H. Wang, A. Wong, and J. Zacher. This work was supported by the Canadian Institutes for Health Research and the Deutsche Forschungsgemeinschaft.

References

Allen, M.J. (1954) The dependence of cyclophoria on convergence, elevation and the system of axes. Am. J. Optom. Arch. Am. Acad. Optom., 31: 297–307.

Carpenter, R.H.S. (1988) Movements of the Eyes. Pion, London.

Crawford, J.D. and Vilis, T. (1991) Axes of eye rotation and Listing's law during rotations of the head. J. Neurophysiol., 65: 407–423.

Garding, J., Porrill, J., Mayhew, J.E. and Frisby, J.P. (1995) Stereopsis, vertical disparity and relief transformations. Vision Res., 35: 703–722.

Guitton, D. and Volle, M. (1987) Gaze control in humans: eye-head coordination during orienting movements to targets within and beyond the oculomotor range. J. Neurophysiol., 58: 427–459.

Hamilton, W. (1853) Lectures on Quaternions: Containing a Systematic Statement of a New Mathematical Method. Hodges and Smith, Dublin.

Helmholtz, H.von. (1867) Handbuch der Physiologischen Optik. L. Voss, Leipzig.

Henriques, D.Y., Klier, E.M., Smith, M.A., Lowy, D. and Crawford, J.D. (1998) Gaze-centered remapping of remembered visual space in an open-loop pointing task. J. Neurosci., 18: 1583–1594.

Hepp, K. (1990) On Listing's law. Commun. Math. Phys. V, 132: 285–292.

Hestenes, D. (1994a) Invariant body kinematics: I. Saccadic and compensatory eye movements. Neural Netw., 7: 65–77.

Hestenes, D. (1994b) Invariant body kinematics: II. Reaching and neurogeometry. Neural Netw., 7: 79–88.

Julesz, B. (1960) Binocular depth perception of computer-generated patterns. BSTJ, 39: 1125–1162.

Kapoula, Z., Bernotas, M. and Haslwanter, T. (1999) Listing's plane rotation with convergence: role of disparity, accommodation, and depth perception. Exp. Brain Res., 126: 175–186.

Klier, E.M., Meng, H. and Angelaki, D.E. (2006) Three-dimensional kinematics at the level of the oculomotor plant. J. Neurosci., 26: 2732–2737.

Laurutis, V.P. and Robinson, D.A. (1986) The vestibulo-ocular reflex during human saccadic eye movements. J. Physiol., 373: 209–233.

McCarthy, J.M. (1990) An Introduction to Theoretical Kinematics. MIT Press, Cambridge, MA.

Minken, A.W. and Van Gisbergen, J.A. (1994) A three-dimensional analysis of vergence movements at various levels of elevation. Exp. Brain Res., 101: 331–345.

Minken, A.W., Van Opstal, A.J. and Van Gisbergen, J.A. (1993) Three-dimensional analysis of strongly curved saccades elicited by double-step stimuli. Exp. Brain Res., 93: 521–533.

Misslisch, H. and Tweed, D. (2000) Torsional dynamics and cross-coupling in the human vestibulo-ocular reflex during active head rotation. J. Vestib. Res., 10: 119–125.

Mok, D., Crawford, J.D. and Vilis, T. (1992) Rotation of Listing's plane during vergence. Vision Res., 32: 2055–2064.

Ogle, K.N. (1950) Researches in Binocular Vision. W.B. Saunders, Philadelphia, PA.

van Opstal, A.J., Hepp, K., Hess, B.J., Straumann, D. and Henn, V. (1991) Two- rather than three-dimensional representation of saccades in monkey superior colliculus. Science, 252: 1313–1315.

Quaia, C. and Optican, L.M. (1998) Commutative saccadic generator is sufficient to control a 3-D ocular plant with pulleys. J. Neurophysiol., 79: 3197–3215.

Raphan, T. (1997) Modelling control of eye orientation in three dimensions. In: Fetter M., Haslwanter T., Misslisch H. and Tweed D. (Eds.), Three-dimensional Kinematics of Eye, Head and Limb Movements. Harwood Academic Publishers, Amsterdam, The Netherlands, pp. 359–376.

Raphan, T. (1998) Modeling control of eye orientation in three dimensions. I. Role of muscle pulleys in determining saccadic trajectory. J. Neurophysiol., 79: 2653–2667.

Rogers, B.J. and Bradshaw, M.F. (1996) Does the visual system use the epipolar constraint for matching binocular images? Invest Ophthalmol. Vis. Sci., 37(Suppl.): 31–25.

Roucoux, A., Crommelinck, M., Guerit, J.M. and Meulders, M. (1980) In: Fuchs A.F. and Becker W.E. (Eds.), Progress in Oculomotor Research. Elsevier, Amsterdam, pp. 309–315.

Schnabolk, C. and Raphan, T. (1994) Modeling three-dimensional velocity-to-position transformation in oculomotor control. J. Neurophysiol., 71: 623–638.

Schreiber, K., Crawford, J.D., Fetter, M. and Tweed, D. (2001) The motor side of depth vision. Nature, 410: 819–822.

Smith, M.A. and Crawford, J.D. (1998) Neural control of rotational kinematics within realistic vestibuloocular coordinate systems. J. Neurophysiol., 80: 2295–2315.

Steffen, H., Walker, M.F. and Zee, D.S. (2000) Rotation of Listing's plane with convergence: independence from eye position. Invest. Ophthalmol. Vis. Sci., 41: 715–721.

Stevenson, S.B. and Schor, C.M. (1997) Human stereo matching is not restricted to epipolar lines. Vision Res., 37: 2717–2723.

Straumann, D., Zee, D.S., Solomon, D., Lasker, A.G. and Roberts, D.C. (1995) Transient torsion during and after saccades. Vision Res., 35: 3321–3334.

Tweed, D. (1997a) Three-dimensional model of the human eye-head saccadic system. J. Neurophysiol., 77: 654–666.

Tweed, D. (1997b) Velocity-to-position transformations in the VOR and the saccadic system. In: Fetter M., Haslwanter T., Misslisch H. and Tweed D. (Eds.), Three-dimensional Kinematics of Eye, Head and Limb movements. Harwood Academic Publishers, Amsterdam, The Netherlands, pp. 375–386.

Tweed, D. (1997c) Visual-motor optimization in binocular control. Vision Res., 37: 1939–1951.

Tweed, D. (2003) Microcosms of the Brain. Oxford University Press, Oxford, UK.

Tweed, D. and Vilis, T. (1987) Implications of rotational kinematics for the oculomotor system in three dimensions. J. Neurophysiol., 58: 832–849.

Tweed, D. and Vilis, T. (1990) Geometric relations of eye position and velocity vectors during saccades. Vision Res., 30: 111–127.

Tweed, D., Glenn, B. and Vilis, T. (1995) Eye-head coordination during large gaze shifts. J. Neurophysiol., 73: 766–779.

Tweed, D., Haslwanter, T. and Fetter, M. (1998) Optimizing gaze control in three dimensions. Science, 281: 1363–1366.

Tweed, D., Misslisch, H. and Fetter, M. (1994) Testing models of the oculomotor velocity-to-position transformation. J. Neurophysiol., 72: 1425–1429.

Tweed, D.B., Haslwanter, T.P., Happe, V. and Fetter, M. (1999) Non-commutativity in the brain. Nature, 399: 261–263.

Van Rijn, L.J. and Van den Berg, A.V. (1993) Binocular eye orientation during fixations: Listing's law extended to include eye vergence. Vision Res., 33: 691–708.

Westheimer, G. (1957) Kinematics of the eye. J. Opt. Soc. Am., 47: 967–974.

P. Cisek, T. Drew & J.F. Kalaska (Eds.)
Progress in Brain Research, Vol. 165
ISSN 0079-6123
Copyright © 2007 Elsevier B.V. All rights reserved

CHAPTER 12

How tightly tuned are network parameters? Insight from computational and experimental studies in small rhythmic motor networks

Eve Marder, Anne-Elise Tobin* and Rachel Grashow

Volen Center MS 013, Brandeis University, 415 South St., Waltham, MA 02454-9110, USA

Abstract: We describe theoretical and experimental studies that demonstrate that a given pattern of neuronal activity can be produced by variable sets of underlying conductances. Experimental work demonstrates that individual identified neurons in different animals may show variations as large as 2–5 fold in the conductance densities of specific ion channels. Theoretical work shows that models with this range of variation in many of their maximal conductances can produce similar activity. Together, these observations suggest that neurons and networks may be less tightly tuned than previously thought. Consequently, we argue that instead of attempting to construct single canonical models of neuronal function, it might be more useful to construct and analyze large families of models that give similar behavior.

Keywords: neuronal models; conductance-based models; Central Pattern Generators; half-center oscillator; neuronal homeostasis

One of the great challenges in neuroscience is to understand how network dynamics depend on the interaction between the intrinsic membrane properties of the network neurons and their synaptic interactions. We know that most neurons have a large number of different kinds of voltage and time dependent ion channels (Marder, 1998). Additionally, synaptic potentials are quite diverse, and can show complex time- and voltage-dependent properties (Zucker and Regehr, 2002). The widespread implementation of synaptic learning rules in neural networks has led to an implicit, almost unconscious, assumption among many neuroscientists that in order for a network to perform well all of its parameters must be quite tightly tuned. This assumption was buttressed by the historical difficulty of tuning complex models to give a desired output. In this chapter we will present a combination of both experimental and computational work that suggests that many solutions can produce similar network performance.

*Corresponding author. Tel.: +1 781 736 3141;
Fax: +1 781 736 3142; E-mail: atobin@brandeis.edu

Compensating conductances in a two-cell network

Two-cell, reciprocally inhibitory networks have been studied for almost 100 years. Early work in motor control defined the concept of a half-center oscillator in which rhythmic bursts of activity would drive alternations in flexor-extensor activity in the spinal cord (Brown, 1911, 1914). Subsequently, this generic circuit has been richly studied computationally (Perkel and Mulloney, 1974;

DOI: 10.1016/S0079-6123(06)65012-7

Wang and Rinzel, 1992; Cymbalyuk et al., 1994; Skinner et al., 1994; Van Vreeswijk et al., 1994; Nadim et al., 1995, 1999; Olsen et al., 1995; White et al., 1998) and experimentally (Friesen, 1994; Sharp et al., 1996; Cymbalyuk et al., 2002; Sorensen et al., 2004).

Reciprocally inhibitory circuits can produce a wide variety of outputs, including alternating spiking, alternating bursting, in-phase spiking, and in-phase bursting (Sharp et al., 1996), depending on a variety of parameters, including the time course of the reciprocal inhibition (Van Vreeswijk et al., 1994). Sharp et al. (1996) used the dynamic clamp (Sharp et al., 1993a, b; Prinz et al., 2004) to construct reciprocally inhibitory networks using two Gastric Mill (GM) neurons from the crab stomatogastric ganglion. They were able to produce alternating bursting in two-cell networks by adding an I_h conductance (also with the dynamic clamp) to each neuron, which provided a slowly activating inward current that sustained slow alternating bursts. The period of the alternating bursts increased as the conductance of the inhibitory synapse was increased and also when the I_h conductance was decreased. In this study, the two conductances were varied individually.

To understand how intrinsic and synaptic parameters may interact, we replicated the essential paradigm used by Sharp et al. (1996); however, we varied the conductances of the inhibitory synapses and I_h simultaneously to construct a 9×9 matrix of 81 different versions of the two-neuron network (Fig. 1). This allows us to examine the behavior of

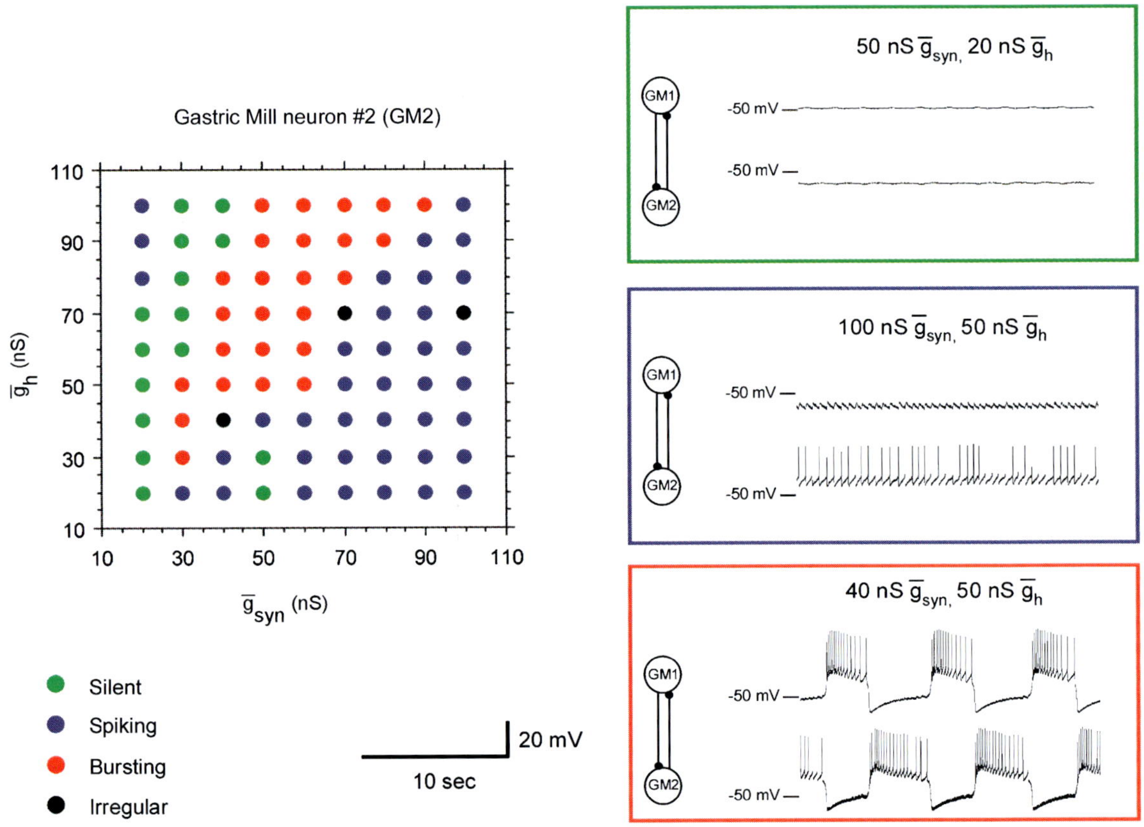

Fig. 1. A range of parameter values can produce similar network output in an artificial reciprocally inhibitory two-cell network. *Left*, map of parameter space (synaptic conductance g_{syn}, and h-conductance g_h) for one GM neuron (GM2) in the network. Green, blue, red, and black mark the parameter values for which GM2 was quiescent, spiking tonically, bursting rhythmically, and exhibiting irregular activity, respectively. *Right*, intracellular recordings exemplifying three characteristic activity patterns of the network for different parameter values. Box colors correspond to activity of GM2 as symbolized in the parameter map legend.

the network as a function of these two conductances. At the top left of the matrix, the green points mark parameter combinations in which GM2 was silent (see recordings shown in the green inset box). The blue points in the lower right of the plot mark the parameter regime that led to tonic spiking in GM2 (inset shown in blue). The points shown in red mark the parameters that gave rise to alternating bursts of activity (recordings shown in red inset box).

It is clear from this experiment that there is a wedge in parameter space in which stable rhythmic alternating bursts were found. The shape and size of the wedge demonstrates that, in this experiment, stable half-center activity was produced over a 3-fold range of both I_h and synaptic conductances, as long as they were covaried (Fig. 1). That is to say, both conductances had to be either large or small, but that within that constraint, they could vary considerably.

Figure 2 shows the frequency of the alternating bursts produced in the same experiment as shown in Fig. 1. The same color code is used: the red points and the recordings show that similar burst frequencies can result from quite different sets of parameters. These dynamic clamp experiments illustrate that a given network output can be produced over an extensive range of parameters, as long as appropriate relationships among some of those parameters are maintained.

Modeling studies of leech heart interneurons have demonstrated similar findings (Cymbalyuk et al., 2002). The leak parameters, E_{leak} and g_{leak} could vary up to 3-fold, as long as the parameters covaried. Within that range, many parameter combinations could produce similar burst frequency, duty cycle, and interburst spike frequency.

Building models to capture the dynamics of real neurons

For many years most researchers wishing to build a conductance-based model of a specific neuron or

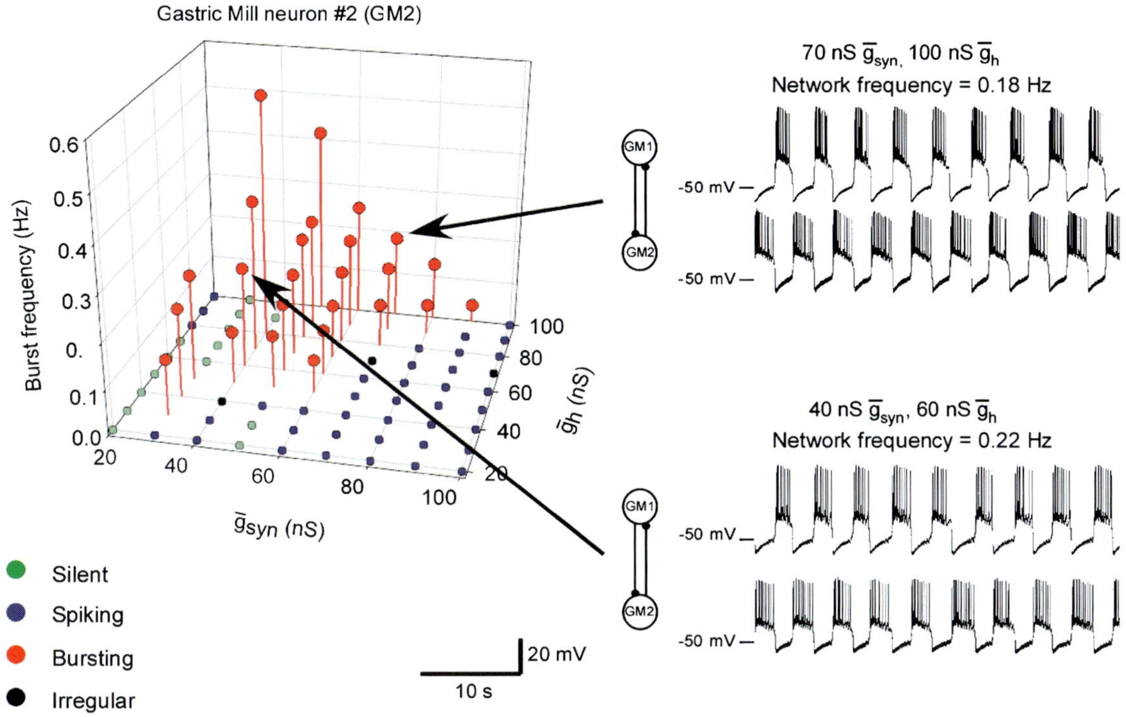

Fig. 2. Similar network frequencies result from different network parameters. *Left*, burst frequency for the artificial half-center network when the values of g_{syn} and g_{h} are varied. *Right*, intracellular recordings from two half-center networks with different parameter values.

neuron type followed the same general strategy. First, all available biophysical data describing the voltage- and time-dependent currents in the neuron were fit with appropriate differential equations (Hodgkin and Huxley, 1952; Buchholtz et al., 1992; Traub et al., 1994). Second, values for currents not measured in the neuron in question were pulled from the literature, either from other cell types in the same species, or from different species. Third, a decision was made to build either a single-compartment model or a multicompartmental model (De Schutter et al., 2005). Fourth, the model was hand-tuned, using some set of measurements of the neuron's firing properties as a criterion for the tuning process (De Schutter et al., 2005).

Inherent in this program are two fundamental assumptions: (a) that all neurons of the same class or type are virtually identical, and (b) that the end-result of the hand-tuning would produce an optimal solution that would capture in detail the parameters of the ideal neuron, and therefore could be used to extract insights into the mechanisms by which interacting currents give rise to specific dynamics. Recent experimental and theoretical work summarized below challenges the validity of both of these assumptions. We now argue that the process of building semi-realistic or realistic model neurons in the future will involve the construction of a family of models that may equally well capture the dynamics and variability of the neurons that are to be modeled.

Biological variability in synaptic and intrinsic conductances

Biophysical measurements of synaptic and intrinsic conductances are conventionally reported as means and standard errors, leading to the assumption that the mean is the "true value." However, we now argue that reporting the range of the underlying data may be as important as reporting means alone. For example, it is now becoming clear that 2–5 fold ranges of both intrinsic and synaptic conductances may be common in many systems (Golowasch et al., 1999a; Swensen and Bean, 2005; Marder and Goaillard, 2006; Schulz et al., 2006). Figure 3a shows electrophysiological recordings from two LP neurons from two different crab stomatogastric ganglia during ongoing activity, and the voltage-clamp measurements of three K currents from those same neurons. Note that while the overall activity pattern is quite similar in the two cases, the outward currents were quite different in the recordings from the two preparations (Schulz et al., 2006). Figure 3b shows data from a larger population of neurons (Schulz et al., 2006) and shows the spread of conductance densities measured in different LP neurons, a result similar to that reported earlier (Golowasch et al., 1999a). Interestingly, the same kind of variability is seen at the level of mRNA expression for these channel genes (Fig. 3c). Note that the mRNA expression and measured conductance are correlated in the same neuron, demonstrating that this variability is not a result of experimental error (Fig. 3d). Thus, the parameter range that we see in the half-center examples in Figs. 1 and 2 may be similar to the parameter ranges found in biological preparations.

This conclusion, that 2–5 fold ranges in conductances can underlie similar activity, flies in the face of years of intuitions we have developed from pharmacological studies in which currents and synapses are modified. These studies often show dramatic alterations in activity from changes in a synaptic or intrinsic conductance of 20–50%. Figure 4A shows the effects of the application of the neuromodulator, dopamine, on the transient outward current, I_A, recorded from the PD neuron of the lobster stomatogastric ganglion (Harris-Warrick et al., 1995a, b; Szucs and Selverston, 2006). Note that while dopamine increased the outward current by ~25%, the PD neuron dramatically changed its firing patterns – much of this change presumably attributable to the effects of dopamine on I_A (Fig. 4B).

While such neuromodulator studies indicate that even moderate changes in a conductance can greatly alter activity, it is extremely important to distinguish between changing one parameter in a neuron or a network at one moment in time and the variance that can occur in a population, when compensatory mechanisms act throughout the lifetime of each animal (Swensen and Bean, 2005; Marder and Goaillard, 2006). This is illustrated by

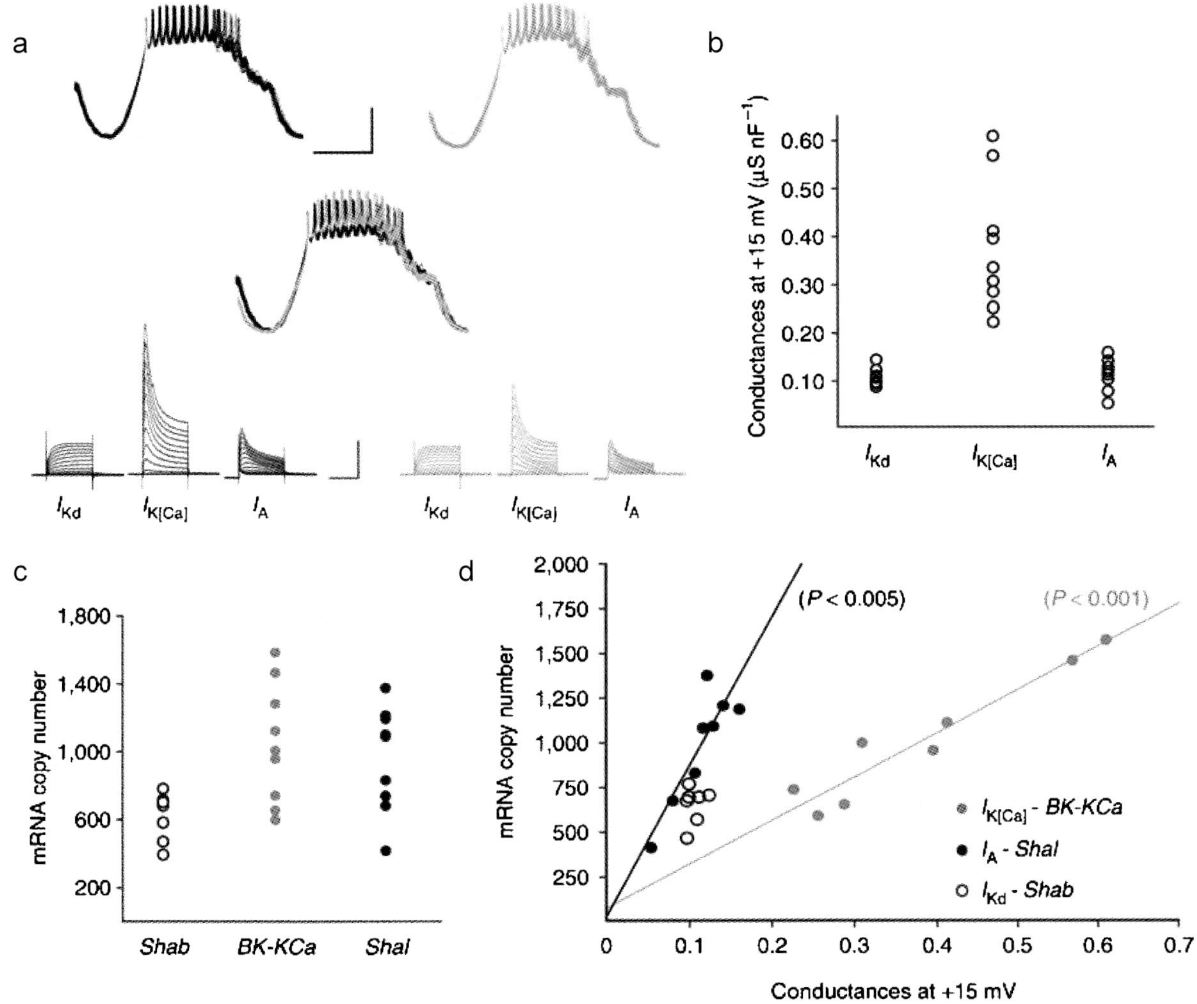

Fig. 3. Variable conductances underlie stereotyped activity in identified neurons. (a) *Top*, 20 burst cycles are overlaid for each of two LP neurons from two different animals, black and gray. Vertical scale is −60 to −50 mV; horizontal scale, 200 ms. *Middle*, traces from the two neurons are overlaid to demonstrate similarity. *Bottom*, voltage-clamp recordings from each neuron of 3 K currents: I_{Kd}, delayed rectifier, $I_{K[Ca]}$, calcium-dependent K, and I_A, fast transient K. Vertical scale is 50 nA; horizontal scale, 100 ms. (b) Normalized conductances for the 3 K currents as measured in voltage clamp. (c) Normalized levels of mRNA of *Shab*, *BK-KCa*, and *Shal* corresponding to the 3 K currents, I_{Kd}, $I_{K[Ca]}$, and I_A, respectively. The mRNA levels are measured in the same neurons for which voltage-clamp measurements are taken. (d) mRNA levels significantly correlate with the conductance measurements for $I_{K[Ca]}$ and I_A. Adapted with permission from Schulz et al. (2006).

the recent experiments from the Harris-Warrick laboratory in which mRNA encoding I_A was injected into single PD neurons. This resulted in 2–3 fold changes in the measured outward current (Fig. 4C), but no change in the neuron's activity (Fig. 4D) because a compensatory upregulation of I_h occurred (MacLean et al., 2003, 2005).

Constructing model families

Hand-tuning detailed conductance-based models is tedious, frustrating, and intellectually unsatisfying. It is intellectually unsatisfying because at the end of a hand-tuning process the modeler has no assurance that the solution is in any way "correct"

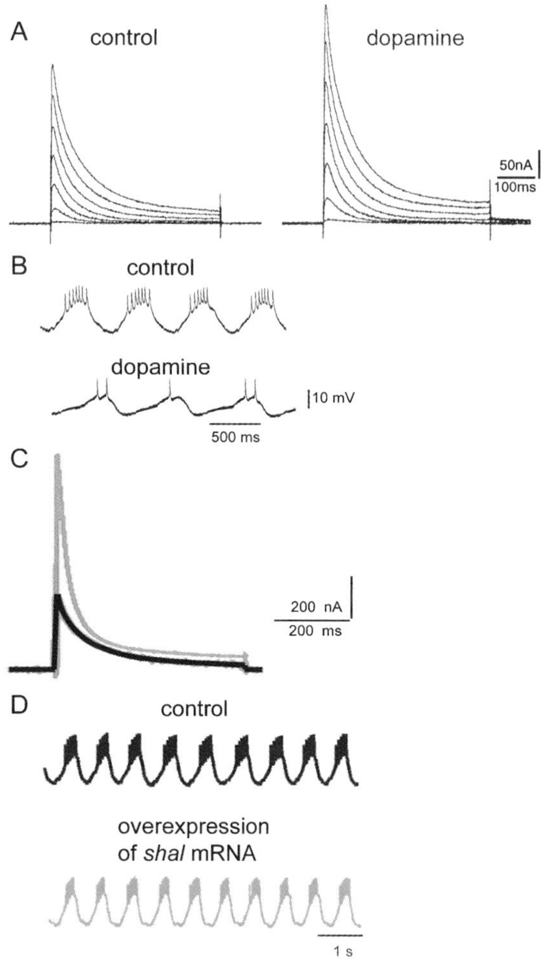

Fig. 4. Effect on PD neuron activity differs whether the K current, I_A, is increased by a modulator or by overexpression of *Shal* mRNA, with a compensatory increase in I_h. (A) Increase in I_A, as measured by a series of voltage-clamp steps, in PD neurons due to dopamine application. (B) Change in PD activity due to dopamine application. (C) I_A is considerably larger, as measured by a single voltage-clamp step, in PD neurons injected with *Shal* mRNA (gray) than in control PD neurons (black). (D) Activity is similar between control PD neurons (black) and those injected with *Shal* mRNA (gray). No vertical scale bar provided. (A) and (B) adapted with permission from Kloppenburg et al. (1999), Figs. 1A and 4A, respectively. (C) and (D) adapted with permission from MacLean et al. (2005), Figs. 1A and 4B, respectively.

or even representative of all of the possible solutions that might be present in a complex, multidimensional parameter space. As computational power has increased, we are seeing the development of multiple strategies to replace hand-tuning. These include randomly generating many potential solutions, and then selecting among them for candidate models (Goldman et al., 2001; Prinz et al., 2003), and using parameter optimization algorithms multiple times to produce multiple solutions (Achard and De Schutter, 2006; Tobin and Calabrese, 2006).

Interestingly, in several recent studies, multiple models that show similar electrophysiological phenotypes are found in connected regions of parameter space (Achard and De Schutter, 2006; Taylor et al., 2006). This latter observation is not to be taken for granted, as there easily could be disconnected islands in multidimensional parameter space that give similar behavior. Nonetheless, if neurons with similar behavior are connected in multidimensional parameter space, a neuron could maintain its overall physiological properties while adjusting ion channel number and distribution using homeostatic tuning rules. More specifically, if activity or other feedback mechanisms are used to control the insertion and deletion of ion channels, then neurons may continuously self-tune to maintain constant activity despite ongoing channel turnover (LeMasson et al., 1993; Liu et al., 1998; Golowasch et al., 1999b). In other words, individual neurons may be "wandering around" in parameter space throughout their lifetime, as long as they stay within the connected region of multidimensional parameter space consistent with a given output pattern of activity. This could then help explain the 2–4 fold range of conductances measured in identified neurons of the same cell type (Fig. 3b).

What can be learned from a population of model neurons with similar behavior that cannot be learned from studying a canonical model? Classically, sensitivity analyses are performed to determine how a model's behavior depends on one or more of its parameters (Guckenheimer et al., 1993, 1997; Nadim et al., 1995; Olsen et al., 1995). These methods are very good at finding the precise location of bifurcations or transitions between states in a model's output. Different information can be obtained by generating a large population of models, all of which mimic well the biological neurons to be studied (Taylor et al., 2006). By looking at the ranges of the values for each of the parameters in the population and at correlations among these parameters, one can begin to discover which combinations of parameters may compensate for each other to maintain a desired output pattern.

Moreover, using a single canonical model to study neuron function is as dangerous as using measurements from a single neuron to represent the neuron population as a whole. Just as the maximal conductances vary from neuron to neuron, neuronal morphology, responses to synaptic inputs, and neuromodulators also vary across preparations. Studies of a single-model neuron run the risk of sampling only a small region of the set of biologically possible examples, and may lead to conclusions that are idiosyncratic to that particular model. In contrast, by studying a population of models, one can look for results that are general, in much the same way that experimentalists look for results that are common to all examples of a single neuron type.

Acknowledgments

This work was supported by MH46742 and the McDonnell Foundation. We thank Dr. Adam Taylor for contributing to all of these ideas.

References

Achard, P. and De Schutter, E. (2006) Complex parameter landscape for a complex neuron model. PLoS Comput. Biol., 2: e94.

Brown, T.G. (1911) The intrinsic factors in the act of progression in the mammal. Proc. R. Soc. Lond. Biol., 84: 308–319.

Brown, T.G. (1914) On the nature of the fundamental activity of the nervous centres, together with an analysis of the conditioning of rhythmic activity in progression, and a theory of the evolution of function in the nervous system. J. Physiol., 48: 18–46.

Buchholtz, F., Golowasch, J., Epstein, I.R. and Marder, E. (1992) Mathematical model of an identified stomatogastric ganglion neuron. J. Neurophysiol., 67: 332–340.

Cymbalyuk, G.S., Gaudry, Q., Masino, M.A. and Calabrese, R.L. (2002) Bursting in leech heart interneurons: cell-autonomous and network-based mechanisms. J. Neurosci., 22: 10580–10592.

Cymbalyuk, G.S., Nikolaev, E.V. and Borisyuk, R.M. (1994) In-phase and antiphase self-oscillations in a model of two electrically coupled pacemakers. Biol. Cybern., 71: 153–160.

De Schutter, E., Ekeberg, O., Kotaleski, J.H., Achard, P. and Lansner, A. (2005) Biophysically detailed modelling of microcircuits and beyond. Trends Neurosci., 28: 562–569.

Friesen, W.O. (1994) Reciprocal inhibition: a mechanism underlying oscillatory animal movements. Neurosci. Biobehav., 18: 547–553.

Goldman, M.S., Golowasch, J., Marder, E. and Abbott, L.F. (2001) Global structure, robustness, and modulation of neuronal models. J. Neurosci., 21: 5229–5238.

Golowasch, J., Abbott, L.F. and Marder, E. (1999a) Activity-dependent regulation of potassium currents in an identified neuron of the stomatogastric ganglion of the crab *Cancer borealis*. J. Neurosci., 19: RC33.

Golowasch, J., Casey, M., Abbott, L.F. and Marder, E. (1999b) Network stability from activity-dependent regulation of neuronal conductances. Neural Comput., 11: 1079–1096.

Guckenheimer, J., Gueron, S. and Harris-Warrick, R.M. (1993) Mapping the dynamics of a bursting neuron. Philos. Trans. R. Soc. Lond. B, 341: 345–359.

Guckenheimer, J., Harris-Warrick, R., Peck, J. and Willms, A. (1997) Bifurcation, bursting, and spike frequency adaptation. J. Comput. Neurosci., 4: 257–277.

Harris-Warrick, R.M., Coniglio, L.M., Barazangi, N., Guckenheimer, J. and Gueron, S. (1995a) Dopamine modulation of transient potassium current evokes phase shifts in a central pattern generator network. J. Neurosci., 15: 342–358.

Harris-Warrick, R.M., Coniglio, L.M., Levini, R.M., Gueron, S. and Guckenheimer, J. (1995b) Dopamine modulation of two subthreshold currents produces phase shifts in activity of an identified motoneuron. J. Neurophysiol., 74: 1404–1420.

Hodgkin, A.L. and Huxley, A.F. (1952) A quantitative description of membrane current and its application to conduction and excitation in nerve. J. Physiol., 117: 500–544.

Kloppenburg, P., Levini, R.M. and Harris-Warrick, R.M. (1999) Dopamine modulates two potassium currents and inhibits the intrinsic firing properties of an identified motor neuron in a central pattern generator network. J. Neurophysiol., 81: 29–38.

LeMasson, G., Marder, E. and Abbott, L.F. (1993) Activity-dependent regulation of conductances in model neurons. Science, 259: 1915–1917.

Liu, Z., Golowasch, J., Marder, E. and Abbott, L.F. (1998) A model neuron with activity-dependent conductances regulated by multiple calcium sensors. J. Neurosci., 18: 2309–2320.

MacLean, J.N., Zhang, Y., Goeritz, M.L., Casey, R., Oliva, R., Guckenheimer, J. and Harris-Warrick, R.M. (2005) Activity-independent coregulation of I_A and I_h in rhythmically active neurons. J. Neurophysiol., 94: 3601–3617.

MacLean, J.N., Zhang, Y., Johnson, B.R. and Harris-Warrick, R.M. (2003) Activity-independent homeostasis in rhythmically active neurons. Neuron, 37: 109–120.

Marder, E. (1998) From biophysics to models of network function. Annu. Rev. Neurosci., 21: 25–45.

Marder, E. and Goaillard, J.M. (2006) Variability, compensation and homeostasis in neuron and network function. Nat. Rev. Neurosci., 7: 563–574.

Nadim, F., Manor, Y., Kopell, N. and Marder, E. (1999) Synaptic depression creates a switch that controls the frequency of an oscillatory circuit. Proc. Natl. Acad. Sci. U.S.A., 96: 8206–8211.

Nadim, F., Olsen, Ø.H., De Schutter, E. and Calabrese, R.L. (1995) Modeling the leech heartbeat elemental oscillator I: interactions of intrinsic and synaptic currents. J. Comput. Neurosci., 2: 215–235.

Olsen, Ø.H., Nadim, F. and Calabrese, R.L. (1995) Modeling the leech heartbeat elemental oscillator II: exploring the parameter space. J. Comput. Neurosci., 2: 237–257.

Perkel, D.H. and Mulloney, B.M. (1974) Motor pattern production in reciprocally inhibitory neurons exhibiting postinhibitory rebound. Science, 185: 181–183.

Prinz, A.A., Abbott, L.F. and Marder, E. (2004) The dynamic clamp comes of age. Trends Neurosci., 27: 218–224.

Prinz, A.A., Billimoria, C.P. and Marder, E. (2003) Alternative to hand-tuning conductance-based models: construction and analysis of databases of model neurons. J. Neurophysiol., 90: 3998–4015.

Schulz, D.J., Goaillard, J.M. and Marder, E. (2006) Variable channel expression in identified single and electrically coupled neurons in different animals. Nat. Neurosci., 9: 356–362.

Sharp, A.A., O'Neil, M.B., Abbott, L.F. and Marder, E. (1993a) The dynamic clamp: artificial conductances in biological neurons. Trends Neurosci., 16: 389–394.

Sharp, A.A., O'Neil, M.B., Abbott, L.F. and Marder, E. (1993b) Dynamic clamp: computer-generated conductances in real neurons. J. Neurophysiol., 69: 992–995.

Sharp, A.A., Skinner, F.K. and Marder, E. (1996) Mechanisms of oscillation in dynamic clamp constructed two-cell half-center circuits. J. Neurophysiol., 76: 867–883.

Skinner, F.K., Kopell, N. and Marder, E. (1994) Mechanisms for oscillation and frequency control in reciprocal inhibitory model neural networks. J. Comput. Neurosci., 1: 69–87.

Sorensen, M., DeWeerth, S., Cymbalyuk, G. and Calabrese, R.L. (2004) Using a hybrid neural system to reveal regulation of neuronal network activity by an intrinsic current. J. Neurosci., 24: 5427–5438.

Swensen, A.M. and Bean, B.P. (2005) Robustness of burst firing in dissociated Purkinje neurons with acute or long-term reductions in sodium conductance. J. Neurosci., 25: 3509–3520.

Szucs, A. and Selverston, A.I. (2006) Consistent dynamics suggests tight regulation of biophysical parameters in a small network of bursting neurons. J. Neurobiol., 66: 1584–1601.

Taylor, A.L., Hickey, T.J., Prinz, A.A. and Marder, E. (2006) Structure and visualization of high-dimensional conductance spaces. J. Neurophysiol., 96: 891–905.

Tobin, A.E. and Calabrese, R.L. (2006) Endogenous and half-center bursting in morphologically inspired models of leech heart interneurons. J. Neurophysiol., 96: 2089–2106.

Traub, R.D., Jefferys, J.G., Miles, R., Whittington, M.A. and Toth, K. (1994) A branching dendritic model of a rodent CA3 pyramidal neurone. J. Physiol., 481(Pt 1): 79–95.

Van Vreeswijk, C., Abbott, L.F. and Ermentrout, G.B. (1994) When inhibition not excitation synchronizes neural firing. J. Comput. Neurosci., 1: 313–321.

Wang, X.-J. and Rinzel, J. (1992) Alternating and synchronous rhythms in reciprocally inhibitory model neurons. Neural Comput., 4: 84–97.

White, J.A., Chow, C.C., Ritt, J., Soto-Trevino, C. and Kopell, N. (1998) Synchronization and oscillatory dynamics in heterogeneous, mutually inhibited neurons. J. Comput. Neurosci., 5: 5–16.

Zucker, R.S. and Regehr, W.G. (2002) Short-term synaptic plasticity. Annu. Rev. Physiol., 64: 355–405.

P. Cisek, T. Drew & J.F. Kalaska (Eds.)
Progress in Brain Research, Vol. 165
ISSN 0079-6123
Copyright © 2007 Elsevier B.V. All rights reserved

CHAPTER 13

Spatial organization and state-dependent mechanisms for respiratory rhythm and pattern generation

Ilya A. Rybak[1,*], Ana P.L. Abdala[2], Sergey N. Markin[1], Julian F.R. Paton[2] and Jeffrey C. Smith[3]

[1]*Department of Neurobiology and Anatomy, Drexel University College of Medicine, Philadelphia, PA 19129, USA*
[2]*Department of Physiology, School of Medical Sciences, University of Bristol, Bristol BS8 1TD, UK*
[3]*Cellular and Systems Neurobiology Section, National Institute of Neurological Disorders and Stroke, National Institutes of Health, Bethesda, MD 20892-4455, USA*

Abstract: The brainstem respiratory network can operate in multiple functional states engaging different state-dependent neural mechanisms. These mechanisms were studied in the in situ perfused rat brainstem–spinal cord preparation using sequential brainstem transections and administration of riluzole, a pharmacological blocker of persistent sodium current (I_{NaP}). Dramatic transformations in the rhythmogenic mechanisms and respiratory motor pattern were observed after removal of the pons and subsequent medullary transactions down to the rostral end of pre-Bötzinger complex (pre-BötC). A computational model of the brainstem respiratory network was developed to reproduce and explain these experimental findings. The model incorporates several interacting neuronal compartments, including the ventral respiratory group (VRG), pre-BötC, Bötzinger complex (BötC), and pons. Simulations mimicking the removal of circuit components following transections closely reproduce the respiratory motor output patterns recorded from the intact and sequentially reduced brainstem preparations. The model suggests that both the operating rhythmogenic mechanism (i.e., network-based or pacemaker-driven) and the respiratory pattern generated (e.g., three-phase, two-phase, or one-phase) depend on the state of the pre-BötC (expression of I_{NaP}-dependent intrinsic rhythmogenic mechanisms) and the BötC (providing expiratory inhibition in the network). At the same time, tonic drives from pons and multiple medullary chemoreceptive sites appear to control the state of these compartments and hence the operating rhythmogenic mechanism and motor pattern. Our results suggest that the brainstem respiratory network has a spatial (rostral-to-caudal) organization extending from the rostral pons to the VRG, in which each functional compartment is controlled by more rostral compartments. The model predicts a continuum of respiratory network states relying on different contributions of intrinsic cellular properties versus synaptic interactions for the generation and control of the respiratory rhythm and pattern.

Keywords: respiratory CPG; brainstem; medulla; pons; pre-Bötzinger complex; computational modeling; respiratory rhythm generation

---

*Corresponding author. Tel.: +1 215 991 8596;
Fax: +1 215 843 9082; E-mail: rybak@drexel.edu

DOI: 10.1016/S0079-6123(06)65013-9

Introduction

Breathing movements in mammals are produced by the respiratory central pattern generator (CPG). The network architecture and neural mechanisms for rhythmic pattern generation in most CPGs in the vertebrate brain are not well understood and are under intense investigation (Grillner et al., 2005). In the mammalian respiratory CPG, rhythm generation appears to involve multiple complex, nonlinear, cross-level interactions of cellular, network and systems-level mechanisms. Because of these nonlinear interactions, the respiratory CPG can operate in multiple functional states engaging and integrating different cellular and network mechanisms. Revealing these complex interactions and state-dependent reorganizations of neural circuits involved in rhythm generation would have a broad impact on our understanding of the key principles of brain/neural control of movements, and especially control of rhythmic movements and processes. Our goal was to investigate the spatial and functional organization of the mammalian respiratory CPG and the neuronal circuits and mechanisms underlying the state-dependency of the rhythm and pattern generation.

The respiratory cycle in mammals consists of two major phases: inspiration (I) and expiration (Cohen, 1979; Euler, 1986; Feldman, 1986). Expiration in turn comprises two phases, post-inspiration (post-I or E1) and active expiration (E2). Therefore, during eupnea (normal breathing), the respiratory motor activity appears to have a three-phase pattern, i.e., contain three phases: I, post-I, and E2 (Richter and Ballantyne, 1983; Richter, 1996), which can be recognized in the integrated activities of the phrenic (PN) and cranial (e.g., laryngeal) nerves. Respiratory neurons are usually classified based on their firing pattern (e.g., decrementing, augmenting) and the phase of activity relative to the breathing cycle, e.g., early-inspiratory (early-I) with decrementing inspiratory pattern; ramp-inspiratory (ramp-I) with augmenting inspiratory pattern; post-inspiratory (post-I) or decrementing expiratory (dec-E); augmenting or stage II expiratory (aug-E or E2); pre-inspiratory (pre-I), etc. (see Richter, 1996, for review).

The location of the respiratory CPG in the lower brainstem was established in vivo using a combination of anatomical and physiological approaches including lesions at different levels of the brainstem and spinal cord. It was shown that the generation of eupnea involves several respiratory regions in the medulla and pons (Lumsden, 1923; Cohen, 1979; Euler, 1986; Feldman, 1986, see Fig. 1A, B). The major concentration of bulbospinal respiratory neurons (projecting to the spinal motoneurons) are found in a region called "the ventral respiratory group" (VRG) located in the ventrolateral medulla. This region is subdivided into the rostral (rVRG) and caudal (cVRG) parts (Fig. 1A, B). The premotor (bulbospinal) inspiratory neurons are dominantly present in rVRG, whereas the bulbospinal expiratory neurons dominate in cVRG. Rostrally to rVRG, there is a region known as the pre-Bötzinger Complex (pre-BötC) that was shown to be a major source of endogenous (inspiratory) rhythmicity in vitro (Smith et al., 1991, 2000; Rekling and Feldman, 1998). More rostrally there is the Bötzinger Complex (BötC), a compact cluster of cells that is considered a principal source of expiratory inhibition within the network (Ezure, 1990; Jiang and Lipski, 1990; Tian et al., 1999). The pontine respiratory regions include the Kölliker–Fuse (KF) nucleus and parabrachial (PB) complex (lateral, LPB, and medial, MPB, nuclei) in the dorsolateral pons and several areas in the ventrolateral pons (see Fig. 1A). The specific role of pontine structures in the generation and control of the respiratory pattern has hitherto not been well defined. However, the pontine structures appear to have specific interactions with multiple medullary compartments, and the pons as a whole provides strong modulation of the medullary respiratory network via tonic and/or respiratory modulated drives (St.-John, 1998; Alheid et al., 2004). In addition, several medullary structures, specifically the retrotrapezoid nucleus (RTN, located rostrally to BötC below the facial nucleus) and the medullary raphe nucleus, both involved in the central chemoreception, also modulate the medullary respiratory network performance via various drives defining the metabolic state of the system (the level of oxygen and carbon dioxide, pH, etc.) (Mulkey et al., 2004; Guyenet et al., 2005).

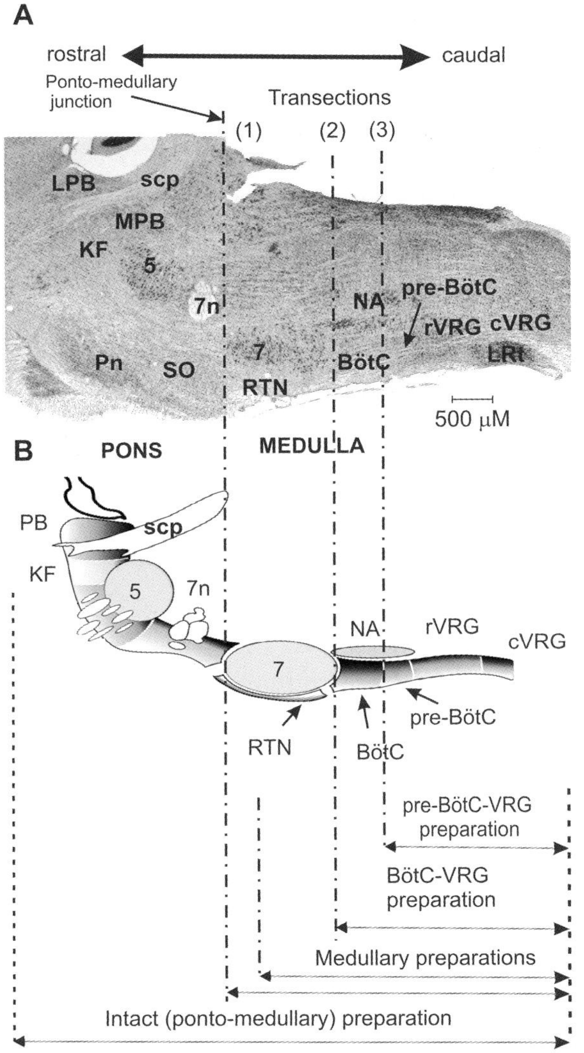

Fig. 1. Parasagittal view of rodent brainstem (section through the level of the compact part of nucleus ambiguus) and spatially arrayed compartments of respiratory CPG network. (A) Respiratory-related ponto-medullary regions in the mature rat brainstem with several transection planes (dot-dashed lines) used in experimental studies. (B) Corresponding schematic diagram of respiratory-related brainstem compartments in parasagittal section of rat brain (created by George Alheid and used with permission) with transactions and resultant reduced preparations indicated at the bottom. Abbreviations: 5 – trigeminal nucleus; 7 — facial nucleus; 7n — facial nerve; BötC — Bötzinger Complex; cVRG — caudal ventral respiratory group; KF — Kölliker–Fuse nucleus; LPB — lateral parabrachial nucleus; LRt: lateral reticular nucleus; MPB — medial parabrachial nucleus; NA — nucleus ambiguus; PB — parabrachial nuclei; Pn — pontine nuclei; pre-BötC — pre-Bötzinger Complex; RTN — retrotrapezoid nucleus; rVRG — rostral ventral respiratory group; scp — superior cerebellar peduncle; SO — superior olive.

Despite a long history of studies there is still no commonly accepted view on the neural mechanisms for respiratory rhythm generation. Proposed mechanisms have been mostly based on particular sets of data obtained from different preparations (decerebrate and/or vagotomized in vivo, arterially perfused in situ brainstem–spinal cord, in vitro isolated brainstem–spinal cord and/or slices from neonatal rodents), which operate under different, often abnormal, metabolic conditions or have significantly reduced circuitry. We believe that the ongoing debate (e.g., Feldman and Smith, 1995; Richter, 1996; Smith et al., 2000; Feldman and Del Negro, 2006) about whether rhythm generation normally is a "pacemaker-driven" or an "emergent" network process has been posed as a mechanistic dichotomy that is largely artificial and requires reframing to take into account the multiple nonlinear state-dependent interactions and potentially different rhythmogenic mechanisms that may emerge and operate in different states. Indeed, we believe that the rhythmogenic mechanism is strongly dependent on the system's state, external inputs, metabolic conditions, etc. Changing these interactions or full elimination of some interactions in reduced preparations by removing part of the network may alter the operating rhythmogenic mechanism and the respiratory pattern generated. For example, the pre-BötC isolated from slice preparations in vitro can intrinsically generate rhythmic inspiratory activity. The rhythm generation in this reduced network involves a persistent sodium current-dependent cellular mechanism operating in the context of an excitatory network (Butera et al., 1999a, b; Koshiya and Smith, 1999; Johnson et al., 2001). However, in the intact system, the pre-BötC is embedded within a wider respiratory network, and its state, operating conditions, and functioning may alter by, and depend on, the excitatory and inhibitory inputs from other parts of the network (Smith et al., 2000; Rybak et al., 2002, 2004a).

Although most respiratory regions are not homogenous and may contain multiple neuron types, a consideration of the neuronal types that predominate in each region leads to the suggestion that the respiratory CPG has a specific functional and spatial organization within the brainstem "respiratory column" that extends in the rostral-to-caudal direction from the pons to the VRG. To test this hypothesis we developed an approach allowing sequential reduction of the respiratory network using highly precise brainstem transactions to remove particular respiratory regions and investigation of the resultant reorganization of the rhythm-generating mechanism by studying alterations in the firing patterns of different neuronal population and motor outputs. Using this approach, we have revealed novel insights into the topographical organization and state-dependency of brainstem mechanisms for respiratory rhythm and pattern generation.

Experimental studies

The experimental studies were performed using the in situ perfused brainstem–spinal cord preparation of the juvenile rat (Paton, 1996). The cranial and spinal nerves in this preparation exhibit discharge patterns similar to those recorded in vivo during eupnea and under different experimental conditions (St.-John and Paton, 2000). A particular advantage of this preparation is that it allows precise control of the perfusion of the brainstem combined with independent control of oxygen and carbon dioxide concentrations in the artificial perfusate. This is crucial when the transections of the brainstem are applied which in vivo would cause hemorrhage leading to brain ischemia. Another advantage is that this preparation allows for administration of pharmacological agents through the perfusate that would be incompatible with viability of in vivo preparations. Juvenile Wistar rats (80–100 g, approximately 4–6 weeks of age) were used. Recordings were obtained from PN, central vagal (cVN), and hypoglossal (XII) motor nerves simultaneously. All procedures were described in detail previously (Paton, 1996; St.-John and Paton, 2000; St.-John et al., 2002; Paton et al., 2006).

The proposed spatial organization of the respiratory CPG defined the experimental approach we adopted. We sequentially reduced the brainstem respiratory network by a series of brainstem transections, starting from a transection at the ponto-medullary junction and continuing with parallel transections though the medulla with the cutting plane sequentially shifted to a caudal direction. The transections were performed by a special, custom-made, piezo cutting and *x-y-z* translation system, allowing precision microsectioning of the brainstem. The precise level of transactions was confirmed histologically post hoc. After each transection, the resultant alterations in the discharge patterns of the PN, cVN, and XII motor nerves were investigated. Riluzole (3–20 μM), a pharmacological blocker of persistent sodium current (I_{NaP}), was applied to the perfusate in the intact and each reduced preparation to determine a role of I_{NaP} for rhythm generation in each given preparation.

The first (most rostral) transection of the brainstem was usually made at the ponto-medullary junction [indicated in Fig. 1 by (1)], which removed the pons. This transection was followed by a series of parallel transections applied by sequential shifting the position of the cutting plane to a caudal direction. These medullary transections went through the facial nucleus (indicated by "7" in Fig. 1) and sequentially removed rostral parts of the medulla. We called the remaining caudal parts the "medullary" preparations (see Fig. 1). The transection at the rostral end of BötC [indicated in Fig. 1 by (2)] reduced the original, intact preparation to a "BötC–VRG" preparation. Finally, a transection was made at the rostral end of pre-BötC [indicated in Fig. 1 by (3)], which produced a "pre-BötC–VRG" preparation. Note that no rhythm could be evoked if a transection was made caudal to pre-BötC.

Figure 2A shows an example of PN, XII, and cVN nerve activities recorded from the intact preparation (containing a complete ponto-medullary circuitry). The discharge patterns of these nerves have the following characteristics, which are also typical for a three-phase eupneic respiratory rhythm recoded in vivo (Duffin, 2003; St.-John and Paton, 2003): (i) the PN bursts have an augmenting shape (the spike frequency increases during the burst); (ii) the onset of the XII bursts usually precedes (50–100 ms) the onset of PN bursts (i.e., XII bursts have a pre-I component); and (iii) the cVN bursts include a prominent decrementing post-I activity.

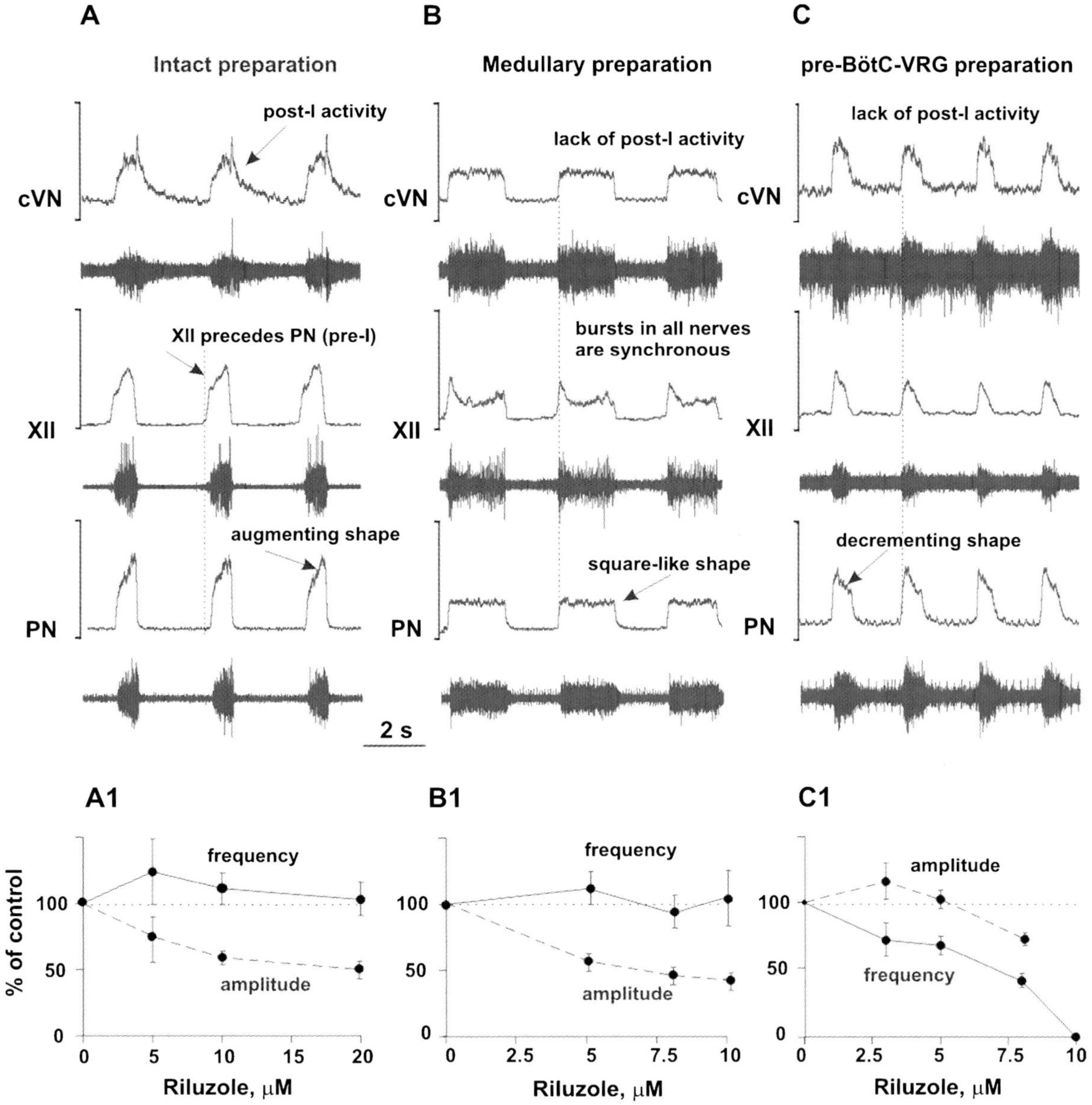

Fig. 2. (A–C) Activity patterns of phrenic (PN), hypoglossal (XII), and central vagus (cVN) nerves from the intact (A), medullary (B), and pre-BötC–VGR (C) preparations. Each diagram shows the recorded (bottom trace) and integrated (upper trace) motor output activities. See text for details. (A1–C1) Dose-dependent effects of I_{NaP} current blocker riluzole on the frequency (solid lines) and amplitude (dashed lines) of PN bursts in the intact (A1), medullary (B1), and pre-BötC–VGR (C1) preparations. Riluzole in the concentrations shown in horizontal axes was added to the perfusate. Note that the frequency of PN bursts does not significantly change in the intact and medullary preparations (A1 and B1), but dramatically decreases with riluzole concentration in the pre-BötC–VGR preparation and, finally, the PN activity was abolished at riluzole concentration of 10 μM (C1).

Brainstem transections between the ponto-medullary junction [indicated in Fig. 1 by (1)] and the rostral border of BötC [indicated in Fig. 1 by (2)] removed the pons and other rostral compartments and hence reduced the intact ponto-medullary preparation to a "medullary" or, in the extreme case, a "BötC–VRG" preparation (see Fig. 1). These transections converted the three-phase rhythm described above into a "two-phase" rhythm (lacking the post-I phase) with the following characteristics: (i) the PN (and XII and cVN) bursts have a "square-like" shape; (ii) bursts in all three nerves are synchronized (i.e., XII burst onset does not precede PN burst); and (iii) the post-I component in cVN bursts disappears. A typical example of this two-phase rhythm is shown in Fig. 2B.

A transection between BötC and pre-BötC [indicated in Fig. 1 by (3)] resulted in further reduction of the preparation to the pre-BötC–VRG (see Fig. 1). This preparation typically generated a respiratory (inspiratory) motor pattern characterized by: (i) a decrementing shape of the burst in all three nerves; (ii) a synchronized activity in all three nerves; and (iii) a lack of post-I activity in vagal nerve (see an example in Fig. 2C). The pattern of this inspiratory activity is very similar to that recorded from pre-BötC and XII nerve in the slice in vitro (e.g., see Koshiya and Smith, 1999; Johnson et al., 2001) and is likely generated due to endogenous bursting mechanisms operating within the pre-BötC without involving inhibitory interactions with other "half-centers". Therefore, we characterize this rhythmic activity as a "one-phase" inspiratory rhythm.

The role of the persistent sodium current for rhythm generation was assessed in each of the three above states (related to the three-, two- and one-phase rhythms, respectively). Figure 2A1–C1 shows the effect of riluzole on the frequency and amplitude of PN discharges. In the intact and the medullary (e.g., BötC–VRG) preparations, riluzole produced a dose-dependent effect on the amplitude of PN discharges but did not affect burst frequency (see Fig. 2A1, A2). In contrast in the pre-BötC–VRG preparation, riluzole had much less effect on the PN amplitude but caused a dose-dependent reduction of PN burst frequency and finally abolished the rhythm at a concentration of about 10 μM (see Fig. 2C1). These data suggest that an intrinsic persistent sodium current-dependent mechanism is essential for the rhythm generation in the pre-BötC–VRG preparation (i.e., for the one-phase rhythm), but its contribution to rhythmogenesis in the BötC–VRG and intact ponto-medullary network (two- and three-phase rhythms, respectively) appears to be less important.

Computational modeling of the brainstem respiratory network

The objectives of our modeling studies were to build a model of the spatially distributed brainstem respiratory network that could reproduce the above experimental findings and suggest an explanation for possible transformations of the rhythm-generating mechanism after sequential reduction of the network. The model has been developed based on a previous model (Rybak et al., 2004a) and represents an extension of the "hybrid pacemaker-network model" proposed by Smith et al. (2000). All neurons were modeled in the Hodgkin–Huxley style (one-compartment models) and incorporated known biophysical properties and channel kinetics characterized in respiratory neurons in vitro. Specifically, the fast sodium (I_{Na}) and the persistent (slowly inactivating, I_{NaP}) sodium currents were described using experimental data obtained from the studies of neurons acutely isolated from the rat's ventrolateral medulla (Rybak et al., 2003a) at the level of the pre-BötC; the high-voltage activated calcium current (I_{CaL}) was described based on the data of Elsen and Ramirez (1998); the intracellular calcium dynamics was described to fit the data of Frermann et al. (1999); the description of potassium rectifier (I_K) and calcium-dependent potassium ($I_{K,Ca}$) currents and all other cellular parameters were the same as in the previous models (Rybak et al., 1997a–c, 2003b, 2004a, b). Each neuronal type was represented by a population of 50 neurons with some parameters and initial conditions randomized within the population. The full description of the model and model parameters can be found in Appendix.

The schematic of the full model is shown in Fig. 3A. Three major medullary regions are considered (in the rostral-to-caudal direction): Bötzinger Complex (BötC), pre-Bötzinger Complex (pre-BötC) and rostral VRG (rVRG). The BötC compartment includes inhibitory populations, aug-E(1) and post-I, each of which serves as a source of expiratory inhibition widely distributed within the medullary respiratory network (Ezure, 1990; Jiang and Lipski, 1990; Tian et al., 1999). In the model, these populations inhibit all populations in the pre-BötC and rVRG compartments and each other (see Fig. 3A). The BötC compartment also includes a second aug-E (aug-E(2)) inhibitory population, providing the additional control of the duration of expiration via inhibition of post-I activity, and an excitatory post-I (post-I(e)) population that provides the expiratory output (e.g., contributes to the cVN motor output). All neurons in the BötC compartment (in the post-I, post-I(e), aug-E(1) and aug-E(2) populations) have intrinsic adapting properties defined by I_{CaL} and $I_{K,Ca}$. Because of this, the post-I neurons exhibit decrementing discharge patterns during expiration. In contrast, the aug-E neurons (under normal conditions) start firing later in expiration and exhibit augmenting patterns because of the slow disinhibition from the adapting inhibitory post-I neurons.

The pre-BötC compartment includes two neural populations: pre-I, and early-I(1) (see Fig. 3A). The pre-I population is the major source of inspiratory activity in the network. It projects to the pre-motor inspiratory ramp-I population of rVRG and also defines the XII motor output. The pre-I population in the model is comprised by excitatory neurons with I_{NaP}-dependent endogenous bursting properties and mutual excitatory synaptic interconnections within the population. Under certain conditions (depending on total tonic drive, phasic inhibition, etc), this population can operate in a bursting mode and intrinsically generate rhythmic inspiratory activity (Butera et al., 1999a, b; Smith et al., 2000; Rybak et al., 2003b, 2004b) similar to that recorded in vitro (Koshiya and Smith, 1999; Johnson et al., 2001). However, in the model under normal conditions, most neurons of this

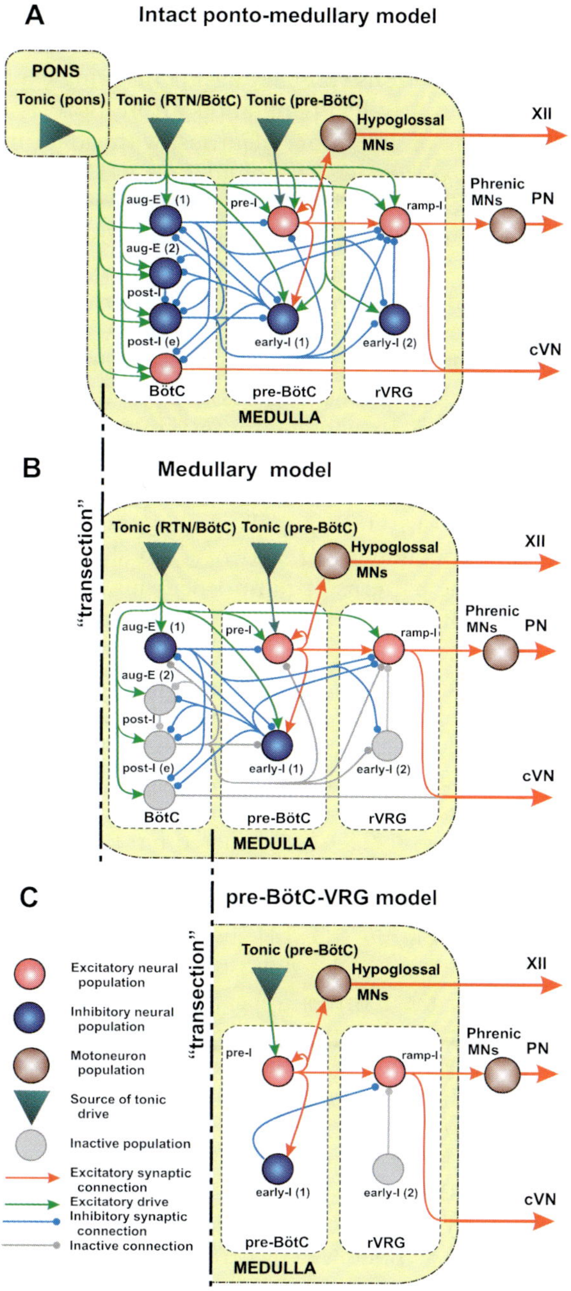

Fig. 3. The schematic of the full (intact) model (A) and the reduced medullary (B) and pre-BötC–VGR (C) models. Neural populations are represented by spheres. Excitatory and inhibitory synaptic connections are shown by arrows and small circles, respectively. Sources of excitatory drives are shown by triangles. All conditional symbols are shown in the left bottom corner. See explanations in the text.

population operate in a tonic spiking mode due to high tonic excitatory input, and are inhibited by expiratory neurons (post-I, aug-E(1)) during expiration. The early-I(1) population of pre-BötC is a population of inhibitory neurons with adapting properties (defined by I_{CaL} and $I_{K,Ca}$). This population receives excitation from the pre-I population and serves as a major source of inspiratory inhibition. In the model, this population inhibits all expiratory neurons during inspiration (see Fig. 3A).

The rVRG compartment contains ramp-I, and early-I(2) populations (Fig. 3A). Ramp-I is a population of excitatory premotor inspiratory neurons that project to PN motor output, and contribute to cVN activity (see Fig. 3A). The major role of the inhibitory early-I(2) population (with adapting neurons containing I_{CaL} and $I_{K,Ca}$) is shaping the augmenting patterns of ramp-I neurons.

The maintenance of normal breathing at the appropriate homeostatic level depends on a variety of afferent inputs to different clusters of respiratory neurons within the brainstem. These inputs are viewed as "excitatory drives" that carry state-characterizing information provided by multiple sources distributed within the brainstem (pons, RTN, raphe, NTS, etc.), including those considered to be major chemoreceptor sites (sensing CO_2/pH), and activated by peripheral chemoreceptors (sensing CO_2/pH and low O_2) (Nattie, 1999; Guyenet et al., 2005). Although currently undefined, these drives appear to have a certain spatial organization with specific mapping on the spatial organization of the brainstem respiratory network. In our model, these drives are conditionally represented by three separate sources located in different compartments (pons, RTN/BötC, and pre-BötC) and providing drives to different respiratory populations (see Fig. 3A).

Figure 4A, B shows the performance of the intact model. The activity of each population is represented by the average spike frequency histogram of population activity. The post-I population of BötC shows decrementing activity during expiration. This population inhibits all other populations (except post-I(e)) during the first half of expiration (post-inspiratory phase). Because of the reduction of post-I inhibition with the adaptation in the post-I activity, the aug-E(1) and then the aug-E(2) population start activity later in expiration forming a late expiratory (E2) phase. At the very end of expiration, the pre-I population of pre-BötC is released from inhibition and activates the early-I(1) population, which inhibits all expiratory populations of BötC. As a result, the ramp-I (and early-I(2)) populations of rVRG release from inhibition (with some delay relative to pre-I) giving the start to the next inspiratory phase (onset of inspiration). During inspiration, the activity of early-I(2) population decreases providing the ramp increase of ramp-I population activity (and PN burst). The activity of early-I(1) population of pre-BötC decreases during inspiration providing a slow disinhibition of the post-I population of BötC. Finally, the post-I population fires and inhibits all inspiratory activity completing inspiratory off-switching. Then the process repeats. In summary, the respiratory rhythm (with a typical three-phase pattern) is generated in the intact model by the neuronal ring comprising the early-I(1), post-I, and aug-E(1) inhibitory populations with the pre-I population participating in the onset of inspiration.

The motor output patterns of the model (PN, XII, and cVN) are shown in Fig. 4B and may be compared with the integrated activities of the corresponding nerves obtained from our experimental studies (Fig. 4C). A comparison clearly demonstrates that the model reproduces all major characteristics of the respiratory pattern recorded under normal conditions from the intact preparation: (i) an augmenting shape of the PN bursts; (ii) a delay in the onset of the PN bursts relative to the XII bursts; and (iii) a decrementing post-I component in cVN bursts. However, the shape of XII busts in the model is slightly different which suggests that the neural organization of the pre-BötC and/or the hypoglossal motor output in the real system is more complicated (more heterogeneous) than that in our model.

Figure 3B shows a schematic of the reduced ("medullary") model used for simulation of a reduced experimental preparation remaining after medullary transections removing the pons or the pons together with an adjacent part of the medulla (e.g., a part of Facial nucleus and RTN).

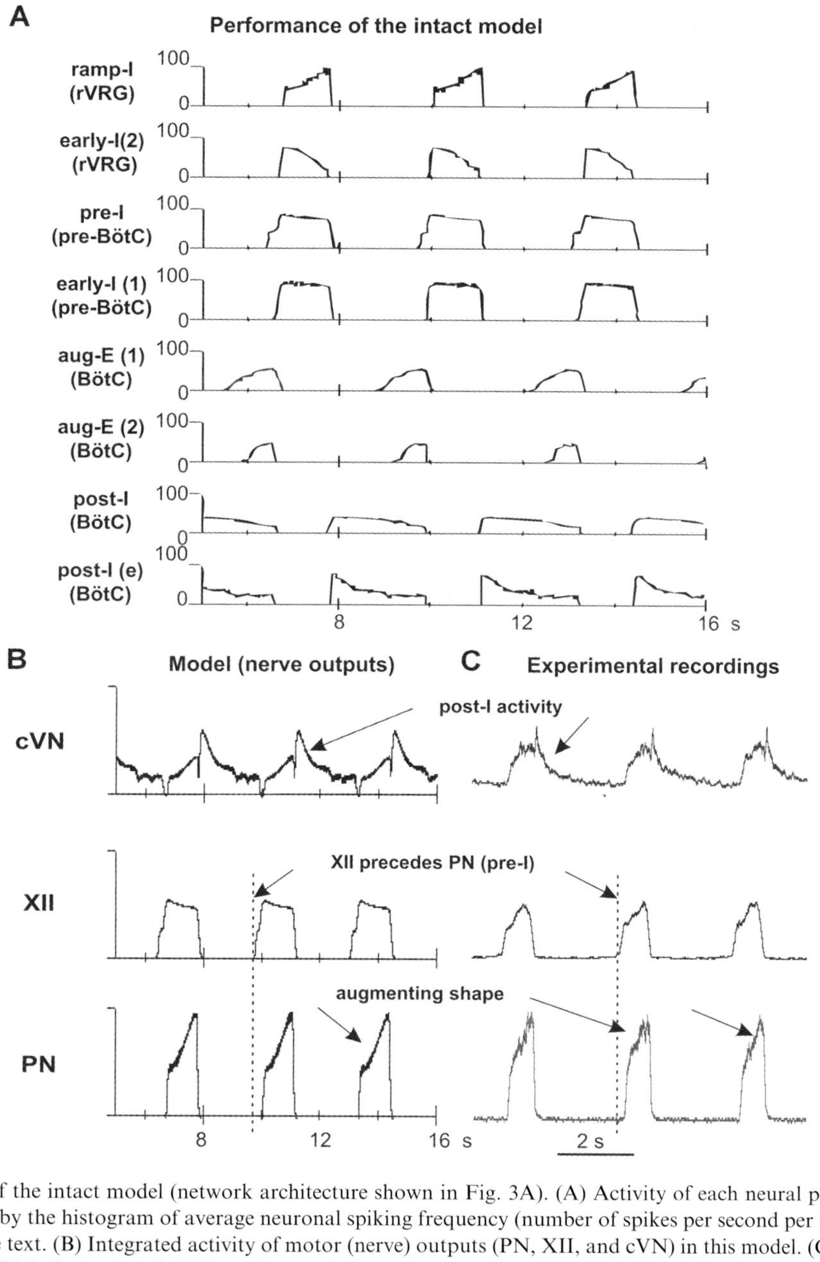

Fig. 4. Performance of the intact model (network architecture shown in Fig. 3A). (A) Activity of each neural population (labeled on the left) is represented by the histogram of average neuronal spiking frequency (number of spikes per second per neuron, bin = 30 ms). See explanations in the text. (B) Integrated activity of motor (nerve) outputs (PN, XII, and cVN) in this model. (C) Integrated patterns of activity of phrenic (PN), hypoglossal (XII), and central vagus (cVN) nerves obtained from the intact preparation (from Fig. 2A) shown for comparison. See all explanations in the text.

The performance of this model is shown in Fig. 5A, B. Based on indirect evidence about a strong excitatory influence of the pons on the post-I neurons (Rybak et al., 2004a; Dutschmann and Herbert, 2006), we have suggested that with the removal of the pons and adjacent medullary regions, all post-I populations of BötC lose a significant portion of the excitatory drive (see Fig. 3B), whereas the drive to aug-E(2) is less dependent on these regions. As a result, a balance

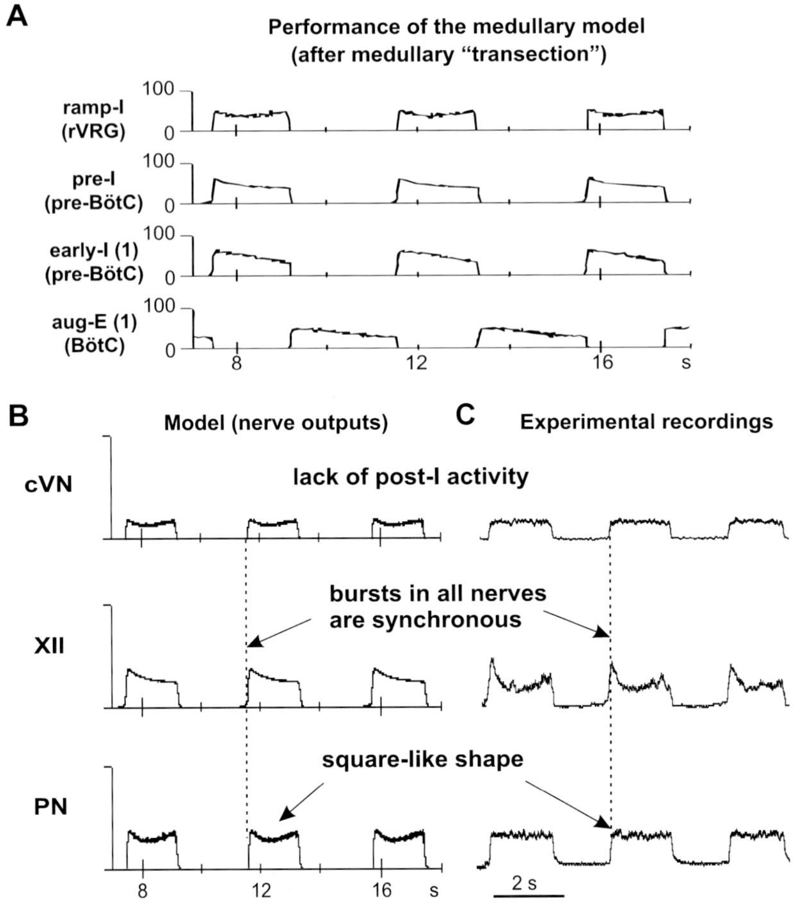

Fig. 5. Neuronal population activities and motor output patterns of the medullary model (shown in Fig. 3B). (A) Activity (spike frequency histograms) of all neural populations (labeled on the left). See explanations in the text. (B) Integrated activity of motor (nerve) outputs (PN, XII, and cVN) in this model. (C) Integrated patterns of activity of phrenic (PN), hypoglossal (XII), and central vagus (cVN) nerves obtained from a medullary preparation (from Fig. 2B) shown for comparison. See explanations in the text.

of mutual inhibitory interactions between aug-E(1) and post-I shifts to the domination of aug-E. The latter now demonstrates a "natural" decrementing pattern (see in Fig. 5A) and completely inhibits all post-inspiratory activity in the network (Figs. 3B and 5A). The respiratory oscillations in this state are generated by a half-center mechanism based on the mutual inhibitory interactions between the adapting early-I(1) and aug-E(1) populations (see Figs. 3B and 5A). The model now generates a typical two-phase rhythm (lacking the post-I phase). In addition, elimination of the drive from more rostral compartments reduces excitability and firing frequency of the pre-I and ramp-I populations, which reduces the amplitudes of all motor outputs of the model (PN, XII, and cVN, see Fig. 5B). Also, because of this drive reduction, the early-I(2) population becomes silent and does not influence the ramp-I population activity, which changes the shape of ramp-I (Fig. 5A) and PN (Fig. 5B) bursts from an augmenting to a square-like pattern. Finally, the patterns of motor outputs in the model (PN, XII, and cVN, Fig. 5B) are very similar to the integrated activities of the corresponding nerves obtained in our experimental studies (Fig. 5C). This reduced model reproduces all major characteristics of the respiratory pattern recorded in the corresponding reduced preparations: (i) an apneustic "square-like" shape of the PN bursts; (ii) a synchronized activities in all three nerves; and (iii) a lack of the post-I component in the cVN bursts.

Figure 3C represents a more reduced model used for simulation of behavior of the reduced pre-BötC–VRG preparation after a transection at the rostral end of pre-BötC. The performance of this model is shown in Fig. 6A, B. As shown in previous modeling studies (Butera et al., 1999a, b; Smith et al., 2000; Rybak et al., 2003b, 2004b) a population of neurons with I_{NaP}-dependent endogenous bursting properties and mutual excitatory connections (as the pre-I population of pre-BötC in the present model) can, under certain conditions, intrinsically generate a population bursting activity similar to that recorded in pre-BötC in vitro (Koshiya and Smith, 1999; Johnson et al., 2001). Specifically, increasing tonic excitatory drive switches the population from a quiescent state to rhythmic population bursting, and then to asynchronous tonic activity (Butera et al., 1999b; Rybak et al., 2003b, 2004b). A relatively strong excitatory drive to this population causes inactivation of NaP channels and maintenance of tonic activity. Alternatively, a reduction of the excitatory drive may hence switch this population to the regime of endogenous bursting activity. In the intact and medullary models above, the pre-I population of pre-BötC receives the total excitatory drive, which is large enough to keep this population in the state of tonic activity. This tonic activity is interrupted by the phasic expiratory inhibition from the post-I (intact model) or aug-E

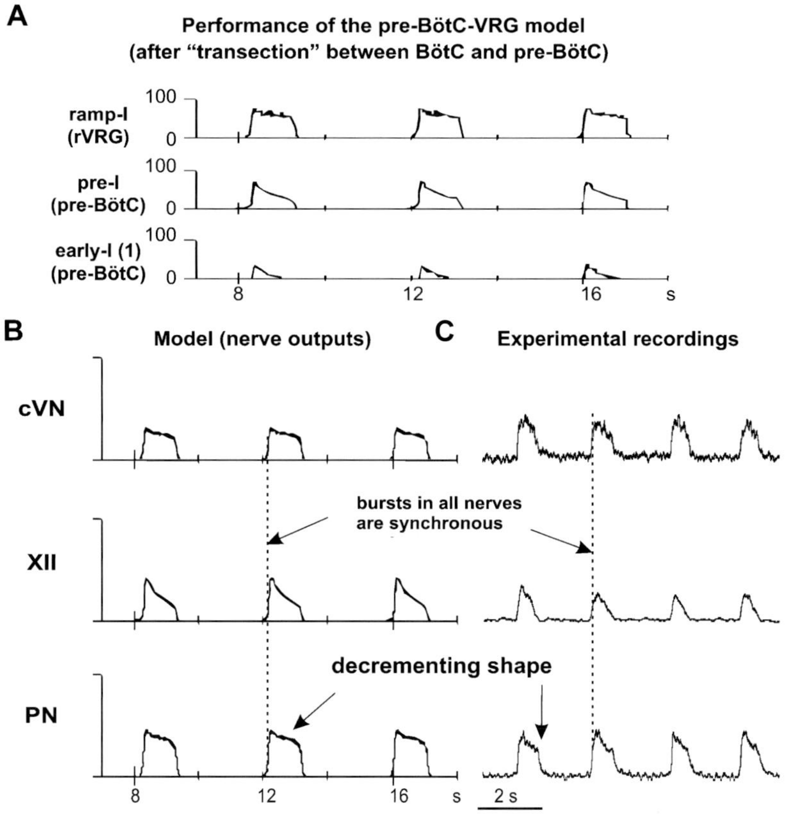

Fig. 6. Performance of the pre-BötC–VGR model (shown in Fig. 3C). (A) Activity of all neural populations (labeled on the left). See explanations in the text. (B) Integrated activity of motor (nerve) outputs (PN, XII, and cVN) in this model. (C) Integrated patterns of activity of phrenic (PN), hypoglossal (XII), and central vagus (cVN) nerves obtained from a pre-BötC–VGR preparation (from Fig. 2C) shown for comparison. See all explanations in the text.

(medullary model) population. Removal of all compartments located caudal to pre-BötC results in further reduction of the drive to the pre-I population (Fig. 3C). In addition, phasic inhibition from expiratory populations of BötC is also eliminated. As described above, with the reduction of tonic excitatory drive and elimination of phasic inhibition, the behavior of pre-I population switches to the regime of endogenous bursting activity. This population now intrinsically generates oscillations with a decrementing burst shape (similar to those recorded in vitro) (see Fig. 6A). Moreover, the bursting activity of the pre-I population now drives the activity of the ramp-I population (Fig. 6A) and all motor outputs that now exhibit one-phase (inspiratory) oscillations with a decrementing burst shape (PN, XII, cVN, see Fig. 6B) similar to that recorded in the pre-BötC–VRG preparation (see Fig. 6C).

In order to investigate the role of the persistent sodium current (I_{NaP}) in the intact (Fig. 3A) and sequentially reduced models (Fig. 3B, C) and compare model behaviors to the corresponding experimental data on the dose-dependent effect of the I_{NaP} blocker riluzole (Fig. 2A1–C1), the maximal conductance of NaP channel ($\bar{g}_{NaP}$) was sequentially reduced in all neurons of the model. The results are shown in Fig. 7A–C. The progressive reduction of $\bar{g}_{NaP}$ (up to zero) in the intact and medullary models does not affect the frequency of respiratory (PN) oscillations and causes only a small reduction of the amplitude of PN bursts (see Fig. 7A, B). This occurs because a relatively high total excitatory tonic drive produces membrane depolarization that holds the (voltage-dependent) I_{NaP} current in a significantly inactivated state In contrast in the pre-BötC–VRG model, with a reduction of the total drive the I_{NaP} essentially contributes to the cellular firing behavior and the reduction of $\bar{g}_{NaP}$ progressively decreases the frequency of PN bursts and finally abolishes the rhythm when $\bar{g}_{NaP}$ becomes less than 2.5 nS (Fig. 7C). These modeling results are fully consistent with our experimental data (Fig. 2A1–C1).

Recent studies in vitro and in vivo (Mellen et al., 2003; Janczewski and Feldman, 2006) have demonstrated that a blockade of activity of inspiratory neurons in the pre-BötC, e.g., by administration of opioids, can produce spontaneous deletions or "quantal" skipping of individual or series of inspiratory bursts in the pre-BötC and/or in PN, while a rhythmic activity persists in a more rostral compartment of the brainstem, the parafacial (pF) region. In this regard, it is interesting to consider the behavior of our model when the activity of the pre-I population of pre-BötC is suppressed. The results are shown in Fig. 8A, B. The pre-I population activity was suppressed by setting

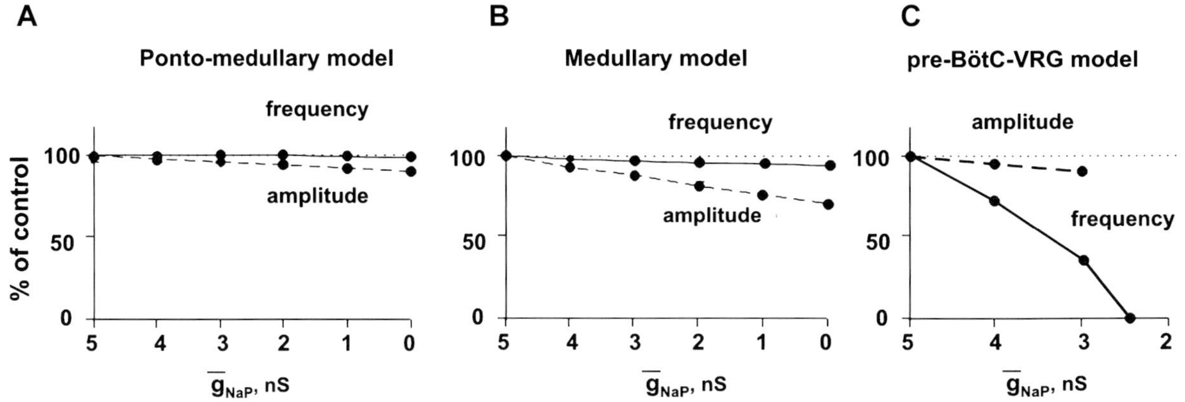

Fig. 7. Effect of reduction of maximum conductance for the persistent sodium channels ($\bar{g}_{NaP}$) in all neurons of the pre-I population of pre-BötC on frequency (solid lines) and amplitude (dashed lines) of PN bursts in the intact (A), medullary (B), and pre-BötC–VGR (C) models. Note that the frequency of PN bursts does not change in the intact and medullary models (A and B), but dramatically decreases with the reduction of $\bar{g}_{NaP}$ in the pre-BötC–VGR model and, finally, the PN activity is abolished at $\bar{g}_{NaP} = 2.5$ nS (C). Compare with the corresponding graphs in Fig. 2A1–C1. See explanations in the text.

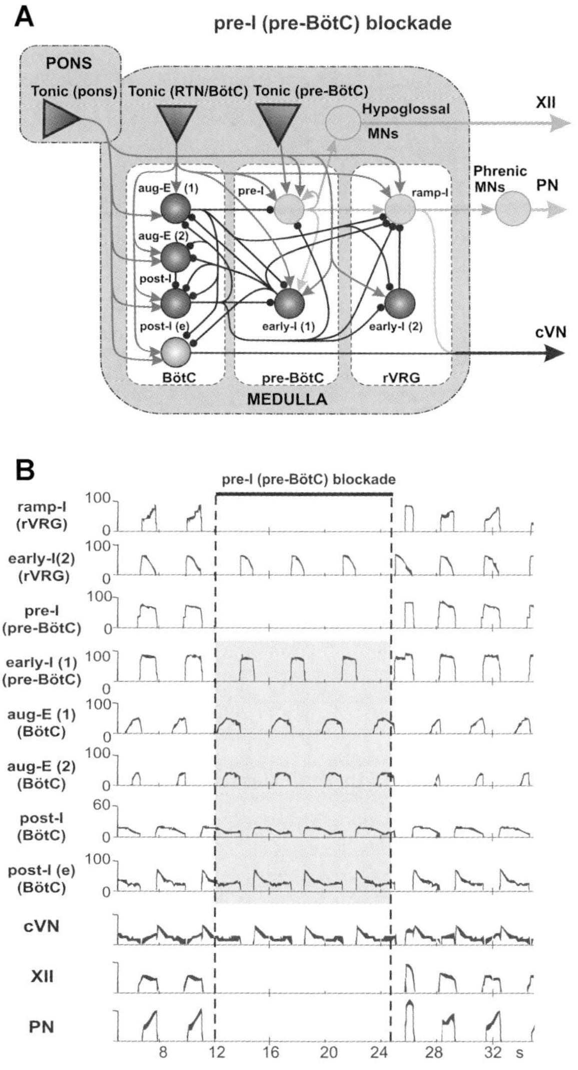

Fig. 8. Neuronal activity patterns in the intact model (Fig. 3A) when the activity of the pre-I population of pre-BötC is suppressed. Activity of pre-I population was suppressed by setting the maximal conductance for fast sodium current to zero for the period shown by a horizontal bar at the top. During this period the pre-I population of pre-BötC as well as the phrenic (PN) and hypoglossal (XII) nerves show no activity. At the same time, an "expiratory" rhythm continues despite of the blockade of inspiration.

the maximal conductance for fast sodium current to zero (Fig. 8A) for some period shown by a horizontal bar at the top of Fig. 8B. During this period, the pre-I population of pre-BötC as well as the PN and XII nerves show no activity. At the same time, despite the complete blockade of the output inspiratory activity, "expiratory" oscillations persist in the network. This expiratory rhythm results from mutual inhibitory interactions between the post-I, aug-E(1) and early-I(1) populations (see marked by gray in Fig. 8B).

Discussion

Our experimental studies have demonstrated that the intact in situ perfused rat brainstem–spinal cord preparation under normal conditions generates a three-phase respiratory rhythm with an augmenting shape of PN discharges, temporal delay between the XII and PN bursts, and prominent post-inspiratory activity (as seen in the cVN bursts). Removal of the pons converts this rhythm into a two-phase rhythm with a square-like shape of PN discharges and lack of the post-I activity. Our data show that administration of riluzole, an I_{NaP} blocker (10–20 μM), does not slow down or abolish these two rhythms (see also Paton et al., 2006). This implicitly supports the conclusion that both these rhythms are generated by network mechanisms without significant contribution of an I_{NaP}-dependent intrinsic mechanism. Another conclusion drawn from these studies is that the input from the pons is necessary for the expression of the post-inspiratory activity in the network and the three-phase rhythm (at least in the absence of pulmonary stretch receptor inputs as in the in situ preparation), which supports previous findings (Rybak et al., 2004a; Dutschmann and Herbert, 2006).

Based on our experimental studies we also conclude that removal of neural circuits rostral to pre-BötC decreases the role of inhibitory network interactions (provided by the expiratory populations of BötC) and increases the role of the endogenous I_{NaP}-dependent rhythmicity in the pre-BötC network. Specifically, the medullary transections between pre-BötC and BötC produce switching to a one-phase inspiratory rhythm that is characterized by a decrementing shape of PN discharges. Application of riluzole produces dose-dependent reduction of the oscillation frequency and finally abolishes this rhythm, as predicted from modeling of heterogeneous populations of excitatory neurons with I_{NaP} (Butera et al., 1999b; Rybak et al., 2003b).

Our experimental and modeling studies reveal a novel spatial and functional organization within the brainstem "respiratory column" that extends from the VRG to the rostral pons. Each functional compartment within this spatial network structure operates under control of more rostral compartments. Specifically, the premotor (bulbospinal) inspiratory neurons of rVRG located caudal to pre-BötC cannot generate rhythmic activity themselves. Their activity is defined by inputs from rostral compartments including excitatory input from pre-BötC and phasic inhibition from expiratory neurons of BötC. In contract, the pre-BötC can, in certain states, intrinsically generate inspiratory bursting activity. However, in the intact system, the pre-BötC is functionally embedded within a more spatially distributed network, and its state and operating conditions are controlled by the more rostral compartments: BötC, inhibiting pre-BötC during expiration, and RTN and pontine nuclei, providing excitatory drives to pre-BötC as well as to BötC and more caudal compartments. Activation of the expiratory populations of BötC (post-I and aug-E), which provide a widely -distributed inhibition within the network during expiration, is critical for the expression of rhythm-generating and pattern-formatting network mechanisms operating in the intact system under normal conditions. In turn, activation of these populations also requires excitatory drives from pons, RTN and other medullary sources as well as a certain balance between these drives. Therefore, both the rhythmogenic mechanism operating in the system under certain conditions (e.g., a network-based, or a pacemaker-driven) and the type of the respiratory pattern generated (e.g., three-phase, two-phase, or one-phase) depend upon the functional states of the system, and specifically, upon the state of BötC (the excitability of post-I and aug-E population, defining the expression of expiratory inhibition within the respiratory network), and the state of the pre-BötC (the expression of the intrinsic I_{NaP}-dependent mechanisms).

We therefore confirm a critical role of the pre-BötC in the generation of respiratory rhythm. In any state of the system, the pre-BötC plays a central role as the major source of inspiratory activity in the network. However, since this complex is embedded in a wider functionally and spatially distributed network, its state, performance, and the expression of endogenous rhythmogenic bursting properties depend upon (and hence are defined by) inputs from other respiratory compartments and external drives (which can be different in different preparations and under different conditions).

Our modeling studies have also demonstrated that a complete suppression of excitatory (pre-I) activity in the pre-BötC does not necessarily eliminate "expiratory" oscillations. Rhythmic expiratory activity may persist due to reciprocal inhibitory interactions within the BötC and between the BötC and inhibitory populations (e.g., early-I) within the pre-BötC region (see Fig. 8). These simulation results may be relevant to the recent findings of the continuing (presumably expiratory) rhythmic activity in the pF region following a suppression (or quantal deletion) of inspiratory activity in pre-BötC (e.g., by administration of opioids) (Mellen et al., 2003; Janczewski and Feldman, 2006). The pF region appears to be adjacent to BötC and functionally may be considered as an extension of BötC. We therefore suggest that similar to our simulations, the expiratory rhythmic activity in pF may be produced by network interactions within the pF/BötC region, and do not necessarily imply the existence of an independent pF expiratory generator normally interacting with a pre-BötC inspiratory oscillator as was recently suggested (Janczewski and Feldman, 2006).

As described above, the pre-I population of pre-BötC and the post-I and aug-E populations of BötC receive multiple drives from the pons, RTN, and other undefined medullary sources. These drives and their balances define a relative excitability and the states of the above key populations and hence implicitly define the operating rhythmogenic mechanism and the respiratory pattern generated. In our experimental and modeling studies, extreme manipulations (transections) were applied to fully eliminate some parts of the network and their influences on the remaining structure in order to uncover the possible states of the system and rhythmogenic mechanisms engaged in each state. Such extreme changes would never happen in real life. At the same time, system states, similar to those uncovered by the transections may occur as a result of alterations in external drives and/or their balances. These drives originate in multiple brainstem regions involved in central chemoreception (such as RTN, raphe, etc.) or come from peripheral chemoreceptors and hence are dependent on the metabolic state of the system. Therefore, specific changes in the metabolic conditions, such as levels of carbon dioxide/pH, or oxygen may alter the balance between the above drives, change interactions between the key respiratory populations, and finally produce transformations to the two- or one-phase rhythms described above. Specifically, hypocapnia (a reduced level of carbon dioxide) can convert the eupneic three-phase rhythm to a two-phase rhythm that appears identical to that obtained by the pontine transection described above (Sun et al., 2001; Abdala et al., 2007). Severe hypoxia (a strong reduction of the oxygen level) can switch the system to a gasping state driven by a one-phase rhythm similar to that obtained by a transaction between pre-BötC and BötC (Paton et al., 2006). In contrast to the intact rhythm and similar to the one-phase rhythm described here, this gasping rhythm is characterized by a decrementing burst shape and can be abolished by the I_{NaP} blocker riluzole (Paton et al., 2006).

In summary our results lead to the conclusion that the neural organization in the respiratory CPG supports multiple rhythmogenic mechanisms. The CPG appears to contain multiple functionally embedded oscillators with rhythmogenic mechanisms ranging from (primarily) inhibitory network-based circuits, resembling classical half-center-type structures, to excitatory networks of neurons with conductance-based endogenous rhythmicity. In the intact system, these circuits are organized in a hierarchy that can be understood in terms of spatially and functionally defined, interacting neuroanatomical compartments. This arrangement makes the respiratory CPG a robust, flexible neural machine for respiratory rhythm generation and control of breathing that can easily adapt to current metabolic demands as well as to various changes in internal and external environment, which would be expected for a rhythmic motor system as vital as the respiratory network.

Acknowledgments

This study was supported by the CRCNS grant R01 NS057815 from the National Institutes of Health (NIH), and in part by the NIH grant R01 NS048844 and the Intramural Research Program of the National Institute of Neurological Disorders and Stroke (NINDS), NIH. JFRP was in receipt of a Royal Society Wolfson Research Merit Award.

Appendix

Single neuron descriptions

All neurons were modeled in the Hodgkin–Huxley style as single-compartment models:

$$C \cdot \frac{dV}{dt} = -I_{Na} - I_{NaP} - I_K - I_{CaL} - I_{K,Ca} - I_L - I_{SynE} - I_{SynI} \quad \text{(A1)}$$

where V is the membrane potential, C the membrane capacitance, and t the time. The terms in the right part of this equation represent ionic currents: I_{Na} — fast sodium (with maximal conductance $\bar{g}_{Na}$); I_{NaP} — persistent (slow inactivating) sodium (with maximal conductance $\bar{g}_{NaP}$); I_K — delayed-rectifier potassium (with maximal conductance $\bar{g}_K$); I_{CaL} — high-voltage activated calcium-L (with maximal conductance $\bar{g}_{CaL}$); $I_{K,Ca}$ — calcium-dependent potassium (with maximal conductance $\bar{g}_{K,Ca}$), I_L — leakage (with constant conductance g_L); I_{SynE} (with conductance g_{SynE}) and I_{SynI} (with conductance g_{SynI}) — excitatory and inhibitory synaptic currents, respectively.

Currents are described as follows:

$$\begin{aligned}
I_{\rm Na} &= \bar{g}_{\rm Na}\cdot m_{\rm Na}^3\cdot h_{\rm Na}\cdot(V-E_{\rm Na});\\
I_{\rm NaP} &= \bar{g}_{\rm NaP}\cdot m_{\rm NaP}\cdot h_{\rm NaP}\cdot(V-E_{\rm Na});\\
I_{\rm K} &= \bar{g}_{\rm K}\cdot m_{\rm K}^4\cdot(V-E_{\rm K});\\
I_{\rm CaL} &= \bar{g}_{\rm CaL}\cdot m_{\rm CaL}\cdot h_{\rm CaL}\cdot(V-E_{\rm Ca});\\
I_{\rm K,Ca} &= \bar{g}_{\rm K,Ca}\cdot m_{\rm K,Ca}^2\cdot(V-E_{\rm K});\\
I_{\rm L} &= g_{\rm L}\cdot(V-E_{\rm L});\\
I_{\rm SynE} &= g_{\rm SynE}\cdot(V-E_{\rm SynE});\\
I_{\rm SynI} &= g_{\rm SynI}\cdot(V-E_{\rm SynI}),
\end{aligned} \tag{A.2}$$

where $E_{\rm Na}$, $E_{\rm K}$, $E_{\rm Ca}$, $E_{\rm L}$, $E_{\rm SynE}$, and $E_{\rm SynI}$ are the reversal potentials for the corresponding channels.

Variables m_i and h_i with indexes indicating ionic currents represent, respectively, the activation and inactivation variables of the corresponding ionic channels. Kinetics of activation and inactivation variables is described as follows:

$$\begin{aligned}
\tau_{m_i}(V)\cdot\frac{d}{dt}m_i &= m_{\infty i}(V)-m_i;\\
\tau_{h_i}(V)\cdot\frac{d}{dt}h_i &= h_{\infty i}(V)-h_i.
\end{aligned} \tag{A.3}$$

The expressions for steady state activation and inactivation variables and time constants are shown in Table 1. The value of maximal conductances for all neuron types are shown in Table 2.

The kinetics of intracellular calcium concentration Ca is described as follows (Rybak et al., 1997a):

$$\frac{d}{dt}{\rm Ca} = k_{\rm Ca}\cdot I_{\rm Ca}\cdot(1-P_B)+\frac{({\rm Ca}_0-{\rm Ca})}{\tau_{\rm Ca}} \tag{A.4}$$

where the first term constitutes influx (with the coefficient $k_{\rm Ca}$) and buffering (with the probability P_B), and the second term describes pump kinetics

Table 1. Steady state activation and inactivation variables and time constants for voltage-dependent ionic channels

Ionic channels	$m_\infty(V)$, V in mV $\tau_m(V)$, ms $h_\infty(V)$, V in mV $\tau_h(V)$, ms
Fast sodium, Na	$m_{\infty\rm Na} = 1/(1+\exp(-(V+43.8)/6))$ $\tau_{m\rm Na} = \tau_{m\rm Na\ max}/\cosh((V+43.8)/14)$, $\tau_{m\rm Na\ max} = 0.252$ $h_{\infty\rm Na} = 1/(1+\exp((V+67.5)/10.8))$ $\tau_{h\rm Na} = \tau_{h\rm Na\ max}/\cosh((V+67.5)/12.8)$, $\tau_{h\rm Na\ max} = 8.456$
Persistent sodium, NaP	$m_{\infty\rm NaP} = 1/(1+\exp(-(V+47.1)/3.1))$ $\tau_{m\rm NaP} = \tau_{m\rm NaP\ max}/\cosh((V+47.1)/6.2)$, $\tau_{m\rm NaP\ max} = 1$ $h_{\infty\rm NaP} = 1/(1+\exp((V+59)/6))$ $\tau_{h\rm NaP} = \tau_{h\rm NaP\ max}/\cosh((V+59)/6)$, $\tau_{h\rm NaP\ max} = 7000$
Delayed rectifier potassium, K	$\alpha_{\infty\rm K} = 0.01\cdot(V+44)/(1-\exp(-(V+44)/5))$ $\beta_{\infty\rm K} = 0.17\cdot\exp(-(V+49)/40)$ $m_{\infty\rm K} = \alpha_{\infty\rm K}/(\alpha_{\infty\rm K}+\beta_{\infty\rm K})$ $\tau_{m\rm K} = \tau_{m\rm K\,max}/(\alpha_{\infty\rm K}+\beta_{\infty\rm K})$, $\tau_{m\rm K\,max} = 1$
High-voltage activated calcium, CaL	$m_{\infty\rm CaL} = 1/(1+\exp(-(V+27.4)/5.7))$ $\tau_{m\rm CaL} = 0.5$ $h_{\infty\rm CaL} = (1+\exp((V+52.4)/5.2))$ $\tau_{h\rm CaL} = 18$
Calcium-dependent potassium, $K(Ca^{2+})$	$\alpha_{\infty\rm K,Ca} = 1.25\times10^8\cdot[{\rm Ca}]_i^2$, $\beta_{\infty\rm K,Ca} = 2.5$ $m_{\infty\rm K,Ca} = \alpha_{\infty\rm K,Ca}/(\alpha_{\infty\rm K,Ca}+\beta_{\infty\rm K,Ca})$ $\tau_{m\rm K,Ca} = \tau_{m\rm K,Ca\ max}\cdot1000/(\alpha_{\infty\rm K,Ca}+\beta_{\infty\rm K,Ca})$, $\tau_{m\rm K,Ca\ max} = 1-8$

Table 2. Maximal conductances of ionic channels in different neuron types

Neuron type	$\bar{g}_{Na}$, nS	$\bar{g}_{NaP}$, nS	$\bar{g}_{K}$, nS	$\bar{g}_{CaL}$, nS	$\bar{g}_{K,Ca}$, nS	g_L, nS
pre-I	300	5.0	180			2.5
ramp-I	400		250			6.0
All others	400		250	0.05	3.0–6.0	6.0

with resting level of calcium concentration Ca_0 and time constant τ_{Ca}.

$$P_B = \frac{B}{(Ca + B + K)} \qquad (A.5)$$

where B is the total buffer concentration and K the rate parameter.

The calcium reversal potential is a function of Ca:

$$E_{Ca} = 13.27 \cdot \ln\left(\frac{4}{Ca}\right)$$
$$\text{(at rest } Ca = Ca_0 = 5 \times 10^{-5}\text{ mM and } E_{Ca} = 150\text{ mV)} \qquad (A.6)$$

The excitatory (g_{SynE}) and inhibitory synaptic (g_{SynI}) conductances are equal to zero at rest and may be activated (opened) by the excitatory or inhibitory inputs, respectively:

$$g_{SynEi}(t) = \bar{g}_E \cdot \sum_j S\{w_{ji}\} \cdot \sum_{t_{kj}<t} \exp\left(-\frac{(t - t_{kj})}{\tau_{SynE}}\right) + \bar{g}_{Ed} \cdot \sum_m S\{w_{dmi}\} \cdot d_{mi};$$
$$g_{SynIi}(t) = \bar{g}_I \cdot \sum_j S\{-w_{ji}\} \cdot \sum_{t_{kj}<t} \exp\left(-\frac{(t - t_{kj})}{\tau_{SynI}}\right) + \bar{g}_{Id} \cdot \sum_m S\{-w_{dmi}\} \cdot d_{mi}, \qquad (A.7)$$

where the function $S\{x\} = x$, if $x \geq 0$, and 0 if $x < 0$. In Eq. (A.7), each of the excitatory and inhibitory synaptic conductances has two terms. The first term describes the integrated effect of inputs from other neurons in the network (excitatory and inhibitory, respectively). The second term describes the integrated effect of inputs from external drives d_{mi}. Each spike arriving to neuron i from neuron j at time t_{kj} increases the excitatory synaptic conductance by $\bar{g}_E \cdot w_{ji}$ if the synaptic weight $w_{ji} > 0$, or increases the inhibitory synaptic conductance by $-\bar{g}_I \cdot w_{ji}$ if the synaptic weight $w_{ji} < 0$. $\bar{g}_E$ and $\bar{g}_I$ are the parameters defining an increase in the excitatory or inhibitory synaptic conductance, respectively, produced by one arriving spike at $|w_{ji}| = 1$. τ_{SynE} and τ_{SynE} are the decay time constants for the excitatory and inhibitory conductances, respectively. In the second terms of equations (A.7), $\bar{g}_{Ed}$ and $\bar{g}_{Id}$ are the parameters defining the increase in the excitatory or inhibitory synaptic conductance, respectively, produced by external input drive $d_{mi} = 1$ with a synaptic weight of $|w_{dmi}| = 1$. All drives were set equal to 1. The relative weights of synaptic connections (w_{ji} and w_{dmi} are shown in Table 3).

Neuronal parameters

Capacitance: $C = 36.2$ pF. Reversal potentials: $E_{Na} = 55$ mV; $E_K = -94$ mV; $E_{SynE} = 0$ mV; $E_{SynI} = E_{Cl} = -75$ mV.

To provide heterogeneity of neurons within neural populations, the value of E_L was randomly assigned from normal distributions using average value $\pm$SD. Leakage reversal potential for all neurons (except for pre-I) $E_L = -60 \pm 1.2$ mV; for pre-I neurons $E_L = -68 \pm 1.36$ mV.

Synaptic parameters: $\bar{g}_E = \bar{g}_I = \bar{g}_{Ed} = \bar{g}_{Id} = 1.0$ nS; $\tau_{SynE} = 5$ ms; $\tau_{SynI} = 15$ ms.

Parameters of calcium kinetics:

$$Ca_0 = 5 \times 10^{-5}\text{mM};\ k_{Ca} = 5.18 \times 10^{-8}\text{mM/C};$$
$$\tau_{Ca} = 500\text{ ms},\ B = 0.030\text{ mM};\ K = 0.001\text{ mM}$$

The motoneuron populations have not been modeled. Integrated activities of the ramp-I and pre-I population were considered as PN and XII motor outputs, respectively. The weighted sum of integrated activities of the ramp-I (1/3) and the post-I(e) (2/3) populations was considered as cVN motor output.

Table 3. Weights of synaptic connections in the network

Target population (location)	Source population (one neuron) or drive {weight of synaptic input}
ramp-I (rVRG)	drive(RTN/BötC) {0.1}; drive(pons) {2.8}; early-I(2) {−0.25}; pre-I {0.12}; early-I(1) {−0.15}; aug-E(1){−1.5}; post-I {−0.5}.
early-I(2) (rVRG)	drive(pons) {1.7}; aug-E(1) {−0.25}; post-I {−0.5}.
pre-I (pre-BötC)	drive(pre-BötC) {0.32}; drive(RTN/BötC) {0.1}; drive(pons) {0.6}; pre-I {0.03}; aug-E(1) {−0.035}; post-I {−0.09}.
early-I(1) (pre-BötC)	drive(RTN/BötC) {0.9}; drive(pons) {1.3}; pre-I {0.026}; aug-E(1) {−0.145}; post-I {−0.185}.
aug-E(1) (BötC)	drive(RTN/BötC) {1.0}; drive(pons) {0.9}; early-I(1) {−0.125}; post-I {−0.16}.
aug-E(2) (BötC)	drive(RTN/BötC) {0.1}; drive(pons) {1.4}; early-I(1) {−0.4}; post-I {−0.16}.
post-I (BötC)	drive(RTN/BötC) {0.1}; drive(pons) {2.9}; early-I(1) {−0.13}; aug-E(1) {−0.03}; aug-E(2) {−0.05}.
post-I (e) (BötC)	drive(RTN/BötC) {0.1}; drive(pons) {2.0}; early-I(1) {−0.2}; aug-E(1) {−0.075}.

Note: Values in brackets represent relative weights of synaptic inputs from the corresponding source populations (w_{ji}) or drives (w_{dmi}), see Eq. (A.7).

Modeling neural populations

In the present model, each functional type of neuron is represented by a population of 50 neurons. Connections between the populations were established so that, if a population A was assigned to receive an excitatory or inhibitory input from a population B or external drive D, then each neuron of population A received the corresponding excitatory or inhibitory synaptic input from each neuron of population B or from drive D, respectively. The heterogeneity of neurons within each population was set by a random distribution of E_L (mean values $\pm$ SD, see above) and initial conditions for values of membrane potential, calcium concentrations and channel conductances. In all simulations, initial conditions were chosen randomly from a uniform distribution for each variable, and a settling period of 20 s was allowed in each simulation before data were collected. Each simulation was repeated 20–30 times, and demonstrated qualitatively similar behavior for particular values of the standard deviation of E_L and initial conditions.

The model was developed using a custom simulation package NSM 2.0, developed at Drexel University by S. N. Markin, I. A. Rybak, and N. A. Shevtsova. Differential equations are solved using the exponential Euler integration method (MacGregor, 1987) with a step of 0.1 ms (for details see Rybak et al., 2003b).

References

Abdala, A.P.L., Koizumi, H., Rybak, I.A., Smith, J.C. and Paton, J.F.R. (2007) The 3-2-1 state respiratory rhythm generator hypothesis revealed by microsectioning, reduced extracellular chloride and alterations in arterial gas tensions in the in situ rat. Exp. Biology Abstr. 610.4.

Alheid, G.F., Milsom, W.K. and McCrimmon, D.R. (2004) Pontine influences on breathing: an overview. Respir. Physiol. Neurobiol., 143: 105–114.

Butera, R.J., Rinzel, J.R. and Smith, J.C. (1999a) Models of respiratory rhythm generation in the pre-Bötzinger complex: I. Bursting pacemaker neurons. J. Neurophysiol., 82: 382–397.

Butera, R.J., Rinzel, J.R. and Smith, J.C. (1999b) Models of respiratory rhythm generation in the pre-Bötzinger complex: II. Populations of coupled pacemaker neurons. J. Neurophysiol., 82: 398–415.

Cohen, M.I. (1979) Neurogenesis of respiratory rhythm in the mammal. Physiol. Rev., 59: 1105–1173.

Duffin, J. (2003) A commentary on eupnoea and gasping. Respir. Physiol. Neurobiol., 139: 105–111.

Dutschmann, M. and Herbert, H. (2006) The Kölliker–Fuse nucleus gates the postinspiratory phase of the respiratory cycle to control inspiratory off-switch and upper airway resistance in rat. Eur. J. Neurosci., 24: 1071–1084.

Elsen, F.P. and Ramirez, J. (1998) Calcium currents of rhythmic neurons recorded in the isolated respiratory network of neonatal mice. J. Neurosci., 18: 10652–10662.

Euler, C.von. (1986) Brainstem mechanism for generation and control of breathing pattern. In: Chernack N.S. and Widdicombe J.G. (Eds.), Handbook of Physiology. The Respiratory System II. American Physiological Society, Washington, DC, pp. 1–67.

Ezure, K. (1990) Synaptic connections between medullary respiratory neurons and consideration on the genesis of respiratory rhythm. Prog. Neurobiol., 35: 429–450.

Feldman, J.L. (1986) Neurophysiology of breathing in mammals. In: Bloom E. (Ed.), Handbook of Physiology, Sec. 1, Vol. 4. American Physiological Society, Bethesda, MD, pp. 463–524.

Feldman, J.L. and Del Negro, C.A. (2006) Looking for inspiration: new perspectives on respiratory rhythm. Nat. Rev. Neurosci., 7: 232–241.

Feldman, J.L. and Smith, J.C. (1995) Neural control of respiratory pattern in mammals: An overview. In: Dempsey J.A. and Pack A.I. (Eds.), Regulation of Breathing. Decker, New York, pp. 39–69.

Frermann, D., Keller, B.U. and Richter, D.W. (1999) Calcium oscillations in rhythmically active respiratory neurones in the brainstem of the mouse. J. Physiol., 515: 119–131.

Grillner, S., Markram, H., DeSchutter, E., Silberberg, G. and LeBeau, F.E.N. (2005) Microcircuits in action — from CPGs to neocortex, TINS, 28: 525–533.

Guyenet, P.G., Stornetta, R.L., Bayliss, D.A. and Mulkey, D.K. (2005) Retrotrapezoid nucleus: a litmus test for the identification of central chemoreceptors. Exp. Physiol., 90: 247–253.

Janczewski, W.A. and Feldman, J.L. (2006) Distinct rhythm generators for inspiration and expiration in the juvenile rat. J. Physiol., 570: 407–420.

Jiang, C. and Lipski, J. (1990) Extensive monosynaptic inhibition of ventral respiratory group neurons by augmenting neurons in the Bötzinger complex in the cat. Exp. Brain Res., 81: 639–648.

Johnson, S.M., Koshiya, N. and Smith, J.C. (2001) Isolation of the kernel for respiratory rhythm generation in a novel preparation: the pre-Bötzinger complex "island." J. Neurophysiol., 85: 1772–1776.

Koshiya, N. and Smith, J.C. (1999) Neuronal pacemaker for breathing visualized in vitro. Nature, 400(6742): 360–363.

Lumsden, T. (1923) Observations on the respiratory centers in the cat. J. Physiol., 57: 153–160.

MacGregor, R.I. (1987) Neural and Brain Modelling. Academic Press, New York.

Mellen, N.M., Janczewski, W.A., Bocchiaro, C.M. and Feldman, J.L. (2003) Opioid-induced quantal slowing reveals dual networks for respiratory rhythm generation. Neuron, 37: 821–826.

Mulkey, D.K., Stornetta, R.L., Weston, M.C., Simmons, J.R., Parker, A., Bayliss, D.A. and Guyenet, PG. (2004) Respiratory control by ventral surface chemoreceptor neurons in rats. Nat. Neurosci., 7: 1360–1369.

Nattie, E. (1999) CO_2, brainstem chemoreceptors and breathing. Prog. Neurobiol., 59: 299–331.

Paton, J.F.R. (1996) A working heart-brainstem preparation of the mouse. J. Neurosci. Meth., 65: 63–68.

Paton, J.F.R., Abdala, A.P.L., Koizumi, H., Smith, J.C. and St.-John, W.M. (2006) Respiratory rhythm generation during gasping depends on persistent sodium current. Nat. Neurosci., 9: 311–313.

Rekling, J.C. and Feldman, J.L. (1998) Pre-Bötzinger complex and pacemaker neurons: hypothesized site and kernel for respiratory rhythm generation. Ann. Rev. Physiol., 60: 385–405.

Richter, D.W. (1996) Neural regulation of respiration: rhythmogenesis and afferent control. In: Gregor R. and Windhorst U. (Eds.), Comprehensive Human Physiology, Vol. II. Springer-Verlag, Berlin, pp. 2079–2095.

Richter, D.W. and Ballantyne, D. (1983) A three phase theory about the basic respiratory pattern generator. In: Schlafke M., Koepchen H. and See W. (Eds.), Central Neurone Environment. Springer, Berlin, pp. 164–174.

Rybak, I.A., Paton, J.F.R. and Schwaber, J.S. (1997a) Modeling neural mechanisms for genesis of respiratory rhythm and pattern: I. Models of respiratory neurons. J. Neurophysiol., 77: 1994–2006.

Rybak, I.A., Paton, J.F.R. and Schwaber, J.S. (1997b) Modeling neural mechanisms for genesis of respiratory rhythm and pattern: II. Network models of the central respiratory pattern generator. J. Neurophysiol., 77: 2007–2026.

Rybak, I.A., Paton, J.F.R. and Schwaber, J.S. (1997c) Modeling neural mechanisms for genesis of respiratory rhythm and pattern: III. Comparison of model performances during afferent nerve stimulation. J. Neurophysiol., 77: 2027–2039.

Rybak, I.A., Paton, J.F.R., Rogers, R.F. and St.-John, W.M. (2002) Generation of the respiratory rhythm: state-dependency and switching. Neurocomputing, 44–46: 603–612.

Rybak, I.A., Ptak, K., Shevtsova, N.A. and McCrimmon, D.R. (2003a) Sodium currents in neurons from the rostroventrolateral medulla of the rat. J. Neurophysiol., 90: 1635–1642.

Rybak, I.A., Shevtsova, N.A., Paton, J.F.R., Dick, T.E., St.-John, W.M., Mörschel, M. and Dutschmann, M. (2004a) Modeling the ponto-medullary respiratory network. Respir. Physiol. Neurobiol., 143: 307–319.

Rybak, I.A., Shevtsova, N.A., Ptak, K. and McCrimmon, D.R. (2004b) Intrinsic bursting activity in the pre-Bötzinger complex: role of persistent sodium and potassium currents. Biol. Cybern., 90: 59–74.

Rybak, I.A., Shevtsova, N.A., St.-John, W.M., Paton, J.F.R. and Pierrefiche, O. (2003b) Endogenous rhythm generation in the pre-Bötzinger complex and ionic currents: modelling and in vitro studies. Eur. J. Neurosci., 18: 239–257.

Smith, J.C., Butera, R.J., Koshiya, N., Del Negro, C., Wilson, C.G. and Johnson, S.M. (2000) Respiratory rhythm generation in neonatal and adult mammals: the hybrid pacemaker-network model. Respir. Physiol., 122: 131–147.

Smith, J.C., Ellenberger, H., Ballanyi, K., Richter, D.W. and Feldman, J.L. (1991) Pre-Bötzinger complex: a brain stem region that may generate respiratory rhythm in mammals. Science, 254: 726–729.

St.-John, W.M. (1998) Neurogenesis of patterns of automatic ventilatory activity. Prog. Neurobiol., 56: 97–117.

St.-John, W.M. and Paton, J.F. (2000) Characterizations of eupnea, apneusis and gasping in a perfused rat preparation. Respir. Physiol., 123: 201–213.

St.-John, W.M. and Paton, J.F.R. (2003) Defining eupnea. Respir. Physiol. Neurobiol., 139: 97–103.

St.-John, W.M., Rybak, I.A. and Paton, J.F.R. (2002) Switch from eupnea to fictive gasping after blockade of inhibitory transmission and potassium channels. Am. J. Physiol. (Regul. Integr. Comp. Physiol.), 283: R721–R731.

Sun, Q.-J., Goodchild, A.K. and Pilowsky, P.M. (2001) Firing patterns of pre-Bötzinger and Bötzinger neurons during hypocapnia in the adult rat. Brain Res., 903: 198–206.

Tian, G.F., Peever, J.H. and Duffin, J. (1999) Bötzinger-complex, bulbospinal expiratory neurones monosynaptically inhibit ventral-group respiratory neurones in the decerebrate rat. Exp. Brain Res., 124: 173–180.

P. Cisek, T. Drew & J.F. Kalaska (Eds.)
Progress in Brain Research, Vol. 165
ISSN 0079-6123
Copyright © 2007 Elsevier B.V. All rights reserved

CHAPTER 14

Modeling a vertebrate motor system: pattern generation, steering and control of body orientation

Sten Grillner[1,*], Alexander Kozlov[1,2], Paolo Dario[3], Cesare Stefanini[3], Arianna Menciassi[3], Anders Lansner[2] and Jeanette Hellgren Kotaleski[1,2]

[1]*Nobel Institute for Neurophysiology, Department of Neuroscience, Karolinska Insititutet, Retzius väg 8, SE-171 77 Stockholm, Sweden*
[2]*Computational Biology and Neurocomputing, School of Computer Science and Communication, Royal Institute of Technology, SE 10044 Stockholm, Sweden*
[3]*CRIM Laboratory, Scuola Superiore Sant'Anna, Viale Rinaldo Piaggio 34, 56025 Pontedera (Pisa), Italy*

Abstract: The lamprey is one of the few vertebrates in which the neural control system for goal-directed locomotion including steering and control of body orientation is well described at a cellular level. In this report we review the modeling of the central pattern-generating network, which has been carried out based on detailed experimentation. In the same way the modeling of the control system for steering and control of body orientation is reviewed, including neuromechanical simulations and robotic devices.

Keywords: locomotor network; lamprey; spinal cord; modeling; steering; robot

Introduction

In order to gain insight into the cellular bases of vertebrate motor behavior, different experimental models are required depending on the type of process explored. For fine control of hand movements, primate models may be a first choice, but for the neural control of goal-directed locomotion or control of body orientation simpler vertebrate models may instead be a better alternative. In this chapter, we will review the extensive knowledge gained on the lamprey nervous system through an interactive process between experiments and modeling (see Grillner, 2003, 2006). The organization of the locomotor system is to a large extent conserved through vertebrate phylogeny, and it is therefore also pertinent to explore to what an extent this knowledge can be applied to the more complex mammalian nervous system.

We will focus on the modeling of the different components of the neural systems underlying goal-directed locomotion, while referring to the detailed experimental evidence. The overall aim is to account for this complex set of behaviors, based on an understanding of the intrinsic cellular mechanisms determining the operation of the different neuronal networks.

The different subsystems involved in the control of locomotion can be represented as follows (see Fig. 1):

- A neural system responsible for selection of the appropriate behavior, in this case

*Corresponding author. Tel.: +46 8 52486900, +46 70 3439900 (cellular); Fax: +46 8 349544; E-mail: sten.grillner@ki.se

DOI: 10.1016/S0079-6123(06)65014-0

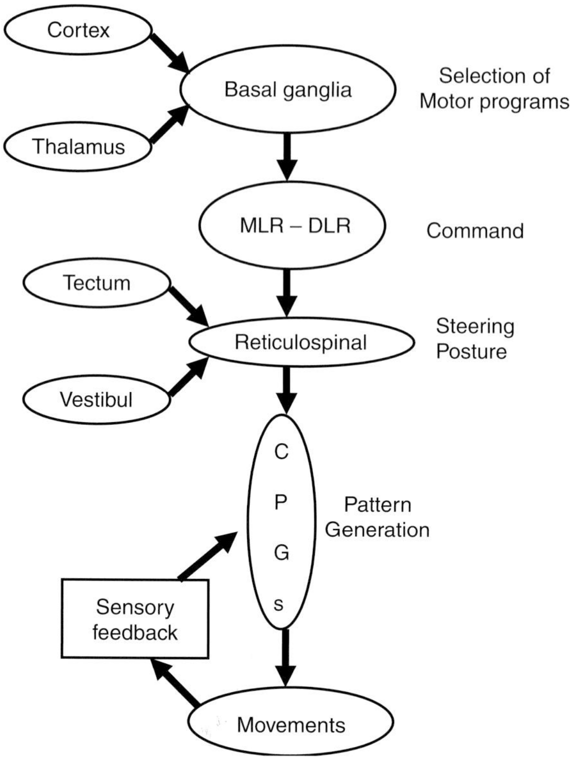

Fig. 1. Subsystems involved in the control of motor behavior, like locomotion. Selection of a certain motor program is performed in the basal ganglia, which receive inputs from cortex (pallium) and thalamus. The basal ganglia output stage (pallidum) inhibits command centers in the diencephalic locomotor region (DLR) and the mesencephalic locomotor region (MLR) during resting conditions. Through a well-controlled inhibition of pallidal regions the spinal CPG for locomotion can be activated via the reticulospinal (RS) neurons. In the brainstem, information is further integrated based on visual, sensory and vestibular inputs to control both steering and posture. As in higher vertebrates, the spinal cord CPG neurons are modulated by local sensory feedback.

locomotion. The striatum, the input layer of the basal ganglia, has an important role in this context. The striatum receives phasic input from pallium (cortex) and thalamus, and a modulatory input from the dopamine system. The GABAergic striatal neurons have a high threshold for activation. When activated they can indirectly release specific motor programs by inhibiting the GABA-ergic output neurons of the basal ganglia (pallidum) that at rest provide tonic inhibition of the different motor programs (see Grillner, 2006).

- A command system that, when released from basal ganglia inhibition, can elicit locomotion by activating the pattern-generating circuits in the spinal cord. Two command systems for locomotion, the mesencephalic (MLR) and the diencephalic (DLR) have been defined. They act via a symmetric activation of reticulospinal neurons that turn on the spinal circuits.
- Segmental and intersegmental networks [central pattern generators (CPGs)] located at the spinal level. The CPGs contain the necessary timing information to activate the different motoneurons (MNs) in the appropriate sequence to produce the propulsive movements.
- The segmental burst-generating network in the lamprey contains excitatory interneurons (EINs) that provide excitation within the pool of interneurons. The alternating pattern between the left and the right sides is provided by reciprocal inhibitory connections between pools of EINs.
- A sensory control system, sensing the locomotor movements, helps to compensate for external perturbations by a feedback action on the spinal CPGs.
- A control system for steering the body toward different goals. The steering commands are superimposed on the basic locomotor activity and will bias the control signals, so as to steer the movements to the left or right side or in other orientations.
- During locomotion the body moves with the dorsal side up, regardless of perturbations. This "postural" control system, for orientation of the body in the gravity field, depends on bilateral vestibular input that detects any deviation from the appropriate orientation of the head, whether tilt to the left or right or changes in pitch angle during locomotion. These vestibular effects are mediated via brainstem interneurons to reticulospinal neurons on the left or right side, respectively, that can elicit compensatory movements that restore the body orientation.

Modeling of the segmental CPG

Basic mechanisms of burst generation

One major problem when studying vertebrate pattern generation has been the intrinsic function of the networks controlling behaviors such as respiration and locomotion. In the case of lower vertebrates like the lamprey and the frog embryo, they are comparatively well understood. At the segmental level recurring locomotor bursts can be generated even in a hemisegment, provided that the excitability is high enough (Cangiano and Grillner, 2003, 2005). The burst generation is primarily dependent on a pool of EINs. Within this pool EINs, excite other EINs (via AMPA and NMDA receptors), although mutual excitation between individual EINs has not been observed [Figs. 2E and 3A (Parker and Grillner, 2000)]. These pools of EINs form burst-generating kernels within the CPG. The excitatory drive to the EIN kernel is provided by the locomotor command regions via glutamatergic reticulospinal neurons, while the burst generation is due primarily to the intrinsic synaptic and membrane properties of the EINs. Although there are inhibitory premotor

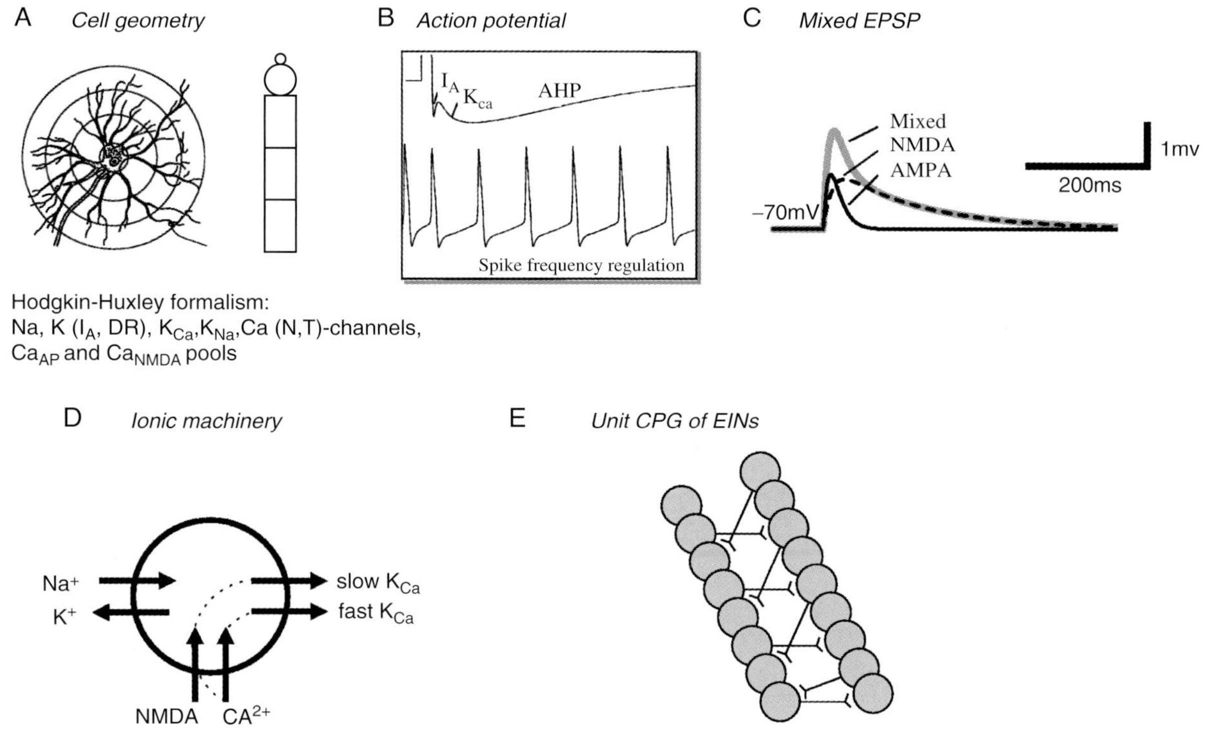

Fig. 2. Cellular properties explored in simulations of the lamprey spinal locomotor networks. (A) The morphology of the CPG neurons is captured using a five-compartment model consisting of an initial segment, a soma and a dendritic tree. Active ion currents are modeled using a Hodgkin-Huxley formalism. Both ion channels involved in spiking behavior (Na^+, K^+) and also slower Ca^{2+} or potassium-dependent processes (K_{Ca}, K_{Na}) are modeled based on available data. (B) Spiking behavior of a CPG neuron. Spike frequency can be regulated by the AHP. (C) Fast synaptic transmission is included in the form of excitatory glutamatergic (AMPA and voltage-dependent NMDA) and inhibitory glycinergic inputs. (D) Main ionic membrane and synaptic currents considered to be important during activation within the CPG network. Slower processes can cause spike frequency adaptation, like Ca^{2+} accumulation during ongoing spiking and resulting activation of K_{Ca}. (E) Unilateral CPG activity can be evoked in local networks of EINs. This basic EIN network can sustain the rhythm seen in ventral roots in vitro during evoked locomotor activity. The left–right alternating activity requires the presence of contralateral inhibition provided by glycinergic interneurons (see below).

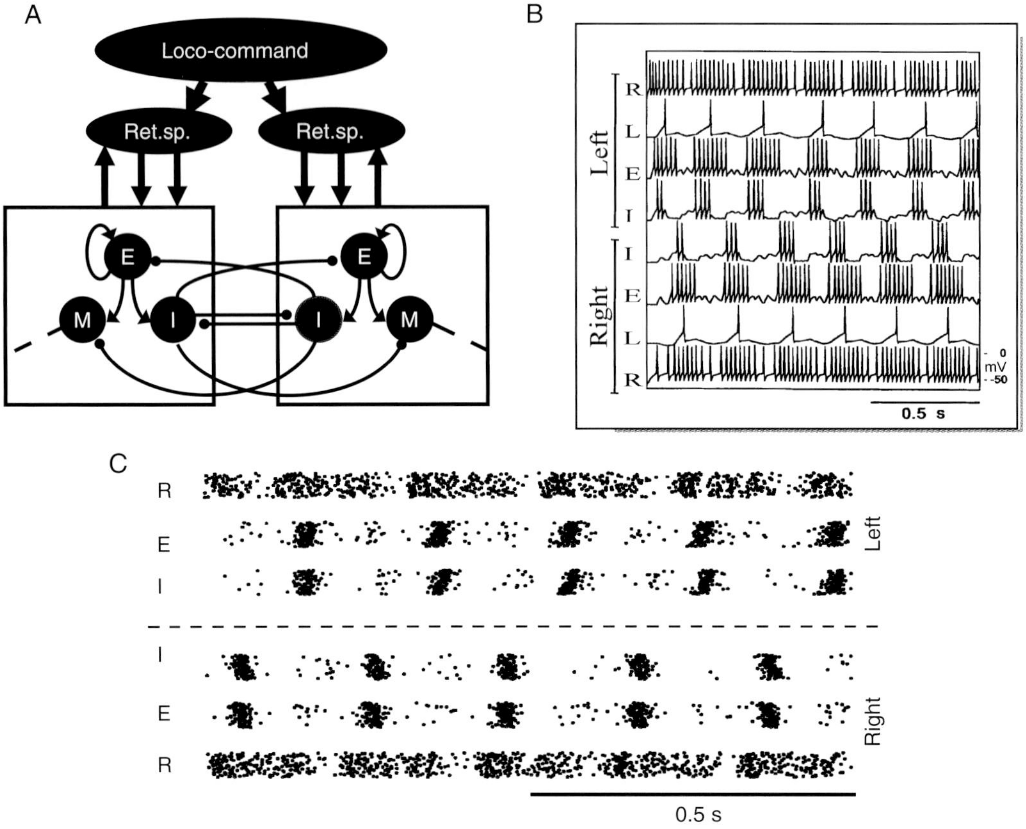

Fig. 3. Generation of locomotor activity in the spinal cord. (A) The CPG network is activated from reticulospinal neurons in the brainstem. EINs (E) excite all types of interneurons including other EINs. The inhibitory glycinergic interneurons (I) provide inhibition to the contralateral side and are also responsible for the left–right alternation seen during normal swimming. M indicates motoneurons. Some phasic feedback modulation (both glutamatergic and glycinergic) from the spinal cord network back to the brainstem also occurs (not indicated). (B) Activation of the spinal CPG by reticulospinal neurons (R) can drive the local network activity, and cause a left–right alternating pattern over the experimentally observed frequency range in the simulations. (C) Similar data are represented as a raster plot with R, E, and I neurons on the right and the left sides (5 segments; 30 neurons/segment) with the connectivity as represented in Fig. 5A, B.

interneurons (Buchanan and Grillner, 1988) that can be active during locomotion, they are not essential for burst generation, since they can be blocked without any effects on burst frequency (Cangiano and Grillner, 2005).

The EINs have overshooting action potentials and a fast and a slow afterhyperpolarization (AHP) similar to that of MNs (Buchanan et al., 1989; Biro et al., 2006). Although, due to their small size, they have been studied in less detail than the larger spinal neurons, their properties appear very similar. They have been modeled using "Hodgkin — Huxley" formalism. Each neuron consists of five compartments (soma, initial segment and three dendritic compartments) and the membrane currents found experimentally in the lamprey spinal cord have been incorporated [Na^+, K^+ (I_A, DR), K_{Ca}, K_{Na}, Ca^{2+} (N, T)-channels and separate pools for the Ca^{2+} entering during the action potential and through NMDA channels; Fig. 2A, B]. These model EINs behave as their biological counterparts, and within the simulated EIN pool the distribution of input impedances is similar to that found experimentally (Ekeberg et al., 1991; Hellgren et al., 1992; Hellgren-Kotaleski et al., 1999b; Kozlov et al., 2007).

Each hemisegment is estimated to have around 30 EINs. In addition to the segmental EIN interaction, EINs have brief ascending (three segments) and longer descending (six to eight segments) axonal branches (Dale, 1986; Buchanan et al., 1989). Each EIN will thus receive excitation not only from segmental EINs but also from EINs located both rostrally and caudally to itself. The connectivity ratio within the pool of EINs is approximated to be 10%. The synaptic interaction between EINS is modeled with conductance increase EPSPs (AMPA) combined with voltage-dependent NMDA receptors (Fig. 2C; Kozlov et al., 2007).

If a pool of model or real EINs (Fig. 2E) is activated from the brainstem or experimentally by bath applying glutamate agonists, burst activity will be produced, the rate of which depends on the excitatory drive. The initial depolarization of EINs will make the EINs with the lowest threshold start firing, and they will in turn activate other EINs synaptically. When the EINs are depolarized, the membrane depolarization can be boosted by the opening of both voltage-dependent NMDA receptors and low voltage-activated Ca^{2+} channels (compare Fig. 2D). This process will be responsible for the initiation of the activity. During the ongoing burst of activity, Ca^{2+} will enter through several different channels, including both NMDA channels and a variety of voltage dependent Ca^{2+} channels. The Ca^{2+} levels will in turn activate Ca^{2+}-dependent K^+ channels (K_{Ca}) that cause adaptation and will pull the membrane potential downward, and thereby terminate the burst (Fig. 2D). In addition to K_{Ca} channels, K_{Na} channels also contribute to the adaptation. Following the termination of the burst, the Ca^{2+} levels will be enhanced and this will lead to a hyperpolarization due to K_{Ca} activation; this will gradually decline as the cell reduces its cytoplasmic Ca^{2+} levels by uptake. When the EIN population has recovered from the K_{Ca} activation, a new burst will be initiated due to the background excitatory drive.

Generation of alternating activity

The generation of alternating activity between two pairs of antagonists like the left and the right sides of a segment is a standard element in many networks. The burst-generating kernels on each side in the lamprey segment can produce recurring burst activity by themselves (Fig. 2E). By connecting them through reciprocal inhibition, alternating activity will occur. In addition to ipsilateral MNs, the EINs also drive inhibitory commissural interneurons that provide inhibition of the different neurons on the contralateral side (Fig. 3A) (Buchanan, 1982; Ohta et al., 1991; Biro et al., 2006). Thus when one side is bursting the other will be inhibited, and as the burst is terminated the other side will initiate a burst due to the background excitatory drive (Hellgren et al., 1992). This side will then in turn inhibit the first side and so forth (Figs. 3B, C).

Modeling of intersegmental coordination

Intersegmental network

The lamprey normally swims by using an undulatory wave propagated from head to tail (Fig. 4A), the faster the wave is propagated backward along the body, the faster the animal will move forward through the water (Grillner, 1974; Wallen and Williams, 1984; Williams et al., 1989). The delay between the activation of each segment is around 1% of the cycle duration regardless of whether the cycle duration is 2 s or one tenth of a second. Since the lamprey has around 100 segments, this means that the delay from the head to the tail fin will be constant at around one full cycle regardless of whether the lamprey swims fast or slow. This rostro-caudal propagation of the wave also occurs during locomotor activity in the isolated spinal cord (Fig. 4B), and is thus basically part of the standard spinal pattern generation.

If caught in a corner, the lamprey can also swim backward (Islam et al., 2006) by reversing the direction of the mechanical wave (Fig. 4A). A reversed phase coupling can be produced in the isolated spinal cord if the caudal part of the spinal cord has higher excitability than more rostral segments (Matsushima and Grillner, 1992). Essentially, the segments with the highest excitability

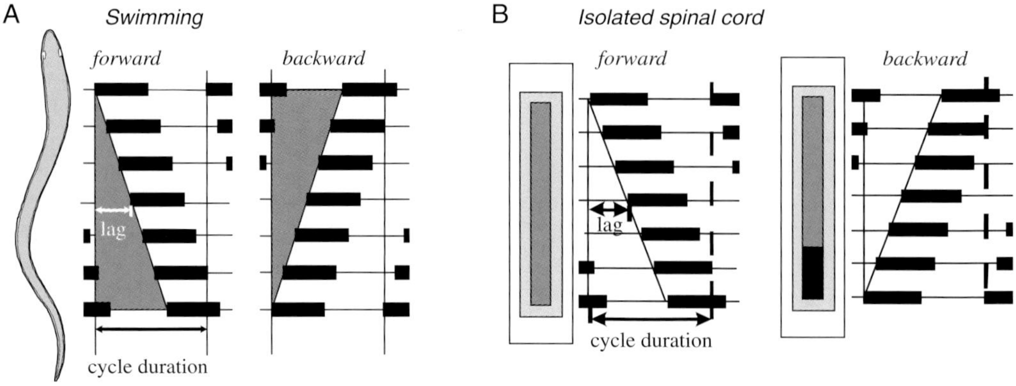

Fig. 4. Lamprey spinal cord intersegmental coordination. (A) During swimming a mechanical wave activating the muscles is transmitted along the spinal cord. When the animal moves forward, there is a lag between consecutive segments in the spinal cord. This lag is always a certain proportion of the cycle duration (i.e. a constant phase lag). It can be reversed into a wave that is propagated from tail to head, as during backward swimming. (B) In the isolated spinal cord preparation, a rostral to caudal lag is also seen. This pattern can be reversed if, e.g., extra excitation is added to the caudal spinal cord. Then the caudal segments receive a higher inherent frequency, and can then entrain the more rostral segmental networks.

will lead the activity and those with lower excitability will follow with a delay (Fig. 4B).

We have modeled the intersegmental coordination (Hellgren-Kotaleski et al., 1999a, b) to test if one can account for the experimental findings with the available knowledge of the segmental pattern generation and the intersegmental connectivity in several different ways. The connectivity of the ipsilateral EINs is represented in Fig. 5A–C. Essentially the connectivity of the EINs extends over an area of around three segments rostral to the cell body and eight segments caudally (Dale, 1986). This applies to all EINs along the spinal cord, and they thus form a continuous network that is not separated by segmental boundaries. Such a hemicord network, only including the EIN population with appropriate synaptic connectivity, can generate not only bursting but also an intersegmental rostro-caudal lag. Figure 5C shows that a rostrocaudal phase lag can be generated in a simulation with 100 segments and 3000 EINs connected as represented in Fig. 5A, B. This number of neurons and connectivity approaches that found biologically. If the relative excitability along the cord is modified so that the rostral part has lower excitability than the caudal one (Fig. 5D), the phase lag is reversed as found experimentally. Figure 5C and D show that the phase lag remains constant along the spinal cord both with a forward and a backward coordination. This finding thus shows that a segmental coupling of exclusively EINs with appropriate cellular and intersegmental connectivity can produce an appropriate phase coupling for forward and backward swimming.

The hemicord coordination is of interest from the analytical point of view, but physiologically there is always a reciprocal activity at the left and the right sides resulting in the propulsive locomotor movements. Segmentally the EINs also activate commissural inhibitory interneurons that ensure that the activity in each segment alternates (see Fig. 3). Figure 6A shows such a simulation with approximately the correct number of neurons in which both the left and the right sides are represented. Blue dots represent neurons that are inhibited, red ones those that fire action potentials and yellow those that are only depolarized but subthreshold. The intersegmental coordination has also been modeled with a more theoretical approach (see Cohen et al., 1992).

Neuromechanical modeling

To extend the modeling from only network activity, a simplified neuromechanical lamprey was developed (Ekeberg and Grillner, 1999). The

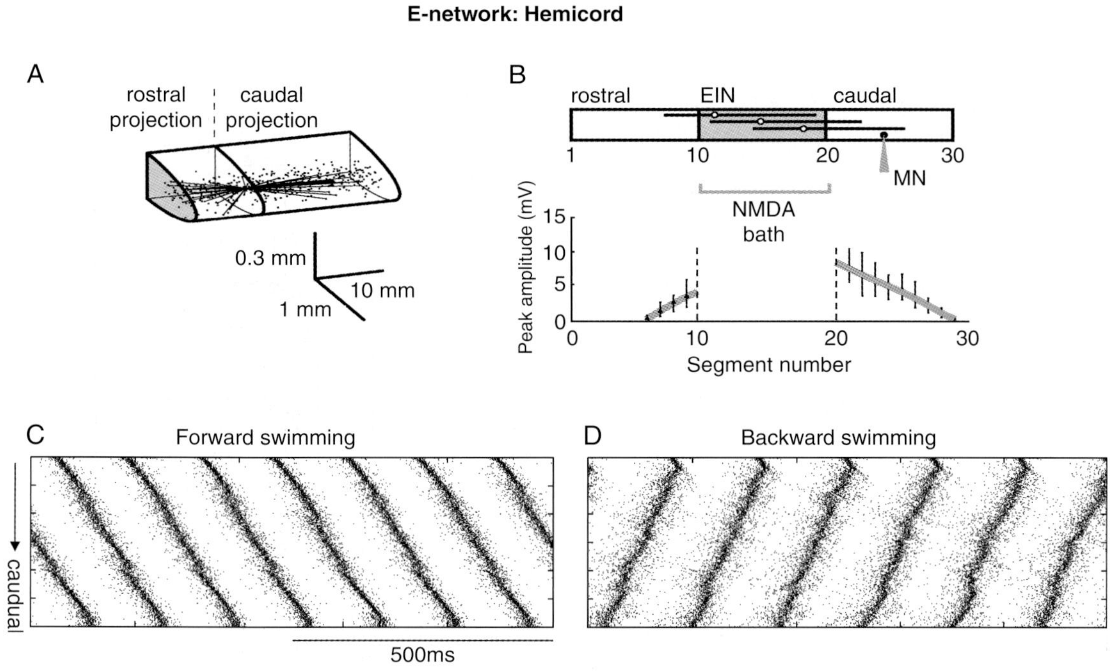

Fig. 5. Simulations of intersegmental coordination in a hemicord (Kozlov et al., 2007). (A) Example of outgoing synaptic connections from a single EIN within a simulated hemisegmental EIN network. Each cell projects up to four segments rostrally and eight segments caudally. (B) The simulated setup for estimating the EIN to motoneuron (MN) synaptic conductances. The trough to peak amplitude of membrane potential in motoneurons (given an input resistance of 14 MΩ) recorded in the rostral and caudal directions of the hemicord decreases when network activity is induced in the middle trunk of the model hemicord (shaded region in upper plot). Error bars show variations of peak amplitude if the simulated MN resistance varies between 10 and 20 MΩ. (C) Raster plots of the activity of each of the 3000 neurons (corresponding to 100 segments) in the model hemicord network during spontaneously forming forward swimming (C), and when a decrease in the excitation to the most rostral segments is simulated (D). As in experiments, a backward swimming pattern can also be produced by increasing the relative excitation to the most caudal segments. Both for forward- and backward-simulated swimming, the phase lag is constant along the cord except at the ends.

model lamprey was reduced to 10 segments and the muscle properties of each myotome with their visco-elastic properties were simulated. The degree of stiffness within the myotome was controlled by the degree of efferent (motoneuronal) activity. A neuronal network operating in a similar way to that discussed above could generate both forward and backward locomotion through the simulated viscous water. Figure 6B shows the outline of the simulated lamprey as it swims through the simulated water with a rostro-caudal phase lag. By changing the excitability in the controlling network, as in Fig. 5D, the model lamprey could even be made to swim backward (not shown). These results provide a confirmation that the interpretation arrived at experimentally can be reproduced and tested by neuromechanical models interacting with the surrounding water. Such models have also allowed us to perform tests that would be difficult to perform experimentally with regard to sensory feedback and the control of steering (see also below).

A lamprey robot

A physical implementation has recently been realized (Stefanini et al., 2006), which is 0.7 m long, based on the principles discussed above (Fig. 7). The different sections are connected through a notochord consisting of a relatively stiff structure that determines the length of the body. Each robot

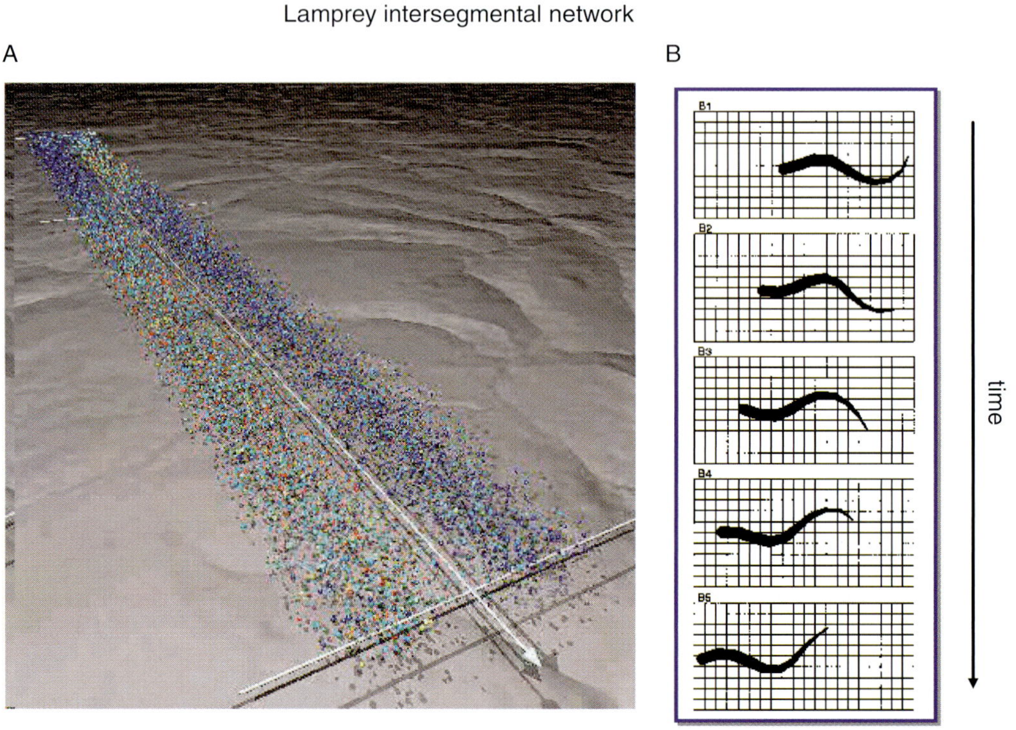

Fig. 6. Intersegmental coordination in a complete spinal cord — full-scale network simulations as well as neuromechanical models. (A) Large-scale network model consisting of both EINs and commissural inhibitory interneurons (Hodgkin-Huxley, five-compartment models). The wave of activity is transmitted from rostral to caudal, with the left and the right sides alternating. Red dots represent actively spiking neurons, yellow ones are depolarized but are not firing and blue dots are inhibited cells. The network extends from the rostral part (lower right) to the caudal aspect (upper left; segment 100), segment 50 is indicated with a hatched line and the perspective is thus compressed in the caudal part. (B) The lamprey swimming in water simulated using a neuromechanical model of the muscles activated by the output from local CPG neurons (Ekeberg and Grillner, 1999).

myotome/segment consists of four actuators, which have biological properties in that they have a length tension curve similar to that of muscle. There are two actuators on each side, one dorsal and one ventral. This allows not only left–right alternation, but also the possibility of a bias on the dorsal or the ventral side, to steer the robot in 3D during swimming. The actuators receive control signals similar to those elicited from the segmental MNs to the different parts of the biological myotome. The resulting movements are sensed by stretch receptors (see Fig. 7) that provide feedback. The lamprey robot swims with a rostro-caudally directed wave with increasing amplitude from head to tail. The speed of locomotion can be controlled with a maintained phase lag. A similar robotics and simulation approach has been taken for salamander swimming and walking (Bem et al., 2003; Ijsbeert et al., 2005).

Steering

What we have modeled so far is the neural bases of symmetric locomotor movements. To make them behaviorally meaningful, we need in addition to steer them. Steering movements to the left or right side are achieved through an asymmetric activation of reticulospinal neurons on the left and the right sides particularly involving the middle and posterior rhombencephalic reticular nuclei (Ohta and Grillner, 1989; Wannier et al., 1998; Deliagina

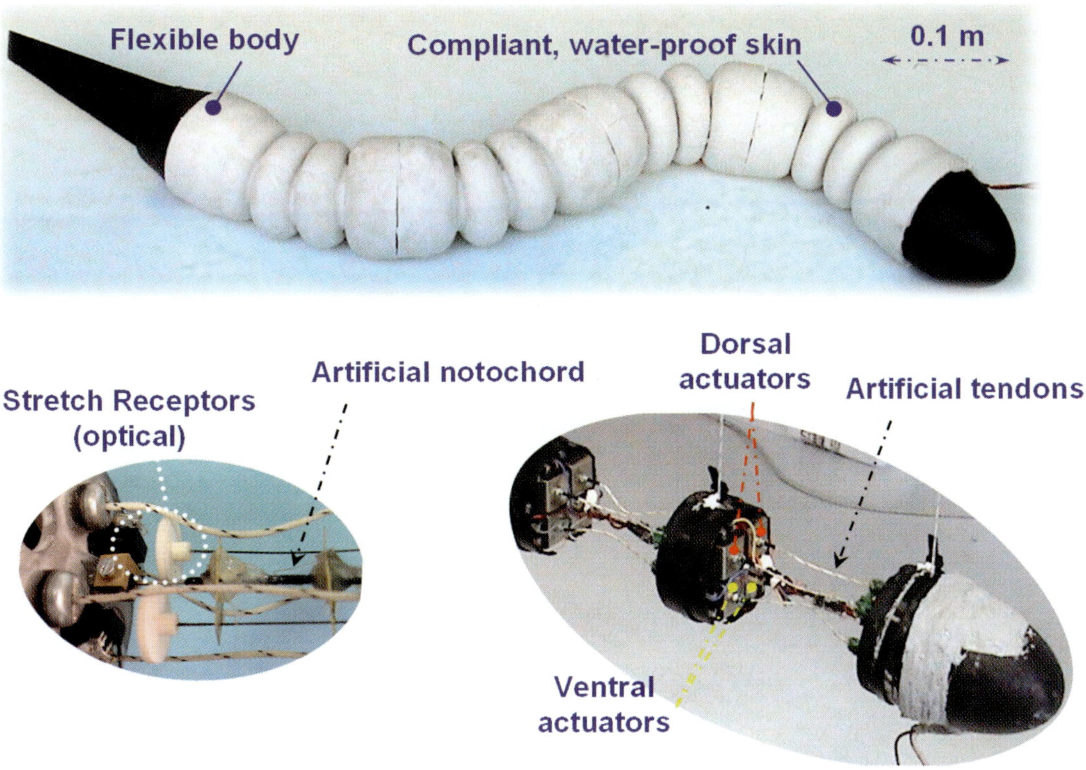

Fig. 7. The lamprey robot built through a collaboration between neuroscientists at the Karolinska Institute and bioengineers of Scuola Superiore Sant'Anna. The segmented body can spatially bend, thanks to a central flexible structure artificially replicating the lamprey's notochord and to four contractile electromagnetic actuators per segment (left/right, ventral/dorsal) acting on adjacent modules through artificial tendons. Proprioceptive receptors based on disturbance-free optical sensors detect body deformation and their signals are used for CPG control. The head is designed to house stereoscopic vision and an artificial vestibular system based on inertial sensors (gyroscopes and accelerometers). Internal components are protected by a compliant, water-proof skin fabricated via silicone rubber molding (Stefanini et al., 2006).

et al., 2000, 2002; Fig. 8A). This results in longer and more intense bursts on the side toward which a turning response occurs (Fig. 8B). Steering responses to the left or right side can be elicited in a reproducible manner in a reduced brainstem–spinal cord in vitro preparation by trigeminal stimuli. Such a stimulus activates the reticulospinal neurons via lower brainstem interneurons (Fagerstedt et al., 2001). The basic brainstem circuitry for turning is thus comparatively simple (Fig. 9A). The reticulospinal neurons impinge on both the EINs and the commissural inhibitory interneurons, and will thereby provide a stronger excitation and thus longer bursts on the same side and concurrently also a longer inhibition of the contralateral side. These results have been analyzed both experimentally and through modeling (Kozlov et al., 2001).

The above results illustrate the basic neural machinery for turning, but they do not unravel the mechanisms by which the animal itself will elicit goal-directed steering toward a particular object or prey. The tectum (superior colliculus) is an important structure in this context. It receives input from the eye, which is organized in a retinotopic fashion. It provides thus a sensory map that is aligned to a motor map, which can elicit eye movements to a target and also orientation movements of the body and finally locomotion (Saitoh et al., 2007), seemingly toward an object that gives rise to the activation of retina (see Fig. 8C). The tectal circuitry in interaction with the basal ganglia

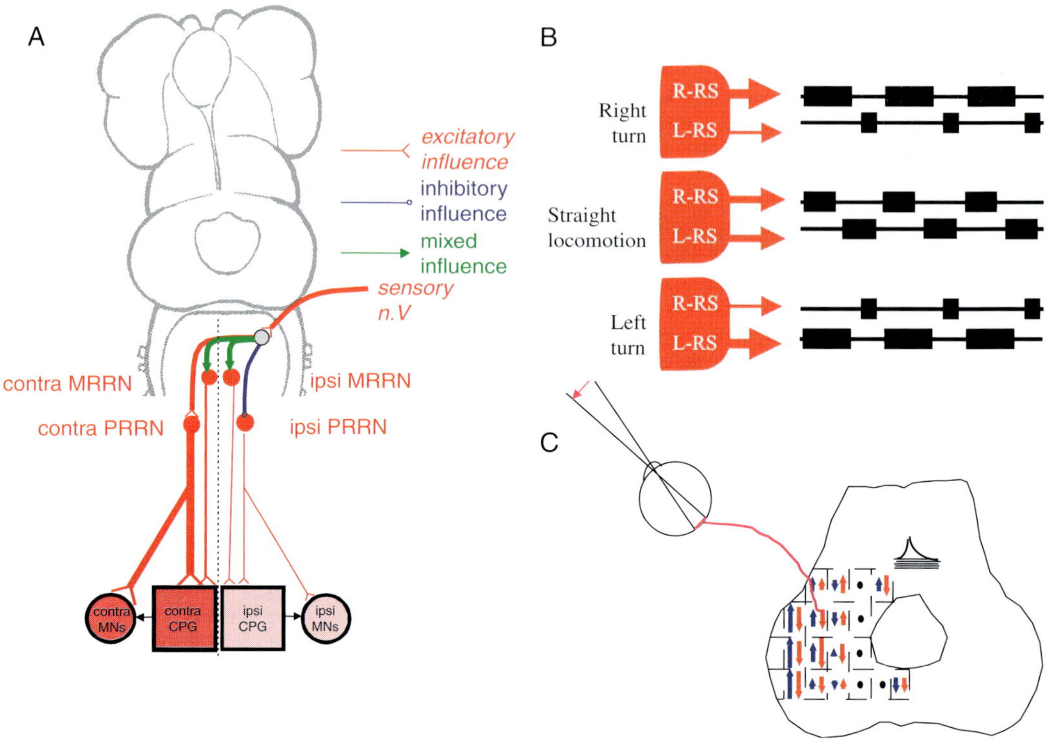

Fig. 8. Exploration of the mechanisms controlling steering in the lamprey. (A) During experimental conditions, steering signals can be produced by trigeminal sensory inputs (sensory n. V.) that via interneurones activate prefentially contralateral medial and posterior rhombencephalic reticulospinal neurons (MRRN and PRRN) that project to the spinal cord (Deliagina et al., 2000). Red indicates excitation, blue inhibition and green mixed effects. The descending signals affect both the CPG and the motoneurons (MNs) directly. (B) An asymmetric activation of RS neurons results in a steering movement toward the more strongly activated side. Behaviorally this will allow the animal to escape, or swim away from, the original stimulus (compare A). (C) A and B illustrate the turning machinery itself, goal-directed steering toward a certain object requires sensory processing. The tectum receives input from the eye in a retinotopic fashion, thus forming a motor map, which can elicit both appropriate eye and orientation movements toward a given target (Saitoh et al., 2007), which will result in steering if the animal is swimming.

presumably is responsible for this action — but this is outside the scope of this brief review.

Control of body orientation

During locomotion the lamprey corrects its body position instantaneously so that the dorsal side is always directed upward. The vestibular organs on the left and the right sides sense any deviation in terms of lateral tilt of the head. The vestibular input is mediated via interneurons to the reticulospinal neurons in the rhombencephalon (Fig. 9), so that a tilt to the left leads to an enhanced activity of reticulospinal neurons on the right side of the brainstem, which elicits a correction of the body position.

When the head is in a "correct" dorsal side-up position, the activity of the reticulospinal neurons on the two sides are at the same level (Fig. 9), any deviation will lead to an enhanced activity on one side and a reduced activity on the other (Deliagina et al., 1992; Orlovsky et al., 1992; Zelenin et al., 2003), which leads to a corrective response. In a hybrid system, we used the recorded tilt elicited signals in lamprey reticulospinal neurons to elicit body corrections via an external motor, which indeed was quite effective (Zelenin et al., 2000). We have studied this control system extensively also by modeling (Kozlov et al., 2001). The lamprey is

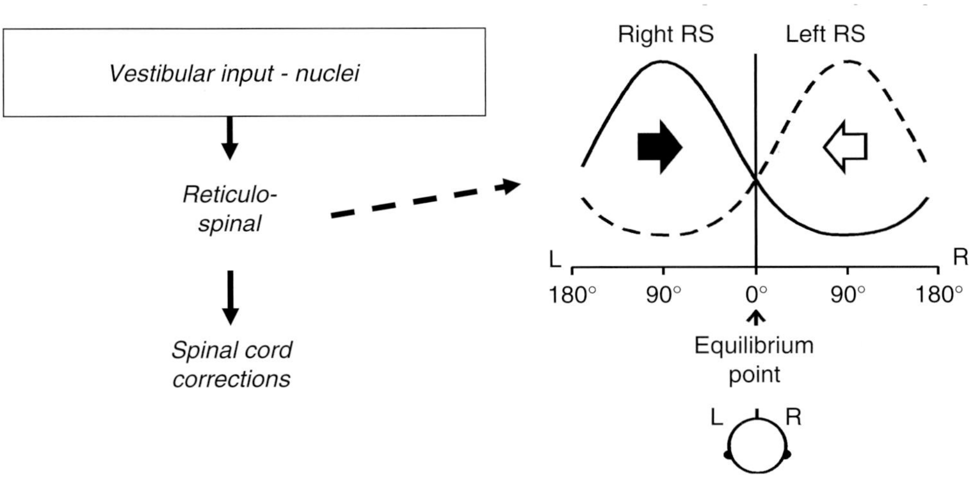

Fig. 9. Conceptual model of roll corrections to maintain body orientation (Deliagina et al., 1992; Orlovsky et al., 1992). (A) The curves represent the activity of the reticulospinal (RS) neurons on the right side (thick line), and correspondingly for the left side (dashed line), as a function of roll angle. Vestibular input causes activation of right RS and left RS, with contralateral tilt via interneurons. Directions of roll caused by right and left RSs are indicated by the black and white arrows, respectively. The system has an equilibrium point at 0 degrees (i.e. dorsal side-up orientation).

not able to stabilize its body orientation without the vestibular system, but the visual system can also provide input to the reticulospinal neurons, and thereby stabilize positions with a certain degree of tilt — the dorsal light response (Ullen et al., 1997). In addition to stabilizing the head position for lateral tilt, the lamprey is also able to stabilize swimming with a certain angle against the horizontal (downward or upward). A similar type of modulation, as unraveled for lateral tilt, has also been found to apply for the stabilization of a certain pitch angle.

Sensory feedback helps compensate for perturbations

The spinal CPG can generate the motor pattern underlying locomotion without sensory feedback as shown with the neuromechanical lamprey model in Fig. 10A (Grillner et al., 1976; Ekeberg and Grillner, 1999; Grillner, 2003). Although the CPG operates without sensory feedback, stretch receptors on the lateral margin of the spinal cord (Grillner et al., 1984) sense the locomotor movements, and have a direct synaptic link to the CPG interneurons (Viana Di Prisco et al., 1990; Fig. 10C). They provide excitation to the ipsilateral side and inhibition to the contralateral side. Artificially imposed movements will be able to entrain the rhythmic activity of the CPG within certain limits, and thus will also be able to compensate for external perturbations. Using the neuromechanical lamprey model in which a perturbation with higher water speed in a limited region was introduced (gray area) the lamprey will not be able to compensate for the perturbations in Fig. 10A but will when sensory feedback is present (Fig. 10B). The sensory feedback allows the lamprey model to compensate for the perturbations so that it can swim through the gray area.

Concluding remarks

In this brief review we have aimed at summarizing the experimentally based modeling of the neural control system for goal-directed locomotion and control of body orientation in the lamprey. Due to the relative simplicity of the lamprey nervous system, it has been possible to address the general principles of the neuronal mechanisms underlying goal-directed motor control. From these, it is possible to infer the control mechanisms underlying

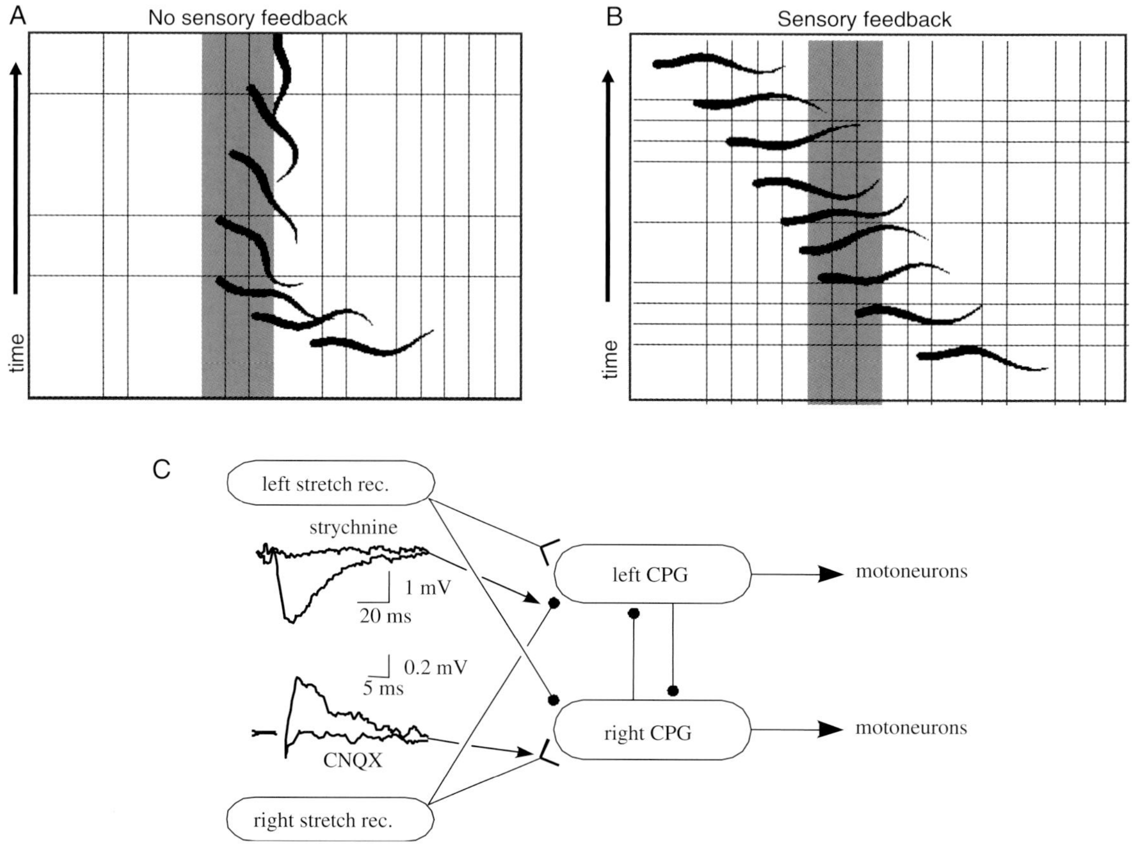

Fig. 10. Role of sensory feedback during lamprey swimming (Ekeberg and Grillner, 1999). (A) Simulation of lamprey swimming in water using a neuromechanical model of the lamprey. The shaded area represents a region with increased water speed. The simulated lamprey will not be able to cross the region. (B) However, in the presence of sensory feedback from spinal cord stretch receptors the simulated lamprey can pass this region successfully, since the stretch receptors stabilize the CPG network by allowing the bending of the body to affect the CPG neuron activity. (C) The connectivity from stretch receptor neurons to the CPGs on the two sides. Stretch receptors with ipsilateral axons provide excitation (glutamatergic EPSPs) to the ipsilateral CPG, and stretch receptors with contralateral axons inhibition (glycinergic IPSPs) of the contralateral CPG. The stretch receptors on one side are activated as the contralateral side is contracting, resulting in an inhibition of the active side and excitation of the side that is going to be active.

goal-directed behavior in more complex vertebrates. Computational approaches are essential tools for analyzing the dynamically interacting processes at the cellular and network levels that underlie motor behavior.

Acknowledgments

We hereby acknowledge the support of the European commission (Neurobotics, PD, SG), The Swedish Research Council (SG, JHK, AL) and The Wallenberg Foundations.

References

Bem, T., Cabelguen, J., Ekeberg, O. and Grillner, S. (2003) From swimming to walking: a single basic networks for two different behaviors. Biol. Cybern., 88: 79–90.

Biro, Z., Hill, R.H. and Grillner, S. (2006) 5-HT modulation of identified segmental premotor interneurons in the lamprey spinal cord. J. Neurophysiol., 96: 931–935.

Buchanan, J.T. (1982) Identification of interneurons with contralateral, caudal axons in the lamprey spinal cord: synaptic interactions and morphology. J. Neurophysiol., 47: 961–975.

Buchanan, J.T. and Grillner, S. (1988) A new class of small inhibitory interneurones in the lamprey spinal cord. Brain Res., 438: 404–407.

Buchanan, J.T., Grillner, S., Cullheim, S. and Risling, M. (1989) Identification of excitatory interneurons contributing to generation of locomotion in lamprey: structure, pharmacology, and function. J. Neurophysiol., 62: 59–69.

Cangiano, L. and Grillner, S. (2003) Fast and slow locomotor burst generation in the hemispinal cord of the lamprey. J. Neurophysiol., 89: 2931–2942.

Cangiano, L. and Grillner, S. (2005) Mechanisms of rhythm generation in a spinal locomotor network deprived of crossed connections: the lamprey hemicord. J. Neurosci., 25: 923–935.

Cohen, A.H., Ermentrout, G.B., Kiemel, T., Kopell, N., Sigvardt, K.A. and Williams, T.L. (1992) Modelling of intersegmental ccordination in the lamprey central pattern generator for locomotion. Trends Neurosci., 15: 434–438.

Dale, N. (1986) Excitatory synaptic drive for swimming mediated by amino acid receptors in the lamprey. J. Neurosci., 6: 2662–2675.

Deliagina, T.G., Orlovsky, G.N., Grillner, S. and Wallen, P. (1992) Vestibular control of swimming in lamprey. II. Characteristics of spatial sensitivity of reticulospinal neurons. Exp. Brain Res., 90: 489–498.

Deliagina, T.G., Zelenin, P.V., Fagerstedt, P., Grillner, S. and Orlovsky, G.N. (2000) Activity of reticulospinal neurons during locomotion in the freely behaving lamprey. J. Neurophysiol., 83: 853–863.

Deliagina, T.G., Zelenin, P.V. and Orlovsky, G.N. (2002) Encoding and decoding of reticulospinal commands. Brain Res. Brain Res. Rev., 40: 166–177.

Ekeberg, O. and Grillner, S. (1999) Simulations of neuromuscular control in lamprey swimming. Philos. Trans. R. Soc. Lond. B. Biol. Sci., 354: 895–902.

Ekeberg, O., Wallen, P., Lansner, A., Traven, H., Brodin, L. and Grillner, S. (1991) A computer based model for realistic simulations of neural networks. I. The single neuron and synaptic interaction. Biol. Cybern., 65: 81–90.

Fagerstedt, P., Orlovsky, G.N., Deliagina, T.G., Grillner, S. and Ullén, F. (2001) Lateral turns in the lamprey. II. Activity of reticulospinal neurons during the generation of fictive turns. J. Neurophysiol., 86: 2257–2267.

Grillner, S. (1974) On the generation of locomotion in the spinal dogfish. Exp. Brain Res., 20: 459–470.

Grillner, S. (2003) The motor infrastructure: from ion channels to neuronal networks. Nat. Rev. Neurosci., 4: 573–586.

Grillner, S. (2006) Biological pattern generation: the cellular and computational logic of networks in motion. Neuron, 52: 751–766.

Grillner, S., Perret, C. and Zangger, P. (1976) Central generation of locomotion in the spinal dogfish. Brain Res., 109: 255–269.

Grillner, S., Williams, T. and Lagerback, P.A. (1984) The edge cell, a possible intraspinal mechanoreceptor. Science, 223: 500–503.

Hellgren, J., Grillner, S. and Lansner, A. (1992) Computer simulation of the segmental neural network generating locomotion in lamprey by using populations of network interneurons. Biol. Cybern., 68: 1–13.

Hellgren-Kotaleski, J., Grillner, S. and Lansner, A. (1999a) Neural mechanisms potentially contributing to the intersegmental phase lag in lamprey. I. Segmental oscillations dependent on reciprocal inhibition. Biol. Cybern., 81: 317–330.

Hellgren-Kotaleski, J., Lansner, A. and Grillner, S. (1999b) Neural mechanisms potentially contributing to the intersegmental phase lag in lamprey. II. Hemisegmental oscillations produced by mutually coupled excitatory neurons. Biol. Cybern., 81: 299–315.

Ijsbeert, A., Crespi, A. and Cabelguen, J.-M. (2005) Simulation and robotics studies of salamander locomotion: applying neurobiological principles to the control of locomotion in robots. Neuroinformatics, 3: 171–196.

Islam, S.S., Zelenin, P.V., Orlovsky, G.N., Grillner, S. and Deliagina, T.G. (2006) Pattern of motor coordination underlying backward swimming in the lamprey. J. Neurophysiol., 96: 451–460.

Kozlov, A.K., Aurell, E., Orlovsky, G., Deliagina, T.G., Zelenin, P.V., Hellgren-Kotaleski, J. and Grillner, S. (2001) Modeling postural control in the lamprey. Biol. Cybern., 84: 323–330.

Kozlov, A.K., Lansner, A., Grillner, S. and Hellgren-Kotaleski, J. (2007) A hemicord locomotor network of excitatory interneurons: a simulation study. Biol. Cybern., 96(2): 229–243.

Matsushima, T. and Grillner, S. (1992) Neural mechanisms of intersegmental coordination in lamprey: local excitability changes modify the phase coupling along the spinal cord. J. Neurophysiol., 67: 373–388.

Ohta, Y., Dubuc, R. and Grillner, S. (1991) A new population of neurons with crossed axons in the lamprey spinal cord. Brain Res., 564(1): 143–148.

Ohta, Y. and Grillner, S. (1989) Monosynaptic excitatory amino acid transmission from the posterior rhombencephalic reticular nucleus to spinal neurons involved in the control of locomotion in lamprey. J. Neurophysiol., 62: 1079–1089.

Orlovsky, G.N., Deliagina, T.G. and Wallen, P. (1992) Vestibular control of swimming in lamprey. I. Responses of reticulospinal neurons to roll and pitch. Exp. Brain Res., 90: 479–488.

Parker, D. and Grillner, S. (2000) The activity-dependent plasticity of segmental and intersegmental synaptic connections in the lamprey spinal cord. Eur. J. Neurosci., 12: 2135–2146.

Saitoh, K., Menard, A. and Grillner, S. (2007) Tectal control of locomotion, steering, and eye movements in lamprey. J. Neurophysiol., 97(4): 3093–3108.

Stefanini, C., Orlandi, G., Menciassi, A., Ravier, Y., La-Spinal, G., Grillner, S. and Dario, P. (2006) A mechanism for biomimetic actuation in lamprey-like robots. In: The First IEEE/RAS-EMBS International Conference on Biomedical Robotics and Biomechatronics, pp. 1–6.

Ullen, F., Deliagina, T.G., Orlovsky, G.N. and Grillner, S. (1997) Visual pathways for postural control and negative phototaxis in lamprey. J. Neurophysiol., 78: 960–976.

Viana Di Prisco, G., Wallén, P. and Grillner, S. (1990) Synaptic effects of intraspinal stretch receptor neurons mediating movement-related feedback during locomotion. Brain Res., 530: 161–166.

Wallen, P. and Williams, T.L. (1984) Fictive locomotion in the lamprey spinal cord in vitro compared with swimming in the intact and spinal animal. J. Physiol. (Lond.), 347: 225–239.
Wannier, T., Deliagina, T.G., Orlovsky, G.N. and Grillner, S. (1998) Differential effects of the reticulospinal system on locomotion in lamprey. J. Neurophysiol., 80: 103–112.
Williams, T., Grillner, S., Smoljaninov, P., Wallén, P., Kashin, S. and Rossignol, S. (1989) Locomotion in lamprey and trout: the relative timing of activation and movement. J. Exp. Biol., 143: 559–566.
Zelenin, P.V., Deliagina, T.G., Grillner, S. and Orlovsky, G.N. (2000) Postural control in the lamprey: a study with a neuromechanical model. J. Neurophysiol., 84: 2880–2887.
Zelenin, P.V., Grillner, S., Orlovsky, G.N. and Deliagina, T.G. (2003) The pattern of motor coordination underlying the roll in the lamprey. J. Exp. Biol., 206: 2557–2566.

P. Cisek, T. Drew & J.F. Kalaska (Eds.)
Progress in Brain Research, Vol. 165
ISSN 0079-6123
Copyright © 2007 Elsevier B.V. All rights reserved

CHAPTER 15

Modeling the mammalian locomotor CPG: insights from mistakes and perturbations

David A. McCrea[1,*] and Ilya A. Rybak[2]

[1]*Spinal Cord Research Centre and Department of Physiology, University of Manitoba, Winnipeg, MB, R3E 3J7, Canada*
[2]*Department of Neurobiology and Anatomy, Drexel University College of Medicine, Philadelphia, PA 19129, USA*

Abstract: A computational model of the mammalian spinal cord circuitry incorporating a two-level central pattern generator (CPG) with separate half-center rhythm generator (RG) and pattern formation (PF) networks is reviewed. The model consists of interacting populations of interneurons and motoneurons described in the Hodgkin-Huxley style. Locomotor rhythm generation is based on a combination of intrinsic (persistent sodium current dependent) properties of excitatory RG neurons and reciprocal inhibition between the two half-centers comprising the RG. The two-level architecture of the CPG was suggested from an analysis of deletions (spontaneous omissions of activity) and the effects of afferent stimulation on the locomotor pattern and rhythm observed during fictive locomotion in the cat. The RG controls the activity of the PF network that in turn defines the rhythmic pattern of motoneuron activity. The model produces realistic firing patterns of two antagonist motoneuron populations and generates locomotor oscillations encompassing the range of cycle periods and phase durations observed during cat locomotion. A number of features of the real CPG operation can be reproduced with separate RG and PF networks, which would be difficult if not impossible to demonstrate with a classical single-level CPG. The two-level architecture allows the CPG to maintain the phase of locomotor oscillations and cycle timing during deletions and during sensory stimulation. The model provides a basis for functional identification of spinal interneurons involved in generation and control of the locomotor pattern.

Keywords: spinal cord; CPG; rhythm generation; locomotion; afferent control

Introduction

Well co-ordinated locomotor activity can be evoked in the mammalian spinal cord in the absence of input from higher brain centers (e.g., in spinalized animals) and rhythmic sensory feedback following neuromuscular blockade, i.e., during fictive locomotion (see Grillner, 1981; Rossignol, 1996; Orlovsky et al., 1999). Such observations have provided evidence for the existence of a central pattern generator (CPG) that generates the locomotor rhythm and pattern of motoneuron activity (Graham Brown, 1914). There appears to be one CPG controlling each limb (see Yamaguchi, 2004; Zehr and Duysens, 2004) since there can be independent rates of left and right stepping in the legs of man (Dietz, 2003; Yang et al., 2004) and in spinal cats (e.g., Forssberg et al., 1980). Cats can also step with independent rates between the fore and hind limbs (Akay et al., 2006). The spinal cord also contains circuitry for inter-limb coordination,

*Corresponding author. Tel.: +1 204 789 3770; Fax: +1 204 789 3930; E-mail: dave@scrc.umanitoba.ca

DOI: 10.1016/S0079-6123(06)65015-2

since coordinated stepping between the fore and the hind limbs is seen in animals with a spinal transection at upper cervical levels (Miller and van der Meche, 1976) and gaits remain matched and coordinated in the hind limbs of cats spinalized at mid-thoracic levels when walking on a treadmill (Forssberg et al., 1980).

The first conceptual scheme of the mammalian locomotor CPG responsible for alternating rhythmic extensor and flexor activity was based on a half-center concept (Graham Brown, 1914). According to the classical half-center architecture and its elaboration by Lundberg and colleagues (see Lundberg, 1981), the locomotor rhythm and the alternating activation of flexor and extensor motoneurons within a limb are produced by a single network consisting of two populations of excitatory interneurons (called the flexor and extensor half-centers) coupled together by reciprocal inhibitory connections such that activity in one half-center inhibits activity in the other. The interplay between tonic excitation of the two half-centers, a fatigue process reducing half-center activity over time, and the reciprocal inhibition between the half-centers results in rhythmic alternating activation of flexor and extensor motoneurons. The advantages of the half-center CPG organization include its relative simplicity and the strict alternation and coupling of flexor and extensor activities. This simple half-center architecture, however, is unable to account for a number of observations including the variety of motoneuron firing patterns observed during locomotion (e.g., Grillner, 1981; Stein and Smith, 1997).

The objective of the present study was to develop a computational model of the neural circuitry in the spinal cord that could provide predictions about the organization of the locomotor CPG and the interactions between the CPG and reflex circuits. We wished to create a model that could reproduce and provide explanations for a series of observations obtained in decerebrate adult cats during fictive locomotion induced by continuous electrical stimulation of the brainstem midbrain locomotor region (MLR) following neuromuscular blockade. One advantage of this preparation is that locomotor activity occurs without descending cortical influences, rhythmic sensory feedback, or the effects of systemic drug administration. Furthermore, the use of an adult preparation avoids developmental issues associated with an immature central and peripheral nervous system. Importantly, the pattern of motoneuron activities recorded in the decerebrate, immobilized cat during fictive locomotion is similar to that in intact preparations (Rossignol, 1996). Our intention was to develop a model in which a variety of simulations could be directly compared to data obtained during fictive locomotion in our laboratory. The simulations to be discussed were limited to creating locomotor-like activity in "pure" flexor and extensor motoneurons. The complex activity of motoneurons innervating muscles spanning more than one joint (bifunctional) is not considered here.

The role of intrinsic neuronal properties and reciprocal inhibition in rhythm generation

The major difficulty in developing a realistic CPG model is that the intrinsic and network mechanisms involved in the generation of the mammalian locomotor rhythm remain largely unknown. It is not yet possible to explicitly model the exact mechanisms operating in the mammalian spinal rhythm generator (RG). Therefore, our approach was to use the available data on spinal CPG operation and to incorporate rhythmogenic mechanisms operating in other mammalian CPGs and vertebrate motor systems. Our goal was to reproduce experimentally observed patterns of motoneuron activity recorded in the cat and their alteration under different conditions. At a minimum the model should be able to generate the locomotor rhythm and reproduce the following characteristics of fictive locomotion: (1) Tonic excitatory drive to the CPG (e.g., excitation mimicking that produced by tonic stimulation of the MLR) should evoke rhythmic activity with two alternating phases ("flexion" and "extension") coupled without intervening quiescent periods. (2) Increasing tonic (MLR) drive to the RG should result in a faster locomotor cadence. (3) Drive-evoked oscillations must encompass the range of step cycle periods and flexor and extensor phase durations observed during fictive locomotion. (4) In the absence of MLR drive, an increase in the excitability of CPG neurons should produce slow locomotor-like activity similar to that evoked by systemic L-DOPA administration in cats. (5) Blocking synaptic inhibition should result in spontaneous synchronized oscillations of flexors and extensors.

There is indirect evidence for the involvement of the persistent (or slowly inactivating) sodium current, I_{NaP}, in rhythmogenesis in different motor systems. For example, this current was shown to play a critical role in respiratory rhythm generation in the pre-Bötzinger complex in vitro and under certain conditions, in vivo (Smith et al., 2000; Rybak et al., 2003, 2004; Paton et al., 2006). Persistent sodium currents have been found in spinal interneurons and motoneurons (e.g., Lee and Heckman, 2001; Darbon et al., 2004; Brocard et al., 2006; Dai and Jordan, 2006; Streit et al., 2006; Theiss et al., 2007), and its blockade (e.g., by riluzole) abolishes the intrinsic cellular oscillations and rhythm generation in cultured rat spinal cord neurons (Darbon et al., 2004; Streit et al., 2006) as well as the NMDA- and 5-HT-evoked fictive locomotor rhythm in the neonatal mouse spinal cord (Zhong et al., 2006). Based on this indirect evidence, we hypothesized that I_{NaP} plays an essential role in the generation of locomotor oscillations in the mammalian spinal cord and incorporated an I_{NaP}-dependent intrinsic oscillatory mechanism in our model of locomotor rhythm generation (Rybak et al., 2006a).

Figure 1A shows the schematic of the spinal locomotor RG implemented in our model. The RG contains a homogenous population of excitatory neurons with intrinsic I_{NaP}-dependent rhythmogenic properties. This homogenous population is subdivided into two half-centers (RG-E and RG-F populations) with excitatory synaptic connections within and between the half-centers. These half-centers reciprocally inhibit each other via corresponding inhibitory interneuron populations (Inrg-E and Inrg-F, see Fig. 1A). Each population in the model contains 20 neurons described as single-compartment, Hodgkin-Huxley type neuron models. Each neuron contains only a minimum set of ionic channels: fast sodium, potassium rectifier, and leakage channels. The excitatory RG and PF neurons also contain the persistent (slowly inactivating) I_{NaP} sodium current. Because voltage- and time-dependent kinetics of activation and inactivation of fast sodium and potassium rectifier channels in mammalian spinal neurons have not been experimentally characterized, generic descriptions of these channels (from Booth et al., 1997) were used in the model. The kinetics of the NaP channel was adapted from previous computational models of neurons in the medullary pre-Bötzinger complex (Butera et al., 1999a, b; Rybak et al., 2003, 2004). The conductance of the NaP channel is characterized by a slow inactivation and is described as the product of three variables: the channel maximum conductance (g_{NaP}), the voltage- and time-dependent activation (m_{NaP}), and inactivation (h_{NaP}). The heterogeneity of neurons within each population was set by a random distribution of neuronal parameters and initial conditions. A full description of the model may be found in Rybak et al. (2006a).

Figure 1B illustrates the results of simulation of MLR-evoked locomotor rhythm in our model. The top three traces show, respectively: the histogram of average neuron activity in the RG-F population, the membrane potential trajectory of one neuron in this population, and the change in the inactivation variable for the NaP channel (h_{NaP}) in the same neuron. The next three traces show the same variables for the RG-E population. Locomotor oscillations in the model are initiated and maintained by a constant excitatory (MLR) drive to both RG populations. This drive is sufficient to depolarize the neurons of the excitatory populations to exceed the threshold for activation of the fast sodium current and to maintain spiking activity even if the persistent sodium current I_{NaP} becomes fully inactivated. In neurons of the currently active RG population, I_{NaP} progressively decreases with time because of the falling h_{NaP} (see third and sixth traces in Fig. 1B). The reduction in I_{NaP} reduces neuronal firing rate and population activity during the burst (see first and fourth traces), but activity remains sufficient to maintain inhibition of the antagonist RG population via the corresponding Inrg population (Fig. 1B, two bottom traces).

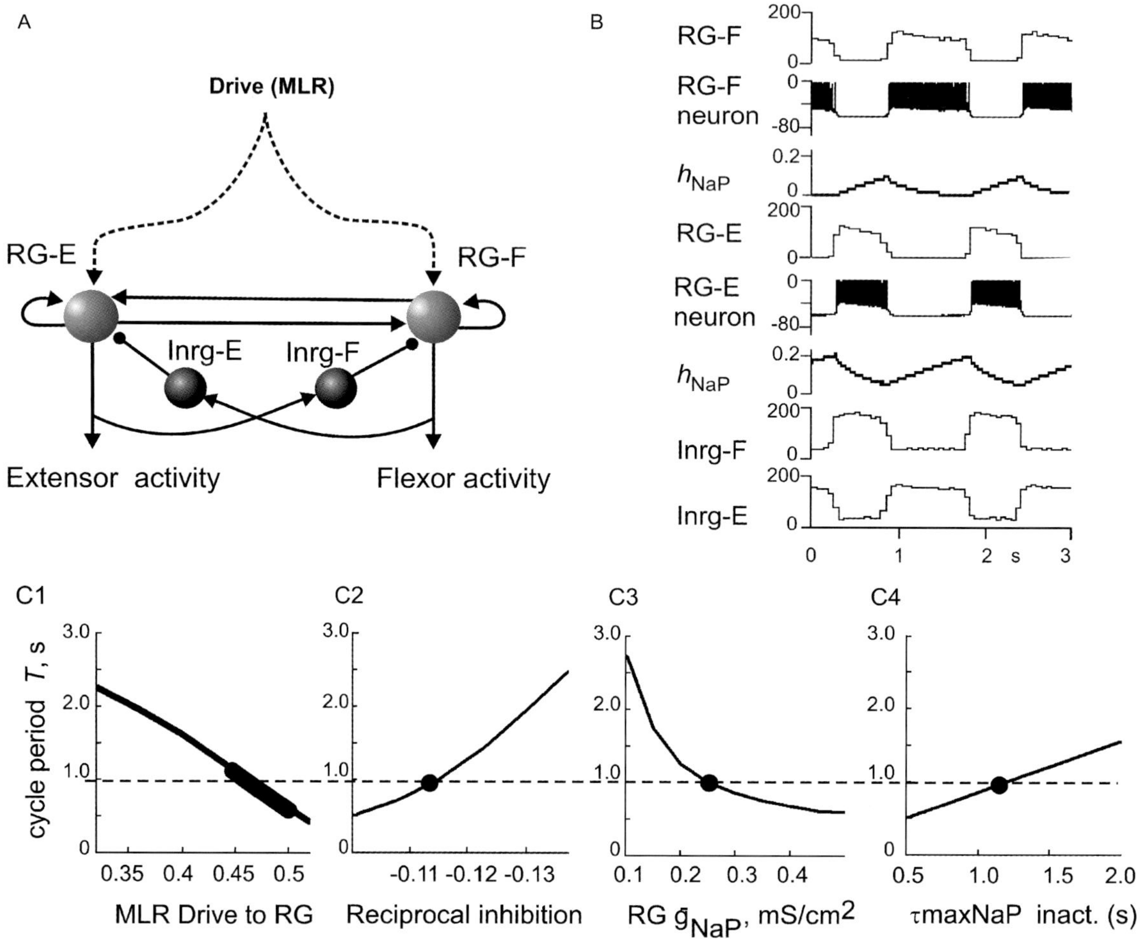

Fig. 1. Operation of the locomotor rhythm generator (RG). (A) Model Schematic. Each sphere represents a 20-neuron population. Excitatory and inhibitory synaptic connections are shown by arrows and small circles, respectively. Excitatory drive (from the MLR) is shown by dashed lines. The RG consists of two populations of excitatory neurons (RG-E and RG-F) interconnected via inhibitory interneuron populations, Inrg-E and Inrg-F, and mutual excitatory connections. (B) Activity of RG populations during two-step cycles. The top three traces for the flexor population show, respectively, the histogram of average RG-F neuron activity, the membrane potential trajectory in one RG-F neuron, and change in the inactivation variable for persistent (slowly inactivating) sodium (NaP) channel (h_{NaP}) in this neuron. The next three traces show the corresponding variables for the extensor portion of the RG. Activities of inhibitory interneuron populations, Inrg-F and Inrg-E, are shown in the two bottom traces. In this and other figures, population activity is represented by a histogram of average firing frequency (number of spikes per second per neuron, bin = 30 ms). (C) Dependence of locomotor step-cycle period on model parameters. (C1) Increased drive to both RG populations (indicated by arbitrary units on in the horizontal axis) reduces step-cycle period (T). (C2) Step-cycle period monotonically increases with an increase in mutual inhibition between the RG populations. Mutual inhibition was increased (see abscissa) by an increase in the weight of inhibitory synaptic input from Inrg-E and Inrg-F to their respective targets. (C3) T monotonically decreases with an increase of the maximal conductance of persistent sodium (NaP) channels in RG neurons (g_{NaP}). (C4) Increasing the time constant for NaP channel inactivation (τ_{maxNaP}) causes a linear increase in T. The dots in C2–C4 and the heavy line in C1 indicate values of these parameters used for most simulations.

At the same time, in neurons of the inhibited RG population (i.e., during the "off" phase of the locomotor cycle) I_{NaP} activation threshold progressively decreases (i.e., h_{NaP} increases). At some point I_{NaP} becomes activated and neuronal firing is initiated. Recurrent excitation within the half-center (see Fig. 1A) synchronizes the onset of firing in the population. The maximal activation of both the fast and the persistent sodium currents results in a high level of RG population activity with the onset of firing. This vigorous activity excites the corresponding Inrg population, which in turn inhibits the previously active (opposite) RG population. Repetition of these processes produces alternating bursts of firing in the RG-E and RG-F populations. In summary, the onset of firing bursts in our model is determined mostly by the activation of the intrinsic excitatory mechanism (I_{NaP}), whereas burst termination is determined by reciprocal inhibition. As a result, the cycle period (T) depends on the external (MLR) drive to RG half-centers, the reciprocal inhibition between them, and the intrinsic characteristics of NaP channels in RG neurons.

The effects of altering these parameters on cycle period (T) are shown in Figs. 1C1–C4. Consistent with the cat fictive locomotor preparation (Sirota and Shik, 1973), T decreases monotonically with increasing MLR drive (see Fig. 1C1). The increase in drive to each RG population in the model mainly affects the inter-burst interval (i.e., the currently silent RG population). Because of the drive-induced increase in excitability, h_{NaP} needs less time to reach the level at which the neuronal excitability overcomes the inhibition provided by the currently active (opposite) half-center. As a result phase switching occurs sooner and cycle period decreases. A decrease in T resulting from increased drive to both RG populations is illustrated in Fig. 1C1. Conversely, increasing the strength of reciprocal inhibition between the RG half-centers (mediated by the inhibitory Inrg populations) increases T. This is because more time is required for h_{NaP} to reach the level at which the neuronal excitability overcomes the increased inhibition (Fig. 1C2). Increasing the maximal conductance of NaP channels (g_{NaP}) in the RG neurons decreases T (Fig. 1C3) since the inactivation variable h_{NaP} needs less time during the inter-burst interval to reach the threshold and produce phase switching. Finally, increasing the maximal time constant for h_{NaP} increases T (Fig. 1C4) because h_{NaP} needs more time to reach the level of I_{NaP} activation during each inter-burst interval.

One important aspect for model evaluation is the ability to reproduce behaviors of the locomotor system observed in other experimental conditions. For this reason, we have used the model to simulate locomotor-like activity evoked without MLR stimulation by the application of monoamine neuromodulators or neuromodulators and to simulate rhythmic activity evoked by blocking spinal synaptic inhibition. We suggest that monoamine cause a net increase in neuron excitability. Such an increase could result, for example, from the 2 to 7 mV reduction in action potential threshold produced by both serotonin and noradrenaline in spinal cord in vitro preparations (Fedirchuk and Dai, 2004) or from a reduction of potassium leak conductance (e.g., Kjaerulff and Kiehn, 2001; Perrier et al., 2003). To imitate a pharmacologically induced increase in excitability in the model, the average leakage reversal potential E_{L} was depolarized in all RG neurons. Figure 2A shows that in the absence of MLR drive, a depolarization of E_{L} by 6 mV evoked slow locomotor-like oscillations with alternating flexor and extensor activities and $T \approx 5$ s. Although the lack of experimental data on neuromodulator mechanisms in the spinal cord precludes more detailed simulations, these oscillations in the model are qualitatively similar to the slower rhythms evoked by neuromodulators such as L-DOPA.

As described above, the inhibitory interneuron populations, Inrg-F and Inrg-E, are responsible for producing the strictly alternating activity between the half-centers. In our model these populations have a low-level background activity, which prevents rhythm generation at rest. This low-level activity is also present between the strong bursts of activity evoked from the corresponding RG populations (bottom traces in Figs. 1B and 2A). As shown experimentally, antagonists of glycinergic and GABAergic inhibition can produce rhythmic activity characterized by synchronized bursting in flexor and extensor motoneurons (e.g., Noga et al., 1993; Cowley and Schmidt, 1995; Kremer and Lev-Tov, 1998; Beato and Nistri, 1999) instead of an alternating, locomotor pattern. To simulate oscillations evoked by blocking inhibitory transmission, the weights of all inhibitory connections in the model were set to zero. Under these conditions, the model generated oscillations in RG-F and RG-E that were synchronized by the

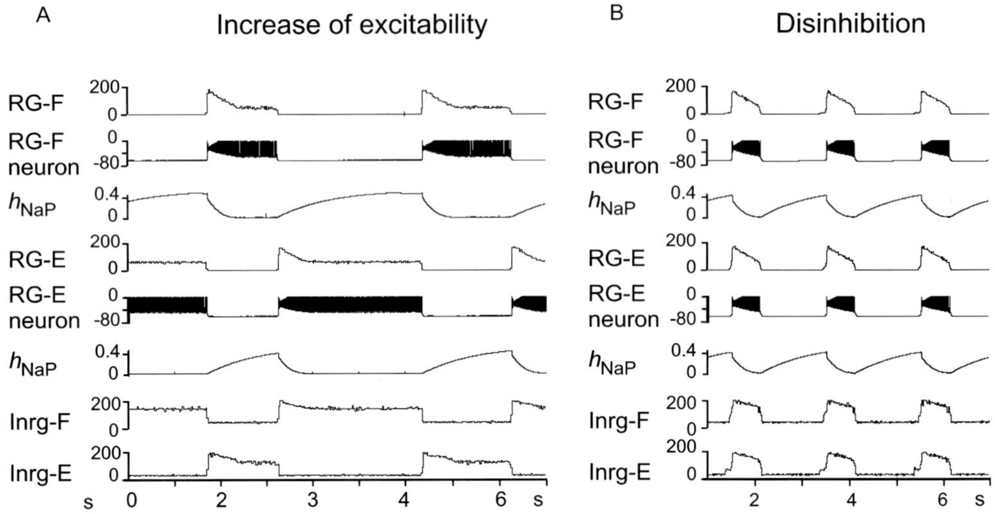

Fig. 2. Rhythms in the absence of MLR drive. (A) Imitation of pharmacologically evoked rhythm in the model. The slow rhythmic activity was produced by an increase in excitability of the excitatory RG populations in the absence of external (MLR) drive. This increased excitability was produced by a 6 mV depolarization of the average leakage reversal potential in all neurons of these populations. (B) Rhythmic activity produced in the model by disinhibiton. To simulate this behavior, the weights of all inhibitory connection in the model were set to zero. Note the synchronized rhythmic bursts of flexors and extensors. See details in the text. Figure adapted with permission from Rybak et al. (2006a).

excitatory connections between the two half-centers (see Fig. 2B) and are qualitatively similar to the synchronized activity evoked experimentally by blocking synaptic inhibition. Without reciprocal inhibition, cycle period is determined mainly by the level of neuronal excitability and the kinetics of I_{NaP} inactivation.

The finding that synchronized oscillations are evoked when synaptic inhibition is blocked in mammalian spinal cord preparations (Noga et al., 1993; Cowley and Schmidt, 1995; Kremer and Lev-Tov, 1998; Beato and Nistri, 1999) has led to the suggestion that the spinal locomotor network has endogenous rhythmogenic properties, which do not require inhibition (Kiehn, 2006; Rossignol et al., 2006). While we agree that a rhythm can be produced in the absence of network inhibition and our model can indeed produce such a rhythm, we do not accept that these oscillations represent the locomotor rhythm. Flexor and extensor bursts are strictly alternating during fictive locomotion, and this coupling is maintained as flexor and extensor phase durations are changed or interrupted by sensory stimulation. Alternating and strict coupling of flexor and extensor activities in a variety of preparations and throughout the entire range of locomotor speeds support the concept that reciprocal inhibition is essentially involved in at least the termination of flexor and extensor discharges and hence in locomotor-phase switching. For this reason, the locomotor pattern generated in our model is strongly dependent on the reciprocal inhibition between the RG half-centers (Fig. 1C2). We believe that regardless of the intrinsic rhythmogenic properties of spinal neurons involved, reciprocal inhibitory interactions are critically important for locomotor rhythm and pattern generation.

Much more needs to be learned about how and which intrinsic cellular mechanisms underlie mammalian locomotion. The locomotor rhythm in our model is generated with an essential contribution from the slowly inactivating sodium current (I_{NaP}). Our hypothesis originates from the role of this current in rhythmogenesis in other mammalian systems including respiratory rhythm generation and the presence of this current in mammalian spinal cord. Without further experimental evidence, it may be premature to claim that the I_{NaP}-dependent mechanism described in our model operates in the real spinal cord during locomotion. Mechanisms described in other preparations (see El Manira et al., 1994; Büschges et al., 2000; Grillner et al., 2001; Butt et al., 2002; Grillner and Wallén, 2002; Grillner, 2003) may also be important for rhythm generation. Our simulations do show, however, that I_{NaP}-dependent cellular properties of RG neurons in combination with reciprocal inhibition between the RG half-centers can reproduce locomotor oscillations as well as rhythmic activities evoked by neuromodulators and disinhibition (Fig. 2A and B).

Structure and operation of the locomotor model

The architecture of the locomotor model shown in Fig. 3A was suggested from two independent lines of experimental evidence obtained during fictive locomotion in cats. One series of experimental studies concerned deletions of motoneuron activity that occur during fictive locomotion in the cat. Deletions are brief periods during locomotion in which the normal alternating activity of flexor and extensor motoneurons is briefly interrupted by a failure to activate a group of synergist motoneurons. Because deletions occur spontaneously without any experimental intervention and simultaneously affect multiple agonist motoneuron pools, it is likely that they result from spontaneous alterations in the excitability of some elements of the CPG and are not simply changes in the excitability of a few motoneurons. For example, the activity of all synergist motoneurons in the limb (e.g., flexors) can fail while activity in antagonist motoneurons (e.g., extensors) becomes tonic for a few step cycles (Lafreniere-Roula and McCrea, 2005). Deletions can be full (i.e., no activity in all synergists) or partial (reduced activity in some synergist motoneuron pools and no activity in others) (Lafreniere-Roula and McCrea, 2005). Examples of deletions are presented in Figs. 5A and B and will be discussed below. The key observation was that during many deletions, rhythmic motoneuron activity returned without phase shift following the deletion. It appeared that some internal structure could "remember" cycle period and the phase of locomotor oscillations when rhythmic motoneuron activity failed. A similar maintenance of cycle period has also been noted during studies of the effects of afferent stimulation on fictive locomotion (Guertin et al., 1995; Perreault et al., 1995; McCrea, 2001; Stecina et al., 2005). In such cases, afferent stimulation delays or causes a premature phase switching within the ongoing step cycle without changing the timing of the following step cycles. Examples of such effects of sensory stimulation and of deletions in which cycle timing is maintained are presented below. In order to reproduce such timing maintenance, we proposed an extension of the classical half-center CPG organization. We suggested that the locomotor CPG has a two-level architecture containing a half-center RG performing a "clock" function, and an intermediate pattern formation (PF) network containing interacting interneuron populations that activate multiple synergist and antagonist motoneuron pools (Lafreniere-Roula and McCrea, 2005; Rybak et al., 2006a, b). Conceptually, similar subdivisions of the mammalian CPG into separate networks for rhythm generation and motoneuron activation have been proposed previously (e.g., Koshland and Smith, 1989; Orsal et al., 1990; Kriellaars et al., 1994; Burke et al., 2001) but have not been formally modeled or considered in the context of non-resetting deletions and sensory stimulation (for discussion of a two-level CPG in the turtle see Lennard, 1985).

Figure 3A shows the schematic of our model of spinal circuitry with a two-level locomotor CPG (Rybak et al., 2006a, b). In the proposed architecture, the RG defines the locomotor rhythm and durations of flexor and extensor phases. It also controls activity in the PF network, which is responsible for activation and inhibition of flexor and extensor motoneurons. Each interneuron or motoneuron population in the model consists of 20 neurons simulated in Hodgkin-Huxley style. Motoneurons are described as two-compartment models (modified from Booth et al., 1997) and interneurons are simulated using single compartment models (Rybak et al., 2006a). The PF network contains excitatory interneuron populations projecting to motoneurons and to inhibitory

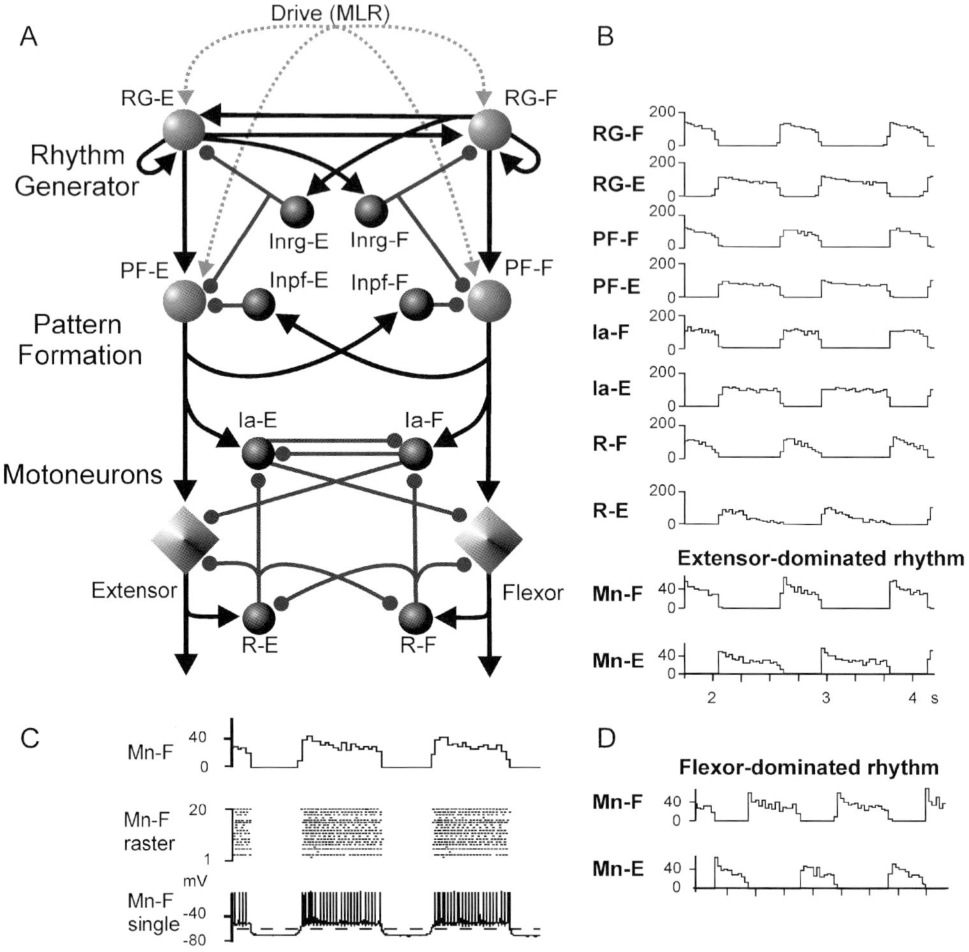

Fig. 3. Model of spinal circuitry with a two-level locomotor CPG without afferent inputs. (A) Schematic of the model. Populations of interneurons are represented by spheres. Excitatory and inhibitory synaptic connections are shown by arrows and small circles, respectively. Populations of flexor (Mn-F) and extensor (Mn-E) motoneurons are shown by diamonds. MLR excitatory drives are shown as dashed lines. See explanations in the text. (B) Locomotor activity patterns generated by the model. Activity of each population is represented by a histogram of average firing frequency. The MLR drive to the RG-E population ($d_{rge} = 0.5$ arbitrary units) is larger than to RG-F population ($d_{rgf} = 0.43$) and the model generates a rhythm with a longer duration extensor phase ($T_E > T_F$). In D, the RG-F population receives a larger drive ($d_{rgf} = 0.51$; $d_{rge} = 0.45$) and the model generates a flexion-dominated rhythm ($T_F > T_E$). Figure adapted with permission from Rybak et al. (2006a). (C) Flexor motoneuron population firing activity obtained in another simulation (upper trace), raster plot of spiking activity in the population (middle) and the trace of the membrane potential from a single flexor motoneuron. Note the rhythmic hyperpolarization of the motoneuron below resting membrane potential (horizontal dashed line) during the extension phase.

interneurons within the PF network. A more complete locomotor model would have multiple PF populations to allow differential control of groups of motoneurons including those activating bifunctional muscles. In the reduced version of the model considered here, the PF network contains only two excitatory neural populations, PF-E and PF-F. Each of these populations receives a weak excitatory input from the homonymous RG population, strong inhibition from the opposite RG population via a corresponding inhibitory population (Inrg-E or Inrg-F) and reciprocal inhibition from the opposite PF population via another inhibitory population (Inpf-F or Inpf-E, see Fig. 3A). Locomotion is initiated in the model by external tonic excitation (from the "MLR") that is distributed to both the RG and PF populations (details in Rybak et al., 2006a).

Figure 3B shows examples of computer simulations of locomotor rhythm generation and flexor- and extensor-motoneuron activities. Alternating busts of activity in the RG-E and RG-F populations (evoked by MLR drive) produce periodic, alternating activity of the PF-E and PF-F populations which leads to rhythmic excitation of extensor (Mn-E) and flexor (Mn-F) motoneurons (bottom traces Fig. 3B). Variations in motoneuron excitability within each population (defined by the random distribution of leakage reversal potential and other neuronal parameters) result in variations in firing rates and recruitment for a given level of PF drive to the population. This is shown in Fig. 3C (middle panel) by the raster plot of the firing in the 20 member flexor motoneuron population obtained from another simulation.

During fictive locomotion and in the absence of sensory activation by muscle spindle afferents, inhibitory Ia interneuron populations are phasically active (e.g., see Feldman and Orlovsky, 1975; McCrea et al., 1980; Pratt and Jordan, 1987) and since they are directly connected to motoneurons (Jankowska, 1992) they must contribute to the rhythmic inhibition of motoneurons. Accordingly, in our model PF activity excites inhibitory Ia-E and Ia-F interneurons (Fig. 3B, 5th and 6th traces from the top). The bottom trace in Fig. 3C is the simulated membrane potential from one flexor motoneuron showing the rhythmic depolarization and hyperpolarization (via Ia interneurons during extension) relative to the resting membrane potential (horizontal dashed line). The synaptic connections between IaINs, Renshaw cells, and motoneurons depicted in Fig. 3A are in accord with the known organization of this network (references in Jankowska, 1992). Thus during the active locomotor phase, motoneurons in the model receive both rhythmic excitation from the PF network and rhythmic inhibition from Renshaw cells (see RC-E and RC-F activities in Fig. 3B). During the opposite phase, they receive inhibition from the antagonist Ia population. Although the rhythmic inhibition of motoneurons is an important part of our model, we do not consider Ia interneurons or Renshaw cells to be part of the CPG. This is because they are not involved in rhythm generation per se and because their connections to motoneurons are organized on a more limited "local" basis (Jankowska, 1992). Thus the inhibition provided by sub-populations of these interneurons would sculpt the firing patterns of individual motoneuron pools during locomotion (Pratt and Jordan, 1987; Orlovsky et al., 1999). Recent evidence suggests that other as yet unidentified interneurons may also contribute to the rhythmic inhibition of motoneurons during locomotion (Gosgnach et al., 2006).

Control of cycle period and phase duration

Most of our simulations were carried out using fixed values for reciprocal inhibition, maximal conductance of NaP channels, and the maximal time constant for NaP channel inactivation (see Rybak et al., 2006a, b). These "standard" values (dots on the respective curves in Fig. 1C2–C4) were chosen to produce a cycle period on the order of 1 s. With these parameters fixed, a wide range of locomotor-cycle periods could be produced by varying the MLR drives to the RG populations (Fig. 1C1). Because of the symmetry of the two RG half-centers, equal MLR drive to both half-centers produced locomotor phases with equal durations. Unequal "flexor" and "extensor" phase durations could be produced by varying the drives to the RG-E and RG-F populations. Thus a stronger MLR drive to RG-E than to RG-F in Fig. 3B resulted in an extensor-phase dominated cycle, while increasing the MLR drive to the flexor circuitry in Fig. 3D produced a flexor phase-dominated rhythm with approximately the same locomotor-cycle period. Because motoneuron excitation is produced in the two-level CPG by the PF network, the level of motoneuron activity is similar in Fig. 3B and D. Although not shown, with a two-level CPG organization excitability changes at the PF level can strongly affect motoneuron activity and recruitment without changing locomotor-phase durations or step-cycle period.

In order to investigate the relative durations of the locomotor phases provided by the model, MLR drive was held constant in one RG population and progressively changed in the other. The thin lines in Fig. 4A1, A2, and B show the durations of the two locomotor phases obtained from the model in such experiments. In Fig. 4A1, reducing the drive to RG-E (d_{rge}) increased the duration of the flexion phase. Drive reduction had relatively little effect on the duration of the extension phase (i.e., the slope of the T_E line was shallow) but substantially prolonged the cycle period. In Fig. 4A2 and B, reducing drive to RG-F significantly prolonged the extension (stance) phase with little effect on the flexor (swing) phase. In all three panels, decreasing excitation to one half-center had only a minor effect on the duration of activity in that half-center. Cycle period increased mainly because of the increased burst duration in the opposite RG population.

During both fictive and real locomotion, changes in cat step-cycle period often involve a disproportionate change in the duration of one of the phases. For example, during treadmill locomotion, faster cycle periods are made primarily at the expense of extensor phase duration (Fig. 4B, thick lines; data replotted from Halbertsma, 1983). As discussed elsewhere (Yakovenko et al., 2005), this asymmetry in phase duration modulation during real walking may result from the influence of particular proprioceptive feedback on the CPG. In the absence of rhythmic sensory feedback during fictive locomotion, some preparations show a similar change in extensor phase duration with cycle period (thick lines, Fig. 4A1, data from Yakovenko et al., 2005) while other preparations show preferential changes in the flexor phase (Fig. 4A2). We consider the fact that fictive locomotion can involve either a dominance of the flexor or extensor phase to be strong evidence for a CPG that is organized symmetrically with regard to its ability to control flexor- and extensor-phase durations

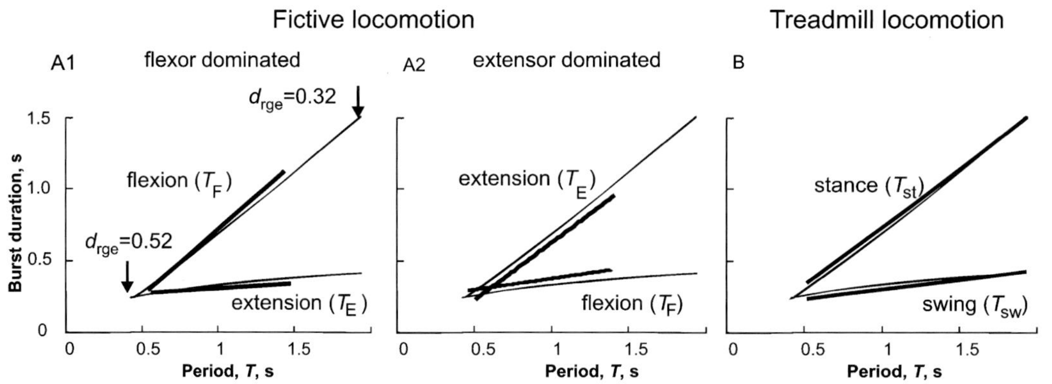

Fig. 4. Comparison of simulated locomotor-phase and step-cycle durations to experimental data. Thin lines in the three panels show the durations of simulated flexor and extensor phases plotted against the step-cycle period on the abscissa. These plots were created by holding the MLR drive to one side of the RG constant (RG-F in A1 and RG-E in A2 and B) and varying the drive to the other side of the RG. In A1, the flexor drive (d_{rgf}) was held constant at 0.52 (arbitrary units) and the extensor drive (d_{rge}) decreased from 0.52 to 0.32. The inverse of this procedure (i.e., holding d_{rge} constant) produced the simulations curves in A2 and B. Note that an increase in d_{rge} speeds up locomotion mainly by decreasing the duration of the opposite (flexor) phase, T_F with little effect on T_E. See text for explanation. The bold lines in the three panels are linear regressions of measurements of cycle phase duration for fictive locomotion in decerebrate cats (A1 and A2, data from Yakovenko et al., 2005) and for treadmill locomotion in intact cats (data from Halbertsma, 1983). The fictive locomotion data is separated into measurements from experiments in which the flexion phase dominated the fictive locomotor pattern (A1), and from those in which extensor activity was longer (A2) (see Yakovenko et al., 2005). Note the close correspondence between actual and modeled phase duration plots for both the duration of extensor (T_E) and flexor (T_F) nerve activity (A1 and A2) during fictive locomotion and for the durations of the stance (T_{st}) and swing phases (T_{sw}) during real locomotion. Adapted with permission from Rybak et al. (2006a).

(discussed in Lafreniere-Roula and McCrea, 2005; Duysens et al., 2006).

Insights into CPG organization from deletions of motoneuron activity during fictive locomotion

As mentioned, the stable alternation of flexor- and extensor-motoneuron activities during fictive locomotion can be briefly interrupted by periods in which motoneuron activity falls silent for a few step cycles and then reappears (e.g., Grillner and Zangger, 1979; Lafreniere-Roula and McCrea, 2005). Such spontaneously occurring errors or "deletions" of motoneuron activity also occur during the scratch reflex in turtles (see Stein, 2005) and during treadmill locomotion in cats (e.g., Duysens, 1977). When spontaneous deletions occur during MLR-evoked fictive locomotion or during fictive scratch, there is usually a failure of activity in multiple synergist motoneuron populations that is accompanied by tonic activity in multiple antagonist motoneuron populations (Lafreniere-Roula and McCrea, 2005). The widespread effect of deletions on the activity of multiple-motoneuron pools is strong evidence that they are produced by failures in the operation of some common spinal circuitry such as the CPG, and not the result of local perturbations affecting only the excitability of particular motoneurons.

An important feature of deletions concerns the timing of the locomotor bursts following a deletion episode. In the classical half-center CPG organization, a single network is responsible both for rhythm generation and motoneuron excitation. Accordingly, one would expect that a spontaneous deletion of motoneuron activity would be accompanied by changes in the timing of the locomotor RG. Since deletion duration could be arbitrary, post-deletion step cycles would often be expected to re-appear with a phase shift relative to the pre-deletion rhythm, i.e., the post-deletion rhythm would be reset. While "resetting" deletions do occur during fictive locomotion, it is more common for the post-deletion rhythm to be re-established without phase shift of the pre-deletion locomotor rhythm (Lafreniere-Roula and McCrea, 2005). The frequent occurrence of "non-resetting" deletions suggests that some internal structure can maintain locomotor period timing during the deletion of motoneuron activity.

Figure 5A and B, show examples of non-resetting deletions of flexor activity obtained during fictive locomotion. In Fig. 5A (from Lafreniere-Roula and McCrea, 2005) the failure of hip (Sart) and ankle (EDL) flexor motoneuron activation was accompanied by continuous activity of extensors operating at the hip (SmAB), knee (Quad), and ankle (Plant). The vertical dashed lines indicate intervals of the mean of the five-step cycle periods preceding the deletion. Note that flexor activity re-appeared at a multiple of this interval. In other words, the rhythm is re-established with a timing that would have been expected had the deletion not occurred. During some of the expected bursts there was also a weak modulation of the sustained extensor-motoneuron activity (marked by *). Figure 5B shows a bout of MLR-evoked fictive locomotion with recordings from ipsi- and contralateral flexor and extensor nerves. During this recording there was a typical non-resetting deletion of ipsilateral flexor activity. Two bursts in the ipsilateral TA trace are omitted and this is accompanied by sustained firing of ipsilateral extensors, AB and LGS. Note that at this particular stage of the experiment, the contralateral nerves were not active (see coTA and coMG traces). This example demonstrates that maintenance of the phase of locomotor oscillations during non-resetting deletions does not require (and cannot be explained by) rhythmic activity in the contralateral hind limb (further discussion in Lafreniere-Roula and McCrea, 2005).

We suggest that deletions result from spontaneously occurring, temporary changes in the excitability of neurons in the RG or PF networks. First consider the effect of a relatively strong additional excitatory drive to one of the RG populations. An increase in excitability or an additional excitatory drive to one RG population can temporarily interrupt rhythm generation by producing sustained activity in this population and suppressing activity in the opposite RG half-center. Increased RG activity will also inhibit the opposite PF population, which then causes a deletion of the corresponding

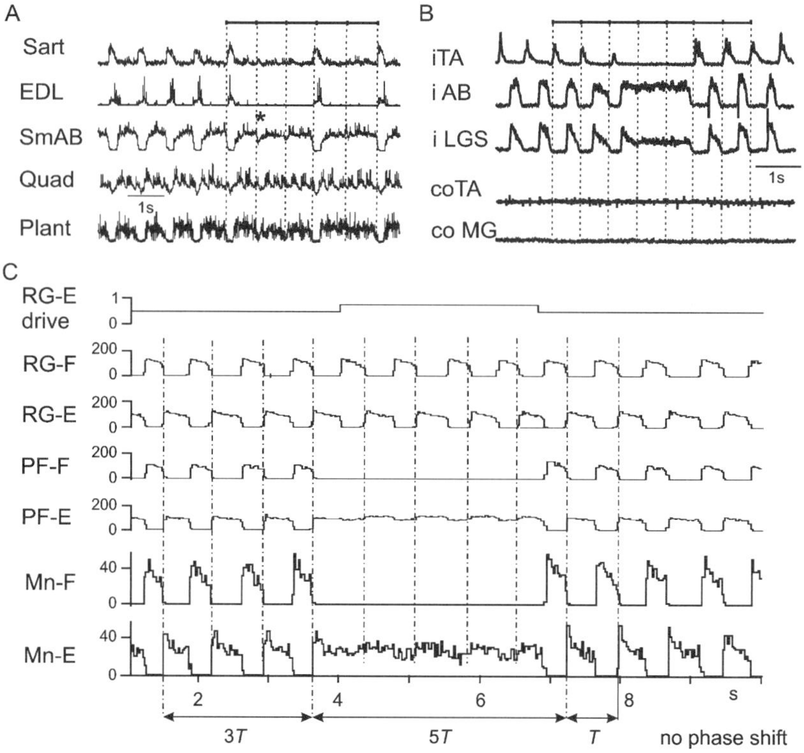

Fig. 5. Examples of deletions occurring during MLR-evoked fictive locomotion and simulation of "non-resetting" deletions. (A) An example of a deletion of flexor activity during fictive locomotion. The traces are rectified-integrated recordings from hindlimb flexors (hip: Sart and ankle: EDL) and extensors (hip: SmAB, knee: Quad, and ankle: Plant). The vertical dashed lines are plotted at intervals of the average cycle period preceding the deletions and indicate where flexor bursts should have occurred. Note the re-emergence of flexor activity at these intervals following the deletion. The * indicates a weak modulation of the sustained extensor motoneuron activity. Adapted with permission from Lafreniere-Roula and McCrea (2005). (B) Another example of deletion of flexor activity in which there was no contralateral flexor and extensor activity (see coTA and coMG traces). The non-resetting deletion of ipsilateral flexor activity (iTA) was accompanied by sustained firing of ipsilateral extensors (iAB and iLGS). Vertical dashed lines show that the phase of the locomotor rhythm is maintained after the deletion despite the absence of contralateral locomotor activity. (C) Simulation of the deletion of flexor activity in A produced by a temporary 90% increase in excitatory drive to the PF-E population (top trace). This drive produced sustained PF-E population activity and consequently sustained activity in the Mn-E population. Inhibition of the PF-F population resulted in a deletion of flexor motoneuron activity. The vertical dashed lines show that the rhythm re-appeared without a phase shift in respect to the pre-deletion rhythm (see arrows at the bottom of each panel). Adapted with permission from Rybak et al. (2006a).

motoneuron activity. The rhythm will restart when the perturbation ends. However since the perturbation can end at an arbitrary time, the post-perturbation rhythm will likely be phase shifted (reset) with respect to the pre-deletion rhythm. Hence perturbations that change the excitability of the RG should generally produce the "resetting" type of deletion observed during fictive locomotion (Lafreniere-Roula and McCrea, 2005).

In contrast to the effects of altering excitability at the RG level, a temporary change in excitability of one of the PF populations can produce a deletion in which the phase of the post-deletion rhythm is maintained. The simulation in Fig. 5C shows the effects of a temporary increase in drive to the PF-E population (see the top trace). This increased drive results in sustained activity in the PF-E population and inhibition of activity in the opposite, PF-F, population. At the motoneuron level (two bottom traces in Fig. 5C), there is a deletion of flexor motoneuron activity with sustained activity of extensor motoneurons. An important feature of this deletion is the lack of resetting of the post-deletion rhythm. Because RG operation is unchanged during the deletion, motoneuron activity re-appears without a phase shift. Note the weak rhythmic modulation of extensor motoneuron activity during the periods where flexor bursts "should" have occurred (vertical dashed lines). This is similar to that observed in experimental records (see Fig. 5A). In the model, the weak modulation of sustained motoneuron activity is the result of continued rhythmic activity at the RG level. In the example shown, RG-F inhibition produces a periodic modulation of PF-E activity that is reflected at the motoneuron level. Simulations of other types of deletions using the model are described in Rybak et al. (2006a). These modeling studies show that the proposed two-level CPG architecture can readily account for both the resetting and non-resetting types of deletions observed during fictive locomotion in the cat.

Afferent control of the CPG at the PF and RG levels

Although the spinal CPG can operate in the absence of sensory feedback (e.g., during fictive locomotion) afferent activity plays a critical role in adjusting the locomotor pattern to the motor task, environment, and biomechanical characteristics of the limbs and body (Pearson, 2004; Rossignol et al., 2006). To illustrate the effects of afferent stimulation on motoneuron activity during locomotion in the context of the two-level CPG organization considered here, we will consider the effects of activation of the extensor group I (Ia muscle spindle and Ib tendon organ) afferents. Modeling the effects of stimulation of other afferents (i.e., flexor and cutaneous) can be found in Rybak et al. (2006b).

A large body of experimental evidence shows that proprioceptive feedback from extensor group I afferents, and particularly those from ankle muscle nerves, results in a strong excitation of extensor motoneurons that contributes to a substantial portion of stance-phase extensor activity in cats during treadmill locomotion (see Donelan and Pearson, 2004; Rossignol et al., 2006), and in man (Sinkjaer et al., 2000). In reduced preparations, activity in extensor group I afferents can also control the transition from stance to swing, regulate the duration of the stance phase, and entrain the step-cycle period (see Guertin et al., 1995; Pearson, 2004; Rossignol et al., 2006).

Figure 6 shows the schematic of the extended model used to simulate control of locomotor pattern by group I extensor afferents. During locomotion, extensor group I afferents can access the spinal circuitry at several levels. The model includes monosynaptic excitation of homonymous motoneurons and disynaptic inhibition of antagonists by group Ia afferents. The weight of Ia monosynaptic excitation has been made low to reflect the presynaptic inhibition of these afferents during locomotion (Gosgnach et al., 2000; McCrea, 2001). The model also includes the locomotor-dependent disynaptic excitation of motoneurons (see Angel et al., 2005) that is mediated by the excitatory population Iab-E. During extension (i.e., when PF-E and Inpf-F are active), the Inpf-F population inhibits In-E thereby removing the In-E inhibition of Iab-E. This disinhibition permits a phase-dependent disynaptic excitation of extensors by group I extensor afferents (Schomburg and Behrends, 1978; McCrea et al., 1995; Angel et al., 1996, 2005). In addition, excitation of the Iab-E population from the PF-E population creates rhythmic extensor-phase activity in Iab-E interneurons in the absence of sensory stimulation that is in accord with that found experimentally (Angel et al., 2005). Thus these Iab-E interneurons also provide a portion of the excitation of extensor motoneurons during locomotion. Nonetheless, we do not consider the Iab-E population to be an integral part of the CPG. This is because they do not participate in rhythm generation and in some locomotor preparations, their excitability is low and group I disynaptic excitation motoneurons cannot be evoked (e.g., low spinal cats, McCrea et al., 1995). A more complete discussion of how group I disynaptic excitation emerges during locomotion and replaces the inhibitory effects evoked at rest is presented elsewhere (Rybak et al., 2006b).

Additional interneuron populations have been added to the model to mediate the effects of

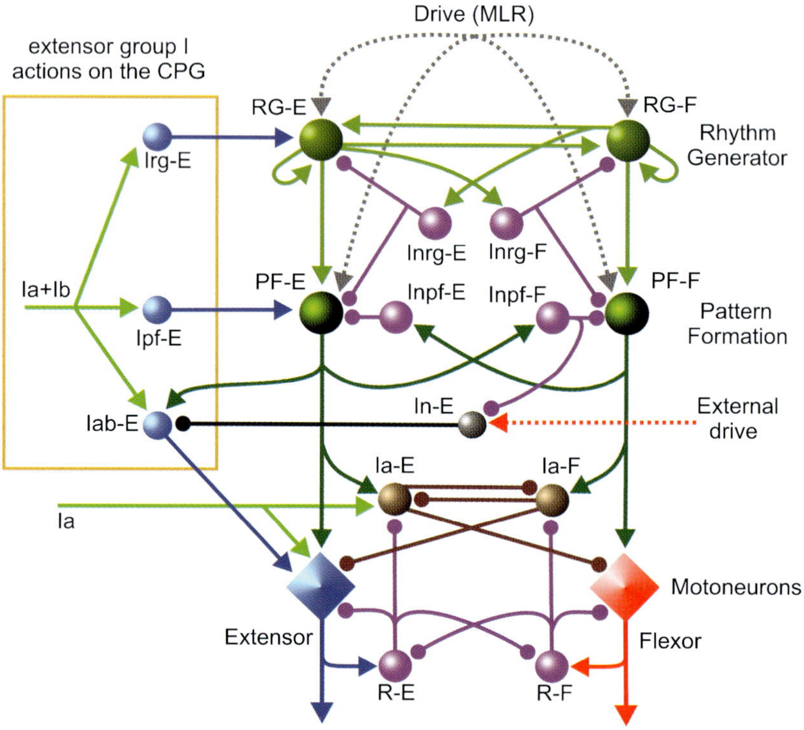

Fig. 6. Model schematic of the spinal cord circuitry integrated with the locomotor CPG used for simulation of the effects of extensor group I afferent stimulation during fictive locomotion. This model extends that shown in Fig. 3A to include pathways for group I extensor afferents. Connections of group I (Ia and Ib) extensor afferents are shown on the left. Interneuron populations Irg-E and Ipf-E mediate the access of extensor group I (Ia and Ib) afferents to the rhythm generator (RG-E) and the pattern formation (PF-E) networks, respectively. The Iab-E and In-E populations provide phase-dependent disynaptic excitation of extensor motoneurons by group I extensor afferents. During the extensor phase of fictive locomotion, the Iab-E population is released from inhibition of In-E and mediates disynaptic excitation of extensor motoneurons. See details in the text.

extensor group I afferents on the CPG during locomotion (references in McCrea, 2001; Pearson, 2004; Rossignol et al., 2006). In the framework of the two-level CPG, we have hypothesized that there are separate pathways for extensor group I excitation through the RG and the PF levels of the CPG via the hypothetical Irg-E and Ipf-E populations, respectively (Fig. 6). According to the suggestion explored here, the synaptic weight of group I input to the PF-E population (controlling extensor activity at the PF level) is stronger than that to RG-E (the extensor half-center of the RG).

Figure 7A1 and A2 shows two examples of simulations of the effects of a short duration stimulus to extensor group I afferents during the extension phase of locomotion, i.e., when both the RG-E and PF-E populations are active. With the moderate intensity stimulation in Fig. 7A1, there was only a small effect on RG population activity. Hence this afferent stimulation did not change the locomotor rhythm generated by the RG (see the second and third traces in Fig. 7A1). The stimulation did, however, enhance and prolong PF-E population activity, which in turn enhanced and prolonged the activity of extensor motoneurons (Fig. 7A1, bottom trace). The prolongation in PF-E activity delayed the switching to the flexion phase at the PF-F level (see fourth and fifth traces in Fig. 7A1). However, because the RG was not affected, the subsequent flexion phase was shortened and the duration of the ongoing step cycle remained constant. This is consistent with the experimental data shown in Fig. 7B1 where plantaris

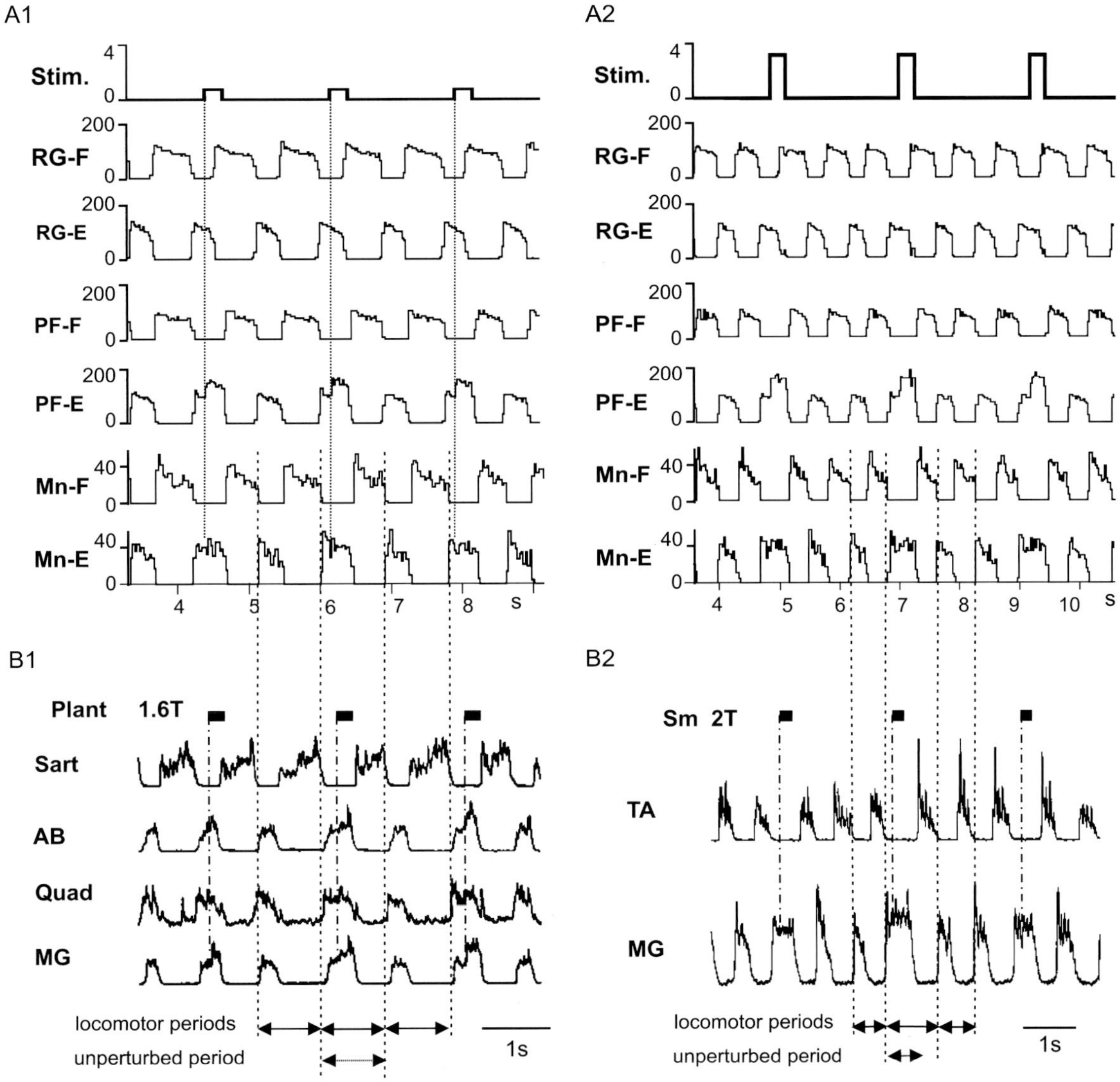

Fig. 7. Modeling the effects of group I extensor afferent stimulation during extension. (A1 and A2) Examples of modeling the effects of stimulation of group I extensor afferents during extension (see text for details). The applied stimuli are shown in the top traces. The stimulus amplitude in A2 was three times than in A1. (B1 and B2) The effects of stimulation of extensor group I afferents during MLR-evoked fictive locomotion. In panel B1, stimulation of plantaris (Plant) group I afferents during extension increased the size and duration of extensor motoneuron activity (MG) and shortened the duration of the following flexor phase as seen in the sartorius (Sart) ENG. Note that in both A1 and B1, the duration of each flexion phase following the prolonged extension phase was shortened so that the locomotor periods did not change. The locomotor rhythm was not reset (see equal length arrows at the bottom). In panel B2, hip extensor (Sm) muscle afferents were electrically stimulated during extension. In contrast to A1 and B1, in both A2 and B2 the flexion phase that follows the stimulus-evoked extension phase prolongation was not shortened and the step cycle period increased with each stimulus delivery (see arrows at the bottom). Adapted with permission from Rybak et al. (2006b).

nerve stimulation enhanced and prolonged extensor motoneuron activity (hip: AB, knee: Quad, and ankle: MG). As in our simulation, there was a corresponding shortening of the subsequent flexor phase such that the ongoing step-cycle period remained unchanged (see arrows at the bottom of Fig. 7B1).

In Fig. 7A2 the intensity of stimulation is three times stronger than that in Fig. 7A1. Unlike the weaker stimulation, this stimulation also increased RG-E activity, which in turn delayed the transition to flexion (see second and third traces in Fig. 7A2). The net effect was an increase and prolongation of extensor motoneuron firing (see the bottom trace in Fig. 7A2). There was, however, no compensatory change in the duration of the subsequent flexor phase. Consequently, each stimulus prolonged the duration of the ongoing locomotor cycle and hence produced a phase shift of the post-stimulation rhythm. An experimental example of a group I extensor stimulation-evoked enhancement and prolongation of extensor motoneuron activity in which the ongoing step-cycle period was increased is shown in Fig. 7B2. In this example, each stimulus applied to group I extensor afferents (Sm, a hip extensor) enhanced and prolonged extensor (MG) bursts. The following flexor phase had an unchanged duration and hence the locomotor rhythm was shifted in time (see arrows at the bottom of the figure).

Based on the results of these simulations, we conclude that stimulation of group I extensor afferents during extension may prolong and enhance activity during the current extension phase, with or without changing the duration of ongoing locomotor cycle and the phase of post-stimulation rhythm. The exact effect in the model depends on how strongly the applied stimulation influences the RG. With separate access of proprioceptive feedback to the RG and the PF networks, the contribution of extensor group I afferents to weight support and the control of stance-swing transitions may be accomplished via separate pathways within the CPG. Specifically we hypothesize that enhanced weight support (i.e., level of extensor motoneuron activity) during stance can be provided by the PF network, while actions on the extensor portion of the RG can control the timing of stance to swing transitions.

Conclusions

Development of the present CPG model began with observations obtained during fictive locomotion in the cat showing that cycle phase could be maintained during deletions and during sensory stimulation. This phase maintenance necessitated consideration of a two-level CPG in which the tasks of rhythm generation and motoneuron activation were separate since such phase maintenance is not easily accommodated within the classical half-center CPG concept. The RG structure and the parameters of RG neurons were selected to be able to reproduce the range of locomotor-cycle periods and phase durations observed during fictive and treadmill locomotion. Modeling shows that this two-level CPG architecture consisting of a half-center RG and a PF network can replicate and explain the existence of resetting and non-resetting deletions. The same model can also realistically reproduce the actions of reflex circuits during locomotion and provide explanations for the effects of afferent stimulation on CPG observed experimentally. We believe that analysis of "mistakes" and sensory perturbations of CPG operation in the fictive locomotion preparation has provided a unique insight into the structure of the CPG operating during real locomotion.

Finally, we must stress that the present model is a work in progress. As more experimental data becomes available, the model will need to be modified to reproduce CPG activity observed under other conditions. Extensions to the model will include incorporating more than two motoneuron pools and adding more sources of sensory and descending control to the CPG.

We also hope that the model may assist in the development of criteria for the functional identification of spinal interneuron classes involved in locomotor pattern and rhythm generation. For example, based on the model, we suggest that excitatory RG neurons should have the following features. These neurons should (i) receive excitation from the MLR region, (ii) demonstrate rhythmic activity during fictive locomotion that persists during non-resetting deletions, and (iii) not be monosynaptically coupled to motoneurons. In contrast, the neurons comprising the PF network should (i) demonstrate rhythmic activity during fictive locomotion that fails during non-resetting deletions and (ii) produce monosynaptic EPSPs in synergist motoneurons. Even a partial experimental identification of some of the classes of CPG neurons postulated in the model would provide opportunities to directly examine the intrinsic cellular properties underlying rhythm generation in the adult mammalian spinal cord.

Acknowledgments

Supported by the NIH (R01 NS048844) and the Canadian Institutes of Health Research (MOP37756).

References

Akay, T., McVea, D.A., Tachibana, A. and Pearson, K.G. (2006) Coordination of fore and hind leg stepping in cats on a transversely-split treadmill. Exp. Brain Res., 175: 211–222.

Angel, M.J., Guertin, P., Jiminez, I. and McCrea, D.A. (1996) Group I extensor afferents evoke disynaptic EPSPs in cat hindlimb extensor motoneurones during fictive locomotion. J. Physiol., 494: 851–861.

Angel, M.J., Jankowska, E. and McCrea, D.A. (2005) Candidate interneurones mediating group I disynaptic EPSPs in extensor motoneurones during fictive locomotion in the cat. J. Physiol., 563(Pt 2): 597–610.

Beato, M. and Nistri, A. (1999) Interaction between disinhibited bursting and fictive locomotor patterns in the rat isolated spinal cord. J. Neurophysiol., 82: 2029–2038.

Booth, V., Rinzel, J. and Kiehn, O. (1997) Compartmental model of vertebrate motoneurons for Ca^{2+}-dependent spiking and plateau potentials under pharmacological treatment. J. Neurophysiol., 78: 3371–3385.

Brocard, F.F., Tazerart, S., Viermari, J.C., Darbon, P. and Vinay, L. (2006) Persistent sodium inward current (INaP) in the neonatal rat lumbar spinal cord and its contribution to locomotor pattern generation. Soc. Neurosci. Abstr. 252.16.

Burke, R.E., Degtyarenko, A.M. and Simon, E.S. (2001) Patterns of locomotor drive to motoneurons and last-order interneurons: clues to the structure of the CPG. J. Neurophysiol., 86: 447–462.

Büschges, A., Wikstroöm, M.A., Grillner, S. and El Manira, A. (2000) Roles of high-voltage activated calcium channel subtypes in a vertebrate spinal locomotor network. J. Neurophysiol., 84: 2758–2766.

Butera, R.J., Rinzel, J.R. and Smith, J.C. (1999a) Models of respiratory rhythm generation in the pre-Bötzinger complex: I. Bursting pacemaker neurons. J. Neurophysiol., 82: 382–397.

Butera, R.J., Rinzel, J.R. and Smith, J.C. (1999b) Models of respiratory rhythm generation in the pre-Bötzinger complex: II. Populations of coupled pacemaker neurons. J. Neurophysiol., 82: 398–415.

Butt, S.J.B., Harris-Warrick, R.M. and Kiehn, O. (2002) Firing properties of identified interneuron populations in the mammalian hindlimb central pattern generator. J. Neurosci., 22: 9961–9971.

Cowley, K.C. and Schmidt, B.J. (1995) Effects of inhibitory amino acid antagonists on reciprocal inhibitory interactions during rhythmic motor activity in the in vitro neonatal rat spinal cord. J. Neurophysiol., 74: 1109–1117.

Dai, Y. and Jordan, L.M. (2006) Characterization of persistent inward currents (PICs) in locomotor activity-related neurons of Cfos-EGFP mice. Soc. Neurosci. Abstr. 130.6.

Darbon, P., Yvon, C., Legrand, J.C. and Streit, J. (2004) INaP underlies intrinsic spiking and rhythm generation in networks of cultured rat spinal cord neurons. Eur. J. Neurosci., 20: 976–988.

Dietz, V. (2003) Spinal cord pattern generators for locomotion. Clin. Neurophysiol., 114: 1379–1389.

Donelan, J.M. and Pearson, K.G. (2004) Contribution of force feedback to ankle extensor activity in decerebrate walking cats. J. Neurophysiol., 92: 2093–2104.

Duysens, J. (1977) Reflex control locomotion as revealed by stimulation of cutaneous afferents in spontaneously walking premammillary cats. J. Neurophysiol., 40: 737–751.

Duysens, J., McCrea, D. and Lafreniere-Roula, M. (2006) How deletions in a model could help explain deletions in the laboratory. J. Neurophysiol., 95: 562–565.

El Manira, A., Tegner, J. and Grillner, S. (1994) Calcium-dependent potassium channels play a critical role for burst termination in the locomotor network in lamprey. J. Neurophysiol., 72: 1852–1861.

Fedirchuk, B. and Dai, Y. (2004) Monoamines increase the excitability of spinal neurones in the neonatal rat by hyperpolarizing the threshold for action potential production. J. Physiol., 557: 355–361.

Feldman, A.G. and Orlovsky, G.N. (1975) Activity of interneurons mediating reciprocal Ia inhibition during locomotion. Brain Res., 84: 181–194.

Forssberg, H., Grillner, S., Halbertsma, J. and Rossignol, S. (1980) The locomotion of the low spinal cat: II. Interlimb coordination. Acta Physiol. Scand., 108: 283–295.

Gosgnach, S., Lanuza, G.M., Butt, S.J., Saueressig, H., Zhang, Y., Velasquez, T., Riethmacher, D., Callaway, E.M., Kiehn, O. and Goulding, M. (2006) V1 spinal neurons regulate the speed of vertebrate locomotor outputs. Nature, 440: 215–219.

Gosgnach, S., Quevedo, J., Fedirchuk, B. and McCrea, D.A. (2000) Depression of group Ia monosynaptic EPSPs in cat hindlimb motoneurones during fictive locomotion. J. Physiol., 526: 639–652.

Graham Brown, T. (1914) On the fundamental activity of the nervous centres: together with an analysis of the conditioning of rhythmic activity in progression, and a theory of the evolution of function in the nervous system. J. Physiol., 48: 18–41.

Grillner, S. (1981) Control of locomotion in bipeds, tetrapods, and fish. In: Brookhart J.M. and Mountcastle V.B. (Eds.), Handbook of Physiology. The Nervous System. Motor Control, Sect. 1, Vol. II. American Physiological Society, Bethesda, MD, pp. 1179–1236.

Grillner, S. (2003) The motor infrastructure: from ion channels to neuronal networks. Nat. Rev. Neurosci., 4: 573–586.

Grillner, S. and Wallén, P. (2002) Cellular bases of a vertebrate locomotor system-steering, intersegmental and segmental co-ordination and sensory control. Brain Res. Rev., 40: 92–106.

Grillner, S., Wallén, P., Hill, R., Cangiano, L. and El Manira, A. (2001) Ion channels of importance for the locomotor pattern generation in the lamprey brainstem-spinal cord. J. Physiol., 533: 23–30.

Grillner, S. and Zangger, P. (1979) On the central generation of locomotion in the low spinal cat. Exp. Brain Res., 34: 241–261.

Guertin, P., Angel, M.J., Perreault, M.-C. and McCrea, D.A. (1995) Ankle extensor group I afferents excite extensors throughout the hindlimb during fictive locomotion in the cat. J. Physiol., 487: 197–209.

Halbertsma, J.M. (1983) The stride cycle of the cat: the modelling of locomotion by computerized analysis of automatic recordings. Acta Physiol. Scand. Suppl., 521: 1–75.

Jankowska, E. (1992) Interneuronal relay in spinal pathways from proprioceptors. Prog. Neurobiol., 38: 335–378.

Kiehn, O. (2006) Locomotor circuits in the mammalian spinal cord. Annu. Rev. Neurosci., 29: 279–306.

Kjaerulff, O. and Kiehn, O. (2001) 5-HT modulation of multiple inward rectifiers in motoneurons in intact preparations of the neonatal rat spinal cord. J. Neurophysiol., 85: 580–593.

Koshland, G.F. and Smith, J.L. (1989) Mutable and immutable features of paw-shake responses after hindlimb deafferentation in the cat. J. Neurophysiol., 62: 162–173.

Kremer, E. and Lev-Tov, A. (1998) GABA-receptor-independent dorsal root afferents depolarization in the neonatal rat spinal cord. J. Neurophysiol., 79: 2581–2592.

Kriellaars, D.J., Brownstone, R.M., Noga, B.R. and Jordan, L.M. (1994) Mechanical entrainment of fictive locomotion in the decerebrate cat. J. Neurophysiol., 71: 2074–2086.

Lafreniere-Roula, M. and McCrea, D.A. (2005) Deletions of rhythmic motoneuron activity during fictive locomotion and scratch provide clues to the organization of the mammalian central pattern generator. J. Neurophysiol., 94: 1120–1132.

Lee, R.H. and Heckman, C.J. (2001) Essential role of a fast persistent inward current in action potential initiation and control of rhythmic firing. J. Neurophysiol., 85: 472–475.

Lennard, P.R. (1985) Afferent perturbations during "monopodal" swimming movements in the turtle: phase-dependent cutaneous modulation and proprioceptive resetting of the locomotor rhythm. J. Neurosci., 5: 1434–1445.

Lundberg, A. (1981) Half-centres revisited. In: Szentagothai J., Palkovits M. and Hamori J. (Eds.), Regulatory Functions of the CNS. Motion and Organization Principles. Pergamon Akadem Kiado, Budapest, Hungary, pp. 155–167.

McCrea, D.A. (2001) Spinal circuitry of sensorimotor control of locomotion. J. Physiol., 533: 41–50.

McCrea, D.A., Pratt, C.A. and Jordan, L.M. (1980) Renshaw cell activity and recurrent effects on motoneurons during fictive locomotion. J. Neurophysiol., 44: 475–488.

McCrea, D.A., Shefchyk, S.J., Stephens, M.J. and Pearson, K.G. (1995) Disynaptic group I excitation of synergist ankle extensor motoneurones during fictive locomotion in the cat. J. Physiol., 487: 527–539.

Miller, S. and van der Meche, F.G. (1976) Coordinated stepping of all four limbs in the high spinal cat. Brain Res., 109: 395–398.

Noga, B.R., Cowley, K.C., Huang, A., Jordan, L.M. and Schmidt, B.J. (1993) Effects of inhibitory amino acid antagonists on locomotor rhythm in the decerebrate cat. Soc. Neurosci. Abstr. 225.4.

Orlovsky, G.N., Deliagina, T. and Grillner, S. (1999) Neuronal control of locomotion: from mollusc to man. Oxford University Press, New York.

Orsal, D., Cabelguen, J.M. and Perret, C. (1990) Interlimb coordination during fictive locomotion in the thalamic cat. Exp. Brain Res., 82: 536–546.

Paton, J.F.R., Abdala, A.P.L., Koizumi, H., Smith, J.C. and St.-John, W.M. (2006) Respiratory rhythm generation during gasping depends on persistent sodium current. Nat. Neurosci., 9: 311–313.

Pearson, K.G. (2004) Generating the walking gait: role of sensory feedback. Prog. Brain Res., 143: 123–129.

Perreault, M.C., Angel, M.J., Guertin, P. and McCrea, D.A. (1995) Effects of stimulation of hindlimb flexor group II afferents during fictive locomotion in the cat. J. Physiol., 487: 211–220.

Perrier, J.F., Alaburda, A. and Hounsgaard, J. (2003) 5-HT1A receptors increase excitability of spinal motoneurons by inhibiting a TASK-1-like K+ current in the adult turtle. J. Physiol., 548: 485–492.

Pratt, C.A. and Jordan, L.M. (1987) Ia inhibitory interneurons and Renshaw cells as contributors to the spinal mechanisms of fictive locomotion. J. Neurophysiol., 57: 56–71.

Rossignol, S. (1996) Neural control of stereotypic limb movements. In: Rowell L.B. and Shepherd J. (Eds.), Handbook of Physiology, Sect. 12. The American Physiological Society, Bethesda, MD, pp. 173–216.

Rossignol, S., Dubuc, R. and Gossard, J.-P. (2006) Dynamic sensorimotor interactions in locomotion. Physiol. Rev., 86: 89–154.

Rybak, I.A., Shevtsova, N.A., Lafreniere-Roula, M. and McCrea, D.A. (2006a) Modelling spinal circuitry involved in locomotor pattern generation: insights from deletions during fictive locomotion. J. Physiol., 577: 617–639.

Rybak, I.A., Shevtsova, N.A., Ptak, K. and McCrimmon, D.R. (2004) Intrinsic bursting activity in the pre-Bötzinger complex: role of persistent sodium and potassium currents. Biol. Cybern., 90: 59–74.

Rybak, I.A., Shevtsova, N.A., St.-John, W.M., Paton, J.F.R. and Pierrefiche, O. (2003) Endogenous rhythm generation in the pre-Bötzinger complex and ionic currents: modelling and in vitro studies. Eur. J. Neurosci., 18: 239–257.

Rybak, I.A., Stecina, K., Shevtsova, N.A. and McCrea, D.A. (2006b) Modelling spinal circuitry involved in locomotor pattern generation: insights from the effects of afferent stimulation. J. Physiol., 577: 641–658.

Schomburg, E.D. and Behrends, H.B. (1978) The possibility of phase-dependent monosynaptic and polysynaptic Ia excitation to homonymous motoneurones during fictive locomotion. Brain Res., 143: 533–537.

Sinkjaer, T., Andersen, J.B., Ladouceur, M., Christensen, L.O.D. and Nielsen, J. (2000) Major role for sensory feedback in soleus EMG activity in the stance phase of walking in man. J. Physiol., 523: 817–827.

Sirota, M.G. and Shik, M.L. (1973) The cat locomotion elicited through the electrode implanted in the mid-brain. Sechenov Physiol. J. U.S.S.R., 59: 1314–1321 (in Russian).

Smith, J.C., Butera, R.J., Koshiya, N., Del Negro, C., Wilson, C.G. and Johnson, S.M. (2000) Respiratory rhythm generation in neonatal and adult mammals: the hybrid pacemaker-network model. Respir. Physiol., 122: 131–147.

Stecina, K., Quevedo, J. and McCrea, D.A. (2005) Parallel reflex pathways from flexor muscle afferents evoking resetting and flexion enhancement during fictive locomotion and scratch in the cat. J. Physiol., 569: 275–290.

Stein, P.S.G. (2005) Neuronal control of turtle hindlimb motor rhythms. J. Comp. Physiol. A, 191: 213–229.

Stein, P.S.G. and Smith, J.L. (1997) Neural and biomechanical control strategies for different forms of vertebrate hindlimb motor tasks. In: Stein P., Grillner S., Selverston A.I. and Stuart D.G. (Eds.), Neurons, Networks, and Motor Behavior. MIT Press, Cambridge, MA, pp. 61–73.

Streit, J., Tscherter, A. and Darbon, P. (2006) Rhythm generation in spinal culture: Is it the neuron or the network? In: Advances in Neural Information Processing Systems. Springer, New York, pp. 377–408.

Theiss, R.D., Kuo, J.J. and Heckman, C.J. (2007) Persistent inward currents in rat ventral horn neurons. J. Physiol., 580: 507–522.

Yakovenko, S., McCrea, D.A., Stecina, K. and Prochazka, A. (2005) Control of locomotor cycle durations. J. Neurophysiol., 94: 1057–1065.

Yamaguchi, T. (2004) The central pattern generator for forelimb locomotion in the cat. Prog. Brain Res., 143: 115–122.

Yang, J.F., Lam, T., Pang, M.Y., Lamont, E., Musselman, K. and Seinen, E. (2004) Infant stepping: a window to the behaviour of the human pattern generator for walking. Can. J. Physiol. Pharmacol., 82: 662–674.

Zehr, E.P. and Duysens, J. (2004) Regulation of arm and leg movement during human locomotion. Neuroscientist, 10: 347–361.

Zhong, G., Masino, M.A. and Harris-Warrick, R.M. (2006) Persistent sodium currents participate in fictive locomotion generation in neonatal mouse spinal cord. Soc. Neurosci. Abstr. 128.17.

P. Cisek, T. Drew & J.F. Kalaska (Eds.)
Progress in Brain Research, Vol. 165
ISSN 0079-6123
Copyright © 2007 Elsevier B.V. All rights reserved

CHAPTER 16

The neuromechanical tuning hypothesis

Arthur Prochazka[1,*] and Sergiy Yakovenko[2]

[1]*Centre for Neuroscience, 507 HMRC University of Alberta, Edmonton, AB T6G 2S2, Canada*
[2]*Departement de Physiologie, Pavillon Paul-G. Desmarais, Universite de Montreal. C.P. 6128, Succ. Centre-ville, Montreal, QC H3C 3J7, Canada*

Abstract: Simulations performed with neuromechanical models are providing insight into the neural control of locomotion that would be hard if not impossible to obtain in any other way. We first discuss the known properties of the neural mechanisms controlling locomotion, with a focus on mammalian systems. The rhythm-generating properties of central pattern generators (CPGs) are discussed in light of results indicating that cycle characteristics may be preset by tonic drive to spinal interneuronal networks. We then describe neuromechanical simulations that have revealed some basic rules of interaction between CPGs, sensory-mediated switching mechanisms and the biomechanics of locomotor movements. We posit that the spinal CPG timer and the sensory-mediated switch operate in parallel, the former being driven primarily by descending inputs and the latter by the kinematics. The CPG timer produces extensor and flexor phase durations, which covary along specific lines in a plot of phase- versus cycle-duration. We coined the term "phase-duration characteristics" to describe such plots. Descending input from higher centers adjusts the operating points on the phase-duration characteristics according to anticipated biomechanical requirements. In well-predicted movements, CPG-generated phase durations closely match those required by the kinematics, minimizing the corrections in phase duration required of the sensory switching mechanism. We propose the term "neuromechanical tuning" to describe this process of matching the CPG to the kinematics.

Keywords: neural control of locomotion; central pattern generators; sensory control of locomotion

Introduction: historical development and overview

The control of animal locomotion was among the first mechanisms of nervous systems to be analyzed in detail (Freusberg, 1874; Magnus, 1909a, b; Sherrington, 1910, 1914). A key contradiction soon arose. It had been shown that in spinally transected dogs, a locomotor rhythm could be initiated in pendent limbs by dropping one of the limbs from a flexed position (Freusberg, 1874). Sherrington found that the rhythm could be halted by holding a limb in mid-cycle (Sherrington, 1910, 1914). He concluded that stepping and walking resulted from a chain of proprioceptive reflexes, the end of one phase triggering the onset of the next. Sensory input was of course crucial to this scheme. The problem was that locomotor-like rhythms were also observed in spinal cats even after sensory input was abolished by extensive deafferentation (Brown, 1911). Brown proposed the existence of what he called the intrinsic factor, located in the spinal cord and capable of producing the locomotor rhythm autonomously, without

*Corresponding author. Tel.: + 1 780 492 3783;
Fax: + 1 780 492 1617; E-mail: arthur.prochazka@ualberta.ca

DOI: 10.1016/S0079-6123(06)65016-4

sensory input or descending control from the brain. Over 60 years later Sten Grillner renamed this mechanism the central pattern generator (CPG) (Grillner and Zangger, 1975).

It is now clear that sensory input interacts with the CPG in at least three ways. It can trigger step cycle phase transitions in a discontinuous, switch-like manner; it can lengthen or shorten phase durations in a more continuous manner and it can provide continuous, proportional control of muscle activation through short reflex pathways (Rossignol et al., 2006).

All of the building blocks of locomotor systems have been separately studied, from the molecular to systems levels. The biomechanical properties of the musculoskeletal actuators and body segments have been determined and modeled (Brown and Loeb, 2000; Zajac, 2002; Zajac et al., 2003). The transducing properties of sensory afferents (Prochazka and Gorassini, 1998b; Prochazka, 1999; Mileusnic and Loeb, 2006; Mileusnic et al., 2006) and their reflex effects on load compensation and locomotor phase transitions have been characterized (Frigon and Rossignol, 2006; Rossignol et al., 2006). The behavior of isolated CPGs in a variety of motor systems has been investigated and network models have been proposed (Selverston, 1993; Arshavsky et al., 1997; Grillner et al., 2000; McCrimmon et al., 2000; Zelenin et al., 2000; Kiehn, 2006; Rybak et al., 2006a, b). In mammals, brainstem, cerebellar, and cortical involvement in adaptive locomotor responses has been explored (Arshavsky et al., 1986; Beloozerova and Sirota, 1998; Orlovsky et al., 1999; Drew et al., 2004). The picture that has emerged from all of this work is that of a multilevel control system (Fig. 1). At the lowest level, the muscles act like damped springs, providing automatic length feedback control of bodily movement. The term "preflex" has been coined to describe this mechanism (Loeb et al., 1999). The second level, which we will call the output level, comprises motoneurons (MNs) that activate muscles. Interneurons that mediate sensory and recurrent feedback, arguably are also included (e.g., Renshaw cells and Ia and Ib interneurons). The third, hypothetical, level comprises neuronal networks that generate individual muscle activation profiles, also without affecting the basic

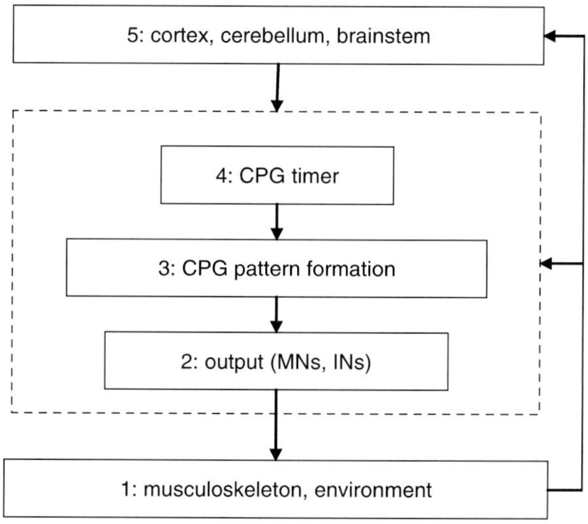

Fig. 1. Hypothesized CNS levels of locomotor control.

rhythm (Perret and Cabelguen, 1980; Perret, 1983; Lafreniere-Roula and McCrea, 2005). This level was recently called the pattern formation layer (Rybak et al., 2006a). The fourth level comprises the CPG rhythm generator or oscillator, whose activity is adjusted or reset by sensory input, as well as by input descending from the fifth and highest level, which comprises the brainstem (Shik et al., 1966; Takakusaki et al., 2004), cerebellum (Arshavsky et al., 1986), and motor cortex (Beloozerova and Sirota, 1993; Drew, 1993; Widajewicz et al., 1994). These higher centers integrate a wide variety of inputs, including motivational (Jordan, 1998), exteroceptive (Drew, 1991; Rossignol, 1996; Patla et al., 1999), and proprioceptive inputs, to define state and intention and to predict upcoming motor requirements.

There have been several excellent reviews of the above mechanisms over the last few years (Pearson, 2004; Frigon and Rossignol, 2006; Rossignol et al., 2006). Yet in spite of all the neurophysiological knowledge, there has been relatively little progress in understanding precisely how the neuronal mechanisms combine with the biomechanics to produce the stability, adaptability and grace of animal movement. Neuromechanical simulations are becoming increasingly useful in this regard (Taga et al., 1991; Taga, 1995; Yakovenko et al., 2004; Ekeberg and Pearson, 2005; Pearson et al., 2006).

In what follows, we will concentrate on the properties of mammalian locomotor systems that are relevant for computational approaches. We will then present the results of recent neuromechanical simulations, followed by some general propositions about locomotor control.

Sensory inputs in mammals

The vast majority of mechanoreceptors are cutaneous or hair follicle receptors. Most of these are only sporadically active during the step cycle, for example upon ground contact in the case of footpad receptors, or during surface airflow for hair follicle receptors (Prochazka, 1996). Although cutaneous afferents have demonstrable reflex actions on MNs and as event detectors, can elicit specific motor programs such as the stumble reaction, most of the *continuous* reflex control during stepping must be attributed to the proprioceptive afferents, muscle spindles and tendon organs. This is not to deny a role for cutaneous input to provide kinesthetic information (Collins and Prochazka, 1996) and affect muscle activation and the timing of locomotor phase transitions (Rossignol et al., 2006). Indeed event-related multiunit activity of cutaneous afferents recorded from nerve cuffs has been used to trigger bursts of functional electrical stimulation (Haugland and Sinkjaer, 1999).

There is much detailed information on the transducing properties of proprioceptors. To a first approximation, tendon organs signal muscle force and muscle spindles signal muscle length and velocity. Models have been developed that predict their responses reasonably accurately. It is of course debatable as to the level of accuracy that is required for biomechanical modeling. Because stretch reflexes probably account for no more than 30% of muscle activation during locomotion (see below), models that account for 80% or more of the variance in afferent ensemble firing rate would introduce 6% or less error in predicted muscle activation.

Models are available that predict the responses of tendon organ ensembles with >80% accuracy (i.e., $r^2 > 0.8$ in linear regressions of predicted versus actual firing rate) (Prochazka, 1996; Mileusnic and Loeb, 2006). Muscle spindles are more problematic, because their responses are modulated by fusimotor action emanating from the CNS. If this action fluctuates substantially, predictions of spindle responses based only on length variations are bound to be inaccurate. Models of varying complexity have been developed and compared with ensemble spindle responses recorded from dorsal roots in freely walking cats (Prochazka and Gorassini, 1998a, b). In some muscles, more than 80% of the variance was accounted for by quite simple models, presumably because fusimotor action does not fluctuate very much in these muscles during walking. In other muscles, for example the ankle extensors, modeling was less satisfactory, even after presumed fusimotor action was added. Recently, a model has been developed that includes not only fusimotor action, but also tendon compliance, muscle pennation and other nonlinear features (Mileusnic and Loeb, 2006; Mileusnic et al., 2006). The main problem here is that fusimotor fluctuations during gait in normal animals have never been established with certainty. They have been inferred from recordings in walking decerebrate cats (Taylor et al., 2000a, b, 2006; Ellaway et al., 2002; Durbaba et al., 2003) but there are discrepancies between the normal and decerebrate data. Because fusimotor activation profiles remain uncertain, the improvement in accuracy the new complex model offers is also uncertain.

Locomotor stretch reflexes

Muscle spindles reflexly excite MNs that innervate their parent muscles, resisting deflecting forces. This reflex action is equivalent to negative length and velocity feedback. During locomotion, tendon organ afferents respond to increments in muscle force by exciting homonymous MNs to produce even more force. This is equivalent to *positive* force feedback, the loop gain of which is evidently less than unity in normal gait, but may transiently exceed unity in bouncing gait (Prochazka et al., 1997; Geyer et al., 2003; Donelan and Pearson, 2004). We have estimated that at most 30% of the activation of extensors in the stance phase of the cat step cycle is attributable to proprioceptive stretch

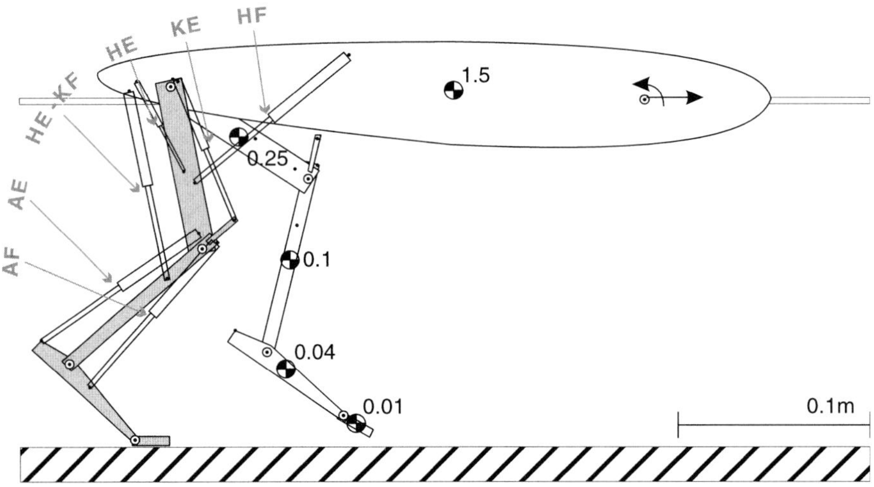

Fig. 2. Neuromechanical model used in locomotor simulations. Reproduced from Yakovenko et al. (2004) with kind permission of Springer Science and Business Media.

reflexes (Prochazka et al., 2002). Furthermore, there is a surprisingly long delay (20–40 ms) before the EMG response to ground contact manifests itself (Gorassini et al., 1994; Gritsenko et al., 2001). Given the modest and delayed contribution of stretch reflexes to muscle activation and the relatively normal locomotion of cats deprived of proprioceptive afferents (Pearson et al., 2003) the importance of stretch reflexes in load compensation came into question (Prochazka and Yakovenko, 2002; Prochazka et al., 2002).

The issue was tackled with a neuromechanical model (Yakovenko et al., 2004) a torso supported at the front by a frictionless horizontal rail and at the back by a pair of hindlimbs (Fig. 2). Each hindlimb comprised four rigid-body segments (thigh, shank, foot, and toes), which were driven by six musculo-tendon actuators. The model was implemented with Matlab and Working Model software. Each actuator had Hill-type muscle properties driven by a CPG that generated muscle activation patterns derived from the literature. Spindle and tendon organ models added reflex components to muscle activations, contributing on average 30% of total activation. The stability of locomotor simulations with and without reflexes was assessed by randomly varying actuator gains and computing the size of stable regions in parametric space.

In the absence of stretch reflexes the CPG, acting through the intrinsic biomechanical properties of the model, could produce stable gait over a surprising range of muscle activation levels. When the activation levels were deliberately set too low to support stable gait (Fig. 3, left panels), stretch reflexes adding 30% to the muscle activation profiles helped "rescue" stability. This is shown by the fact that when the reflexes were suddenly withdrawn, the model fell within two step cycles. When the CPG activation profiles were set to levels that produced stable gait (Fig. 3, right panels), the addition of stretch reflexes made the kinematics slightly more vigorous and slightly increased the stable range of activation in parametric space. In some cases when CPG activation was high, the addition of stretch reflexes resulted in a fall after a few steps (not shown in Fig. 3). We concluded that stretch reflexes could rescue locomotion when CPG activation levels were low and improve overall stability by a modest amount (Yakovenko et al., 2004).

Phase switching with If–Then sensory rules increases stability

Some years ago it was suggested that although stretch reflexes may contribute to load compensation

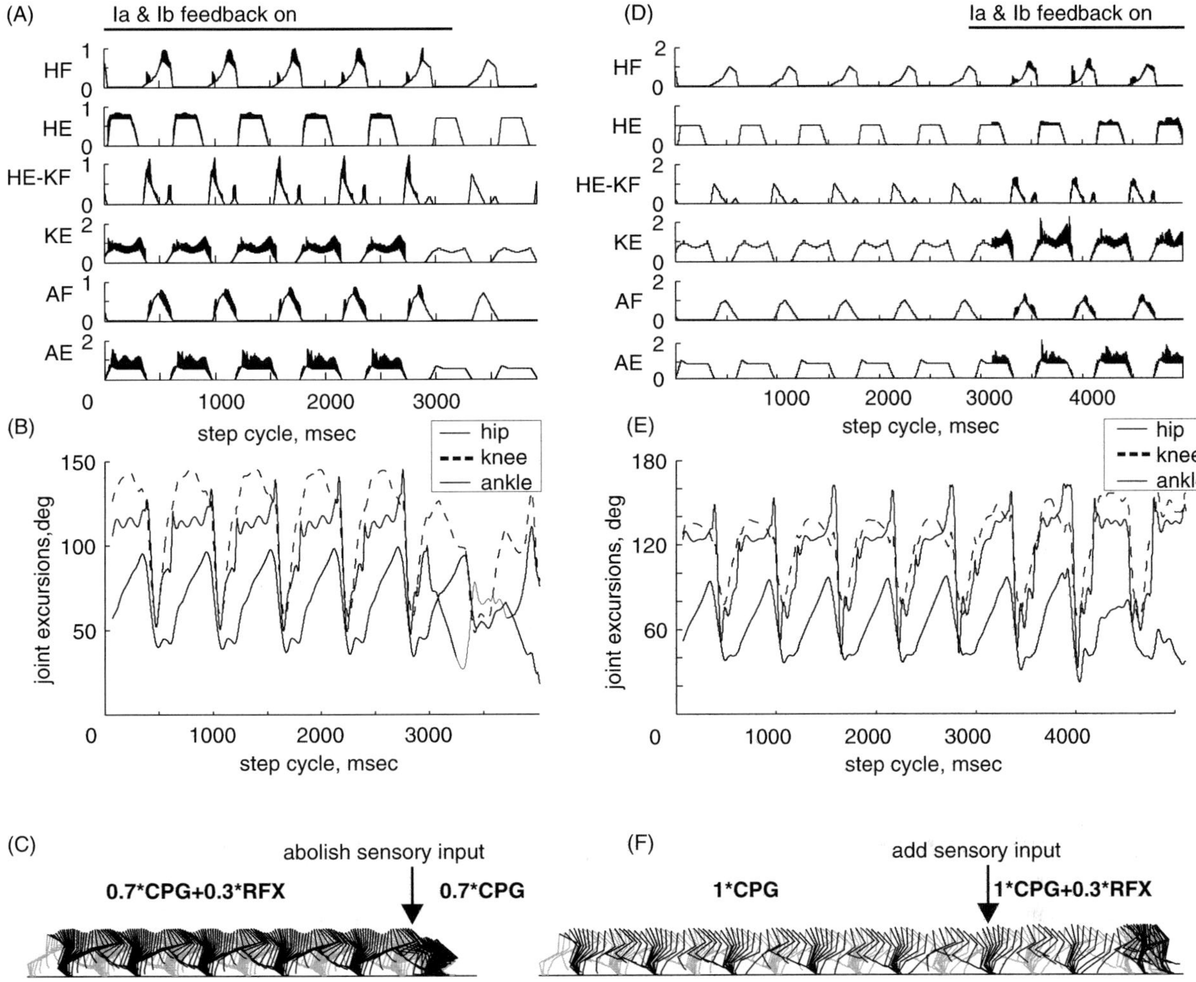

Fig. 3. Contribution of stretch reflexes during modeled locomotion. Left panels: centrally generated levels of muscle activation were insufficient of themselves to support stable locomotion. Stretch reflex contributions comprising 30% of the total activation (black portions of the activation profiles) were initially added and then suddenly removed. This caused a rapid collapse (stick figures). Right panels: centrally generated muscle activation levels were sufficient to support locomotion. Initially stretch reflexes were absent and then suddenly added. The only effect was a slight increase in gait velocity and a more vigorous gait. Reproduced in adapted form from Yakovenko et al. (2004) with kind permission of Springer Science and Business Media.

at the output level defined in Fig. 1, the sensory control of cycle duration was mediated by some other mechanism, most likely through the timing and patterning elements of the CPG at the third or fourth levels. By analogy with robotic systems, it was proposed that finite state ("If–Then") sensory rules underlay this higher-level interaction (Cruse, 1990; Prochazka, 1993). In our modeling study we therefore also explored the effect on stability of this form of control (Yakovenko et al., 2004). The sensory rules were of the type:

1. Stance to swing transition: IF stance AND ipsilateral hip is extended AND contralateral leg is loaded THEN swing;
2. Swing to stance transition: IF swing AND ipsilateral hip is flexed AND ipsilateral knee is extended THEN stance.

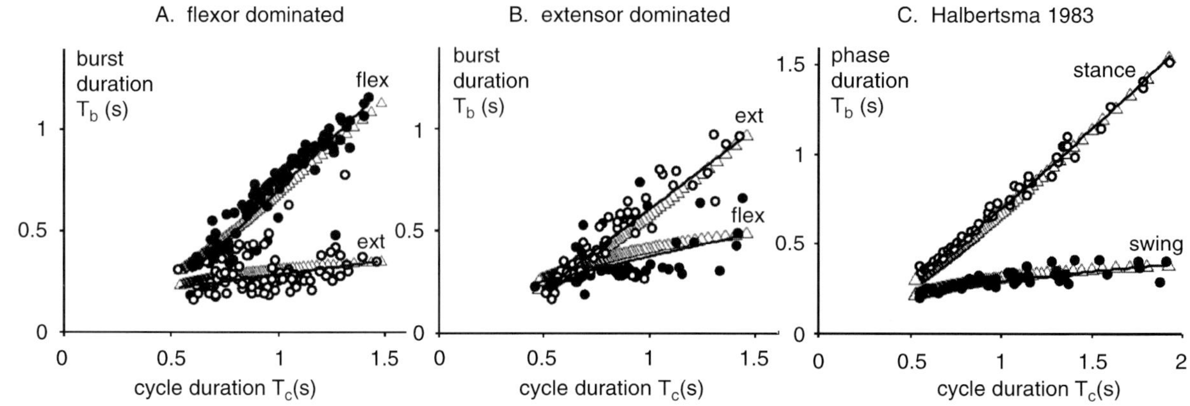

Fig. 4. Phase-duration versus cycle-duration plots in cat fictive locomotion (A and B) and normal cats (C). Filled circles: flexor phase (swing) durations, open circles: extensor phase (stance) durations. Triangles show the phase durations obtained from a simple oscillator model (Fig. 5), having adjusted the gain and offset parameters to fit the start and endpoints of the regression lines (solid) fitted to the data points. Reproduced from Yakovenko et al. (2005) with kind permission from the American Physiological Society.

A large number of simulations were performed with and without these If–Then rules. Analysis showed that the rules provided dramatic improvements in flexibility and stability of level overground locomotion in our model. The key to the improvement was that each step cycle was adjusted to the prevailing kinematic state. Similar conclusions were reached in a later neuromechanical study in which locomotion was generated entirely by If–Then rules, in the absence of a modulated CPG pattern (Ekeberg and Pearson, 2005).

Control of locomotor phase durations within the CPG

In the next sections we will discuss new findings that indicate that CPG oscillators are "set" to generate phase durations best suited for the biomechanics of locomotion. In normal locomotion in most animals, cycle duration varies mainly as a result of changes in extensor phase duration (Halbertsma, 1983; Fig. 4C). However, recently it was found that in fictive locomotion in decerebrate cats elicited by stimulation of the midbrain locomotor region (MLR), in which the locomotor rhythm is generated almost exclusively by the CPG, flexor phase durations varied more than extensor phase durations in over half of the sequences observed (Fig. 4A). The phase (flexion or extension) showing the larger variation was termed the "dominant" phase (Yakovenko et al., 2005). In a given animal, phase-duration plots were similar from one sequence to the next, suggesting that in a given preparation MLR stimulation produced specific descending signals that determined the phase-duration characteristics. We concluded that the locomotor CPG is not inherently extensor- or flexor-dominant, but depending on the balance of descending drives, it can show a continuum between the two. All three phase-duration plots in Fig. 4 were fitted remarkably well with a simple oscillator model (Fig. 5) by adjusting just two pairs of parameters that corresponded to "bias" and "gain" of the oscillator's timing elements. This suggested that in real CPGs the phase-duration characteristics could be the consequence of particular set levels of drive to neural timing elements in the CPG. The half-center receiving the lower set level would respond to additional drive with the larger variation in phase duration. On this view, the set level, or background drive, would determine which half-center was dominant.

Interestingly, it is known that neurons in which persistent inward currents (PICs) have been activated show an inverse relationship between PIC level and sensitivity to synaptic inputs (Lee et al., 2003; Li et al., 2004). This raises the intriguing possibility that interneurons in the extensor timing element may receive less PIC-generating input and

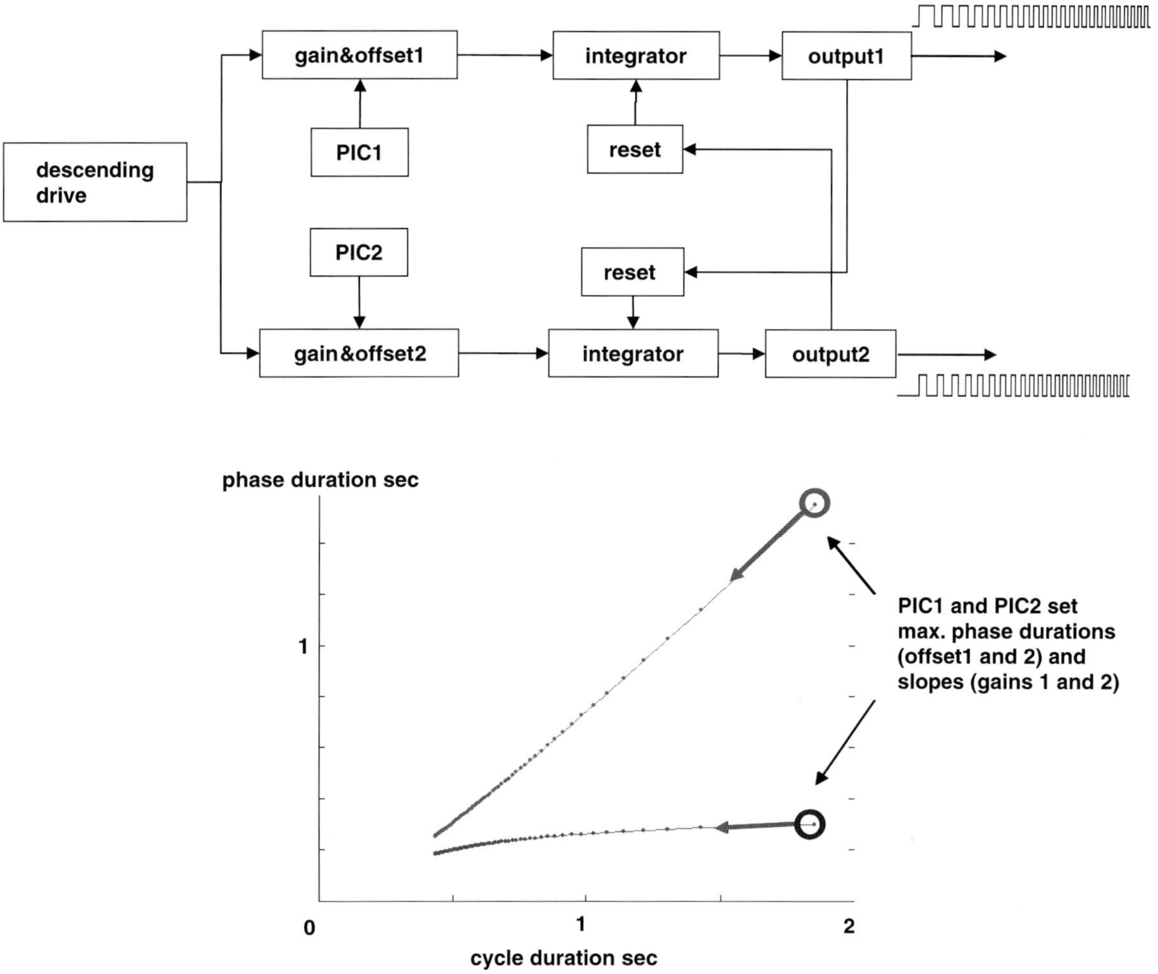

Fig. 5. Top: CPG oscillator model. PIC1 and PIC2 represent hypothesized persistent inward currents that determine the phase-duration characteristics (bottom). Open circles represent phase durations set by PICs when descending drive is zero. As descending drive increases, phase and cycle durations decline according to the sensitivity set by the PIC inputs. Note that the PIC inputs are hypothesized to set both gain and offset.

therefore as a network they are not only set to have longer half-cycle durations, but also to be more sensitive to synaptic commands for higher or lower cycle rates descending from supraspinal areas. The descending control of cadence could then reduce to a single signal driving these flexor and extensor timing interneurons (Fig. 5). As we will see next, in locomotion controlled by finite state rules, extensor-dominant phase-duration characteristics produce stable gait. We speculate that the appropriate drives to the extensor and flexor half-centers to produce these characteristics are set at birth and are subsequently adjusted during the years of motor learning and body growth.

Sensory control of phase durations during locomotion

The section above discusses phase-duration control in the absence of sensory input. The spinal rhythm generator is effectively blind to the unfolding kinematics, except when supraspinal areas provide descending commands based on exteroceptive inputs. We saw that in fictive locomotion in decerebrate MLR-stimulated cats the spinal CPG could generate cycles ranging from extensor- to flexor-dominant (Fig. 4A, B), presumably because the balance of descending drives to the half-centers ranged from normal to abnormal.

We wondered whether our neuromechanical model, provided with If–Then rules (see above), would exhibit phase-duration plots such as those in Fig. 4C. If it did, this would suggest that the biomechanics of locomotion require extensor-dominant phase-duration characteristics. It would also suggest that to harmonize with the kinematics and therefore the sensory input, the CPG oscillator should not only have an extensor-dominant phase-duration characteristic, but its operating points on this characteristic should be matched as closely as possible to the upcoming biomechanical requirements.

Figure 6 shows phase-duration plots computed from 20 simulations, each involving a minimum of five sequential step cycles. In a given simulation, the amplitudes of CPG activation profiles and durations as well as the trigger levels for If–Then rules were set to a variety of different levels, in order to generate gait of varying velocity and cadence. In two of the simulations, on–off activation profiles were used rather than the modulated EMG profiles obtained from the literature. In spite of all these parametric differences, the phase durations in all stable sequences (five steps or more without falling) were constrained along extensor-dominant phase-duration characteristics.

This result supports the idea that phase-duration characteristics are dictated by biomechanical attributes. The structure of the body (segment lengths and masses), as well as the disposition and force-generating properties of the muscles would be among the important determining factors. In future simulations we will explore the effect of limb segment length, body mass and other factors, including for example physical properties of the support surface. At this stage the results are sufficiently persuasive for us to formulate the conclusions below.

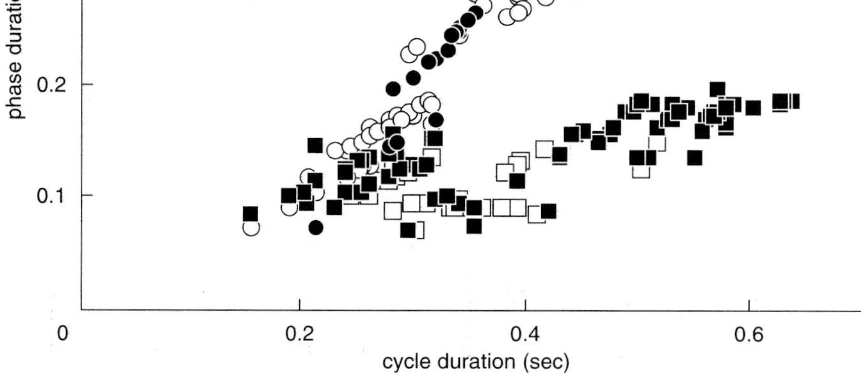

Fig. 6. Phase-duration plots computed from 20 simulations, each involving a minimum of 5 sequential step cycles. Circles: stance phase, squares: swing phase. Filled symbols indicate phases terminated by an If–Then rule, open symbols indicate phases terminated by completion of the CPG profile for that phase.

Conclusions: general propositions

1. For gait to be stable, swing and stance phase durations are constrained to characteristic values, described by two lines in a plot of phase- versus cycle-duration.
2. The phase-duration characteristics were modeled surprisingly well by setting just two pairs of parameters in a simple oscillator. This suggests that a particular phase-duration characteristic is preset by tonic drive to spinal interneuronal networks forming the timing elements of the CPG.
3. Stable gait was also achieved without specific CPG activation profiles, by switching muscles on and off according to sensory-mediated rules. The key new finding is that the phase-duration characteristic in such simulations was similar to the extensor-dominant characteristics in purely CPG-generated rhythms. This suggests that the constraints on phase durations in normal behavior are predetermined by the biomechanics.
4. We posit a spinal CPG timer and a sensory-mediated switch that operate in parallel, the former being driven primarily by descending inputs and the latter by the kinematics. We suggest that the system works best when the CPG timer is preset to produce an extensor-dominant phase-duration characteristic. Descending input from higher centers then adjusts the operating point on this preset phase-duration characteristic according to anticipated biomechanical requirements. In well-predicted movements, CPG-generated phase durations closely match those required by the kinematics. Residual errors are corrected by the sensory switching mechanism. We propose the term "neuromechanical tuning" to describe this process.

Epilogue

It is always humbling to discover that conclusions derived from complicated mathematical analyses were anticipated many years ago. Here is an extract from T. Graham Brown (1911).

"*A purely central mechanism of progression ungraded by proprioceptive stimuli would clearly be inefficient in determining the passage of an animal through an uneven environment. Across a plain of perfect evenness the central mechanism of itself might drive an animal with precision. Or it might be efficient for instance in the case of an elephant charging over ground of moderate unevenness. But it alone would make impossible the fine stalking of a cat over rough ground. In such a case each step may be somewhat different to all others, and each must be graded to its conditions if the whole progression of the animal is to be efficient. The hind limb which at one time is somewhat more extended in its posture as it is in contact with the ground, in another step may be more flexed. But the forward thrust it gives as its contribution to the passage of the animal must be of a comparatively uniform degree in each consecutive step. It may only be so if it is graded by the posture of the limb when in contact with the ground, and by the duration of its contact with the ground. This grading can only be brought about by peripheral stimuli. Of these we must regard the proprioceptive stimuli from the muscles themselves as the most important, and the part which they play is essentially the regulative — not the causative.*"

Acknowledgments

This work was supported by the Canadian Institutes of Health Research (CIHR), the Alberta Heritage Foundation for Medical Research (AHFMR), and the Fonds de Recherche en Santé du Québec (FRSQ).

References

Arshavsky, Y.I., Deliagina, T.G. and Orlovsky, G.N. (1997) Pattern generation. Curr. Opin. Neurobiol., 7: 781–789.

Arshavsky, Y.I., Gelfand, I.M. and Orlovsky, G.N. (1986) Cerebellum and Rhythmical Movements. Springer, Berlin.

Beloozerova, I.N. and Sirota, M.G. (1993) The role of the motor cortex in the control of accuracy of locomotor movements in the cat. J. Physiol., 461: 1–25.

Beloozerova, I.N. and Sirota, M.G. (1998) Cortically controlled gait adjustments in the cat. Ann. N.Y. Acad. Sci., 860: 550–553.

Brown, I.E. and Loeb, G.E. (2000) Measured and modeled properties of mammalian skeletal muscle: IV. dynamics of activation and deactivation. J. Muscle Res. Cell Motil., 21: 33–47.

Brown, T.G. (1911) The intrinsic factors in the act of progression in the mammal. Proc. R. Soc. Lond. Ser. B, 84: 308–319.

Collins, D.F. and Prochazka, A. (1996) Movement illusions evoked by ensemble cutaneous input from the dorsum of the human hand. J. Physiol., 496: 857–871.

Cruse, H. (1990) What mechanisms coordinate leg movement in walking arthropods? Trends Neurosci., 13: 15–21.

Donelan, J.M. and Pearson, K.G. (2004) Contribution of force feedback to ankle extensor activity in decerebrate walking cats. J. Neurophysiol., 92: 2093–2104.

Drew, T. (1991) Visuomotor coordination in locomotion. Curr. Opin. Neurobiol., 1: 652–657.

Drew, T. (1993) Motor cortical activity during voluntary gait modifications in the cat. I. Cells related to the forelimbs. J. Neurophysiol., 70: 179–199.

Drew, T., Prentice, S. and Schepens, B. (2004) Cortical and brainstem control of locomotion. Prog. Brain Res., 143: 251–261.

Durbaba, A., Taylor, R., Rawlinson, S.R. and Ellaway, P.H. (2003) Static fusimotor action during locomotion in the decerebrated cat revealed by cross-correlation of spindle afferent activity. Exp. Physiol., 88: 285–296.

Ekeberg, O. and Pearson, K. (2005) Computer simulation of stepping in the hind legs of the cat: an examination of mechanisms regulating the stance-to-swing transition. J. Neurophysiol., 94: 4256–4268.

Ellaway, P., Taylor, A., Durbaba, R. and Rawlinson, S. (2002) Role of the fusimotor system in locomotion. Adv. Exp. Med. Biol., 508: 335–342.

Freusberg, A. (1874) Reflexbewegungen beim Hunde. Pflueger's Archiv fuer die gesamte Physiologie, 9: 358–391.

Frigon, A. and Rossignol, S. (2006) Experiments and models of sensorimotor interactions during locomotion. Biol. Cybern., 95: 607–627.

Geyer, H., Seyfarth, A. and Blickhan, R. (2003) Positive force feedback in bouncing gaits? Proc. Biol. Sci., 270: 2173–2183.

Gorassini, M.A., Prochazka, A., Hiebert, G.W. and Gauthier, M.J. (1994) Corrective responses to loss of ground support during walking. I. Intact cats. J. Neurophysiol., 71: 603–610.

Grillner, S., Cangiano, L., Hu, G., Thompson, R., Hill, R. and Wallen, P. (2000) The intrinsic function of a motor system—from ion channels to networks and behavior. Brain Res., 886: 224–236.

Grillner, S. and Zangger, P. (1975) How detailed is the central pattern generation for locomotion? Brain Res., 88: 367–371.

Gritsenko, V., Mushahwar, V. and Prochazka, A. (2001) Adaptive changes in locomotor control after partial denervation of triceps surae muscles in the cat. J. Physiol., 533: 299–311.

Halbertsma, J.M. (1983) The stride cycle of the cat: the modelling of locomotion by computerized analysis of automatic recordings. Acta Physiol. Scand. Suppl., 521: 1–75.

Haugland, M. and Sinkjaer, T. (1999) Interfacing the body's own sensing receptors into neural prosthesis devices. Technol. Health Care, 7: 393–399.

Jordan, L.M. (1998) Initiation of locomotion in mammals. In: Kiehn O., Harris-Warrick R.M., Jordan L.M., Hultborn H. and Kudo N. (Eds.), Neuronal Mechanisms for Generating Locomotor Activity. New York Academy of Sciences, New York, pp. 83–93.

Kiehn, O. (2006) Locomotor circuits in the mammalian spinal cord. Annu. Rev. Neurosci., 29: 279–306.

Lafreniere-Roula, M. and McCrea, D.A. (2005) Deletions of rhythmic motoneuron activity during fictive locomotion and scratch provide clues to the organization of the mammalian central pattern generator. J. Neurophysiol., 94: 1120–1132.

Lee, R.H., Kuo, J.J., Jiang, M.C. and Heckman, C.J. (2003) Influence of active dendritic currents on input-output processing in spinal motoneurons in vivo. J. Neurophysiol., 89: 27–39.

Li, Y., Gorassini, M.A. and Bennett, D.J. (2004) Role of persistent sodium and calcium currents in motoneuron firing and spasticity in chronic spinal rats. J. Neurophysiol., 91: 767–783.

Loeb, G.E., Brown, I.E. and Cheng, E.J. (1999) A hierarchical foundation for models of sensorimotor control. Exp. Brain Res., 126: 1–18.

Magnus, R. (1909a) Zur Regelung der Bewegungen durch das Zentralnervensystem. Mitteilung I. Pflügers Archiv für die gesamte Physiologie des Menschen und der Tiere, 130: 219–252.

Magnus, R. (1909b) Zur Regelung der Bewegungen durch das Zentralnervensystem. Mitteilung II. Pflügers Archiv für die gesamte Physiologie des Menschen und der Tiere, 130: 253–269.

McCrimmon, D.R., Ramirez, J.M., Alford, S. and Zuperku, E.J. (2000) Unraveling the mechanism for respiratory rhythm generation. Bioessays, 22: 6–9.

Mileusnic, M.P., Brown, I.E., Lan, N. and Loeb, G.E. (2006) Mathematical models of proprioceptors: I. Control and transduction in the muscle spindle. J. Neurophysiol., 96: 1772–1788.

Mileusnic, M.P. and Loeb, G.E. (2006) Mathematical models of proprioceptors: II. Structure and function of the Golgi tendon organ. J. Neurophysiol., 96(4): 1789–1802.

Orlovsky, G.N., Deliagina, T.G. and Grillner, S. (1999) Neuronal control of locomotion (1st ed.). Oxford University Press, Oxford.

Patla, A.E., Prentice, S.D., Rietdyk, S., Allard, F. and Martin, C. (1999) What guides the selection of alternate foot placement during locomotion in humans. Exp. Brain Res., 128: 441–450.

Pearson, K., Ekeberg, O. and Buschges, A. (2006) Assessing sensory function in locomotor systems using neuromechanical simulations. Trends Neurosci., 29: 625–631.

Pearson, K.G. (2004) Generating the walking gait: role of sensory feedback. Prog. Brain Res., 143: 123–129.

Pearson, K.G., Misiaszek, J.E. and Hulliger, M. (2003) Chemical ablation of sensory afferents in the walking system of the cat abolishes the capacity for functional recovery after peripheral nerve lesions. Exp. Brain Res., 150: 50–60.

Perret, C. (1983) Centrally generated pattern of motoneuron activity during locomotion in the cat. Symp. Soc. Exp. Biol., 37: 405–422.

Perret, C. and Cabelguen, J.M. (1980) Main characteristics of the hindlimb locomotor cycle in the decorticate cat with special reference to bifunctional muscles. Brain Res., 187: 333–352.

Prochazka, A. (1993) Comparison of natural and artificial control of movement. IEEE Trans. Rehabil. Eng., 1: 7–17.

Prochazka, A. (1996) Proprioceptive feedback and movement regulation. In: Rowell L. and Sheperd J.T. (Eds.), Exercise: Regulation and Integration of Multiple Systems. American Physiological Society, New York, pp. 89–127.

Prochazka, A. (1999) Quantifying proprioception. Prog. Brain Res., 123: 133–142.

Prochazka, A., Gillard, D. and Bennett, D.J. (1997) Implications of positive feedback in the control of movement. J. Neurophysiol., 77: 3237–3251.

Prochazka, A. and Gorassini, M. (1998a) Ensemble firing of muscle afferents recorded during normal locomotion in cats. J. Physiol., 507: 293–304.

Prochazka, A. and Gorassini, M. (1998b) Models of ensemble firing of muscle spindle afferents recorded during normal locomotion in cats. J. Physiol., 507: 277–291.

Prochazka, A., Gritsenko, V. and Yakovenko, S. (2002) Sensory control of locomotion: reflexes versus higher-level control. Adv. Exp. Med. Biol., 508: 357–367.

Prochazka, A. and Yakovenko, S. (2002) Locomotor control: from spring-like reactions of muscles to neural prediction. In: Nelson R.J. (Ed.), The Somatosensory System. Deciphering the Brain's Own Body Image. CRC Press, Boca Raton, FL, pp. 141–181.

Rossignol, S. (1996) Visuomotor regulation of locomotion. Can. J. Physiol. Pharmacol., 74: 418–425.

Rossignol, S., Dubuc, R. and Gossard, J.P. (2006) Dynamic sensorimotor interactions in locomotion. Physiol. Rev., 86: 89–154.

Rybak, I.A., Shevtsova, N.A., Lafreniere-Roula, M. and McCrea, D.A. (2006a) Modelling spinal circuitry involved in locomotor pattern generation: insights from deletions during fictive locomotion. J. Physiol., 577: 617–639.

Rybak, I.A., Stecina, K., Shevtsova, N.A. and McCrea, D.A. (2006b) Modelling spinal circuitry involved in locomotor pattern generation: insights from the effects of afferent stimulation. J. Physiol., 577: 641–658.

Selverston, A.I. (1993) Modeling of neural circuits: what have we learned? Ann. Rev. Neurosci., 16: 531–546.

Sherrington, C.S. (1910) Flexion-reflex of the limb, crossed extension-reflex, and reflex stepping and standing. J. Physiol. (Lond.), 40: 28–121.

Sherrington, C.S. (1914) Further observations on the production of reflex stepping by combination of reflex excitation with reflex inhibition. J. Physiol., 47: 196–214.

Shik, M.L., Severin, F.V. and Orlovsky, G.N. (1966) Control of walking and running by means of electrical stimulation of the mid-brain. Biophysics, 11: 756–765.

Taga, G. (1995) A model of the neuro-musculo-skeletal system for human locomotion. II. Real-time adaptability under various constraints. Biol. Cybern., 73: 113–121.

Taga, G., Yamaguchi, Y. and Shimizu, H. (1991) Self-organized control of bipedal locomotion by neural oscillators in unpredictable environment. Biol. Cybern., 65: 147–159.

Takakusaki, K., Oohinata-Sugimoto, J., Saitoh, K. and Habaguchi, T. (2004) Role of basal ganglia-brainstem systems in the control of postural muscle tone and locomotion. Prog. Brain Res., 143: 231–237.

Taylor, A., Durbaba, R., Ellaway, P.H. and Rawlinson, S. (2000a) Patterns of fusimotor activity during locomotion in the decerebrate cat deduced from recordings from hindlimb muscle spindles. J. Physiol., 522: 515–532.

Taylor, A., Durbaba, R., Ellaway, P.H. and Rawlinson, S. (2006) Static and dynamic gamma-motor output to ankle flexor muscles during locomotion in the decerebrate cat. J. Physiol., 571: 711–723.

Taylor, A., Ellaway, P.H., Durbaba, R. and Rawlinson, S. (2000b) Distinctive patterns of static and dynamic gamma motor activity during locomotion in the decerebrate cat. J. Physiol., 529: 825–836.

Widajewicz, W., Kably, B. and Drew, T. (1994) Motor cortical activity during voluntary gait modifications in the cat. II. Cells related to the hindlimbs. J. Neurophysiol., 72: 2070–2089.

Yakovenko, S., Gritsenko, V. and Prochazka, A. (2004) Contribution of stretch reflexes to locomotor control: a modeling study. Biol. Cybern., 90: 146–155.

Yakovenko, S., McCrea, D.A., Stecina, K. and Prochazka, A. (2005) Control of locomotor cycle durations. J. Neurophysiol., 94: 1057–1065.

Zajac, F.E. (2002) Understanding muscle coordination of the human leg with dynamical simulations. J. Biomech., 35: 1011–1018.

Zajac, F.E., Neptune, R.R. and Kautz, S.A. (2003) Biomechanics and muscle coordination of human walking: part II: lessons from dynamical simulations and clinical implications. Gait Posture, 17: 1–17.

Zelenin, P.V., Deliagina, T.G., Grillner, S. and Orlovsky, G.N. (2000) Postural control in the lamprey: a study with a neuromechanical model. J. Neurophysiol., 84: 2880–2887.

P. Cisek, T. Drew & J.F. Kalaska (Eds.)
Progress in Brain Research, Vol. 165
ISSN 0079-6123
Copyright © 2007 Elsevier B.V. All rights reserved

CHAPTER 17

Threshold position control and the principle of minimal interaction in motor actions

Anatol G. Feldman[1,2,*], Valeri Goussev[2], Archana Sangole[2,3] and Mindy F. Levin[2,3]

[1]*Department of Physiology, Neurological Science Research Center, Institute of Biomedical Engineering, University of Montreal, Montreal, QC, Canada*
[2]*Center for Interdisciplinary Research in Rehabilitation, Rehabilitation Institute of Montreal and Jewish Rehabilitation Hospital, Laval, QC, Canada*
[3]*School of Physical and Occupational Therapy, McGill University, Montreal, QC, Canada*

Abstract: The answer to the question of how the nervous system controls multiple muscles and body segments while solving the redundancy problem in choosing a unique action from the set of many possible actions is still a matter of controversy. In an attempt to clarify the answer, we review data showing that motor actions emerge from central resetting of the threshold position of appropriate body segments, i.e. the virtual position at which muscles are silent but deviations from it will elicit activity and resistive forces (threshold position control). The difference between the centrally-set threshold position and the sensory-signaled actual position is responsible for the activation of neuromuscular elements and interactions between them and the environment. These elements tend to diminish the evoked activity and interactions by minimizing the gap between the actual position and the threshold position (the principle of minimal interaction). Threshold control per se does not solve the redundancy problem: it only limits the set of possible actions. The principle of minimal interaction implies that the system relies on the natural capacity of neuromuscular elements to interact between themselves and with the environment to reduce this already restricted set to a unique action, thus solving the redundancy problem in motor control. This theoretical framework appears to be helpful in the explanation of the control and production of a variety of actions (reaching movements, specification of different hand configurations, grip force generation, and whole-body movements such as sit-to-stand or walking). Experimental tests of this theory are provided. The prediction that several types of neurons specify referent control variables for motor actions may be tested in future studies. The theory may also be advanced by applying the notion of threshold control to perception and cognition.

Keywords: motor control; lambda model; equilibrium-point hypothesis; redundancy problem; referent body configuration; posture and movement; sit-to-stand movement; modeling

Introduction

Threshold position control is a well-established empirical phenomenon that shows that motor actions emerge following resetting the threshold (referent) position of appropriate body segments, i.e. the position at which muscles are silent but ready to generate activity and forces in response to deviations from this position (Asatryan and Feldman, 1965; Archambault et al., 2005; Foisy and Feldman, 2006). Descending systems (cortico-,

*Corresponding author. Tel.: +1 (514) 340-2111 Ext. 2192; Fax: +1 (514) 340-2154; E-mail: feldman@med.umontreal.ca

DOI: 10.1016/S0079-6123(06)65017-6

reticulo-, rubro- and vestibulo-spinal) can reset the threshold limb position (Feldman and Orlovsky, 1972; Nichols and Steeves, 1986). Such resetting is mediated by pre-, post-, mono- or poly-synaptic inputs to α- and/or γ-motoneurons (Matthews, 1959; Feldman and Orlovsky, 1972; Capaday, 1995). These control influences can be conveyed to motoneurons by all spinal neurons, including interneurons of reflex loops, for example, those influenced by group I and II muscle spindle afferents acting both mono- or polysynaptically on motoneurons.

Any deviation of the body from the threshold position, elicited either by external forces acting on the body or by central resetting of the threshold position, results in a change in proprioceptive signals to motoneurons. These signals facilitate motoneurons of those muscles that resist the deviation of the body from the threshold position. The proprioceptive response to the deviation also elicits activation of interneurons of reflex loops, some of which mediate interactions between muscles. Gelfand and Tsetlin (1971) suggested that the response of neuromuscular elements to any imposed activity and interactions, elicited either by reflexes or descending central influences, is guided by the *principle of minimal interaction*. According to this principle, the system tries to diminish the imposed activity by bringing the body to a state in which all-possible interactions in the system are minimized, in the limits defined by external forces. Reformulated in the framework of threshold position control (Feldman, 2007), this principle implies that neuromuscular elements act individually or collectively to diminish the imposed activity and interactions by minimizing the gap between the actual and threshold positions. Not only are the interactions between neuromuscular elements themselves minimized but also those between these elements and hierarchically higher, control levels, as well as the interactions of these elements and levels with the environment. This process continues until the gap is either fully eliminated or diminished to a degree such that the residual muscle activity gives rise to forces just sufficient to counterbalance external forces. Neurophysiologically, manifestations of this principle can be seen from typical mechanical and reflex responses of muscles to activation. Specifically, as suggested below, the activity of motoneurons depends on the difference $x-\lambda^*$ between the actual (x) and the threshold (λ^*) length of the muscle they innervate. The motoneurons can be activated by muscle stretch (i.e., due to an increase in x) or centrally (due to a decrease in λ^*), or in both ways. In response, the muscle contracts, thus diminishing the difference $x-\lambda^*$. Following muscle shortening, the autogenic facilitation of motoneurons from muscle spindle afferents decreases thus creating conditions for de-recruitment of motoneurons. Recurrent inhibition of motoneurons contributes to this process. Reciprocal inhibition mediated by Ia interneurons helps agonist muscles and motoneurons to diminish the resistance of antagonist muscles to the minimization of activity of agonist muscles. The process of minimization of overall activity of the neuromuscular system may not be completed with the de-recruitment of motoneurons of agonist muscles: gradually released from Ia inhibition and facilitated by stretching, motoneurons of antagonist muscles take their turn in bringing the body to a position at which a global minimum of the activity in the neuromuscular system is reached. Thus the minimization process may be non-monotonous. Also note that the two rules — threshold position control and the principle of minimal interaction — go together in the production of motor actions.

It is important to note that threshold control per se does not solve the redundancy problem of how the nervous system chooses a unique action from many possible actions that may or may not meet the motor goal: it only limits the set of possible motor actions. After that, the controller relies on the principle of minimal interaction, i.e. on the natural capacity of neuromuscular elements to interact between themselves and with the environment to reduce the already restricted set of motor actions to a unique action (Feldman, 2007). Due to variability in the properties of neuromuscular elements, the emerging action varies with repetition, even if the control pattern of threshold shifts and the external conditions remain unchanged — an illustration of the suggestion that "movement does not repeat itself" (Bernstein, 1967).

The threshold pattern may either be successful in eliciting an adequate action or it may require corrections to modify the previous or to initiate a new action. The learned adequate patterns of threshold shifts can be stored in motor memory and reproduced, if necessary, in similar behavioural situations.

The principle of minimal interaction may underlie the functioning of not only executive but also control levels responsible for resetting the threshold position: the difference between the position of effectors from the desired position defined by the motor goal may force the control levels to adjust the threshold positions at these and subordinate levels in order to diminish the difference.

The effectiveness of the two rules has been demonstrated in simulations of several single- and multi-joint motor actions, including locomotion and those that rely on adaptation, anticipation and learning (Flanagan et al., 1993; Feldman and Levin, 1995; Laboissière et al., 1996; Weeks et al., 1996; St.-Onge et al., 1997; Gribble et al., 1998; Günther and Ruder, 2003; Foisy and Feldman, 2006; Pilon and Feldman, 2006; Pilon et al., 2007).

In this chapter, we will elaborate on the basic notions outlined above. We first review the neurophysiological mechanisms underlying threshold position control. Then we consider different forms of such control at corresponding neuromuscular levels (neurons, muscles, whole body, and task effectors) and review experimental data that are consistent with the existence of such forms. This analysis will be combined with explanations of how threshold control and the principle of minimal interaction operate and elicit different motor actions involving the limbs or the whole body. We postulate the existence of neurons that are responsible for different forms of threshold position control, thus offering the possibility of further testing of the theory by means of neural analysis of motor behaviour.

Physiological origin of threshold position control

Thresholds of motoneurons and other neurons are usually measured in electrical units (threshold membrane potentials or currents). The observations that the nervous system can modify the *threshold position* at which muscles become active imply that the electrical thresholds are somehow transformed into positional variables, thus *placing our actions in a spatial frame of reference associated with the body or with the environment*. Such a transformation can be explained by considering how proprioceptive and other sensory inputs are combined with independent, control inputs at the level of the membrane of motoneurons or neurons, a process called *sensory-control integration* (Fig. 1A). We will avoid the commonly used term "sensory-motor integration" since in the present context it cannot be defined in a consistent way.

Consider some properties of a motoneuron when descending central influences on the neuromuscular system remain unchanged or absent. Due to proprioceptive feedback from muscle spindle afferents, the membrane potential of the motoneuron depends on the current muscle length. This means that a slow, quasi-static stretch of this muscle results in a gradual increase in the membrane potential of the motoneuron (Fig. 1B, lower diagonal line). The motoneuron begins to generate spikes when the current membrane potential starts to exceed the threshold potential (V_+). In the presence of length-dependent feedback, the same event becomes associated with spatial variables: motoneuronal recruitment occurs when the muscle length (x) reaches a specific, threshold length (λ_+). Now consider the case when a constant control input is added by descending systems, thus changing the membrane potential of the motoneuron (Fig. 1B, vertical arrows). If the net effect of such an input is facilitatory, the same muscle stretch elicits motoneuronal recruitment at a shorter threshold length, λ. The electrical effect of control inputs is thus transformed into a spatial variable — a change in the threshold muscle length. The firing frequency of active motoneurons and their number increases when the muscle is stretched beyond the threshold. This process resembles the size principle for motor unit recruitment (Henneman, 1981) except that in the context of threshold control the recruitment order for motoneurons is defined by threshold lengths, rather than sizes of motoneuronal bodies (Feldman, 1986). Although

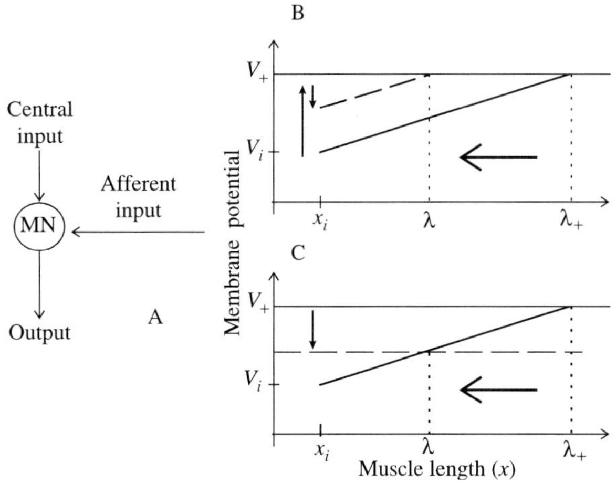

Fig. 1. Physiological origin of threshold position control. (A) Basic components of sensory-control integration underlying threshold position control at the level of motoneurons (MN). Each MN receives afferent influences that depend on the muscle length as well as on central control influences that are independent of muscle length. The MN is recruited when the membrane potential exceeds the electrical threshold (V_+). (B) When the muscle innervated by the MN is stretched quasi-statically from an initial length (x_i) the motoneuronal membrane potential increases from its initial value (V_i) according to afferent length-dependent feedback from the muscle (solid diagonal line). The electrical threshold (V_+) is eventually reached at length λ_+, at which the motoneuron begins to be recruited. When independent control inputs are added (↑: depolarization, ↓: hyper-polarization), the same stretch elicits motoneuronal recruitment at a shorter threshold length (λ). (C) Shifts in the spatial threshold (horizontal arrow) can also result from changes in the electrical threshold (vertical arrow). In both cases (B or C), shifts in the membrane potentials and respective changes in the threshold position are initiated prior to the onset of EMG activity and force generation (a feed-forward process). Thereby, the activity of motoneurons and muscle force emerge depending on the difference between the actual (x) and the threshold (λ) muscle length. Adapted with permission from Pilon et al. (2007).

the two definitions of the recruitment order are likely consistent, its definition in neurophysiological terms of threshold lengths goes beyond the anatomical context in which the size principle was originally formulated. As mentioned earlier, the threshold muscle length can be changed by descending systems that influence the membrane potential of α-motoneurons either directly or indirectly via interneurons or γ-motoneurons. A similar effect may result from shifting the threshold membrane potential (V_+) of motoneurons directly by descending systems (Fig. 1C; see Fedirchuk and Dai, 2004).

Physiological data also indicate that the threshold length is comprised of several additive components with only one component controlled centrally (Matthews, 1959; Feldman and Orlovsky, 1972; Feldman, 2007). To reflect these findings, we use the symbol λ^* for the composite (net) threshold whereas symbol λ is reserved for its central component:

$$\lambda^* = \lambda - \mu\omega - \rho + \varepsilon(t) \tag{1}$$

where λ and μ are controllable parameters; μ a temporal parameter related to the dynamic sensitivity of muscle spindle afferents (Feldman and Levin, 1995); ω the velocity of change in the muscle length ($\omega = \mathrm{d}x/\mathrm{d}t$); ρ the shift in the threshold resulting from reflex inputs such as those responsible for the inter-muscular interaction and cutaneous stimuli (e.g., from pressure-sensitive receptors in the finger pads during grasping); $\varepsilon(t)$ represents temporal changes in the threshold resulting, in particular, from intrinsic properties of motoneurons.

Let the net threshold, λ^*, be the threshold muscle length for the first motoneuron from which recruitments of motor units of a muscle starts. Then the muscle begins to be activated if the difference between the actual and the net threshold length is not negative, i.e. when $x - \lambda^* \geq 0$. Otherwise the motoneuron and the whole muscle are silent. In a supra-threshold state, the frequency and number of recruited motoneurons increase with the increasing difference between the actual and the threshold muscle length, so that the activity of the muscle (electromyographic, EMG, magnitude) is proportional to A, where

$$A = [x - \lambda^*]^+ \tag{2}$$

Here $[u]^+ = u$ if $u \geq 0$ and 0 otherwise.

The sensory-control integration described for motoneurons has several important implications:

(1) It is the motoneuronal membrane, i.e. the site where electrical control inputs are transformed into a spatial quantity — shifts in the

threshold muscle length (λ) — that the length-dependent afferent signals should exceed to begin motoneuronal recruitment.

(2) Proprioceptive feedback is critically important in this transformation. This point is illustrated by the observation of sensory and motor deficits in deafferented subjects. In the absence of proprioceptive feedback in these subjects, the electrical control signals issued by descending systems cannot be transformed into spatio-dimensional variables, resulting in the inability to recognize, specify and stabilize appropriate limb positions when the eyes are closed, not to mention that these subjects cannot stand or walk without assistance (Forget and Lamarre, 1995). Note that threshold position control does not exclude the possibility of motor actions when the normal sensory-control integration is damaged by deafferentation — it only suggests that these actions will be deficient. The importance of threshold position control is also emphasised by findings that lesions of different brain structures in stroke survivors limit the range of threshold regulation, resulting in numerous motor deficits such as muscle weakness, spasticity, as well as impaired coordination (Levin et al., 2000; Mihaltchev et al., 2005).

(3) By switching from a silent to an active state or vice versa, motoneurons signify that the values of the actual and threshold muscle lengths match each other. This *matching* between a physical variable and its central referent counterpart may be specific not only for motoneurons but also other neurons involved in *pattern recognition*, an essential aspect of cognitive processes.

(4) Once the difference between the current and threshold lengths becomes positive, the emerging activity of the muscle tends to shorten the muscle and thus diminish the gap between the actual and the threshold lengths. Other mechanisms (e.g., the stretch reflex, recurrent and reciprocal inhibition of antagonist muscles) apparently contribute to this process that represents a manifestation of the principle of minimal interaction at the neuromuscular level (see above).

(5) The right-hand side of Eq. (2) represents a strongly non-linear function of $x-\lambda^*$, implying that contrary to the servo-assistance hypothesis, muscle activation cannot be decomposed into two additive components, one resulting from central and the other from reflex influences on motoneurons. Furthermore, the same central shift in the threshold length may or may not elicit muscle activation — the choice between the two events equally depends on the current muscle length, velocity and other variables signalled by sensory inputs. The EMG activity thus represents an indivisible whole that was "born" by motoneurons with the equal help of both "parents" — sensory and control signals.

(6) Shifts in threshold positions result from changes in the membrane potentials and/or electrical thresholds of motoneurons, and thus *precede* changes in muscle activity and forces. This suggests a *forward nature* of threshold control. As has been shown (Pilon and Feldman, 2006), the forward nature of threshold control allows the system to overcome destabilizing effects of reflex and electromechanical delays. Most important, the forward nature of threshold control implies that essential changes in the state of the neuromuscular system start before any changes in EMG activity so that the latter emerges with a substantial contribution of proprioceptive feedback. Indeed, the forward nature of threshold position control may underlie anticipatory and predictive control strategies employed in many motor actions. For example, threshold control is helpful in producing arm movement while simultaneously increasing the grip force acting on the object to prevent it from sliding off the fingers (Pilon et al., 2007).

Neural basis for other forms of threshold position control

Motoneurons are not unique in having a membrane with an electrical threshold, gradual sensory inputs and independent central inputs (Fig. 1A): these features are characteristic of many, if not all, neurons. Therefore, the membrane of any such neuron is the site where central control inputs acquire the dimensionality of the sensory input. Decoded in this way, the control signal defines the threshold (referent) value that the sensory signals should exceed to activate the neuron. By assuming that the principle of minimal interaction is valid for all neurons related to motor control, one can say that when activated, each neuron from this group forces other neurons, motoneurons and muscles to diminish, if possible, the gap between the referent and the actual value of the sensory input, thus eliciting a motor action. Sensory inputs may be different for different neurons, implying the existence of different forms of threshold control. Consider several examples (Fig. 2).

Aperture neurons

Suppose there are neurons that receive facilitation in proportion to the distance (aperture) between the thumb and the index finger, for example, during a pinch type of grasp. Then, similar to the scheme described for motoneurons, independent, central inputs can be interpreted as setting the threshold position (threshold aperture) for activation of the muscles involved in the grip. When a rigid object is held between the fingers, the actual aperture (Q_a) is determined by the size of the object. In contrast, the centrally specified referent aperture (R_a) can be smaller than Q_a, as if the fingers virtually penetrate the object (Fig. 2A). Due to the deviation from the threshold aperture, the hand muscles generate activity and resistive forces that tend to diminish the gap between Q_a and R_a. In other words, the resistive (grip) force emerges since the object prevents the fingers from reaching the referent aperture. This idea was used to simulate arm movements with anticipatory changes in the grip force (Pilon et al., 2007).

Neurons controlling single joints

It is assumed that a population of neurons receive mono- or/and poly-synaptic afferent signals from

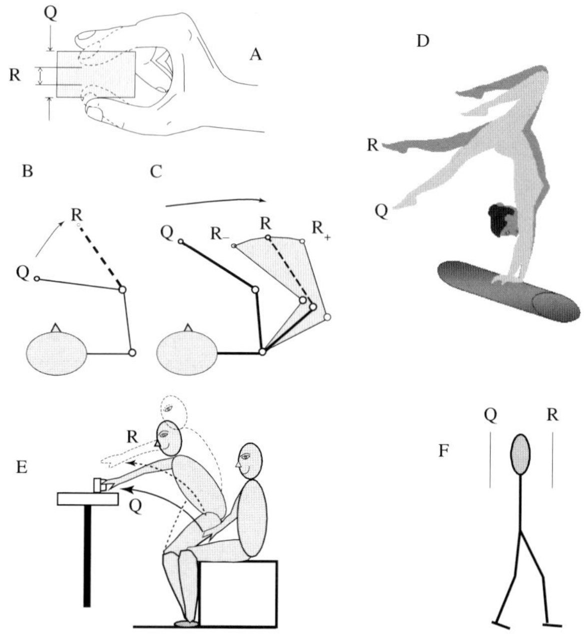

Fig. 2. Motor actions are elicited by the nervous system by shifting the referent position (R) of the body or its segments at which muscles reach their activation thresholds. The muscle activity and forces emerge due to the difference between the system's actual (Q) and the referent position. The specific form of R is chosen depending on the desired action. (A) In precision grip force control, R is the referent aperture that defines a virtual distance between the index finger and the thumb. In the presence of the object, the actual aperture (Q) is constrained by the size of the object held between the fingers whereas, in the referent position, the fingers virtually penetrate the object. Deviated by the object from their thresholds of activation, hand muscles generate activity and grip forces in proportion to the gap between the Q and R. (B) In motor tasks involving a single joint, R is the referent joint angle. (C) In tasks involving the whole arm, R is the referent arm configuration that defines a common threshold position for all arm muscles, except that the system may set thresholds for agonist and antagonist muscle groups differently such that these groups will be co-active at the R configuration and in the range (R_-, R_+) of adjacent configurations (co-activation zone or C command; grey area). (D) In tasks involving skeletal muscles of the whole body, the R is the referent body configuration. During the gymnastic exercise, the athlete presumably specifies an R configuration at which the net joint torques are zero and cannot compensate the weight torques of body segments. The body will move until the difference between Q and R become sufficient to elicit muscle activation and torques that balance the weight torques. (E) To reach for a cup, the nervous system shifts a referent hand position in the direction of the cup whereas the changes in the actual hand position and other body segments emerge following the principle of minimal interaction. (F) A single step or continuous walking are produced by a discrete or, respectively, continuous shifts in the referent position of the whole body in space.

muscle, joint and skin afferents such that the membrane potential of each neuron monotonically depends on a joint angle, say, the elbow angle. We define the elbow angle (Q_e) as increasing with elbow extension, i.e., with lengthening of elbow flexors. We assume that with a passive elbow extension, the membrane potential of some of these *elbow neurons* will increase and at a certain angle, called the referent angle (R_e), elbow motoneuronal recruitment will begin. The principle of minimal interaction suggests that the neuromuscular system will try to diminish the evoked activity by bringing the arm to the referent position. This will be possible if the neurons have appropriate descending projections — they should facilitate flexor motoneurons and disfacilitate or inhibit extensor motoneurons. Another population of such neurons will be activated during passive flexion from angle R_e and should also have reciprocal descending projections that facilitate extensor and inhibit flexor motoneurons, possibly via Ia interneurons of reciprocal inhibition. The principle of minimal interaction may thus be helpful in predicting some connections between neuromuscular elements. It is possible that the same principle guides not only the motor behaviour after neural connections have been maturated but also during proliferation of these connections in the process of development, when it is necessary to eliminate those connections that are not helpful in the formation of a coherent neural network that tends to minimize evoked activity. It is further assumed that, by influencing the whole population of elbow neurons, the nervous system can set a new referent angle (Fig. 2B). The motor response to shifts in the referent elbow angle may vary depending on the initial elbow angle and external condition (load). For example, if the elbow angle (Q_e) is fixed (isometric condition) and the final $R_e \neq Q_e$, then the appropriate group of elbow muscles will be activated in proportion to the difference between the R_e and Q_e and eventually produce a constant isometric torque, like in the case of grip force production (see above). In contrast, in the absence of an external load, the emerging muscle activation will eventually bring the system to threshold position $Q_e = R_e$. By reaching the common threshold angle, the activity of all muscles of the joint, regardless of their biomechanical function, will be reduced to a near-zero level. Such a manifestation of the principle of minimal interaction has been observed in experimental studies (Ostry and Feldman, 2003; Foisy and Feldman, 2006, see also Fig. 3). In the case of movement against a constant or spring-like load, the emergent muscle activity will first elicit a torque moving the arm towards the final threshold angle, R_e. Because of the resistive load, reduction of muscle activity to zero will not be achieved in this case — the movement will terminate at an intermediate position at which the muscle activity is just sufficient to generate the net torque that balances the load torque.

The fact that the same shift in the referent position may give rise to different actions does not prevent reaching the desired motor goal. For example, if the goal is to produce a specific isometric torque at the elbow joint, the subject may push the arm against an immovable object and change the referent angle until the desired elbow torque is reached. If the goal is to reach a desired elbow angle Q_e, with or without a load, the subject can also change the referent angle until the desired angle Q_e is attained. Indeed, in isometric conditions, the subject should first remove the obstacle to arm motion to reach the desired angle. Note that the control strategy suggested here does not resemble a servo-control scheme: the R_e is not an internal representation of the desired position, it is just a tool for reaching the desired arm position (Q_e) usually defined in external space (in relation to the body or to objects in the environment). As long as the desired angle is reached, the movement error is zero even in the cases when the R_e is substantially different from Q_e, contrary to what is postulated in servo-control schemes.

Neurons specifying the referent configuration of the body or its segments

Note that R_a and R_e described above resemble reciprocal (R) commands defined in the earlier version of the threshold control theory. Here we generalize these notions to multi-muscle and multi-joint systems.

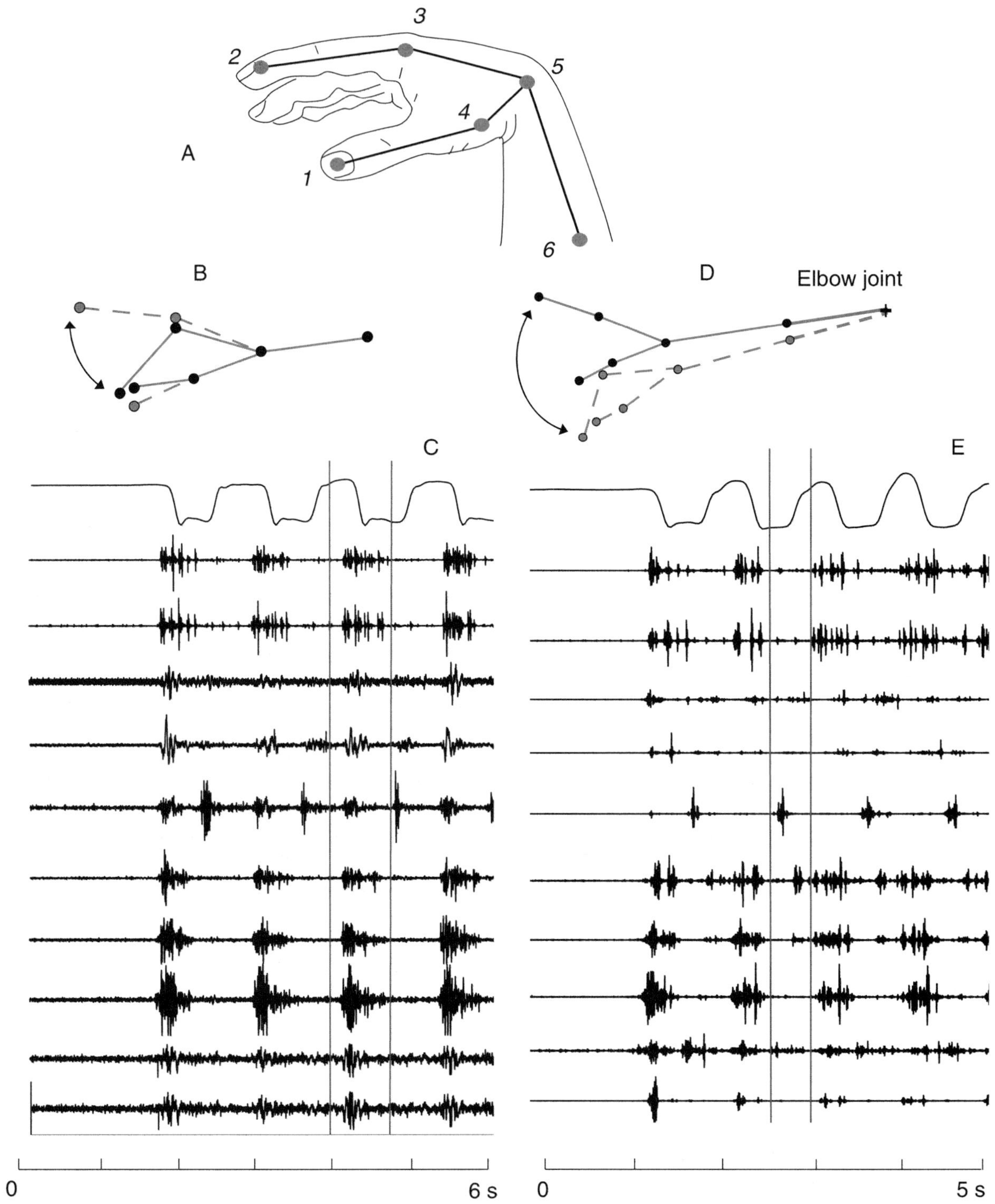
3
2
5
4
A
1
6
B
D
Elbow joint
C
E
0
6 s
0
5 s

Suppose there are neurons that receive synaptic influences which monotonically depend on sensory signals resulting from some coordinated changes in multiple degrees of freedom (DFs) of the body — a gradual increase or decrease in the height of the body, leaning the body forward or backward or more complex changes in the body configuration. We postulate that a network consisting of such neurons is used to set and reset the threshold position (configuration) of the whole body. Like for motoneurons and other neurons described above, the threshold body configuration can be considered as a referent (R) configuration that may or may not coincide with the actual body configuration (Q). According to the principle of minimal interaction, the projections of these neurons to motoneurons are organized in a manner that makes it possible to diminish the difference between Q and R in the limits defined by internal and external constraints. Thereby, this difference appears to be a global factor that guides the activity of all skeletal muscles without redundancy problems (see Fig. 2C–E). By changing the referent body configuration, the system may elicit, for example, jumping, walking or dancing.

Single-joint referent positions (see above) can be considered as components of referent body configurations. The notion of referent configuration can be applied to several segments of the body, for example, of the arm and/or hand so that one can consider the activity of all muscles spanning these segments as dependent on the difference between the actual and the referent configuration of respective segments (Fig. 2D, E).

In many actions, the system may also change muscle activation thresholds in order to create a range of body configurations centred around and including the R configuration at which agonist and antagonist muscles are co-active (spatial co-activation zone; for details, see Levin and Dimov, 1997; Feldman, 2007; Fig. 2C). Such a C command is used when it is necessary to increase the muscle forces resulting from shifts in the R configuration, enhance the movement speed and stability of posture and movement (Feldman and Levin, 1995).

Effector neurons

The list of possible threshold positions and neurons associated with the specification of these positions can be extended to everyday motor actions that are associated with reaching or approaching objects in an external space, for example, picking a tea cup from the table, grasping a ball or walking to a door. In these cases, the position of end-effectors (e.g., the hand) or the whole body should be associated with the environment, rather than with body parts. Qualitatively, threshold control and the principle of minimal interaction guide, for example, the hand to a tea cup in the following way (Fig. 2E). The movement is produced by shifting the referent coordinates (R) of the hand by influencing neurons that receive afferent inputs related to the actual coordinates (Q) of the hand in space. We assume that the projections of these neurons to motoneurons of multiple muscles are organized according to the principle of minimal interaction so that the neurons drive the hand until the difference between its referent and actual positions becomes minimal. This occurs when neurons of subordinate levels, including motoneurons, also minimize their activity and interactions. Based on the previous experience, the system takes into account gravity by shifting the referent position somewhat higher than the cup. The weight of the arm will deviate the hand from the referent position to a position convenient for grasping the cup. The minimization process will thus be finished when the residual muscle activity produces net joint torques that balance the weight torques of

Fig. 3. Gobal EMG minima and referent arm-hand configurations during rhythmical movement involving only hand (B, C) or also the forearm (D, E) in a horizontal plane. (A) Coordinates of markers (1–6) were reordered during the movements. (B, D) Referent configurations at the time when the global EMG minima occurred (vertical lines in C and E, respectively). (C, E) Displacement of the tip of the index finger (upper curve, amplitude of 17–18 cm) and EMG activity of 10 hand, wrist and elbow muscles (from top to bottom: thenar opponens, abductor pollicis brevis, flexor brevis digiti minimi, abductor digit minimi, lumbrical II, extensor pollicis longus, extensor indicis proprius, extensor carpi ulnaris, extensor digitorum longus, flexor carpi ulnaris).

arm segments. According to the principle of minimal interaction, whether or not a DF will be involved in the action depends on its capacity to contribute to the minimization process. This capacity depends not only on the difference between *R* and *Q* but also on mechanical and internal constraints. For example, when the subject is sitting and the cup is located within the arm's reach, the legs and trunk cannot contribute to the minimization and usually remain motionless, although they can be moved intentionally. In contrast, the capacity of leg and trunk DFs to contribute to the minimization process increases when the cup is beyond the arm's reach: depending on the distance to the cup, subjects only lean the trunk or also raise the body or even make a step to reach the cup. Similarly, the capacity of hand DFs to contribute to its transport to an object is minimal but their capacity to pre-shape the hand in order to grasp the object increases with decreasing distance to it. Thereby the transport and grasp components of movement emerge without pre-planning or internal modelling of their sequence (cf. Smeets and Brenner, 1999).

In each trial, there will be no uncertainty in choosing one pattern of coordination between DFs of the body of the set of all possible patterns — each time the minimizing process will produce a unique coordination pattern. If necessary, with additional corrective shifts in *R*, the target will be reached. The coordination pattern can, indeed, vary with task repetitions, time-dependent changes in the system (e.g., due to fatigue), intentional involvements or restrictions (e.g., following pain) of some DFs of the body. When the motor action is controlled at the level of task effectors, the system may not be involved directly in the specification of referent body configurations — these emerge automatically, following the minimization process initiated at a higher level of effector neurons.

Effector neurons can be involved in the production of a single step or continuous walking — these actions supposedly result from discrete or, respectively, continuous shifts in the referent position of the whole body in space (Fig. 2F). Again, they emerge following the tendency of the system to minimize the difference between the actual and the centrally-specified referent position of the body. As a result, after each step the body will restore approximately the same configuration, but in another part of space. Modulations of the referent body position at the level of appropriate effector neurons may also underlie changes in the direction ("steering"), stepping over, jumping or avoiding obstacles during locomotion (Drew et al., 2004; cf. Fajen and Warren, 2003; Warren, 2006). Thus, task-specific changes in the referent position of the body in the environment, which lead to a smooth transition from one equilibrium state of the body to another, may be a major function of the central pattern generator for locomotion (see Ustinova et al., 2006).

Experimental identification of task-specific referent body configurations and other tests

Many ideas formulated above conflict with conventional views on motor control (Ostry and Feldman, 2003). This has been the basis for some misunderstanding leading to false rejections of the whole theory (for recent review, see Feldman and Latash, 2005). However, predictions of theory have undergone vigorous testing and many have been confirmed. In particular, this was the case for the prediction that the control strategy relying on effector neurons should guide the hand along the same trajectory regardless of the number of DFs involved in pointing movements (Adamovich et al., 2001).

Here we describe some additional findings that are consistent with the referent body configuration concept. Such findings also help to experimentally identify the referent configurations underlying motor actions, including those involving multiple muscles.

Global EMG minima

Biomechanical factors, such as the inertia of body segments and external forces (the weights of body segments), may prevent the actual and the referent configurations from matching and thus establishing a zero global EMG minimum. Matching is still possible in some cases. For example, by reversing the changes in the *R* configuration, the nervous system may reverse the movement direction. Due to inertia, however, body segments may continue to move in the same direction before yielding to reversal in the changes of the *R* configuration. During this period, the actual (*Q*) and the *R* configurations may approach and temporarily match each other. Since activity of each muscle depends on the difference between *Q* and *R*, matching between *Q* and *R* should result in minimization of the EMG activity of all muscles involved, regardless of their biomechanical function. The depth of the global EMG minima may be limited by the level of co-activation of agonist and antagonist muscles. In the absence of co-activation, all skeletal muscles will be deactivated when a global EMG minimum occurs. The referent body configuration(s) employed by the nervous system for a given motor action can thus be identified experimentally as coinciding with those actual body configurations at which global EMG minima occur.

Knowing that skeletal muscles of the body are biomechanically and functionally diverse, it seemed unlikely that global EMG minima may actually occur and, at this point, the theory was at the risk of being rejected. It appeared, however, that such minima are regular phenomena in many movements involving reversals of direction. Minima in the EMG activity of numerous, functionally diverse muscles of the body have been found during vertical jumps in humans, rhythmical horizontal head rotations in monkeys, jaw movements in rabbits, arm pointing movement with reversals, some ballet movements in dancers, sit-to-stand (STS) and hammering movements in humans (Archambault et al., 1998; Feldman et al., 1998; Weijs et al., 1999; Lestienne et al., 2000; St-Onge and Feldman, 2004; Feldman and Balasubramaniam, 2004; Lepelley et al., 2006). Figure 3 shows global minima in the EMG activity of arm and hand muscles and the respective arm-hand referent configurations obtained during rhythmical hand opening and closing movements with simultaneous swinging of the arm in a horizontal plane.

Although global EMG minima can be observed in many movements with reversals, it is important to emphasize that there are movements in which such minima cannot occur despite reversals. For example, rhythmical chin-ups on a bar can be produced by shifting the referent body position up and down, and thus eliciting changes in the actual position of the body. The necessity to deal with the high body weight in this task makes matching between these positions and thus global minima impossible: the referent position should be kept higher than the actual body position so that the difference between them will cause a high level of arm muscle activation required for at least partial compensation of the body weight in this task.

An unexpected aspect of threshold position control is the implication that synchronization in the activity of multiple muscles (EMG minima in all of them) can result from a spatial event — matching between the central referent (*R*) and actual (*Q*) configurations. One can assume that synchronization of activity of neuronal ensembles in the brain may also result from matching of centrally specified referent variables with respective variables delivered by sensory signals, not only in motor action but also in perception and cognition.

Simulation of sit-to-stand movement

Although the concept of referent body configuration has been verified empirically (see above), it seems important to test the feasibility of this concept and the principle of minimal interaction in simulations of whole body movements involving multiple muscles. This has been done for locomotion (Günther and Ruder, 2003). Here we briefly describe simulation of STS movements in order to illustrate that multiple muscles can actually be guided as a coherent unit and that postural stability during the body transition to standing can be achieved, contrary to standard biomechanical approaches, without internal computations or monitoring of shifts in the centre of foot pressure or body mass.

The simulations of STS movements is complicated by the necessity to realistically represent the properties of neuromuscular elements to derive appropriate equations and responses to the selected control pattern. In contrast, the control pattern can be described in a simple way. We assume that a set of appropriate patterns of changes in the referent body configuration is selected by the controller, based on previous experience (i.e., by trial and error). The changes in the referent configuration have two phases: virtually leaning the trunk forward to an appropriate position while sitting on a stool and raising the trunk above the stool to the desired height by extending the ankle, knee and hip joint angles (Fig. 4). Each phase is defined by two parameters: the rate and duration of the changes in the referent body configuration (ramp-shaped patterns) as well as by the time between the offset of one and the onset of the other referent phase. Changes in the *R* configurations were combined with a *C* command that, as mentioned before, enhances the muscle torques resulting from changes in the *R* configuration and increases stability of the transition from sitting to standing.

In simulations of responses to selected control patterns, the body is considered as having three DFs: a planar model with one DF at each of the three (ankle, knee and hip) joints. For simplicity, the trunk, arms and head are considered as a single unit. Lagrange formalism and Mathematika software were used to obtain an analytical form of equations of motion of the system. Twenty-three anatomically defined muscles spanning the three joints on each side of the body and the dependency of their length on joint angles with appropriate moment arms were included in the model. Formula

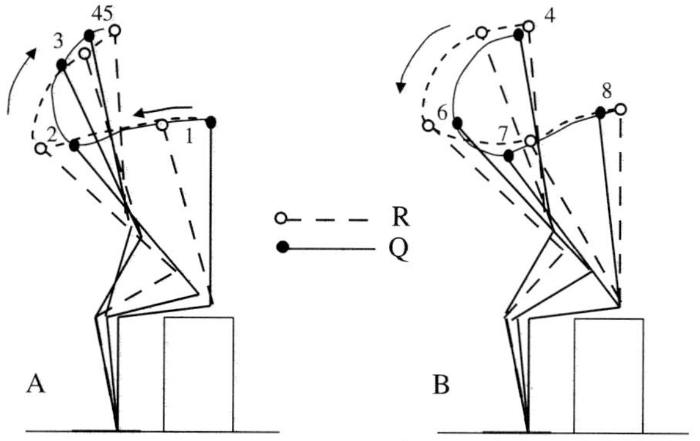

Fig. 4. Simulation of sit-to-stand movements (left panel) with reversal (right panel). The referent and actual body configurations (1–8) are shown at time 0.4, 1.0, 1.4, 1.8, 2.43, 2.6, 3.0, and 3.5 s after the onset of movement, respectively. The total movement time is ~4 s.

2 was used to compute the muscle activation by taking individual values of variables for each muscle. Muscle torques resulting from muscle activation were obtained based on equations described by Pilon and Feldman (2006). These equations take into account that muscle activation elicits gradual (time-dependent) muscle torque development. This torque also depends on the muscle length and velocity coming from two sources: (1) The properties of the contractile apparatus (Huxley and Hanson, 1954) that elicits muscle forces depending on these kinematic variables practically *without delay*. (2) The properties of proprioceptive inputs that depend on the same kinematic variables but with *some delay* (50 ms in our model). In addition, muscle activation elicits muscle torques after electromechanical delay (20 ms in our model). Each muscle also includes a series elastic component that plays the role of a buffer in the transmission of muscle force to body segments.

According to threshold control theory, the values of individual muscle thresholds (λ–s) of numerous muscles emerge following minimization of the neuromuscular activity evoked by changes in the referent body configurations. In simulations, these can be derived by two methods. First, by designing a neuronal network that produces such minimization and yields the respective muscle thresholds. Second, by assuming that such a network exists and extrapolating the values of thresholds that such a network could yield: mathematically, the minimal value(s) of a function can often be obtained without knowing the function in detail (e.g., by finding the lowest point on its graph). Physiologically, the first method is desirable but since the second method yields about the same threshold values (see below) it appeared more practical in a computational sense and therefore was used in the present simulation. Thereby, we do not want to suggest that the nervous system uses this method to specify muscle activation thresholds: this method just symbolically imitates what the nervous system specifies.

The second method takes into account that in the absence of co-activation of agonist and antagonist muscles, each referent body configuration represents, by definition, a position of the body at which all muscles reached their threshold lengths (i.e., $\lambda^* = x$; see Eq. (2)). We also assumed that each threshold configuration corresponds to a steady state at which the reflex inter-muscular interaction and intrinsic variations of the threshold are negligible (i.e., $\omega = \rho = \varepsilon = 0$ in Eq. (1)). In this case, $\lambda = x$ for each muscle. In other words, at any referent body configuration, the central component of the threshold length of each muscle coincides with its actual muscle length at this configuration. This implies that the relationship between the control variables for multiple muscles reflects the geometric relationship between actual muscle lengths at the respective referent configuration (the *rule of biomechanical correspondence in motor control*). This rule was used to derive the threshold muscle lengths for each referent body configuration. Indeed, this method does not take into account some temporal dynamics in the specification of threshold muscle lengths corresponding to a given referent configuration.

The model also employs a generalized C command that defines a range (R_-, R_+) of body configurations centred around the R configuration within which all muscles can be co-active (see Fig. 2C). The parameters in the model were subdivided into two groups: (1) constant parameters in the equations that describe the properties of muscle, reflex and other components of the system; (2) parameters characterizing the ramp-shaped patterns of changes in the R and C commands. The command patterns were chosen from the set of patterns that do not lead to the body falling. This did not require computation of the position of the centre of mass in relation to the area of the feet. When possible, the values of constant parameters were chosen from the ranges reported in other studies. Specific values of such parameters were identified by simulating real STS movements from a representative trial, individually for each of the nine subjects participating in the study (kinematic recording of STS movements with an optoelectronic system, Optotrak). Once identified, these values remained unchanged in the simulation of data from the remaining trials of the same subject and only a comparatively limited number of parameters related to the control variables (R and C) could be modified to reproduce all STS movements.

The resemblance between simulated and experimental curves (goodness of fit) was evaluated by the coefficient of correlation (r^2) and by the root mean square error. All simulations were made using Matlab and Mathematika software.

The model was robust in reproducing experimental kinematics of STS movements (Fig. 4; $r^2 = 0.85–0.99$ for the group of subjects) (see Animation).

Conclusions

Two rules — threshold position control and the principle of minimal interaction — seem to play a fundamental role in the control and production of motor actions. These rules also suggest how multiple muscles and DFs can be guided without redundancy problems. The prediction of several types of neurons that provide referent control variables for motor actions may be tested in future studies. The idea that the nervous system compares centrally specified referent with respective sensory signals may be relevant not only to action but also to perception and cognition (Feldman and Latash, 1982, 2005).

Abbreviations

DF	degree of freedom
EMG	electromyographic
STS	sit-to-stand

Acknowledgments

Supported by NSERC, FQRNT and CIHR (Canada).

Appendix A. Supplementary data

Supplementary data (Animation) associated with this article can be found in the online version at doi:10.1016/S0079-6123(06)65017-6.

References

Adamovich, S.V., Archambault, P.S., Ghafouri, M., Levin, M.F., Poizner, H. and Feldman, A.G. (2001) Hand trajectory invariance in reaching movements involving the trunk. Exp. Brain Res., 138: 288–303.

Archambault, P.S., Levin, M.F., Mitnitski, A. and Feldman, A.G. (1998) Multiple muscle control may be guided by a referent body configuration. Neurosci. Abstr., 24: 1158.

Archambault, P.S., Mihaltchev, P., Levin, M.F. and Feldman, A.G. (2005) Basic elements of arm postural control analyzed by unloading. Exp. Brain Res., 164: 225–241.

Asatryan, D.G. and Feldman, A.G. (1965) Functional tuning of the nervous system with control of movements or maintenance of a steady posture: I. Mechanographic analysis of the work of the joint on execution of a postural task. Biophysics, 10: 925–935.

Bernstein, N.A. (1967) The Coordination and Regulation of Movements. Pergamon Press, London.

Capaday, C. (1995) The effects of Baclofen on the stretch reflex parameters of the cat. Exp. Brain Res., 104: 287–296.

Drew, T., Prentice, S. and Schepens, B. (2004) Cortical and brainstem control of locomotion. Prog. Brain Res., 143: 251–261.

Fajen, B.R. and Warren, W.H. (2003) Behavioral dynamics of steering, obstacle avoidance, and route selection. J. Exp. Psychol. Hum. Percept. Perform., 29: 343–362.

Fedirchuk, B. and Dai, Y. (2004) Monoamines increase the excitability of spinal neurones in the neonatal rat by hyperpolarizing the threshold for action potential production. J. Physiol., 557: 355–561.

Feldman, A.G. (1986) Once more on the equilibrium point hypothesis (λ Model) for motor control. J. Mot. Behav., 18: 17–54.

Feldman, A.G. (2007) Equilibrium point control (an essay). In: Karniel, A. (Ed.), Encyclopedic Reference of Neuroscience. Field: Computational Motor Control (in press).

Feldman, A.G. and Balasubramaniam, R. (2004) Guiding movements without redundancy problems. In: Kelso J.K.S. (Ed.), Coordination Dynamics. Springer, Berlin, pp. 155–176.

Feldman, A.G. and Latash, M.L. (1982) Afferent and efferent components of joint position sense; interpretation of kinaesthetic illusion. Biol. Cybern., 42: 205–214.

Feldman, A.G. and Latash, M.L. (2005) Testing hypotheses and the advancement of sciences: recent attempts to falsify the equilibrium point hypothesis. Exp. Brain Res., 161: 91–103.

Feldman, A.G. and Levin, F.M. (1995) The origin and use of positional frames of reference in motor control. Behav. Brain Sci., 18: 723–806.

Feldman, A.G., Levin, M.F., Mitnitski, A.M. and Archambault, P. (1998) 1998 ISEK congress keynote lecture: multi-muscle control in human movements. J. Electromyogr. Kin., 8: 383–390.

Feldman, A.G. and Orlovsky, G.N. (1972) The influence of different descending systems on the tonic stretch reflex in the cat. Exp. Neurol., 37: 481–494.

Flanagan, J.R., Ostry, D.J. and Feldman, A.G. (1993) Control of trajectory modifications in target-directed reaching. J. Mot. Behav., 25(3): 140–152.

Foisy, M. and Feldman, A.G. (2006) Threshold control of arm posture and movement adaptation to load. Exp. Brain Res., 175: 726–744.

Forget, R. and Lamarre, Y. (1995) Postural adjustments associated with different unloadings of the forearm: effects of proprioceptive and cutaneous afferent deprivation. Can. J. Physiol. Pharmacol., 73: 285–294.

Gelfand, I.M. and Tsetlin, M.L. (1971) Some methods of controlling complex system. In: Gelfand I.M., Gurfinkel V.S., Fomin S.V. and Tsetlin M.L. (Eds.), Models of structural-functional organization of certain biological systems. MIT Press, Cambridge, pp. 329–345.

Gribble, P.L., Ostry, D.J., Sanguineti, V. and Laboissière, R. (1998) Are complex control signals required for human arm movement? J. Neurophysiol., 79: 1409–1424.

Günther, M. and Ruder, H. (2003) Synthesis of two-dimensional human walking: a test of the λ-model. Biol. Cybern., 89: 89–106.

Henneman, E. (1981) Recruitment of motor neurons: the size principle. In: Desmedt J.E. (Ed.), Progress in clinical neurophysiology: motor units types, recruitment and plasticity in health and disease, Vol. 9, Basel, Karger.

Huxley, H. and Hanson, J. (1954) Changes in the cross-striations of muscle during contraction and stretch and their structural interpretation. Nature, 173(4412): 973–976.

Laboissière, R., Ostry, D.J. and Feldman, A.G. (1996) The control of multi-muscle systems: human jaw and hyoid movements. Biol. Cybern., 74: 373–384.

Lepelley, M.C., Thullier, F., Koral, J. and Lestienne, F.G. (2006) Muscle coordination in complex movements during Jete in skilled ballet dancers. Exp. Brain Res., 175: 321–331.

Lestienne, F.G., Thullier, F., Archambault, P., Levin, M.F. and Feldman, A.G. (2000) Multi-muscle control of head movements in monkeys: the referent configuration hypothesis. Neurosci. Lett., 283(1): 65–68.

Levin, M.F. and Dimov, M. (1997) Spatial zones for muscle coactivation and the control of postural stability. Brain Res., 757: 43–59.

Levin, M.F., Selles, R.W., Verheul, M.H. and Meijer, O.G. (2000) Deficits in the coordination of agonist and antagonist muscles in stroke patients: implications for normal motor control. Brain Res., 853: 352–369.

Matthews, P.B.C. (1959) A study of certain factors influencing the stretch reflex of the decerebrate cat. J. Physiol., 147: 547–564.

Mihaltchev, P., Archambault, P.S., Feldman, A.G. and Levin, M.F. (2005) Control of double-joint arm posture in adults with unilateral brain damage. Exp. Brain Res., 163: 468–486.

Nichols, T.R. and Steeves, J.D. (1986) Resetting of resultant stiffness in ankle flexor and extensor muscles in the decerebrated cat. Exp. Brain Res., 62: 401–410.

Ostry, D.A. and Feldman, A.G. (2003) A critical evaluation of the force control hypothesis in motor control. Exp. Brain Res., 153: 275–288.

Pilon, J.-F. and Feldman, A.G. (2006) Threshold control of motor actions prevents destabilizing effects of proprioceptive delays. Exp. Brain Res., 174: 229–239.

Pilon, J.F., De Serres, S.J. and Feldman, A.G. (2007) Threshold position control of arm movement with anticipatory increase in grip force. Exp. Brain Res., 181: 49–67.

Smeets, J.B.J. and Brenner, E. (1999) A new view on grasping. Motor Control, 3: 237–271.

St-Onge, N., Adamovich, S.V. and Feldman, A.G. (1997) Control processes underlying elbow flexion movements may be independent of kinematic and electromyographic patterns: experimental study and modelling. Neuroscience, 79: 295–316.

St-Onge, N. and Feldman, A.G. (2004) Referent configuration of the body: a global factor in the control of multiple skeletal muscles. Exp. Brain Res., 155: 291–300.

Ustinova, K.I., Feldman, A.G. and Levin, M.F. (2006) Central resetting of neuromuscular steady states may underlie rhythmical arm movements. J. Neurophysiol., 96: 1124–1134.

Warren, W.H. (2006) The dynamics of perception and action. Psychol. Rev., 113: 358–389.

Weeks, D.L., Aubert, M.P., Feldman, A.G. and Levin, M.F. (1996) One-trial adaptation of movement to changes in load. J. Neurophysiol., 75(1): 60–74.

Weijs, W.A., Sugimura, T. and van Ruijven, L.J. (1999) Motor coordination in a multi-muscle system as revealed by principal components analysis of electromyographic variation. Exp. Brain Res., 127: 233–243.

P. Cisek, T. Drew & J.F. Kalaska (Eds.)
Progress in Brain Research, Vol. 165
ISSN 0079-6123
Copyright © 2007 Elsevier B.V. All rights reserved

CHAPTER 18

Modeling sensorimotor control of human upright stance

Thomas Mergner*

Neurological University Clinic, Neurocenter, Breisacher Street 64, 79106 Freiburg, Germany

Abstract: We model human postural control of upright stance during external disturbances and voluntary lean. Our focus is on how data from various sensors are combined to estimate these disturbances. Whereas most current engineering models of multisensory estimation rely on "internal observers" and complex processing, we compute our estimates by simple sensor fusion mechanisms, i.e., weighted sums of sensory signals combined with thresholds. We show with simulations that this simple device mimics humanlike postural behavior in a wide range of situations and diseases. We have now embodied our mechanism in a biped humanoid robot to show that it works in the real world with complex, noisy, and imperfectly known sensors and effectors. On the other hand, we find that the more complex, internal-observer approach, when applied to bipedal posture, can also yield human-like behavior. We suggest that humans use both mechanisms: simple, fast sensor fusions with thresholding for automatic reactions (default mechanism), and more complex methods for voluntary movements. We suggest also that the fusion with thresholding mechanisms are optimized during phylogenesis but are mainly hardwired in any one organism, whereas sensorimotor learning and optimization is mainly a domain of the internal observers.

Keywords: human; posture; sensor fusion; model; sensorimotor control; top-down approach; robot

Introduction

How can organisms control their behavior when their knowledge of themselves and the world is based on noisy, variable, imperfect sensors? We are studying this question in the context of the upright posture of humans, which is a prerequisite for most of their voluntary movements. Upright stance is subject to many disturbances, including gravity, Coriolis and centrifugal forces, the body's own inertia, pushes and pulls exerted from the outside, and motion of the support surface. In view of all these hazards, one might assume that postural control must be very complicated. But we have shown that human-like behavior can be achieved by remarkably simple mechanisms (Mergner et al., 2002, 2003, 2005; Maurer et al., 2006). Implemented in simulations and in a real, biped humanoid robot ("PostuRob"), these mechanisms mimic the behavior of healthy humans and neurological patients in a wide range of experiments.

The central question is how the brain combines data from multiple imperfect sensors to estimate the disturbances. Nowadays, the usual approach in engineering is to use "internal observers" and this method has been used also in engineering-inspired biological models (e.g., Oman, 1982; Borah et al., 1988; Merfeld et al., 1993; van der Kooij et al., 2001). To explain this point, we refer

*Corresponding author. Tel.: +49 (0) 761 2705313;
Fax: +49 (0) 761 2705203; E-mail: mergner@uni-freiburg.de

DOI: 10.1016/S0079-6123(06)65018-8

to an experience one often makes during voluntary behavior. One has the intuitive notion that one compares expected and actually occurring sensory and motor effects and normally finds that they match. This notion was discussed by von Holst and Mittelstaedt (1950). In their "reafference principle," sensory input is decomposed into self-produced sensory input (reafferent input) and input stemming from external disturbances (exafferent) with the help of knowledge about one's own activity (efference copy). An internal observer makes a similar comparison. It uses an "internal model" of the body, its dynamics, etc., to simulate its behavior in parallel to the actual outside behavior. The difference between the simulated behavior and the actual (monitored by sensors) is iteratively minimized to yield the observer's best guess as to what is really going on. This kind of model is powerful and widely used in engineering. But in what follows we explore the capabilities of a much simpler mechanism, which computes its estimates by summing and weighting sensor signals and putting them through thresholds. The difference in establishing disturbances estimates between the simple sensor fusion concept and the "observer" concept is highlighted in Fig. 1.

There have been a number of earlier attempts to model human postural control by multisensory feedback models (Nashner, 1972; Johansson and Magnusson, 1991; Fitzpatrick et al., 1996). But it remained for Peterka (2002) to validate this approach by demonstrating a close correspondence between human experimental and model simulation data (he did not address, however, the modeling of experimentally observed response nonlinearities and sensory re-weightings). In these previous studies, one may still recognize the classical textbook concept of postural reflexes with essentially direct sensor-actuator couplings. An exception is the model of van der Kooij et al. (1999, 2001) that is based on the internal observer concept with Kalman filter for sensory re-weighting and noise minimization (but still feeds back sensor signals). In contrast, our approach ignores the reflex concept. Instead, it focuses in an "inverse" (top-down) approach on implementing intersensory interaction principles derived from perception studies into sensorimotor control.

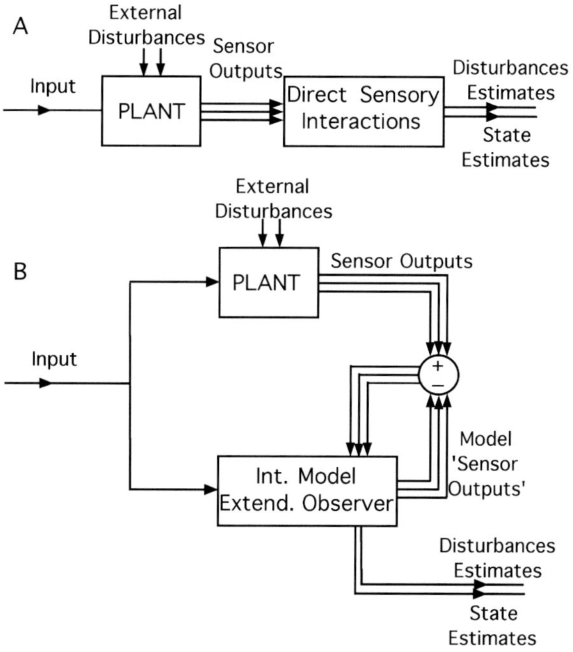

Fig. 1. Schematic presentation of two different concepts of how the brain may derive internal estimates of the external disturbances. (A) Simple "hardwired sensor fusion" concept. The estimates are obtained by direct (forward) intersensory interactions (i.e., without iterations). (B) "Observer" concept. An internal model of the body, its dynamics, etc., in the brain is used to "predict" sensor signals. They are compared to the actual sensor signals (delivered through the real body; Plant). The difference (error) is used in an iterative way to improve the predictions until it essentially becomes zero. Then the brain mechanism is able to deliver estimates of the external disturbances (Extended Observer) for use in sensorimotor control.

Sensors and sensory interaction principles in spatially oriented behavior

Physiological and clinical evidences indicate that humans use mainly *four different sensory inputs* for their spatially oriented behavior (e.g., Horak and Macpherson, 1996). In the following list, we include the technical sensors that are used by PostuRob.

(a) *Vestibular system — Inertial motion sensor.* The vestibular system measures 3D linear and angular body motion. It receives input from two receptor organs, the macular and semicircular canal organs. Both are encapsulated deep within the bone of the skull and thus are advantageous over other force-sensitive receptors in that they are not directly affected by internal or external tissue deformation (by reaction forces). Processing in the brain is thought to yield neural estimates of three quantities: 3D body-in-space rotational velocity, 2D body angle with respect to the gravitational vector, and 3D body-in-space linear velocity (Mergner and Glasauer, 1999; Zupan et al., 2002). Technical equivalents for the macular and canal organs would be a 3D accelerometer and a 3D gyrometer, respectively. They were combined in PostuRob in a biologically inspired vestibular system (see below).

(b) *Joint angle proprioception/sensor*. The human sense of joint position and velocity stems mainly from stretch receptors in muscles. But it may also be influenced by skin and joint-capsule receptors. PostuRob's analog is a goniometer that delivers both position and velocity signals.

(c) *Joint torque proprioception/sensor*. Joint torque is measured by force receptors in the tendons of the muscles that actuate the joint (Duysens et al., 2000). Another source is receptors deep in the foot arch which measure center of pressure (COP) shifts or other aspects of ground reaction forces (compare van der Kooij et al., 2005). The evidence comes from human posture control studies (e.g., Maurer et al., 2000, 2001). In PostuRob the analogous organs were sensors in the insertions of the artificial muscles and a COP monitor measuring pressure on forefoot and heel.

(d) *Visual motion and orientation sensors*. Visual motion stimuli induce body-sway responses (e.g., Mergner et al., 2005). However, the contribution of visual cues to posture control is not a crucial one in the present context. We therefore do not consider it here.

In psychophysical work on human self-motion perception, we have disclosed a number of sensor fusion principles that have been summarized elsewhere (Mergner, 2002). The main finding is that humans combine sensory signals to estimate the external events that evoke the motion. For instance, subjects perceive a passive head rotation on a stationary trunk when signals from vestibular and neck proprioceptive inputs are equal and opposite. When vestibular signals are matched by equal and opposite signals from leg proprioceptors, subjects perceive a body rotation on stationary feet. When vestibular signals are not balanced by any proprioception, subjects perceive that their support surface is rotating (given that haptic cues are indicating contact with a surface).

In our psychophysical studies of vestibular-neck and vestibular-leg interaction in human self-motion perception in the horizontal rotational plane (Mergner et al., 1991, 1993), we used sinusoidal rotation stimuli and compared gain and phase curves of the vestibular responses with simulations of the canal afferents' transfer characteristics known from sensory physiology and the vestibulo-ocular reflex (VOR; showing some prolongation of time constant). We found that, in contrast to these, the gain of the perceptual response is affected by the magnitude of the stimulus, unlike the phase. Alternating between experimental and modeling approaches suggested a central mechanism with the effect of a velocity-detection threshold, whose existence and characteristics could be experimentally determined (see Mergner et al., 1991). The approach and model elements were subsequently extended to cover the neck proprioceptive stimulus and finally various combinations of the two stimuli, until all experimental data could be reproduced in simulations of a "simplest-possible" model. The modeling process also took into consideration certain constraints. For instance, care was taken that it could be applied to other body segments such as the legs (see above) and extended to more than two body segments ("modularity" constraint). Indeed, it could be used to construct a broad and general framework for intersensory interactions in human spatially oriented behavior (see Mergner, 2002). Some of them will be addressed below (e.g., an automatic sensory re-weighting with noise reduction).

Sensor fusions in the sensorimotor control model

As a first step to modeling postural control, we note that under most common conditions, the kinetic equations of the system can be simplified. Body sway is usually small ($<4°$), which allows a small-angle approximation, ANGLE $\approx$ sin(ANGLE). Excursions can be mimicked as occurring predominantly about just one joint, the ankle, so the geometry can be approximated by an "inverted pendulum."

Accordingly, we built our principles for sensor fusion into an inverted-pendulum model of human stance (Fig. 2; simulation software: Simulink, Matlab®). For simplicity, the model is restricted to the sagittal plane. Kinetic aspects are added (impacts of gravity and external contact forces). Fusion-derived estimates of external disturbances are fed into the feedback loop together with a "voluntary lean" set point signal (assuming that volition normally deals with the body's orientation in space). The difference between feedback and set point is fed into a PID controller (P, proportional, I, integrative, and D, differential factors) to produce ankle joint torque by means of actuators (here taken to be ideal; we conceived that mechanisms in the human spinal cord account for non-optimal actuator characteristics such as nonlinearities and establish a single controller mechanism to simplify and unify feedback signals

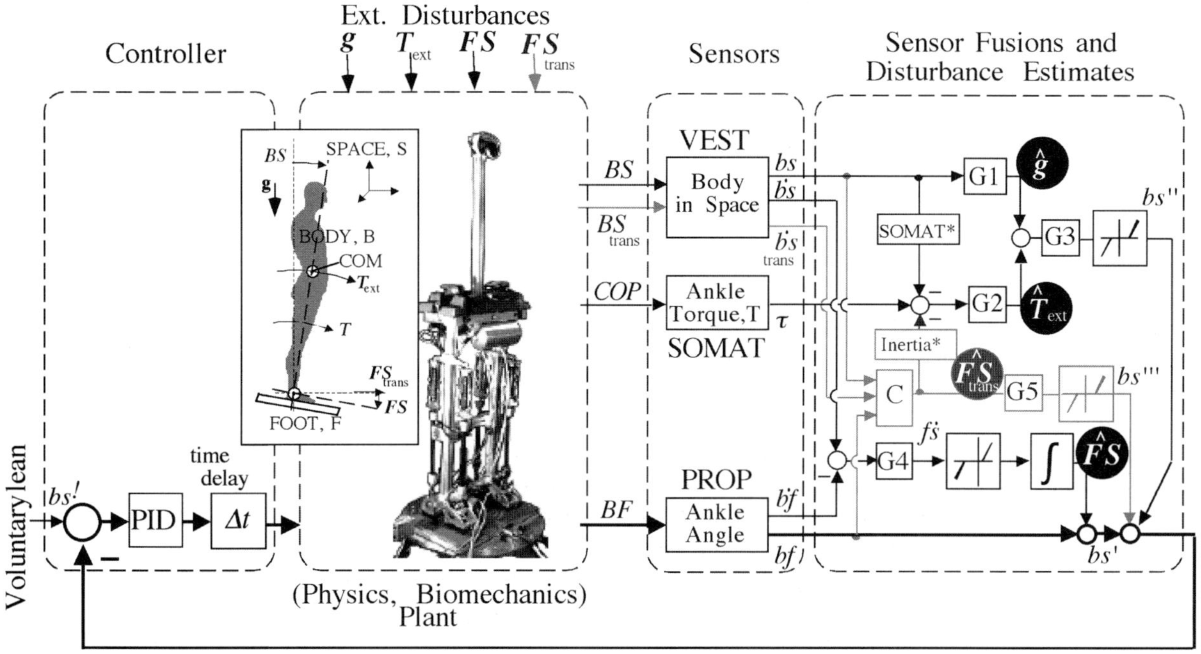

Fig. 2. Model of human posture control in sagittal plane. The model was derived from experimental data in humans and describes these in the form of a sensory feedback stabilization of an inverted pendulum (inset). Its sensorimotor control modules (boxes "Sensor Fusions and Disturbance Estimates" and "Controller") were furthermore used to control a biped humanoid robot (PostuRob; box Plant), where for the sensors (box Sensors) and actuators (not shown) corresponding "biologically inspired" technical devices were used. There are two main principles of the control. First, "*hardwired*" *sensor fusions* (intersensory interactions) yield *internal estimates* of the *external disturbances* (i.e., $\hat{g}$ for gravitational vector g, $\hat{T}_{ext}$ for external contact force T_{ext}, and $\hat{FS}$ for foot-space angle FS upon support-surface tilt; highlighted in circles). In this article we complete the model by adding the fourth external disturbance and its estimate ($\hat{FS}_{trans}$ for foot-space translation FS_{trans} upon support-surface translation; gray lines and symbols). The second main principle is that the estimates, after giving them a body-space angle dimension (bs', bs'', bs'''), are acting in a local proprioceptive feedback loop (bold lines) that receives a body-space angle "Voluntary lean" signal ($bs!$) as set point signal. This yields an "*internal feed forward disturbance compensation.*" Sensors (transfer characteristics omitted): VEST, vestibular sensor, transforming "Body-Space" angle BS into corresponding angular position and velocity signals, bs and $\dot{b}s$ ("Body-Space" translation BS_{trans} into $\dot{b}s_{trans}$); SOMAT, yielding from somatosensory force receptors in the foot, a measure of center of pressure shift, COP, and transforming it into a measure of ankle torque τ; PROP, ankle angle sensor providing measures of body-foot angle and its velocity, bf and $\dot{b}f$. G1–G5, gain and transforming factors (further details in text).

and supra-spinal control signals from many different sources). As evaluated in a number of experiments in humans, it describes normal subjects' and vestibular loss patients' postural responses during various external disturbances (Mergner et al., 2003, 2005; Maurer et al., 2006). For instance, it describes the response gain and phase curves in Bode diagrams obtained during platform tilts and pull stimuli, both with and without body-sway referencing of the platform. It is also the control mechanism used in PostuRob (Mergner et al., 2006).

The main control principle is that the external disturbances are compensated for by (i) creating internal estimates of each disturbance, and (ii) feeding these estimates into a "local" body-foot (*bf*) proprioceptive feedback loop ("local loop," bold lines in Fig. 2). Kinetic disturbances are gravity, g, and external force, T_{ext}. Kinematic disturbances are a change in foot-space angle, *FS*, generated by support-surface tilt about the ankle joint, and translation of the foot in space at the ankle joint, FS_{trans}, upon support-surface translation. Before feeding the estimates into the feedback loop, they are referred to the same coordinates as the set point signal (body orientation in space, bs', in °). The feedback concept may be viewed as an "internal feed forward disturbance compensation". The loop gain per se, determined by the local proprioceptive loop, is very low (compare Fitzpatrick et al., 1996, who first reported a low-loop gain in postural control). But it increases to the extent that a given external disturbance has impact. The low-loop gain allows for considerable delay in the system (in the model, all delays are lumped together in one, 150 ms, taken from experimental data; Maurer et al., 2006).

Although the feedback loop combines both kinematic and kinetic estimates, a kinematic disturbance (e.g., support tilt) leads to a kinematic compensation, i.e., a change in body-foot angle that tends to keep the body upright in space. Correspondingly, a kinetic disturbance (e.g., external pull) leads to a kinetic compensation in terms of a corresponding counter torque in the ankle joint, while the body-space angle is maintained. (Note that the kinematic and kinetic parameters are linked to each other in the form that the body-space angle *BS* determines the gravitational ankle torque, T_{ank}, in the form of $T_{ank} = m \times g \times h \times \sin(BS)$; m, body mass; h, height of m above ankle joint.) Conceivably, the performance of this simple control mechanism depends on the quality of the disturbance estimates (see below). Yet, there is redundancy in the system. Let us assume, for instance, that during an external force (pull) stimulus the ankle torque signal τ and thus the external force estimate $\hat{}T_{ext}$ is too low or too high, for some reason. This leads to an insufficient compensation, and this, in turn, to a body-space excursion with corresponding change of the gravitational ankle torque and its internal estimate ($\hat{}g$; derived from a vestibular signal of body-space angle, *bs*). Since $\hat{}g$ tends to reorient the body upright, it helps to cope with the $\hat{}T_{ext}$ error, so that balancing is nevertheless performed quite well. But the estimation errors remain.

The internal estimate of the support/foot tilt, $\hat{}FS$, in the model is derived from a combination of a vestibular body-space velocity signal and a proprioceptive body-foot velocity signal (via gain factor, velocity threshold, and mathematical integration). Since we are dealing here with signals of coplanar rotations, their combination is by simple vector summation, while in the 3D case it would require a coordinate transformation (see Mergner et al., 1997). This sensor fusion entails an automatic *sensory re-weighting* with two interesting aspects. This is explained using an enlarged section of the model (Fig. 3). There, the proprioceptive sensor measures body-foot position and velocity (*bf*, $b^{\bullet}f$). The vestibular sensor, upon body-space acceleration (second derivative, $\delta\ \delta$, of *BS*) delivers, after an acceleration-to-velocity integration (integration symbol), a measure of body-space velocity ($b^{\bullet}s$). It is assumed that this vestibular signal is very noisy and that the integration accentuates its low-frequency components. In contrast, noise in the *bf* and $b^{\bullet}f$ signals is considered to be small. In the subsequent central processing, the $b^{\bullet}f$ and $b^{\bullet}s$ signals are summed ($f^{\bullet}s = b^{\bullet}s - b^{\bullet}f$) and passed through a velocity threshold that is slightly larger than the vestibular noise, before an integration yields the estimate of foot-space tilt, $\hat{}FS$. Combining the estimate then with the proprioceptive *bf* signal yields a bs' signal

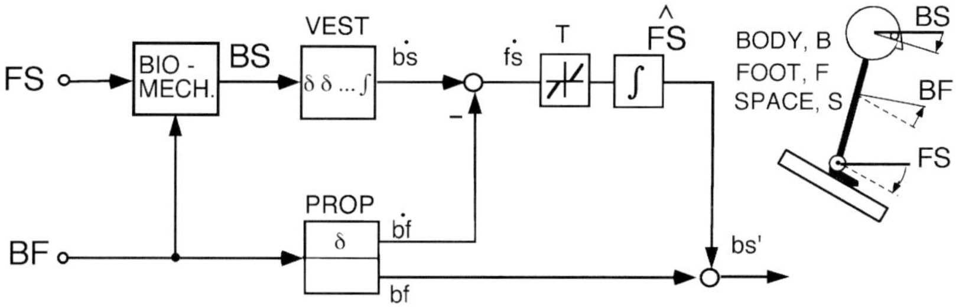

Fig. 3. Automatic sensory re-weighting mechanism (detail from model in Fig. 1; same abbreviations). The body-space output signal *bs'*, which is used for the feedback, results from combining the foot-space angle estimate $\hat{}FS$ with the body-foot angle signal *bf* (by which the local body-foot control becomes upgraded into a body-space control). This mechanism yields that *bs'* is determined by proprioception (*bf*) when the support surface is stationary (since with $FS = 0°$ also $\hat{}FS$ becomes 0°), while *bs'* becomes mainly determined by the vestibular signal $b^{\bullet}s$ during support tilt (since the $b^{\bullet}f$ and *bf* contributions to *bs'* then tend to cancel each other). Furthermore, with the vestibular signal $b^{\bullet}s$ carrying large noise and the proprioceptive signals small noise, the threshold (*T*; in the order of the vestibular noise) yields that the *bs'* signal shows small noise with stationary support and large noise with tilting support. Note that *T* is a velocity threshold and therefore especially effective in reducing low-frequency noise contributions to *bs'*.

(used for the feedback). Note that this *bs'* signal is determined solely by the low-noise proprioceptive input *bf* during body motion on stationary support (since $\hat{}FS = 0°$; note that support stationarity corresponds to the most common behavioral situation). The *bs'* signal involves the noisy vestibular signal only to the extent that there occurs a support tilt. The effect can be viewed as a noise "optimization".

To appreciate the other aspect of the sensory re-weighting in Fig. 3, let us start with the general consideration that a proprioceptive dominated *bs'* signal would be appropriate only with stationary support; since it tends to align the body with respect to the support rather than in space, it would be inappropriate with support tilt. In contrast, a vestibular derived *bs'* feedback signal would be appropriate with either stationary or tilting support, because it always tries to orient the body upright. The sensory re-weighting mechanism in Fig. 3 accounts for this by making the signal *bs'* proprioceptive-determined whenever the support is stationary ($bs' = bf$; $\hat{}FS = 0°$), and it makes it more and more vestibular-determined with increasing support tilt ($bs' \approx \int b^{\bullet}s$). Note from Fig. 3 that the threshold plays an important role in the sensory re-weighting. The thresholds in Fig. 2 were originally introduced to account for considerable amplitude nonlinearities of the stimulus responses (note that they are clearly lower than in the self-motion perception). For a quantitative assessment of their sensory re-weighting effects in humans upon changes in the external stimulus conditions, see Maurer et al. (2006).

In order to stress once more the robustness of the control, we refer to the estimate of the external contact force in Fig. 2, $\hat{}T_{ext}$, which is extracted from a measure of ankle torque, τ (here derived from foot-force receptors and the measure of COP shift; see above). The estimate is obtained by decomposing τ into its constituents, removing from it with the help of vestibular signals the dynamic and static torque components that arise with body excursions (box "Somat*;" see Maurer et al., 2006, for details). $\hat{}T_{ext}$ is then summed with the other kinetic estimate $\hat{}g$ and transformed by a factor (G3) into a kinematic equivalent, *bs''*. Above, we have considered that $\hat{}g$ helps to cope with errors of $\hat{}T_{ext}$. Here we consider the reverse, which is also true. The signal *bs''* receives a contribution from the vestibular *bs* signal not only via $\hat{}g$, but also via the box "Somat*," yet with sign reversal, so that both essentially cancel each other out. Therefore, balancing on stationary support is not affected considerably by an inaccurate vestibular signal. Also upon complete loss of the vestibular sensor, where the signal paths via G1, box "Somat*," and G4 and G5 are set to zero (by some still-to-be-defined mechanisms in the patients), balancing on stationary support is not impaired. In contrast, balancing on tilting support is clearly affected (in the absence of vision), and in a way that is predicted by the model (Maurer et al., 2000, 2006).

In previous versions of the model, support translation as an external disturbance was not included. Yet, the previous model does compensate for this stimulus to the extent that it generates an ankle torque that tends to compensate inertial torque. In this form, however, it does not provide an estimate of the translation stimulus, but rather produces an error of the external force estimate. In the model of Fig. 2, we now have included a foot/support translation (FS_{trans}) and its internal estimate ($\hat{}FS_{trans}$), shown as gray print. The estimation is derived from vestibular signals of body-space translatory velocity and body-space angle, a body-foot angle proprioceptive signal, and known parameters (such as height of vestibular system and center of mass (COM) above joint, etc.; box "C"), and is taken to correct the external torque estimate (via box "Inertia*").

Embodiment of control principles into humanoid

The simple model of Fig. 2, conceivably, does not account for all currently known experimental findings in the human posture control literature, but does account for the types of sensors involved, the major aspects of the sensor fusions, and the control principles. To test this notion, the model was implemented into PostuRob, while making actuator performance essentially ideal. The robot's hardware (Fig. 2, box "Plant") consists of an aluminum frame with two feet, two rigid legs fixed to a pelvic girdle, and a spine, with the main "body" mass being represented by lead weights on the pelvis. It is freely standing on a motion platform that can be tilted or translated about the ankle joints. Each leg carries a front and back pneumatic "muscle" with "tendon" (spring) to move the body with respect to the foot and its support about the "ankle joint". To realize our notion of essentially "ideal actuators", we control the muscle-tendon system using tendon force feedback. The above-described sensors were implemented using appropriate electronics. The boxes "Sensor Fusions and Disturbance Estimates" and "Controller" in Fig. 2 were implemented in a Simulink version on an embedded PC under the control of a host PC. (Further technical specifications are given under www.uniklinik-freiburg.de/neurologie/live/forschung/sensorfusion/PostuRob.html.)

PostuRob was presented with the same experiments as in humans using similar PID factors. It is able to perform "voluntary" body lean movements while, at the same time, support tilt and external force stimuli are applied ("superposition criterion" of Mergner, 2004). Furthermore, it copes with compliant support surface (i.e., standing on foam rubber) and body sway referenced platform by which the proprioceptive feedback loop is opened (compare Maurer et al., 2006). Gain and phase curves of the postural responses closely mimic those of humans (i.e., with some under-compensation; see Tahboub and Mergner, 2007). The robot shows a low-loop gain that allows for delays well above 100 ms and makes the control very "soft" and compliant. Inaccurate sensor and control signals remain without major degradation of performance.

Thus, PostuRob's control architecture with "disturbance compensation" yields a robust and human-like performance. This may well be of interest for engineers who construct multipurpose humanoid robots (a "bionics" aspect). Furthermore, the explicit representation of the disturbances estimates create a meta-level that would allow PostuRob to communicate its knowledge about the outside world with other robots, or to store this information, etc. Future developments of PostuRob will focus on further and alternative control features, additional sensory re-weighting mechanisms, and on the modularity of the system by adding further body segments.

The work with PostuRob alerted us to a number of interesting aspects that did not per se arise from the computer simulations (for instance, unforeseen nonideal sensor performance, signal noise, offsets, drifts, etc., all often slightly changing "spontaneously" over time). Furthermore, we conceive that PostuRob's human-like performance will allow us to use it for developing neurological diagnostic and therapeutic tools using it in a hardware-in-the-loop simulation approach (see Mergner et al., 2006). Furthermore, the approach allows for a direct technical realization of medical sensorimotor aids such as protheses and exoskeletons. The biological background of the control principles likely increases the chance to win patients' compliance.

The fact that PostuRob's simple control concept (based on "hardwired" sensor fusions) differs considerably from contemporary engineering approaches ("observer" concept; see Introduction) led us to compare the two approaches.

Control engineering approach

In collaboration with a colleague from the robotics field, we started a closer inspection and analysis of PostuRob's structure and control architecture and parameters using mathematical models and modern control theory techniques (Tahboub and Mergner, 2007). We designed a state and disturbance estimator and a controller that allows voluntary motion in the presence of the above mentioned external disturbances (support tilt, pull). The goal is again to create a humanlike stance performance. Thus, PostuRob serves as a linkage between the robotics and the biological approach. To ease comparison, the main features of the biological control structure were used for the modeling framework: the architecture was structured as a combination of a tracking controller, sensor fusions, disturbance estimations, and disturbance compensation. However, instead of using the "inverted-pendulum" simplification, we modeled the humanoid fully with its two rigid bodies. The same sensors as before were available. Furthermore, we took into account the friction (or shear) force between the foot and the platform, which we also measured.

For the model, equations of motion were linearized and expressed in state-space form. Geometric and mass parameters were identified and measured in experiments which comprised inclinations of the robot's body and/or tilts of the platform with known angles (1–4°), while measuring platform normal forces and applied torques. The mass moment of inertia was approximated. For the estimation of the external disturbances, measures of body-foot angle (and angular velocity), body acceleration at a known point, and reaction forces were obtained. Using a linearized dynamics model, the estimation was performed by the means of an extended observer (which later may ease application of these methods to higher dimensional systems including more body segments and more degrees of freedom). Thus, in addition to the use of the regular states body-foot angle and its velocity, the support tilt angle and the external pull force were taken as the third and fourth states. Since the latter states are extraneous to the system, a solution with quasi static linear state-space representation was chosen and observability was proven. A set of full-order observers were tested, with the difference between the real measurements and the observer-generated measurements being fed back to force the estimation error to converge to zero.

The control strategy was geared to that of a PID controller as before. However, a different interpretation was given to this controller. It was viewed as a classical robust tracking and disturbance rejection mechanism where the desired input to be tracked is a step input requiring an integral internal model. The PD part is seen as the required state-feedback control inner loop. The closed-loop feedback gains were chosen such that they were similar to those identified in human control.

For the simulations (in a Simulink environment), the eigenvalues of the extended estimator were chosen to be faster than those of the closed-loop controlled system to guaranty "real-time" estimation convergence. The same eigenvalues were used for a set of observers using different measurements. Stability and convergence were demonstrated in all experiments even in the presence of measurement noise and a 100 ms time delay. Employing, in addition to the body-foot angle and angular velocity, measures of the normal force, shear force, or acceleration yielded essentially similar estimation results and similar tracking errors for voluntary motion. Employing all measures at the same time did not improve the estimation considerably. However, when white noise was added (10% of signal amplitude), the use of all measures did yield a clear improvement.

In the robot, the "engineering" control yielded qualitatively similar responses as before the "biological" one (preliminary results). Before reaching detailed conclusions, however, we have to await further extensive testing. An interesting, though preliminary, finding was that the use of PostuRob's artificial vestibular system considerably improves its control, especially if the disturbances include support translation. This sensor system provides measures of body-space angle, angular velocity, and translatory velocity, derived by means of a sensor fusion of gyrometer and accelerometer signals (see above, Sensors). In the following discussion we present some "biologically inspired" steps in the development of the technical sensor system, which vice versa inspired interesting inferences on the biological sensor.

Artificial vestibular system

In higher animals and man, the oculomotor system and the skelotomotor postural system show spontaneous movements in terms of slow fluctuations (low-frequency noise), which appear to stem mainly from vestibular input. In the absence of visual stabilization, the eyes show slow spontaneous drifts. Also the body shows spontaneous sway with a low-frequency preponderance (Carpenter et al., 2001). Furthermore, the aforementioned vestibular self-motion perception shows considerable slow variations over time, unlike its proprioceptive counterpart (Mergner et al., 2001). This perceptual phenomenon is particular in so far that the variability is observed only during motion, whereas rest (i.e., stationarity of the body) is associated with subjective stability of the self and surroundings. We had explained this stability by a relatively high central velocity threshold in the vestibular perception (Mergner et al., 1991). We postulated that the threshold copes for an increase of low-frequency noise of the canal afferent signal, which results from a central prolongation of the time constant (from $T \approx 5$ s to $T \approx 20$ s). When designing PostuRob's artificial vestibular system, we reconsidered this problem.

With angular acceleration as input to a gyrometer (or to a vestibular canal system) and with a further processing of the gyrometer's angular velocity signal to yield angular displacement as output for control purposes, one has overall two integrations (in the mathematical sense). The first (acceleration-to-velocity integration) in the sensor is mechanical (in the canals considered to be "leaky" with 5 s time constant). Theoretically, at least, noise arising at the input site somewhere before the integration is transformed into a $1/f$ noise (meaning that noise amplitude increases, the lower the frequency is). The second integration (velocity-to-position) transforms this then into $1/f^2$ noise, and it transforms the noise that arises after the first integration into $1/f$ noise. Low-frequency noise is especially detrimental to function, because the motor behavior occurs mainly in the mid- to low-frequency range. How to deal with this problem? In the engineering field, one performs traditionally a high-pass filtering of the gyrometer signal. Our "biologically inspired" approach is similar, but goes beyond it.

In PostuRob's artificial vestibular system, the gyrometer output at rest shows noise with a clear preponderance at low frequency, albeit not exactly with a $1/f$ characteristics (Fig. 4A; $1/f^2$ in power spectrum histograms). The low-frequency preponderance becomes stronger after the velocity-to-position integration (Fig. 4B, 1). The latter signal showed very slow drifts that were eliminated by high-pass filtering the gyrometer signal with a 100 s time constant (which is in the order of the vestibular adaptation time constant) (Fig. 4B, 2). When further adding after this a high-pass filtering with $T = 5$ s (to mimic the "leak" in the canal's acceleration-to-velocity integration), a pronounced further reduction in noise resulted (Fig. 4B, 3). Introducing finally a velocity threshold led to still further noise reduction (Fig. 4B, 4).

Conceivably, a gyrometer response to a 0.2 Hz rotation stimulus is not affected considerably by these procedures (apart from a slight reduction through the velocity threshold). In contrast, a response at 0.02 Hz would be clearly affected. This was accounted for concerning the vertical planes by a sensor fusion with accelerometer signals, which provided the missing low frequency response components. The method we used is based again simply on a "hardwired" fusion (described elsewhere; Mergner and Glasauer, 1999). The fusion also decomposes gravitational and translational (inertial) components of the accelerometer

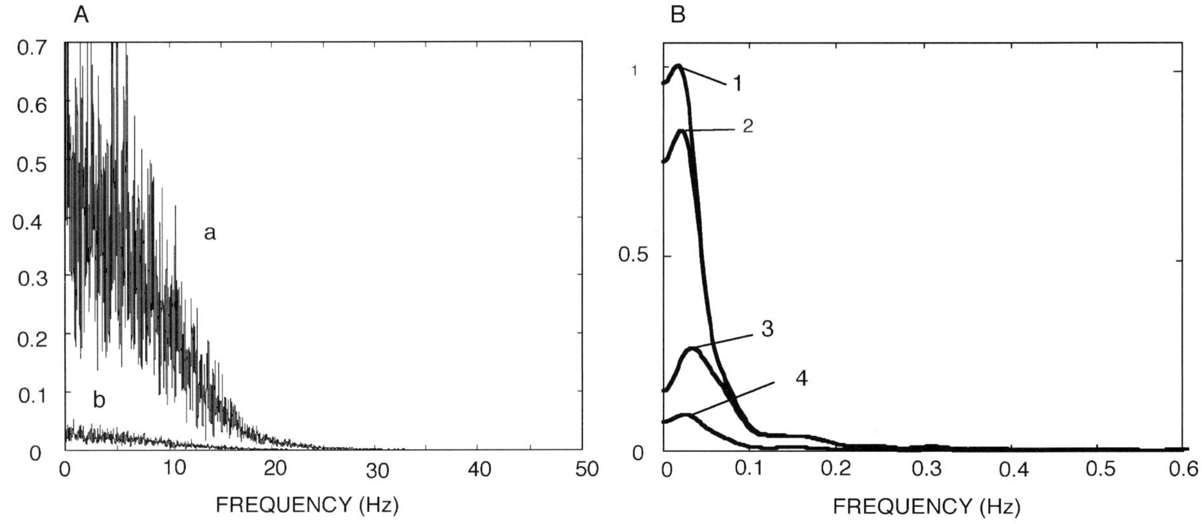

Fig. 4. "Biologically inspired" noise reduction of gyrometer readings (taken to represent vestibular canal signals) for PostuRob's artificial vestibular system. Plots of power spectrum histograms, PSH, of 400 s time series. (A) Superposition of PSP derived from one gyrometer signal (a) (ADXRS401, Analog Devices) and of a PSP derived from averaging 16 gyrometer signals (b). Note the low-frequency preponderance of the noise, presumed to be related to the acceleration-to-velocity integration of the sensor's operation. The clear reduction of the low-frequency noise by the averaging demonstrates that the noise sources were essentially independent of each other. (B) Effect of subjecting the averaged gyrometer signal to a velocity-to-position integration (1) and introducing, in addition, an adaptation time constant (2; high-pass filtering, $T = 100$ s, which is similar to the human canal adaptation time constant) which entails some noise reduction. More noise reduction is obtained by adding a further high-pass filtering, mimicking the "leak" in the acceleration-to-velocity integration with $T = 5$ s (3) and a velocity threshold (4; 0.5°/s). We conceive that nature uses such a bundle of "hardwired" noise reduction measures for the use of the canal signal in a canal-otolith sensor fusion (see text).

signals. Overall, it yields the bs, $b^{\bullet}s$, and $b^{\bullet}s_{\mathrm{trans}}$ signals shown in Fig. 2. Noticeably, there are comparable canal-otolith interaction models in the literature, but again these use or build upon the iterative "observer" concepts inspired by contemporary engineering methods (Merfeld et al., 1993; Zupan et al., 2002).

Discussion

In our "inverse" approach, we evaluated sensorimotor control of human spatially oriented behavior in a "top-down" rather than in a "bottom-up" way. We proceeded from the general notion that, in the metaphor of "the whole and its elements," different ways of combining the elements in a synthetic way would be possible, with differing results as to the quality of the whole (meaning here behavior). However, as long as the quality of the whole is not defined, the choice of the bottom-up synthetic way is arbitrary. To overcome this problem, we first performed behavioral experiments and studied the interaction between sensory signals. As already mentioned, we proceeded from the assumption that, within the sensorimotor framework chosen, the motor aspects are correctly performed.

One could object to this approach because the choice of the synthetic way is not arbitrary as, actually, demonstrated by our studies. They show that the disclosed sensor fusion mechanisms reconstruct the external physics. The internal estimates of the external disturbances are then used to neutralize the disturbances and this allows for undisturbed volition. From this viewpoint, the solution represents the most straight-forward way of dealing with sensorimotor control and thus is trivial. However, as described elsewhere (Mergner, 2002), the internal reconstruction of the physics is not necessarily straightforward. For instance, vestibular information that is used in a retrospective spatial arm-pointing task is not directly transformed from head via trunk to arm coordinates, but the transformation is via an estimate of the kinematics of the body support surface (Mergner et al., 2001).

Furthermore, this objection represents a post hoc view and does not acknowledge the novelty of the present concept as compared to the postural reflex concept in current textbooks where the sensor signals are coupled directly to the motor system instead of using them for disturbance estimation (compare Mergner, 2004). And this view does not consider the problem that arises from information loss due to non-ideal sensors and from noise, drifts, etc. It is mainly this problem that led engineers to establish elaborate mechanisms that can deal with inaccurate and changing sensory information. These include the aforementioned iterative "observer concept" (see Introduction). The fact that also the simple "hardwired sensor fusion" concept successfully deals with the problem (by way of the described "automatic sensory re-weighting with noise optimization") is certainly not trivial. The problem is dealt with, again in a "hardwired" way, already at the level of the sensors (see above, Artificial vestibular sensor).

Similarly, the use of the disturbance estimates for an "internal feed forward disturbance compensation" (by feeding them into the local proprioceptive feedback loop) is not straightforward. This implementation creates a number of noteworthy features. One consequence, for instance, is that loop gain is low, thus allowing for considerable feedback delay. Because of this, PostuRob shows a soft and humanlike response "behavior." Furthermore, the resulting network yielded the described remarkable robustness of the control (and, as to be shown in the future, eases modularity of the concept). Finally, future applications of the concept may profit from the fact that the disturbance estimates provide a meta-level for communicating or storing the robot's estimates about the outside world and, vice versa, at this level stored or communicated knowledge can be fed into the control for correcting or improving the estimates.

Another objection could be that the simple "hardwired sensor fusion" control is actually the result of our modeling attempt to simplify the model as far as possible (following "Occam's razor" rule; Gibbs and Sugihara, 1996/1997) rather than of a simplification performed by nature during phylogenesis. Therefore, one cannot exclude that the mechanism actually is more complex (which, indeed, is true). For instance, one could conceive that the same functionality is achieved with control mechanisms that extract the disturbance estimates by way of the "observer concept." It has been shown that such models can describe different types of sensorimotor control and that, with certain extensions, they show adaptive properties, which is one of the most outstanding features of brain function (Tin and Poon, 2005). A postural control model of this kind with adaptive Kalman filter for sensory re-weighting and noise minimization has, indeed, been suggested by van der Kooij et al. (2001), as we mentioned before.

However, since we have not investigated adaptation in our experimental studies and have no data from which we might derive its rules, we should not include it into our model. Yet, we conceive that adaptive properties play an important role, together with cognitive mechanisms that have been shown to also contribute to posture control (e.g., Blumle et al., 2006). We like to point out, however, that the "hardwired sensor fusion" model would be prepared to incorporate adaptive and cognitive mechanisms (via the internal estimates as interface or further sensory re-weighting mechanisms).

It also contains internal models (internal representations) of the outside world and the body, some of which have to be parameterized and can be adjusted. These models relate, however, solely to the information pick-up by the sensors (e.g., the topology of the wiring relates to the way in which the physical stimuli interact, and certain parameters account for body mass, gravity, etc.) and are not used to internally create a "virtual body motion" as in the "observer" concept.

We see in the above described "hardwired sensor fusion" concept a simple, fast, and non-iterative alternative to the "observer" concept. It is possible that the latter shows advantages as concerns mathematical treatment and engineering applicability (this appears to relate primarily to research and not to praxis, however). Yet, these aspects tell nothing about the biological plausibility. We try to resolve this confrontation in terms of two alternatives by hypothesizing their coexistence. In this framework, the simple "hardwired" concept would represent the fast, basic mechanism (optimized during phylogenesis and possibly already "pre-wired" at birth), on which a more complex "observer" concept is superimposed. A fawn, for example, might use the basic simple mechanism for posture control when forced to flee shortly after birth (i.e., before it could learn much). The adult individual may still automatically resort to this as a default mechanism in response to unforeseen external disturbances. The superimposed "observer" mechanism, in contrast, might be used mainly during voluntary ("proactive") movements for learning and optimization and not so much during compensatory ("reactive") movements that aim to maintain posture and are more stereotype.

This framework might explain certain differences between the "proactive" and "reactive" types of movements described in the literature. For example, the control of a voluntary targeting of the eyes or the head in space tends to be restricted to two degrees of freedom, which may be learned in order to minimize computational effort, for instance, whereas 3D external disturbances of these movements are compensated for in 3D (compare Tweed, 2003).

It is to say, however, that evidence for the "efference copy" (EC) concept, at least so far, is primarily related to the oculomotor system. There, for some unknown reasons, kinematic and kinetic proprioception appears to contribute little to ongoing sensorimotor control, so that most researchers use in modeling an EC signal in substitution for the proprioceptive eye-head position or velocity signal. Otherwise, there is not much evidence for the EC concept in sensorimotor control, to my knowledge. Reported evidence relates mainly to higher levels of brain function that include perception, such as in the aforementioned intuitive example where we said that one tends to compare expected and actually occurring sensory inputs.

The simple "hardwired" concept has had its "baptism of fire" already. When we "embodied" it into PostuRob, we found that it works and that the robot shows the expected humanlike response behavior. It is robust against noise, signal drifts, etc., and fulfils the aforementioned "superposition criteria" when combining different stimuli and volitional lean. We conceive that its medical use in the form of "hardware-in-the-loop simulations" may help to better understand normal stance control, its impairment in patients, and the effects of therapies and may help to design new therapies and medical sensorimotor aids (see Mergner et al., 2006). Furthermore, it may give inspirations to colleagues from the robotics field (where the engineering attempts to create humanlike biped walking have not been very successful, to date). The embodiment of the "observer" concept into PostuRob still has to prove its validity. A preliminary finding is that it can do a similar job, but a detailed comparison is still missing. We expect from the comparison also considerable insights into presumed biological sensorimotor control concepts, which brings us back to our idea of an "inverse" approach (Introduction).

Given that both concepts turn out to be functionally equivalent in most respects, one may ask which is then the one that is used by humans for a given task. Waiting for evidence from electrophysiological and functional imaging research is, in the short term, not very promising. We speculate that some hint may be obtained from behavioral observations. Both the "hardwired sensor fusion" concept and the "observer" concept allow one to predict postural responses after loss of vestibular sensor function. The "observer" concept, as currently realized in our approach at least, would predict little functional impairment. The prediction of the "hardwired sensor fusion" concept is similar for situations where the support surface is stationary. However, it predicts inadequate postural responses upon platform tilts (with eyes closed). This is in line with our experimental findings. Blindfolded vestibular loss patients, unlike normal subjects, fall with fast platform tilts in a way as predicted (Maurer et al., 2006). It therefore appears that non-vestibular graviceptive sensors, for instance visceral ones (Mittelstaedt, 1996) or force/torque related ones (see above), can make only a partial substitution (essentially static or low frequency). Thus, an elaborate iterative disturbance estimation is here apparently not performed, at least not to the extent that the disturbance responses become normal. This may depend, conceivably, on the currently available processing resources. Vestibular loss patients' ability to consciously detect and to correctly interpret horizontal body rotations is impaired as compared to intact subjects, and this clearly is more when they are involved in performing some task (Schweigart et al., 1993).

Last but not least, we discuss our observations made with the "biologically inspired" vestibular system. The situation here is essentially parallel to the above considerations, in that we realized a "hardwired sensor fusion" concept, while others favor corresponding "observer" concepts (Merfeld et al., 1993; Zupan et al., 2002). However, we like to focus here on a different aspect, which is our notion that an embodiment of the biological concepts into a simple technical device may have an impact on our understanding of these concepts. We considered that a gyrometer is, at least to some extent, an equivalent to the vestibular canal system. We mimicked the "leak" of the acceleration-to-velocity integration ($T \approx 5$ s), known from the afferent canal signal, in the gyrometer signal (technically specified as having DC sensitivity) by applying a corresponding high-pass filtering. The measured noise of the gyrometer's velocity signal, which showed a clear low-frequency preponderance, was clearly reduced by this filtering, as especially evident in the corresponding angular displacement signal.

Our inference back for the biological sensor would be that the "leak" in the canal signal processing does not so much reflect a functional deficit, but rather a useful "noise reduction measure," similar as the other two "biological" steps we applied, i.e., the inclusion of an adaptation time constant and a velocity detection threshold. Future research may readdress this aspect in relation to the canal's geometry. The sensor has essentially not changed form and dimensions since its invention early in phylogenesis in fishes and across many species with vastly different body sizes. This feature was taken in a recent study (Squires, 2004) to ask whether the canal's dimensions reflect a design by nature for optimal sensitivity. Taking into account the basic building materials and physical and physiological operation constraints, the author calculated the optimal canal geometry and concludes that the result of his calculation corresponds largely to the biological realization. We would hold that such a consideration of the canal's optimal sensitivity should include the "leak" versus noise aspect.

In summary, implementation of simple sensor fusion principles derived from perception studies into a sensorimotor control model of human stance led to a powerful solution that could successfully be embodied into a humanoid robot and bears well comparison with much more complex engineering solutions. Our interplays between biological and engineering approaches yielded interesting and novel vistas from which both approaches profit. As a hypothesis for the future, we conceive that a simple and fast fusion-with-thresholding control coexists in parallel to a more complex observer-based control in humans. From this we like to suggest such a dual control also for humanoid robots, with a fast and essentially hardwired sensor fusion control (mostly passive, electronics-based) and a more complex and flexible software-based (active) control, such that the former takes over when the latter is too slow or fails.

Abbreviations

$BF/bf/b^{\bullet}f$	body-foot angle/internal proprioceptive signal of body-foot angle/velocity
bs'	voluntary lean command signal (a body-space signal)
$BS/bs/b^{\bullet}s$	body-space angle/vestibular signal of body-space angle/velocity
$BS_{trans}/b^{\bullet}s_{trans}$	body-space translation/vestibular signal of body-space translatory velocity
COM	center of mass
COP	center of pressure
EC	efference copy

$FS/fs/\dot{f}s$	foot-space angle/internal signal of foot-space angle/velocity
$FS_{trans}/\hat{}FS_{trans}$	foot-space translation/internal estimate of it
$g/\hat{}g$	gravity/estimate of gravitational contribution to ankle torque
G1–G5	internal gain and transforming factors
PROP	proprioceptive sensor
SOMAT/"Somat*"	somatosensory sensor/internal model of it
$T_{ext}/\hat{}T_{ext}$	external force/internal estimate of it
τ	ankle torque signal
VEST	vestibular sensor
VOR	vestibulo-ocular reflex

Acknowledgments

Supported by DFG Me 715/5-3. I like to thank W. Becker and D. Tweed for helping me to improve the manuscript.

References

Blumle, A., Maurer, C., Schweigart, G. and Mergner, T. (2006) A cognitive intersensory interaction mechanism in human postural control. Exp. Brain Res., 173: 357–363.

Borah, J., Young, L.R. and Curry, R.E. (1988) Optimal estimator model for human spatial orientation. Ann. N.Y. Acad. Sci., 545: 51–73.

Carpenter, M.G., Frank, J.S., Winter, D.A. and Peysar, G.W. (2001) Sampling duration effects on centre of pressure summary measures. Gait Posture, 13: 35–40.

Duysens, J., Clarac, F. and Cruse, H. (2000) Load-regulating mechanisms in gait and posture: comparative aspects. Physiol. Rev., 80: 83–133.

Fitzpatrick, R., Burke, D. and Gandevia, S.C. (1996) Loop gain of reflexes controlling human standing measured with the use of postural and vestibular disturbances. J. Neurophysiol., 76: 3994–4008.

Gibbs, P. and Sugihara, H. (1996/1997) What is Occam's Razor? Usenet Physics FAQ, http://math.ucr.edu/home/baez/physics/General/occam.html

von Holst, E. and Mittelstaedt, H. (1950) Das Reafferenzprinzip (Wechselwirkungen zwischen Zentral nervensystem und Peripherie). Naturwissenschaften, 37: 464–476.

Horak, F.B. and Macpherson, J.M. (1996) Postural orientation and equilibrium. In: Rowell L. and Shepherd J. (Eds.), Handbook of Physiology I: Excercise, Regulation and Integration of Multiple Systems. Oxford University Press, New York, pp. 255–292.

Johansson, R. and Magnusson, M. (1991) Human postural dynamics. Biomed. Engin., 18: 413–437.

van der Kooij, H., van Asseldonk, E. and van der Helm, F.C.T. (2005) Comparison of different methods to identify and quantify balance control. J. Neurosci. Methods, 145: 175–203.

van der Kooij, H., Jacobs, R., Koopman, B. and Grootenboer, H. (1999) A multisensory integration model of human stance control. Biol. Cybern., 80: 299–308.

van der Kooij, H., Jacobs, R., Koopman, B. and van der Helm, F. (2001) An adaptive model of sensory integration in a dynamic environment applied to human stance control. Biol. Cybern., 84: 103–115.

Maurer, C., Mergner, T., Bolha, B. and Hlavacka, F. (2000) Vestibular, visual, and somatosenory contributions to human control of upright stance. Neurosci. Lett., 281: 99–102.

Maurer, C., Mergner, T., Bolha, B. and Hlavacka, F. (2001) Human balance control during cutaneous stimulation of the plantar soles. Neurosci. Lett., 302: 45–48.

Maurer, C., Mergner, T. and Peterka, R.J. (2006) Multisensory control of human upright stance. Exp. Brain Res., 171: 231–250.

Merfeld, D.M., Young, L., Oman, C. and Shelhamer, M. (1993) A multi-dimensional model of the effect of gravity on the spatial orientation of the monkey. J. Vestib. Res., 3: 141–161.

Mergner, T. (2002) The Matryoshka Dolls principle in human dynamic behavior in space: a theory of linked references for multisensory perception and control of action. Curr. Psychol. Cogn., 21: 129–212.

Mergner, T. (2004) Meta level concept versus classic reflex concept for the control of posture and movement (in honour of O. Pompeiano). Arch. Ital. Biol., 142: 175–198.

Mergner, T. and Glasauer, S. (1999) A simple model of vestibular canal-otolith signal fusion. Ann. N.Y. Acad. Sci., 871: 430–434.

Mergner, T., Hlavacka, F. and Schweigart, G. (1993) Interaction of vestibular and proprioceptive inputs. J. Vestib. Res., 3: 41–57.

Mergner, T., Huber, W. and Becker, W. (1997) Vestibular-neck interaction and transformations of sensory coordinates. J. Vestib. Res., 7: 119–135.

Mergner, T., Huethe, F., Maurer, C. and Ament, C. (2006) Human equilibrium control principles implemented into a biped robot. In: Zielinska, T. and Zielinski, C. (Eds.), Robot Design, Dynamics, and Control (Romansy 16, Proceedings of the sixteenth CISM-IFToMM symposium), CISM courses and lectures No. 487, Springer, Wien, NY, pp. 271–279.

Mergner, T., Maurer, C. and Peterka, R.J. (2002) Sensory contributions to the control of stance: a posture control model. Adv. Exp. Med. Biol., 508: 147–152.

Mergner, T., Maurer, C. and Peterka, R.J. (2003) A multisensory posture control model of human upright stance. Prog. Brain Res., 142: 189–201.

Mergner, T., Nasios, G., Maurer, C. and Becker, W. (2001) Visual object localization in space: interaction of retinal, eye position, vestibular and neck proprioceptive information. Exp. Brain Res., 141: 33–51.

Mergner, T., Schweigart, G., Maurer, C. and Blümle, A. (2005) Human postural responses to motion of real and virtual visual environments under different support base conditions. Exp. Brain Res., 167: 535–556.

Mergner, T., Siebold, C., Schweigart, G. and Becker, W. (1991) Human perception of horizontal head and trunk rotation in space during vestibular and neck stimulation. Exp. Brain Res., 85: 389–404.

Mittelstaedt, H. (1996) Somatic graviception. Biol. Psychol., 42: 53–74.

Nashner, L.M. (1972) Vestibular postural control model. Kybernetik, 10: 106–110.

Oman, C. (1982) A heuristic mathematical model for the dynamics of sensory conflict and motion sickness. Acta Otolaryngol. Suppl., 392: 1–44.

Peterka, R.J. (2002) Sensorimotor integration in human postural control. J. Neurophysiol., 88: 1097–1118.

Schweigart, G., Heimbrand, S., Mergner, T. and Becker, W. (1993) Role of neck input for the perception of horizontal head and trunk rotation in patients with loss of vestibular function. Exp. Brain Res., 95: 533–546.

Squires, T.M. (2004) Optimizing the vertebrate vestibular semicircular canal: could we balance any better? Phys. Rev. Lett., 93: 1981061–1981064.

Tahboub, K. and Mergner, T. (2007) Biological and engineering approaches to human postural control. Integrated Comput. Aided Engin., 14: 15–31.

Tin, C. and Poon, C.-S. (2005) Internal models in sensorimotor integration: perspectives from adaptive control theory. J. Neural Eng., 2: S147–S163.

Tweed, D. (2003) Microcosmos of the Brain: What Sensorimotor Systems Reveal about the Mind. Oxford University Press, Oxford, NY.

Zupan, L., Merfeld, D.M. and Darlot, C. (2002) Using sensory weighting to model the influence of canal, otolith and visual cues on spatial orientation and eye movements. Biol. Cybern., 86: 209–230.

P. Cisek, T. Drew & J.F. Kalaska (Eds.)
Progress in Brain Research, Vol. 165
ISSN 0079-6123
Copyright © 2007 Elsevier B.V. All rights reserved

CHAPTER 19

Dimensional reduction in sensorimotor systems: a framework for understanding muscle coordination of posture

Lena H. Ting*

The Wallace H. Coulter Department of Biomedical Engineering at Georgia Tech and Emory University, 313 Ferst Drive, Atlanta, GA 30332-0535, USA

Abstract: The simple act of standing up is an important and essential motor behavior that most humans and animals achieve with ease. Yet, maintaining standing balance involves complex sensorimotor transformations that must continually integrate a large array of sensory inputs and coordinate multiple motor outputs to muscles throughout the body. Multiple, redundant local sensory signals are integrated to form an estimate of a few global, task-level variables important to postural control, such as body center of mass (CoM) position and body orientation with respect to Earth-vertical. Evidence suggests that a limited set of muscle synergies, reflecting preferential sets of muscle activation patterns, are used to move task-variables such as CoM position in a predictable direction following postural perturbations. We propose a hierarchal feedback control system that allows the nervous system the simplicity of performing goal-directed computations in task-variable space, while maintaining the robustness afforded by redundant sensory and motor systems. We predict that modulation of postural actions occurs in task-variable space, and in the associated transformations between the low-dimensional task-space and high-dimensional sensor and muscle spaces. Development of neuromechanical models that reflect these neural transformations between low- and high-dimensional representations will reveal the organizational principles and constraints underlying sensorimotor transformations for balance control, and perhaps motor tasks in general. This framework and accompanying computational models could be used to formulate specific hypotheses about how specific sensory inputs and motor outputs are generated and altered following neural injury, sensory loss, or rehabilitation.

Keywords: muscle; balance; EMG; muscle synergy; motor control; biomechanics; feedback; sensorimotor integration

Postural control is a fundamental motor task ideally suited for investigating questions of sensorimotor integration and redundancy. The ability to maintain posture and balance are precursors to other voluntary movements such as reaching or walking over uneven terrain. Moreover, loss of balance is a clinically important problem, as falls are a primary cause of injury and accidental death in older adults (Minino et al., 2002). Yet, we currently have little understanding of the underlying neuromechanical principles that govern patterns of muscle activation during postural control or other basic motor behaviors.

*Corresponding author. Tel.: +1 404 894 5216; Fax: +1 404 385 5044; E-mail: lting@emory.edu

DOI: 10.1016/S0079-6123(06)65019-X

Although technological advances allow the simultaneous measurement of multiple kinematic, kinetic, and electromyographic (EMG) data channels during behavioral experiments, we lack a framework for understanding how all of these measured variables are related to the control and performance of a functional task. Without this basic understanding, we cannot begin to understand or predict how patterns of muscle activation *should* be altered to perform novel tasks, nor can we understand the functional impact of disordered patterns of muscle activation in neurologically impaired individuals. In addition to making multiple measurements, advances in motor control must reveal *why* a particular pattern of muscle activation is chosen by the nervous system to achieve task goals. A quantitative framework for understanding the neuromechanical interactions and sensorimotor transformations during standing postural control is critical to unraveling the underpinnings of this important behavior.

A general framework will be presented that can be used to formulate specific hypotheses about sensorimotor transformations, and provide an organizational scheme for formulating computational and experimental studies of postural control. Computer simulations of the neuromechanical transformations from sensory input to motor output are critical for understanding the neural mechanisms underlying spatial and temporal patterns of muscle activation. Such simulations can also serve as a virtual test-bed for quantifying the functional impact of neurological disorders on postural control. Moreover, the framework has significant implications for understanding and evaluating experimentally measured changes in muscle activation patterns due to learning, adaptation, injury, or disease. The general principle of dimensional reduction may be common to many motor control processes and can guide our approaches to understanding and improving motor dysfunction.

First, the fundamental "degrees of freedom" problem and its relevance to postural control will be reviewed. Then, experimental evidence that establishes the critical role of task-level sensorimotor integration processes during standing balance will be presented. Next, findings demonstrating that muscle activation patterns used during postural control can be simplified to combinations of a few muscle synergies — patterns of muscle activity used to control task-level biomechanical variables — will be discussed. Finally, a framework that integrates these observations of dimensional reduction in sensorimotor signals during postural control will be presented.

Degrees of freedom problem

To maintain standing balance, the nervous system must confront the classic "degrees of freedom" problem posed by Nikolai Bernstein (1967), where many different solutions to a task are available due to the large number of elements that need to be controlled, or degrees of freedom, in the system. In postural control, muscles and joints across the limbs, trunk, and neck must be coordinated to maintain the body's center of mass (CoM) over the base of support, typically formed by the feet. The many degrees of freedom afforded by the joints and muscles allows for multiple (i.e., redundant) solutions, allowing the nervous system flexibility in performing the postural task. This redundancy poses a problem to the nervous system: it must choose from a large set of possible solutions because the task requirements are not sufficient to uniquely specify how each muscle and joint must be controlled.

Bernstein proposed a neural strategy for simplifying the control of multiple degrees of freedom by coupling, or grouping, output variables at the kinematic level (Bernstein, 1967). This scheme was based on experimental observations that multiple joint angles appear to be controlled together, rather than independently, during motor tasks. For example, during running, the hip, knee, and ankle joints all flex and extend at the same time, suggesting that they are not controlled independently. This covariation of joint angles has the effect of moving the CoM vertically in a simple motion that mimics the bouncing of a spring and mass system (Blickhan, 1989; McMahon and Cheng, 1990; Farley et al., 1993). In walking, the lower limb joint angles covary in a different pattern such that the overall motion of the CoM resembles that of an inverted pendulum (Cavagna et al., 1977; Minetti, 2001). Therefore, the overall effect of such joint angle covariations, or *kinematic synergies*, may be to produce a predictable and simple motion of the task-variable at hand — in the case of locomotion, the trajectory of the CoM.

Yet, the strict covariation of joint angles *themselves* does not appear to be the end-goal of the nervous system computation. Task-variables such as the CoM trajectory in postural control, or the finger trajectory in pointing movements, appear to be more precisely controlled by the nervous system than individual joint angles (Scholz and Schoner, 1999; Scholz et al., 2000), suggesting that the task-variables have special significance to the nervous system.

The neural principles and mechanisms underlying our ability to control task-variables within a high degree of freedom system are still unknown and require further investigation into the specific sensorimotor transformations that facilitate these behaviors. While kinematic observations can identify important correlations and task-variables controlled by the nervous system, they do not directly specify which muscle activation patterns should be used to produce the movements. This problem arises because Bernstein's degrees of freedom problem also exists in the transformation between muscle activation patterns and kinematic patterns of movement. This additional redundancy in the musculature results not only because multiple muscles cross each joint, but also because the biomechanical equations of motion are such that different temporal patterns of muscle activation can lead to similar joint trajectories (Gottlieb, 1998). Therefore both spatial and temporal muscle activation patterns have a degree of redundancy that must be managed by the nervous system. To gain a deeper understanding of the underlying neural mechanisms for controlling task-variables, the complex spatiotemporal coordination of multiple muscles and their effect on biomechanical task outputs must be considered.

We propose that the reduction of degrees of freedom observed at the biomechanical level reflects a reduction in degrees of freedom at the level of the neural circuits that activate muscles. *Muscle synergies* could be a mechanism through which the nervous system achieves repeatable and correlated multi-joint coordination. We define muscle synergies to be a specific pattern of muscle coactivation. Each muscle synergy is presumed to be controlled by a single neural command signal, which modulates the overall magnitude of the patterns specified by a muscle synergy. Although muscle synergies simplify spatial coordination of muscles, temporal variations of the neural command signal must still be specified to achieve a motor task. We will address possible mechanisms for simplification of both spatial and temporal muscle activation patterns.

Postural responses to perturbations

The overall goal in standing equilibrium can be simply defined as maintaining the CoM over the base of support; however, there are multiple strategies for accomplishing this goal. For example, it is possible to extend the base of support by taking a step or using the hands to hold on to a stable object (Horak and Macpherson, 1996; Maki et al., 2003). The ability to choose an appropriate postural control strategy reflects complex and integrative sensorimotor processes. Successful balance control depends on having accurate knowledge of the entire body configuration in space, as well as the location of the body CoM relative to the line of gravity and the base of support. The activation of muscles in response to a perturbation results from integration of multiple sensory signals to properly estimate CoM displacement and Earth-vertical. This role of local sensory signals in estimation of critical task-level variables is also illustrated by psychophysical experiments whereby perturbations to a single sensory channel create illusions of shifting Earth-vertical, or the entire body orientation (Mergner and Rosemeier, 1998; Hlavacka et al., 2001; Scinicariello et al., 2002; Mergner et al., 2003; Hatzitaki et al., 2004). While postural responses themselves are thought to be integrated in the brainstem and cannot be voluntarily suppressed following a perturbation, postural strategy selection and postural response amplitude can depend on many different descending, cognitive, and emotional influences, such as habituation, divided attention, or fear (Keshner et al., 1987; Woollacott and Shumway-Cook, 2002; Carpenter et al., 2006).

Even within a specific postural response strategy, the rapid activation of muscles to stabilize the body CoM is finely tuned to the biomechanics of the perturbation, in particular, direction. When the support surface is translated in each of many directions in the horizontal plane, simulating a "slip," multiple muscles across the body are activated by an amount related to the direction of the perturbation (Nashner, 1976; Macpherson, 1988; Horak and Macpherson, 1996; Henry et al., 1998). For example, a different set of muscles is activated in response to a forward perturbation versus a backward perturbation (Fig. 1A). The directional sensitivity of the postural response can be represented as muscle tuning curves that illustrates the response amplitude as a function of direction and demonstrate that little co-contraction of antagonist muscles occurs (Fig. 1B). Each muscle also has a unique tuning curve, demonstrating that postural responses are not fixed response patterns (Macpherson, 1988; Horak and Macpherson, 1996; Ting and Macpherson, 2004, 2005). The tuning of muscle activation is already evident in the initial *automatic postural response* (APR) that

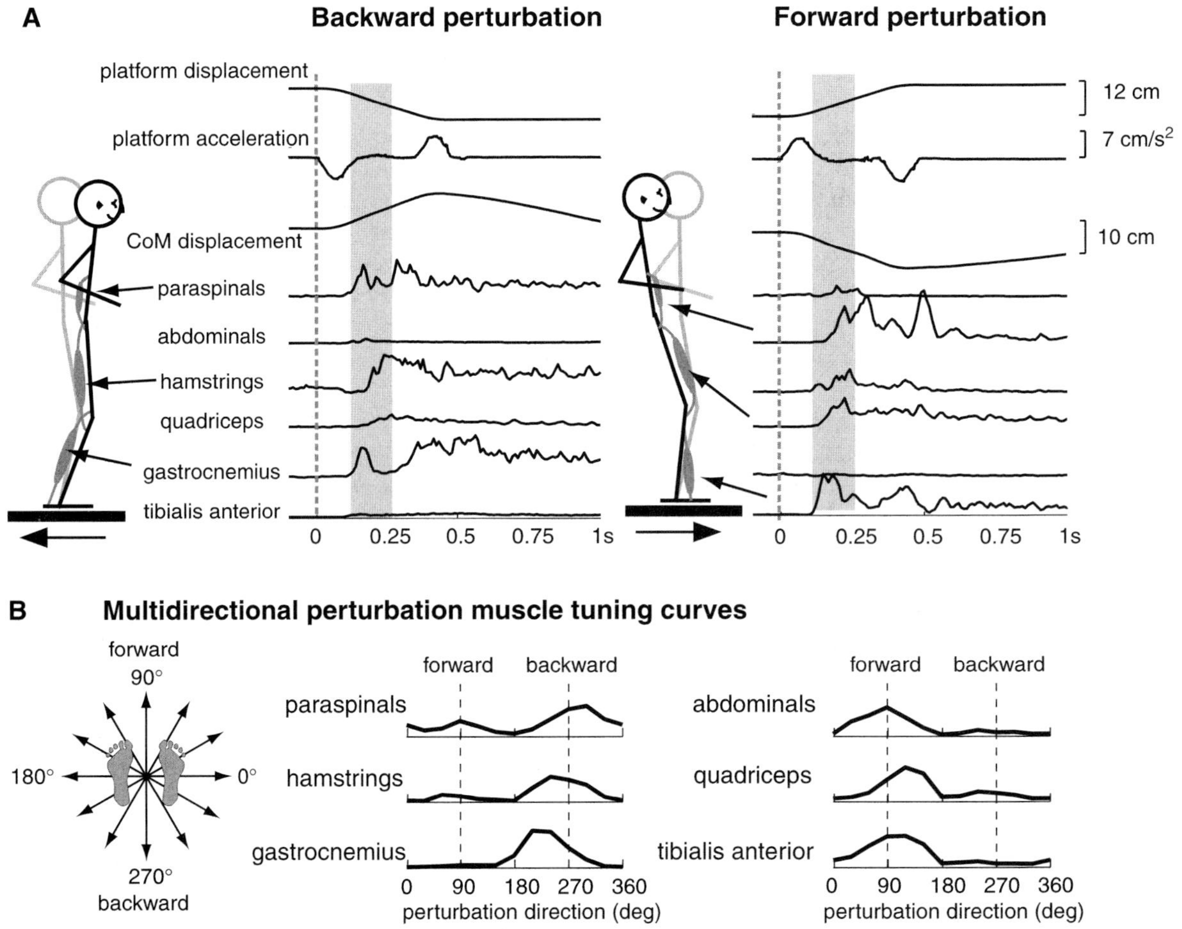

Fig. 1. Muscle activity evoked following perturbations to the support surface. (A) Backward perturbations of the support surface elicit activity in muscles on the posterior side of the body. Forward perturbations elicit activity in muscles on the anterior side of the body. The gray area represents the initial muscular response to perturbation, called the automatic postural response (APR). Note that at the onset of the APR, the amplitude of platform and center of mass displacement are quite small. (B) The magnitude of the response during the APR varies as a function of direction and can be plotted as a tuning curve. Each muscle has a unique tuning curve, suggesting that each muscle is activated by a separate neural command signal.

begins 100 ms after the onset of platform motion in humans (Fig. 1A), well before the CoM displacement reaches its maximum (Horak and Macpherson, 1996).

Multisensory integration for postural control

Directional tuning in postural responses reflects integration of multiple sensory inputs to arrive at an estimate of the CoM displacement — the task-variable that must be controlled by the nervous system. This has been deduced from several studies showing that the spatial patterns of muscle activation during the postural response cannot be consistently correlated to any single sensory signal (Keshner et al., 1988; Inglis and Macpherson, 1995; Horak and Macpherson, 1996; Allum et al., 1998; Runge et al., 1998; Carpenter et al., 1999; Ting and Macpherson, 2004; Allum and Carpenter, 2005). Only the direction of CoM displacement can accurately predict the muscle activation patterns used in response to a perturbation (Nashner, 1977; Gollhofer et al., 1989; Carpenter et al., 1999; Ting and Macpherson, 2004). Since CoM motion is due to movement of all body segments, a combination of multiple sensory signals, including visual, vestibular, proprioceptive, and cutaneous signals, which encode local variables such as head motion in space or joint angle, must be considered to explain accurate directional tuning of postural responses. This means that no single sensory modality can accurately predict the direction of CoM movement caused by a perturbation.

As an example, support surface rotations in pitch and roll can elicit *similar* patterns of muscle activation to support surface translations in the horizontal plane as long as the net motion of the CoM induced by the perturbation is the same (compare Fig. 2A and C). However, these conditions impose *opposite* changes in local joint angle displacements and head acceleration, and therefore *opposite* proprioceptive, vestibular, as well as visual cues (Diener et al., 1983; Nardone et al., 1990; Carpenter et al., 1999; Ting and Macpherson, 2004). Depending upon postural perturbation characteristics, short-latency responses can be evoked, but they do not provide any appreciable torque for stabilizing the

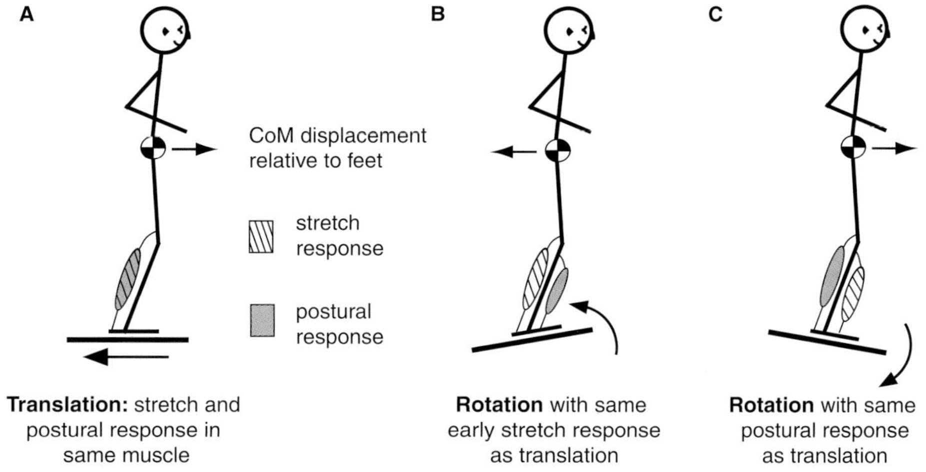

Fig. 2. Illustration demonstrating that local muscle stretch cannot predict postural responses. (A) In translation perturbations of the support surface, the muscle stretch and postural response occur in the *same* muscles — in the case of backward perturbations, the triceps surae. (B) In rotation perturbations of the support surface, the muscle stretch and the postural response occur in the *opposite* muscles. In the case of a toes-up rotation, the triceps surae is stretched, but the postural response occurs in the antagonist, the tibialis anterior. Therefore, the short-latency stretch response is possibly destabilizing. (C) In toes-down perturbations of the support surface, the postural response occurs in the triceps surae muscle, the same as in the backward translation in A. In both cases, the center of mass is displaced in the forward direction relative to the base of support, requiring triceps surae activation. The direction of this more global, task-level variable that is not directly detected by any one sensory modality is the best predictor of muscle activation patterns during postural responses. (Modified from Ting and Macpherson, 2004, used with permission.)

body, and in some cases may be destabilizing (Nashner, 1976; Carpenter et al., 1999). The short-latency responses occur in muscles that are stretched and reflect local monosynaptic spinal circuits, whereas the long-latency APR is multisynaptic and reflects both the direction of impending CoM destabilization and the postural strategy selected. In translation perturbations, short-latency responses occur in the *same* muscles as do the long-latency postural response (Fig. 2A), but in rotation perturbations, the short- and long-latency responses occur in opposite muscles (Fig. 2B and C). Thus, if local variables such as muscle stretch were used to generate postural responses an incorrect response would occur in the case of rotations (Fig. 2B).

Neurophysiological, psychophysical, and biomechanical studies all demonstrate that estimates of body position and motion in space are achieved by combining multiple sensory information through an internal model and not through a simple summation of sensory inputs (Merfeld et al., 1999; Zupan et al., 2002; Mergner et al., 2003; Kuo, 2005). Supporting this idea, perturbations to a particular sensory organ through experimental manipulations tend to alter the global perception of vertical rather than the more local variables of head orientation in space, or ankle angle, respectively (Popov et al., 1986; Gurfinkel and Levick, 1991; Bisdorff et al., 1996; Tardy-Gervet and Severac-Cauquil, 1998; Lackner et al., 2000; Maurer et al., 2001; Sorensen et al., 2002; Park et al., 2006). It is not clear where in the nervous system multisensory integration occurs. There is evidence that task-level variables such as leg length, orientation, velocity, and end-point force are already represented in ascending afferent pathways originating from the spinal cord (Bosco et al., 1996; Bosco and Poppele, 1997, 2001; Lemay and Grill, 2004).

Synergy organization of motor outputs in posture

While each muscle's directional tuning curve is unique, the control of muscles for posture appears to be simplified by the activation of a limited set of muscle synergies. We define a muscle synergy to be a muscle activation pattern with consistent spatial and temporal characteristics. Each muscle synergy is presumed to be controlled by a single neural command signal that modulates the magnitude of the muscle activation pattern specified by the synergy.

Older concepts of muscle synergies were restrictive in specifying a small set of fixed postural response patterns. Clinically, the term synergy sometimes refers to an inflexibility in motor patterns, such as the abnormal coactivation of flexors or of extensors seen in hemiplegia associated with stroke (Bourbonnais et al., 1989). In postural control, the idea of muscle synergies arose from the observation of distinct and mutually exclusive muscle activation patterns in response to two opposite directions of perturbation of the support surface, forward and backward (Fig. 1A, after Nashner, 1977). In this early conception, the two identified muscle synergies define just two possible muscle activation patterns that specify strict correlations across multiple muscles (Figs. 3A and 4). Moreover, each muscle belongs to only one synergy, and only a single synergy can be activated during any given postural response. However, when the experimental paradigms for investigating postural control were expanded to include multiple perturbation directions in the horizontal plane, muscle activation patterns were not found to be strictly correlated across all directions in both cats and humans (e.g., Figs. 1B and 3B, Macpherson, 1988; Henry et al., 1998), and the question of whether muscle synergies were a useful or physiological concept was debated (Macpherson, 1991).

In more recent formulations, it has been recognized that a limited number of muscle synergies can give rise to a continuum of postural responses. Although muscle activation patterns may not be strictly correlated across all perturbation directions, the set of postural responses has a lower dimension than the number of perturbation directions or muscles controlled, and can be accounted for by the flexible "mixing" of a limited set of muscle synergies (Fig. 3B), as well as the fact that muscles can participate in more than one muscle synergy. In the example shown, the amplitudes of neural commands C_1 and C_2 can be varied

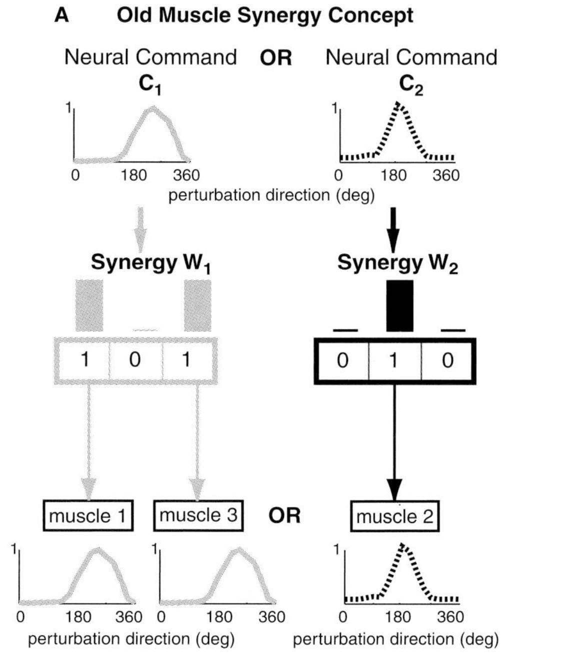

$$M = \begin{bmatrix} m_1(\theta) \\ m_2(\theta) \\ m_3(\theta) \end{bmatrix} = \begin{bmatrix} 1 \\ 0 \\ 1 \end{bmatrix} C_1(\theta) \quad \text{OR} \quad M = \begin{bmatrix} m_1(\theta) \\ m_2(\theta) \\ m_3(\theta) \end{bmatrix} = \begin{bmatrix} 0 \\ 1 \\ 0 \end{bmatrix} C_2(\theta)$$

$$M = \begin{bmatrix} m_1(\theta) \\ m_2(\theta) \\ m_3(\theta) \end{bmatrix} = \begin{bmatrix} w_{11} & w_{21} \\ w_{12} & w_{22} \\ w_{13} & w_{23} \end{bmatrix} \begin{bmatrix} C_1(\theta) \\ C_2(\theta) \end{bmatrix}$$

Fig. 3. Comparison of "fixed" versus "flexible" muscle synergy concepts. (A) In the original muscle synergy concept, only one muscle synergy was elicited at a time, and muscles could only be activated by one synergy. Therefore, all muscles activated by the same synergy would have the same tuning curve, determined by the neural command, C_1, that activated it. (B) In the new concept, more than one synergy can be activated at a time. Further, muscles can participate in multiple synergies, and have different weightings in each synergy. Each muscle's tuning curve is a weighted average of the two tuning curves of each muscle synergy. This allows flexibility in muscle tuning curves while also reducing the dimension of the neural control task.

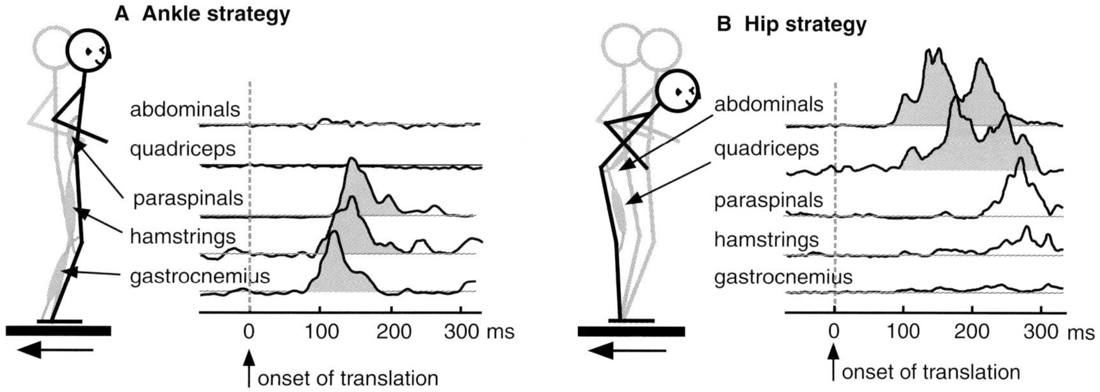

Fig. 4. Two postural strategies for controlling the center of mass in response to backward perturbations of the support surface. The two postural strategies are characterized by different joint motions and muscle activation patterns. (A) In the ankle strategy, motion is restricted to the ankle joint, and muscles on the posterior side of the body are activated. (B) In the hip strategy, the hip is flexed and muscles on the anterior side of the body are activated, but at longer latencies than in the ankle strategy. These two strategies represent extremes of a continuum, and a mixture of the two strategies can be observed in most postural responses (Runge et al., 1999; Creath et al., 2005).

independently, resulting in three different muscle tuning curves (Fig. 3B).

When analyzing a set of muscle activation patterns measured experimentally, the muscle synergies used to generate that set can be identified using matrix factorization techniques (Tresch et al., 1999, 2006). Mathematically, each muscle activation pattern is hypothesized to be a linear combination of a few (n) muscle synergies $\mathbf{W_i}$, whose elements, w_{ij}, specify the pattern of muscle activity defined by that muscle synergy (Fig. 3B, bar plots). Each muscle synergy is activated by one neural command, c_i, which can vary as a function of experimental condition such as perturbation direction, θ. $\mathbf{C_i}(\theta)$ is a vector where each element specifies the level of the neural command over a range of perturbation directions, θ (Fig. 3B, top). The net muscle activation pattern vector for any muscle over a range of perturbation direction, $\mathbf{m_i}$ is therefore hypothesized to take the form:

$$\mathbf{m_i}(\theta) = \mathbf{C_1}(\theta)w_{1i} + \mathbf{C_2}(\theta)w_{2i} + \cdots + \mathbf{C_n}(\theta)w_{ni} \quad (1)$$

where w_{1i} is the i'th element of synergy 1, $\mathbf{W_1}$ and so on. Similarly, the overall muscle activation pattern for any given perturbation direction can be expressed a vector where each element is the resulting level of activation in each muscle:

$$\mathbf{m}(\theta_k) = c_{1k}\mathbf{W_1} + c_{2k}\mathbf{W}_2 + \cdots + c_{nk}\mathbf{W_n} \quad (2)$$

where c_{1k} represents the k'th element of $C_1(\theta)$, corresponding to the particular perturbation θ_k.

The matrix $\mathbf{M}$ is a concatenation of responses in all muscles across different experimental conditions, where each row represents a muscle, and each column an experimental condition such that:

$$\mathbf{M} = \mathbf{C_1}(\theta)\mathbf{W}_1 + \mathbf{C_2}(\theta)\mathbf{W_2} + \cdots + \mathbf{C_n}(\theta)\mathbf{W_n} \quad (3)$$

Each element of $\mathbf{W_i}$ takes a value between 0 and 1, representing the relative contribution of each muscle to that muscle synergy. In postural responses, this analysis has been used to investigate the initial response in a single time window, where the columns of $\mathbf{M}$ represent different perturbation directions. However, the muscle synergies can also be viewed as being modulated by a set of independent time-varying neural commands, $c_i(t)$, where each time point is treated as a condition in the columns of $\mathbf{M}$ (Ivanenko et al., 2003, 2004, 2005). Several mathematical analysis techniques such as principal components analysis (PCA), independent components analysis (ICA), and factor analysis (FA) can be used to find muscle synergies (Tresch et al., 2006). Another such technique, non-negative matrix factorization (NMF), allows complex data sets to be more successfully partitioned into meaningful parts (Lee and Seung, 1999; Ting and Macpherson, 2005; Tresch et al., 2006).

Because the number of muscle synergies is smaller than the number of muscles, the spectrum of muscle activation patterns that can be generated using muscle synergies is still more limited than the case where muscles are controlled independently. Multiple muscle synergies may exist even for a single postural perturbation. In backward translation of the support surface in humans, two types of responses can be elicited (Nashner, 1976). One is called the "ankle strategy," where the body remains upright and most of the motion occurs around the ankle joint. The other is called the "hip strategy," where the trunk tilts forward and the hip angle motion is most predominant (Fig. 4). Each strategy can be defined by a distinct spatiotemporal pattern of muscle activation (Fig. 4) and a specific pattern of joint torques (Runge et al., 1999; Alexandrov et al., 2001a, b, 2005). Muscle synergy analysis of human postural responses demonstrates that each strategy corresponds to an independently modulated muscle synergy (Torres-Oviedo and Ting, 2007). This is consistent with studies at the joint torque and joint motion level suggesting that hip and ankle strategies represent two biomechanical response modes which are combined to form a continuum of postural responses (Runge et al., 1999; Creath et al., 2005). The flexible combination of two different synergies may underlie variations in the APR that have been shown to occur with perturbation amplitude, prior experience, and anticipation (Keshner et al., 1987; Maki et al., 1991; Brown et al., 2002; Woollacott and Shumway-Cook, 2002; Carpenter et al., 2004, 2006). Muscle synergy analysis might therefore provide a method to quantitatively compare postural responses with variable contributions from the two strategies. If so, this would suggest that muscle synergies are mechanisms by which descending influences can affect postural strategy selection (Kuo, 1995; Horak et al., 1997; Park et al., 2004).

The robustness of muscle synergies has been most thoroughly demonstrated in cat postural responses using different experimental conditions (Torres-Oviedo et al., 2006). The muscle synergies extracted from translation perturbation responses represent coordinated patterns of muscle activity across the entire limb (Fig. 5A). These patterns are roughly grouped by anatomical function, but the patterns are not strictly predictable from muscle moment arms alone. In quiet standing, only one muscle synergy, dominated by extensor muscles, is active (Fig. 5B, red). Following a perturbation, each muscle synergy has a distinctive tuning curve, which represents the purported neural command, c_i, to each muscle synergy for each perturbation direction (Fig. 5B, lower). These muscle synergies are robust in that they are used under multiple biomechanical configurations that produce changes in the muscle tuning curves (Fig. 5) as well as the active forces for stabilization. For example, when stance distance is changed, the neural commands to some of the muscle synergies change, while others remain relatively constant (Fig. 5B). These changes can explain the variations in muscle tuning curves in the hindlimb muscles (Fig. 6, black lines). Each muscle tuning curve is the sum of all the synergies activating a given muscle. Therefore, a change in any of the synergy commands will result in a change in the overall muscle tuning curve. Thus, the contribution of each muscle synergy to each tuning curve can explain the different tuning curve changes with stance distance in each muscle (Fig. 6, colored lines). The robustness of the synergy structure is further demonstrated by the fact that the same muscle synergies can account for the postural responses to two very different types of perturbation. For example, the same set of synergies account for muscle tuning curves from postural responses to both translations and rotations, where, as explained in the previous section, the sensory inputs from the two types of perturbation vary dramatically. This further demonstrates that the muscle synergies reflect a motor output mechanism that is distinct from local or central processing of afferent information.

Muscle synergies have a direct functional effect, as demonstrated in cats where modulation of the neural commands to each muscle synergy changes the biomechanical output produced during postural responses. Each muscle synergy was correlated to the production of a specific active force vector at the endpoint of the hindlimb (Fig. 5C); this relationship is consistent across all of the different postural conditions discussed above. Therefore changes in muscle activation *and* forces produced during postural response in different stance configurations (Macpherson, 1994; Torres-Oviedo et al., 2006) can be explained by simply changing the proportion of contribution of each muscle synergy.

These findings suggest that muscle synergies could be a functional mechanism by which descending neural commands related to the desired control of task-level variables are transformed into specific patterns of muscle activation that affect those task-level variables. This type of direct mechanism for motor coordination is appealing in that high-level computations in the nervous system can occur in the context of task-level variables rather than in local afferent or efferent signals. Moreover, selection of muscle synergies would not require online forward or inverse computation of the muscle activation to motor output transformation. The muscle synergy pattern itself can be thought of as an element in a look-up table of the muscle activation to motor output transformation. Therefore each person's motor outputs would be defined by the repertoire of muscle synergies available, which could exceed the number of muscles or degrees of freedom in the body.

Hierarchal feedback model of postural control

How can the many experimental and theoretical findings in postural control be unified in a coherent framework? By studying the nature of the dimensional reduction in sensorimotor systems, it may be possible to explain the apparent tension between the commonalities and variations in postural behaviors observed across individuals. Frequently, individual variations deviate from general characterizations of muscle activation patterns during postural responses to perturbation. Are there common

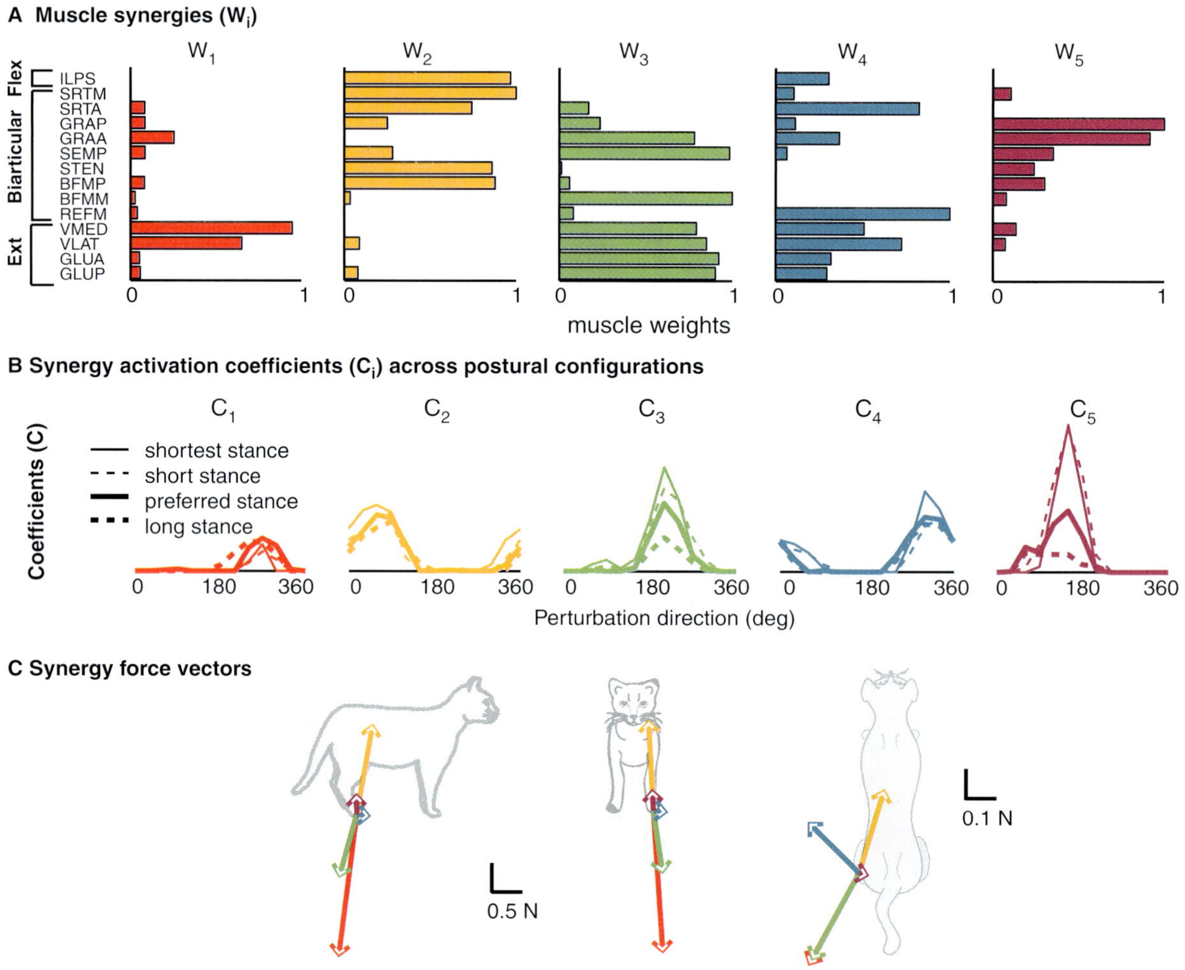

Fig. 5. Muscle synergies robustly produce endpoint forces in cat postural control. (A) Five muscle synergy vectors, $\mathbf{W_i}$, extracted from postural responses to support surface translation at the preferred stance distance in a cat. These five muscle synergies account for over 96% of the total variability accounted for in the preferred stance. Each bar represents the relative level of activation for each muscle within the synergy. Note that muscles can contribute to multiple muscle synergies. (B) Activation coefficients, $\mathbf{C_i}$, representing the purported neural commands to each muscle synergy during postural responses in four different postural configurations. Upper traces show background activity of each muscle synergy during the quiet stance period before perturbations. Lower traces show the synergy tuning curves in response to support surface translations. Changes in muscle tuning curves at different stance distances are due to variations in the amplitude of the neural commands to the various muscle synergies. Some muscle synergies (e.g., red, yellow) are relatively constant amplitude across all conditions, whereas others (e.g., green, purple) are highly modulated. (C) Endpoint force vectors are produced by each muscle synergy (same color coding), in the sagittal, frontal, and horizontal planes. Vectors are expressed as forces applied by the limb against the support surface. The amplitude of each the force vectors in any postural response is directly modulated by the amplitude of the neural command to each muscle synergy. (Modified from Torres-Oviedo et al., 2006, used with permission.)

organizational themes underlying muscle activation patterns across tasks and individuals that can be used to guide our approaches to understanding postural control?

Rather than using the muscle activation patterns themselves as the primary determinant of postural strategies, we propose that the nature of the dimensional reduction within an experimental data set

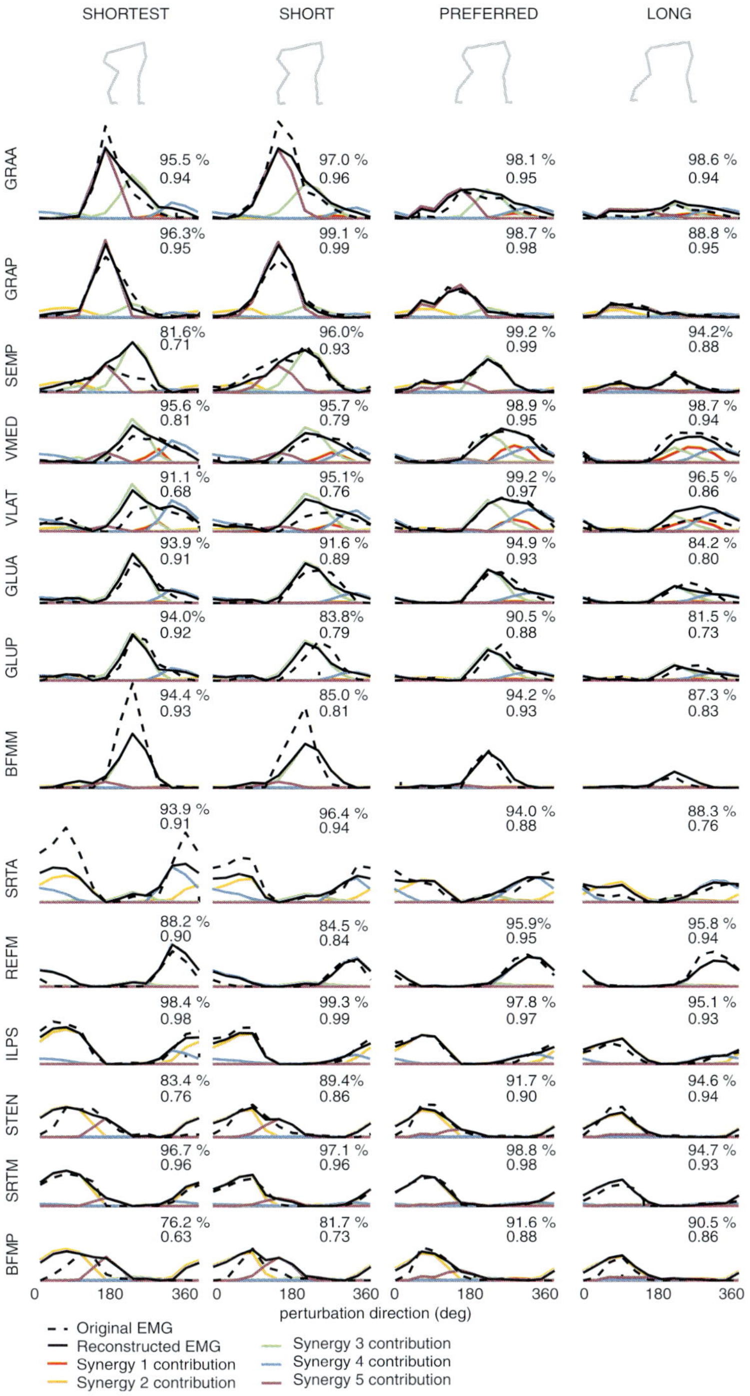

Fig. 6. Muscle tuning curves reconstructed using the same set of muscle synergies in four different postural configurations. Muscle tuning curves vary across postural responses to support surface perturbation when postural configuration is varied. These variations can be reconstructed using the set of muscle synergies extracted from the preferred stance configuration. The original data are shown by the dashed black line, and the reconstructed data by the solid black line. The contribution from each synergy to the reconstruction is shown by the corresponding colored line. This is computed by multiplying each functional synergy vector **W** by its activation coefficient **C**. (From Torres-Oviedo et al., 2006, used with permission.)

can be better used to characterize and compare motor outputs across trials and across individuals. Because of redundancy in the musculoskeletal system, even muscle activation patterns associated with a common task-variable can differ across individuals. For example, in cat postural control, the same number of muscle synergies were found across several cats (Torres-Oviedo et al., 2006). These muscle synergies were found to have similar tuning curves and produce similar force directions in different animals (Fig. 7). This suggests that the neural commands to the muscle synergies and the motor output generated by the muscle synergies in response to perturbations are similar across cats. Postural responses in different individuals are probably modulated by disturbances in similar task-level variables, which by definition are independent of individual variations in morphometry, or postural configuration. But, muscle synergy composition differs significantly across individuals (Fig. 7). Therefore, the exact muscle synergy mapping from task-variable to muscle activation patterns appears to be specific to each individual. Likewise, the sensory mappings leading to the estimation of the relevant task-variables are also likely to be individual-specific. Therefore, individuals are more similar in terms of the task-variables that are controlled, and not to specific sensorimotor patterns.

In this view, the number of muscle synergies and their corresponding neural commands carry more information than the activation pattern of individual muscles because they reflect the task-variables that are sensed and regulated by the nervous system. Changes in muscle activation patterns might then be thought of in terms of either changes in the activation patterns of a consistent set of muscle synergies, changes in the number of muscle synergies recruited, or changes in the composition of muscle synergies. This leads to a general framework in which processes causing variability due to influences at all levels of the nervous system can be explained using concepts of dimensional reduction in sensorimotor transformations during postural control (Fig. 8).

The general framework for understanding muscle coordination is presented in terms of understanding sensorimotor transformations in postural responses, but can also be applied to most sensorimotor processes. The framework consists of a nested set of hierarchal feedback loops with much lower dimensionality at the higher levels than the lower levels. At the highest level, the relevant task-variables depend on the goal-level decisions in the nervous system (Fig. 8A and B). For successful task performance, these goals must also be nominally matched to biomechanical constraints, and can be regulated in a feedback manner (Fig. 8B). The estimate of the relevant task-variables depends upon sensory transformations that integrate high-dimensional multisensory signals; these transformations can also be influenced by behavioral goals (Fig. 8C). Muscle synergies perform the symmetric function of transforming the desired control of task-variables into high-dimensional multiple muscle activations (Fig. 8C). Muscle synergy activation patterns ultimately interact with spinal circuits and intrinsic muscle properties to produce biomechanical outputs (Fig. 8D). Simultaneously, these biomechanical outputs induce sensory signals in afferents across the body that are then mapped onto task-variable estimates in the nervous system (Fig. 8D and C). While the role of descending influences is primarily at the level of the relatively low-dimensional task-variable space, they can also affect the state of low-level circuits in the spinal cord that ultimately will affect task performance (Fig. 8D). This framework can be used to make predictions about how changes in muscle activation patterns are regulated by the nervous system and then suggest how computational studies can be used to substantiate the hypothesis.

In this framework, descending influences primarily modulate the relatively low-dimensional task-variable space. It has been hypothesized that a simple, linear neural control mechanism might sit at the top of a complex hierarchy of sensorimot or feedback loops used for movement control q(Todorov, 2000; Scott, 2004). The control of task-level variables ultimately has to be considered and implemented in a high-dimensional space. It is possible that the role of the motor cortex is to perform this translation between task-level and local variables, perhaps by selecting the appropriate muscle synergies. Neuronal populations in the motor cortex reflect a wide array of both task-level and local variables (Scott, 2003). Moreover, long

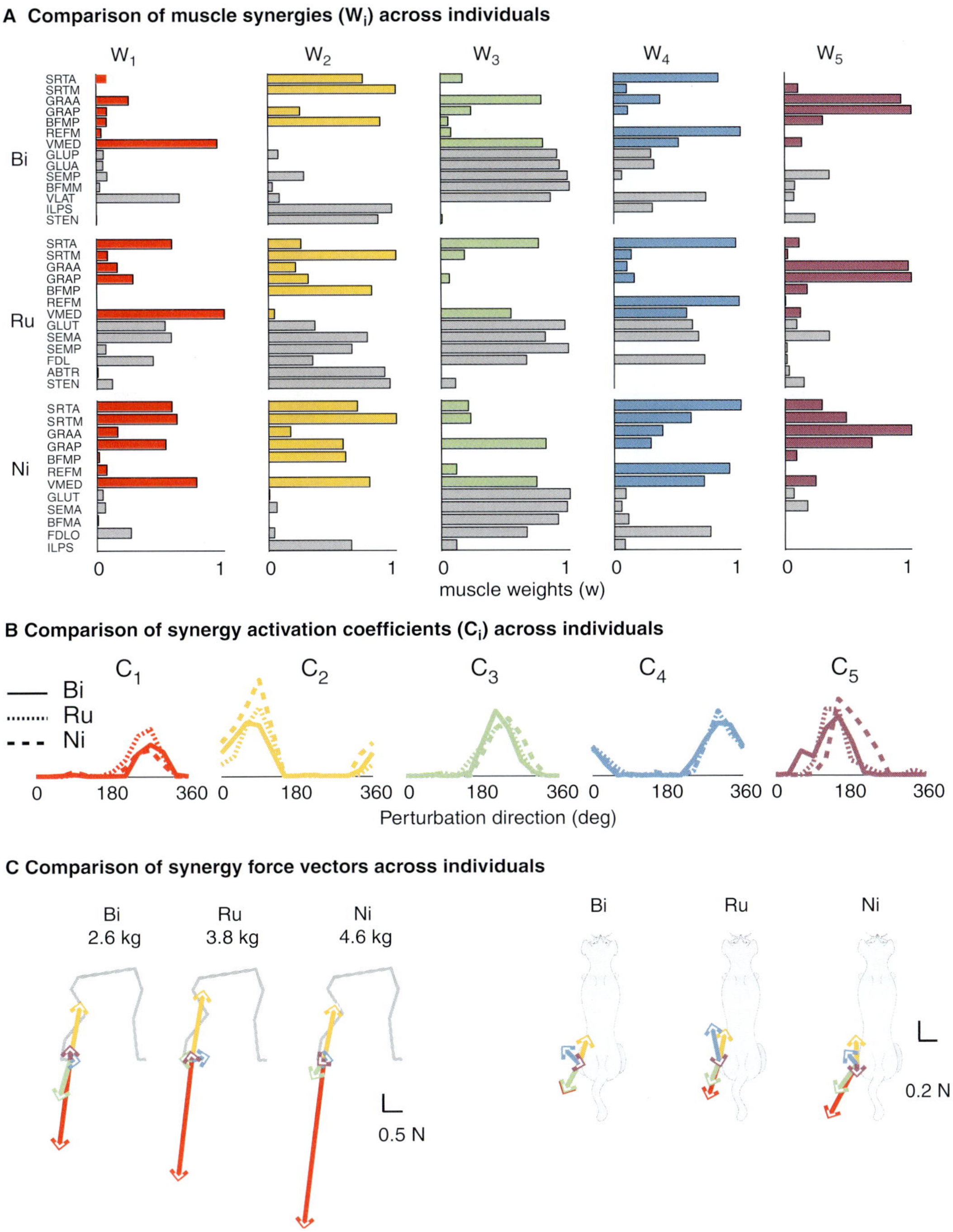

Fig. 7. Example of similar dimensional-reduction and task-variable encoding across individuals. In all cats, five synergies accounted for >96% of the variability in response to translation at the preferred stance. (A) Muscle synergies for each individual. Colored bars indicate muscles that were measured across all individuals. Gray bars indicate the remaining muscles collected for each individual. While there are general similarities in the most highly activated muscles in each synergy, substantial variation in muscles contributing to the synergies exist across individuals. (B) Activation coefficients across animals are similar, indicating that they are activated in similar perturbation directions. (C) Force vectors produced by each synergy are also quite similar. Taken together, this data demonstrates that neural commands encoding force-vector directions are quite similar across individuals, but the specific muscle synergy mapping used can vary. (Modified from Torres-Oviedo et al., 2006, used with permission.)

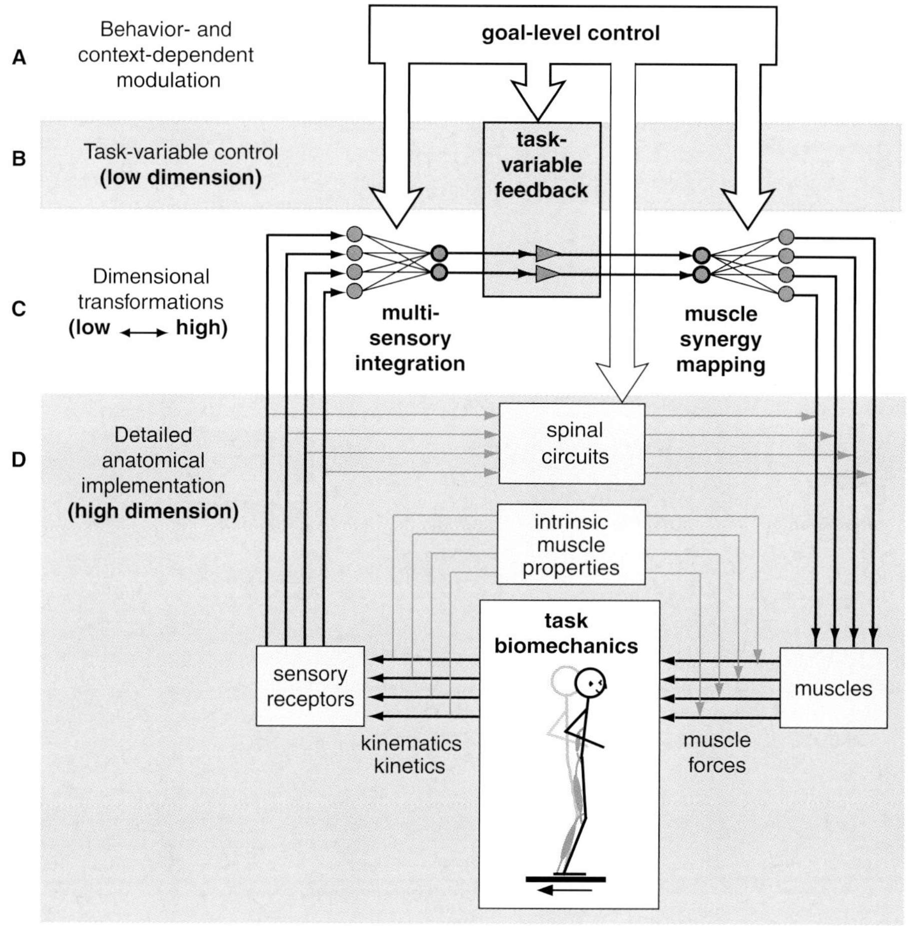

Fig. 8. General framework for understanding dimensional reduction in muscle coordination of posture. The framework consists of a nested set of hierarchal feedback loops with much lower dimensionality at the higher levels than the lower levels. (A) Goal-level modulation of postural responses occurs in task-variable space. Therefore behavioral or cognitive-level modulation can alter the task-variables attended to, as well as the way they are estimated and regulated by (B) low-dimensional feedback in the case of postural control. (C) Mappings between low- and high-dimensional spaces are necessary for estimation and control of task-level variables. A dimensional reduction occurs in the multisensory integration mappings that use multiple afferent signals to estimate task-variables. Once the desired effect on the task-level variable is determined, a dimensional expansion occurs via muscle synergy mappings, allowing the action to be implemented in (D) specific anatomical details. At this level there are many nonlinearities and state-dependent effects that can influence the eventual biomechanical output produced through the activation of a muscle synergy. However, some of these factors, such as spinal circuits, may be used to make the system more controllable by the reduced-dimension controller, and are also influenced by high-level centers. This general framework can be used to make specific hypotheses about the characteristics of changes in muscle activation patterns in postural responses due to changes at all levels in the nervous system. In addition, it can be used to guide computational studies focused on understanding mappings to and from the low-dimensional space where movement is controlled by the nervous system.

train stimulation of sites in the motor cortex generates coordinated movement of the limb to a common, final posture regardless of initial position, suggesting encoding of higher order movement parameters (Graziano, 2006).

Changes in motor output due to descending, goal-level control affecting postural strategy selection can be thought of as changes in muscle synergy selection. For example, behavior-dependent modulation could influence the selection and modulation of appropriate sensory and motor mappings, consistent with the "strategy selection centers" proposed for postural control (Kuo, 1995, 2005; Horak et al., 1997; Park et al., 2004). Different postural strategies could be preferentially selected by altering the selection of muscle synergies. Changes in postural responses due to descending influences could therefore be due to changing the threshold for selection of an ankle or hip synergy. Such changes would be represented in terms of variations in the neural commands to muscle synergies, or preferential activation of particular muscle synergies within a given set, rather than changes in the muscle synergy patterns themselves. It has been shown that variability in locomotor behaviors can be explained by differences in the activation of muscle synergies and not random variability in the activation of each muscle (Tresch et al., 1999; d'Avella and Bizzi, 2005). We predict that variability in postural response also arises due to variations in the relatively low-dimensional set of neural commands.

The dimensional reductions occurring in the sensory and motor systems appear to be symmetric processes that serve the function of controlling task-level variables for motor behaviors. The ability to sense and to control relevant task-variables must match in order for feedback regulation of task-variables to occur. The sensory and motor mappings are independent but related processes in that both must resolve similar issues of sensorimotor redundancy in mapping between low-dimensional task-variables and high-dimensional anatomical details (Fig. 8D). The redundancy allows flexibility in the mappings, ensuring that the control of task-level variables is not directly or immutably linked to any particular afferent or efferent pathway. This is consistent with recent studies demonstrating substantial non-uniqueness in neural circuits mapping sensory inputs into sparse neural representations in the production of motor outputs (Prinz et al., 2004; Leonardo, 2005).

In postural control, this concept is consistent with the fact that CoM kinematics cannot be reliably derived from any single sensory afferent population (Nashner, 1977) and that direct feedback of sensory signals does not explain human postural behaviors, particularly in situations of sensory conflict where an erroneous sensory signal must be ignored (Nashner, 1977; Kuo, 2005). The sense of verticality can be derived from multiple sensory systems, and the contributions of each sensory organ can be dynamically reweighted under various experimental conditions (Peterka, 2002; Mergner et al., 2003; Peterka and Loughlin, 2004; Carver et al., 2006; Jeka et al., 2006).

Conversely, the control of task-variables is achieved through the inverse transformation to muscle activations encoded by the muscle synergies. In this case the muscle synergy defines one of many possible muscle activation patterns that has the desired effect on the task-variable at hand. Despite the numerous possibilities for affecting task-variables, it has been shown that individuals use the same muscle synergies across a wide range of tasks across days and weeks (Torres-Oviedo et al., 2006). Therefore, while the sensory and motor transformations are roughly optimal in the sense that they are comparable to solutions derived from optimization techniques (Kuo, 1995, 2005; Todorov and Jordan, 2002; Scott, 2004; Todorov, 2004; Kurtzer et al., 2006), these mappings do not appear to be updated on a rapid trial-by-trial time scale. In tasks that are relatively uncommon in the experience of an individual, the nominal set of muscle synergies may not be optimal, and can result in less effective biomechanical outputs. This appears to be the case in the relaxation of the force constraint strategy for postural control at shorter stance distances (Macpherson, 1994; McKay et al., 2007; Torres-Oviedo et al., 2006).

Thus, the formation of muscle synergies and sensory transformations is considered to be a separate process from the goal-level decisions influencing their regulation and selection. In this framework, the sensory transformations and muscle synergies represent a relatively fixed set of preferred mappings, influenced by each individual's experiences and motor training as well as the biomechanics of the task. If the formation of these mappings is influenced by experience, it may not be possible to directly compare muscle synergies across individuals in terms of their exact composition, but only on the task-level variables they encode. Similarities are inevitable because of the constraints imposed by task biomechanical constraints; however, redundancies in the sensory and motor systems allow for substantial individual variation. It is possible that such variations give rise to individual movement characteristics, as movement styles that nonetheless conform to physical constraints can be encapsulated through patterns of joint torque weightings in computer simulations for animation (Liu and Popovic, 2002; Liu et al., 2005).

Most individuals have reasonably consistent movement patterns, but muscle synergy number or composition could be altered through experience and training, in particular due to neural or musculoskeletal injury or disease. New muscle synergies might form after extended experience with a new motor task — such as with skiing or bicycling. It has also been shown that dancers have postural responses that tend to emphasize the orientation and alignment of the body, as compared to non-dancers who simply maintain the CoM within the base of support (Mouchnino et al., 1992, 1993). Likewise, the environment in which an individual is raised also affects sensory integration mappings (Wallace and Stein, 2006; Wallace et al., 2006). Orienting responses to a stimulus are enhanced when visual and auditory cues are congruent. However, animals raised in the dark experience no such enhancement (Wallace et al., 2004).

In speech, an instructive example can be found that demonstrates a process of dimensional reduction in sensorimotor systems that can be thought of as an experience-dependent "interpretation" of the relevant task-variables (Kuhl, 1994, 2004). A similar phenomenon of matched dimensional reduction in sensory and motor processing occurs in the perception and production of speech sounds. The native language of each individual shapes his or her ability to both distinguish and produce speech sounds (Kuhl et al., 1997; Zhang et al., 2005). Essentially, a reduction in dimension occurs that is based on the native language of an individual. Idealized templates of speech sounds are formed in the nervous system that can be thought of as "sensorimotor synergies" — these synergies underlie the characteristic accents of individuals speaking a foreign language. Sensorimotor synergies in language are so strong that sounds considered to be very distinct in one language may not be perceivable, much less producible by native speakers of a different language. While there are similar characteristics of these synergies across a native-language population, they are also specific to each individual and can change through experience-dependent processes like intensive speech training. Therefore, sensorimotor transformations that map between low-dimensional task-variables and high-dimensional anatomical variables underlie individual speech or movement characteristics allowing us to recognize distinctive features of a person even when performing a novel task because of each individual's distinctive set of "building blocks," or sensorimotor synergies. Clearly we have dedicated circuits for language production, as well as motor behaviors, and yet these structures do not specify the exact synergy patterns in individuals, but facilitate the formation and general applicability of sensorimotor synergies in sensorimotor processes.

While the use of reduced dimension task-level control is appealing, the reality is that movements must be implemented in complex, nonlinear dynamic systems that are not easily controlled. The long latencies that exist between descending commands and peripheral action add further challenges to task performance, particularly for standing postural control in an unstable, bipedal postural configuration. It has been proposed that physiological "linearization" mechanisms may exist that allow a low-dimensional hierarchal feedback architecture to work, but the nature of this mechanism has not been discussed. However, there are many candidate components of the neuromuscular system that are modifiable through descending and neuromodulatory influences. Intrinsic muscle properties provide instantaneous stabilizing influences, which can be influenced by activation level, or the motion history of the muscles. Spinal heterogenic stretch reflex circuits coordinate the limb (Nichols, 1994; Nichols et al., 1999; Wilmink and Nichols, 2003), but more importantly their strength can be altered by the state of the spinal network such that the online processing of afferent and efferent signals is altered. Neuromodulatory effects on motoneuron excitability can be affected by joint angle (Hyngstrom et al., 2007). But, influence of these state-dependent changes in the spinal cord extends far beyond mono- or poly-synaptic reflex loops, and can alter the influence of descending commands on the activation of single muscles, as well as the strength of ascending afferent signals. It is therefore likely that the context-dependent modulation of spinal circuits through descending control as well as neuromodulator release works in tandem with descending muscle synergy commands in order to produce predictable, stable movements. Therefore the spinal circuitry is an essential component of the implementation of the "simple" hierarchal control architecture, although it may not be responsible for specifying the muscle synergies used in postural control.

Future directions

This framework that links low- and high-dimensional representations of movement is an overarching hypothesis that lends itself to testing through computer simulations. Our philosophy is that neither a simple conceptual model nor a complex anatomical model in isolation can effectively elucidate principles of motor coordination. Current models of posture and movement are formulated either in the low-dimensional task-space, or in the high-dimensional anatomical details where individual muscles and joints are considered. Each has its strengths and weaknesses that cannot alone be used to understand neural mechanisms of movement. The neural mechanisms through which musculoskeletal systems exhibit "collapses in dimension" must be explicitly studied (Holmes et al., 2006). But to date, the sensorimotor transformations between low- and high-dimensional spaces have only been addressed by a few studies demonstrating that a muscle synergy organization is sufficient to control the task-variables (Valero-Cuevas et al., 1998; Raasch and Zajac, 1999; Loeb et al., 2000; Valero-Cuevas, 2000).

In posture, simplified feedback control models of posture have been used to explain how task-level variables are regulated by sensorimotor mechanisms (Kuo, 1995; van der Kooij et al., 1999; Peterka, 2000, 2002; Bortolami et al., 2003). These models have been instructive in understanding the importance of various sensory channels on postural control (Kuo et al., 1998; van der Kooij et al., 2001; Peterka, 2002) but are not sufficient for understanding muscle activation patterns. On the other hand, current musculoskeletal models can explain individual muscle activations in a specific motor task (Neptune, 2000; Pandy, 2001; Zajac et al., 2003). But, in the absence of feedback loops, a small change in the pattern of muscle activation can completely destabilize the simulated system (Risher et al., 1997), allowing only the analysis of explicitly modeled conditions. Because these models lack sensorimotor mechanisms that allow them to respond to perturbations, they cannot yet be used to understand the neuromechanical principles coordinating muscles.

To bridge the gap between concepts about task-variable control and its implementation at the level of individual muscle activation patterns, novel methods for complementary and parallel development of simple and complex musculoskeletal models of posture must be developed (Full and Koditschek, 1999). The framework presented demonstrates why both low-dimensional and high-dimensional models alone can be used to produce reasonable simulations of movement. However, the functional relevance of a simple model of postural control to multiple joint motions depends critically on its integration with more complex musculoskeletal models. The future challenges in computational studies will be to incorporate relevant dimensional reduction mechanisms in the control of multiple muscles. As an example, six muscle synergies can be used to produce a range of natural pedaling behaviors in simulations, such as slow, fast, smooth, jerky, and backwards pedaling (Raasch et al., 1997; Raasch and Zajac, 1999; Zajac et al., 2003). These simulations were found to predict phase changes in muscle activation patterns that were unexpected based on prior hypotheses (Ting et al., 1999). Moreover, when the model used only flexor and extensor synergies it was unable to advance the limb through the transition from extension to flexion (Raasch and Zajac, 1999). Similarly, stroke patients limited to flexion and extension synergies (Bourbonnais et al., 1989) have difficulties through the same phase transition (Brown et al., 1997). A similar model for understanding postural control would be critical to understanding the functional consequences of neural impairments that lead to balance disorders. Recent steps in this direction include models demonstrating reduced dimension in feedback control of posture. "Eigenmovements" that couple joint motions in multiarticular models of standing posture, can be used to reduce the dimension of the feedback parameters required for postural control (Alexandrov et al., 2005). Consistent with the eigenmovement hypothesis, at the level of muscle activation patterns, simulations demonstrate the need for coordinated control of multiple muscles to achieve task-variable control (Bunderson et al., 2007; Van Antwerp et al., 2007). Moreover, it has been shown that multiple muscle activation patterns in both cats and humans are regulated by simple feedback control laws (Lockhart, 2005; Welch and Ting, 2005), suggesting that a feedback control system might act at the level of the neural commands to muscle synergies.

The integration of simple and complex models may also be clinically relevant. The ability of patients to conform to overall control principles may be more important than the enforcement of specific synergies or detailed movement patterns. In cerebral palsy subjects with hemiplegia, different patterns of joint angle changes and EMGs are observed in each leg. These differences are difficult to interpret through direct comparison of the multiple variables. However, the overall mechanics and energy exchange mechanisms in the unaffected and affected limbs can be characterized by two simple models of gait: an inverted pendulum and a spring-mass model, respectively (Fonseca et al., 2001, 2004). Therefore, more insight is gained from understanding the control of task-level variables versus local variables. These conceptual frameworks can inform the analysis of data and design of new experiments and complex models, and may explain why prior attempts to enforce specific muscle activation patterns in clinical rehabilitation were unsuccessful. Development of computational models that can predict the functional consequences of muscle activation patterns in postural control may be more effective at predicting how postural function could improve in individuals with specific impairments. The resulting muscle activation patterns may not resemble a "normal" pattern, but take advantage of the capabilities of the individual.

Finally, the framework presented calls for more computationally sophisticated methods of data analysis that reflect the hypothesized neural organization principles. The framework suggests that a relatively low number of parameters can be used to describe complex changes in muscle activation patterns. Therefore, understanding low-dimensional task-level variables can lead to a better understanding of changes in local variables. For example, cats with large-fiber peripheral sensory neuropathy that destroys afferents from muscle spindles and Golgi tendon organs exhibit postural instability and delayed postural responses (Stapley et al., 2002). Application of a simple feedback model demonstrates that changes in the entire timecourse of multiple muscle activations can be described as a decrease in the feedback gain associated with CoM acceleration (Lockhart et al., 2005; Ting et al., 2005). This change can explain the apparent delay in the response through a change in only one of four feedback parameters. Further, since muscle directional tuning remains intact (Stapley et al., 2002), it is likely that the muscle synergy patterns in these animals is unaffected by the sensory loss. The framework further predicts that nominal changes in postural behaviors from goal- or task-level control due to changes in mental state, such as anticipation, adaptation, fear, or divided attention (Keshner et al., 1987; Maki et al., 1991; Brown et al., 2002; Woollacott and Shumway-Cook, 2002; Carpenter et al., 2004, 2006), would occur only in the modulation and selection of postural synergies. However, more changes due to disease or injury might result in the inability to activate particular muscle synergies, such as in Parkinson's disease (Dimitrova et al., 2004), or an inappropriate activation of muscle synergies, as in cerebellar loss (Timmann and Horak, 1997), or a reorganization of muscle synergies themselves. The ability to differentiate these different mechanisms of changes may lead to greater insight into the neurological underpinnings of motor dysfunctions and the development of potential interventions.

Abbreviations

APR	automatic postural response
CoM	center of mass
EMG	electromyographic

Muscle abbreviations used in figures

BFMA	anterior biceps femoris
BFMM	medial biceps femoris
BFMP	posterior biceps femoris
EDL	extensor digitorum longus
FDL	flexor digitorum longus
GLUA	anterior gluteus medius
GLUP	posterior gluteus medius
GLUT	gluteus medius
GRAA	anterior gracilis
GRAP	posterior gracilis
ILPS	iliopsoas
LGAS	lateral gastrocnemius
MGAS	medial gastrocnemius
PLAN	plantaris
REFM	rectus femoris
SEMA	anterior semimembranosus
SEMP	posterior semimembranosus
SOL	soleus
SRTA	anterior sartorius
SRTM	medial sartorius
STEN	semitendinosus
TFL	tensor fasciae lata
TIBA	tibialis anterior
VLAT	vastus lateralis
VMED	vastus medialis

Acknowledgments

I would like to thank Jane Macpherson for many interesting discussions about multiple aspects of postural control. Thanks to everyone who helped critique and prepare this chapter: J.L. McKay, K.W. van Antwerp, S. Chvatal, J. Gottschall, G. Torres-Oviedo. Supported by NIH HD46922.

References

Alexandrov, A.V., Frolov, A.A., Horak, F.B., Carlson-Kuhta, P. and Park, S. (2005) Feedback equilibrium control during human standing. Biol. Cybern.: 1–14.

Alexandrov, A.V., Frolov, A.A. and Massion, J. (2001a) Biomechanical analysis of movement strategies in human forward trunk bending I: modeling. Biol. Cybern., 84: 425–434.

Alexandrov, A.V., Frolov, A.A. and Massion, J. (2001b) Biomechanical analysis of movement strategies in human forward trunk bending II: experimental study. Biol. Cybern., 84: 435–443.

Allum, J.H., Bloem, B.R., Carpenter, M.G., Hulliger, M. and Hadders-Algra, M. (1998) Proprioceptive control of posture: a review of new concepts. Gait Posture, 8: 214–242.

Allum, J.H. and Carpenter, M.G. (2005) A speedy solution for balance and gait analysis: angular velocity measured at the centre of body mass. Curr. Opin. Neurol., 18: 15–21.

d'Avella, A. and Bizzi, E. (2005) Shared and specific muscle synergies in natural motor behaviors. Proc. Natl. Acad. Sci. U.S.A., 102: 3076–3081.

Bernstein, N. (1967) The Coordination and Regulation of Movements. Pergamon Press, New York.

Bisdorff, A.R., Wolsley, C.J., Anastasopoulos, D., Bronstein, A.M. and Gresty, M.A. (1996) The perception of body verticality (subjective postural vertical) in peripheral and central vestibular disorders. Brain, 119(Pt 5): 1523–1534.

Blickhan, R. (1989) The spring-mass model for running and hopping. J. Biomech., 22: 1217–1227.

Bortolami, S.B., Dizio, P., Rabin, E. and Lackner, J.R. (2003) Analysis of human postural responses to recoverable falls. Exp. Brain Res., 151: 387–404.

Bosco, G. and Poppele, R.E. (1997) Representation of multiple kinematic parameters of the cat hindlimb in spinocerebellar activity. J. Neurophysiol., 78: 1421–1432.

Bosco, G. and Poppele, R.E. (2001) Proprioception from a spinocerebellar perspective. Physiol. Rev., 81: 539–568.

Bosco, G., Rankin, A.M. and Poppele, R.E. (1996) Representation of passive hindlimb postures in cat spinocerebellar activity. J. Neurophysiol., 76: 715–726.

Bourbonnais, D., Vanden Noven, S., Carey, K.M. and Rymer, W.Z. (1989) Abnormal spatial patterns of elbow muscle activation in hemiparetic human subjects. Brain, 112(Pt 1): 85–102.

Brown, L.A., Gage, W.H., Polych, M.A., Sleik, R.J. and Winder, T.R. (2002) Central set influences on gait: age-dependent effects of postural threat. Exp. Brain Res., 145: 286–296.

Brown, D.A., Kautz, S.A. and Dairaghi, C.A. (1997) Muscle activity adapts to anti-gravity posture during pedalling in persons with post-stroke hemiplegia. Brain, 120(Pt 5): 825–837.

Bunderson, N.E., Ting, L.H. and Burkholder, T.J. (2007) Asymmetric interjoint feedback contributes to postural control of redundant multi-link systems. J. Neural Eng., 4: 234–245.

Carpenter, M.G., Adkin, A.L., Brawley, L.R. and Frank, J.S. (2006) Postural, physiological and psychological reactions to challenging balance: does age make a difference? Age Ageing, 35: 298–303.

Carpenter, M.G., Allum, J.H.J. and Honegger, F. (1999) Directional sensitivity of stretch reflexes and balance corrections for normal subjects in the roll and pitch planes. Exp. Brain Res., 129: 93–113.

Carpenter, M.G., Frank, J.S., Adkin, A.L., Paton, A. and Allum, J.H. (2004) Influence of postural anxiety on postural reactions to multi-directional surface rotations. J. Neurophysiol., 92: 3255–3265.

Carver, S., Kiemel, T. and Jeka, J.J. (2006) Modeling the dynamics of sensory reweighting. Biol. Cybern., 95: 123–134.

Cavagna, G.A., Heglund, N.C. and Taylor, C.R. (1977) Mechanical work in terrestrial locomotion: two basic mechanisms for minimizing energy expenditure. Am. J. Physiol., 233: R243–R261.

Creath, R., Kiemel, T., Horak, F., Peterka, R. and Jeka, J. (2005) A unified view of quiet and perturbed stance: simultaneous co-existing excitable modes. Neurosci. Lett., 377: 75–80.

Diener, H.C., Bootz, F., Dichgans, J. and Bruzek, W. (1983) Variability of postural "reflexes" in humans. Exp. Brain Res., 52: 423–428.

Dimitrova, D., Horak, F.B. and Nutt, J.G. (2004) Postural muscle responses to multidirectional translations in patients with Parkinson's disease. J. Neurophysiol., 91: 489–501.

Farley, C.T., Glasheen, J. and Mcmahon, T.A. (1993) Running springs: speed and animal size. J. Exp. Biol., 185: 71–86.

Fonseca, S.T., Holt, K.G., Fetters, L. and Saltzman, E. (2004) Dynamic resources used in ambulation by children with spastic hemiplegic cerebral palsy: relationship to kinematics, energetics, and asymmetries. Phys. Ther., 84: 344–354 Discussion 355–358.

Fonseca, S.T., Holt, K.G., Saltzman, E. and Fetters, L. (2001) A dynamical model of locomotion in spastic hemiplegic cerebral palsy: influence of walking speed. Clin. Biomech. (Bristol, Avon), 16: 793–805.

Full, R.J. and Koditschek, D.E. (1999) Templates and anchors: neuromechanical hypotheses of legged locomotion on land. J. Exp. Biol., 202(Pt 23): 3325–3332.

Gollhofer, A., Horstmann, G.A., Berger, W. and Dietz, V. (1989) Compensation of translational and rotational perturbations in human posture: stabilization of the centre of gravity. Neurosci. Lett., 105: 73–78.

Gottlieb, G.L. (1998) Muscle activation patterns during two types of voluntary single-joint movement. J. Neurophysiol., 80: 1860–1867.

Graziano, M. (2006) The organization of behavioral repertoire in motor cortex. Annu. Rev. Neurosci., 29: 105–134.

Gurfinkel, V.S. and Levick, Y.S. (1991) Perceptual and automatic aspects of the postural body scheme. In: Paillard J. (Ed.), Brain and Space. Oxford University Press, Oxford.

Hatzitaki, V., Pavlou, M. and Bronstein, A.M. (2004) The integration of multiple proprioceptive information: effect of ankle tendon vibration on postural responses to platform tilt. Exp. Brain Res., 154: 345–354.

Henry, S.M., Fung, J. and Horak, F.B. (1998) EMG responses to maintain stance during multidirectional surface translations. J. Neurophysiol., 80: 1939–1950.

Hlavacka, F., Dzurkova, O. and Kornilova, L.N. (2001) Vestibular and somatosensory interaction during recovery of balance instability after spaceflight. J. Gravit. Physiol., 8: P89–P92.

Holmes, P., Full, R., Koditschek, D.E. and Guckenheimer, J. (2006) The dynamics of legged locomotion: models, analyses, and challenges. Sci. Am. Rev., 48: 207–304.

Horak, F.B., Henry, S.M. and Shumway-Cook, A. (1997) Postural perturbations: new insights for treatment of balance disorders. Phys. Ther., 77: 517–533.

Horak, F.B. and Macpherson, J.M. (1996) Postural orientation and equilibrium. In: Handbook of Physiology. American Physiological Society, New York Section 12.

Hyngstrom, A.S., Johnson, M.D., Miller, J.M. and Heckman, C.J. (2007) Intrinsic electrical properties of spinal motoneurons vary with joint angle. Nat. Neurosci., 10(3): 363–369.

Inglis, J.T. and Macpherson, J.M. (1995) Bilateral labyrinthectomy in the cat: effects on the postural response to translation. J. Neurophysiol., 73: 1181–1191.

Ivanenko, Y.P., Cappellini, G., Dominici, N., Poppele, R.E. and Lacquaniti, F. (2005) Coordination of locomotion with voluntary movements in humans. J. Neurosci., 25: 7238–7253.

Ivanenko, Y.P., Grasso, R., Zago, M., Molinari, M., Scivoletto, G., Castellano, V., Macellari, V. and Lacquaniti, F. (2003) Temporal components of the motor patterns expressed by the human spinal cord reflect foot kinematics. J. Neurophysiol., 90: 3555–3565.

Ivanenko, Y.P., Poppele, R.E. and Lacquaniti, E. (2004) Five basic muscle activation patterns account for muscle activity during human locomotion. J. Physiol. Lond., 556: 267–282.

Jeka, J., Allison, L., Saffer, M., Zhang, Y., Carver, S. and Kiemel, T. (2006) Sensory reweighting with translational visual stimuli in young and elderly adults: the role of state-dependent noise. Exp. Brain Res., 174: 517–527.

Keshner, E.A., Allum, J.H. and Pfaltz, C.R. (1987) Postural coactivation and adaptation in the sway stabilizing responses of normals and patients with bilateral vestibular deficit. Exp. Brain Res., 69: 77–92.

Keshner, E.A., Woollacott, M.H. and Debu, B. (1988) Neck, trunk and limb muscle responses during postural perturbations in humans. Exp. Brain Res., 71: 455–466.

van der Kooij, H., Jacobs, R., Koopman, B. and Grootenboer, H. (1999) A multisensory integration model of human stance control. Biol. Cybern., 80: 299–308.

van der Kooij, H., Jacobs, R., Koopman, B. and Van Der Helm, F. (2001) An adaptive model of sensory integration in a dynamic environment applied to human stance control. Biol. Cybern., 84: 103–115.

Kuhl, P.K. (1994) Learning and representation in speech and language. Curr. Opin. Neurobiol., 4: 812–822.

Kuhl, P.K. (2004) Early language acquisition: cracking the speech code. Nat. Rev. Neurosci., 5: 831–843.

Kuhl, P.K., Andruski, J.E., Chistovich, I.A., Chistovich, L.A., Kozhevnikova, E.V., Ryskina, V.L., Stolyarova, E.I., Sundberg, U. and Lacerda, F. (1997) Cross-language analysis of phonetic units in language addressed to infants. Science, 277: 684–686.

Kuo, A.D. (1995) An optimal control model for analyzing human postural balance. IEEE Trans. Biomed. Eng., 42: 87–101.

Kuo, A.D. (2005) An optimal state estimation model of sensory integration in human postural balance. J. Neural Eng., 2: S235–S249.

Kuo, A.D., Speers, R.A., Peterka, R.J. and Horak, F.B. (1998) Effect of altered sensory conditions on multivariate descriptors of human postural sway. Exp. Brain Res., 122: 185–195.

Kurtzer, I., Pruszynski, J.A., Herter, T.M. and Scott, S.H. (2006) Primate upper limb muscles exhibit activity patterns that differ from their anatomical action during a postural task. J. Neurophysiol., 95: 493–504.

Lackner, J.R., Rabin, E. and Dizio, P. (2000) Fingertip contact suppresses the destabilizing influence of leg muscle vibration. J. Neurophysiol., 84: 2217–2224.

Lee, D.D. and Seung, H.S. (1999) Learning the parts of objects by non-negative matrix factorization. Nature, 401: 788–791.

Lemay, M.A. and Grill, W.M. (2004) Modularity of motor output evoked by intraspinal microstimulation in cats. J. Neurophysiol., 91: 502–514.

Leonardo, A. (2005) Degenerate coding in neural systems. J. Comp. Physiol. A Neuroethol. Sens. Neural Behav. Physiol., 191: 995–1010.

Liu, C.K., Hertzmann, A. and Popovic, Z. (2005) Learning physics-based motion style with nonlinear inverse optimization. ACM Transactions on Graphics, 24(3): 1071–1081.

Liu, C.K. and Popovic, Z. (2002) Synthesis of complex dynamic character motion from simple animations. ACM Transactions on Graphics, 21(3): 408–416.

Lockhart, D.B. (2005) Prediction of muscle activation patterns during postural control using a feedback control model. In: Woodruff School of Mechanical Engineering. Georgia Institute of Technology, Atlanta, GA.

Lockhart, D.B., Stapley, P.J., Macpherson, J.M. and Ting, L.H. (2005) Prediction of muscle activation patterns during postural perturbation before and after peripheral neuropathy in cats. Program #868.2 2005, Abstract Viewer/Itinerary Planner. Washington, DC, Society for Neuroscience. Online.

Loeb, E.P., Giszter, S.F., Saltiel, P., Bizzi, E. and Mussa-Ivaldi, F.A. (2000) Output units of motor behavior: an experimental and modeling study. J. Cogn. Neurosci., 12: 78–97.

Macpherson, J.M. (1988) Strategies that simplify the control of quadrupedal stance II: electromyographic activity. J. Neurophysiol., 60: 218–231.

Macpherson, J.M. (1991) How flexible are muscle synergies? In: Humphrey D.R. and Freund H.-J. (Eds.), Motor Control: Concepts and Issues. Wiley, New York.

Macpherson, J.M. (1994) Changes in a postural strategy with inter-paw distance. J. Neurophysiol., 71: 931–940.

Maki, B.E., Holliday, P.J. and Topper, A.K. (1991) Fear of falling and postural performance in the elderly. J. Gerontol., 46: M123–M131.

Maki, B.E., Mcilroy, W.E. and Fernie, G.R. (2003) Change-in-support reactions for balance recovery. IEEE Eng. Med. Biol. Mag., 22: 20–26.

Maurer, C., Mergner, T., Bolha, B. and Hlavacka, F. (2001) Human balance control during cutaneous stimulation of the plantar soles. Neurosci. Lett., 302: 45–48.

Mckay, J.L., Burkholder, T.J. and Ting, L.H. (2007) Biomechanical capabilities influence postural control strategies in the cat hindlimb. J. Biomech., 40(10): 2254–2260.

McMahon, T.A. and Cheng, G.C. (1990) The mechanics of running: how does stiffness couple with speed? J. Biomech., 23(Suppl 1): 65–78.

Merfeld, D.M., Zupan, L. and Peterka, R.J. (1999) Humans use internal models to estimate gravity and linear acceleration. Nature, 398: 615–618.

Mergner, T., Maurer, C. and Peterka, R.J. (2003) A multisensory posture control model of human upright stance. Prog. Brain Res., 142: 189–201.

Mergner, T. and Rosemeier, T. (1998) Interaction of vestibular, somatosensory and visual signals for postural control and motion perception under terrestrial and microgravity conditions: a conceptual model. Brain Res. Brain Res. Rev., 28: 118–135.

Minetti, A.E. (2001) Walking on other planets. Nature, 409: 467–469.

Minino, A.M., Arias, E., Kochanek, K.D., Murphy, S.L. and Smith, B.L. (2002) Deaths: final data for 2000. Natl. Vital Stat. Rep., 50: 1–119.

Mouchnino, L., Aurenty, R., Massion, J. and Pedotti, A. (1992) Coordination between equilibrium and head-trunk orientation during leg movement: a new strategy build up by training. J. Neurophysiol., 67: 1587–1598.

Mouchnino, L., Aurenty, R., Massion, J. and Pedotti, A. (1993) Is the trunk a reference frame for calculating leg position? Neuroreport, 4: 125–127.

Nardone, A., Giordano, A., Corra, T. and Schieppati, M. (1990) Responses of leg muscles in humans displaced while standing: effects of types of perturbation and of postural set. Brain, 113(Pt 1): 65–84.

Nashner, L.M. (1976) Adapting reflexes controlling the human posture. Exp. Brain Res., 26: 59–72.

Nashner, L.M. (1977) Fixed patterns of rapid postural responses among leg muscles during stance. Exp. Brain Res., 30: 13–24.

Neptune, R.R. (2000) Computer modeling and simulation of human movement: applications in sport and rehabilitation. Phys. Med. Rehabil. Clin. N. Am., 11: 417–434 viii.

Nichols, T.R. (1994) A biomechanical perspective on spinal mechanisms of coordinated muscular action: an architecture principle. Acta Anat. (Basel), 151: 1–13.

Nichols, T.R., Cope, T.C. and Abelew, T.A. (1999) Rapid spinal mechanisms of motor coordination. Exerc. Sport Sci. Rev., 27: 255–284.

Pandy, M.G. (2001) Computer modeling and simulation of human movement. Annu. Rev. Biomed. Eng., 3: 245–273.

Park, S., Gianna-Poulin, C., Black, F.O., Wood, S. and Merfeld, D.M. (2006) Roll rotation cues influence roll tilt perception assayed using a somatosensory technique. J. Neurophysiol., 96: 486–491.

Park, S., Horak, F.B. and Kuo, A.D. (2004) Postural feedback responses scale with biomechanical constraints in human standing. Exp. Brain Res., 154: 417–427.

Peterka, R.J. (2000) Postural control model interpretation of stabilogram diffusion analysis. Biol. Cybern., 82: 335–343.

Peterka, R.J. (2002) Sensorimotor integration in human postural control. J. Neurophysiol., 88: 1097–1118.

Peterka, R.J. and Loughlin, P.J. (2004) Dynamic regulation of sensorimotor integration in human postural control. J. Neurophysiol., 91: 410–423.

Popov, K.E., Smetanin, B.N., Gurfinkel, V.S., Kudinova, M.P. and Shlykov, V. (1986) Spatial perception and vestibulomotor reactions in man. Neirofiziologiia, 18: 779–787.

Prinz, A.A., Bucher, D. and Marder, E. (2004) Similar network activity from disparate circuit parameters. Nat. Neurosci., 7: 1345–1352.

Raasch, C.C. and Zajac, F.E. (1999) Locomotor strategy for pedaling: muscle groups and biomechanical functions. J. Neurophysiol., 82: 515–525.

Raasch, C.C., Zajac, F.E., Ma, B. and Levine, W.S. (1997) Muscle coordination of maximum-speed pedaling. J. Biomech., 30: 595–602.

Risher, D.W., Schutte, L.M. and Runge, C.F. (1997) The use of inverse dynamics solutions in direct dynamics simulations. J. Biomech. Eng., 119: 417–422.

Runge, C.F., Shupert, C.L., Horak, F.B. and Zajac, F.E. (1998) Role of vestibular information in initiation of rapid postural responses. Exp. Brain Res., 122: 403–412.

Runge, C.F., Shupert, C.L., Horak, F.B. and Zajac, F.E. (1999) Ankle and hip postural strategies defined by joint torques. Gait Posture, 10: 161–170.

Scholz, J.P. and Schoner, G. (1999) The uncontrolled manifold concept: identifying control variables for a functional task. Exp. Brain Res., 126: 289–306.

Scholz, J.P., Schoner, G. and Latash, M.L. (2000) Identifying the control structure of multijoint coordination during pistol shooting. Exp. Brain Res., 135: 382–404.

Scinicariello, A.P., Inglis, J.T. and Collins, J.J. (2002) The effects of stochastic monopolar galvanic vestibular stimulation on human postural sway. J. Vestib. Res., 12: 77–85.

Scott, S.H. (2003) The role of primary motor cortex in goal-directed movements: insights from neurophysiological studies on non-human primates. Curr. Opin. Neurobiol., 13: 671–677.

Scott, S.H. (2004) Optimal feedback control and the neural basis of volitional motor control. Nat. Rev. Neurosci., 5: 534–546.

Sorensen, K.L., Hollands, M.A. and Patla, E. (2002) The effects of human ankle muscle vibration on posture and balance during adaptive locomotion. Exp. Brain Res., 143: 24–34.

Stapley, P.J., Ting, L.H., Hulliger, M. and Macpherson, J.M. (2002) Automatic postural responses are delayed by pyridoxine-induced somatosensory loss. J. Neurosci., 22: 5803–5807.

Tardy-Gervet, M.F. and Severac-Cauquil, A. (1998) Effect of galvanic vestibular stimulation on perception of subjective vertical in standing humans. Percept. Mot. Skills, 86: 1155–1161.

Timmann, D. and Horak, F.B. (1997) Prediction and set-dependent scaling of early postural responses in cerebellar patients. Brain, 120(Pt 2): 327–337.

Ting, L.H., Kautz, S.A., Brown, D.A. and Zajac, F.E. (1999) Phase reversal of biomechanical functions and muscle activity in backward pedaling. J. Neurophysiol., 81: 544–551.

Ting, L.H., Lockhart, D.B., Stapley, P.J. and Macpherson, J.M. (2005) Feedback regulation of temporal muscle activation patterns for postural control before and after peripheral neuropathy. Gait Posture, 21: S62–S63.

Ting, L.H. and Macpherson, J.M. (2004) Ratio of shear to load ground-reaction force may underlie the directional tuning of the automatic postural response to rotation and translation. J. Neurophysiol., 92: 808–823.

Ting, L.H. and Macpherson, J.M. (2005) A limited set of muscle synergies for force control during a postural task. J. Neurophysiol., 93: 609–613.

Todorov, E. (2000) Direct cortical control of muscle activation in voluntary arm movements: a model. Nat. Neurosci., 3: 391–398.

Todorov, E. (2004) Optimality principles in sensorimotor control. Nat. Neurosci., 7: 907–915.

Todorov, E. and Jordan, M.I. (2002) Optimal feedback control as a theory of motor coordination. Nat. Neurosci., 5: 1226–1235.

Torres-Oviedo, G., Macpherson, J.M. and Ting, L.H. (2006) Muscle synergy organization is robust across a variety of postural perturbations. J. Neurophysiol., 96: 1530–1546.

Torres-Oviedo, G. and Ting, L.H. (2007) Muscle synergies characterizing human postural responses. J. Neurophysiol. [Epub ahead of print].

Tresch, M.C., Cheung, V.C. and d'Avella, A. (2006) Matrix factorization algorithms for the identification of muscle synergies: evaluation on simulated and experimental data sets. J. Neurophysiol., 95: 2199–2212.

Tresch, M.C., Saltiel, P. and Bizzi, E. (1999) The construction of movement by the spinal cord. Nat. Neurosci., 2: 162–167.

Valero-Cuevas, F.J. (2000) Predictive modulation of muscle coordination pattern magnitude scales fingertip force magnitude over the voluntary range. J. Neurophysiol., 83: 1469–1479.

Valero-Cuevas, F.J., Zajac, F.E. and Burgar, C.G. (1998) Large index-fingertip forces are produced by subject-independent patterns of muscle excitation. J. Biomech., 31: 693–703.

Van Antwerp, K.V., Burkholder, T.J. and Ting, L.H. (2007) Interjoint coupling effects on muscle contributions to endpoint force and acceleration in a musculoskeletal model of the cat hindlimb. J. Biomech., doi:10.1016/j.jbiochem.2007.06.001

Wallace, M.T., Carriere, B.N., Perrault Jr., T.J., Vaughan, J.W. and Stein, B.E. (2006) The development of cortical multisensory integration. J. Neurosci., 26: 11844–11849.

Wallace, M.T., Perrault Jr., T.J., Hairston, W.D. and Stein, B.E. (2004) Visual experience is necessary for the development of multisensory integration. J. Neurosci., 24: 9580–9584.

Wallace, M.T. and Stein, B.E. (2006) Early experience determines how the senses will interact. J. Neurophysiol., 97(1): 921–926.

Welch, T.D.J. and Ting, L.H. (2005) The initial burst of the human automatic postural response scales with the perturbation acceleration and velocity during quiet stance. Society for Neuroscience, Program #56.11 2005 Abstract Viewer/Itinerary Planner. Washington, DC, Society for Neuroscience. Online.

Wilmink, R.J. and Nichols, T.R. (2003) Distribution of heterogenic reflexes among the quadriceps and triceps surae muscles of the cat hind limb. J. Neurophysiol., 90(4): 2310–2324.

Woollacott, M.H. and Shumway-Cook, A. (2002) Attention and the control of posture and gait: a review of an emerging area of research. Gait Posture, 16: 1–14.

Zajac, F.E., Neptune, R.R. and Kautz, S.A. (2003) Biomechanics and muscle coordination of human walking. Part II: lessons from dynamical simulations and clinical implications. Gait Posture, 17: 1–17.

Zhang, Y., Kuhl, P.K., Imada, T., Kotani, M. and Tohkura, Y. (2005) Effects of language experience: neural commitment to language-specific auditory patterns. NeuroImage, 26: 703–720.

Zupan, L.H., Merfeld, D.M. and Darlot, C. (2002) Using sensory weighting to model the influence of canal, otolith and visual cues on spatial orientation and eye movements. Biol. Cybern., 86: 209–230.

P. Cisek, T. Drew & J.F. Kalaska (Eds.)
Progress in Brain Research, Vol. 165
ISSN 0079-6123
Copyright © 2007 Elsevier B.V. All rights reserved

CHAPTER 20

Primitives, premotor drives, and pattern generation: a combined computational and neuroethological perspective

Simon Giszter*, Vidyangi Patil and Corey Hart

Neurobiology and Anatomy, Drexel University College of Medicine, 2900 Queen Lane, Philadelphia, PA 19129, USA

Abstract: A modular motor organization may be needed to solve the degrees of freedom problem in biological motor control. Reflex elements, kinematic primitives, muscle synergies, force-field primitives and/or pattern generators all have experimental support as modular elements. We discuss the possible relations of force-field primitives, spinal feedback systems, and pattern generation and shaping systems in detail, and review methods for examining underlying motor pattern structure in intact or semi-intact behaving animals. The divisions of systems into primitives, synergies, and rhythmic elements or oscillators suggest specific functions and methods of construction of movement. We briefly discuss the limitations and caveats needed in these interpretations given current knowledge, together with some of the hypotheses arising from these frameworks.

Keywords: primitives; motor synergies; force-fields; modularity; feedback; motor pattern analysis; decomposition; rhythm generation; pattern shaping

Introduction

The goal of this paper is to present a perspective on spinal cord and lower motor system modularity in the context of both comparative/neuroethological (see e.g., Hoyle, 1970) and computational neuroscience. We briefly review a neuroethological perspective on modularity, and the degrees of freedom problem that motivates much research on modularity. We then analyze force-field primitives as modules and computational elements and the hypotheses arising from such primitives. Finally, we examine how best to detect and analyze primitives and oscillators in the behavior and electromyographic (EMG) activity patterns of awake animals, and how primitives may fit into a general pattern generator framework.

Modularity from a neuroethological perspective

From a Darwinian perspective the objective of all organisms is to maximize inclusive fitness (Hamilton, 1964; McFarland and Houston, 1981; Wilson, 2000). For this objective various strategies can succeed. Control policies range from those of sessile unicellular prokaryotes (Kurata et al., 2006) to organisms with complex nervous systems and motor capabilities. All of these incorporate predictions about the best interactions of the organism with its niche, honed on an evolutionary timescale.

*Corresponding author. Tel.: +1 215 991 8412;
Fax: +1 215 843 9082; E-mail: simon.giszter@drexel.edu

DOI: 10.1016/S0079-6123(06)65020-6

The special advantage enjoyed by animals with complex nervous systems is the use of more complex behaviors in rapid and flexible anticipatory responses. Such motor skills are often elaborated both during ontogeny and as adults. An ecological niche is a moving target, as climatic changes demonstrate. Both ontogenetic and adult adaptations can assist in tracking niche changes. These enable the "Baldwin effect" in the natural selection process (Baldwin, 1896; see Bateson, 2004). The lower level motor apparatus is the substrate for building complex behaviors. How these are constructed is one of the topics of neuroethology and our focus here.

Statisticians are philosophically divided into Bayesian or Frequentist groups. Crudely, the former are willing to assign probability values to beliefs and events, and the latter assign probability only to observable events from a sample space. Despite these major differences in their philosophy, for most statisticians the process of prediction has a Bayesian form. Thus Bayesian analysis can likely provide major insight into how animals organize adaptive processes. However, it is unknown to what extent the near-Bayesian predictions represented by many skilled behaviors and motor functions should be attributed to the products of evolution, of ontogeny, or of rapid learning processes in the adult. This is an issue of broad importance. Each possibility implies differing degrees of flexibility and computational sophistication, and differing origins for modularity (see e.g., Callebaut and Rasskin-Gutman, 2005). One area in which this issue comes into particularly sharp contrast is in the degrees of freedom problem in the motor system.

Degrees of freedom problem and evolution

The first clear statement of the degrees of freedom problem in motor control is probably due to Bernstein (1967). Excess of degrees of freedom are available at the joint-level for positioning the limb. There also is a redundancy of muscles available to generate torques at these joints. This embarrassment of riches makes the issue of choosing the best limb postures, the best trajectories, and the best muscle activation patterns potentially very complex (the so-called "curse of dimensionality"). For example, Wolpert et al. (2001) point out that human "muscle space" comprises ~600 muscles. Considering only binary activation of these muscles, there are 2^{600} patterns, which is a number larger than the estimated count of atoms in the known universe! The available degrees of freedom are powerful if engaged appropriately. They enable immense flexibility, and afford rapid use of novel tools, and new motor strategies. These may include new types of movement perhaps never employed previously in evolutionary history (e.g., the Fosbury flop strategy in the high jump). However, the high flexibility available with these degrees of freedom comes at a cost. Some of this cost can be alleviated by inclusion of additional constraints (e.g., Mussa-Ivaldi and Hogan, 1991), but this problem can nevertheless be severe.

To make the potential costs of the flexibility of the motor apparatus concrete, consider a wildebeest calf born on the plains of Africa, or a hatchling turtle on a Caribbean beach. Both are subject to rapid predation, and they are likely to soon be devoured if they do not rapidly organize directed locomotion. The wildebeest must move with the herd within a few hours. The hatched turtle must rapidly leave the beach. Natural selection will thus favor rapid movement development and management of motor complexity, and degrees of freedom issues.

Modularity may provide a solution to the degrees of freedom problem for the developing animal. Modularity is also an obvious way to solve several other problems in motor control. It can help to (1) manage complexity, (2) provide a basis to build more complex motions, and (3) provide a "seed" or bootstrap to motor learning, based on expectations about the world. Again, modularity like that needed by the wildebeest calf could either be built-in by evolution directly, or it could arise as a result of an evolved computational strategy, which possesses a strong predisposition to modularize. Additional finer adjustments of the motor nervous system to the specifics of the individual motor apparatus, and novel motor adaptations or inventions, can then be added through experience.

Forms of Modularity in the motor system

The many forms of modularity currently identified in motor control are largely operationally defined elements. These can be broadly classified as reflexes, primitives (kinematic strokes, force-fields, and/or synergies) and central pattern generators (CPGs), and component oscillators. Marr's decomposition of a computational problem solved by the central nervous system (CNS) breaks a problem into a task specification, the possible algorithms, and the implementation details (Marr, 1982). This framework can be useful in thinking about these differing modular descriptions and integrating them.

Reflexes represent rapid and relatively fixed responses adapted to a single stimulus or several stimuli. The Sherrington classification (see Sherrington, 1961), refined by others, provides a fundamental conceptual framework. Refinement of the basic notions of Sherrington remains part of the business of contemporary motor control. Reflexes are fundamental aspects of neural implementation details. Polysynaptic protective reflexes (e.g., scratch reflex) can in addition encompass goal-directed features, spanning from task level, through algorithm, to implementation.

Kinematic primitives (also termed kinematic strokes) are unitary elements or patterns in trajectory formation. These were first introduced as a basis for analysis of multiple and figured movements (Viviani and Terzuolo, 1982). Strokes are usually considered to conform to some kinematic objective function (usually considered to be minimum jerk, Hogan, 1984). The Flash, Milner, Hogan, and other laboratories have developed careful decompositions in terms of such individual trajectory strokes (e.g., Flash and Hogan, 1985; Burdet and Milner, 1998; Sanger, 2000; Rohrer et al., 2002; Sosnik et al., 2004; Flash and Hochner, 2005). These provide a strong account of planning of motion. It is also possible that some features of kinematic primitives may be broadly conserved across evolution (e.g., see Sumbre et al., 2006).

Kinematic oscillations or rhythmic kinematic primitives have also been developed. They form a computational framework to capture imitation learning and planning due to Schaal, Ijspeert, Sternad, Kawato, and colleagues (Ijspeert et al., 2003; Schaal et al., 2003). Formal development partly relates to earlier Gibsonian frameworks (see e.g., Kelso et al., 1981; Balasubramaniam and Turvey, 2004). Rhythmic primitives form compact dynamical representations of multi-joint task or planning kinematics that guarantee task achieval and stability. Their modules include: (1) discrete point-to-point kinematic motions, with static attractors (similar to strokes); and (2) limit-cycle oscillators at joints. The scheme supports imitation learning in biomorphic robots and gives a plausible account of kinematic imitation learning in man. Currently, execution details are managed downstream from these kinematic representations (by a separated inverse dynamics or execution capacity). These oscillation primitives resemble the earlier motion planning and kinematic decompositions, and the unit oscillators and pattern generators that follow, but defined at a task level.

CPGs represent the intrinsic capacity of the CNS for generating rhythmic ordered pattern in the absence of ordered and patterned input either centrally or peripherally (Wilson, 1961; Grillner et al., 1976). In the operational definition of a CPG, sequencing or rhythmicity of muscle use are central, and amplitude relations are usually considered less crucial. Dynamical systems or limit cycle oscillators driving motor patterns are then usually inferred. There is an enormous wealth of research in this area that other chapters will address with significantly more authority. Perhaps the most powerful result of the CPG operational definition has been the frequent use and active development of fictive preparations (e.g., Kiehn et al., 1997).

Motor synergies represent co-occuring or, more strongly, covarying joint or muscle use. Bernstein emphasized this idea. Today there are sometimes quite different usages (Gottlieb, 1998; d'Avella et al., 2003; Cappellini et al., 2006; Torres-Oviedo et al., 2006). A central reduction in degrees of freedom is inferred. The modularity of muscle effectors as synergies are closer to algorithm and implementation than to task in Marr's scheme, thus linking these two together.

Force-field primitives and the broader motor primitives were introduced as execution elements. These support modular execution and organization of the biomechanics of movement construction. Force-field primitives as execution elements represent a specific structure of synergies and reflexes. Force-field structures may imply modular linearly covarying muscle use (Bizzi et al., 1991; Giszter et al., 1993; Mussa-Ivaldi et al., 1994; Giszter et al., 2000). Primitives are often considered to be recruited in fixed duration or bounded duration bursts (Kargo and Giszter, 2000; Hart and Giszter, 2004). The force-field description thus implies specific quantitative and testable hypotheses about premotor drive, feedback organization, and the construction of movement.

Muscle synergy burst, premotor drive, and motor primitive have been used as largely synonymous terms by several researchers in this area. The term "primitive" may best indicate the idea of a set of building blocks or developmental bootstrap elements used in a constructive or compositional fashion. A few types of force-field "primitives" have been found in spinal cord in the frog, rat (Tresch and Bizzi, 1999) and cat (Lemay and Grill, 2004). Combination of these by vector superposition can be shown (Mussa-Ivaldi et al., 1994, see Fig. 1), and compositional frameworks have been developed (Mussa-Ivaldi, 1992; Mussa-Ivaldi and Giszter, 1992). Force-field primitives imply specific classes of relationships among the motor pools, muscle synergies and reflexes in implementation, and specific limb kinematics and kinetics. Without these, their compositional use is less tenable, as described next.

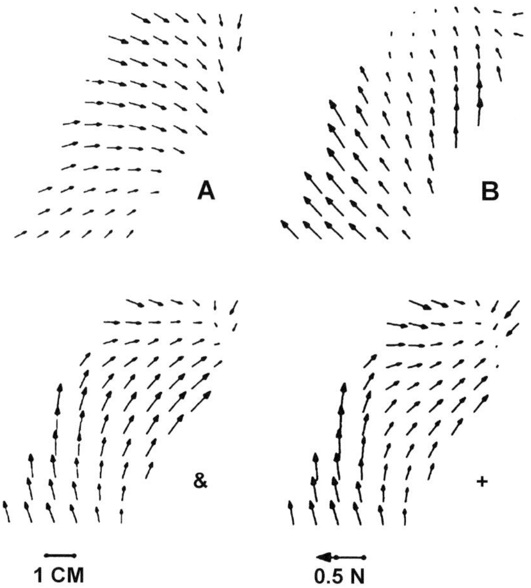

Fig. 1. Vector superposition of force-fields from spinal motor primitives. Adapted with permission from Mussa-Ivaldi et al., 1994. Results of costimulation of the lumbar gray matter in a frog. Force-fields were constructed to represent stimulation results. Each force-vector in the field represents the results of stimulation with the limb held in a configuration with the ankle at that location, and is the force generated at the ankle. It thus takes into account the effects of muscle moment arms, the skeletal linkage force transmission, and any feedback modulation of the muscle recruitment by stimulation. Two results are found: (a) vector summation (~85% of cases) and (b) winner-take-all (~15% of cases). An example of vector summation. (Upper) Fields obtained by separate stimulation of two sites, A and B. (Lower) Costimulation fields (&) and summation field (+). This was by far (87.8% of cases) the most common outcome. The other outcome was winner-take-all. The quality of fit is determined as the inner product derived Cosine. Cos(a/b) close to 1.0 indicate near identity of two tested fields, a and b. In this panel Cos (&/+) = 0.967.

Force-field primitives, premotor drives, and reflex effects: a computational framework

Force-fields may form a natural basis for organizing motor control. Examining how this basis can be used, and under what circumstances it may become unusable can generate testable hypotheses and experiments. This section is, of necessity, equation laden with this goal in mind.

A time varying force pattern F drives motion or interaction with the environment from which we can write limb or body dynamics:

$$M(q)\ddot{q} + G(q, \dot{q}) + E(q, \dot{q}, t) = F(q, \dot{q}, t) \quad (1)$$

where $M(q)$ represents inertial terms, G interaction terms, E environmental forces, and F the torque generation by the musculo-skeletal plant. The problem of movement generation and interaction is then to deliver an appropriate set of forces F to satisfy the task constraints. These constraints may be purely kinematic (e.g., pointing, or gesture), the action of counteracting an environmental force/load (e.g., stationary support of a book), or a combination of target motions and interaction forces with the environment (e.g., martial arts or manipulation of a paint brush, see Mussa-Ivaldi and Bizzi, 2000). Following the approach of Mussa-Ivaldi, activity of muscles and feedback pathways as a group can be generically represented as a multidimensional time varying force-field in joint space:

$$F(q, \dot{q}, t) = C(q, \dot{q}, u(t)) \tag{2}$$

where q_i, $\dot{q}$, are joint angles and angular velocities, t time, F the field expressed as joint torques, in general joint coordinates, $u(t)$ the applied control, and C a (noninvertible) function transforming muscle activations to F. Spinal force-field motor primitives can provide a modular basis for constructing this potentially arbitrary field and designing $u(t)$. Mussa-Ivaldi (1992) showed that an arbitrary smooth static field may be approximated with a combination of equal numbers of static circulating and conservative fields:

$$F(q) = \sum_i \Phi_i(q) + \sum_i \Theta_i(q) \tag{3}$$

where the Θ are circulating fields and Φ conservative fields. The fields found in practice in experimental work show no significant circulation (usually under 5%), and thus emulate passive conservative systems. Accordingly, we can eliminate $\Theta(q)$ in the approximation in Eq. (3). This is consistent with constraints that guarantee passive stability in interaction (see Colgate and Hogan, 1988). Muscles have viscous properties and including these and extending (3) leads to an approximation of Eq. (2) as follows:

$$F(q, \dot{q}, t) = \sum_i A_i a_i(t) \Phi_i(q, \dot{q}) \tag{4}$$

where a viscous term is added in each Φ_i and the $a_i(t)$ represents the normalized (unit amplitude) time-courses of the force-field primitives needed for the approximation, and A_i the amplitude scalings of each force-field primitive Φ_i. In the spinal and semi-intact frog, experimental data support a fixed duration for all primitives. The task is then to find the best combination of constant-duration pulses of the force-field elements (see below). Experiments show that the pulsed force-fields can potentially be initiated at arbitrary times or phases (Kargo and Giszter, 2000). We can then express trajectory formation as:

$$F(q, \dot{q}, t) = \sum_i A_i a(t + \tau_i) \phi_i(q, \dot{q}) \tag{5}$$

where A_i is the amplitude of the pulse, τ_i the timing of the pulse, and $a(t)$ the time course of the pulse which is similar for all primitives. A set of predictable pulsed force-field structures $\Phi_i(q, \dot{q}, t)$ arise from synchronous activation of muscles and appropriately balanced feedback. A scheme for this construction of movement combining rhythm generation and pulsed primitives is shown in Fig. 2. However, some feedback could dissolve the fields.

To relate these pulses to muscle activation and feedback and see this, we note that each force-field pulse must arise as the coordinated action of multiple muscles. This can be expressed as:

$$\Phi(q, \dot{q}, t) = \sum_j^N B_j(q, \dot{q}) a(t) \Psi_j(q, \dot{q}) \tag{6}$$

where the measured forces arise as the sum of the endpoint forces $\Psi(q_i, \dot{q})$ produced by the N different component muscles at limb state $q_i, \dot{q}$ where $\Psi(q_i, \dot{q})$ is the normalized force-field effect of the muscle. In isometric conditions, with configuration $(q, 0)$, the overall force direction is fixed and can be expressed as a scaling of $\Phi(q_i, \dot{q})$ (obtained from the summation of the individual muscle $\Psi(q_i, \dot{q})$ scaled by scalar functions B and a). $B_i(q_i, \dot{q})$ is the activation driven peak force amplitude achieved by the muscle in the synergy at location and velocity $(q_i, \dot{q})$, and $a(t)$ the normalized activation. (This can be further reduced to a set of Hill-type or other muscle model form incorporating sarcomere length and muscle moment arm data (Kargo and Rome, 2002).) If $B_i(q_i, \dot{q})$ and $a_i(t)$ are consistent in form across the muscles then the musculo-skeletal plant generates a field which remains a scaled version of itself at all time points. In this case the force-field primitives have a set of properties that can be tested and established experimentally. (Otherwise force vectors rotate over time.) First, the inner product of the direction vectors in each instance of a field (n, m) at the same location are close to 1:

$$\langle \phi_n(q_i, \dot{q}_i) \rangle \cdot \langle \phi_m(q_i, \dot{q}_i) \rangle \sim= 1 \ \forall i, n, m \tag{7}$$

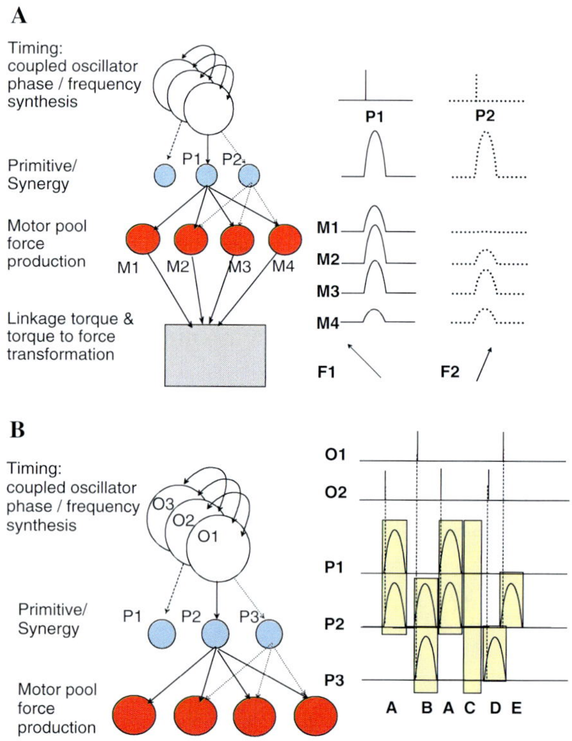

Fig. 2. Superposition of motor primitives in spinal cord: a working hypothesis. Panel A: Our working hypothesis is that spinal mechanisms for trajectory formation, protective reflexes, and locomotion comprise a timing or rhythm generation system driving a pattern-shaping layer with a set of unitary synergies or primitives. The rhythm generating system of one or more oscillator/timers organizes timing and switching of pulses to recruit primitives. Primitives (P1, P2) are thus elicited at specific phases and states of the rhythm generator, or by external cues. Each primitive produces a fixed duration pulse in several motor pools. The balance of drive to the pools (M1–M4) determines torque balance at the joints in a given limb configuration and thereby endpoint forces F1 and F2. F1 and F2 differ in direction and magnitude as a result of the differing balances of M1–M4. Panel B: An example of a hypothetical behavior generated by superposition of primitives. The panel shows the several levels of analysis implied in this scheme. Coupled oscillators O1–O3 generate a rhythm of pulses and switchings (pulse series O1 and O2 on right). These recruit specific combinations of primitives (P1–P3) in sequence (the equivalent of a pattern-shaping layer). This recruitment and switching represents a "clutching" of the rhythm system to the (pattern shaping) layer of primitives, and can differ among tasks. The combinations and overlaps of primitives in the task can be treated as a symbol string, or a chain of state transitions. Thus repeating and branching sequences (measured as symbol strings/ motor pattern), their timing among effectors (rhythm generation), and the drive pulses (primitives) are separately controllable, and identifiable, in this scheme.

where $\langle\ \rangle$ denotes unit direction vector. Second, force amplitude ratios between locations q_i and q_j are also similar across instances of force-field primitives:

$$\sqrt{\left(\frac{\phi_m(q_i,\dot{q}_i).\phi_m(q_i,\dot{q}_i)}{\phi_m(q_j,\dot{q}_j).\phi_m(q_j,\dot{q}_j)}\right)} = \sqrt{\left(\frac{\phi_n(q_i,\dot{q}_i).\phi_n(q_i,\dot{q}_i)}{\phi_n(q_j,\dot{q}_j).\phi_n(q_j,\dot{q}_j)}\right)} = K_{q_i,q_j}\ \forall i,\ j,\ n,\ m \quad (8)$$

This is observed experimentally. Conserving these features during feedback from different reflex systems requires several sets of neural and reflex constraints (see Fig. 3). First, the ratios of muscle amplitudes ($B(q,\dot{q})$ as expressed in Eq. (6)) must either be fixed in ratio across the whole field under the influence of sensory feedback (Eq. (9)) or must have a simple proportionality between fields at each configuration, although different proportionality between configurations q_k and q_m, is allowed (Eq. (10)), i.e.:

$$\frac{B_i(q,\dot{q})}{B_j(q,\dot{q})} = K_{i,j}\ \forall i,j,q,t \quad (9)$$

or

$$\frac{B_i(q_k,\dot{q}_k)}{B_j(q_k,\dot{q}_k)} = K_{i,j,k} \neq \frac{B_i(q_m,\dot{q}_m)}{B_j(q_m,\dot{q}_m)} = K_{i,j,m} \quad (10)$$

The first (fixed ratios of recruitment across the workspace, Eq. (9)), or mild versions of the second, also imply that there is a muscle synergy, and that this could be extracted by statistical means (see following sections). The second form of relation among muscles need not lead to a fixed synergy, but nonetheless preserves the structure of a compositional force-field primitive. Feedback effects could also alter timing of muscles throughout a burst or simply at onset.

Feedback preserving force-field properties

A feedback driven scaling u for each muscle is consistent with Eqs. (7)–(10) and would be:

$$\Phi(q,\dot{q}) = \sum_i^N u\cdot[B_{i\cdot}\cdot a(t)\cdot\Psi_i(q,\dot{q})] \quad (11)$$

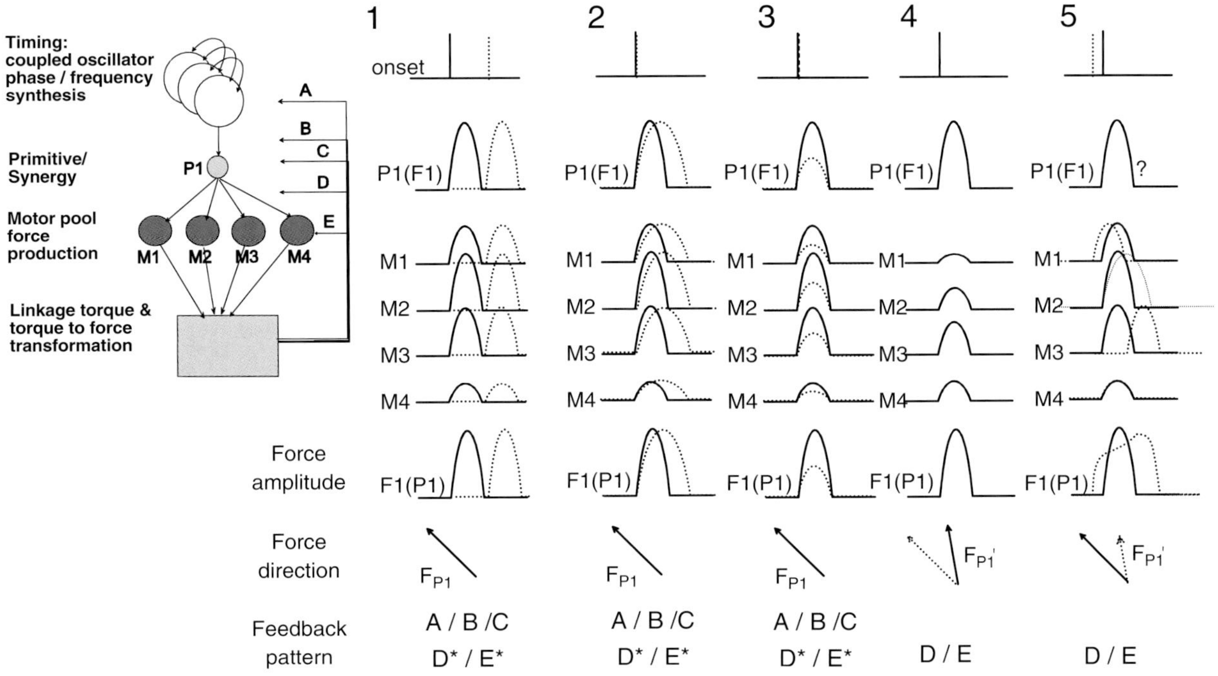

Fig. 3. Types of feedback influences on motor pools and their impact on organization of primitives. We distinguish five levels at which feedback can act in our scheme (A–E on left). These can potentially either alter or preserve the structure and balance of the premotor drive from P1 (i.e., the force-field primitive structure) depending on balance and site of action. Thus if the spinal cord output routinely uses a conserved set of primitives, this places specific constraints on how feedback systems must operate. A represents feedback resetting the oscillators or rhythm synthesis layer; B represents feedback modulating recruitment of a primitive by the rhythm/timing layer; C represents modulation of a whole primitive directly; D represents modulation of the connections of primitive drive to motor pools (e.g., by primary afferent depolarization); and E represents heteronymous or homonymous feedback to motor pools. D* and E* represent balanced feedback within a primitive. The onset timing (top), primitive activation P1, muscle activation M1 through M4, and force Pulse F1 are shown for five patterns of feedback effect. Patterns 1, 2, and 3 represent the groups of feedback effects that can preserve the force pattern structure if balanced (Type A feedback in "Relations among primitives and pattern generators" of the text). Pattern 1 uniformly shifts phase. This could be achieved in several ways. Pattern 2 modulates duration of the pulse uniformly. Pattern 3 modulates amplitude proportionately in all motor pool drives. Pattern 4 and 5 disrupt the predicable force and torque balance of the primitive: Pattern 4 alters balance of muscle amplitudes independently and thereby alters force direction; Pattern 5 shifts recruitment timing of pools independently and thereby alters direction and duration patterns.

where u is a feedback based amplitude scaling. Similarly, a uniform phase shift could be applied based on feedback u and preserves properties in Eqs. (7)–(10):

$$\Phi(q,\dot{q}) = \sum_{i}^{N} [B_{i\cdot} \cdot a(t+u\tau) \cdot \Psi_i(q,\dot{q})] \qquad (12)$$

where τ is a uniform phasing parameter that preserves both field structure and pulse duration.

These patterns of adjustment by feedback are fully consistent with the kind of movement construction observed experimentally in the frog (Eq. (5)). However, there are several alternate feedback effects possible.

Feedback modulated pulse changes

Dilation of the pulse by a feedback parameter u preserves force-field structure alone, but not pulse properties:

$$\Phi(q,\dot{q}) = \sum_{i}^{N} [B_{i\cdot} \cdot a(ut) \cdot \Psi_i(q,\dot{q})] \qquad (13)$$

where u is a feedback based time dilation or frequency parameter. Eq. (13) conserves field structure (i.e., satisfies Eqs. (7)–(10)) but approximation in the form of Eq. (5), as combination of fixed duration pulses, is no longer possible.

Modulation of the basis set

Feedback that causes each component muscle contributing to a force-field to scale based on the configuration will violate only Eq. (9). However, this alters the basis set with feedback compared to that with no feedback:

$$\Phi(q,\dot{q}) = \sum_{i}^{N} u_i(q,\dot{q}) \cdot [B_{i\cdot} \cdot a(t) \cdot \Psi_i(q,\dot{q})] \quad (14)$$

Each $u_i(q_i, \dot{q})$ is a scalar function of configuration. Eq. (14) generates a new set or basis of force-fields. Depending on offset and depth of modulation in the functions $u_i(q_i, \dot{q})$ these can be divided into a set of unmodulated basis fields (offset or feedforward force-fields satisfying Eqs. (8) and (9)) and a set of feedback modulated basis fields. These can continue to satisfy Eqs. (7) and (10) (directional invariance at a location), and preserve $a(t)$ but will violate the condition of fixed amplitude ratios of muscle recruitment across configurations (Eqs. (8) and (9)). The feedback alters the basis set of force-fields but combinations and approximations will still be feasible with the larger but still modular feedback-formed basis. These effects in combination still preserve a predictable, albeit more complex, force-field basis set:

$$\Phi(q,\dot{q}) = \sum_{i}^{N} u_i(q,\dot{q}) \cdot [B_{i\cdot} \cdot a(wt+\tau) \cdot \Psi_i(q,\dot{q})] \quad (15)$$

Thus preserving both force pattern and pulse form will be satisfied only by specific balanced feedback regulations and patterns through homonymous and heteronymous pathways.

Feedback destruction of modular organization

Strong feedback can severely challenge the idea of a compositional basis of force-fields if feedback alters force-direction at a given configuration (i.e., violates Eq. (7)). A necessary (but not sufficient) condition for violating Eq. (7), is that any individual force-field primitives' component muscles are altered independently of one another. In all except a minority of special cases (those involving the trading of joint torque contributions between several muscles) the endpoint force direction is altered, if adjustments by feedback take any of the following forms:

A. New activation source or addition of a muscle, $u_i(q, \dot{q})$

$$\Phi(q,\dot{q}) = \sum_{i}^{N} (B_i(q,\dot{q}) + u_i(q,\dot{q}))[a(t)] \cdot \Psi_i(q,\dot{q}) \quad (16)$$

B. Separate phasing of muscle activations, ($u\tau_i$ does not equal $u\tau_k$)

$$\Phi(q,\dot{q}) = \sum_{i}^{N} B_i(q,\dot{q})[a(t+u_i\tau_i)] \cdot \Psi_i(q,\dot{q}) \quad (17)$$

C. Separate time dilation of individual muscle activations (by w_j)

$$\Phi(q,\dot{q}) = \sum_{i}^{N} B_i(q,\dot{q})[a(w_i t)] \cdot \Psi_i(q,\dot{q}) \quad (18)$$

$$\Phi(q,\dot{q}) = \sum_{j}^{M} (B_j(q,\dot{q}) + v_j(q,\dot{q}))[a(w_j t + \tau_j)] \cdot \Psi_j(q,\dot{q}) \quad (19)$$

Each of these feedback patterns would reorganize the basis set and timing and make approximation using the form of Eq. (5) impossible. This framework of basis force-field primitives thus leads to several strong experimentally testable hypotheses: Feedback patterns that independently alter the onset, time course, or strength of recruitment of single muscles in a primitive will not occur, if the compositional scheme in Eq. (5) is to hold. We have been looking for such violations, caused by altered feedback or load in the frog, but to date we have found none. It is also interesting in this regard that feedback gains often drop precipitously in complex tasks such as cats' pedestal walking and speculatively might thereby better preserve the basis set.

Experimental competence of a basis of force-field primitives

Competence of some of the controls that can be formulated according to Eq. (5) to capture rich behavior have now been examined in number of studies based on theoretical analyses (e.g., Mussa-Ivaldi and Bizzi, 2000), on data-driven simulation in 2D (e.g., Giszter and Kargo, 2001), or in 3D models (Kargo et al., unpublished data).

Early support for this general framework has come primarily from the frog. Vector summation (Mussa-Ivaldi et al., 1994; Lemay et al., 2001) and modular EMG decompositions (d'Avella et al., 2003; Hart and Giszter, 2004) have been shown. A trajectory correction response using a primitive has been identified (Kargo and Giszter, 2000). These results are similar to blending in turtle scratch (Earhart and Stein, 2000). Deletions of primitives have also been shown (Giszter and Kargo, 2000), similar to deletions in turtle and in fictive cat locomotion (Lafreniere-Roula and McCrea, 2005). Correction of hindlimb wipes in frogs is an interesting demonstration of the uses of primitives in a real behavior and does not require any statistical decomposition. The correction element represents one of the sets of six primitives identified that is added to the motor pattern de-novo following a collision. This addition acts to modify trajectory generation so as to circumvent the impacted obstacle in a spinalized frog. Remarkably, as a result of the correction primitive addition, the limb reaches the target within 50 ms of its unperturbed arrival time. The primitive is only recruited following a collision during the period of motion when its superposition can successfully achieve this with a hip flexion. The spinal cord thus appears to assemble primitives based on their potential utility in achieving the task goal. In some way it represents the contingent relationships among the primitives and task. The degree to which this result for scratch reflexes should be thought of as representing the existence of a predictive internal model of the task is currently unclear, but is an interesting issue.

An alternate scheme: time-varying synergies as primitives

It is possible to combine rhythm and pattern generator frameworks with pulsed primitives in a hierarchical fashion as discussed here. However, an alternative scheme combines these together, into *time-varying synergies* (d'Avella and Bizzi, 2005). In these, the element of construction is a sequence of waveforms of muscle use. These waveforms need not represent unitary and common premotor drives to muscle groups. For example, they could take the form described by Eqs. (17)–(19) above, and be inconsistent with Eq. (5). They are a combination of sequencings of individually phased muscle drives, treated as a unit. From a perspective of statistical decomposition, joint estimation of muscle activation and its timing is attractive. Indeed, if the time-varying patterns are actually built from synchronous pulses, the two formulations differ very little. However, estimating more complex waveform structures may require significantly more data than the equivalent successive identification and estimation of synergies and then pulses, in a timing and synergy cascade (e.g., see Westwick and Kearney, 2003). Extending the use of more complex waveforms to rapid on-line adaptations such as frog corrections may also present computational challenges. These are presented graphically in Fig. 4. Time-varying synergies as a descriptive scheme must ultimately be judged based on parameter parsimony, quality of fit achieved, adaptive flexibility, and simplicity of the compositional scheme. Here we focus on the hierarchical organization of pulses and synchronous premotor drives.

Relations among primitives and pattern generators

It is well established that various reflexes can be recruited and modulated by the CPG or rhythm generator (e.g., in locomotion). However, much more problematic is the relation of primitives, synergies, and pattern generation. There are a

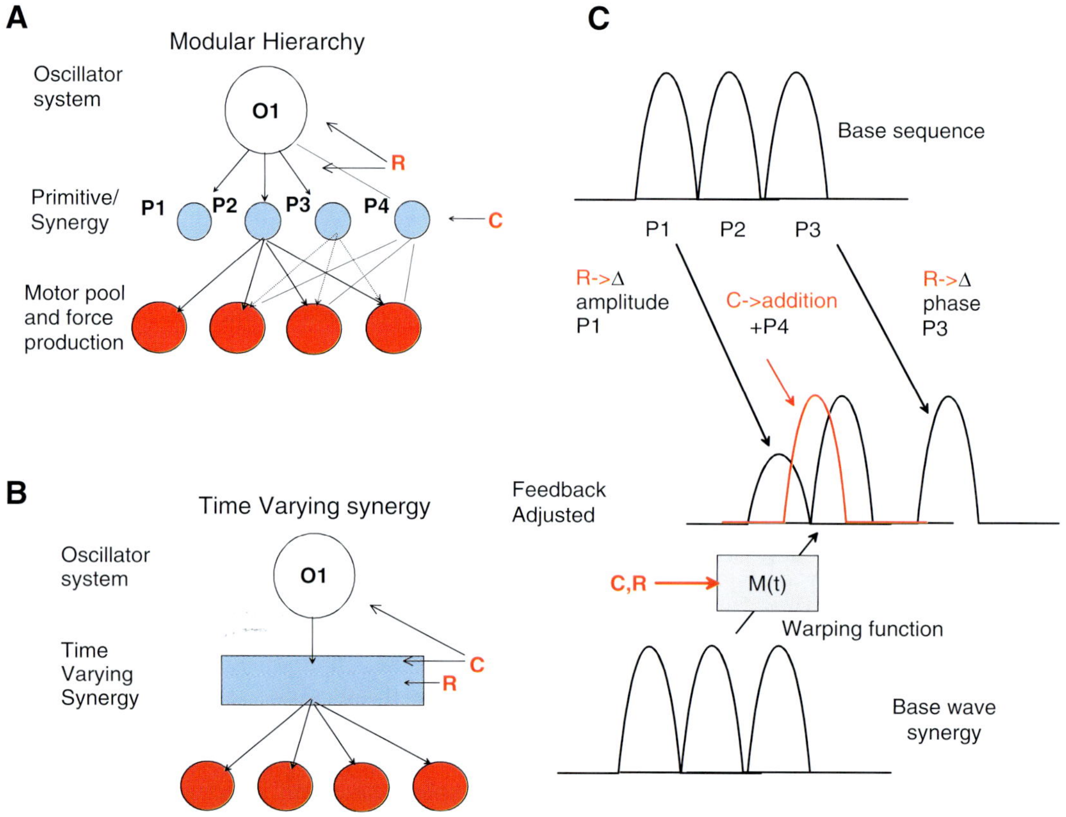

Fig. 4. Adapting motor patterns: comparing modular primitives, with time-varying synergies as flexible basis sets. Panel A: The simple responses to feedback and perturbations possible in the modular hierarchy scheme shown in Fig. 3. This is contrasted in Panel B with generation of the same adjustments using a system of time-varying synergies. Panel C shows the target changes in the pattern. Reflex set up that is based on initial configuration R and a perturbation collision C, such as seen in the frog, cause a feedback-adjusted pattern. Three relatively simple adjustments, all observed in our experiments in frog (not shown here), accomplish this in the modular hierarchy (amplitude adjustment of P1 based on R, addition of P4 based on occurrence of event C, and delay or phase adjustment of P3 based on R). For the time varying synergy framework, these same effects must be obtained by a warping or distorting (local amplifying, local frequency altering, and/or phase shifting) function M(t)1000 or two interacting functions $M_R(t)$ and $M_C(t)$ for each input. The process of constructing the M(t) function and the complexity of M functions that will be needed for such adjustments have not yet been addressed in detail by proponents of unitary time varying synergy ideas.

range of perspectives. Primitives could be fundamental building blocks recruited by CPGs (e.g., Figs. 2 and 8). Synergies or primitives could be emergent properties of the pattern generator. Synergies or primitives could simply be the result of an optimal control (Todorov, 2004), not requiring conventional ideas of circuit modularity, rhythmogenesis, and CPGs, and even perhaps thoroughly divorced from them. However, in the most dismissive perspectives and critiques, they have been seen either as an epiphenomenon of the structure naturally occurring in highly stereotyped movements, or as artifacts of microstimulation of a CPG, or of some random neural assembly actually physiologically irrelevant to real CNS operation.

Whether we can explain the control and richness of intact behavior, and gain insight into the neural underpinnings are the ultimate acid-tests of these ideas. Decomposition approaches to intact behaviors, and gradual integration of these analyses with the various experimental frameworks may help toward these ends.

Detecting primitives and pattern generators in EMG behaviors

Different decomposition approaches have been proposed to examine motor behavior (Tresch et al., 2006). These include principal components analysis (PCA) and factor analysis, independent components analysis (ICA), non-negative matrix factorization (NNMF), and direct components analysis (DCA). These all can be very powerful in addressing modularity primarily because they allow an analysis of modularity across a range of experimental conditions (Hart and Giszter, 2004; Cappellini et al., 2006; d'Avella et al., 2006; Krouchev et al., 2006; Torres-Oviedo et al., 2006). Each emphasizes slightly different aspects of motor pattern data. Key points of difference between the different schemes for the purpose of these techniques are (1) systems that drive covariation of muscles, compared to (2) systems that control primarily timing but may drive co-occurrence of independently modulated muscles, and finally (3) systems that organize relatively fixed but phased timing sequences (e.g., time-varying synergies). Figure 2 shows the hypothesized scheme of rhythm generation projecting to primitives used here.

In the following examples we present two decompositions and analyses. Both are of rhythmic patterns, one in the rat and one in the insect. Together, these bring up many of the issues of detection of modularity and detection of oscillator activity and interaction using EMG data and statistical methods.

Rhythmic motor patterns in rat locomotion

In the data presented here, four rats instrumented with 12–16 differential EMG leads walked on a treadmill at a stepping rate of ~2 Hz. Analysis of EMG from intact rats walking on the treadmill was performed using ICA (Bell and Sejnowski, 1995; Brown et al., 2001). This was followed by Best Basis or Matching Pursuit Wavepacket decomposition (see Hart and Giszter, 2004). The 16 pairs of leads were placed primarily in hindleg muscles or trunk muscles (Fig. 5). We picked proximal hip through ankle muscles in the hindlimb. In the trunk we selected muscles acting at the pelvis and thorax, and representative hip and shoulder muscles. Both types of sets of myograms show well-ordered bursting locomotor patterns in the raw recordings (Fig. 5A and B). Decomposition of the leg muscles in the rat matched the results in frog and cat: a small number of synergies captured 85–90% of variance (as evidenced by the elbow in the curve in Fig. 5C).

We examined whether phasing of onsets in the leg musculature could be well reconstructed and captured by a description in terms of modular drives derived from ICA. We reconstructed the EMG using only the few synergies obtained by ICA that together captured 85% of variance and each individually had more than ~7% of total variance (1/16), while treating the other components as noise. We then examined the onset and offset timing correspondences of individual EMG bursts, comparing between the original and the reconstructed EMG signals. The two signal types (original and reconstructed) onset and offset timings differed by very little, with a standard deviation of ~15 ms. Reconstructions of timing based on covarying synergies were thus within ~30 ms of that in the original data. One concern with the use of blind separation methods applied to locomotor movements, where phase may be an important parameter, was thus allayed.

Patterns of unitary pulses can be fitted to the drive patterns using wavepacket decomposition (Hart and Giszter, 2004) or other pulse fitting methods. These results also fit well with the analyses of Krouchev et al., 2006, in cat hindlimb using onset/offset times as a clustering basis in their DCA method. The notion of a set of modules recruited by a pattern generator is thus consistent with the data here and with the fairly precise capture of onset timings of pulsed premotor drives with ICA (e.g., see Fig. 6).

A second question or concern about ICA was addressed serendipitously. This was whether a decomposition of rhythmic behaviors using the ICA

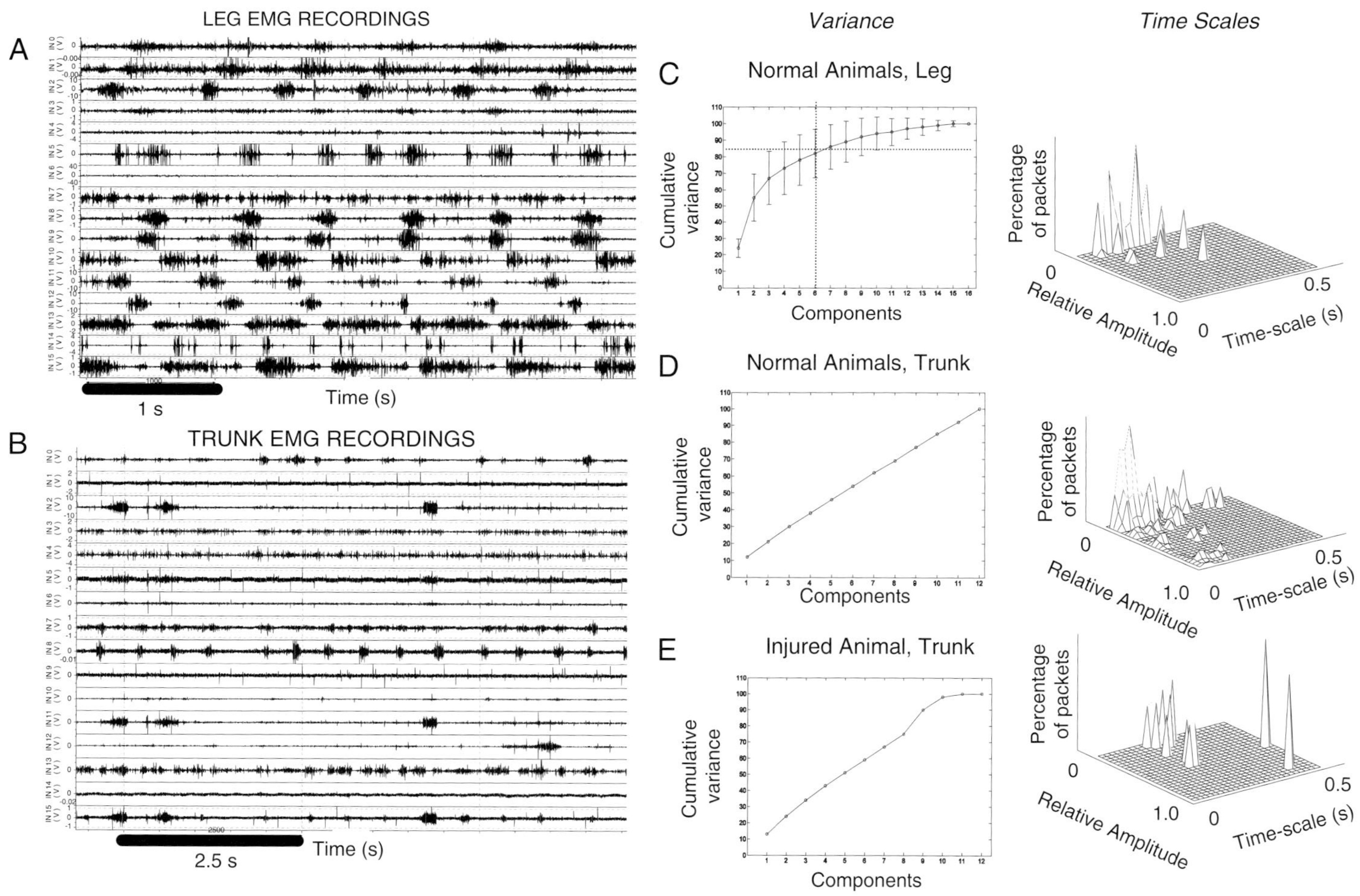

Fig. 5. Modular primitives for leg but not trunk in rat locomotion EMG patterns. 16 EMG recordings were collected in several rats, focused either in leg muscles (A) or in trunk (B). The recordings were rectified, filtered identically, and subject to ICA (see Fig. 6, Panel A) to discover modular organization. In C the presence of an elbow in the cumulative variance for the leg activity is seen. About six components account for 85% of variance (dotted lines). In contrast, in both normal D and neonatal spinal transected rats (E, "injured") the trunk EMG cumulative variance is close to linear in number of components, and the mixing matrix is the identity matrix (not shown). This analysis shows that for ICA a modularization based on information need not occur. Inspection of trunk EMG in B shows covariation of EMG timing occurs. DCA would capitalize on this timing to find modules. However, the ICA shows the bursts are independently modulated despite similar onset timing.

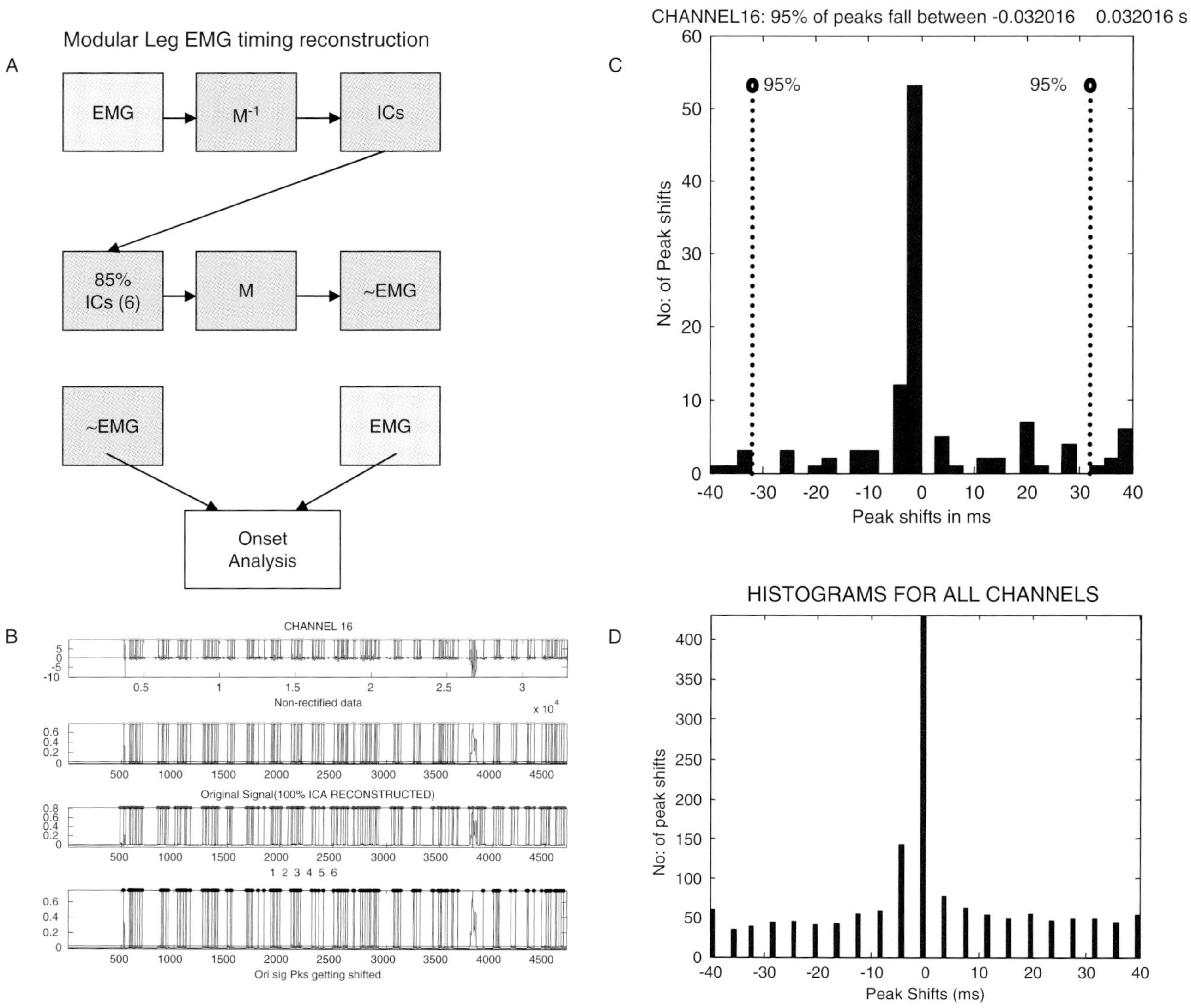

Fig. 6. Leg EMG modularity reconstructions preserve timing information in rat locomotion. (A) The process used to test preservation of timing information. EMG from the leg of a locomoting rat is subject to ICA and the six ICs accounting for 85% of variance are used to reconstruct EMG alone. (B) The timing of onsets in reconstructed EMG is compared to raw EMG using an onset detection and matching process. (C) In a single channel most peaks fall under 10 ms from 0 shift. (D) In all 16 channels the statistics indicate a standard deviation of peaks of ~15 ms. Reconstruction of detailed timing from the reduced set of components was thus reasonable. This result is in keeping with the similar outcomes of ICA and DCA in cat locomotion (Krouchev et al., 2006).

method was bound to discover a lower dimensional modularity in an EMG set, especially during rhythmic behaviors like locomotion. This is a criticism often leveled at decomposition methods. ICA was applied to trunk EMG collected from rats locomoting as described above. We found that although it had been processed identically to the leg EMG, ICA did not extract synergies (Fig. 5D and E). Rather, the individual muscles were obtained as the best independent components. This was initially surprising, given the systematic treadmill locomotion recorded. However, if the timing of the bursts covaried but the individual amplitudes of muscle bursts were adjusted independently of one another cycle to cycle, then this result makes sense. ICA would not separate such signals based simply on onsets. This result also suggests that the trunk motor behavior may exhibit very rich amplitude adjustments on a cycle-to-cycle basis. Synchronous bursting onset of bursts clearly occurred at least occasionally and usually much more often in the trunk data. This suggested that the DCA analysis of Krouchev et al. (2006) was very appropriate to detect modular onset behaviors. We have applied this technique, using only onset timings, and intermuscle delays on a cycle-by-cycle basis, and referenced the activity and intermuscle phase using iliopsoas onset as a cycle marker. The analysis reveals the covariation of muscle timing modularity, presumably as a result of the rhythmic pattern generator drive evidenced at onset.

Taken together, these data add confidence to these decomposition approaches as tools in analyzing modularity. First, the ICA analysis of relatively stereotyped behaviors is not "bound to succeed." Second, the failure of ICA, coupled to a DCA analysis, indicates particular rhythmic structure of trunk control, separate from amplitude, which differs when compared to the hindlimb.

Using pulses, which have amplitude and phase variations within short sequences, may be a basis for adapting a modular pattern to configuration or load conditions as suggested by these analyses. Onset variations among co-occuring groups or modules may also be used to examine rhythm generation structure in coupled oscillator systems as described the next section.

Detecting oscillator interactions and timing

Examining oscillator interactions using DCA or other timing events from recordings of EMG in intact behavior is feasible. This type of analysis is shown here in a "simpler system" analysis of insect breathing. The system is relatively unique in that a single pair of muscles with single motoneurons generate almost all active inspiratory behavior and these motoneurons can be characterized from single unit EMG. This feature makes this system a very good model for timing-based analysis of rhythmic interactions.

The breathing rhythm of grasshoppers is almost fully synchronous. However, isolated abdominal ganglia can oscillate. Using CPG operational criteria of deafferentation and isolation of individual segments the presence of pattern generators or oscillators in each segment can be shown. For example, cycle duration drifts can be revealed by autocorrelation of stationary time series from the EMG recordings in the isolated abdomen. These show that all the segmental abdominal oscillators are active in the adult (Fig. 8D). These oscillators can play roles in specialized behaviors of the adult, e.g., phase synthesis in digging for oviposition (Thompson, 1986a, b). How do they operate in unstressed breathing?

Figure 7 shows the patterns of firing of a pair of inspiratory myograms (i.e., motoneurons) from adjacent segments in the abdomen, innervated by two different abdominal ganglia. These patterns represent the total drive to these two muscles.

Burst onsets across segments are all driven by identical descending information carried in paired descending ventilatory interneurons. Segmental timing differences must represent variations in local abdominal state. If these variations in phase persist cycle to cycle, or show phase resetting effects, the operation and connectivity of underlying oscillators can be assessed (Giszter, 1984). These effects can be examined using conventional time series analysis methods if the behavior is statistically stationary for more than 200 cycles.

By examining the auto and cross-correlation patterns of expiratory muscle onset delays or estimated phases across a stationary time series of more than 200 ventilatory events in multiple

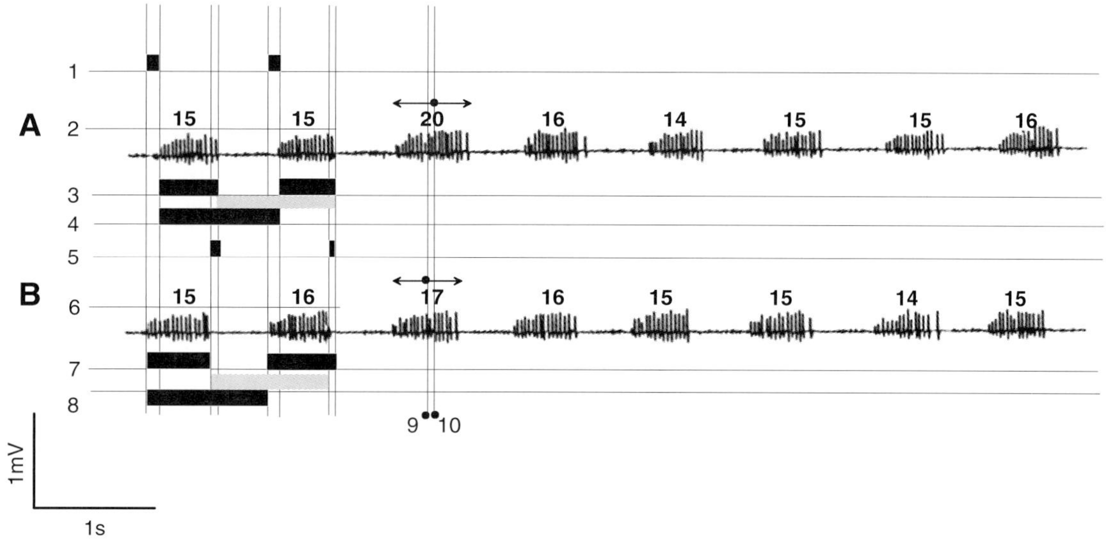

Fig. 7. Multiple segment recordings of single unit EMG to detect oscillator couplings. Single unit EMG traces of inspiratory muscles in two adjacent segments (A and B) are shown. In each single unit EMG the onset, offset, and spike number and pattern are collected. Several coordination parameters can then be employed in subsequent analysis: (1) Segment A/B burst onset delay or estimated phase (delay/period); (2) Segment A spike number; (3) Segment A burst duration; (4) Segment A period (referenced to onset in black or offset in gray); (5) Segment A/B burst termination delay or estimated phase (delay/period); (6) Segment B spike number; (7) Segment B burst duration; and (8) Segment B period (referenced to onset in black or offset in gray). In addition, interburst can be constructed from these. Finally, one approach not favored in this article references timing to the burst peak or center (9 and 10). This method can average separate phase variations at burst onset and termination and occasionally obscures biomechanically important timing.

segments, a standard linear recursive model of the coupling mechanisms can be developed. (Pikovsky et al., 2000). This is possible under the assumption of phaselocked behavior across the series (e.g., see Kiemel and Cohen, 1998; Kiemel et al., 2003; Boothe et al., 2006).

Expressing relative timing of segments as onset delays and perturbations in an ARMAX model in ordinal (cycle) time:

$$A(q^{-1})d(t) = B(q^{-1})u(t) + C(q^{-1})e(t) \qquad (20)$$

where q is the delay or backshift operator, $d(t)$ the vector of relative onset delays among oscillators on cycle t, $u(t)$ the variation of the synchronous driving input to each oscillator in cycle t, and $e(t)$ the vector of cycle-by-cycle drift or shocks to the individual oscillators. In stationary conditions of breathing $u(t)$ and $e(t)$ are approximately independent and identically distributed (i.i.d.). [Note: There are auto- and cross-correlations that arise out of necessity in the system of delays, periods, and burst durations when these are expressed in ordinal (oscillator cycle) time.] Expression of timing as relative phases can also be used, by division by cycle period.

This analysis is equivalent to small signal analysis of phase variations at the dynamical system fixed points, with linearization near the fixed point. Phase can be estimated in several ways from discrete events: burst onset, burst termination, burst center. It should be noted that burst centers may average phase variations at onset and termination that may not be ideal in this analysis.

The matrices A, B, C in Eq. (20) can be identified and estimated from the autocorrelation, partial autocorrelation (ACF), and cross-correlation (CCF) matrices of the time series (see Box and Jenkins, 1994; Ljung and Soderstrom, 1983) using least-squares or maximum likelihood approaches. Similar analyses of coupled biological oscillator systems are now well established, though not broadly employed (see Boothe et al., 2006).

Several possible patterns of control of the motoneurons in the locust can be tested. Different three way oscillator couplings in linear phaselocking models generate specific cross-correlation and auto-correlation patterns in simple ARMA models. Similar patterns can be obtained from the data in the locust (Fig. 8). (Muscles driven by the unitary metathoracic oscillator show the pattern predicted by a null hypothesis of a single oscillator with random phase variations.) Muscles in other segments (e.g., segments 4 and 5 expiratory muscles) show evidence of autonomous abdominal oscillations and of oscillator phase couplings in a chain, driven by the master oscillator. Examples of delay and estimated phase based ACF and CCF analyses of segments 4 and 5 are shown in Fig. 8. For comparison the patterns of the null hypothesis, observed in segments 3 and 4 inspiratory muscles, are shown in red. The upshot of this analysis is to show that the local abdominal oscillators appear to be functionally coupled unidirectionally, caudal to rostral, with a small expiratory delay running caudal to rostral during the period of stationary quiet ventilation.

Functionally, the local oscillators and their connecting circuitry build a local prediction of the next arrival time of the descending information. This improves the reception and transmission to the musculature of the inspiratory onset event that is crucial to the insect due to the brevity of inspiratory spiracle opening. The local oscillators can be thought of as generating an internal model of the expected behavior of the upstream neural activity. They form a pattern reception system in at least equal measure to a system of pattern generation. The structure of connectivity and operation is reminiscent of a delay line multiplier phase-locked loop arrangement used in early digital tape recording systems to manage tape speed wobble, and predict bit transition times (Gardner, 2005), which is an analogous problem to the locusts.

In summary, using event timing throughout a motor pattern, like that extracted by DCA, ICA, or wavepacket decompositions, it is feasible to extract coupling patterns and information flow among different potentially autonomous oscillators from intact or semi-intact behavior records and ENG or EMG.

Fig. 8. Time series auto and cross-correlation signatures of phaselocked coupled oscillators in grasshopper abdomen. (A) The recorded segments (bold): two adjacent segments innervated by abdominal ganglia but also driven by the descending control from the metathoracic were used in the analysis presented. (B) The null hypothesis model: drive from a metathoracic master with local oscillators that are either suppressed or tightly locked (with type 0 resetting of Winfree) so that there is effectively no phase history. By progressive model fitting using standard approaches, the least complex model capturing the data and accounting for high variance can be obtained. Here we show correlation patterns of the original time series data, and the residual delay shocks from the simple model in B. These support additional local coupling patterns, as shown in C, and weaker (Winfree type 1) phase couplings. D shows initial CCF structures for the burst delay and features of the simple time series. The paired fine horizontal lines above and below zero in each plot indicate the levels above and below which the CCF values are significant at $p<0.05$ for these data. (D1) CCF of segment 4 interburst with burst delay, no significant correlations. (D2) CCF of segment 5 interburst and burst delay: significant at lag -1 but not lag 0, compared to Null hypothesis CCF pattern of negative correlation at lag 0 (red). (D3) CCF of segment 4 burst and burst delay: significant peak at lag 0. (D4) CCF of segment 5 burst and burst delay: no significant features. (D5) CCF of interburst and burst delays: significant at -1 lag under null hypothesis (red) but 0 in observed data. (D6) CCF of segment 4 period duration and burst delay: significant correlation at lag 0 here but not under null hypothesis (red). (D7) CCF of segment 5 period duration and burst delay with significant correlation at lag -1 but not 0 compared to null hypothesis pattern (red). Panel E shows the matching cross-correlation patterns of the estimated residual delays (i.e., the noise process or shocks that were extracted in a time-series fitting of the model in 8B, hypothesizing no local couplings) and the same features of the simple time series as in Panel D. These CCFs indicate the need for additional model fitting. The pattern predicted from model, which has no local coupling is shown in red. E1–E7 are cross-correlations to the same time series features used in D1–D7. In addition, E8 shows the CCF of the actual burst delay and the residual shocks. This CCF shows a persistent memory or contribution of shocks at several preceding lags. This "memory" is one that should be completely absent under the null-hypothesis control structure fitted. Panels F and G show the same CCF structures, constructed for phase delay and phase residuals (also obtained under the model in 8B), instead of time delay and duration. Ordinal time series of the phase measures were constructed. Delays in D and E become phase lags in F and G. Time durations in D and E become phase durations (from 0 to 1) in the matching plots in F and G. Note the changed Null hypothesis and data fit in F7 and G7 compared with D7 and E7, and the smoother residual to phase delay curve in G8, still with significant phase memory.

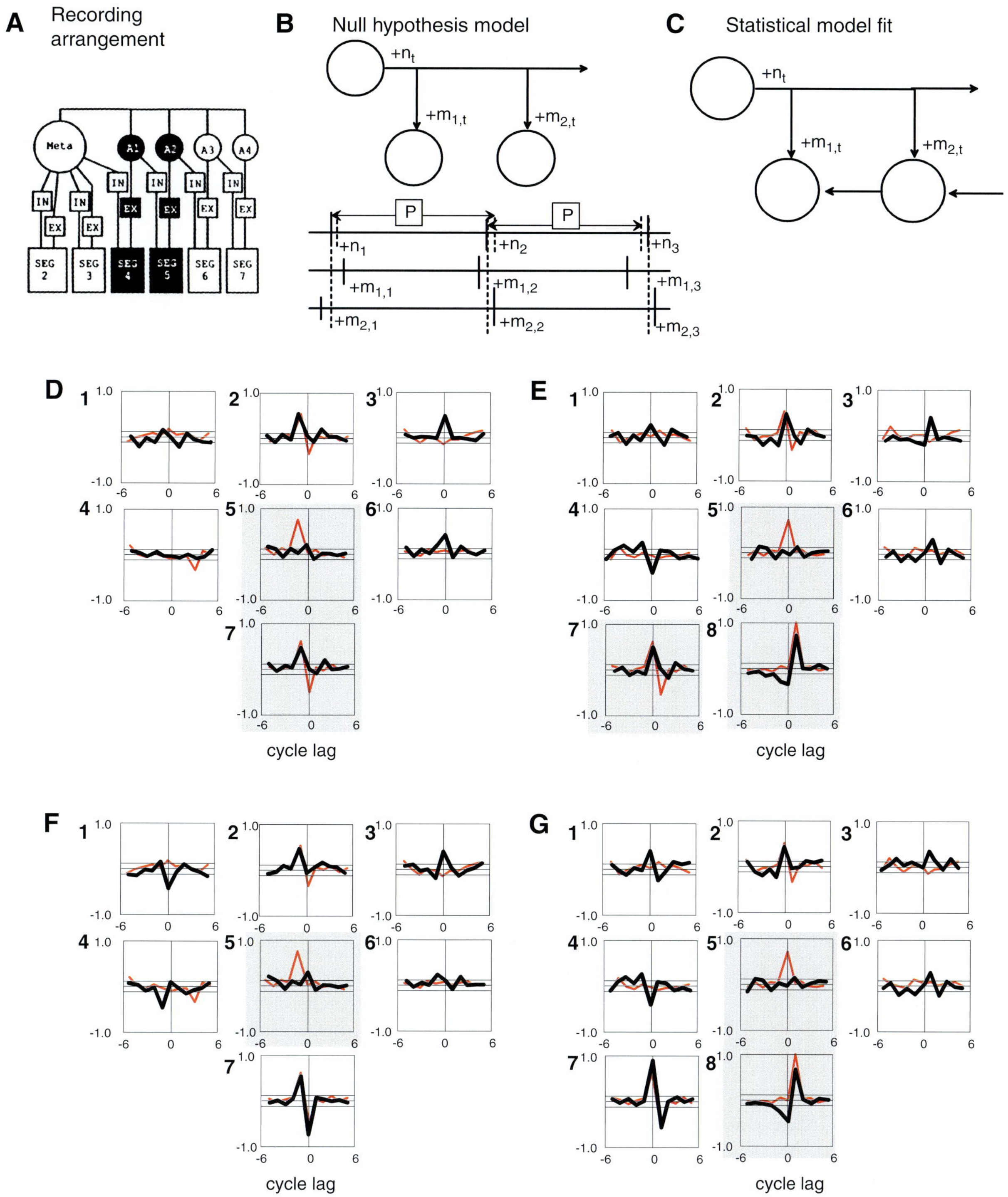
A Recording arrangement
Meta
A1
A2
A3
A4
IN
EX
SEG 2
SEG 3
SEG 4
SEG 5
SEG 6
SEG 7
B Null hypothesis model
$+n_t$
$+m_{1,t}$
$+m_{2,t}$
P
$+n_1$
$+n_2$
$+n_3$
$+m_{1,1}$
$+m_{1,2}$
$+m_{1,3}$
$+m_{2,1}$
$+m_{2,2}$
$+m_{2,3}$
C Statistical model fit
$+n_t$
$+m_{1,t}$
$+m_{2,t}$
D
E
F
G
1.0
-1.0
-6
0
6
cycle lag

Discussion

We have reviewed different approaches to modularity in the motor system. A number of topics bear further discussion.

Are modular descriptions adequate for motor tasks?

A range of different evolutionary task constraints can impact motor tasks and modularity. For example, in control of terrestrial locomotion, an animal might be most concerned with mechanical efficiency, specific patterns of energy delivery and recovery, motion symmetry, foot placement, joint and muscle wear and tear, balance and stability to perturbation, terrain tracking, steering, or efficacy of pursuit or avoidance strategies. Several of these task constraints may be handled as much by physical design as by active control (e.g., Kuo et al., 2005). Many of these may also be near-optimized simultaneously (Collins, 1995). Simplified descriptions of multi-limb coordination and dynamics arise from algorithms that describe the locomotor task in a simplified virtual leg and spring-loaded inverted pendulum formalism. These may help interpret neural controls as well as aiding robot designs (Raibert, 1986; Schmitt and Holmes, 2000). However, despite these consolidating task descriptions, producing seamless working robots with performance comparable to the biological "gold-standard" has remained challenging.

Animals perform well across significant body size changes, in a range of terrains and orientations to gravity and in very rough and cluttered terrain across a range of speeds. These performance ranges can radically alter strategies of control, effector use, and roles of musculature. For example the cockroach can run at speeds that preclude the use of feedback in the same way as at slower speeds (Full and Tu, 1991; Ting et al., 1994), and can climb vertical walls and crawl upside down with radically altered body loading (Duch and Pflüger, 1995; Larsen et al., 1995). Cats alter locomotor patterns significantly on inclines (Smith and Carlson-Kuhta, 1995). These observations underscore the need for neural analyses of intact animals in complex environments, to complement neural analyses of fictive patterns. The requirements and input issues associated with real-world behaviors cannot be easily sidestepped. The mechanisms in play in the CNS must support the full range of motor adaptations used, and perhaps anticipate them in their structural organization. Indeed, much very interesting recent research in locomotion has focused on examining such mechanisms and adding back real-world perturbations to otherwise highly predictable behaviors in intact or fictive preparations (Gorassini et al., 1994; Pang et al., 2003; Lafreniere-Roula and McCrea, 2005; Quevedo et al., 2005; Akay et al., 2006; Cappellini et al., 2006). Fitting pattern generator, oscillator, and primitive-based modularity in this framework is essential.

Oscillators and pattern generation

Oscillations in motor behaviors are found repeatedly across the many species tested. These seem to be a natural modular element. Many experiments demonstrate the operational criteria for pattern generation. However, it is also useful to consider the range of ways oscillators could be employed functionally in computation. Some of these might be important in the rapid and broad adaptability of organisms like the cockroach. In electrical engineering, circuit oscillators provide at least six functions in circuitry when they are incorporated in phaselocked loops:

(1) Phase and frequency synthesis (analogous to pattern and gait synthesis).
(2) Frequency multipliers and dividers (analogous to gait and syncopation).
(3) Demodulation and phase measurements from carrier signals (sensory functions).
(4) Data synchronization and synchronization of computation (clocking computations, predicting state, and timing (and reversing) gains reflexes).
(5) Coherent transponders (predictive relays, for boosting and reception of signals).
(6) Time-out clocks/counters (backup for events controlled in other ways).

Although the last four mechanisms and functions are not strictly pattern generation, these mechanisms could nonetheless satisfy current operational CPG tests. Similarly, as discussed more below, several classes of computational and engineering models, that are far from classic pattern generator ideas, will appear to be oscillators, or act as network oscillators and drive patterned output when deprived of inputs. By the same token, dynamical oscillators can also be used in an engineering framework as simple predictive internal models (Kuo, 2002). Indeed this aspect features in several of the possible functions listed above. In the grasshopper, the local oscillators in quiet breathing may be primarily transponders and backup systems for the accurate relay of descending signals generated by a master oscillator for breathing. Modular organizations may usefully subserve sensory as well as motor functions.

Unitary pulses and primitives

There is strong support for sets of unitary drives across species (e.g., Hart and Giszter, 2004; Cappellini et al., 2006; Torres-Oviedo et al., 2006). These drives usually appear to be pulsed. Pulses may provide a simple compositional framework for movement. Pulses also lead to intrinsically time-limited behavior, compared to continuous oscillatory drives. Oscillators may operate close to the margins of stability. Motion can be synthesized by modulating amplitude, duration, phasing, and sequencing of pulses. There are strong a-priori arguments for a separable hierarchy of rhythm and pattern shaping in pattern generation systems and for a pre-specified basis or modularity in the pattern-shaping layer (see Lafreniere-Roula and McCrea, 2005). Pattern generator/rhythm generators recruiting pulsed primitives from a modular pattern-shaping layer may allow flexible "clutching" of a gait rhythm to motor production in response to the different mechanical needs that occur in different environments and loadings (e.g., in cockroach locomotion). In addition, the amplitude modulation of pulses (occurring downstream and decoupled from the system of phase and frequency synthesis) may free the timing synthesis system from a range of potential dynamical and control problems. Such problems might arise from the relationships of amplitude, phase, and frequency in nonlinear limit cycles and dynamical systems. Pattern-shaping layers without any a-priori structure reopen the Pandora's box of the degrees of freedom problem. At the level of kinematic planning a separation of actions into rhythmic and pulsed elements of composition may also exist. However, it is important to note that each task-level kinematic cycle, action, or stroke is likely to be organized using sequencing of several execution level primitives and drive pulses, not a single unitary pulse of muscle activity. Separating rhythm generation from force-production allows complex clutching and syncopation between actions in the world and motion timing that represent what we often consider peak performance in music or sport.

Engineering, dynamical, and ecological perspectives on pattern generation and primitives

The process of organizing the many computations needed for motor tasks such as locomotion in real-world conditions, and thus modularity, can be viewed from several different competing perspectives.

In some views, the presence of timing systems or oscillators in computation, that largely dominate ideas of pattern generation, have been seen as unnecessary or misleading. In principle, appropriate internal models, the representation of state and needed transitions should be a sufficient description. State machines have been used in running/hopping robots and work well. These merely require an occasional time-out backup at some points in their operation (Raibert, 1986). Passive walking in principle requires no active drive or pattern generation, and merely an occasional tweak (see Kuo et al., 2005). Although none of the authors cited here may dismiss pattern generation or timing representations, their work shows there are clearly different perspectives possible.

Contrasting these perspectives is the "ecological psychology" perspective, in which any computations in the CNS should be entirely implicit and emergent from a highly distributed shallow ensemble. This line of thought was pioneered by Gibsonians (e.g., Kelso et al., 1981; Balasubramaniam and Turvey, 2004). Proponents historically have eschewed both engineering-based formulations and CPG-based interpretation of experiments as being highly misleading.

Some of these strong differences in perspective must clearly be artificial. Opposing arguments may arise from taking differing task, algorithm, or implementation based research perspectives, and Marr's framework (Marr, 1982) may go a long way toward resolving some of the differences. The middle ground and synthesis is shown in a number of published studies and discussions (e.g., Schaal et al., 2003; Tin and Poon, 2005; Prinz, 2006). Oscillators can be structured together with feedback pathways to provide simplified "implicit" internal models (Kuo, 2002). Specific types of oscillating dynamical systems can be organized as "contracting systems" with strong stability guarantees that again may unify perspectives (Slotine and Lohmiller, 2001; Wang and Slotine, 2005).

From a "neuroethological perspective," it is generally agreed that shallow and rapid computations in lower motor structures are probably evolutionarily "de rigeur." If there are simplified representations and fast computations ("evolutionary hacks," or matched filters) achievable with specific neural mechanisms, the expectation is that they will be employed by evolution. This may be for reasons of the phylogenetic history of biological systems, if nothing else. If such approximation mechanisms and dynamics can be cheaply and easily constructed in "wetware," and if they simplify descending controls, minimize control effort, and provide sufficiently accurate information for control with less computational effort, what engineer or Gibsonian would quibble?

Were primitives constructed on an evolutionary time-scale and are they hard-wired? How modules or primitives arise during development is an interesting and unresolved issue. There are arguments both for and against hard-wired primitives. Results from stomatogastric ganglia of crustacea show that pattern generation may range widely within a contiguous region of the adjustable neural parameter space in adults (see Marder and Goaillard, 2006; Prinz, 2006). However, we do not know if developmental tasks faced by most animals allow use of a similarly contiguous domain in neural parameter space through development. The stomatogastric environment is somewhat constrained. We do not know if solutions for movement arising early in development can seamlessly be modified and carried forward as behavior becomes complex for limbs and whole body control. Our speculation is that after choices of appropriate bases they can be (see Bradley et al., 2005). In the development of the rat (see Clarac et al., 2004), crawling skills may transfer continuously to waddling and finally to parasagittal limb use in walking, trotting, and galloping. However, the modular bases needed for a development like the rat's might also have to be predictive of future needs across the task sets, from the outset. We speculate that some of the predictive components of spinal bases or modules may be built on an evolutionary time scale and hardwired. It is conceivable that the structures of the bases needed in adulthood cannot be fully anticipated or smoothly extrapolated through ontogeny, so as to provide the survival tools needed, e.g., in a wildebeest calf. Many adult motor tasks and skills do have "knacks" and special strategies (e.g., the Fosbury flop high jump). If such discontinuous jumps or nonoverlapping tasks changes happen in motor skills through development, CNS may need to set up and sequester appropriate modularity, and resources in anticipation of future, as yet unused, actions. Evolution could ensure this.

Theoretical work by Todorov and Ghahramani (2003) and by Chabra and Jacobs (2006) indicate that both simple neural and computational mechanisms to locate a common shared modularity are feasible. In simplified model conditions these may be learned or constructed de-novo. Extension to physical plants as complex as real-world whole organisms remains to be shown. Again, from the neuroethological perspective, whether such ontogenetic learning can be made sufficiently rapid and efficient to be employed in biological systems is unknown.

There is strong evidence for some hardwired "scaffolding" of motor systems in spinal cord. We know that motor pools are segregated anatomically, with numbers proportional to their ultimate needs, into columns in stereotypic locations during development (Dasen et al., 2003), and a relatively uniform fraction of cells are then pruned by the subsequent cell death. Primary afferent projections to motor pools are also relatively hard-wired. These observations suggest there are quite orderly genetic controls of several spinal cord structures that are "built-in" by evolution before plastic tuning. To suppose that this orderly structuring of circuitry extends to at least a few cells and synapses further into the premotor spinal cord is not too far-fetched. Projections are precisely ordered (e.g., Lundberg et al., 1987a, b; Cabaj et al., 2006). Some of these circuits may relate to described primitives (e.g., possibly those of Lundberg et al., 1987b). Further, evolution of precise modular sensory mechanisms is believed to have occurred repeatedly (e.g., amphibian tongue strike mechanisms, Nishikawa et al., 1992). Behaviorally relevant modular and hardwired spinal circuit organizations may be fairly direct targets of natural selection.

Descending systems must build an appropriate connectivity with the spinal apparatus of pattern generators and primitives. They both face the problem of where to connect, of how to modulate pathways, and of making credit assignments for control outcomes (e.g., see Abbott, 2006). Mechanisms to tune these projections are likely to work best in the framework of some initial ordered connectivity and modularity of their targets. Throughout life, further tuning, plasticity, and adaptation to ontogenetic variations and life history mishaps are also clearly needed and used in most animals (e.g., see Wolpaw and Carp, 1993; Martin et al., 2004; Schouenborg, 2004; Rossignol, 2006). A strong speculation is that employing relatively structured hardwiring of modularity at the low levels of motor control simplifies many of these learning and plasticity issues and creates an appropriate scaffold and bootstrap for building motor function as efficiently as possible.

Conclusions

In summary, we believe that current data, computational and comparative neuroscience all favor modularity in low-level motor organization. Separating rhythm and sequence formation from a set of lower-level primitives may provide segmental mechanisms with motor building blocks or words. It is speculated that some modules may have a hardwired basis from early phylogeny. These modular primitives can be assembled in a task-dependent manner by the rhythmogenic and pattern synthesis systems, and by reflex pathways. This structure is a flexible, experimentally testable, and plausible framework. It may also potentially support complex motor adaptation.

Acknowledgments

Manuscript preparation supported by NIH NS40412. The authors wish to thanks William Kargo, Andrea d'Avella, Matt Tresch, Terry Sanger and Michel Lemay for extensive discussion, and the editors for their very helpful comments.

References

Abbott, L.F. (2006) Where are the switches on this thing? In: van Hemmen J.L. and Sejnowski T.J. (Eds.), 23 Problems in Systems Neuroscience. Oxford University Press, USA, pp. 423–431.

Akay, T., McVea, D.A., Tachibana, A. and Pearson, K.G. (2006) Coordination of fore and hind leg stepping in cats on a transversely-split treadmill. Exp. Brain Res., 175(2): 211–222.

d'Avella, A. and Bizzi, E. (2005) Shared and specific muscle synergies in natural motor behaviors. Proc. Natl. Acad. Sci., 102(8): 3076–3081.

d'Avella, A., Portone, A., Fernandez, L. and Lacquaniti, F. (2006) Control of fast-reaching movements by muscle synergy combinations. J. Neurosci., 26(30): 7791–7810.

d'Avella, A., Saltiel, P. and Bizzi, E. (2003) Combinations of muscle synergies in the construction of a natural motor behavior. Nat. Neurosci., 6(3): 300–308.

Balasubramaniam, R. and Turvey, M.T. (2004) Coordination modes in the multisegmental dynamics of hula hooping. Biol. Cybern., 90(3): 176–190.

Baldwin, M.J. (1896) A new factor in evolution. Am. Nat., 30(354): 441–451.

Bateson, P.J. (2004) The active role of behaviour in evolution. Biol. Philos., 19: 283–298.

Bell, A.J. and Sejnowski, T.J. (1995) An information-maximization approach to blind separation and blind deconvolution. Neural Comput., 7: 1129–1159.

Bernstein, N. (1967) The Co-ordination and Regulation of Movements. Pergamon Press, Oxford.

Bizzi, E., Mussa-Ivaldi, F.A. and Giszter, S. (1991) Computations underlying the execution of movement: a biological perspective. Science, 253(5017): 287–291.

Boothe, D.L., Cohen, A.H. and Troyer, T.W. (2006) Temporal correlations in stochastic models of double bursting during simulated locomotion. J. Neurophysiol., 95(3): 1556–1570.

Box, G.E.P., Jenkins, G.M. and Reinsel, G. (1994) Time Series Analysis: Forecasting and Control (3rd ed.). Prentice Hall, USA.

Bradley, N.S., Solanki, D. and Zhao, D. (2005) Limb movements during embryonic development in the chick: evidence for a continuum in limb motor control antecedent to locomotion. J. Neurophysiol., 94(6): 4401–4411.

Brown, G.D., Yamada, S. and Sejnowski, T.J. (2001) Independent components analysis (ICA) at the neural cocktail party. Trends Neurosci., 24: 54–63.

Burdet, E. and Milner, T.E. (1998) Quantization of human motions and learning of accurate movements. Biol. Cybern., 78: 307–318.

Cabaj, A., Stecina, K. and Jankowska, E. (2006) Same spinal interneurons mediate reflex actions of group Ib and group II afferents and crossed reticulospinal actions. J. Neurophysiol., 95(6): 3911–3922.

Callebaut, W. and Rasskin-Gutman, D. (Eds.). (2005) Modularity: Understanding the Development and Evolution of Natural Complex Systems. MIT Press, Cambridge, MA.

Cappellini, G., Ivanenko, Y.P., Poppele, R.E. and Lacquaniti, F. (2006) Motor patterns in human walking and running. J. Neurophysiol., 95(6): 3426–3437.

Chabra, M. and Jacobs, R.A. (2006) Properties of synergies arising from a theory of optimal motor behavior. Neural Comput., 18: 2320–2342.

Clarac, F., Brocard, F. and Vinay, L. (2004) The maturation of locomotor networks. Prog. Brain Res., 143: 57–66.

Colgate, J.E. and Hogan, N. (1988) Robust control of dynamically interacting systems. Int. J. Control, 48(1): 65–88.

Collins, J.J. (1995) The redundant nature of locomotor optimization laws. J. Biomech., 28: 251–267.

Dasen, J.S., Liu, J.-P. and Jessell, T.M. (2003) Motor neuron columnar fate imposed by sequential phases of Hox-c activity. Nature, 425: 926–933.

Duch, C. and Pflüger, H.-J. (1995) Motor patterns for horizontal and upside walking and vertical climbing in the locust. J. Exp. Biol., 198: 1963–1976.

Earhart, G.M. and Stein, P.S. (2000) Scratch-swim hybrids in the spinal turtle: blending of rostral scratch and forward swim. J. Neurophysiol., 83(1): 156–165.

Flash, T. and Hochner, B. (2005) Motor primitives in vertebrates and invertebrates. Curr. Opin. Neurobiol., 15(6): 660–666.

Flash, T. and Hogan, N. (1985) The coordination of arm movements: an experimentally confirmed mathematical model. J. Neurosci., 5(7): 1688–1703.

Full, R.J. and Tu, M.S. (1991) Mechanics of rapid running insects: two-, four- and six-legged locomotion. J. Exp. Biol., 156: 215–231.

Gardner, F.M. (2005) Phaselock Techniques (3rd ed.). Wiley, New York, NY.

Giszter, S.F. (1984) Maintenance of temporal patterning in stochastic environments: ventilation in the locust. PhD Thesis, Department of Biology and Institute of Neuroscience, University of Oregon.

Giszter, S.F. and Kargo, W.J. (2000) Conserved temporal dynamics and vector superposition of primitives in frog wiping reflexes during spontaneous extensor deletions. Neurocomputing, 32–33: 775–783.

Giszter, S.F. and Kargo, W.J. (2001) Modeling of dynamic controls in the frog wiping reflex: force-field level controls. Neurocomputing, 38–40: 1239–1247.

Giszter, S.F., Moxon, K.A., Rybak, I. and Chapin, J.K. (2000) A neurobiological perspective on design of humanoid robots and their components. IEEE Intell. Syst., 15(4): 64–69.

Giszter, S.F., Mussa-Ivaldi, F.A. and Bizzi, E. (1993) Convergent force fields organized in the frog spinal cord. J. Neurosci., 13: 467–491.

Gorassini, M.A., Prochazka, A., Hiebert, G.W. and Gauthier, M.J. (1994) Corrective responses to loss of ground support during walking I: intact cats. J. Neurophysiol., 71(2): 603–610.

Gottlieb, G.L. (1998) Muscle activation patterns during two types of voluntary single-joint movement. J. Neurophysiol., 80: 1860–1867.

Grillner, S., Perret, C. and Zangger, P. (1976) Central generation of locomotion in the spinal dogfish. Brain Res., 109(2): 255–269.

Hamilton, W.D. (1964) The genetical evolution of social behaviour. J. Theor. Biol., 7: 1–52.

Hart, C.B. and Giszter, S.F. (2004) Modular premotor drives and unit bursts as primitives for frog motor behaviors. J. Neurosci., 24(22): 5269–5282.

Hogan, N. (1984) An organizing principle for a class of voluntary movements. J. Neurosci., 4(11): 2745–2754.

Hoyle, G. (1970) Cellular mechanisms underlying behavior: neuroethology. Adv. Insect Physiol., 7: 349–444.

Ijspeert, A., Nakanishi, J. and Schaal, S. (2003) Learning attractor landscapes for learning motor primitives. In: Becker S., Thrun S. and Obermayer K. (Eds.), Advances in Neural Information Processing Systems 15. MIT Press, Cambridge, MA, pp. 1547–1554.

Kargo, W. and Rome, L. (2002) Functional morphology of proximal hindlimb muscles in the frog rana pipiens. J. Exp. Biol., 205(Pt 14): 1987–2004.

Kargo, W.J. and Giszter, S.F. (2000) Rapid corrections of aimed movements by combination of force-field primitives. J. Neurosci., 20: 409–426.

Kelso, J.A., Holt, K.G., Rubin, P. and Kugler, P.N. (1981) Patterns of human interlimb coordination emerge from the properties of non-linear, limit cycle oscillatory processes: theory and data. J. Mot. Behav., 13(4): 226–261.

Kiehn, O., Hounsgard, J. and Sillar, K.T. (1997) Basic building blocks of vertebrate CPGs. In: Stein P.S.G., Grillner S., Selverston A.I. and Stuart D.G. (Eds.), Neurons, Networks and Motor Behavior. MIT press, Cambridge, MA, pp. 47–60.

Kiemel, T. and Cohen, A.H. (1998) Estimation of coupling strength in regenerated lamprey spinal cords based on a stochastic phase model. J. Comput. Neurosci., 5(3): 267–284.

Kiemel, T., Gormley, K.M., Guan, L., Williams, T.L. and Cohen, A.H. (2003) Estimating the strength and direction of functional coupling in the lamprey spinal cord. J. Comput. Neurosci., 15(2): 233–245.

Krouchev, N., Kalaska, J.F. and Drew, T. (2006) Sequential activation of muscle synergies during locomotion in the intact cat as revealed by cluster analysis and direct decomposition. J. Neurophysiol., 96(4): 1991–2010.

Kuo, A.D. (2002) The relative roles of feedforward and feedback in the control of rhythmic movements. Motor Control, 6(2): 129–145.

Kuo, A.D., Donelan, J.M. and Ruina, A. (2005) Energetic consequences of walking like an inverted pendulum: step-to-step transitions. Exerc. Sport Sci. Rev., 33(2): 88–97.

Kurata, H., El-Samad, H., Iwasaki, R., Ohtake, H., Doyle, J.C., Grigorova, I., Gross, C.A. and Khammash, M. (2006) Module-based analysis of robustness tradeoffs in the heat shock response system. PLoS Comput. Biol., 2(7): e59.

Lafreniere-Roula, M. and McCrea, D.A. (2005) Deletions of rhythmic motoneuron activity during fictive locomotion and scratch provide clues to the organization of the mammalian central pattern generator. J. Neurophysiol., 94(2): 1120–1132.

Larsen, G.S., Frazier, S.F., Fish, S.E. and Zill, S.N. (1995) Effects of load inversion in cockroach walking. J. Comp. Physiol. A, 176: 229–238.

Lemay, M.A., Galagan, J.E., Hogan, N. and Bizzi, E. (2001) Modulation and vectorial summation of the spinalized frog's hindlimb end-point force produced by intraspinal electrical stimulation of the cord. IEEE Trans. Neural. Syst. Rehabil. Eng., 9(1): 12–23.

Lemay, M.A. and Grill, W.M. (2004) Modularity of motor output evoked by intraspinal microstimulation in cats. J. Neurophysiol., 91(1): 502–514.

Ljung, L.J. and Soderstrom, T. (1983) Theory and Practice of Recursive Identification. MIT Press, Cambridge, MA.

Lundberg, A., Malmgren, K. and Schomburg, E.D. (1987a) Reflex pathways from group II muscle afferents: 2. Functional characteristics of reflex pathways to α-motoneurones. Exp. Brain Res., 65: 282–293.

Lundberg, A., Malmgren, K. and Schomburg, E.D. (1987b) Reflex pathways from group II muscle afferents: 3. Secondary spindle afferents and the FRA: a new hypothesis. Exp. Brain Res., 65: 294–306.

Marder, E. and Goaillard, J.M. (2006) Variability, compensation and homeostasis in neuron and network function. Nat. Rev. Neurosci., 7(7): 563–574.

Marr, D. (1982) Vision: A Computational Investigation into the Human Representation and Processing of Visual Information. WH Freeman, USA.

Martin, J.H., Choy, M., Pullman, S. and Meng, Z. (2004) Corticospinal system development depends on motor experience. J. Neurosci., 24(9): 2122–2132.

McFarland, D.J. and Houston, A. (1981) Quantitative Ethology: The State Space Approach. Pitman, Boston.

Mussa-Ivaldi, F.A. (1992) From basis functions to basis fields: using vector primitives to capture vector patterns. Biol. Cybern., 67: 479–489.

Mussa-Ivaldi, F.A. and Bizzi, E. (2000) Motor learning through the combination of primitives. Philos. Trans. R. Soc. Lond. B Biol. Sci., 355(1404): 1755–1769.

Mussa-Ivaldi, F.A. and Giszter, S.F. (1992) Vector field approximation: a computational paradigm for motor control and learning. Biol. Cybern., 67: 491–500.

Mussa-Ivaldi, F.A., Giszter, S.F. and Bizzi, E. (1994) Linear combination of primitives in vertebrate motor control. Proc. Natl. Acad. Sci., 91: 7534–7538.

Mussa-Ivaldi, F.A. and Hogan, N. (1991) Integrable solutions of kinematic redundancy via impedance control. Int. J. Robot. Res., 10: 481–491.

Nishikawa, K.C., Anderson, C.W., Deban, S.M. and O'Reilly, J.C. (1992) The evolution of neural circuits controlling feeding behavior in frogs. Brain Behav. Evol., 40(2–3): 125–140.

Pang, M.Y., Lam, T. and Yang, J.F. (2003) Infants adapt their stepping to repeated trip-inducing stimuli. J. Neurophysiol., 90(4): 2731–2740.

Pikovsky, A., Rosenblum, M. and Kurths, J. (2000) Phase synchronization in regular and chaotic systems. Int. J. Bifurcat. Chaos, 10(10): 2291–2305.

Prinz, A.A. (2006) Insights from models of rhythmic motor systems. Curr. Opin. Neurobiol., 16(6): 615–620.

Quevedo, J., Stecina, K., Gosgnach, S. and McCrea, D.A. (2005) Stumbling corrective reaction during fictive locomotion in the cat. J. Neurophysiol., 94(3): 2045–2052.

Raibert, M.H. (1986) Legged Robots that Balance. MIT Press, Cambridge, MA.

Rohrer, B., Fasoli, S., Krebs, H.I., Hughes, R., Volpe, B., Frontera, W.R., Stein, J. and Hogan, N. (2002) Smoothness during stroke recovery. J. Neurosci., 22(18): 8297–8304.

Rossignol, S. (2006) Plasticity of connections underlying locomotor recovery after central and/or peripheral lesions in the adult mammals. Philos. Trans. R. Soc. Lond. B Biol. Sci., 361(1473): 1647–1671.

Sanger, T.D. (2000) Human arm movements described by a low-dimensional superposition of principal components. J. Neurosci., 20(3): 1066–1072.

Schaal, S., Ijspeert, A. and Billard, A. (2003) Computational approaches to motor learning by imitation. Philos. Trans. R. Soc. Lond. B Biol. Sci., 358(1431): 537–547.

Schmitt, J. and Holmes, P. (2000) Mechanical models for insect locomotion: dynamics and stability in the horizontal plane I. Theory. Biol. Cybern., 83(6): 501–515.

Schouenborg, J. (2004) Learning in sensorimotor circuits. Curr. Opin. Neurobiol., 14(6): 693–697.

Sherrington, C.S. (1961) The Integrative Action of the Nervous System. Yale University Press, New Haven, CT.

Slotine, J.J. and Lohmiller, W. (2001) Modularity, evolution, and the binding problem: a view from stability theory. Neural Netw., 14(2): 137–145.

Smith, J.L. and Carlson-Kuhta, P. (1995) Unexpected motor patterns for hindlimb muscles during slope walking in the cat. J. Neurophysiol., 74(5): 2211–2215.

Sosnik, R., Hauptmann, B., Karni, A. and Flash, T. (2004) When practice leads to co-articulation: the evolution of geometrically defined movement primitives. Exp. Brain Res., 156: 422–438.

Sumbre, G., Fiorito, G., Flash, T. and Hochner, B. (2006) Octopuses use a human-like strategy to control precise point-to-point arm movements. Curr. Biol., 16(8): 767–772.

Thompson, K.J. (1986a) Oviposition digging in the grasshopper I: functional anatomy and the motor programme. J. Exp. Biol., 122: 387–411.

Thompson, K.J. (1986b) Oviposition digging in the grasshopper II: descending neural control. J. Exp. Biol., 122: 413–425.

Tin, C. and Poon, C.-S. (2005) Internal models in sensorimotor integration: perspectives from adaptive control theory. J. Neural Eng., 2: S147–S163.

Ting, L.H., Blickhan, R. and Full, R.J. (1994) Dynamic and static stability in hexapedal runners. J. Exp. Biol., 197: 251–269.

Todorov, E. (2004) Optimality principles in sensorimotor control. Nat. Neurosci., 7(9): 907–915.

Todorov, E. and Ghahramani, Z. (2003) Unsupervised learning of sensory-motor primitives. Proceedings of the 25th Annual International Conference of the IEEE, Engineering in Medicine and Biology Society, IEEE, Cancun Mexico.

Torres-Oviedo, G., Macpherson, J.M. and Ting, L.H. (2006) Muscle synergy organization is robust across a variety of postural perturbations. J. Neurophysiol., 96(3): 1530–1546.

Tresch, M.C. and Bizzi, E. (1999) Responses to spinal microstimulation in the chronically spinalized rat and their relationship to spinal systems activated by low threshold cutaneous stimulation. Exp. Brain Res., 129(3): 401–416.

Tresch, M.C., Cheung, V.C. and d'Avella, A. (2006) Matrix factorization algorithms for the identification of muscle synergies: evaluation on simulated and experimental data sets. J. Neurophysiol., 5(4): 2199–2212.

Viviani, P. and Terzuolo, C. (1982) Trajectory determines movement dynamics. Neuroscience, 7(2): 431–437.

Wang, W. and Slotine, J.J. (2005) On partial contraction analysis for coupled nonlinear oscillators. Biol. Cybern., 92(1): 38–53.

Westwick, D.T. and Kearney, R.E. (2003) Identification of Nonlinear Physiological Systems. IEEE Book series in Biomedical engineering, IEEE Press series on Biomedical engineering, ISBN 0-471-27456-9, Wiley.

Wilson, D.M. (1961) Central nervous control of flight in a locust. J. Exp. Biol., 38: 471–490.

Wilson, E.O. (2000) Sociobiology: The New Synthesis (25th anniv. ed.). Belknap Press, Cambridge, MA.

Wolpaw, J.R. and Carp, J.S. (1993) Adaptive plasticity in the spinal cord. Adv. Neurol., 59: 163–174.

Wolpert, D.M., Ghahramani, Z. and Flanagan, J.R. (2001) Perspectives and problems in motor learning. Trends Cogn. Sci., 5(11): 487–494.

P. Cisek, T. Drew & J.F. Kalaska (Eds.)
Progress in Brain Research, Vol. 165
ISSN 0079-6123
Copyright © 2007 Elsevier B.V. All rights reserved

CHAPTER 21

A multi-level approach to understanding upper limb function

Isaac Kurtzer[3] and Stephen H. Scott[1,2,3,*]

[1]*Department of Anatomy & Cell Biology, Queen's University, Center for Neuroscience, Botterell Hall, Kingston, ON K7L 3N6, Canada*
[2]*CIHR Group in Sensory-Motor Systems, Queen's University, Kingston, ON K7L 3N6, Canada*
[3]*Centre for Neuroscience Studies, Queen's University, Kingston, ON K7L 3N6, Canada*

Abstract: Here we describe a multi-level approach to study upper limb control. By using non-human primates we were able to examine several different levels of motor organization within the same individual including their voluntary behavior, musculoskeletal plant, and neural activity. This approach revealed several parallels in the global patterns of activity of upper arm muscles and neurons in primary motor cortex (M1). For example, during postural maintenance both arm muscles and arm-related M1 neurons exhibit a bias in torque-related activity towards whole-limb flexion and whole-limb extension torque. A similar bias could be reproduced with a mathematical model of muscle recruitment that minimized the effects of motor noise suggesting a common constraint for the population activation of muscles and cortical neurons. That said, M1 neurons were not merely "upper motor neurons" as they exhibited substantial context-dependency in torque-related activity compared to arm muscles. This flexible association with low-level processing is consistent with M1 having a pivotal role in an optimal feedback controller.

Keywords: primary motor cortex; posture; reaching; monkey; muscle activity

Introduction

A truism of motor control is that it is highly complex and involves multiple levels of organization (Fig. 1). One basic level of motor organization is the behavioral goal or the target state of our actions. Notably, behavioral goals are often composed of a constellation of more specific subgoals as "driving a car" is subserved by lane and speed control which are likewise subserved by coordinated arm and leg movements. A second level is the neural circuitry that supports behavioral goals. The neural network underlying voluntary behavioral goals is highly distributed and includes cerebral and subcortical structures. The last basic level is the peripheral apparatus of muscles, tendons, and skeletal structure. Importantly, an animal's peripheral apparatus has evolved to support particular behavioral goals as exemplified by the specialization of the human hand and kangaroo hindlimb for prehension and hopping, respectively. In sum, behavioral goals, neural circuits, the peripheral apparatus are three basic and interrelated levels of motor organization.

A central problem that confronts motor control researchers is how to fruitfully study this multi-leveled complexity (Scott, 2003). We feel it is

*Corresponding author. Tel.: +1 613 533 2855;
Fax: +1 613 533 6840; E-mail: steve@biomed.queensu.ca

DOI: 10.1016/S0079-6123(06)65021-8

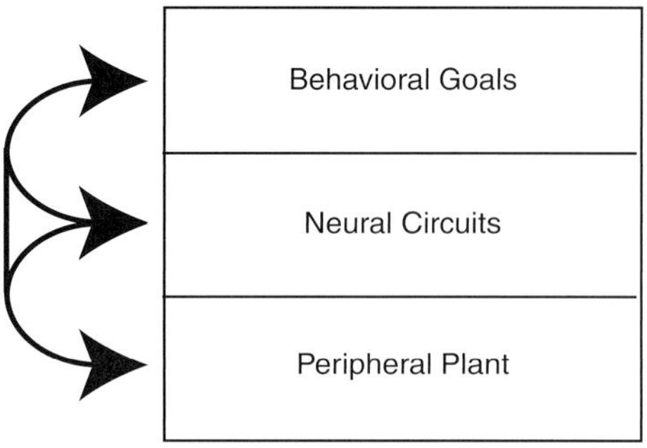

Fig. 1. Framework for examining upper limb function. Cartoon of a putative multi-level organization and the interactions between different levels.

essential to understand each level in order to understand the whole and, consequently, have employed a multi-level approach to study upper limb function. The use of behaving macaque monkeys (in our case, *Macaca mulatta*) is central to this paradigm and allows us to utilize simple behavioral tasks, monitor task-related muscular/neural activity, and examine the musculoskeletal properties in the same individuals. Monkeys are also an invaluable animal-model of human sensori-motor function given their close evolutionary kinship, trainability on arbitrary/complex tasks, dominant use of vision for guiding action, and similar motor repertoire to humans, e.g., ability to reach and grasp objects. Hence, the mechanisms identified from these studies can be readily extended to human studies of motor performance and learning.

Throughout our studies we constrain the upper limb to motion in the horizontal plane as a compromise between single-joint and unconstrained tasks (Fig. 2A) (Scott, 1999). Single-joint studies can be rigorously controlled (Evarts, 1968; Thach, 1978; Cheney and Fetz, 1980) but cannot examine the rich pattern of multi-joint coordination (Hollerbach and Flash, 1982). In contrast, unconstrained tasks involve multiple degrees of freedom but exert far less experimental control (Georgopoulos et al., 1982; Moran and Schwartz, 1999). Our approach limits the limb's motion to flexion and extension at the elbow and shoulder so that we can readily identify the limb's dynamics and examine multi-joint coordination. Although similar paradigms have been extensively used in human studies (Morasso, 1981; Hollerbach and Flash, 1982; Karst and Hasan, 1991; Gordon et al., 1994; Shadmehr and Mussa-Ivaldi, 1994; Sainburg et al., 1999; Flanagan and Lolley, 2001; Singh and Scott, 2003), they are surprisingly rare in monkey studies. The following three sections summarize our results on musculoskeletal mechanics, task-related activity in muscles, and the parallel/unique aspects of cortical processing related to the primate upper limb.

Section 1: Global features of upper limb mechanics

The non-neural "plant" of the motor system is comprised of a complex musculoskeletal system. Since the intrinsic properties of this plant have co-evolved with a species' behavioral goals, its influence on motor function is undoubtedly deep and qualifies any cross-species comparison of motor function. To address this issue our lab undertook several studies on the segmental dynamics and muscle properties of the macaque upper limb.

Limb dynamics

The relation between joint motion and joint torque is qualitatively different between single- and multi-joint systems. Whereas torque and motion are linearly related in a single-joint system, inter-joint coupling allows single-joint torque to induce multi-joint motion and multi-joint torque to induce single-joint motion. Such multi-joint dynamics can be formally represented by a coupled set of complex non-linear equations (Hollerbach and Flash, 1982) that allows us to recognize important component terms including centripetal forces, the combined mass of both segments, and position-dependencies. A more heuristic understanding can be gained by examining the co-varying patterns of joint torque, joint motion, and hand motion in a familiar behavioral task (Buneo et al., 1995; Gottlieb et al., 1997). Note that movement-dependent torque cannot be directly measured but can be calculated with linked-segment models (Scott, 1999) and morphometric tables to scale the size dimensions of individual segments into inertial estimates (Cheng and Scott, 2000). Applying the following procedures to center-out movements revealed

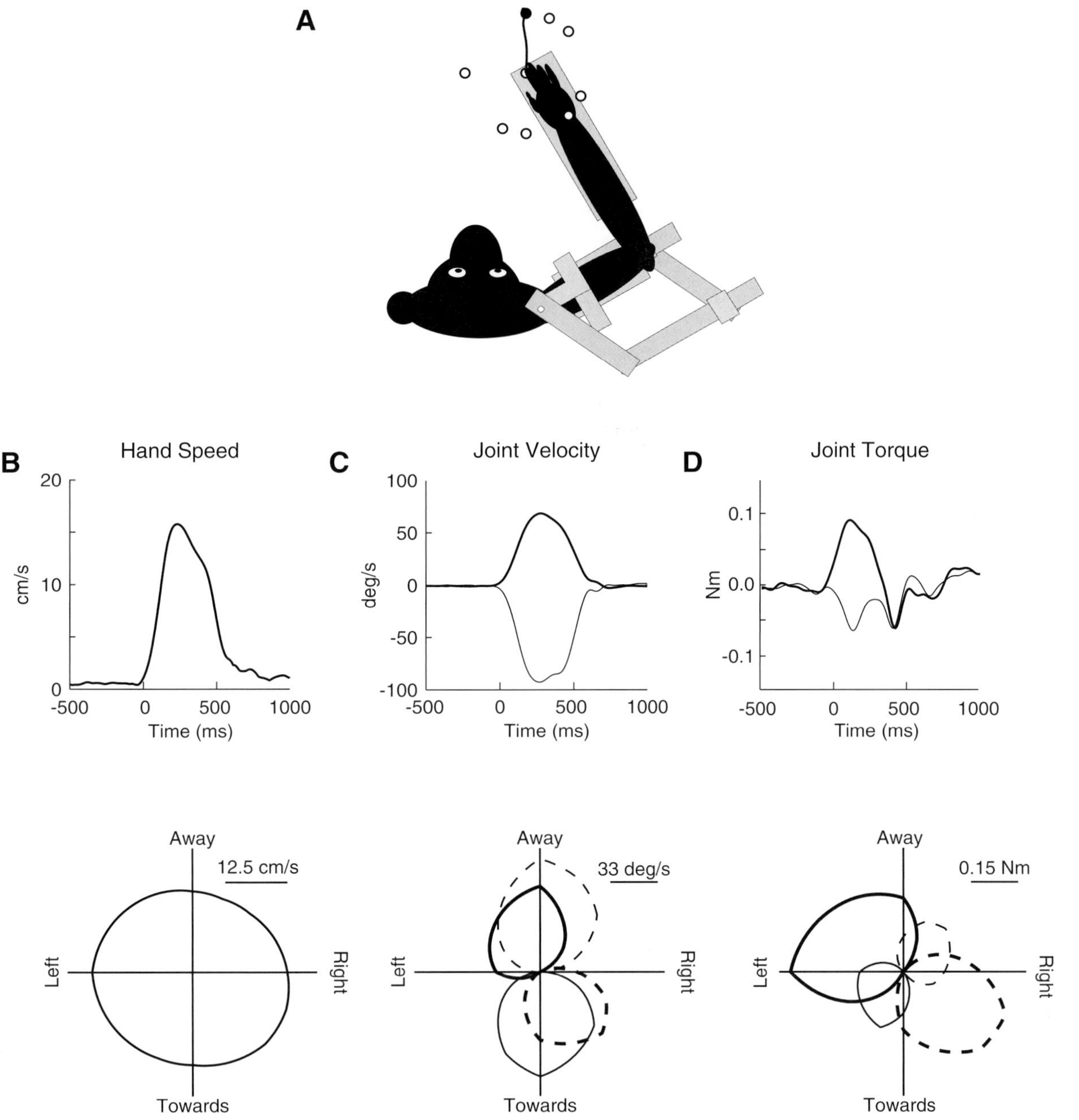

Fig. 2. Limb mechanics during center-out reaching. (A) Schematic of a monkey in the KINARM device to scale and target arrangement during the center-out reaching task. The targets had a non-uniform distribution in hand-space to allow a more uniform distribution in joint space. The thick trace shows an exemplar hand path to the target directly in front of the starting position, filled circle. (B–D) Measured limb mechanics: hand velocity, joint motion, and joint torque. Top panels show kinematics/kinetics from exemplar reach. Bottom panels show data across all targets depicted in polar coordinates of magnitude and angular position; magnitude indicates the average value between movement onset and peak hand velocity whereas angular position indicates the direction of hand motion. Thick and thin lines denote the shoulder and elbow joint, respectively. Flexion and extension are denoted by solid and dashed lines, respectively. All data is from one representative monkey.

several significant patterns in the monkey forelimb (Graham et al., 2003a).

Monkeys were trained to move their hand from a central start position to a peripheral target (Fig. 2A). Since reinforcement depended on endpoint timing and accuracy the monkeys adopted similar peak hand speeds and movement times across target direction (Fig. 2B). In contrast to the hand-based kinematics, joint motions changed substantially with target direction (Fig. 2C). Maximum joint velocities and joint displacements occurred near the fore-aft axis with elbow velocity being ~50% greater than shoulder velocity. Elbow flexion and extension was greatest towards and away from the body whereas shoulder flexion and extension was greatest away/slightly-left and towards/slightly-right of the body. The combined shoulder-elbow motion is about four times greater for movements towards or away from the body as compared to movements to the left or right. This mapping reflects the "inefficiency" of opposing flexion and extension at the shoulder and elbow to cause a net hand displacement towards or away from the body (Graham et al., 2003a).

The muscular torque generated at the shoulder and elbow also exhibited significant non-uniformities across target direction (Fig. 2D). The combined muscular torque of the two joints is about two times greater for movements to the left or right compared to movements towards or away from the body. The largest shoulder torque occurs for movements away and to the left, and movements towards and to the right. Elbow torque is maximal in the opposite directions (away and to the right, and towards and to the left) at about half of the magnitude observed for the shoulder. And as previously mentioned, joint motion does not necessarily equal joint torque in a multi-joint system. During movements to the left and away from the body, elbow extension occurs from purely shoulder-flexion torque whereas movements directly right involve negligible elbow motion but significant elbow torque, approximately half that of the shoulder.

This task presents several similarities to human studies. First, the monkeys reached with gently curved hand paths and single-peaked velocity profiles as observed in human studies (Morasso, 1981). Second, their small movement amplitudes (6 cm) were appropriate for the monkeys' smaller size (shoulder-elbow length ~30 cm) (Cheng and Scott, 2000) when compared to the movement amplitudes in human studies (between 4 and 30 cm); a typical length for the shoulder and elbow in the adult human is ~55 cm (Diffrient et al., 1978). Moreover, the similar arm geometry of monkeys led to comparable patterns of kinematics and kinetics to those in human studies (Buneo et al., 1995; Gottlieb et al., 1997). The main difference in limb dynamics is the smaller mass of the monkey's arm (~1/7th of human) leading to a much lower moment of inertia, ~1/40th of humans!

Muscle morphometry

All mammalian skeletal muscle exhibits dynamic properties that constrain the efficacy of motor commands. These include a time-dependent transformation of neural excitation into activation of the molecular motors (activation dynamics) and from the molecular motors into muscle force (contraction dynamics) (Zajac, 1989). Muscle force also reflects a complex function of the muscle's length and velocity (Rack and Westbury, 1969; Scott et al., 1996). In brief, the peak active force of a muscle occurs at a single length whereas shortening and lengthening velocities result in decreases and increases in muscle force, respectively.

The close evolutionary kinship of humans and old-world monkeys ensures several additional similarities in their proximal arm musculature (Graham and Scott, 2003b). Both possess the same 14 arm muscles for flexion-extension at the elbow-shoulder, excepting a biarticular (dorsoepitrochlearis) that non-human primates use for arboreal ambulation. Also, the tendons of both monkey and human arm muscles are relatively short precluding significant energy storage as in the kangaroo hindlimb (Zajac, 1989).

When comparing between human and monkey muscles the most significant difference is their relative strength. Monkey muscles have a much larger physiological cross-sectional area (PSCA) when scaled to body-weight than humans, between 8 and 10 times larger (Graham and Scott, 2003b).

These muscles are also composed of a large percentage of fast fibers (50%) (Singh et al., 2002), which allows efficient force production at high shortening velocities. The greater relative strength and lower limb inertia of monkeys allows them to generate much faster and more powerful movements than humans.

We observed several broad patterns across the monkey's upper limb muscles (Graham and Scott, 2003b; Graham et al., 2003a). The passive torques of the elbow and shoulder (which includes moment arms, tendon slack length, and optimum fascicle length of all spanning muscles) are well described by simple cubic functions. Second, the zero torque-intercept of these cubic functions — approximately 90° elbow angle and 30° shoulder angle — is near the muscles' optimum fascicle angle — approximately 100° elbow angle and 0° shoulder angle. A third general pattern is that the maximum torque that can be produced at the shoulder and elbow varies systematically and symmetrically with its joint angle: greater flexion/extension elbow torque occurs with greater flexion, and greater flexion/extension shoulder torque occurs with greater extension.

The broad patterns just described strongly contrast with the significant diversity among the individual arm muscles (Graham and Scott, 2003b). Biceps brachii longus and triceps brachii longus both span the shoulder and elbow joint but closer examination reveals that biceps has a greater effect on the elbow (~3:1) whereas triceps has far greater affect on the shoulder (~3:1). The moment arms of the limb's muscles also include increases, decreases, minimal change, and even non-monotonic changes with elbow and/or shoulder angle. Finally, the diverse properties of muscles can even lead to similar effects through reciprocal contributions that balance each other out. Brachioradialis and brachialis (both elbow flexors) have significant differences in their PSCA and moment arm but possess a similar torque capability since these differences are reciprocal, small cross-section/large moment arm and large cross-section/small moment arm.

In sum, the upper arms of monkeys and humans possess similar mechanical and muscular properties. The following section will describe how the most salient properties of the peripheral plant (including the non-uniform joint mechanics and muscle dynamics) shape the patterns of muscle activity during voluntary motor tasks.

Section 2: Task-related activity of upper limb muscles

A thorough understanding of an organism's peripheral apparatus is insufficient for predicting its pattern of motor behavior. This gap reflects, in part, the well-known "degrees of freedom" problem (Bernstein, 1967). A single motor task such as "reaching for a cup" can be achieved by many different joint trajectories. The fact that multiple muscles have a similar mechanical action also ensures that a particular joint torque can be achieved by many different patterns of muscle activity. To address this issue we examined muscle activation of the monkey upper limb during two basic tasks: postural maintenance and point-to-point reaching. Importantly, we applied loads to the shoulder and elbow joints via a custom robotic exo-skeleton (KINARM, BKIN Technologies, Kingston) allowing us to examine muscular (and neural) activity associated with load compensation at each joint and during multi-joint loads.

Posture task

During the postural task the animal maintained a fixed hand position while countering flexion and extension loading of their shoulder and/or elbow joints (Cabel et al., 2001; Herter et al., 2007). Since the limb kinematics were effectively fixed changes in muscular activity could be related specifically to the change in joint torque. The most significant finding was that a muscle's activation pattern could differ from its anatomical action (Kurtzer et al., 2006b). For example, brachioradialis varied its activity with torque at the shoulder even though it lacked any direct mechanical action at that joint (Fig. 3A). This anatomical elbow flexor was maximally activated by a combination of elbow flexor and shoulder extensor torque. A similar bias in preferred torque direction (PTD) was observed for the entire sample of monoarticulars — shoulder extensors and elbow flexors were maximally active with shoulder-extension/elbow-flexion and shoulder flexors and elbow extensors were maximally active with shoulder-flexion/elbow-extension torque. Moreover, the PTDs of biarticular extensor and flexor muscles were not directed to extension–extension and flexion–flexion torques (as one might expect from their moment arms) but towards

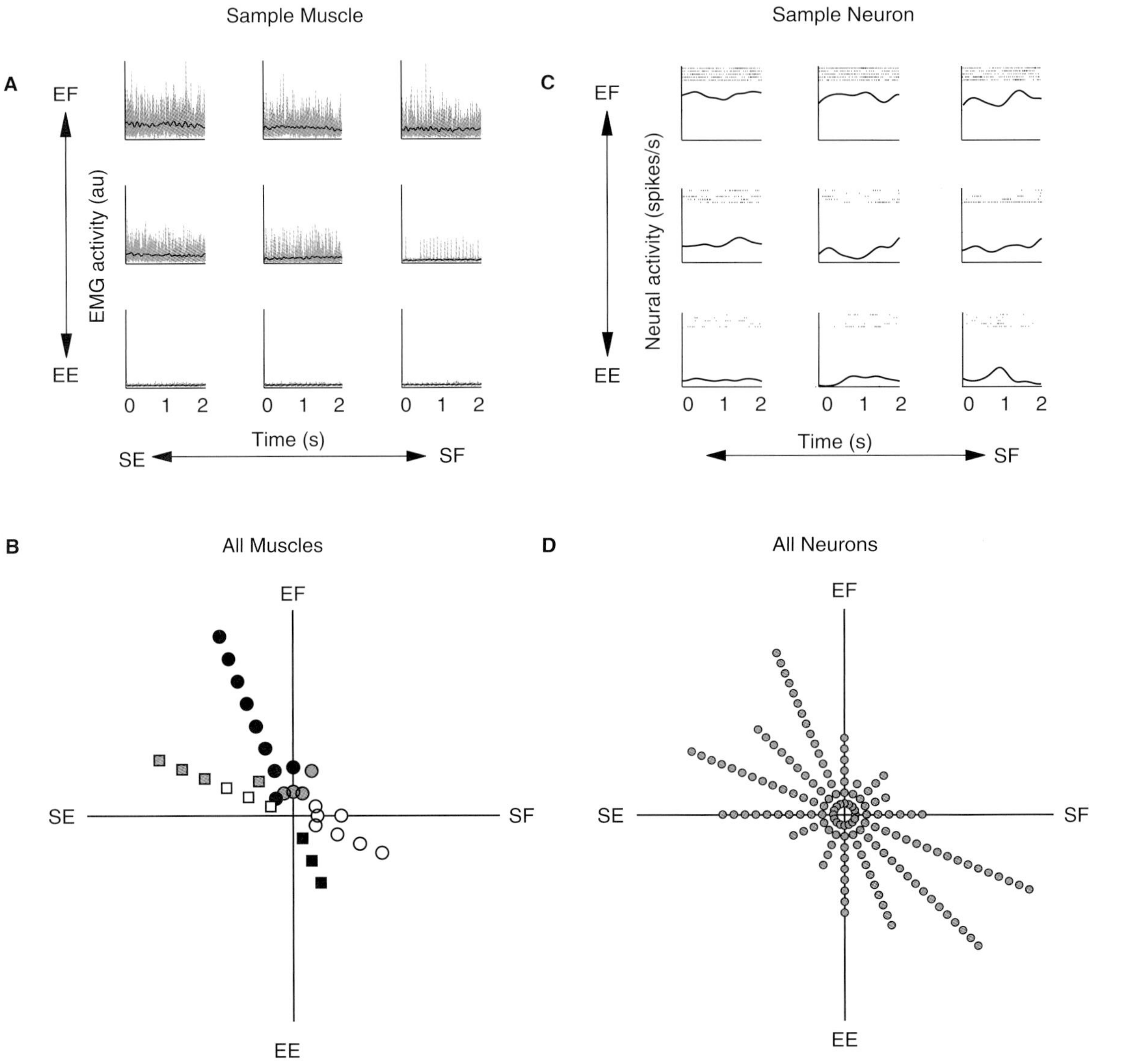

Fig. 3. Torque-related activity during the posture task. (A) Activity of a sample muscle (elbow flexor/brachioradialis) during the nine load conditions of shoulder-elbow torque. The load conditions are presented in a particular coordinate system: shoulder flexion is 0°, elbow flexion is 90°, shoulder extension is 180°, and elbow extension is 270°. Note that the muscle's maximal activity is directed towards elbow-flexion and shoulder-extension. (B) Polar histogram summarizes the preferred torque direction (PTD) from all sampled muscles during the posture task. Each muscle's PTD is represented by a single icon within an angular bin (16 bins of 22.5° used throughout figures). White, black, and gray symbols indicate shoulder monoarticulars, elbow monoarticulars, and biarticulars; circles and squares indicate flexors and extensors. (C) Sample neuron during the posture task. Same format as A. (D) Polar histogram summarizes the PTDs from all sampled neurons during the posture task. Same format as B. Gray circle denote neurons. (Panels A, B adapted from Kurtzer et al., 2006b; Panel C adapted with permission from Kurtzer et al., 2005.)

shoulder-extension/elbow-flexion torques. The resulting global pattern of upper limb muscle activity was a bimodal distribution of PTDs to shoulder-extension/elbow-flexion and shoulder-flexion/elbow-extension torque (Fig. 3B).

Our results are consistent with earlier dissociations of muscle function and action in the monkey wrist (Hoffman and Strick, 1999), human leg (Nozaki et al., 2005), and human arm (van Zuylen et al., 1988; Buchanan et al., 1989). Several of these previous studies (such as Fagg et al., 2002; Nozaki et al., 2005) have suggested that this dissociation reflects an optimization process for muscle coordination. We tested whether a similar process could account for our results. In brief, a lumped representation for the six muscle groups — shoulder extensor, shoulder flexor, elbow extensor, elbow flexor, biarticular flexor, biarticular extensor — was employed using known values for the monkey moment arms, fascicle length and PCSA (Cheng and Scott, 2000; Graham and Scott, 2003b). The model was further constrained so that conditions were isometric and muscles could only pull. Finally, an iterative procedure scaled each muscle group's activity to achieve a target torque with the minimal muscle noise $\sum(\text{force}_i^*/\text{PCSA}_i)^2$, (]Harris and Wolpert, 1998; van Bolhuis and Gielen, 1999; Hamilton et al., 2004) also equivalent to muscle stress (van Bolhuis and Gielen, 1999).

The optimization model successfully predicted PTD rotations towards shoulder-flexion/elbow-extension and shoulder-extension/elbow-flexion torque (Fig. 4). In fact, a similar PTD bias will result from minimizing any number of different variables related to muscle activity such as metabolic energy and muscle force (Kurtzer et al., 2006b). Consider if several muscles can contribute to a particular agonist torque then each can be activated at relatively low levels. But if fewer muscles can contribute then each must have a higher relative activity. And since biarticular muscles in the primate proximal forelimb generate extension–extension and flexion–flexion torques this leads to a PTD bias towards the "gaps" in torque space, i.e., shoulder-extension/elbow-flexion and shoulder-flexion/elbow-extension torque. It should be noted that although the model qualitatively accounts for

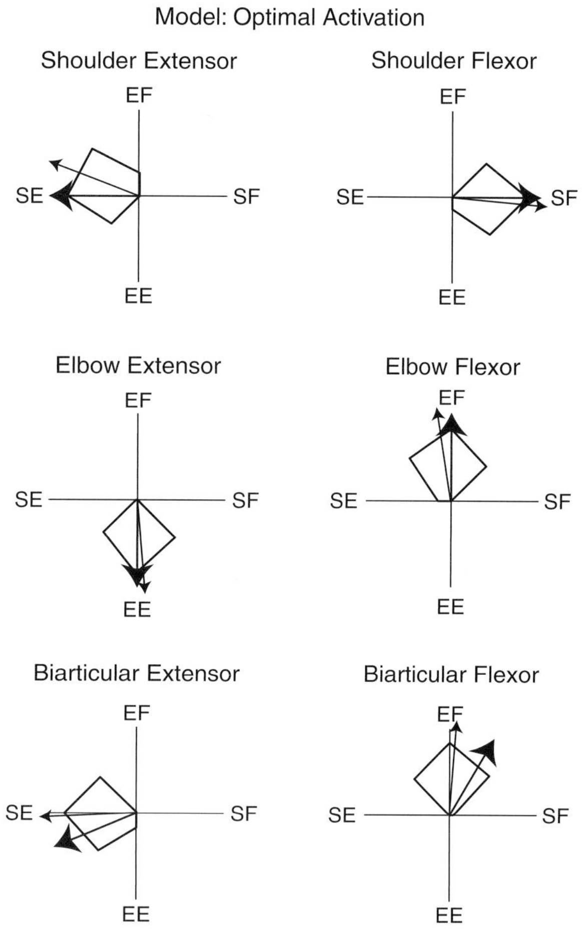

Fig. 4. Modeled activation of muscles during posture task. (Adapted with permission from Kurtzer, 2006b.) Panels illustrate the predicted activation when minimizing net muscle noise. Large and small arrows indicate anatomical action and PTD of each "muscle group": shoulder flexors/extensors, elbow flexors/extensors, and biarticular flexors/extensors.

PTD rotations, (in our hands) it underestimated the magnitude of rotation and did not predict the relative amount of rotation on a muscle group by muscle group basis.

Reaching task

Inferring the function of muscular activity during multi-joint movement is complicated by intersegmental and muscular dynamics (Zajac and Gordon, 1989). However, one can "factor out"

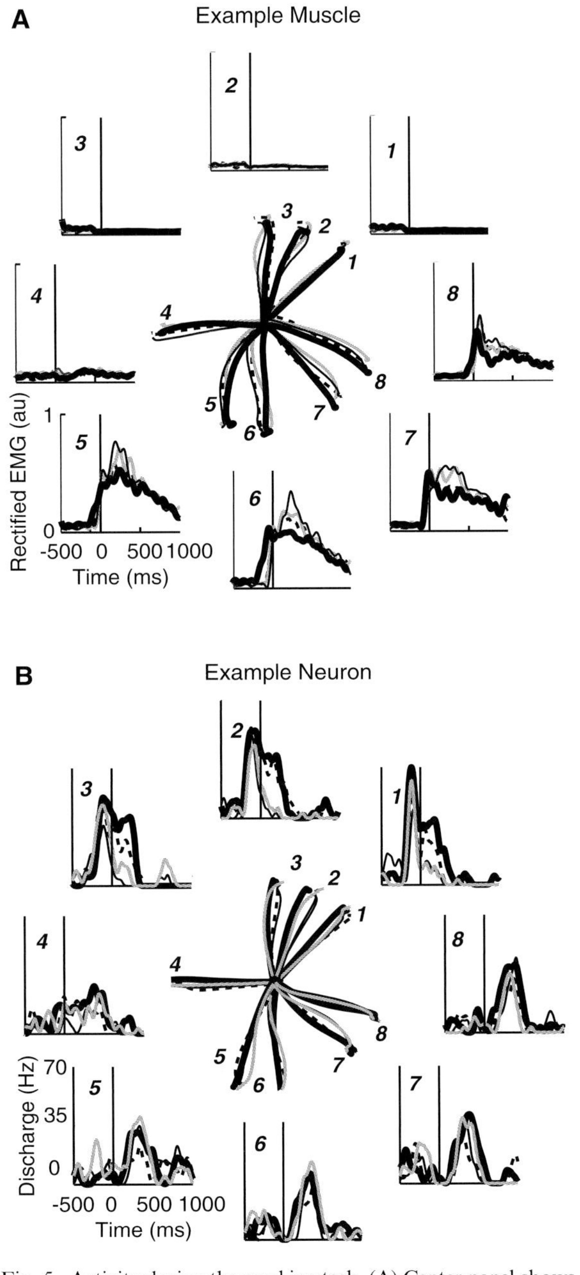

Fig. 5. Activity during the reaching task. (A) Center panel shows the hand path during reaching movements to the eight targets depicted in Fig. 2A. Different lines denote the different load conditions: unloaded (thick black), viscous shoulder (dashed), viscous elbow (gray), and viscous both (thin black). Peripheral panels show associated activity of a sample muscle (elbow flexor/brachioradialis). Here the greatest activity during unloaded reaching occurs towards the body whereas the different load conditions lead to further increases in activation near this preferred hand direction. (B) Center panel shows the hand path from a different session. Peripheral panels show associated activity of a sample neuron. Its greatest (initial) activity during unloaded reaching occurs away from the body and the different load conditions lead to decreased activation to targets near the PHD. (Panel A adapted with permission from Kurtzer et al., 2006a.)

the complex transformation of muscle activity to joint torque by enforcing similar movement patterns under different load conditions. Any change in activity between conditions can then be related specifically to the changes in load across conditions (Gribble and Scott, 2002). This novel paradigm revealed that upper arm muscles often exhibited changes in activity across the different load conditions during a reaching task. An example muscle is showed in Fig. 5A. Moreover, comparing the change in activity versus the change in joint torque revealed that the muscles' PTDs deviated from their anatomical action towards shoulder-flexion/elbow-extension and shoulder-extension/elbow-flexion torques (Fig. 6A) (Kurtzer et al., 2006a). Hence, the muscles exhibited the same torque bias during reaching as seen during in the posture task (Fig. 3B).

Another striking result was that during unloaded reaching the monoarticular and biarticular arm muscles were mostly active for movements towards or away from the body (Kurtzer et al., 2006a). Single-joint elbow flexors and extensors had preferred hand directions (PHDs) towards and away from the body, respectively (Figs. 5A and 7A). In contrast, the single-joint shoulder flexors and extensors had PHDs away/slightly-left and towards/slightly-right, respectively. Importantly, the muscles' PHDs better mirrored the fore-aft orientation of joint motion than the left–right orientation of the joint torque (Fig. 2B, C). Without knowing that these responses reflected the activity of force generators one could wrongly conclude that muscles were encoding joint velocity!

So how did the muscles' PHDs come to match joint velocity better than joint torque? This bias likely reflects the impact of several factors already identified. Reaching movements in the fore-aft axis require the fastest joint velocities and largest joint displacements. Since greater activity is necessary to

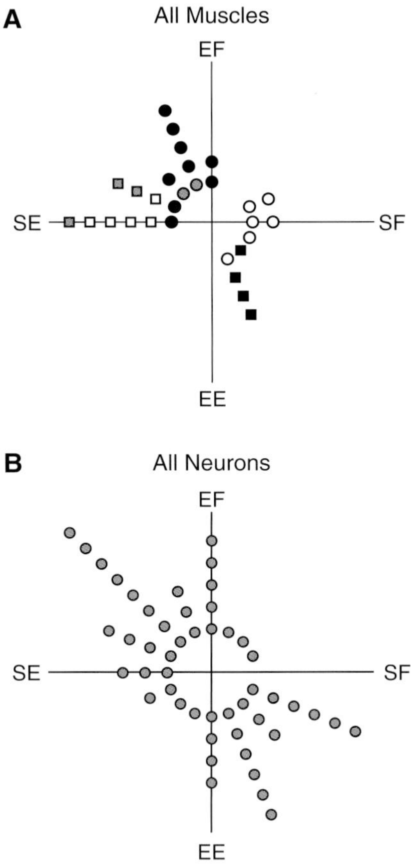

Fig. 6. Summary of observed and simulated preferred torque directions during reaching. Polar histograms of preferred torque directions determined by comparing change in activity to change in torque between unloaded and load reaching movements. (A, B) PTDs from sampled muscles and sampled neurons. Same icons and format as Fig. 3B. (Panel A adapted with permission from Kurtzer et al., 2006a.)

compensate the velocity-dependent drop in force and length-dependent resistance to center-out displacements, the additional activity in these directions is measured as a PHD bias to the fore-aft axis. Another likely factor for the fore-aft bias of muscles' PHDs is their bias in load-related activity towards shoulder-flexion/elbow-extension and shoulder-extension/elbow-flexion torques. Since these torque combinations occur with movements towards and away from the body (Fig. 2B, C), then the torque bias again induces a spatial bias to the fore-aft axis.

In fact, we could reproduce the fore-aft bias of PHDs (Fig. 7C) with a model of muscle activation that included length- and velocity-dependent properties and a recruitment strategy that minimizes the total muscular activity (Kurtzer et al., 2006a), i.e., an expansion of the model used for the posture task. Without accounting for the known muscle properties or interactions among muscles the PHDs simply mirror the left–right bias of the joint torque (Fig. 7D).

In sum, we observed that the salient features of the peripheral plant — included non-uniformities in joint mechanics, muscle dynamics, and muscle redundancy — shaped the activity of muscle during voluntary tasks. The following section will compare and contrast these patterns of activity with those of primary motor cortex.

Section 3: Relation to motor cortical processing

Over a century of research has established a significant role for primary motor cortex in voluntary motor control (Hepp-Reymond, 1988; Porter and Lemon, 1993; Scott, 2003). Anatomically, M1 possesses an intimate link with the motor periphery including somatosensory inputs (Wong et al., 1978; Asanuma et al., 1979) and a dense output to intermediate spinal lamina and a small direct input to motor neurons (Fetz and Cheney, 1987; Dum and Strick, 1991; Lemon and Griffiths, 2005). The specific functional role of M1 is less apparent due to feedback effects, the complexity of the musculoskeletal properties, and the many roles for spinal processing. However, many M1 neurons clearly exhibit muscle-related activity. Cortico–muscular associations include frequency coherence between EMG and cortical oscillations (Baker et al., 1999), steady-state activity in M1 linked to static load requirements (Evarts, 1968; Smith et al., 1975; Thach, 1978; Fromm, 1983), and time-varying activation linked to the dynamic load requirements and muscle activity (Morrow and Miller, 2003; Sergio and Kalaska, 2005). Here we add to this rich story by demonstrating parallels and contrasts in their global patterns of reach- and torque-related activity.

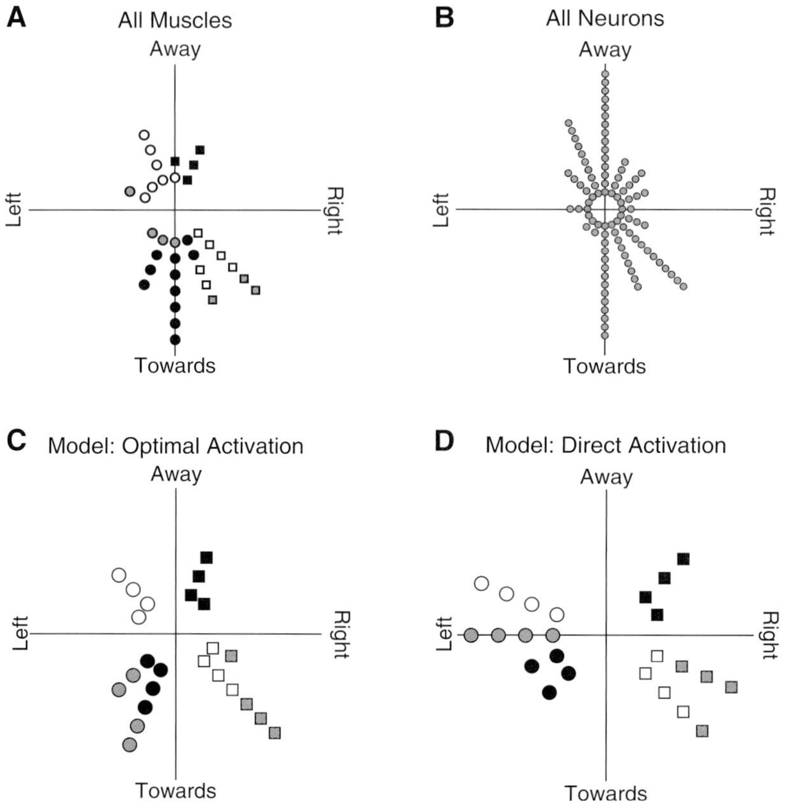

Fig. 7. Summary of preferred hand directions during reaching. Polar histograms of preferred hand directions during unloaded reaching movements. (A, B) PHDs from sampled muscles and sampled neurons. Same icons as Fig. 3B (Panel A adapted with permission from Kurtzer et al., 2006a). (C, D) PHDs of simulated muscles when including both muscle properties and a minimization criterion (Optimal Activation) or directly matching the required torque (Direct Activation).

Similarities in global activity

As described in the previous section, arm muscles exhibited several distinctive patterns of activity during postural maintenance and reaching tasks. Upper arm muscles were maximally activated with shoulder-flexion/elbow-extension and shoulder-extension/elbow-flexion torques while countering joint loads in a fixed limb posture (Kurtzer et al., 2006b). This pattern was recapitulated in the population activity of M1 neurons (Fig. 3C, D) from the same animals (Cabel et al., 2001; Herter et al., 2007). It should be emphasized that this similarity was not an entirely foregone conclusion. In principle, the complex connectivity inherent in M1 could allow it to preferentially represent single-joint torques, or equally represent all torque combinations, or offset the whole-limb extensor bias of brainstem structures controlling quadrapedal stance.

Further parallels were observed during visually guided reaching. In the unloaded condition, M1 neurons typically had a preferred hand direction near the fore-aft axis (Figs. 5B and 7B) (Scott et al., 2001) as previously shown for arm muscles (Figs. 5A and 7A) (Kurtzer et al., 2006a). Moreover, many of these M1 neurons exhibited changes in activity when reaching across different load conditions with an overall trend of PTD biased to shoulder-flexion/elbow-extension and shoulder-extension/elbow-flexion torques (Kurtzer et al., 2006a). Hence, the global pattern of reach- and torque-related activity was similar for M1 neurons and arm muscles.

Note that this mirroring relation during posture and movement tasks does not imply a point-to-point mapping where M1 neurons are merely upper motor neurons. Rather, it suggests that there are similar constraints on the activation of muscles and cortical neurons — such as minimizing the effects of motor noise, the motion-dependent aspects of muscle force, and the mechanics of multi-joint movements.

Context-dependent cortical activity

While several parallels exist between muscular and (some) M1 activity it should be emphasized that M1 neurons express a diverse range of activity patterns and there is no ideal M1 neuron (Scott, 2003). Some, but not all, M1 neurons receive strong somatosensory inputs. Some, but not all, are modulated with muscle force. And some M1 neurons are sensitive to events at a single joint whereas others reflect events across multiple joints. This diversity likely reflects the many possible roles for M1 in spinal processing — an afferent template to fusimotor neurons, gating of sensory input, and setting muscle tone — in addition to providing patterned input to alpha motor neurons. In addition, M1 neurons show an impressive degree of plasticity not present in limb muscles (Sanes and Donoghue, 2000; Li et al., 2001; Scott, 2003; Paz et al., 2004); short-term adaptation paradigms (~5–15 min) can focally alter a neuron's tuning orientation and shape whereas long-term practice or trauma (days to months) can even affect how much cortical area subserves a particular effector.

Our studies have extended this diverse range in M1 processing to include how neurons process torque-related information across different behavioral tasks. A direct comparison of different behavioral tasks is typically compromised because of the intrinsic differences in kinematics such as between posture and movement. We addressed this issue by examining how neurons express torque-related activity, a salient non-movement feature. Interestingly, we observed a broad range of responses. Some neurons showed significant torque-related activity in both tasks (Fig. 8, left column). Other neurons only showed torque-related activity during the reaching task and were wholly irresponsive to loads during posture (Fig. 8, center column). And still others only expressed torque-related activity during the posture task; their reach-related activity was unaffected by loads (Fig. 8, right column). In all, ~20% of M1 neurons were unaffected by load conditions (even though they expressed strong reach-related activity) whereas ~30% had torque-related activity in both tasks. The arm muscles showed a much more uniform pattern with almost all having load effects (97%) in at least one task and most in both tasks (64%). Further, neurons expressed torque-related decreases in activity during reaching equally often as increases whereas the muscular responses were dominated by increases in torque-related activity (Fig. 5A, B).

Another across-task comparison involved comparing different aspects of torque-related activity (Kurtzer et al., 2005). One aspect is the PTD, an index of the relative sensitivity to shoulder and elbow loads. Interestingly, PTDs of neurons and muscles were conserved across tasks with a mean absolute difference of 52° and 28°, respectively; a random paring would have a mean of ~90°. [Note that larger inter-task correlations were observed when the animals countered a constant load during both posture and movement tasks (Kurtzer et al., 2005).] This implies that both neurons and muscles had a consistent relative effect at the motor periphery (Fig. 9A, C). In contrast, the absolute sensitivity to torque magnitude, or torque gain, was completely unpredictable across tasks for M1 neurons ($r = 0.13$) but highly similar for muscles ($r = 0.81$) (Fig. 9B, D); the greater load-sensitivity of muscles during reaching presumably reflects the force–velocity relation.

Summary and interpretation

The previous three sections elaborated on a multi-level approach to motor function whereby we could examine the peripheral apparatus, muscular activation patterns, and cortical processing within the same organism. This approach allowed us to compare and contrast the global properties of each level. Briefly, the limb's dynamics during unloaded reaching involved a significant non-uniformity at

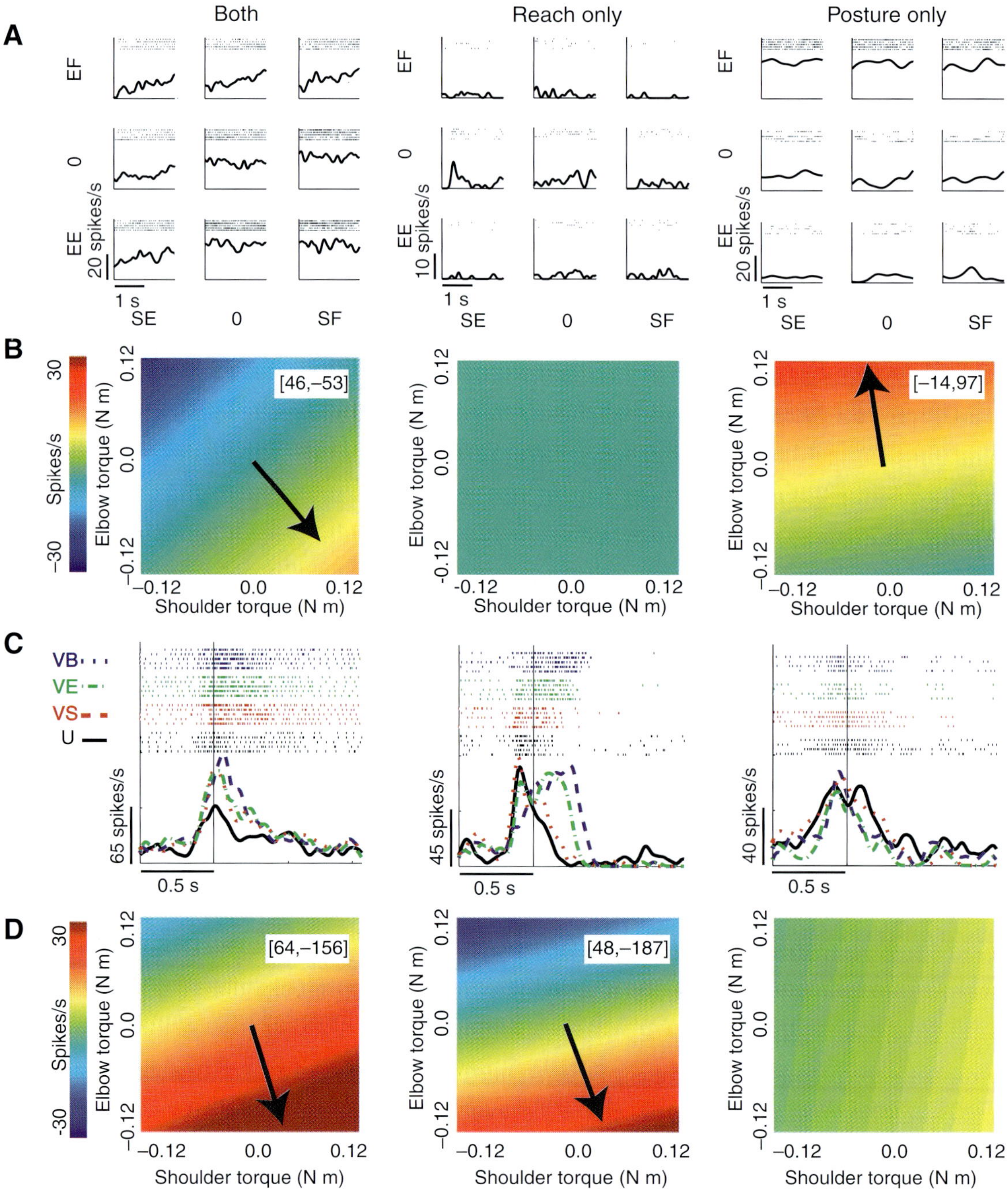

Fig. 8. Exemplar neurons exhibiting torque-related activity. Three sample neurons showing torque-related activity during both the posture and movement tasks (left column), only the movement task (center column), and only during the posture task (right column). (A) Activity during the posture task. (B) Planar regressions of torque-related activity during the posture task indicated by gradient. Arrows denote the PTD for significant torque-related activity. Numerical inset shows the change in activity versus the change in torque, large numbers indicate a high sensitivity to torque magnitude. (C) Activity during the reaching task to the target nearest the neuron's PHD during the viscous shoulder (VS), viscous elbow (VE), viscous both (VB), and unloaded (U) conditions. (D) Planar regressions of torque-related activity during the reaching task. (Adapted with permission from Kurtzer et al., 2005.)

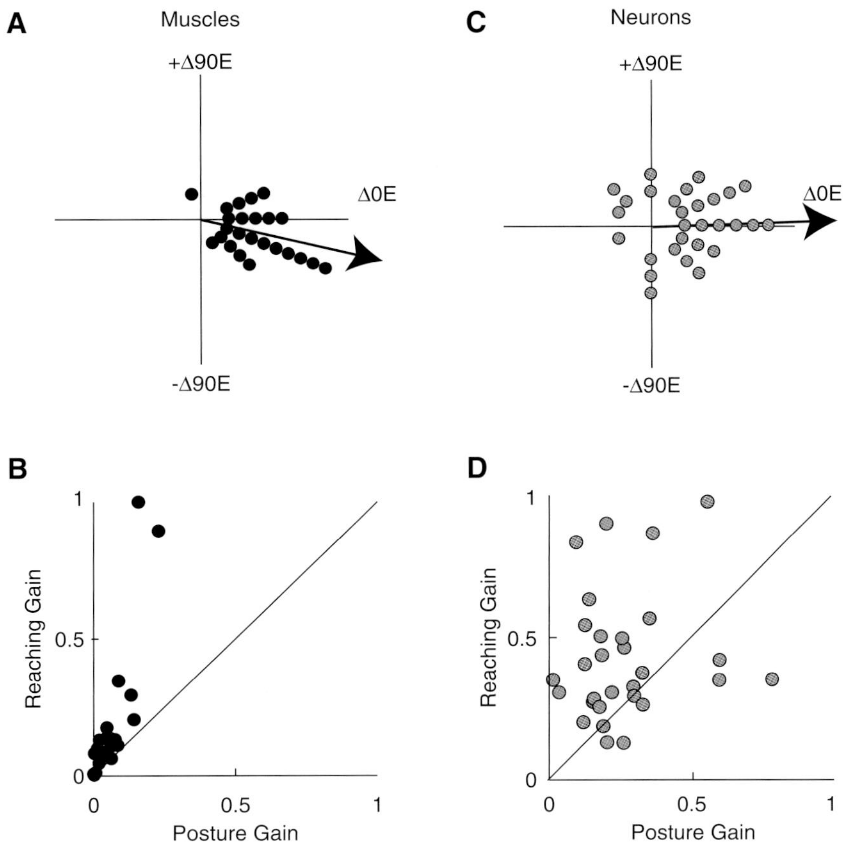

Fig. 9. Comparing torque-related activity across posture and reaching tasks. (A, B) Polar histogram of angular difference in preferred torque direction across tasks. Identical PTDs would be aligned to the right at Δ0°. Data are shown separately for muscles and neurons (Panel B is adapted with permission from Kurtzer et al., 2005). (C, D) Linear regression of torque gain across tasks; unity line indicates an identical torque gain. Torque gain is the sensitivity to absolute magnitude of torque. Data is shown separately for muscles and neurons and is normalized to the peak gain. (Panel D is adapted with permission from Kurtzer et al., 2005.)

the joint level with larger combined joint torque with movement towards/right and away/left and larger joint motion in the fore-aft axis. This pattern was mirrored in the global pattern of both muscular and M1 activity. Second, the torque-related activity of muscles and M1 neurons tasks had an overall bias towards whole-limb flexor and whole-limb extensor torques which could be modeled as an optimization of muscle activity. And finally, M1 neurons (but not muscles) exhibited substantial changes in their torque-related activity between posture and movement contexts including specificity to a particular task.

On a more cautionary note, without adopting a multi-level approach to study the motor system one is simply unable to identify (and experimentally control) the potential co-variations among these levels. For example, Georgopoulos and colleagues recently confirmed our observation of a fore-aft bias of preferred reaching directions by using an unconstrained 3D task though this only emerged when a very large number of neurons were sampled ($n > 1000$) (Naselaris et al., 2006). Without knowing how the limb behaves in its generative details, the authors could only interpret these findings as an augmented representation or "hyperacuity" of particular spatial directions. We suspect that limb mechanics remains an important factor regardless of whether movements are performed in the plane or throughout 3D space and that the factors we have identified provide a simpler explanation for their observation. However, unraveling the link between neural processing and limb mechanics for 3D movements is fraught with many caveats and pitfalls due to the rapid escalation in mechanical complexity.

While our observations elaborate the significant parallels between multiple levels of motor organization, they do not provide a theoretic framework. Such a theory would have to comfortably address several (apparently) disparate features. Why would M1 reflect the global features of the motor periphery? Why would M1 represent such a diverse range of "low-level" features? Why would these representations exhibit plasticity on so many time scales from behavioral context to long-term changes? And how do M1's substantial (and generally neglected) somatosensory inputs figure into this organization?

We are currently exploring the merits of optimal feedback control as theory of motor function (Scott, 2004). In broad strokes, optimal feedback control involves a modifiable sensory-motor mapping (or feedback law) that is tailor-made for the task at hand by balancing multiple conflicting demands such as speed, accuracy, and effort (Todorov and Jordan, 2002). The sensory inputs to this controller can also include multiple sources of information that are flexibly integrated with motor commands for predictive estimates. Thereby, optimal control can provide rapid "intelligent" responses that reflect "higher-level" goals using "low-level" representations. These aspects are broadly consistent with the known anatomy and physiology of M1 including the observations we've described throughout this chapter. Determining whether optimal feedback control is a heuristic framework requires careful experimentation and is the focus of our current research.

Abbreviations

M1	primary motor cortex
PHD	preferred hand direction
PSCA	physiological cross-sectional area
PTD	preferred torque direction

Acknowledgments

We thank K. Moore for expert technical help and T. Herter for help with the figures. This research was supported by grants from the Canadian Institutes of Health Research (CIHR) and National Science and Engineering Research Council (NSERC) to SHS as well CIHR salary awards to SHS and IK.

References

Asanuma, H., Larsen, K.D. and Zarzecki, P. (1979) Peripheral input pathways projecting to the motor cortex in the cat. Brain Res., 172: 197–208.

Baker, S.N., Kilner, J.M., Pinches, E.M. and Lemon, R.N. (1999) The role of synchrony and oscillations in the motor output. Exp. Brain Res., 128: 109–117.

Bernstein, N.A. (1967) The Coordination and Regulation of Movements. Pergamon Press, Oxford.

van Bolhuis, B.M. and Gielen, C.C. (1999) A comparison of models explaining muscle activation patterns for isometric contractions. Biol. Cybern., 81: 249–261.

Buchanan, T.S., Rovai, G.P. and Rymer, W.Z. (1989) Strategies for muscle activation during isometric torque generation at the human elbow. J. Neurophysiol., 62: 1201–1212.

Buneo, C.A., Boline, J., Soechting, J.F. and Poppele, R.E. (1995) On the form of the internal model for reaching. Exp. Brain Res., 104: 467–479.

Cabel, D.W., Cisek, P. and Scott, S.H. (2001) Neural activity in primary motor cortex related to mechanical loads applied to the shoulder and elbow during a postural task. J. Neurophysiol., 86: 2102–2108.

Cheney, P.D. and Fetz, E.E. (1980) Functional classes of primate corticomotoneuronal cells and their relation to active force. J. Neurophysiol., 44: 773–791.

Cheng, E.J. and Scott, S.H. (2000) Morphometry of *Macaca mulatta* forelimb. I. Shoulder and elbow muscles and segment inertial parameters. J. Morphol., 245: 206–224.

Diffrient, N., Tillery, A.R. and Bardagjy, J.C. (1978) Humanscale. MIT Press, Cambridge, MA.

Dum, R.P. and Strick, P.L. (1991) The origin of corticospinal projections from the premotor areas in the frontal lobe. J. Neurosci., 11: 667–689.

Evarts, E.V. (1968) Relation of pyramidal tract activity to force exerted during voluntary movement. J. Neurophysiol., 31: 14–27.

Fagg, A.H., Shah, A. and Barto, A.G. (2002) A computational model of muscle recruitment for wrist movements. J. Neurophysiol., 88: 3348–3358.

Fetz, E.E. and Cheney, P.D. (1987) Functional relations between primate motor cortex cells and muscles: fixed and flexible. Ciba. Found. Symp., 132: 98–117.

Flanagan, J.R. and Lolley, S. (2001) The inertial anisotropy of the arm is accurately predicted during movement planning. J. Neurosci., 21: 1361–1369.

Fromm, C. (1983) Changes of steady state activity in motor cortex consistent with the length-tension relation of muscle. Pflugers Arch., 398: 318–323.

Georgopoulos, A.P., Kalaska, J.F., Caminiti, R. and Massey, J.T. (1982) On the relations between the direction of two-dimensional arm movements and cell discharge in primate motor cortex. J. Neurosci., 2: 1527–1537.

Gordon, J., Ghilardi, M.F. and Ghez, C. (1994) Accuracy of planar reaching movements. I. Independence of direction and extent variability. Exp. Brain Res., 99: 97–111.

Gottlieb, G.L., Song, Q., Almeida, G.L., Hong, D.A. and Corcos, D. (1997) Directional control of planar human arm movement. J. Neurophysiol., 78: 2985–2998.

Graham, K.M., Moore, K.D., Cabel, D.W., Gribble, P.L., Cisek, P. and Scott, S.H. (2003a) Kinematics and kinetics of multijoint reaching in nonhuman primates. J. Neurophysiol., 89: 2667–2677.

Graham, K.M. and Scott, S.H. (2003b) Morphometry of *Macaca mulatta* forelimb. III. Moment arm of shoulder and elbow muscles. J. Morphol., 255: 301–314.

Gribble, P.L. and Scott, S.H. (2002) Overlap of internal models in motor cortex for mechanical loads during reaching. Nature, 417: 938–941.

Hamilton, A.F., Jones, K.E. and Wolpert, D.M. (2004) The scaling of motor noise with muscle strength and motor unit number in humans. Exp. Brain. Res., 157: 417–430.

Harris, C.M. and Wolpert, D.M. (1998) Signal-dependent noise determines motor planning. Nature, 394: 780–784.

Hepp-Reymond, M. (1988) Functional organization of motor cortex and its participation in voluntary movements. In: Alan R. (Ed.), Comparative Primate Biology. Liss, New York, pp. 501–624.

Herter, T.M., Kurtzer, I., Cabel, D.W., Haunts, K.A. and Scott, S.H. (2007) Characterization of torque-related activity in primary motor cortex during a multijoint postural task. J. Neurophysiol., 97: 2887–2899.

Hoffman, D.S. and Strick, P.L. (1999) Step-tracking movements of the wrist. IV. Muscle activity associated with movements in different directions. J. Neurophysiol., 81: 319–333.

Hollerbach, J.M. and Flash, T. (1982) Dynamic interactions between limb segments during planar arm movement. Biol. Cybern., 44: 67–77.

Karst, G.M. and Hasan, Z. (1991) Initiation rules for planar, two-joint arm movements: agonist selection for movements throughout the work space. J. Neurophysiol., 66: 1579–1593.

Kurtzer, I., Herter, T.M. and Scott, S.H. (2005) Random change in cortical load representation suggests distinct control of posture and movement. Nat. Neurosci., 8: 498–504.

Kurtzer, I., Herter, T.M. and Scott, S.H. (2006a) Non-uniform distribution of reach-related and torque-related activity in upper arm muscles and neurons of primary motor cortex. J. Neurophysiol., 96: 3220–3230.

Kurtzer, I., Pruszynski, J.A., Herter, T.M. and Scott, S.H. (2006b) Primate upper limb muscles exhibit activity patterns that differ from their anatomical action during a postural task. J. Neurophysiol., 95: 493–504.

Lemon, R.N. and Griffiths, J. (2005) Comparing the function of the corticospinal system in different species: organizational differences for motor specialization? Muscle Nerve., 32: 261–279.

Li, C.S., Padoa-Schioppa, C. and Bizzi, E. (2001) Neuronal correlates of motor performance and motor learning in the primary motor cortex of monkeys adapting to an external force field. Neuron, 30: 593–607.

Moran, D.W. and Schwartz, A.B. (1999) Motor cortical representation of speed and direction during reaching. J. Neurophysiol., 82: 2676–2692.

Morasso, P. (1981) Spatial control of arm movements. Exp. Brain Res., 42: 223–227.

Morrow, M.M. and Miller, L.E. (2003) Prediction of muscle activity by populations of sequentially recorded primary motor cortex neurons. J. Neurophysiol., 89: 2279–2288.

Naselaris, T., Merchant, H., Amirikian, B. and Georgopoulos, A.P. (2006) Large-scale organization of preferred directions in the motor cortex. I. Motor cortical hyperacuity for forward reaching. J. Neurophysiol., 96: 3231–3236.

Nozaki, D., Nakazawa, K. and Akai, M. (2005) Muscle activity determined by cosine tuning with a nontrivial preferred direction during isometric force exertion by lower limb. J. Neurophysiol., 93: 2614–2624.

Paz, R., Wise, S.P. and Vaadia, E. (2004) Viewing and doing: similar cortical mechanisms for perceptual and motor learning. Trends Neurosci., 27: 496–503.

Porter, R. and Lemon, R.N. (1993) Corticospinal Function and Voluntary Movement. Oxford University Press, New York.

Rack, P.M. and Westbury, D.R. (1969) The effects of length and stimulus rate on tension in the isometric cat soleus muscle. J. Physiol., 204: 443–460.

Sainburg, R.L., Ghez, C. and Kalakanis, D. (1999) Intersegmental dynamics are controlled by sequential anticipatory, error correction, and postural mechanisms. J. Neurophysiol., 81: 1045–1056.

Sanes, J.N. and Donoghue, J.P. (2000) Plasticity and primary motor cortex. Annu. Rev. Neurosci., 23: 393–415.

Scott, S.H. (1999) Apparatus for measuring and perturbing shoulder and elbow joint positions and torques during reaching. J. Neurosci. Methods, 89: 119–127.

Scott, S.H. (2003) The role of primary motor cortex in goal-directed movements: insights from neurophysiological studies on non-human primates. Curr. Opin. Neurobiol., 13: 671–677.

Scott, S.H. (2004) Optimal feedback control and the neural basis of volitional motor control. Nat. Rev. Neurosci., 5: 532–546.

Scott, S.H., Brown, I.E. and Loeb, G.E. (1996) Mechanics of feline soleus: I. Effect of fascicle length and velocity on force output. J. Muscle Res. Cell Motil., 17: 207–219.

Scott, S.H., Gribble, P.L., Graham, K.M. and Cabel, D.W. (2001) Dissociation between hand motion and population vectors from neural activity in motor cortex. Nature, 413: 161–165.

Sergio, L.E. and Kalaska, J.F. (2005) Motor cortex neural correlates of output kinematics and kinetics during isometric force and arm-reaching tasks. J. Neurophysiol., 94: 2353–2378.

Shadmehr, R. and Mussa-Ivaldi, F.A. (1994) Adaptive representation of dynamics during learning of a motor task. J. Neurosci., 14: 3208–3224.

Singh, K., Melis, E.H., Richmond, F.J. and Scott, S.H. (2002) Morphometry of *Macaca mulatta* forelimb. II. Fiber-type composition in shoulder and elbow muscles. J. Morphol., 251: 323–332.

Singh, K. and Scott, S.H. (2003) A motor learning strategy reflects neural circuitry for limb control. Nat. Neurosci., 6: 399–403.

Smith, A.M., Hepp-Reymond, M.C. and Wyss, U.R. (1975) Relation of activity in precentral cortical neurons to force and rate of force change during isometric contractions of finger muscles. Exp. Brain Res., 23: 315–332.

Thach, W.T. (1978) Correlation of neural discharge with pattern and force of muscular activity, joint position, and direction of intended next movement in motor cortex and cerebellum. J. Neurophysiol., 41: 654–676.

Todorov, E. and Jordan, M.I. (2002) Optimal feedback control as a theory of motor coordination. Nat. Neurosci., 5: 1226–1235.

Wong, Y.C., Kwan, H.C., MacKay, W.A. and Murphy, J.T. (1978) Spatial organization of precentral cortex in awake primates. I. Somatosensory inputs. J. Neurophysiol., 41: 1107–1119.

Zajac, F.E. (1989) Muscle and tendon: properties, models, scaling, and application to biomechanics and motor control. Crit. Rev. Biomed. Eng., 17: 359–411.

Zajac, F.E. and Gordon, M.E. (1989) Determining muscle's force and action in multi-articular movement. Exerc. Sport Sci. Rev., 17: 187–230.

van Zuylen, E.J., Gielen, C.C. and Denier van der Gon, J.J. (1988) Coordination and inhomogeneous activation of human arm muscles during isometric torques. J. Neurophysiol., 60: 1523–1548.

P. Cisek, T. Drew & J.F. Kalaska (Eds.)
Progress in Brain Research, Vol. 165
ISSN 0079-6123
Copyright © 2007 Elsevier B.V. All rights reserved

CHAPTER 22

How is somatosensory information used to adapt to changes in the mechanical environment?

Theodore E. Milner[1,*], Mark R. Hinder[2] and David W. Franklin[3,4]

[1]*School of Kinesiology, Simon Fraser University, Burnaby, BC, V5A 1S6, Canada*
[2]*Perception and Motor Systems Laboratory, School of Human Movement Studies, University of Queensland, Brisbane, Queensland 4072, Australia*
[3]*Kobe Advanced ICT Research Center, NiCT, 2-2-2 Hikaridai, Keihanna Science City, Kyoto, 619-0288, Japan*
[4]*ATR Computational Neuroscience Laboratories, 2-2-2 Hikaridai, Keihanna Science City, Kyoto, 619-0288, Japan*

Abstract: Recent studies examining adaptation to unexpected changes in the mechanical environment highlight the use of position error in the adaptation process. However, force information is also available. In this chapter, we examine adaptation processes in three separate studies where the mechanical environment was changed intermittently. We compare the expected consequences of using position error and force information in the changes to motor commands following a change in the mechanical environment. In general, our results support the use of position error over force information and are consistent with current computational models of motor learning. However, in situations where the change in the mechanical environment eliminates position error the central nervous system does not necessarily respond as would be predicted by these models. We suggest that it is necessary to take into account the statistics of prior experience to account for our observations. Another deficiency in these models is the absence of a mechanism for modulating limb mechanical impedance during adaptation. We propose a relatively simple computational model based on reflex responses to perturbations which is capable of accounting for iterative changes in temporal patterns of muscle co-activation.

Keywords: motor learning; error feedback; internal model; mechanical impedance

Introduction

One of the fundamental questions of motor learning is how adaptation to a changing mechanical environment occurs. By mechanical environment we mean the mechanical properties of any physical system with which a human interacts. This includes properties such as the stability of a support surface, the rigidity, dimensions and mass of manipulated objects and the dynamic characteristics of forces applied by the support surface or the manipulated object. If the mechanical environment changes in some way, performance of an activity will deteriorate unless motor commands to muscles are modified to compensate for these changes. Clearly, deterioration in performance is perceived through feedback from sensory receptors. Therefore, modification of motor commands must be linked to use of sensory information by the central nervous system. Since performance tends to improve incrementally with training, it is

*Corresponding author. Tel.: +1-604-291-3499; Fax: +1-604-291-3040; E-mail: tmilner@sfu.ca

DOI: 10.1016/S0079-6123(06)65022-X

likely that perception of past poor performance (sensory error) is used to modify motor commands to muscles in a way that is expected to reduce sensory error on the next performance. Modification of motor commands results in modification of the forces applied to the environment and modification of the mechanical impedance presented to the environment. By mechanical impedance we mean properties which resist imposed motion, which in the case of muscles refers primarily to their viscoelastic properties. Mechanical impedance increases primarily by means of co-activation of antagonistic muscles. Any increase in the mechanical impedance of a limb will act to reduce the effect of perturbing forces applied by the environment to the limb. In particular, any generalized increase in the activation of elbow and shoulder muscles during reaching movements will reduce performance (sensory) error arising from a change in the mechanical environment. However, a change in the force applied to the environment will only reduce performance error if its magnitude and direction are appropriate. Inappropriate changes in the force will increase performance error. Consequently, the question posed initially should be reformulated as two questions. First, to what extent does the central nervous system increase mechanical impedance as opposed to modifying the applied force to adapt to changes in the mechanical environment? Second, when the applied force is modified what sensory information is used to compute the magnitude and direction of the change? The question marks in Fig. 1 identify where these processes would occur in a feedforward learning scheme.

Until recently, models of motor learning required both sensory (performance) error signals and motor (command) error signals (Marr, 1969; Albus, 1971; Wolpert et al., 1998; Kawato, 1999). However, a recent computational model of the cerebellum (Porrill et al., 2004) suggests that the recurrent architecture of projections between motor cortex and cerebellum can be exploited in such a way that only sensory error information is required.

Sensory information about hand position and force are both available to the central nervous system and theoretically both could be used to represent sensory error (Fig. 1). It is possible to compare a sensory representation (visual or proprioceptive) of actual hand position with desired position, providing a position error signal. Indeed, there is convincing evidence that hand position error is critical for rapid adaptation to changes in the mechanical environment (Scheidt et al., 2000). One of the most prominent theories of motor learning involves the formation and refinement of an internal model of the interaction dynamics between the arm and the environment. The internal model is a neural representation of these dynamics that is used to compute the neural commands to control the movement. In the scheme shown in Fig. 1, these commands are implemented by an impedance controller and a force controller. In current models of motor learning under the internal model rubric, hand force rather than hand

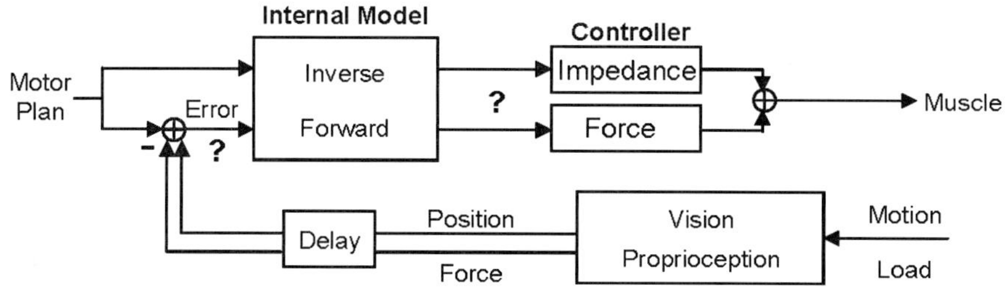

Fig. 1. Schematic representation of elements involved in modification of feedforward commands during learning. The motor plan is converted to feedforward commands to the impedance and force controllers by means of an internal model. The internal model also receives sensory information about performance error, which may include both position and force error. The internal model may include both inverse and forward models. The controllers issue activation commands to muscles which sum with feedback control signals (not shown) farther downstream. Question marks highlight the focus of our experimental investigations.

position is the output (Shadmehr, 2004). Consequently, learning involves the reduction of force error. Thoroughman and Shadmehr (2000) and Donchin et al. (2003) have proposed that hand position error can be used as a proxy for hand force error by transforming position error to force error. However, it is also possible that information about the force applied to the hand can be used to reduce performance error. We have recently examined the extent to which the central nervous system uses the strategy of increasing mechanical impedance to adapt to changes in the mechanical environment (Hinder and Milner, 2005; Milner and Franklin, 2005; Milner and Hinder, 2006). These studies also address the question of what type of sensory information is used to reduce performance error.

Methods

A total of 21 subjects participated in three experiments. Ten male subjects participated in the first experiment, six male and two female subjects participated in the second experiment and five male and four female subjects participated in the third experiment. Six of the subjects who participated in the second experiment also participated in the third experiment. All subjects gave informed consent prior to participating in the study. The protocol was approved by the institutional ethics review committee and conformed to the ethical standards set down in the Declaration of Helsinki.

In all three experiments, subjects adapted to changes in the dynamics of a robotic manipulandum during reaching movements away from the body. Movements began from a position about 30 cm in front of the shoulder and ended at a target 25 cm farther forward. We investigated both position-dependent and velocity-dependent changes in dynamics. More details of the methods can be found in previously published studies (Hinder and Milner, 2005; Milner and Franklin, 2005; Milner and Hinder, 2006).

In the first experiment presented here (Milner and Hinder, 2006), subjects adapted to a position-dependent force field for the first 125 trials. The force field acted purely in the x-direction and pushed the subject's hand to the left ($-x$)

$$F_x = K(x + 0.032(y - y_s)(y - y_e)) \tag{1}$$

with $K = 1.5$ N/cm on most trials. Occasionally, K was doubled for one (five times) or two trials (six times). Subjects were not informed that this would occur. When K was doubled for one trial, on the following trial only, a velocity-dependent force field was instituted, which also acted purely in the x-direction, but pushed the hand to the right ($+x$)

$$F_x = B\dot{y} \tag{2}$$

with $B = 0.15\ \mathrm{N \cdot s/cm}$. On three occasions during the training period (trials 3, 61 and 120), the force field was unexpectedly replaced by a virtual wall, i.e., the manipulandum acted like a stiff (10 N/cm) damped spring to lateral (x) displacement to the right. The wall effectively eliminated any rightward lateral error in hand position.

Following the 125 trials in the position-dependent force field, subjects adapted to the velocity-dependent force field for 50 additional trials. Hand position and force applied to the manipulandum were recorded at 1 kHz. We evaluated error by comparing the hand trajectory and force to the mean hand path and force at the end of the training period. Modification of the feedforward motor command during adaptation to the velocity-dependent force field was quantified by computing the lateral force impulse applied to the manipulandum between movement onset and peak tangential velocity.

In the second experiment presented here (Milner and Franklin, 2005), subjects intermittently adapted to a velocity-dependent force field given by

$$\begin{bmatrix} F_x \\ F_y \end{bmatrix} = \begin{bmatrix} -B & B \\ B & B \end{bmatrix} \begin{bmatrix} \dot{x} \\ \dot{y} \end{bmatrix} \tag{3}$$

with $0.09 \leq B \leq 0.15\ \mathrm{N \cdot s/cm}$, dependent on the subject's capacity to adapt. This force field tended to push the subject's hand to the right and towards the target with the perturbing force increasing as a function of velocity. Subjects performed 27–28 sets of three consecutive trials in the velocity-dependent force field, each separated by a random number of between 4 and 8 null field (no force) trials. On several occasions (4–6), the third of the three force field trials was unexpectedly replaced by a virtual channel, which created stiff elastic walls with a stiffness of 40 N/cm to lateral (x) displacement in either direction. On a similar number of occasions (4–5), the third of the three force field trials was omitted so that subjects performed only two force field trials before returning to the null field. Modification of the feedforward command undergone during the two or three force field trials was evaluated by comparing hand trajectory and force.

In the third experiment presented here (Hinder and Milner, 2005), subjects performed 150 trials in a position-dependent force field, that also pushed the hand to the right with the force increasing in a parabolic fashion during the first 5 cm of the movement and then decreasing symmetrically during the next 5 cm. The following equation describes the perturbing force, which was purely in the x-direction

$$F_x = -0.32(y_s - y)(y - y_e), \quad -(y_s - y)(y - y_e) \geq 0$$
$$F_x = 0, \quad -(y_s - y)(y - y_e) < 0 \tag{4}$$

where $y_s = 0$, $y_e = 10$ cm, y is the current location of the hand, giving a maximum force of 8 N when $y = 5$ cm. On every fifth trial, the force field was predictably replaced by a virtual channel (40 N/cm). Changes in the force measured on these trials indicated how subjects progressively adapted their lateral force to compensate for the perturbing effect of the force field. The effect of the virtual channel on the feedforward command was evaluated in terms of change in force applied to the channel and change in hand trajectory on the trial which followed the channel trial.

Results

The focus of the first study is on the performance error after a change in the mechanical environment and the type of sensory information used to modify the feedforward motor command on the following trial to reduce performance error. In the first study, the position-dependent force field produced a large deviation (~20 cm) to the left of the straight line joining the targets the first time it was experienced, following a series of null field training trials. On the next trial (trial 2), the lateral deviation was dramatically reduced and eventually stabilized at $\sim\pm0.5$ cm after extensive training (Milner and Hinder, 2006). Trials 3, 61 and 120 were trials on which the force field was replaced by a virtual wall. Comparison of the trajectories on pre-wall trials 2 and 119 (Fig. 2A) and the lateral forces on wall trials 3 and 120 (Fig. 2B), provides insight into the adaptation process. There are only minor differences between the trajectories until after the midpoint of the movement (> 13 cm). Yet, comparing the subsequent wall trials, the lateral force applied to the wall begins to differ shortly after movement onset and marked differences are already apparent 8 cm into the movement. This suggests that limb stiffness made an important

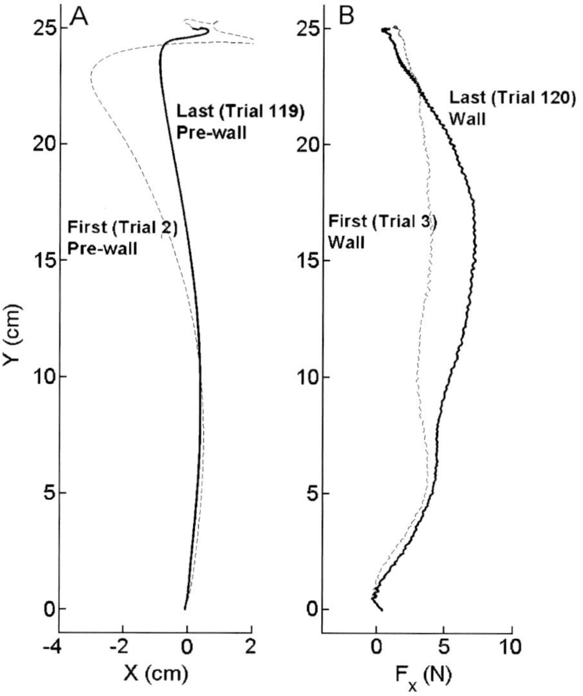

Fig. 2. (A) Mean hand paths across subjects ($N = 10$) early (dashed line) and late (solid line) during learning of the position-dependent force field. There is very little difference in the hand path until near the midpoint of the movement. (B) Mean lateral force across subjects on first wall trial (dashed line) and last wall trial (solid line) plotted as a function of forward position. Forces begin to differ well before the midpoint of the movement, indicating that the dynamics of the force field are not yet well compensated by trial 3.

contribution to reducing the performance error between the first and second trials otherwise greater initial deviation in the trajectories on trials 2 and 119 would have been expected based on differences in the profile of the applied lateral force recorded on the wall trials. Nevertheless, from Fig. 2B there is good evidence that subjects were already applying a substantial force to the right by trial 3. It seems clear that information about the direction and approximate magnitude of the force error must have been extracted from sensory information during the first two force field trials even though the profile of the force error was still inaccurate.

Observations from trials where the force field strength was doubled support this interpretation. Unexpected doubling of the force field strength from 1.5 N/cm to 3 N/cm for two trials in succession occurred on 5 occasions during training. This again produced a large lateral (x) deviation to the left on the first trial followed by a substantial reduction on the second trial, very similar to the first two force field trials (Milner and Hinder, 2006). On the following trial, the force field returned to its initial strength. Any increase in lateral force to compensate for doubling of the force field strength should have been evident as an aftereffect, i.e., a noticeable trajectory deviation to the right. Although the expected aftereffect was observed (Fig. 3A), the force applied to the manipulandum indicated that it was the result of a relatively small increase in force over the first 5 cm of the movement, relative to the force field trial which preceded doubling of the force field strength (Fig. 3B). This again suggests that although information about the direction and magnitude of the change in the mechanical environment was captured from sensory information when the force field strength was doubled, an important contribution to resisting the perturbing effect of the force field must have been derived from an increase in arm stiffness. It should be noted that the maximum force applied to the manipulandum on the aftereffect trial was substantially less than on the trial which preceded doubling of the force field strength. Because the force field drops to the right of the line joining the targets, subjects likely reduced their force by a combination of stretch reflexes in antagonist muscles and voluntary corrective action. It is also possible that high arm stiffness would limit the perturbing effect of the decline in force field magnitude to the right.

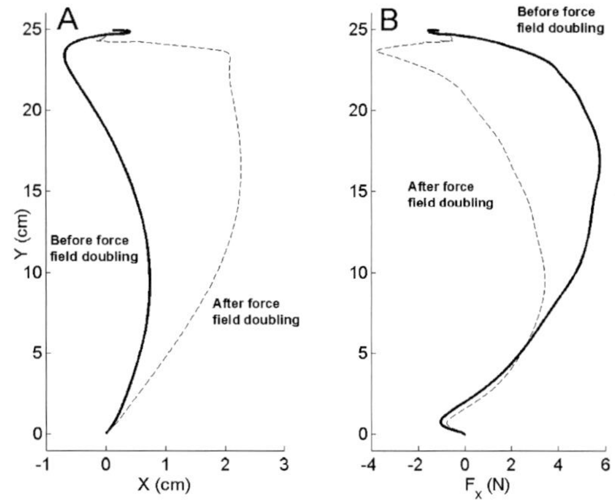

Fig. 3. (A) Mean hand paths across subjects ($N = 10$) for force field trials before (solid line) and after (dashed line) trials where the force field strength was doubled for two consecutive trials. There is a large aftereffect to the right after doubling of the force field strength, indicating that subjects increased their lateral force to the right. (B) Corresponding forces recorded at the hand, indicating that the aftereffect is created by a relatively small increase in rightward force applied to the manipulandum shortly after movement onset, during the initial 5 cm of the reach. The large negative force difference later occurs because deviation of the hand to the right results in lower forces due to the nature of the position-dependent force field (Milner and Hinder, 2006).

When a velocity-dependent force field (VF), which pushed subjects to the right, followed doubling of the force field strength, there was a large lateral deviation (~10 cm) to the right (Milner and Hinder, 2006). However, there was no evidence that subjects modified the direction of their lateral force. When the position-dependent dynamics of the initial force field was restored on the next trial all subjects initially applied a rightward force to the manipulandum, i.e., in the *same* direction as the perturbing effect of the VF. Although the trajectory of some subjects did deviate slightly to the left this occurred because they applied a lateral force to the right, which was now less than the magnitude of the leftward position-dependent force which they were opposing (Milner and Hinder, 2006).

The failure to use information about the direction of the change in the mechanical environment to reduce performance error was not a phenomenon related to one-trial learning. Following trial 125, the VF dynamics were maintained for 50 consecutive trials. We determined how soon after switching to the VF subjects changed the direction of lateral force applied to the manipulandum by comparing the force measured by the load cell attached to the manipulandum handle with the theoretical VF force calculated from Eq. (2). As long as the load cell force was less than the VF force, it indicated that subjects applied a force in the same direction as the VF, i.e., a rightward force, thereby unloading the load cell. A net rightward force over the acceleration phase of the movement (movement onset to peak y-velocity) was recorded even after 10 consecutive VF trials (Fig. 4A). Nonetheless, performance error was incrementally reduced over the first 30 VF trials (Fig. 4B). What this indicates is that subjects did not integrate information about the change in direction of force applied by the environment into the adaptation process. Otherwise, a change in the direction of initial force compensation would have been expected after one or two trials in the VF.

Previously published results of the second study (Milner and Franklin, 2005) confirmed that subjects stiffen the arm to reduce the perturbing effects of a change in the mechanical environment. Subjects normally performed movements in a null field, but intermittently, without prior warning, the null field changed to a VF [Eq. (3)] for the three consecutive trials. Increased activation of all recorded muscles, starting before movement onset, was found on the second VF trial. However, the changes in muscle activation patterns on the second VF trial also increased lateral force to partially counteract the VF force as inferred from aftereffects and channel trials. Occasionally, the third VF trial was omitted, i.e., it was replaced by a null field trial or by a virtual channel, to determine whether subjects were adapting by compensating the force produced by the force field. On trials where the third VF was replaced by the null field, there was a clear aftereffect, evident as deviation in the hand path, soon after movement onset, opposite to the direction of the lateral force

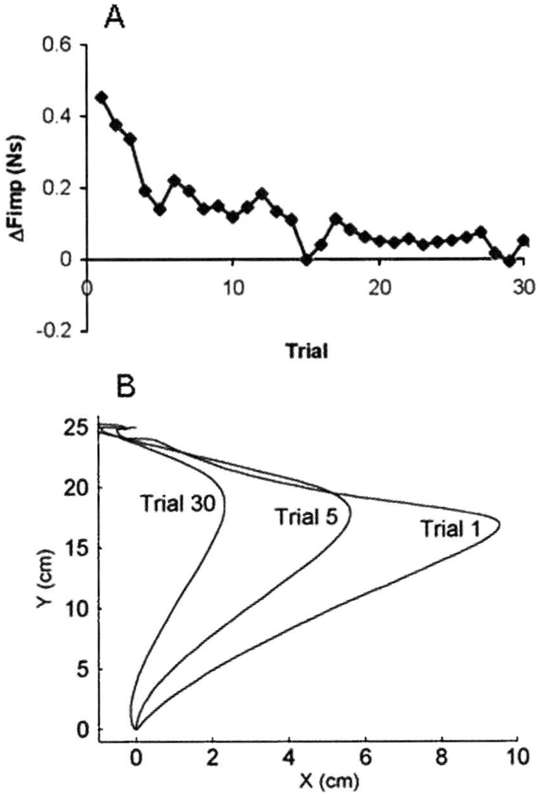

Fig. 4. (A) Difference between the force impulse generated by the velocity-dependent force field and the force impulse applied to the manipulandum by the subject, computed from movement onset to peak y-velocity, for the first 30 VF trials. Mean values for the 10 subjects are shown. (B) Mean hand paths across subjects ($N = 10$) during adaptation to the velocity-dependent force field, showing a gradual reduction in lateral error from trial 1 to trial 30.

created by the force field, i.e., to the left (Fig. 5A). This aftereffect was produced by a relatively small difference between the lateral force applied to the manipulandum on the null field trials preceding and following the three VF trials (Fig. 5B). Channel trials were also preceded by two VF trials and followed by a null field trial. The aftereffect of the VF trials following an intervening channel trial was only about half the size of the aftereffect observed when the null field trial occurred immediately after the second VF trial, despite a very small difference in the lateral force applied to the manipulandum (Fig. 5B). From this we can conclude that only small lateral forces are required to

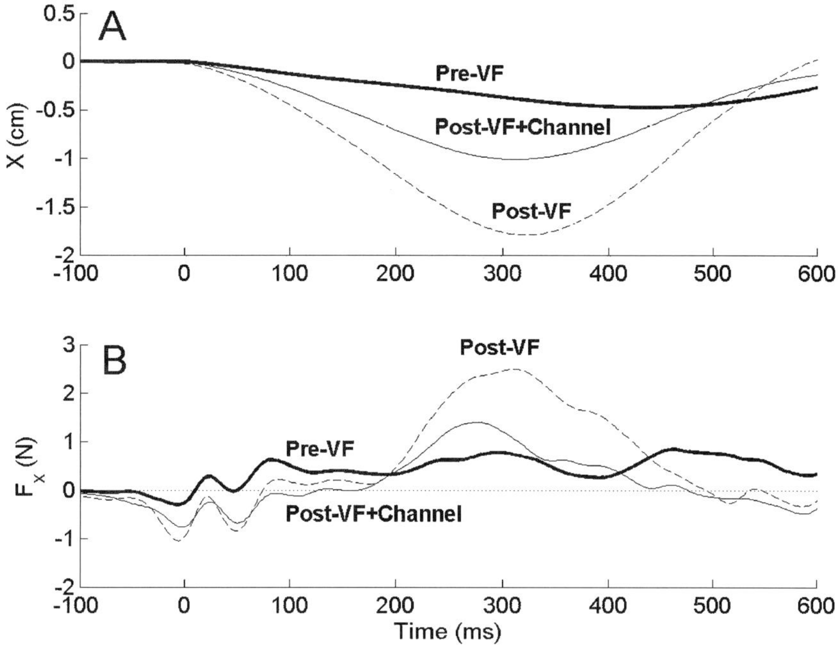

Fig. 5. (A) Mean lateral displacement from a straight line movement ($N = 4$) for null field trials prior to VF trials (Pre-VF, thick solid line), null field trials after two VF trials (Post-VF, dashed line) and null field trials after two VF trials and a channel trial, (Post-VF + Channel, thin solid line). (B) Corresponding forces applied to the manipulandum. A relatively large difference in lateral displacement is produced by relatively small differences in force during the first 200 ms of the movement (compare the thin solid line and dashed line).

produce aftereffects and that small differences in the lateral force can result in relatively large differences in aftereffect size.

In the third study, subjects adapted to a position-dependent force field (PF), which produced a lateral force to the right with a parabolic profile that peaked 5 cm from the start position and returned to zero 10 cm from the start position. The final 15 cm of each movement was performed in a null field. On the initial trial, the force field produced a lateral deviation of several centimeters to the right that reached its maximum during the null field portion of the movement. The lateral deviation was reduced to about a third of its initial value within five trials (Hinder and Milner, 2005). The strategy which subjects employed was to rapidly increase their lateral force at movement onset, pushing the manipulandum in the opposite direction to the force field, i.e., to the left. This resulted in a hand path that initially curved to the left, reaching maximum leftward deviation when approximately a third the force field region had been traversed. During the portion of the movement beyond the boundary of the force field (> 10 cm), the hand path curved to the right of the straight line joining the targets before eventually coming back to the final target position (Fig. 6). The curvature to the right was the consequence of applying a smaller lateral force impulse to the manipulandum in the force field region than the force impulse created by the force field. On every fifth trial, the force field was replaced by a virtual channel. The focus of the current analysis is on what effect the channel trials during the last half of the training session had on the motor command. By this time, subjects had effectively adapted completely to the force field, as judged from the absence of changes in hand paths, lateral force profiles or muscle activation patterns (Hinder and Milner, 2005).

The effect of the channel trial was to increase the lateral deviation of the hand path on the force field

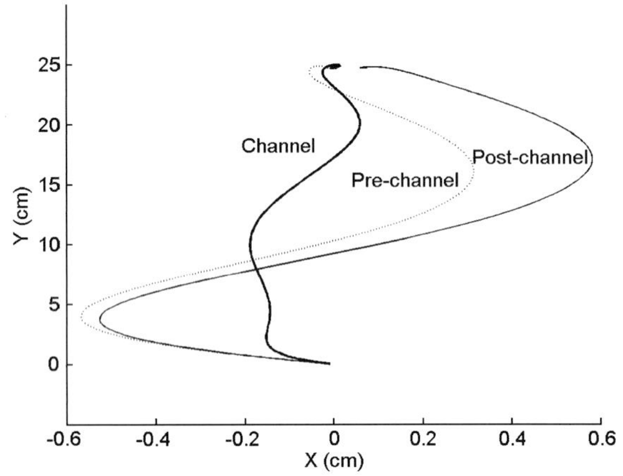

Fig. 6. Mean hand paths across subjects ($N = 9$) for position-dependent force field trials preceding (dotted line) and following (solid line) channel trials (thick solid line). Lateral displacement to the left is decreased and lateral displacement to the right is increased following the channel trial, indicating the subjects applied less leftward lateral force following the channel trial.

trial, which followed the channel trial, relative to the force field trial, which preceded the channel trial (Fig. 6). There were two distinct changes in the features of the hand path following the channel trial. Subjects produced less initial leftward curvature in the force field region and more rightward curvature beyond the force field boundary compared to the trial which preceded the channel trial. On channel trials, the manipulandum applied a lower peak (reaction) force to the hand than on force field trials. Position error on channel trials was reduced in both directions, i.e., both the initial curvature to the left and the later curvature to the right were effectively eliminated. Clearly, initial force was reduced on the trial following the channel trial compared to the trial preceding the channel trial because there was less initial displacement opposite to the direction of the force field. A reduction in force would also lead to a reduction in the stiffness of the arm, although stiffness may have also been reduced by a decrease in co-contraction of antagonistic muscles.

Discussion

Taken together the results of these studies suggest that any change to the mechanical environment, which increases lateral perturbing force, is resisted by a subsequent increase in the stiffness of the arm. However, the central nervous system also increments the lateral force opposing the perturbation if no lateral perturbing force existed previously or if the previously existing lateral force did not change direction. We have also demonstrated that muscle co-contraction is gradually reduced during training, decreasing arm stiffness if the lateral perturbing force does not change (Hinder and Milner, 2005; Milner and Franklin, 2005). This decrement in co-contraction apparently occurs in conjunction with incremental modification of the subject's lateral force profile such that the lateral force applied by the subject straightens the hand path. However, even after extensive training, the hand path is generally slightly curved, with the details of the curvature depending on the characteristics of the perturbing force.

In general, computational models of motor learning predict that if a reaching movement is repeatedly executed in the same environment, performance will improve incrementally, provided that the interaction between the arm and the environment is mechanically stable. Optimal control models such as those of Harris and Wolpert (1998) or Todorov and Jordan (2002) can be used to make general predictions about movement trajectories and movement variability at the end of learning, but are not structured to predict incremental changes from one trial to the next during learning. The most relevant learning models are those of Thoroughman and Shadmehr (2000) and Donchin et al. (2003). They predict that applied hand force will be iteratively modified in proportion to the position error, progressively reducing the error. Because they iteratively modify the internal model based on position error these models are consistent with the observation that reversing the direction of the environmental force does not immediately result in a reversal of the direction of the applied force, i.e., the models predict that the applied force will progressively decrement on successive trials until it eventually changes direction. However, they do not unequivocally predict an increase in performance error following a channel trial.

On trials following channel trials during the early phase of adaptation when position error was expected (second study), the lateral force applied by the subject at movement onset (and possibly the amount of muscle co-contraction) was less than expected had there been no channel trial. The channel trial followed a trial with a relatively large performance error. If the central nervous system interprets somatosensory information during the channel trial as indicating that this performance error has been eliminated then the learning models of Thoroughman and Shadmehr (2000) and Donchin et al. (2003) would not predict any change in the motor command on the subsequent trial; in particular, no reduction in lateral force, which is inconsistent with our observations. However, this interpretation ignores prior expectation. Since it is likely that the force field was expected rather than the channel, elimination of performance error by the channel created a discrepancy. At this early stage of learning, the discrepancy would have most likely been interpreted as a decrease in force field strength. The appropriate response to a decrease in force field strength would be to reduce lateral force, as observed. This could be accommodated in the learning models by including a term in the computation of the update to the internal model which incorporates the statistics of prior experience, e.g., Bayesian statistics (Kording et al., 2004; Krakauer et al., 2006).

Once subjects have adapted completely to a persistent change in the mechanical environment position error may be very small so the constrained trajectory imposed by the channel may not be perceived as a perturbation. Indeed, the work of Scheidt et al. (2000) suggests this to be the case for a velocity-dependent force field. However, if hand paths are sufficiently curved after adaptation, as in the case of adaptation to the parabolic force field of our third study (Hinder and Milner, 2005), the channel might be perceived as a perturbing force opposite to the direction of curvature. This would be the case if the desired trajectory changed during adaptation, a possibility raised by Donchin et al. (2003). This might explain why subjects reduced their leftward lateral force after channel trials in our third study, resulting in greater curvature to the right during the latter portion of the movement. As is evident from Fig. 6, the channel effectively eliminated the normal rightward curvature during the latter portion of the movement. Assuming that the desired trajectory includes rightward curvature, the channel creates a position error to the left. Since adaptation to the force field involved applying force to the left, the learning models would predict that following a channel trial this force should be reduced which would increase curvature to the right, consistent with our observations. Therefore, it seems that the desired trajectory can change during adaptation, depending on the nature of the change in the mechanical environment.

Because a reversal of the direction of the perturbing force does not evoke an immediate reversal of the direction of the applied force the question arises as to whether sensory receptors sensitive to force provide unambiguous information about force direction. Force-sensitive sensory receptors in muscle, Golgi tendon organs, probably cannot unambiguously signal the direction of the external force. A reduction in force field strength or a change in force field direction would both unload the Golgi tendons organs of contracting muscles, resulting in a similar change in their output. Signals from other force-sensitive receptors, cutaneous mechanoreceptors in the hand, could resolve the ambiguity, but only if the assisting force exerted by the subject was less than the external force. In this case, pressure sensitive mechanoreceptors in the hand would detect a force being applied by the manipulandum in the direction of the force field due to the decelerating effect of the inertia of the arm. On the other hand, if the subject exerted an assisting force that was larger than the external force, these mechanoreceptors would experience a force due to the decelerating effect of the inertia of the manipulandum, which would not indicate a change in force field direction. To avoid making potentially larger errors by misinterpreting such ambiguous information, the central nervous system may rely only on information about position error for incrementally reducing performance error. Therefore, computational models of motor learning that use position error to drive learning are likely to more closely reflect physiological reality than models which use force error.

There is one aspect of control which none of the current learning models addresses and that is the modulation of mechanical impedance. There is no mechanism in any of these models that can explain the large feedforward increase in muscle co-contraction that is frequently observed after introducing a perturbing force nor the iterative reduction in co-contraction that occurs during adaptation to the perturbing force. We have developed a computational model that can account for these processes (Milner et al., 2006). The model uses reflex responses to the perturbation as a template for changes to feedforward motor commands. By taking a proportion of the reflex response, i.e., scaling it by a learning factor, shifting the result forward in time and adding it to the previous motor command, iterative changes in temporal patterns of muscle activation observed during learning can be reproduced. Once performance error is reduced below some threshold, a second learning factor is applied which incrementally reduces the muscle activation, reproducing the gradual reduction in co-contraction until a steady-state is achieved around the performance error threshold.

In summary, our results provide support for the computation models of motor learning developed by Shadmehr and Thoroughman (2000) and Donchin et al. (2003). In addition, they suggest that position error is the primary driving signal for modifying the internal model. We also have evidence for the speculation by Donchin et al. (2003) that the desired trajectory can be altered during adaptation to certain types of changes in the mechanical environment. These models do have several limitations though, that need to be addressed. In particular, incorporating the statistics of prior experience may be necessary, as Krakauer et al. (2006) suggest. As well, a mechanism for the modulation of limb mechanical impedance through adjustments in co-contraction levels of antagonistic muscles is required. To rectify this limitation, we propose a model in which reflex responses to perturbations serve as time-delayed templates for iterative adjustments to motor commands that result in improved performance (Milner et al., 2006).

Acknowledgments

This work was supported by NSERC. DWF was supported by funding from NiCT, Japan.

References

Albus, J.S. (1971) A theory of cerebellar function. Math. Biosci., 10: 25–61.

Donchin, O., Francis, J.T. and Shadmehr, R. (2003) Quantifying generalization from trial-by-trial behavior of adaptive systems that learn with basis functions: theory and experiments in human motor control. J. Neurosci., 23: 9032–9045.

Harris, C.M. and Wolpert, D.M. (1998) Signal-dependent noise determines motor planning. Nature, 394: 780–784.

Hinder, M.R. and Milner, T.E. (2005) Novel strategies in feedforward adaptation to a position dependent perturbation. Exp. Brain Res., 165: 239–249.

Kawato, M. (1999) Internal models for motor control and trajectory planning. Curr. Opin. Neurobiol., 9: 718–727.

Kording, K.P., Ku, S. and Wolpert, D.M. (2004) Bayesian integration in force estimation. J. Neurophysiol., 92: 3161–3165.

Krakauer, J.W., Mazzoni, P., Ghazizadeh, A., Ravindran, R. and Shadmehr, R. (2006) Generalization of motor learning depends on the history of prior action. PloS Biol., 4: e316.

Marr, D. (1969) A theory of cerebellar cortex. J. Physiol., 202: 437–470.

Milner, T.E. and Franklin, D.W. (2005) Impedance control and internal model use during the initial stage of adaptation to novel dynamics. J. Physiol., 567: 651–664.

Milner, T.E. and Hinder, M.R. (2006) Position informa tion but not force information is used in adapting to changes in environmental dynamics. J. Neurophysiol., 96: 526–534.

Milner, T.E., Ng, B. and Franklin, D.W. (2006) Learning feedforward commands to muscles using time-shifted sensory feedback. In: Ishii K., Natsume K. and Hanazawa A. (Eds.), Brain-Inspired IT II Decision and Behavioral Choice Organized by Natural and Artificial Brains. Elsevier, Amsterdam, pp. 113–116.

Porrill, J., Dean, P. and Stone, J.V. (2004) Recurrent cerebellar architecture solves the motor-error problem. Proc. R. Soc. Lond. B, 271: 789–796.

Scheidt, R.A., Reinkensmeyer, D.J., Conditt, M.A., Rymer, W.Z. and Mussa-Ivaldi, F.A. (2000) Persistence of motor adaptation during constrained, multi-joint, arm movements. J. Neurophysiol., 84: 853–862.

Shadmehr, R. (2004) Generalization as a behavioral window to the neural mechanisms of learning internal models. Hum. Mov. Sci., 23: 543–568.

Thoroughman, K.A. and Shadmehr, R. (2000) Learning of action through adaptive combination of motor primitives. Nature, 407: 742–747.

Todorov, E. and Jordan, M.I. (2002) Optimal feedback control as a theory of motor coordination. Nat. Neurosci., 5: 1226–1235.

Wolpert, D.M., Miall, R.C. and Kawato, M. (1998) Internal models in the cerebellum. Trends Cogn. Sci., 2: 338–347.

P. Cisek, T. Drew & J.F. Kalaska (Eds.)
Progress in Brain Research, Vol. 165
ISSN 0079-6123
Copyright © 2007 Elsevier B.V. All rights reserved

CHAPTER 23

Trial-by-trial motor adaptation: a window into elemental neural computation

Kurt A. Thoroughman*, Michael S. Fine and Jordan A. Taylor

Department of Biomedical Engineering, Washington University, 1 Brookings Dr., Saint Louis, MO 63130, USA

Abstract: How does the brain compute? To address this question, mathematical modelers, neurophysiologists, and psychophysicists have sought behaviors that provide evidence of specific neural computations. Human motor behavior consists of several such computations [Shadmehr, R., Wise, S.P. (2005). MIT Press: Cambridge, MA], such as the transformation of a sensory input to a motor output. The motor system is also capable of learning new transformations to produce novel outputs; humans have the remarkable ability to alter their motor output to adapt to changes in their own bodies and the environment [Wolpert, D.M., Ghahramani, Z. (2000). Nat. Neurosci., 3: 1212–1217]. These changes can be long term, through growth and changing body proportions, or short term, through changes in the external environment. Here we focus on trial-by-trial adaptation, the transformation of individually sensed movements into incremental updates of adaptive control. These investigations have the promise of revealing important basic principles of motor control and ultimately guiding a new understanding of the neuronal correlates of motor behaviors.

Keywords: motor learning; motor control; psychophysics; reaching; neural network; generalization

Introduction

Despite the centrality of motor learning to basic and clinical neuroscience, we know very little about the quantitative role neural systems play in human motor behavior (Flash and Sejnowski, 2001). Motor behavior, of the hand and arm in particular, consists of many mathematical problems that the central nervous system solves effortlessly. Many reaching tasks seem only positionally dependent; bringing a glass to one's mouth, shaking a friend's hand, and writing with a pen all require the identification and acquisition of arm and hand placements. But this oversimplified positional description belies the complexity of these movements, because desired positions must be related to the velocities, torques, and forces necessary to execute the movement (Atkeson, 1989). For every desired movement of the arm, the central nervous system somewhere needs to calculate the motor neuronal activity appropriate to generate the muscle forces necessary to actuate the movement. The dynamic equations necessary to move an unencumbered arm are very complex (Chan and Moran, 2006); adding the interaction forces of manipulated objects makes the calculations even more daunting.

Motor learning of external dynamics

In the mid 1990s two research groups (Brandeis and MIT) discovered that important properties of

*Corresponding author. Tel.: +1 314 935 9094;
Fax: +1 314 935 7448;
E-mail: thoroughman@biomed.wustl.edu

DOI: 10.1016/S0079-6123(06)65023-1

human motor behavior could be discovered by bypassing the complexities of natural arm movements and adding externally generated, novel dynamic demands. The Brandeis group (Lackner and Dizio, 1994) generated these novel forces by seating participants in the center of a rotating room. When the room rotated at constant speed and was dark, participants had no sensory perception that they were moving. When making outward reaches, however, they experienced novel Coriolis forces due to the interaction of the rotation and the arm movement. Lackner and Dizio discovered that even without visual feedback, participants could adapt to the Coriolis forces to move straight to the target. When the room stopped its rotation, participants generated an after-effect, making pointing errors opposing the direction of the initial error. These after-effects demonstrated that people could build a lasting adaptation to predictive force control over mere minutes of training in a novel dynamic environment.

The MIT group (Shadmehr and Mussa-Ivaldi, 1994) achieved similar results with a very different paradigm. Shadmehr and Mussa-Ivaldi trained participants to make reaching movements while holding a robotic arm (termed a manipulandum) that generated unusual velocity-dependent (viscous) forces. As with the Brandeis findings, Shadmehr and Mussa-Ivaldi observed that participants could learn to counter the novel forces, and generated an after-effect when the forces were removed. The manipulandum, however, offered the flexibility of removing the perturbation in single trials within blocks of viscous force generation. Later, Shadmehr and Brashers-Krug (1997) would use the growth of after-effect magnitude as a novel metric of force field learning and as evidence of interference across training paradigms.

Shadmehr and Mussa-Ivaldi initially trained participants to reach with their hands in front of their torsos. They then asked if and how participants transferred this motor memory when controlling the robot with their arms outstretched, lateral to their shoulders. Shadmehr and Mussa-Ivaldi discovered that participants did transfer learned dynamics to this new posture, as evidenced by better-than-naïve performance with the viscous forces on, and lingering after-effects with the forces off. The experimenters also altered the dependence of the forces such that the viscosity was related to either hand velocity or joint velocity; upon changing postures, experienced forces therefore remained consistent in either hand coordinates or joint coordinates. Participants performed better when the forces were consistent in joint coordinates, providing evidence that the motor memory was represented in terms of joint velocity.

The Brandeis and MIT experiments ushered in a new subdiscipline of exploring motor learning via the psychophysics of adapting to external dynamics. Other studies have considered the transfer of motor learning across postures (Shadmehr and Moussavi, 2000), speeds (Goodbody and Wolpert, 1998), movement directions (Gandolfo et al., 1996), tasks (Conditt et al., 1997), or from one hand to the other (Criscimagna-Hemminger et al., 2003). Studies also considered possible interference across learning multiple dynamic (force-generating) environments (Brashers-Krug et al., 1996; Caithness et al., 2004) or between learning visual and dynamic perturbations (Tong et al., 2002). All these studies provided behavioral evidence of motor memories built over one or several training sessions.

Trial-by-trial adaptation in dynamic environments

This body of work characterized the temporal and spatial properties of motor memories, but did not consider the influence of individual movements on motor adaptation. Motor adaptation was usually investigated with blocks of movements in which the dynamic perturbation was held fixed. This design enabled a measure of the rise time and asymptote of learning over many trials, as well as the transfer of that memory to different times, different movements, and different environments. Since most paradigms kept the movement environment fixed, they could not identify how errors in a single movement led to adaptation in the very next movement. Some experiments interspersed catch trials, single movements in which forces are unexpectedly removed, during training sets (Shadmehr and Mussa-Ivaldi, 1994; Shadmehr and Brashers-Krug, 1997), but did not examine how those catch trials altered adaptation. The consensus was that since the catch trials were rare, they only subtly affected adaptation over hundreds of movements and could therefore be ignored in their contribution to adaptation.

Two groups, at Northwestern and Johns Hopkins, provided novel analyses to begin the quantitative investigation of motor adaptation across individual trials. These new experiments focused on protocols and analyses that shifted the focus away from blocks of movements to the contribution of individual experiences. The block design of previous protocols provided stable environments from which to abstract relevant metrics. The trial-by-trial approaches of the Northwestern and Johns Hopkins groups altered the environmental dynamics within training, therefore preventing the settling of behavior into easily identifiable time constants and asymptotes. At first glance this design would seem to obscure rather than illuminate adaptive processes. The experimenter, however, controlled the trial-by-trial sequence of experienced forces. This sequence was later used as a template to identify subtle changes in control following each force presentation. The totality of the complete sequence of presentations and responses gave the trial-by-trial approach the same interrogative power as block designs, but with the additional vantage point of identifying the immediate transfer of sensory information into action.

The Northwestern group (Scheidt et al., 2001) trained subjects to reach in a single direction, directly away from their bodies, while experiencing forces whose strength was randomly drawn from a distribution in each trial. Scheidt et al. found that participants' performance in a trial, as quantified by maximal hand deviation, linearly depended on forces experienced in that movement, and also on forces and error experienced in the previous movement. The overall learning over a training session was found to consist simply of the summation of these increments of adaptation.

The Johns Hopkins group (Thoroughman and Shadmehr, 2000) did not draw forces from a distribution but rather reanalyzed previous data in which participants learned a viscous environment with occasional catch trials. New analyses shifted the focus to adaptation across single trials, enabled by the catch trial occurrences. Participants trained in eight directions of movement, so these analyses could newly investigate if and how incremental adaptation generalized across movement directions. The analysis of Thoroughman and Shadmehr revealed that participants generalized sensed error to improve subsequent control across many movement directions.

A neural network model mimicked this generalization if the force estimate relied upon neural units broadly tuned to movement direction and speed. This model exemplified a broader theory of neural computation, that the weighted linear combination of broadly tuned neurons allows for generalization and that learning may occur solely in changing the weights (Poggio, 1990). Further studies have indicated that this broadly tuned neural network model mimicked motor learning in several viscous environments (Donchin et al., 2003) as well as position-dependent environments (Hwang et al., 2003). The constancy of this broad trial-by-trial generalization supports the theory that fixed neuronal tuning simplifies learning and, in motor adaptation, provides a simple, consistent expectation of environmental complexity (Pouget and Snyder, 2000).

In the aforementioned experiments, people demonstrated trial-by-trial adaptation that depended co-linearly on two factors: a signal proportional to error, and a generalization of that error across the movement space. We have recently reported that both of these dependencies are not fixed, but rather change as a function of short-term experience; adaptation can be markedly disproportional and generalization can be narrow or broad depending on short-term environmental experience.

Experience-dependent flexibility of error generalization

Previous research has suggested that the extent of trial-by-trial generalization remains fixed during learning, which provides a stable adaptive platform through which people can process sensed signals into motor output. These experiments have generated forces that mimicked natural environments in their low spatial complexity, where complexity describes the change in force direction across movement space. However, if forces change rapidly across movement space and the human motor system generalizes broadly and remains fixed, then generalization itself can be maladaptive. To directly test this theory, we (Thoroughman and Taylor, 2005) trained human participants in dynamics with low, medium, and high complexity, as determined by the equations:

$$F = -15\sqrt{\dot{x}^2 + \dot{y}^2}\begin{bmatrix} -\sin(m\phi) \\ \cos(m\phi) \end{bmatrix}$$
$$\phi = \arctan\left(\frac{\dot{y}}{\dot{x}}\right) \qquad (1)$$

where $\dot{x}$, $\dot{y}$, and ϕ are the Cartesian components and direction of hand velocity. By increasing m, the spatial complexity of the function mapping velocity direction into force direction could be varied (Fig. 1A–C). On different days, the force direction was changed as fast ($m = 1$; Field One), twice as fast ($m = 2$; Field Two), or four times as fast ($m = 4$; Field Four) as the velocity direction. We found that people could learn all three environments over training. Participants reached the same degree of learning in Fields 1 and 2, while slightly less learning occurred in Field 4.

Surprisingly, participants quickly changed the way they adapted in each environment (Thoroughman and Taylor, 2005). We quantified the trial-by-trial generalization of error into adaptation (Fig. 1D) by using a state-space model. A vector parameter B determined the strength of adaptation dependent on the angular separation (θ) between the direction of the movement in which error is sensed and the direction of the next controlled movement. Our parameterization did not capture a significant trial-by-trial adaptation in Field Four; this may be because learning plateaued early in this environment leaving little trial-by-trial signal throughout the training sets. Trial-by-trial generalization in Field One exhibited the strongest and broadest generalization. The generalization function was always positive, such that an error sensed in one direction generated the same sign (positive or negative) of adaptation in all subsequent movement directions. For example, a rightward force experienced while reaching away from the body would generalize to an expectation of a rightward force, even while reaching toward the right or toward the body. In contrast, Field Two had a remarkably different shape; it was narrower above the X-axis and featured negative components in directions far away from sensed error. A rightward force experienced while reaching away from the body would not generalize to movements toward the right, but would generalize to an expectation of a leftward force for movements toward the body (as $B \approx 0$ for $\theta = 90°$ and $B < 0$ for $\theta = 180°$).

When the forces changed rapidly across movement space, people changed the way they learned the environment. This reshaping of motor adaptation suggests that people can rapidly change their internal mapping of the movement space that informs transformations of sense into action. Coupled with our previous and current computational models and with prominent theories linking neuronal tuning to generalization, we suggest that the functional tuning of motor space may be plastic, to induce either appropriately broad or appropriately narrow trial-by-trial generalization.

Trial-by-trial adaptation in response to pulsatile perturbations

While these studies quantified motor adaptation across single trials, the continuous nature of the perturbing force was ill suited to investigate how different time points within individual movements informed subsequent adaptation. By perturbing participants with brief pulses of force, we recently (Fine and Thoroughman, 2006) explored if and how the feedback experienced at the beginning, middle, or end of a single movement differentially affected the control of the next movement.

Pulses were applied in 20% of movements, either to the left or right, 2, 3, 4, 5, 6, or 7 cm into a 10 cm movement. Pulses were never experienced in two consecutive movements. Prepulse movements (movements immediately before a pulsed movement) were relatively straight, curved only a couple of millimeters. The pulse significantly perturbed these prepulse movements. Participants adapted in movements immediately after pulsed movements (termed postpulse movements). Participants adapted to the left after rightward pulses

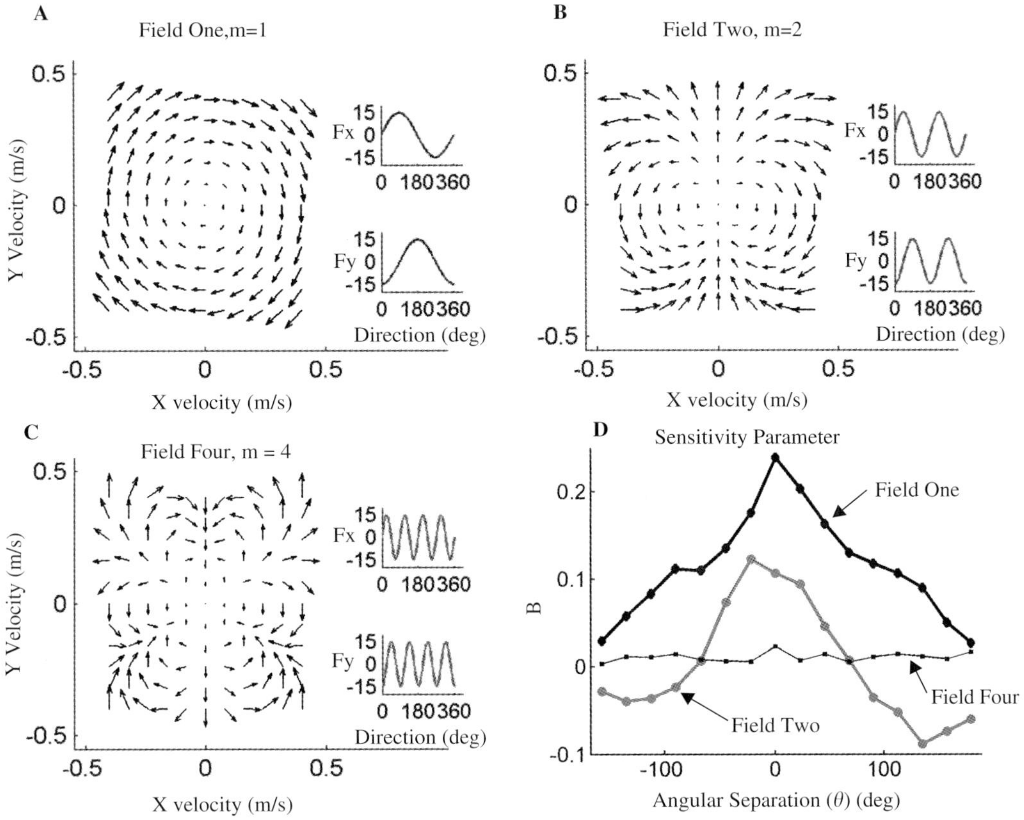

Fig. 1. (A–C) Trial-by-trial sensitivity was estimated in Fields One (A), Two (B), and Four (C). The main subfigures show the magnitude and direction of the force (arrows) as a function of the *X*- and *Y*-components of hand velocity. Insets show the dependence of the *X*- and *Y*-components of force on hand velocity direction. (D) The sensitivity function (*B*) plotted against the angular separation (θ) between sensed and adapted movement directions for each environment. Adapted with permission from Thoroughman and Taylor (2005), © 2005 Society for Neuroscience.

and to the right after leftward pulses. Surprisingly, the timing of postpulse adaptation did not depend significantly on the position of the pulse; pulses early, middle, and late in movement all altered subsequent control from the very beginning of the next movement. Postpulse trajectories also followed similar paths; adaptation to leftward and rightward pulses was statistically indistinguishable regardless of the position of the pulse, whether displacements were measured early, in the middle of, or late in the postpulse movement.

We also investigated the dependence of postpulse adaptation on the magnitude of the pulse. Arm position error (Jordan and Rumelhart, 1992), stiff and viscous feedback (Wolpert and Ghahramani, 2004), and predictive torque error (Jordan and Wolpert, 1999) have all been hypothesized to drive adaptation. All three of these hypotheses have, at their core, an assumption that adaptation strives to regress across experience to minimize the overall magnitude of error. Even reinforcement learning (Sutton and Barto, 1998), which lacks an explicit training signal, presumes regression to drive performance to an optimum. Each of these hypotheses, then, requires the magnitude of adaptation to vary proportionally with the magnitude of error. Current state-space models have included this presumption and have successfully reproduced human behavior when people experienced novel force perturbations throughout the extent of a movement.

We found that the human response to pulsatile force perturbations, in contrast, contained no proportionality to any error signal. We tested participants with force pulses of 70 ms duration and 6, 12, or 18 N peak force (Fig. 2A–C). Pulses were again experienced 3, 5, or 7 cm into a 10 cm movement, either to the left or the right. The error induced by the forces sensibly scaled with the magnitude of the perturbation. If the pulse pushed participants to the left, they adapted to the right in the very next movement; if the forces pushed participants to the right, they adapted by moving to the left. The size of the response, however, was constant, regardless of the magnitude of the force pulse (Fig. 2D–F).

Sensitivity to pulse direction, but insensitivity to pulse magnitude, is most apparent by averaging across pulse position, and plotting adaptation against the direction and magnitude of the force pulse (Fig. 3). This result was surprising, given the proportional responses measured in novel dynamic environments. This component of adaptation therefore cannot scale with any previously hypothesized error signal, nor can it depend on a real-valued critical metric. The magnitude of the force perturbations and positional errors were well within the ranges experienced in previous studies (e.g., Scheidt et al., 2001), so our categorical result is very unlikely to arise due to saturation effects. These results, when combined with previous studies, suggest that people can adopt different modes of adaptation that either can or cannot scale with error magnitude (Fine and Thoroughman, 2007). Both the transformation of error to adaptation and the generalization of error across movement space can therefore change with the environmental demands of a task.

Discussion

Monkey neurophysiologists have correlated the activity of neurons in specific brain areas to many different features of arm movement. Foundational work identified a strong correlation between the activity of primary motor neurons and the direction of arm movement (Georgopoulos et al., 1986). More recent work has investigated how cortical neurons predict both hand speed and movement direction (Moran and Schwartz, 1999) and may also encode several other movement parameters, such as posture of the arm (Scott and Kalaska, 1997) or muscle force (Li et al., 2001; Sergio et al., 2005). Activity of Purkinje cells in the cerebellum, meanwhile, can not only correlate to hand movement speed and direction (Coltz et al., 1999), but also to joint position and muscle force (Thach, 1978). The specific role of cerebral or cerebellar activity in normal motor behavior remains elusive largely due to the correlative nature of these experiments, the high number of neurons that participate in each movement, and the continuous time series of a large number of movement states (hand and joint position and velocity; joint torque; muscle force) that can all contribute to positive post-hoc correlations.

The goal of trial-by-trial analyses is to identify the particular computations used by participants to adapt across individual trials. These and further studies will determine specific signal processing computations that will provide neurophysiologists with protocols with which they can identify the specific contribution a particular brain area makes to the trial-by-trial transformation of sense into incremental adaptation.

Our two new results provide two examples of possible experimental correlates. The observed experience-dependent narrowing or broadening of trial-by-trial adaptation, when compared to theories of generalization, suggests that the neuronal activities underlying adaptation change their tuning with environmental demands. Neurophysiologists could train monkeys in similar environments first to confirm that non-human primates replicate the flexibility in trial-by-trial behavior. Neuronal recordings could then seek to identify areas of activity that change their tuning to movement direction when the environmental complexity changes.

Our observed categorical adaptation to force pulses, in turn, could be used to differentiate areas that encode error from areas that encode adaptation. The bevy of theories that posited a linear relation between sensed error and incremental adaptation could never, because of the linearity, dissociate error and adaptation encoding. If monkeys replicate our human psychophysical response to force pulses, mid-movement feedback

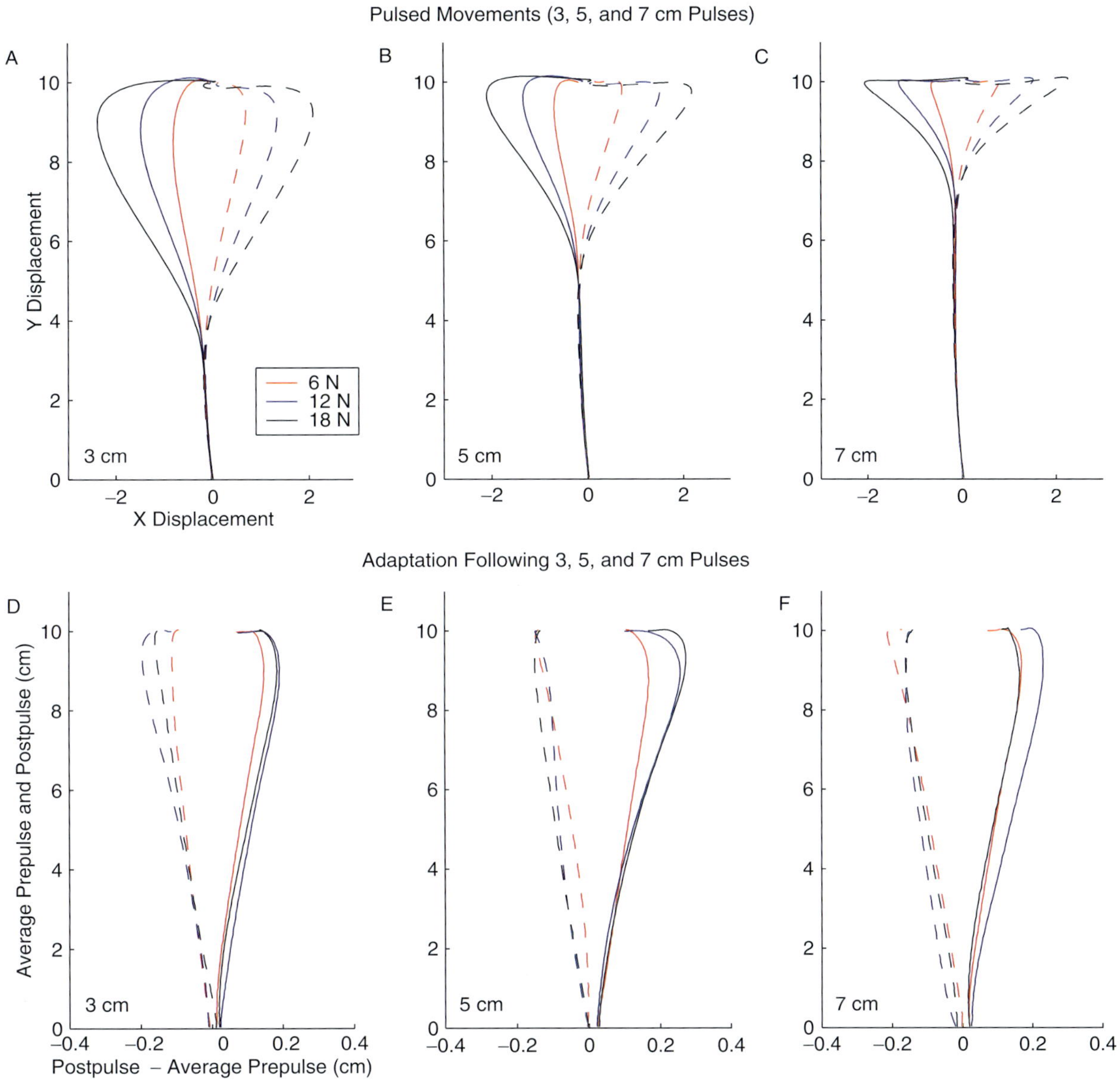

Fig. 2. (A–C) Movement trajectories, averaged across movements and participants, when pulses (C, inset) were applied 3, 5, or 7 cm into a 10 cm movement. The positional error and mid-movement correction were proportional to the magnitude of the pulse (indicated by color). (D–F) To quantify adaptation, we subtracted movements before a pulse (prepulse) from movements after a pulse (postpulse), and averaged across all participants within a particular pulsetype. Notice the *X*-axis is magnified. Adaptation counters the pulse direction but does not scale with the pulse magnitude. Adapted with permission from Fine and Thoroughman (2006), © 2006 American Physiological Society.

responses to the pulses will scale with pulse amplitude, but adaptation in the next movement will depend on the direction and not the magnitude of the pulse. Neuronal recordings could then determine whether a particular brain region's activity responded in proportion to pulse magnitude (as in Fig. 2A–C) or in response to direction but not magnitude (as in Figs. 2D–F and 3). Activity

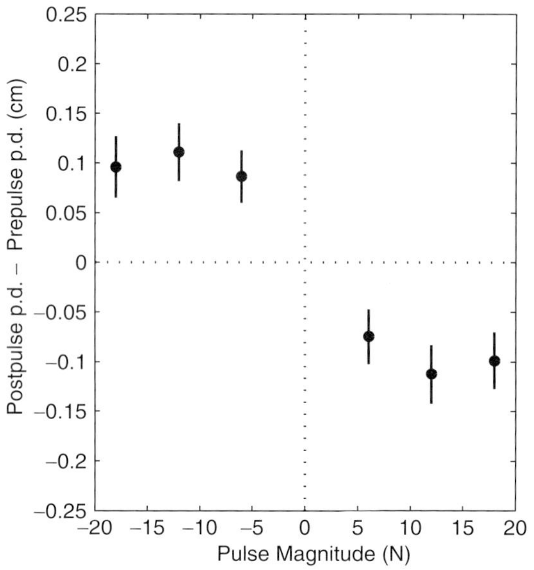

Fig. 3. The magnitude, quantified as the change in mid-movement perpendicular displacement, of aftereffects following force pulses of varying magnitude and direction. Data are averaged across pulse position and participants. Dots in the upper-left quadrant reflect the average aftereffect magnitude after leftward pulses; in the lower-right, after rightward pulses. If the average participant response was proportional, aftereffects following 18 N pulses would be three times the size of those following 6 N pulses. Adapted with permission from Fine and Thoroughman (2006), © 2006 American Physiological Society.

that scaled with pulse magnitude would indicate encoding of error; activity that responded to direction but not magnitude would indicate encoding of the incremental adaptation of subsequent control.

In particular, the trial-by-trial approach correlates well with established models of cerebellar motor adaptation, models that to date have not been tested with trial-by-trial analyses of arm movements. These models have their origins in the Marr notion of adaptation of parallel fiber synapses onto Purkinje cells (Marr, 1969). The foundational Marr hypothesis posits that synaptic efficacy changes as a function of two terms: an error signal, carried by the climbing fibers and encoded as a difference in complex firing rates from baseline, and the tuning curves of the parallel fiber input. The climbing fiber input forms a global error signal that might serve to change synaptic efficacy in many parallel fiber synapses; synaptic specificity comes via the dependency on the parallel input itself. Similar to a backpropagation framework (Jordan and Rumelhart, 1992), the synapse specificity enables the fine tuning of synapses in proportion to their contribution to the neuronal causation of the error. Recent neurophysiology has corroborated these theories; the complex spikes of the Purkinje cells seem to encode sensory error signals in motor command coordinates (Kawato et al., 1987; Kawato and Gomi, 1992). Neural correlates of this model have been identified in ocular following response (Shidara and Kawano, 1993; Gomi et al., 1998; Kobayashi et al., 1998) and in arm movements (Gilbert and Thach, 1977; Kitazawa et al., 1998).

The trial-by-trial approach enables a highly specific computational identification as it relies upon individual errors as inputs and resultant movement modifications as outputs. The careful analysis of current and emerging trial-by-trial adaptation protocols will make possible the construction of very precise hypotheses of neuronal processing on four levels: the tuning of the sensory representation, the representation of mid-movement corrective control, trial-by-trial adaptation, and the across-trial reshaping of that adaptation. The specificity of the model predictions, trial-by-trial monkey behavior, and the corresponding neural activity will foster a careful identification of the computations underlying sensorimotor adaptation and will enable a richly supported, biologically detailed integration of human psychophysics, computation, and primate neurophysiology.

References

Atkeson, C.G. (1989) Learning arm kinematics and dynamics. Annu. Rev. Neurosci., 12: 157–183.

Brashers-Krug, T., Shadmehr, R. and Bizzi, E. (1996) Consolidation in human motor memory. Nature, 382: 252–255.

Caithness, G., Osu, R., Bays, P., Chase, H., Klassen, J., Kawato, M., Wolpert, D.M. and Flanagan, J.R. (2004) Failure to consolidate the consolidation theory of learning for sensorimotor adaptation tasks. J. Neurosci., 24: 8662–8671.

Chan, S.S. and Moran, D.W. (2006) Computational model of a primate arm: from hand position to joint angles, joint torques and muscle forces. J. Neural Eng., 3: 327–337.

Coltz, J.D., Johnson, M.T. and Ebner, T.J. (1999) Cerebellar Purkinje cell simple spike discharge encodes movement velocity in primates during visuomotor arm tracking. J. Neurosci., 19: 1782–1803.

Conditt, M.A., Gandolfo, F. and Mussa-Ivaldi, F.A. (1997) The motor system does not learn the dynamics of the arm by rote memorization of past experience. J. Neurophysiol., 78: 554–560.

Criscimagna-Hemminger, S.E., Donchin, O., Gazzaniga, M.S. and Shadmehr, R. (2003) Learned dynamics of reaching movements generalize from dominant to nondominant arm. J. Neurophysiol., 89: 168–176.

Donchin, O., Francis, J.T. and Shadmehr, R. (2003) Quantifying generalization from trial-by-trial behavior of adaptive systems that learn with basis functions: theory and experiments in human motor control. J. Neurosci., 23: 9032–9045.

Fine, M.S. and Thoroughman, K.A. (2006) Motor adaptation to single force pulses: sensitive to direction but insensitive to within-movement pulse placement and magnitude. J. Neurophysiol., 96: 710–720.

Fine, M.S. and Thoroughman, K.A. (2007) The trial-by-trial transformation of error into sensorimotor adaptation changes with environmental dynamics. J. Neurophysiol., (published online July 5, 2007). doi:10.1152/jn.00196.2007.

Flash, T. and Sejnowski, T.J. (2001) Computational approaches to motor control. Curr. Opin. Neurobiol., 11: 655–662.

Gandolfo, F., Mussa-Ivaldi, F.A. and Bizzi, E. (1996) Motor learning by field approximation. Proc. Natl. Acad. Sci. U.S.A., 93: 3843–3846.

Georgopoulos, A.P., Schwartz, A.B. and Kettner, R.E. (1986) Neuronal population coding of movement direction. Science, 233: 1416–1419.

Gilbert, P.F. and Thach, W.T. (1977) Purkinje cell activity during motor learning. Brain Res., 128: 309–328.

Gomi, H., Shidara, M., Takemura, A., Inoue, Y., Kawano, K. and Kawato, M. (1998) Temporal firing patterns of Purkinje cells in the cerebellar ventral paraflocculus during ocular following responses in monkeys I: simple spikes. J. Neurophysiol., 80: 818–831.

Goodbody, S.J. and Wolpert, D.M. (1998) Temporal and amplitude generalization in motor learning. J. Neurophysiol., 79: 1825–1838.

Hwang, E.J., Donchin, O., Smith, M.A. and Shadmehr, R. (2003) A gain-field encoding of limb position and velocity in the internal model of arm dynamics. PLoS Biol., 1: E25.

Jordan, M.I. and Rumelhart, D.E. (1992) Forward models: supervised learning with a distal teacher. Cogn. Sci. Multidisciplinary J., 17: 463–496.

Jordan, M.I. and Wolpert, D.M. (1999) Computational motor control. In: Gazzaniga M.S. (Ed.), The Cognitive Neurosciences. MIT Press, Cambridge, MA, pp. 601–620.

Kawato, M., Furukawa, K. and Suzuki, R. (1987) A hierarchical neural-network model for control and learning of voluntary movement. Biol. Cybern., 57: 169–185.

Kawato, M. and Gomi, H. (1992) The cerebellum and VOR/OKR learning models. Trends Neurosci., 15: 445–453.

Kitazawa, S., Kimura, T. and Yin, P.B. (1998) Cerebellar complex spikes encode both destinations and errors in arm movements. Nature, 392: 494–497.

Kobayashi, Y., Kawano, K., Takemura, A., Inoue, Y., Kitama, T., Gomi, H. and Kawato, M. (1998) Temporal firing patterns of Purkinje cells in the cerebellar ventral paraflocculus during ocular following responses in monkeys II: complex spikes. J. Neurophysiol., 80: 832–848.

Lackner, J.R. and Dizio, P. (1994) Rapid adaptation to Coriolis force perturbations of arm trajectory. J. Neurophysiol., 72: 299–313.

Li, C.S., Padoa-Schioppa, C. and Bizzi, E. (2001) Neuronal correlates of motor performance and motor learning in the primary motor cortex of monkeys adapting to an external force field. Neuron, 30: 593–607.

Marr, D.A. (1969) Theory of cerebellar cortex. J. Physiol., 202: 437–470.

Moran, D.W. and Schwartz, A.B. (1999) Motor cortical representation of speed and direction during reaching. J. Neurophysiol., 82: 2676–2692.

Poggio, T. (1990) A theory of how the brain might work. Cold Spring Harb. Symp. Quant. Biol., 55: 899–910.

Pouget, A. and Snyder, L.H. (2000) Computational approaches to sensorimotor transformations. Nat. Neurosci., 3(Suppl): 1192–1198.

Scheidt, R.A., Dingwell, J.B. and Mussa-Ivaldi, F.A. (2001) Learning to move amid uncertainty. J. Neurophysiol., 86: 971–985.

Scott, S.H. and Kalaska, J.F. (1997) Reaching movements with similar hand paths but different arm orientations I: activity of individual cells in motor cortex. J. Neurophysiol., 77: 826–852.

Sergio, L.E., Hamel-Paquet, C. and Kalaska, J.F. (2005) Motor cortex neural correlates of output kinematics and kinetics during isometric-force and arm-reaching tasks. J. Neurophysiol., 94: 2353–2378.

Shadmehr, R. and Brashers-Krug, T. (1997) Functional stages in the formation of human long-term motor memory. J. Neurosci., 17: 409–419.

Shadmehr, R. and Moussavi, Z.M. (2000) Spatial generalization from learning dynamics of reaching movements. J. Neurosci., 20: 7807–7815.

Shadmehr, R. and Mussa-Ivaldi, F.A. (1994) Adaptive representation of dynamics during learning of a motor task. J. Neurosci., 14: 3208–3224.

Shidara, M. and Kawano, K. (1993) Role of Purkinje cells in the ventral paraflocculus in short-latency ocular following responses. Exp. Brain Res., 93: 185–195.

Sutton, R.S. and Barto, A.G. (1998) Reinforcement Learning: An Introduction. MIT Press, Cambridge, MA.

Thach, W.T. (1978) Correlation of neural discharge with pattern and force of muscular activity, joint position, and

direction of intended next movement in motor cortex and cerebellum. J. Neurophysiol., 41: 654–676.
Thoroughman, K.A. and Shadmehr, R. (2000) Learning of action through adaptive combination of motor primitives. Nature, 407: 742–747.
Thoroughman, K.A. and Taylor, J.A. (2005) Rapid reshaping of human motor generalization. J. Neurosci., 25: 8948–8953.
Tong, C., Wolpert, D.M. and Flanagan, J.R. (2002) Kinematics and dynamics are not represented independently in motor working memory: evidence from an interference study. J. Neurosci., 22: 1108–1113.
Wolpert, D.M. and Ghahramani, Z. (2004) Computational motor control. In: Gazzaniga M.S. (Ed.), The Cognitive Neurosciences (3rd Ed.). MIT Press, Cambridge, MA, pp. 485–494.

P. Cisek, T. Drew & J.F. Kalaska (Eds.)
Progress in Brain Research, Vol. 165
ISSN 0079-6123
Copyright © 2007 Elsevier B.V. All rights reserved

CHAPTER 24

Towards a computational neuropsychology of action

John W. Krakauer[1] and Reza Shadmehr[2,*]

[1]*The Motor Performance Laboratory, Department of Neurology, Columbia University College of Physicians and Surgeons, New York, NY 10032, USA*
[2]*Laboratory for Computational Motor Control, Department of Biomedical Engineering, Johns Hopkins School of Medicine, Baltimore, MD 21205, USA*

Abstract: From a computational perspective, the act of using a tool and making a movement involves solving three kinds of problems: we need to learn the costs that are associated with our actions as well as the rewards that we may experience at various sensory states. We need to learn how our motor commands produce changes in things that we can sense. Finally, we must learn how to actually produce the motor commands that are needed so that we minimize the costs and maximize the rewards. The various computational problems appear to require different kinds of error signals that guide their learning, and might rely on different kinds of contextual cues that allow their recall. Indeed, there may be different neural structures that compute these functions. Here we use this computational framework to review the motor control capabilities of two important patients who have been studied extensively from the neuropsychological perspective: HM, who suffered from severe amnesia; and BG, who suffered from apraxia. When viewed from a computational perspective, the capabilities and deficits of these patients provide insights into the neural basis of our ability to willfully move our limbs and interact with the objects around us.

Keywords: motor control; computational models; forward models; optimal control

Introduction

In 1997, one of us performed a 2-day experiment on the severely amnesic patient HM to see how well he could learn a new motor task and retain it in memory (Shadmehr et al., 1998). The task was the standard reach adaptation task where subjects hold the handle of a robotic arm and learn to use it to guide a cursor to a sequence of targets (Shadmehr and Mussa-Ivaldi, 1994). When seated in front of the robot, HM, like all naïve volunteers, sat quietly and waited for instructions. We asked him to put his hand on the robot's handle and move it around a bit. Naturally, he kept his gaze on his hand as he moved the robot's handle. He was instructed to not look at his hand, but rather at the video monitor, where a cursor was present. After a minute or so of moving the cursor around, a center target was presented and he was asked to move the cursor to that location. Subsequently another target was shown and he was encouraged to move the cursor there. After a few minutes of practice in reaching to targets, the robot began to impose forces on HM's hand, perturbing the path of the cursor. With more practice, he altered his motor commands so to predicatively compensate for the forces, as evidenced by large errors in catch trials during which the field was turned off. We then thanked him for his time and he left to have lunch.

*Corresponding author. Tel.: +1 410-614-2458; Fax: +1 410-502-2826; E-mail: reza@bme.jhu.edu

DOI: 10.1016/S0079-6123(06)65024-3

When he came back to the experiment room 4 h later, he claimed that he had never seen the robotic device or knew what it was for. We pushed the robotic arm aside and asked him to sit down. He sat down, but then voluntarily reached and grabbed the robot's handle, brought it toward him, and looked at the video monitor, where the cursor was present. It was clear that despite having no conscious recollection of having done the task before, some part of HM's brain recognized that the contraption was a tool that had a particular purpose: to manipulate cursors on a screen. Furthermore, his brain knew that in order to operate this tool, he had to hold its handle. When a target was presented, he showed strong after-effects of the previous training. That is, his brain expected the robot to perturb his movements, and so he generated motor commands that attempted to compensate for these forces. His brain knew all this, yet he was consciously unaware of this knowledge.

This experiment suggested to us that the brain could solve two general problems without conscious awareness and without the medial temporal lobes. First, the brain could learn how to use a tool in order to achieve a purpose. Second, the visual sight of the tool at a later time was sufficient to allow recall of both the purpose of the tool and the motor commands needed to achieve that purpose, although the latter may have required kinesthetic cues from the handle. This is despite the fact that the same visual information was not sufficient to recall conscious memory of having done the task.

Brenda Milner had of course made a similar observation in HM some 30 years earlier in a task where he drew on a piece of paper while looking in a mirror (Milner, 1968). In the novel visual feedback setting, HM adapted his motor output and learned to draw accurately. When he returned the next day, the visual and/or tactile cues associated with the experimental setup were sufficient to allow him to recall the motor skill that he had learned before. Over the years, a number of other investigators made similar observations in other amnesic patients (Gabrieli et al., 1993; Yamashita, 1993; Tranel et al., 1994), culminating in the theory that formation of motor memories are independent of the medial temporal lobe (Mishkin et al., 1984).

Subsequent research has made great strides in understanding the anatomical, physiological, and behavioral characteristics of motor memory. For example, many studies experimentally characterized and computationally modeled adaptation and re-adaptation of reaching movements in force-fields (Shadmehr and Brashers-Krug, 1997; Hwang et al., 2003; Krouchev and Kalaska, 2003; Caithness et al., 2004; Hwang and Shadmehr, 2005; Wainscott et al., 2005) and altered visuomotor environments (Krakauer et al., 1999, 2000, 2005; Wigmore et al., 2002; Caithness et al., 2004). In addition, single unit recording (Li et al., 2001; Padoa-Schioppa et al., 2002), lesion analysis (Maschke et al., 2004; Smith and Shadmehr, 2005), functional imaging (Krakauer et al., 2004; Diedrichsen et al., 2005), and some computational models (Donchin et al., 2003) successfully mapped adaptation processes onto the anatomy and physiology of the cerebral cortex, the basal ganglia, and the cerebellum. However, it appears to us that some of the fundamental implications of the HM experiments have been largely overlooked. For example, what cued him to recall the motor memory that he had acquired in the initial training session with the novel tool? In the original mirror drawing experiments, was it the visual cues associated with seeing the pencil and the mirror, or the act of holding the pencil? In the robot reaching task, was it the sight of the robot, or was it the act of holding the handle? What motivated him to learn these tasks and how did he sustain that motivation in subsequent sessions? Did he need to have a conscious desire to reduce errors for some perceived reward or did the motor system have its own implicit reward system?

Our aim here is to first pose the problem of motor control in a broad computational framework and then show that this framework sheds light on the neuropsychological basis of motor learning and motor memory.

Seeking rewards and observing the consequences of action

Our decisions to perform a task are guided by measures of costs and rewards. For example, in a task when volunteers are asked to control a robotic handle in order to move a cursor to a target, they are often told to try to get to the target in a given amount of time. If they do so, they are "rewarded" by a target explosion, whereupon they hear the experimenter make an encouraging comment. They are also often paid for their time, and this payment may depend on performance.

Was HM's behavior driven by reward? For HM, the target explosion triggered a childhood memory of going bird hunting. As he was performing the task and was able to get a target explosion, he would spend the next few minutes describing the memory in detail: the type of gun that he used, the porch in the rear of his childhood home, the terrain of the woods in his backyard, and the kinds of birds that he hunted. He repeated these details to us many times during the 2-day experiment. Sue Corkin (our colleague who had examined HM for many decades) mentioned that she had not heard HM talk about this before. Although we did not record the conversation, the re-telling of the memory appeared to be related to the target explosions.

When HM came back on his second and third sessions with the robot, he did not need instructions on what the task was about; he sat down, reached for the robot, looked at the video screen and waited for the targets to appear. Perhaps the earlier experience with the task was such that it left a memory of expected reward: upon return, the sight and touch of the robot was sufficient to encourage a motor act that would be expected to be rewarding. If on the other hand, use of the robot in the first session had been paired with a shock or another noxious stimulus, it seems likely that he would have been reluctant to use the device again.

In a broader sense, experience with a tool provides us with information about how its use is associated with costs and rewards. In the robot reaching task, the reward is achieved only if the target explodes, and the cost is the energy spent doing the reaching. (For a typical volunteer, there are also costs associated with time away from their normal routine, but for simplicity let us ignore these realities.) Assuming that cursor position is $\mathbf{y}$ and target position is $\mathbf{r}$, then through experience we learn that the objective of the task is to minimize $(\mathbf{y}-\mathbf{r})^T(\mathbf{y}-\mathbf{r})$ at time N after the reach starts (this is the time that the target will explode if we are near it). Superscript T is the transpose operator. To denote the fact that this cost is zero except for time N, we write this cost as:

$$\sum_{n=1}^{N}(\mathbf{y}^{(n)}-\mathbf{r})^T T^{(n)}(\mathbf{y}^{(n)}-\mathbf{r})$$

where the superscript (n) refers to a discrete measure of time, and matrix T is a measure of our cost at each time step (which may be zero except at time N). We also have a cost associated with our motor commands $\mathbf{u}$, which here we assume grows as a quadratic function. Now the total cost becomes:

$$\sum_{n=1}^{N}(\mathbf{y}^{(n)}-\mathbf{r})^T T^{(n)}(\mathbf{y}^{(n)}-\mathbf{r})+\mathbf{u}^{(n)T}L^{(n)}\mathbf{u}^{n} \qquad (1)$$

where matrix L is a time-dependent measure of the weighted costs associated with the motor commands. To get reward (explode the target), we need to find the motor commands that will minimize this cost. We learn this by observing that moving the robot handle will move the cursor. In particular, we learn that pushing on the robot will result in a specific proprioceptive and visual feedback regarding the state of our body and the state of the cursor. These are the sensory consequences of our actions. Grossly simplifying the problem, here we write these consequences as a linear function of our motor commands:

$$\begin{aligned}\mathbf{x}^{(n+1)} &= A\mathbf{x}^{(n)}+B\mathbf{u}^{(n)}+\varepsilon_u^{(n)}\\ \mathbf{y}^{(n)} &= C\mathbf{x}^{(n)}+\varepsilon_y^{(n)}\end{aligned} \qquad (2)$$

where $\mathbf{x}^{(n)}$represents the state of the body and the world that we interact with, $\mathbf{u}^{(n)}$ is the motor command, $\varepsilon_u^{(n)}$ a stochastic variable representing motor noise, and $\varepsilon_y^{(n)}$ a stochastic variable representing sensory noise. If Eq. (2) is an accurate model of how motor commands to our arm muscles produce changes in the state of our body and the cursor, then we can use it as a set of constraints with which to minimize Eq. (1). This is the classic linear quadratic problem in optimal control. Solving this problem under the assumption that the noise variables are Gaussian yields a linear feedback control law that specifies the state-dependent motor commands that we should produce at each time step:

$$\mathbf{u}^{(n)} = (L^{(n)} + B^T W^{(n+1)} B)^{-1} B^T (\boldsymbol{\theta}^{(n+1)} - W^{(n+1)} A \mathbf{x}^{(n)}) \quad (3)$$

The two new variables in this equation, W and $\mathbf{q}$, are time dependent quantities that depend on Lagrange multipliers that reflect the constraints in Eq. (2) in minimizing Eq. (1).

To summarize, the computational problem of learning motor control may be described as having three components:

1. To perform any action, we need to know the costs that are associated with our actions as well as the sensory states that are rewarding [Eq. (1)]. In the reaching task, through instruction or observation we might learn that target explosions are a rewarding act and they occur only when the cursor reaches the target at a specific time. The relative benefit of this reward with respect to the cost of the motor commands will dictate an internal value of this act.
2. We need to know how our motor commands produce changes in things that we can observe [Eq. (2)]. That is, through experience we must learn that when we are holding the robot in hand, our motor commands will result in a specific change in the state of our arm and the state of the cursor. This is a system identification problem associated with the particular tool. Learning of this map is called forming a forward model.
3. Finally, we must learn how to actually produce the motor commands that are needed so that we minimize the costs and maximize the reward [Eq. (3)]. That is, we need to figure out the "best" motor commands that bring the cursor to the target and get it to explode. This is the constrained minimization problem [minimize Eq. (1) under the constraints of Eq. (2)]. The result of the minimization is a feedback control law that specifies the motor response to the sensory states that we observe in our body and the environment. Learning of this feedback control law is called forming an inverse model.

This computational framework for representing the problem of biological motor control is largely due to the pioneering work of Todorov and Jordan (2002). At the heart of the approach is a departure from a framework in adaptive control where agents are provided with desired trajectories. Indeed, here we have no one to tell us what actions to follow. Rather, the goal is to acquire rewards, and the means is through observing the consequences of our actions.

What motivated HM to learn the reach adaptation task?

We expected that a severely amnesic individual who was performing a novel task would have to be regularly reminded of the task's instructions: "try to move the cursor to the target fast enough so it explodes." However, after HM had exploded a few targets, he no longer needed verbal reminders. Strikingly, when he returned the next day he voluntarily reached for the robot handle and began preparing for onset of targets by moving the cursor to the center location. This remarkable behavior suggests that during the first session, he learned the reward basis of the task [Eq. (1)] implicitly, while during the later sessions, the visual appearance of the machine, and the act of holding its handle, was sufficient to trigger a recall of this reward structure.

Why should target explosions be an implicitly rewarding action for HM? Perhaps because he associated them with his earlier experience with bird hunting, a memory that he very much enjoyed retelling. It is entirely possible that if we had chosen another mode of task performance feedback, say a numeric score, the intrinsic value of the task for HM might have been much lower. This might result in a subject who is ambivalent or even reluctant to perform a task.

We would conjecture that the reward mechanism that is a part of Eq. (1) is implemented in the brain through the release of dopamine. We have recently shown (Mazzoni et al., 2007) that although patients with Parkinson's disease (PD) reach more slowly than age-matched controls, they are nevertheless able to make fast movements without a loss in accuracy if forced to move fast by the experimental task. However, the patients take longer (require more trials) to accumulate a set number of movements at the required speed. Thus, bradykinesia (slowness of movements), one of the hallmarks of PD, may represent decreased intrinsic value of a given motor task because of an imbalance between an estimate of an implicitly-determined cost for making a fast movement and the expected rewards.

Learning sensory consequences of motor commands vs. learning optimum control policies

In order for HM to explode targets, he had to learn how motor commands to his arm produced changes in the proprioceptive state of his limb and the visual state of the cursor [Eq. (2)]. This equation represents a forward model, an association between actions and sensory consequences. We now know that patients with cerebellar dysfunction are often severely impaired in learning this association, whereas basal ganglia damage appears to spare this ability (Maschke et al., 2004; Smith and Shadmehr, 2005). If the cerebellum is the crucial site for this learning, then the error signals that form forward models are likely delivered through climbing fiber activity.

With an accurate forward model [Eq. (2)], one can produce actions that achieve fair performance levels even without learning an optimal inverse model [Eq. (3)]. This is because imperfect motor commands can be rapidly compensated through internal feedback of the forward model. Indeed, for control of saccadic eye movements, the cerebellum appears to fit the role of a forward model that "steers" ongoing oculomotor commands rather well. However, optimal performance requires forming both an accurate forward model and an optimal inverse model. For example, if the robot is producing a force field, then learning an accurate forward model without an optimal inverse model will result in nearly straight trajectories. In theory, however, the optimum cursor trajectory that will result in the maximum probability of exploding the target is not a straight trajectory. Rather, the optimum cursor path is one that is slightly curved, resulting in an overcompensation of the forces early in the reach and an under compensation near the end. The reason for this is that because of noise associated with motor commands, producing larger commands early in the movement is a better policy than late in the movement because feedback allows one to correct the early errors but not the late errors. Interestingly, these curved cursor paths are similar to those that we have recorded in healthy volunteers (Thoroughman and Shadmehr, 2000).

In case of the reaching/robot task, theory predicts that the curved movements that appear to be a signature of optimal inverse models will not occur if training includes catch trials (randomly interleaved trials in which the force field is turned off). The over-compensation, which results from the large motor commands early in the trajectory, and the resulting curved movements, are optimal only in the case where one is certain that the force field will be present. If there are catch trials, then the optimal feedback control law is a path that is fairly close to a straight line.

Catch trials were part of the protocol with HM, and therefore we do not know if learning of optimal inverse models was intact in him. However, the error signals that guide formation of optimal inverse models need to link changes in expectations of cost and reward with the control policy that links sensory states to motor commands. As dopamine appears to be closely linked to reward prediction, it may play a central role in this learning. An interesting prediction of this hypothesis is that patients with basal ganglia damage should learn forward models normally, but be unable to learn optimal inverse models. To test this idea, current theoretical models of action need to advance so that they can predict tasks and conditions where subtle modifications of task characteristics can alter probability of success, resulting in clear changes in behavior in the optimal (presumably normal) learner.

The dissociation that we have made between learning motor commands that can produce rewarding states (optimal inverse models) and learning to predict sensory consequences of those commands (accurate forward models) can help explain a recent experimental result (Mazzoni and Krakauer, 2006). In our task, subjects reached to a target but could not see their hand. Instead, a cursor was presented on a vertical screen. The screen displayed eight targets, arranged around a circle. On each trial one of these targets would be highlighted and the subject was instructed to take the cursor to that target. The novel part of this experiment was that the subjects were told that there is a 45° counter-clockwise (CCW) rotation of visual feedback during their reaching movements. Importantly, they were given a cognitive strategy to counter this perturbation: aim for the target 45° clockwise (CW) from the desired target in order to ensure that the cursor enters the desired target. The prediction was that successful implementation of the strategy would result in an abrupt stepwise cancellation of errors and the absence of after effects. The term aftereffect, seen in catch trials, refers to a trajectory deviation in the direction opposite to the imposed perturbation and indicates learning of an internal model.

Subjects were indeed initially effective in canceling the rotation with errors returning immediately to near zero. Surprisingly however, as subjects continued to make movements they made increasingly large directional errors, leading the cursor away from the desired target. This indicates that the verbal instructions provided them with the means to produce motor commands that resulted in rewarding states. However, because the rotation altered the relationship between the motor commands and the implicitly expected cursor path, the forward model began to adapt. This adaptation resulted in internal feedback during the reach, altering the trajectory and producing consistent errors. When subjects were asked about what they thought was happening, they often expressed frustration at the fact that they became progressively worse at hitting the target (the desired reward) and were unaware that they were adapting to the rotation. We interpret this result as evidence for the idea that despite the fact that they initially produced the optimum motor commands, never the less the forward model began to adapt, resulting in gradually incorrect movement. We think that the process of learning forward models may be distinctly separate from learning optimum control policies.

One way to view this is to assume that the motor system has two independent reward systems, one that evaluates the reward basis of a task and determines the success or failures of a control policy, and another that compares predicted sensory consequences of motor commands with their measured values, resulting in forward models. Our evidence from patients with PD who show a reluctance to make fast accurate reaching movements even though they are capable of making them, and from healthy subjects who adapt to a visuomotor rotation despite this being contrary to the explicit goal of the task, appears consistent with this framework.

Apraxia and the case of patient BG

With the exception of the idea that the motor system might have its own implicit reward system, the computational framework described thus far has dealt with relatively low-level aspects of motor control, namely execution and adaptation of reaching trajectories. However, patients with focal lesions of the CNS, especially of premotor and parietal regions, display a variety of higher-order motor disorders that have not been addressed to any great degree by this framework. Here we will focus on apraxia as an example of a syndrome where neuropsychological observations might be informed by a motor control perspective and vice versa.

Apraxia is a foreboding topic for a non-clinically oriented motor physiologist. Dictionary definitions of apraxia are usually a variation on the following: the inability of a person to perform voluntary and skillful movements of one or more body parts to command even though there is no muscle weakness, incoordination, sensory loss, aphasia or dementia. The unsatisfactory nature of this definition is apparent in the seemingly innumerable overlapping definitions and subtypes of apraxia that abound in the literature. This is undoubtedly because attempts to ascertain a core behavioral manifestation for apraxia have proven frustrating. The extant literature is largely descriptive and non-quantitative with few attempts to incorporate clinical phenomena into the emerging framework provided by basic research on sensorimotor integration and the parietal lobe (Leiguarda and Marsden, 2000). In contrast, recent research on motor control has used restrained, and arguably oversimplified and unnatural laboratory based tasks to study motor learning, motor generalization and the role of context (Shadmehr and Wise, 2005). Thus, these tasks might be inadequate to elicit apraxic deficits. A caveat, however, is that patients with ideomotor apraxia have been shown to have kinematic abnormalities even for simple natural movements (Clark et al., 1994; Poizner et al., 1995). These kinematic abnormalities tend to be ignored by the neuropsychological and neurological literature (although see Ietswaart et al., 2006). Thus, there is an apparent divide between the approach to motor control taken by physiologists, in which the emphasis is on sensorimotor integration for simple movements, and the work on apraxia undertaken by clinical neuropsychologists, where the emphasis is on "higher order" motor deficits (see Table 1).

However, if there is to be progress it is critical for the two fields to learn from one another. In particular, it will allow us to determine whether there are categorical differences between sensorimotor integration seen, for example, during prism adaptation and sensorimotor integration when observed behavior has to be imitated. The case of HM attests to this need. Detailed neuropsychological description of this single patient over 50 years has yielded innumerable insights and avenues of investigation for neurophysiologists interested in the medial temporal lobe system (Corkin, 2002). Analogously, we would argue that motor control scientists stand to benefit greatly from taking a more careful look at apraxia.

Patients with "ideomotor" apraxia show all or a subset of the following motor execution abnormalities (see Leiguarda and Marsden, 2000; Koski et al., 2002, for review):

1. Spatiotemporal errors in the production of both over-learned and novel motor acts. For example, Poizner and colleagues have shown that patients asked to make a slicing movement as though cutting bread fail to show the correct phase relationships between pairs of joint angles (Poizner et al., 1995).
2. Impaired ability to pantomime symbolic gestures or tool-use to command or when given the tool itself. For example, a patient asked to imitate how they would use a pair of scissors may oppose their forefinger and index finger together as though they are themselves the blades of the scissors.
3. Impaired ability to imitate meaningful and meaningless motor acts. For example, a patient may not be able to copy a teeth-brushing motion or copy a random dance move.
4. An inability to adopt complex hand postures. For example, a dog-shaped shadow puppet made with the hands cannot be copied by a patient.

Patients with "ideational' and "conceptual" apraxia make higher-level errors compared to patients with ideomotor apraxia. They seem to have lost the semantic meaning (concept) of an action and can have trouble performing an action in the correct sequence. For example, while making a cup of tea they might pour the water into the cup before it has boiled and forget the tea bag. They can show choice of the wrong movement for a given

Table 1. An apparent divide between the approach to motor behavior taken by motor control physiologists versus that taken by clinical neuropsychologists

	Motor control approach	Neuropsychological approach
Questions	Trajectory control, adaptation, precision grip	Higher order motor deficits, e.g., apraxia, optic ataxia, and neglect
Experimental population	Healthy college students	Patients
Tasks	Restrained movements, e.g., planar reaching, sequential finger movements	3D everyday natural tasks, e.g., using knife and fork, mailing a letter, writing
Planning	Spatial accuracy, reaction time, scaling of early trajectory variables	Perform a task in the right sequence, e.g., making cup of tea
Execution	Continuous kinematic measures	Nominal scales, goal accomplishment
Conceptual framework	Quantitative: control theory, biomechanics, computational	Qualitative: neuroanatomical, descriptive, diagrammatic

transitive act. For example, they might make a chopping action when asked to pantomime brushing their teeth. They may also be impaired in recognizing a gesture or the purpose of a tool.

The majority of patients with apraxia have parietal lobe damage in the hemisphere contralateral to their dominant hand, but cases have been described with lesions in the premotor cortex bilaterally or lesions in the non-dominant parietal cortex (Halsband et al., 2001). There is currently no unifying explanatory framework for apraxic phenomena, which is perhaps not surprising given the multiplicity of parietal functions in motor control (Fogassi and Luppino, 2005; Culham and Valyear, 2006). We would argue that it would be more fruitful to map identified and putative physiological and computational processes in motor control onto specific apraxic phenomena. The goal is to have the clinical phenomena inform future experiments and models of motor control and parietal function.

Let us focus on two central features of ideomotor apraxia, both of which were present and rigorously described in patient BG, an important case study described by Buxbaum et al. (2000). We use BG as an illustrative case whilst recognizing that unlike HM, she did not have an identifiable circumscribed lesion. She had a primary progressive ideomotor apraxia, resulting in an inability to pantomime tool use either by verbal command, by viewing the object, or through imitation. However, when she held the tool in her hand, she was able to gesture and demonstrate its use near normally. That is, holding the tool allowed BG to recall the purpose of that tool and demonstrate its use through correct motor commands. This showed that the motor memory was present but not retrievable by verbal command or by looking at the object. When we compare this ability with HM, we see that it is rather important that with visual cues alone, HM demonstrated an ability to recall that the tool was associated with a rewarding behavior, resulting in his voluntary act of reaching for it, holding the handle in hand, and waiting for the task to begin. Furthermore, recall that once he held the tool, he was able to recall the motor commands necessary to control it.

For BG, why was semantic knowledge or visual observation of the tool insufficient to recall the memory of its purpose or how to hold it? As we described in the first section, knowledge of the behavior of a rotated cursor tool did not allow subjects to learn how to use it and in fact the motor system overrode healthy subjects' explicit knowledge (Mazzoni and Krakauer, 2006). Another clue to the failure of verbal commands or vision to call up the correct gesture comes from a large number of experiments that have tried to use visual cues to recall a particular internal model of a tool, either a robotic force field or rotated cursor (Cunningham and Welch, 1994; Gandolfo et al., 1996; Miall et al., 2004; Shadmehr et al., 2005). In these experiments, subjects learn to associate a force field or visuomotor rotation with a color or some other arbitrary symbolic cue. For example, blue for a CW rotation and red for a CCW rotation. Unexpectedly, subjects are either unable or require extensive training [monkeys had to be trained on a color-force field association for 12 months (Krouchev and Kalaska, 2003)] to use color as a cue to call up the associated internal model. This is strikingly reminiscent of the inability of BG to use visual cues to recall the appropriate gesture.

What was it about holding the tool itself, rather than the vision of the tool, which allowed BG to recall the appropriate gesture? An experiment we have recently performed in healthy subjects sheds light on this (Krakauer et al., 2006). In this experiment we hypothesized that the contextual cues subjects use to recall internal models are kinesthetic and implicit rather than visual and explicit. Specifically, we hypothesized that the relevant contextual cue is an implicit memory of action with a particular body part. To test this hypothesis we had subjects to learn a visuomotor rotation (the screen cursor's trajectory was rotated 30° CCW to the hand's trajectory) in what we conjectured would be two different contexts: by moving their hand through motion of their shoulder and elbow, or through motion of their wrist.

Our hypothesis was confirmed by the demonstration that subjects could recall opposite visuomotor rotations when each rotation was associated with a different body part. We would argue that the kinesthetic memory of the body part used to learn each rotation operates in a mechanistically related manner to how kinesthetic information from handling of a tool allowed BG to recall the appropriate gesture. For example, we would predict that BG would be able to pantomine tool use if she first held the tool in her other hand. In the case of HM, we think that viewing the robot allowed him to remember the costs and rewards associated with the task, while the act of holding the robot allowed him to remember the forward model associated with control of the machine.

What remains to be explained in this framework is how healthy subjects use visual cues or semantic knowledge to recall internal models of everyday tools. We do not know the answer to this but suggest that it is a serial process in which multiple iterations with kinesthetic cues are required first before transitioning to visual cues. This makes intuitive sense when one considers the overshoot that occurs when we lift a cup that we think is full but is actually empty. We overshoot because we have lifted full cups in the past and retained the kinesthetic memory of the act.

A second prominent abnormality was BG's inability to imitate meaningless gestures. This finding, seen in some patients with ideomotor apraxia, has been a puzzle to neuropsychologists because a popular view of apraxia is that patients are impaired in their ability to recall stored motor programs. However, imitation of a meaningless gesture clearly cannot be related to impaired retrieval. Therefore, the explanation is likely separate from the contextual cuing effects discussed above.

In the study of BG, the authors conjectured that the difficulty with imitating gestures, nonsense gestures in particular, is due to a deficit in representing the relative positions of body parts (Buxbaum et al., 2000). In support of this notion is the observation that patients with posterior parietal lesions, unlike normal subjects or patients with motor cortex damage, show poor correlation between imagined and executed sequential finger movements, which suggests that they are not good at simulating movements. The hand laterality task is often used to assess such simulation. In this task one measures the time it takes for a subject to determine whether the picture is of a right hand or a left hand, has been shown to be a function of the position of the participant's own hand. Therefore, when one sees a picture of a hand, our ability to determine if it is a left or right hand is based on our ability to imagine translating and rotating our hand from its current posture to the viewed posture. Interestingly, in a recent study of 55 patients with left hemispheric stroke, there was a correlation between performance on the hand laterality task and the ability to imitate meaningless gestures (Schwoebel et al., 2004). This type of finding is consistent with the conjecture that the basis for BG's problem with meaningless gesture imitation may be due to a problem with her notion of body schema, which is defined as a representation of the relative positions of body parts derived from multiple sensory inputs and, perhaps, efference copy (Schwoebel and Coslett, 2005).

There is a problem with this viewpoint, however. The problem can be appreciated by considering patient PJ, an individual with an extra-axial cyst encroaching upon the left superior parietal lobule (Wolpert et al., 1998). Without vision of her right arm, PJ was unsure of where it was in space and in fact would feel as though it had disappeared. The unique aspect to her symptoms, as compared to patients who are deafferented from either central or peripheral lesions, was that she had normal tactile sensation and proprioception but that these proprioceptive sensations faded without vision of her right arm. The interpretation was that without vision, PJ had an inability to store a proprioceptively derived internal estimate of the state of her right arm and thus accumulated error over time. Notably, however, PJ was not apraxic. Thus, a deficit in limb state estimation that is correctible by vision is not likely to be the explanation for ideomotor apraxia in which viewing of tool use does not help imitation. Thus, notions of body schema (or estimation of limb state) are not sufficient explanation for the inability of patients with ideomotor apraxia to imitate transitive and nonsense gestures. Instead, it seems more plausible to relate these apraxic abnormalities in imitation of hand-held tool use and hand postures to the specific role of the IPL (and ventral premotor cortex) in visuomotor transformations for grasping (Sakata, 2003).

In their seminal work, Sakata and colleagues found that there were object-specific visual and visuomotor cells in the anterior intraparietal area

(AIP) of the IPL that coded for object shape and hand preshaping, respectively (Sakata, 2003). However, patients with ideomotor apraxia show a clear dissociation between impaired gesture imitation and intact reaching and grasping (Ietswaart et al., 2006). Thus, the ability to perform simple visuomotor transformations for grasping is relatively intact in patients with ideomotor apraxia. This suggests that visuomotor transformations related to actions *with* an object are distinct from visuomotor transformations needed to merely grasp an object. In retrospect this distinction is perhaps not surprising given that many of these patients perform much better with the tool itself and therefore will need to have grasped it in the first place. A recent fMRI experiment found that AIP was activated during both object-manipulation and observation of object manipulation by others (Shmuelof and Zohary, 2006). Their findings map well onto a recent study that showed a strong correlation between impairments in the imitation of transitive gestures and the recognition of object-related gestures and hand postures (Buxbaum et al., 2005). Finally, an fMRI study showed that left IPL activation was associated specifically with the somatic perception of hand-object interactions (Naito and Ehrsson, 2006), which suggests a particular form of sensorimotor integration between limb and object states. Damage to this area might explain why some patients with ideomotor apraxia have trouble making transitive gestures even when holding a tool.

Thus, the ability to improve ideomotor apraxia by holding a tool suggests that motor memories of tool use can be contextually triggered by implicit kinesthetic cues. However, the inability to imitate transitive and nonsense gestures, and the correlation of this inability with impairment in hand posture recognition, may suggest a specific role of the IPL (and ventral premotor cortex) in visuomotor transformations for object-related actions.

It should be apparent that the concepts derived from the current computational framework for motor control could only partially explain components of the ideomotor apraxia syndrome. These patients can be considered to have the implicit procedural analogs of the declarative abnormalities in HM. The study of higher order aspects of motor behavior in healthy subjects is difficult in the laboratory setting because complex perturbations are likely to be required. A productive synergy can therefore be envisaged whereby the computational models and experimental approaches derived from studies in healthy subject are applied and adapted to patients with higher-order motor abnormalities. Such an approach is a necessary complement to functional imaging studies in healthy subjects, which rely on activation differences to gain insight into higher order motor processes. Transcranial magnetic stimulation may prove effective in inducing higher-order errors in healthy subjects but as of yet has not yielded insights comparable to those obtained from patients. A rigorous computational approach in patients with motor disorders has been and will continue to be a fruitful avenue of investigation.

Acknowledgments

The work was supported by National Institutes of Health (NIH) grant K02 NS-048099 (JWK), NIH R01-037422 (RS), and a grant from the Human Frontiers Science Foundation (RS).

References

Buxbaum, L.J., Giovannetti, T. and Libon, D. (2000) The role of the dynamic body schema in praxis: evidence from primary progressive apraxia. Brain Cogn., 44: 166–191.

Buxbaum, L.J., Kyle, K.M. and Menon, R. (2005) On beyond mirror neurons: internal representations subserving imitation and recognition of skilled object-related actions in humans. Brain Res. Cogn. Brain Res., 25: 226–239.

Caithness, G., Osu, R., Bays, P., Chase, H., Klassen, J., Kawato, M., Wolpert, D.M. and Flanagan, J.R. (2004) Failure to consolidate the consolidation theory of learning for sensorimotor adaptation tasks. J. Neurosci., 24: 8662–8671.

Clark, M.A., Merians, A.S., Kothari, A., Poizner, H., Macauley, B., Gonzalez Rothi, L.J. and Heilman, K.M. (1994) Spatial planning deficits in limb apraxia. Brain, 117(Pt 5): 1093–1106.

Corkin, S. (2002) What's new with the amnesic patient H.M.? Nat. Rev. Neurosci., 3: 153–160.

Culham, J.C. and Valyear, K.F. (2006) Human parietal cortex in action. Curr. Opin. Neurobiol., 16: 205–212.

Cunningham, H.A. and Welch, R.B. (1994) Multiple concurrent visual-motor mappings: implications for models of adaptation. J. Exp. Psychol. Hum. Percept. Perform., 20: 987–999.

Diedrichsen, J., Hashambhoy, Y., Rane, T. and Shadmehr, R. (2005) Neural correlates of reach errors. J. Neurosci., 25: 9919–9931.

Donchin, O., Francis, J.T. and Shadmehr, R. (2003) Quantifying generalization from trial-by-trial behavior of adaptive systems that learn with basis functions: theory and experiments in human motor control. J. Neurosci., 23: 9032–9045.

Fogassi, L. and Luppino, G. (2005) Motor functions of the parietal lobe. Curr. Opin. Neurobiol., 15: 626–631.

Gabrieli, J.D., Corkin, S., Mickel, S.F. and Growdon, J.H. (1993) Intact acquisition and long-term retention of mirror-tracing skill in Alzheimer's disease and in global amnesia. Behav. Neurosci., 107: 899–910.

Gandolfo, F., Mussa-Ivaldi, F.A. and Bizzi, E. (1996) Motor learning by field approximation. PNAS, 93: 3843–3846.

Halsband, U., Schmitt, J., Weyers, M., Binkofski, F., Grutzner, G. and Freund, H.J. (2001) Recognition and imitation of pantomimed motor acts after unilateral parietal and premotor lesions: a perspective on apraxia. Neuropsychologia, 39: 200–216.

Hwang, E.J., Donchin, O., Smith, M.A. and Shadmehr, R. (2003) A gain-field encoding of limb position and velocity in the internal model of arm dynamics. PLoS Biol., 1: E25.

Hwang, E.J. and Shadmehr, R. (2005) Internal models of limb dynamics and the encoding of limb state. J. Neural Eng., 2: S266–S278.

Ietswaart, M., Carey, D.P. and Della Sala, S. (2006) Tapping, grasping and aiming in ideomotor apraxia. Neuropsychologia, 44: 1175–1184.

Koski, L., Iacoboni, M. and Mazziotta, J.C. (2002) Deconstructing apraxia: understanding disorders of intentional movement after stroke. Curr. Opin. Neurol., 15: 71–77.

Krakauer, J.W., Ghez, C. and Ghilardi, M.F. (2005) Adaptation to visuomotor transformations: consolidation, interference, and forgetting. J. Neurosci., 25: 473–478.

Krakauer, J.W., Ghilardi, M.F. and Ghez, C. (1999) Independent learning of internal models for kinematic and dynamic control of reaching. Nat. Neurosci., 2: 1026–1031.

Krakauer, J.W., Ghilardi, M.F., Mentis, M., Barnes, A., Veytsman, M., Eidelberg, D. and Ghez, C. (2004) Differential cortical and subcortical activations in learning rotations and gains for reaching: a PET study. J. Neurophysiol., 91: 924–933.

Krakauer, J.W., Mazzoni, P., Ghazizadeh, A., Ravindran, R. and Shadmehr, R. (2006) Generalization of motor learning depends on the history of prior action. PLoS Biol., 4.

Krakauer, J.W., Pine, Z.M., Ghilardi, M.F. and Ghez, C. (2000) Learning of visuomotor transformations for vectorial planning of reaching trajectories. J. Neurosci., 20: 8916–8924.

Krouchev, N.I. and Kalaska, J.F. (2003) Context-dependent anticipation of different task dynamics: rapid recall of appropriate motor skills using visual cues. J. Neurophysiol., 89: 1165–1175.

Leiguarda, R.C. and Marsden, C.D. (2000) Limb apraxias: higher-order disorders of sensorimotor integration. Brain, 123(Pt 5): 860–879.

Li, C.S., Padoa-Schioppa, C. and Bizzi, E. (2001) Neuronal correlates of motor performance and motor learning in the primary motor cortex of monkeys adapting to an external force field. Neuron, 30: 593–607.

Maschke, M., Gomez, C.M., Ebner, T.J. and Konczak, J. (2004) Hereditary cerebellar ataxia progressively impairs force adaptation during goal-directed arm movements. J. Neurophysiol., 91: 230–238.

Mazzoni, P., Hristova, A. and Krakauer, J.W. (2007) Why don't we move faster? Parkinson's disease, movement vigor, and implicit motivation. J. Neurosci., 27: 7105–7116.

Mazzoni, P. and Krakauer, J.W. (2006) An implicit plan overrides an explicit strategy during visuomotor adaptation. J. Neurosci., 26: 3642–3645.

Miall, R.C., Jenkinson, N. and Kulkarni, K. (2004) Adaptation to rotated visual feedback: a re-examination of motor interference. Exp. Brain Res., 154: 201–210.

Milner, B. (1968) Pathologie de la memoire. La Memoire, Geneve, pp. 185–212.

Mishkin, M., Malamut, B. and Bachevalier, J. (1984) Memories and habits: two neural systems. In: Lynch G. and MacGaugh J. (Eds.), Neurobiology of Learning and Memory. Guilford Press, pp. 65–77.

Naito, E. and Ehrsson, H.H. (2006) Somatic sensation of hand-object interactive movement is associated with activity in the left inferior parietal cortex. J. Neurosci., 26: 3783–3790.

Padoa-Schioppa, C., Li, C.S. and Bizzi, E. (2002) Neuronal correlates of kinematics-to-dynamics transformation in the supplementary motor area. Neuron, 36: 751–765.

Poizner, H., Clark, M.A., Merians, A.S., Macauley, B., Gonzalez Rothi, L.J. and Heilman, K.M. (1995) Joint coordination deficits in limb apraxia. Brain, 118(Pt 1): 227–242.

Sakata, H. (2003) The role of the parietal cortex in grasping. In: Siegel A.M., Andersen R.A., Freund H.-J. and Spencer D.D. (Eds.), Advances in Neurology, Vol. 93. Lippincott Williams & Wilkins, Philadelphia, PA, pp. 121–139.

Schwoebel, J., Buxbaum, L.J. and Coslett, H.B. (2004) Representations of the human body in the production and imitation of complex movements. Cogn. Neuropsychol., 21: 285–298.

Schwoebel, J. and Coslett, H.B. (2005) Evidence for multiple, distinct representations of the human body. J. Cogn. Neurosci., 17: 543–553.

Shadmehr, R., Brandt, J. and Corkin, S. (1998) Time-dependent motor memory processes in amnesic subjects. J. Neurophysiol., 80: 1590–1597.

Shadmehr, R. and Brashers-Krug, T. (1997) Functional stages in the formation of human long-term motor memory. J. Neurosci., 17: 409–419.

Shadmehr, R., Donchin, O., Hwang, E., Hemminger, S. and Rao, A. (2005) Learning dynamics of reaching. In: Riehle A. and Vaadia E. (Eds.), Motor Cortex in Voluntary Movements: A Distributed System for Distributed Functions. CRC Press, Boca Raton, FL, pp. 297–328.

Shadmehr, R. and Mussa-Ivaldi, F.A. (1994) Adaptive representation of dynamics during learning of a motor task. J. Neurosci., 14: 3208–3224.

Shadmehr, R. and Wise, S.P. (2005) The Computational Neurobiology of Reaching and Pointing: A Foundation for Motor Learning. The MIT press, Cambridge, MA.

Shmuelof, L. and Zohary, E. (2006) A mirror representation of others' actions in the human anterior parietal cortex. J. Neurosci., 26: 9736–9742.

Smith, M.A. and Shadmehr, R. (2005) Intact ability to learn internal models of arm dynamics in Huntington's disease but not cerebellar degeneration. J. Neurophysiol., 93: 2809–2821.

Thoroughman, K.A. and Shadmehr, R. (2000) Learning of action through adaptive combination of motor primitives. Nature, 407: 742–747.

Todorov, E. and Jordan, M.I. (2002) Optimal feedback control as a theory of motor coordination. Nature Neurosci., 5: 1226–1235.

Tranel, D., Damasio, A.R., Damasio, H. and Brandt, J.P. (1994) Sensorimotor skill learning in amnesia: additional evidence for the neural basis of nondeclarative memory. Learn. Mem., 1: 165–179.

Wainscott, S.K., Donchin, O. and Shadmehr, R. (2005) Internal models and contextual cues: encoding serial order and direction of movement. J. Neurophysiol., 93: 786–800.

Wigmore, V., Tong, C. and Flanagan, J.R. (2002) Visuomotor rotations of varying size and direction compete for a single internal model in motor working memory. J. Exp. Psychol. Hum. Percept. Perform., 28: 447–457.

Wolpert, D.M., Goodbody, S.J. and Husain, M. (1998) Maintaining internal representations: the role of the human superior parietal lobe. Nat. Neurosci., 1: 529–533.

Yamashita, H. (1993) Perceptual-motor learning in amnesic patients with medial temporal lobe lesions. Percept. Mot. Skills, 77: 1311–1314.

P. Cisek, T. Drew & J.F. Kalaska (Eds.)
Progress in Brain Research, Vol. 165
ISSN 0079-6123
Copyright © 2007 Elsevier B.V. All rights reserved

CHAPTER 25

Motor control in a meta-network with attractor dynamics

N.I. Krouchev* and J.F. Kalaska

GRSNC, Département de Physiologie, Faculté de Médecine, Pavillon Paul-G. Desmarais, Université de Montréal, C.P. 6128, Succursale Centre-ville, Montréal, QC, H3C 3J7, Canada

Abstract: A neural-network module with attractor dynamics has been shown recently to be robust to stochastic noise in external and internal signals, and to converge rapidly onto an output signal that is an unbiased and efficient representation of the environment. We suggest here a modular network architecture with attractor dynamics that can compute the time-varying signals that are presumably required to control arm movements. The architecture is composed of several linked modules and implements the joint torque control of a planar biomechanical model of the arm, in the presence of external dynamic perturbations. The meta-network is robust to noise and to the unreliable availability of some signals and can provide feedback correction for unexpected external perturbations.

Keywords: recurrent neural networks; attractor dynamics; dynamic attractor networks; sensorimotor transformations; internal models; dynamical computations

Introduction

An important insight from early neural-network studies was the idea that complex dynamic computations and behaviors can be mimicked by artificial neural networks (NNs) as they relax to low-energy attractor points in activity space. Pioneering examples are *Hopfield nets* (Hopfield, 1982), constructed of binary computational units interconnected by symmetric weight matrices between network layers. Since Hopfield published his seminal work, network models have become more elaborate, e.g. with continuous-activation units. However Hopfield's original concept of reciprocal connectivity patterns with equal connection weights between layers is still in wide use in many recent computational models (Boucheny et al., 2005; Brunel, 1996; Brunel and Latham, 2003; Brunel and Nadal, 1998; Hinton and Salakhutdinov, 2006; Toulouse, 1987; Xie et al., 2002), because it guarantees convergence of network dynamics to a stable steady-state. The latter is achieved by the network as it rolls downhill to a low-energy valley in the landscape of its energy function. Symmetric connectivity is a sufficient condition for the existence of a Lyapunov-stability (or energy) function (Hertz et al., 1991). Interestingly, the first energy function introduced by Hopfield for his model turned out to be a mathematically formal way to express the Hebbian principle of unsupervised learning, in which only effective connections between neurones are strengthened.

Recurrent neural nets (RNNs) were introduced to allow processing and generation of time-varying

*Corresponding author. Tel.: +1 514 343 6111 Ext. 4361; Fax: +1 514 343 6113; E-mail: Krouchen@physio.umontreal.ca

DOI: 10.1016/S0079-6123(06)65025-5

signals (Hertz et al., 1991). However, these traditional multi-layer NNs have a number of drawbacks. For instance, despite the enthusiasm originally accompanying the back-propagation algorithm, it provides no guarantee of convergence by gradient descent to a global extremum. Additionally, the training results are adversely affected by increasing the length of the patterns that the network needs to reproduce (Bengio and Frasconi, 1994a, b).

Dynamic attractor nets (DANs) (e.g., Latham et al., 2003) are a form of stochastic RNNs using radial basis functions (RBFs). They have selectively distributed symmetric weight patterns between layers, which are closer to the CNS blueprint, whose distribution are described by specific functions which reflect the experimentally observed broad tuning of single cortical neurones, as opposed to the full random connectivity of traditional RNNs. Their symmetric connectivity guarantees that DANs will converge on specific sub-manifolds in neural state space.

DANs have a number of properties and neural processing principles which are reminiscent of key results in experimental neurophysiology. Among these are population-level encoding and population-vector computation (Georgopoulos et al., 1983; Schwartz and Moran, 1999), and the emergence of gain fields as a computational mechanism to achieve sensorimotor coordinate transformations (Redish and Touretzky, 1994; Andersen et al., 1997; Salinas and Thier, 2000). In addition, this class of computations capture essential aspects of functional neural interaction. These models include interaction of several layers that represent sensory stimuli and their internal representations at various levels of abstraction. Starting from basic primitives of the current sensory input, such as the angular orientation of linear components of a visual image, subsequent model layers become less and less anchored in the immediate context and extract increasingly general and global features of the input. The functional dynamic interplay of input/output and hidden layers resembles the analysis/synthesis interactions within and between primary and secondary cortical areas.

Computational challenges in motor control

To use the arm to interact successfully with the environment, the motor system must successfully solve a number of well-known and non-trivial computational problems. The ultimate goal of those computations is to produce the time-varying patterns of muscle forces necessary to cope with the complex dynamical mechanical properties of both the multi-articular, multi-muscle arm motor plant and the environment. To achieve this goal, it must possess robust computational mechanisms to process noisy sensory inputs, to integrate information from different sensory sources and to perform a series of sensorimotor coordinate transformations. Among those transformations, the motor system must find some form of solution to the *inverse kinematics problem* between a desired spatio-temporal trajectory of hand displacement through space and the required sequence of shoulder, elbow and wrist joint rotations. Similarly, the motor system must find a solution to the *inverse dynamics problem*, i.e. how to generate the temporal sequence of muscle forces required to produce the desired movement trajectory. Furthermore, successful motor performance often requires the ability to anticipate the future state of the motor apparatus and the environment, such as during prey capture or its more civilized contemporary equivalents like catching a ball. Such motor behavior requires carefully coordinated integration of proactive *feedforward* and reactive *feedback* mechanisms. For instance, to perform a skilled movement, the motor system must know not only what forces to produce, but equally importantly, must know how the most recently applied forces affected the limb's position and motion. This knowledge permits feedback-mediated online error correction and feedback error-based adaptation and motor skill acquisition. Both feedforward and feedback processes depend on multi-modal sensory signals that are updated at different latencies.

Given these computational considerations and the highly parallel, distributed architecture of the sensorimotor control system, with its multiple convergent, divergent and reciprocal connectivity patterns, it is simplistic to regard motor control as a purely serial and unidirectional forward computation. We propose here a mechanism incorporating predictions and actual sensory inputs in a seamless bi-directional computation.

Recurrent attractor-dynamics models allow rapid output convergence to unbiased and efficient sensory estimates (Latham et al., 2003) and generate dynamic intermediate response patterns despite not explicitly dealing with the representation of time. Here we propose to use similar ideas to build a biologically inspired functional model of the time-varying sensory–motor transformations required in the skilled control of limb movement.

The CNS cannot provide instantaneous real-time processing, because neural information transmission occurs at a finite rate determined by inherent factors such as axonal conduction times, synaptic delays and the biophysical and bioelectric properties of neurons. This chapter proposes a solution by which the CNS can approximate the computation of time-varying sequences of sensory estimates and motor output control signals. DANs provide the basic building-blocks (Fig. 1). They are compatible with cortical anatomy and physiology, and perform a well-defined class of computations that are sufficient to provide for many input–output transformations required in motor control. Here, we extend the DAN architecture to encode and predict time-varying signals. We demonstrate that it is possible to cope with this problem in quasi-statics, i.e. to implement motor task-related dynamics via a time-varying sequence of transient equilibria of neural activity driven by attractor dynamics.

Methods

The basic building unit

The model presented here is built using several identical basic DAN modules like those in Deneve et al. (2001) and Latham et al. (2003). For the convenience of the reader, an overview of the architecture of the basic module is presented here, and they are referred to the original articles for a detailed description.

In Deneve et al. (2001), the DAN consisted of a hidden layer and three visible input/output layers that encoded eye position, and the retinal position and head-centered position of a visual stimulus, respectively (Fig. 1). The positions were defined as three angular (circle-symmetric) input/output signals whose range was $0 \ldots 2\pi$. The positions were represented by N neural units in each visible layer, indexed by their preferred direction (PD) expressed as the circular angle x_j:

$$x_j = 0,\ dx,\ 2 \cdot dx,\ \ldots j \cdot dx\ \ldots\ (N-1) \cdot dx$$

where $dx = 2\pi/N$.

Each visible layer was symmetrically connected with the hidden layer. Connection weights were such that activation of a single unit in a single visible layer would activate a set of corresponding columns of units across the hidden layer at a level that diminished with the distance of each column of hidden-layer units from the position of the activated unit in the visible layer. The spread of activation across the hidden-layer columns exhibited a bell-shaped (e.g., Gaussian, or truncated cosine) activity profile (Fig. 1).

The DAN internal dynamics were described by a set of coupled non-linear equations, and a non-linear divisive normalization function of the "softmax" type. Detailed analysis of this intrinsic dynamics is beyond the scope of this chapter (Latham et al., 2003).

We used the parameter values given in Deneve et al. (2001). Latham et al. (2003) showed that with such an appropriate choice of parameters, the DAN would be attracted to and settle at a smooth hill activation pattern, determined by the set of connection strengths. As in Hopfield's original work, symmetric synaptic weights guarantee that at each subsequent iteration the network activity is closer to the asymptotic fixed point. Latham et al. (2003) provide a theoretical confirmation of model convergence. Convergence is also rapid, and provides viable estimates of encoded network input stimuli and motor output within two to three iterations.

Deneve et al. (2001) demonstrated that DAN population codes are unbiased and efficient in the

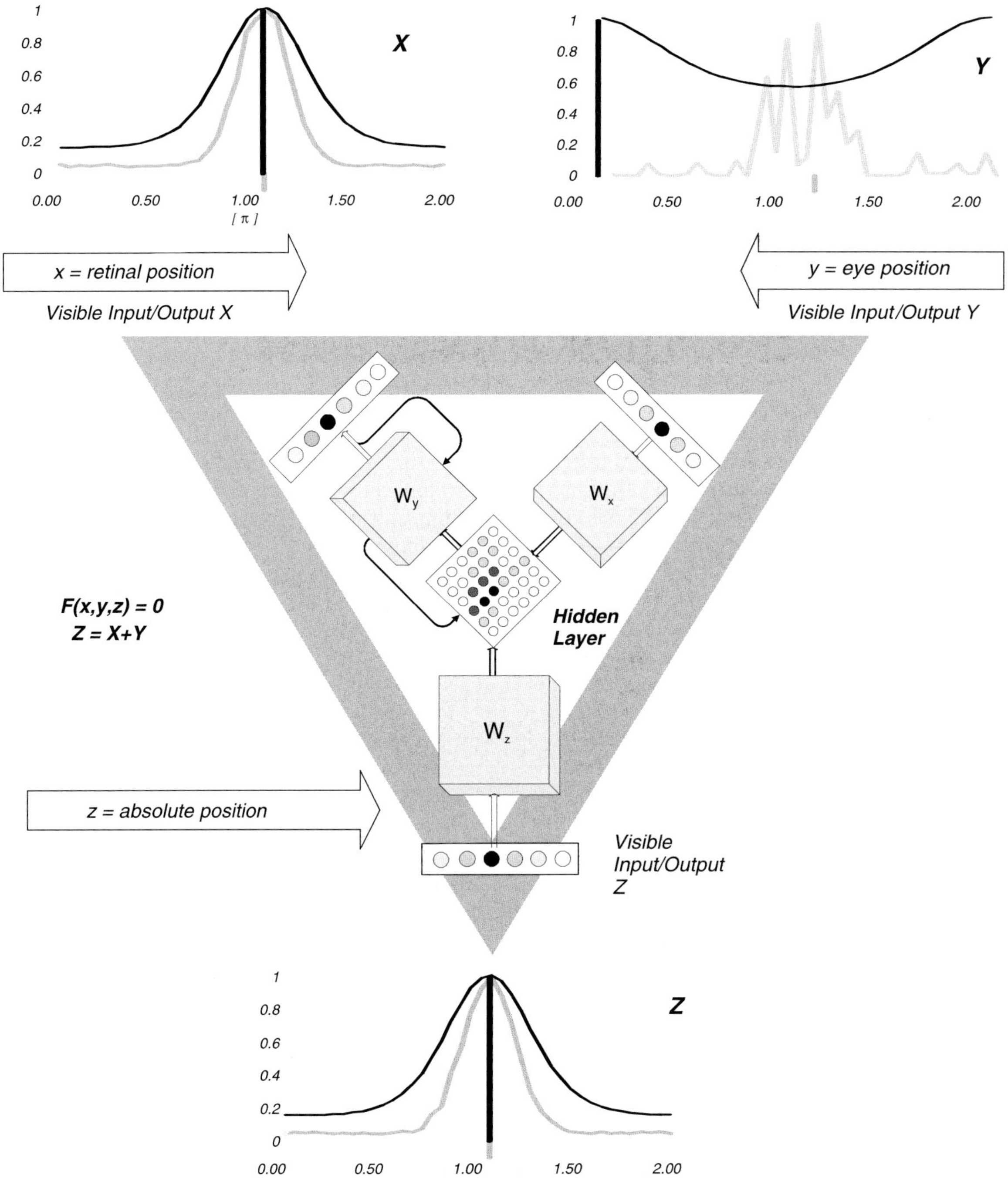
X
Y
Z
[π]
x = retinal position
Visible Input/Output X
y = eye position
Visible Input/Output Y
W_y
W_x
W_z
F(x,y,z) = 0
Z = X+Y
Hidden Layer
z = absolute position
Visible Input/Output Z

presence of encoding noise. They initiated the module with noisy population-coded signals on the eye-position and retinal-position layers and random noise on the head-centered layer. Each neuron in the eye-position and retinal-position layers encoded the angular position relative to its own PD in a stochastic fashion. Each neuron's discharge rate was perturbed by identically distributed Poisson or Gaussian noise terms with mean levels determined by the tuning curve of each unit for a given value of the encoded stimuli. The module rapidly relaxed from this initial state to the attractor state and extracted a smooth best estimate of all three visible-layer signals, including head-centered position, within three computational iterations.

Latham et al. (2003) derived the general DAN conditions that assured that network estimates would be *unbiased and efficient*. The latter means that the network preserves the first-order Fisher information conveyed by the stochastic population code of the environmental stimuli. The variability of the network estimates is *not much higher than the minimal possible variance for unbiased estimates* and a given statistical characterization (distribution type, variance, correlations etc.) of noise terms. The latter minimum is provided by the Cramér-Rao bound.

Bi-directional computation: what determines the direction (and result) of the computation that a module performs at any given moment?

An important feature of the DAN module architecture is that its symmetric connectivity permits bi-directional computation between each visible layer and the hidden layer. The module in Fig. 1 computes the transformation $y = z-x$ which is similar to Deneve et al. (2001). The visible population X encodes the retinal position r, the population Y encodes the eye position e, and the population Z encodes the absolute head-centered position a. The X and Y visible populations serve as input layers that encode simple features of the primary sensory signals and project them orthogonally onto the hidden layer. Conversely, the Z population is "synthetic" in that it extracts its own associated feature from the activity projected onto the hidden layer by the X and Y populations, by virtue of its connectivity pattern with the hidden layer.

However, since the network updates *all* input/output layers via symmetric connectivity matrices between the three visible layers and the hidden layer, the module could, in principle, compute any of the three transformations:

$$z = x + y \quad y = z - x \quad x = z - y$$

Consider a simple example, in which the initial state of all three visible layers is set so that $x = y = z = 2$. If that initial state does not lie along the attractor basin for the DAN, the outcome of the network computation could be either one of: ($x = 2$, $y = 2$, $z = 4$), ($x = 2$, $y = 0$, $z = 2$) or ($x = 0$, $y = 2$, $z = 2$).

This implies that although the X and Y populations may often serve as input signals to the module, they could also potentially extract those primary features from the module by virtue of their symmetrical connections with the hidden layer. A critical issue therefore concerns the factors that could influence the direction in which the network computes.

Fig. 1. Dynamic attractor recurrent neural network (DAN) — the basic building unit. Our model is built using basic modules of the same architecture as in Deneve et al. (2001), where the DAN consisted of a hidden layer and three visible layers that encoded vision- and proprioception-related angular positions. The network does not explicitly produce these values. It encodes them implicitly by the fact that units in the visible populations that have preferred stimulus orientations (PD) close to the encoded values are maximally active. The visible population X encodes the retinal position r, the population Y encodes the eye position e, and finally the population Z encodes the absolute head-centered position a. The circles in each visible layer represent individual neurons, whose activity level is illustrated by different shades of gray (e.g., maximal in black, minimal — white). The activity patterns in the hidden layer are originally due to the interplay of activity in the visible inputs. The hidden layer then determines the activity in the visible output layer(s) according to the network's dynamics equation (see text). Any of three transformations: $z = x+y$, $y = z-x$, $x = z-y$ could be computed (see text); in the insets, the module computes the difference $y = z-x$, achieved by setting the initial activity levels in Y substantially more noisy than in the X and Z populations (see text). Abscissa on all insets: unit activation in the visible layers.

One simple factor that will affect the overall direction of computation of the module is to continually update the signals on one or two of the visible layers, leading to a time series of changes in the state of the layer(s) that are not updated from external signal sources. For instance, if the retinal and eye position signals in layers x and y respectively in the original network configuration of Deneve et al. (2001) had been changed in a series of time steps, the result would have been a time series of $z = x + y$ transformations that yielded a spatio-temporal trajectory of estimates of the head-centered location of the visual stimulus. By providing external signals to update the state of one or more visible layers, the module will continually update the state of the other visible layers, making them the de facto outputs and establishing a net direction of computation across time.

However, a more fundamental and important factor is the reliability (quality and quantity) of the input information initially encoded in each visible layer. This determines the precise shape and location of the equilibrium attractor state of the module. Each visible layer is a population estimate of the true nature of a particular signal. If the reliability of that estimate is significantly poorer in one layer than the others (i.e., if it is noisier or weaker), the attractor dynamics of the network will result in a greater change in the state of that layer compared to the others once the network relaxes to its attractor.

A concrete example of that process is illustrated in the insets in Fig. 1. All three visible layers were initialized so that $x = y = z = \pi$ (Fig. 1, gray lines). However, the initial signal in layer Y was considerably noisier than in layers X and Z, even though the total strength of each signal (area under the curves) was approximately equal. The greater noise in layer Y reflects greater uncertainty about the true value of that input signal. After only three iterations however (black lines), the activity in the Y population changed substantially from a noisy ensemble that encoded a stimulus estimate around π radians to a smooth hill of activity encoding a value of 0 radians. The X and Z population activity also became somewhat smoother. In contrast to the Y population however, there was little change between their respective initial and final estimates, which remained at values around π radians. The network, based on that set of initial conditions, had computed the difference transform $y = z - x$ to extract the best estimate of eye position from the more reliable signals for retinal and head-centered position. The network would also compute the same transform if the initial signal on layer Y had the same relative noise level as on layers X and Z but a much smaller peak amplitude. If, in a different set of conditions, the initial states of layers Y and Z had been reversed because the current estimate of Y was more reliable than of Z, the network would tend to relax toward the solution $z = x + y$. Those were exactly the conditions in the simulations by Deneve et al. (2001).

This simple didactic example illustrates the important point that the overall direction of a module's computation at any given moment will depend on the reliability of the available signals in each visible layer, which determines the attractor state and which visible layer will show the greatest change at equilibrium, i.e., which layer will be the de facto output layer at a given moment. The network always evolves toward the best estimate of the nature of the three visible-layer signals, determined by the relative absolute amplitude, depth of tuning, noise and signal-to-noise ratio of each signal. All of these factors influence the confidence that the CNS can have in the encoded value of the signal on each visible layer, and also inherently influence the direction of overall network computation. This indicates that the DAN module architecture provides a substrate for a truly bi-directional computation in accord with the information available at a given moment.

What are the broader implications of the DAN module architecture for CNS function, especially the cerebral cortex? The cerebral cortex is organized into networks of functional regions that are extensively interconnected by convergent, divergent and reciprocal axonal projections (whether the reciprocal connections are truly computationally symmetrical is not yet clear). At any given time, different combinations of external and centrally generated signals are available to each functional node in the network to process, due to transmission delays, variable availability of external inputs and central processes. However noisy or unavailable some signals may be at a given moment, the CNS must often continue to perform its functions as smoothly as possible in real time. This can be achieved by a highly distributed and redundant system, whose nodes (i.e., modules) can function in one direction or another depending on local signal availability and reliability at each node within the network. For instance, a given modular representation of the environment (i.e., an 'internal model') could potentially perform either a forward or an inverse computation (Wolpert and Kawato, 1998) depending on the information available on its visible layers, according to the current demands of the task.

In much the same way as in the example of Deneve et al. (2001), DAN modules could also process and interpret noisy and ambiguous sensory input. For example, neurons in area MT encode the local component motion in noisy visual flow stimuli (Shadlen and Newsome, 2001; Ditterich et al., 2003). This noisy local information projects to parietal cortex area LIP, where it appears to be integrated across time and across the visual scene until sufficient information is accumulated to permit a perceptual decision based on a thresholding mechanism (Gold and Shadlen, 2002; Mazurek et al., 2003; Hanks et al., 2006; Lo and Wang, 2006). The DAN architecture provides an alternative implementation of such a mechanism. In a seamless way, the DAN dynamics would shift from one attractor point to a different one, determined by the moment to moment quality of the sensory estimates converging onto its input populations. The new attractor point would then be propagated further in the cortical circuitry to determine a perceptual decision or action.

Extension of the model's architecture to compute time-varying signals

We now extend the architecture in Deneve et al. (2001) to code and compute time-varying signals in motor control such as position, velocity or applied torques. To apply DAN architecture to motor control, we introduce some algorithmic extensions.

Most importantly, calculating movement trajectories involves the production of a sequence of estimates of all signals that are correlated as they refer to adjacent steps in time. This is achieved by reinitializing the activity in each module at each time step as in the original static model, but only in the populations coding for input signals. This is a step toward achieving continuity of computed trajectories. Moreover, Deneve et al. (2001) encoded angular signals. Time-varying signals in motor control, such as joint position, can also be angular. However, in general to achieve the continuity of a computed signal which is neither strictly positive, nor angular or periodic, we applied the scaling:

$$\bar{x} = x \cdot K$$

$$K = \frac{\pi}{\max(x) - \min(x)}$$

where the original range of values x is converted to that of an angle-like signal. The outputs were scaled back. Note that this allowed us to achieve two important goals:

(a) to keep the connection strengths W from Deneve et al. (2001), and hence to avoid confounding the analysis by reparameterization of the model;
(b) to achieve a straightforward normalized representation of any signal range by populations of the same size N.

In the original Hopfield network, connection weights required no training. The network was set up in such a way that, for every initial state, it would rapidly converge to the corresponding attractor with prescribed properties (e.g., a given input–output transformation). They were mathematically expressed, which turned the memory patterns stored by the network into fixed points in the network's dynamics. Hence, weight patterns could also be seen as the implicit result of training which may reflect either biological or physical assumptions as to which were the underlying principles (e.g., The Hebb rule) or which cost (energy) function was minimized.

In the DAN modules used here, weights are also mathematically expressed. The module in Deneve et al. (2001) implements a linear transformation between signals encoded over linear ranges. This is sufficient for the purposes of the present work. Indeed, transformations using linear combinations of visible environmental variables can be applied to many situations. However, it is noteworthy that we have also been able to obtain analytic solutions for synaptic weight functions that can implement non-linear input/output ranges or transformations, such as the logarithmic representation of auditory stimulus frequency in the auditory cortex, or muscle output force as a non-linear function of length/velocity (data not provided here).

Meta-network simulation

In this section, we will show how time-varying signals can be computed from their initial conditions and model input (e.g., desired trajectory).

We simulated a simple single-joint motor task involving flexion and extension movements of the elbow joint from a central starting angle (Fig. 2A), either without an external perturbing force field (null field) or with a velocity-dependent viscous torque applied to the elbow joint. We assumed an ideal bell-shaped velocity profile (Fig. 2B). A desired joint angle position trajectory, sampled in a series of time-step via-points, was formed by integrating the ideal velocity profile (Fig. 2B).

The computation was implemented by a recurrent meta-network composed of four functionally identical DAN modules. Each module included three visible layers of 40 units each, and a 40×40 unit square-matrix hidden layer. Starting with initial conditions for position and velocity (Fig. 3), the modules provided estimates of current velocity error, current acceleration error and current torque, respectively, at each time step. In each time step, each module in the sequence was allowed to relax to its attractor during three computational cycles before activating the next

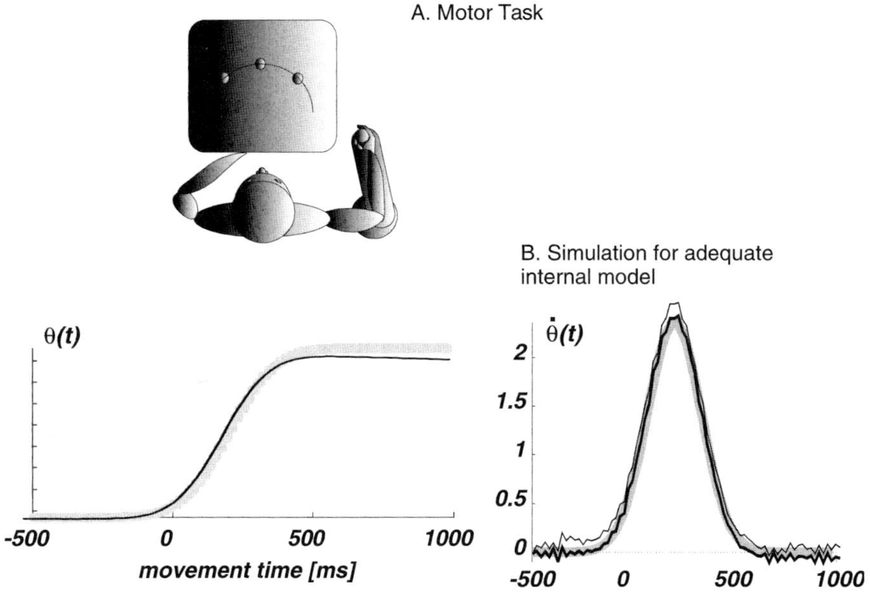

Fig. 2. Motor psychophysics experiment. (A) Task description. Human subjects performed a one degree-of-freedom (flexion/extension) elbow movement task, using the handle of a robotic manipulandum. Elbow angle trajectory was acquired and mapped onto PC screen positions along an arc. A subject was instructed to attain a desired elbow position within a specified time-window and hold steady for a certain period. Arm control was challenged with external force perturbations generated by the manipulandum. Perturbations were of viscous nature (manipulandum torque was proportional to the angular velocity of the subject's elbow joint). (B) Simulation of a skilled subject's performance in a null field (no external force field). Left: Desired elbow angular position (θ) profile (gray) and elbow angular trajectory achieved by the meta-network (black; average of 10 simulations). Time 0 is the time of movement onset. Right: Desired elbow angular velocity ($\dot{\theta}$) profile (gray), and the elbow angular velocity achieved by the meta-network (black). The thick black line is the mean velocity profile from 10 simulations, and the thin black line is the s.d.

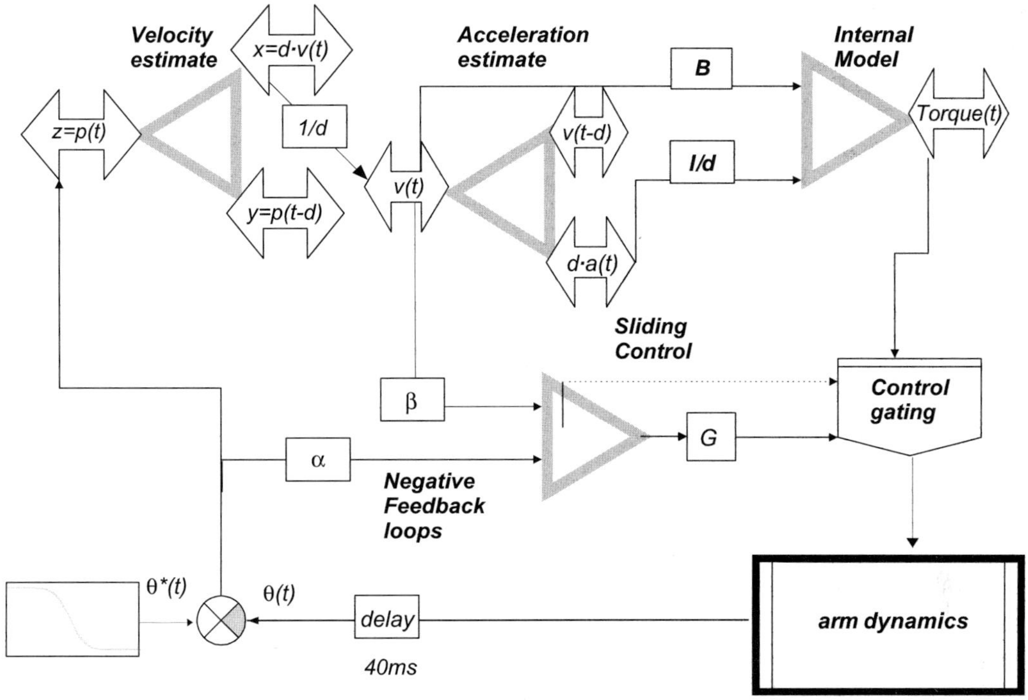

Fig. 3. Functional architecture: the modular control system and the limb plant. The modular neural control model is a meta-network built using four basic modules of the same class as in Fig. 1. At each time step, the first module (upper left) converts a position error signal p (the difference between the current and desired angular positions at the start of each time step) into an estimate of the current velocity error. The position error at the previous time step is coded by the y layer. The next module converts this to an estimate of current acceleration error. The third module is an internal model of plant and environmental dynamics that estimates the current joint torque error. A fourth module implements sliding-mode control to provide feedback error correction when the internal model is inadequate for the dynamical conditions of a given trial (see text). In the comparative sum operator at the bottom left one of the pie wedges is gray to mean that a subtraction is taking place.

sequential module in the meta-network. Successive recurrent iteration of the entire meta-network was performed in a series of discrete time steps from the beginning to the end of the desired trajectory. This yielded time-varying estimates of output signals that correspond to movement position, velocity, acceleration and required joint torques.

According to Newtonian mechanics the second order equation of the plant, i.e. the subject's elbow joint angular motion, is described as a system of first-order (state-space) ordinary differential equations (ODEs) ("Arm Dynamics" in Fig. 3):

$$\dot{x}_1(t) = \dot{\theta}(t) \equiv x_2(t)$$
$$\dot{x}_2(t) = \ddot{\theta}(t) = \frac{1}{I}[u(t) - B \cdot x_2(t)] \qquad (1)$$
$$\frac{dx}{dt} \equiv x, \quad \theta(t) \equiv x_1$$

where $\theta(t)$ is actual subject/manipulandum elbow angular position at a given time t, I the plant's total inertia (of the manipulandum handle and subject's forearm) and B the viscous perturbation torque gain which is negative ($B = -1$) in the assistive and positive ($B = 5$) in the resistive condition.

To estimate the torque required to achieve a desired trajectory the CNS is hypothesized to build and maintain *internal models* (IMs) of arm dynamics, internal and external loads applied at the joint during movement and static postures. According to the second equation in (1) the IM provides an estimate of the required torque $u(t)$ as:

$$u(t) = I \cdot a(t) + B \cdot v(t) \qquad (2)$$

where $a(t)$ is the estimated error between actual and desired angular acceleration, and $v(t)$ is the estimated error between actual and desired angular velocity. Equation (2) depends only on the estimates of the elbow joint's acceleration and velocity. The IM is also applicable for the more general case of the pursuit of a non-stationary target. Note that Eq. (2) is in essence a transform of the form: $z = x + y$, where: $z = u$, $x = I \cdot a(t)$ and $y = B \cdot v(t)$.

In this simulation, we assume that a subject is well trained in the described tasks and therefore our model network implemented the dynamics (2) using the best available estimates of the I and B parameters. The latter were also used in the plant simulation. However, there were also substantial differences between the IM and the plant simulation. The limb plant was simulated by a system of conventional first-order ODEs, integrated by the Runge-Kutta method. In contrast, the IM of the limb plant was computed by a DAN, based on either a centrally generated *estimate* or a *delayed* peripheral signal (via proprioception or vision) of angular kinematics.

The IM has a central place and role in the modular *meta-network* neural control circuitry in Fig. 3. We built the meta-network using three serially connected basic modules of the same class as in Fig. 1. For simplicity, the computation is assumed to occur in discrete time steps of length d, which was set to 20 ms in this simulation.

The desired joint angle position trajectory $\theta^*(t)$, $t = 0 \ldots T$, expressed as a series of scalar time-step via-points, is compared at every 20 ms time step t with the actual joint angle position of the plant $\theta(t)$ to yield position error $p(t)$ as:

$$p(t) = \theta^*(t) - \theta(t) \tag{3}$$

Simulation is started at zero initial conditions for all unknown quantities. A first module (top left on Fig. 3) uses the estimates of current position error $p(t)$ and previous position error $p(t\text{–}d)$ to produce an estimate of the difference $v(t)$ of current velocity to the desired angular velocity, according to:

$$d \cdot v(t) = p(t) - p(t - d) \tag{4}$$

Notice that Eq. (4) is a first-order approximation to differentiation.

The second module (top middle on Fig. 3) uses the output of the first module and the previous estimate of velocity error to estimate acceleration error $a(t)$, once more using approximate differentiation:

$$d \cdot a(t) = v(t) - v(t - d) \tag{5}$$

The estimated velocity and acceleration errors are used in a third module (top right — "Internal Model" in Fig. 3) that computes the estimated torque as specified by Eq. (2).

A key aspect of the operation by the neural assembly is reliance on predictions of the sensory reafference that would result from the computed control output. The motor behavior in the task thus can be achieved, at each time step, in essentially *open loop* until the actual consequences of actions onto the environment become known. In the simulation done here, delayed feedback starts to arrive after an initial time period of 40 ms. Since position changes in a monotonic fashion, the presence of feedback delay is less of a problem. With an adequate IM, the torque output computed on the basis of estimates of kinematics terms that are slightly biased is still adequate to produce a successful movement.

Finally, a fourth module (bottom middle — "Sliding control" in Fig. 3) computes sliding-mode control torques (Hanneton et al., 1997). The sliding (composite) variable $z(t)$ is also proportional to control error terms:

$$z(t) = \alpha \cdot p(t) + \beta \cdot v(t) \tag{6}$$

Once again, Eq. (6) takes the familiar form $z = x + y$.

This sliding-control module provides a *feedback* error compensation mechanism. The generation of motor output using the IM torque estimates operates in open-loop or *feedforward* mode, based on its own best estimate of the dynamics of the plant. In contrast, the sliding-control module *reacts* to the delayed sensory processing that provides direct knowledge of the actual performance of the motor plant of the environment and generates a torque signal proportional to the sensed error in position and velocity. When performance is skilled, the sensed error is small and the sliding-control module has no impact on motor output. However, when the internal model of plant dynamics is inadequate, such as when a subject encounters an external perturbing force field for the first time, the IM will not produce the appropriate torques. In the present meta-network, when the sensed error exceeds an arbitrary threshold, the output from the IM is gated off and control is mediated via the torque output signals from the sliding-control module. In the present simulations, the subject is presumed to be skilled and the IM generates torques adequate for the perturbing torques confronted by the plant. However, we will demonstrate the effectiveness of the sliding-control module in compensating for the unexpected absence of the perturbing torque in 'catch' trials.

Behavioral data

To assess the performance of the extended network, we compared the results from its simulations to the behavior of human subjects in a motor task.

A single-joint movement task was used (flexion-extension movements of the elbow) in the presence of perturbing external loads. Human subjects performed elbow movements with one degree of freedom using the handle of a robotic manipulandum (Fig. 2). Elbow angle trajectory was acquired and mapped onto monitor screen positions along an arc. The subjects were instructed to attain a desired elbow position within a specified time-window and to hold that target angle steady for a certain period of time. Arm control was challenged by external torques generated by the manipulandum, which were proportional to the instantaneous angular velocity of the subject's elbow joint rotation. They were applied either in the direction of elbow rotation (assistive viscous torque) or in the opposite direction (resistive viscous torque).

Results

The described extended network was first validated by reproducing the results from the unperturbed behavior of the subjects in the 1 DOF elbow rotation motor task. Figure 2B shows the results of simulation for an internal model for $B = 0$ (i.e., null field). The network generates a time series of torque outputs to displace the elbow along a sequence of quasi-equilibria as the cascade of DAN modules iteratively compute the best estimates of the velocity and acceleration error signals at each time step. When there was no noise on any visible layer, the meta-network reproduced the desired joint angle trajectory nearly perfectly, with minimal variability or error. The introduction of noise resulted in variability of the motor output of the network (Fig. 2B). Nevertheless, the network succeeded in generating elbow movements with a range of variability reminiscent of normal human motor performance. By substituting $B = 0$ by $B = 5$ or $B = -1$ in the internal model module, the network is equally capable of generating desired elbow movements, thereby simulating the performance of a skilled subject in the resistive and assistive viscous force fields (data not shown).

After practice in the described tasks with external force fields, the human subjects learned to produce a temporal pattern of elbow torque output to counteract the external force perturbations. Interspersed among the sequence of trials with the expected external force field, we also presented 'catch' trials in which there was no external force perturbation (i.e., $B = 0$). In such trials the IMs applied by the subjects in anticipation of external forces, according to the previously conditioned context, were inadequate. The task dynamics unexpectedly changed from assistive or resistive viscous loads to a zero load. Figure 4 illustrates the simulated and actual human performance in 'catch' trials.

Figure 4A illustrates the simulated torque profiles. The thick gray lines in Fig. 4A indicate the torques estimated by the IM to produce the desired elbow rotations in the resistive (left) and assistive (right) viscous fields. When the elbow encountered a resistive viscous torque, the network must generate a large torque pulse in the direction of movement to overcome the velocity-dependent external force field. In contrast, in the assistive field, the network must generate a small initial torque pulse to initiate the elbow rotation but then must generate a torque in the opposite direction to decelerate and stop the movement at the target against the assistive external force field. The torque estimates produced by the IM module

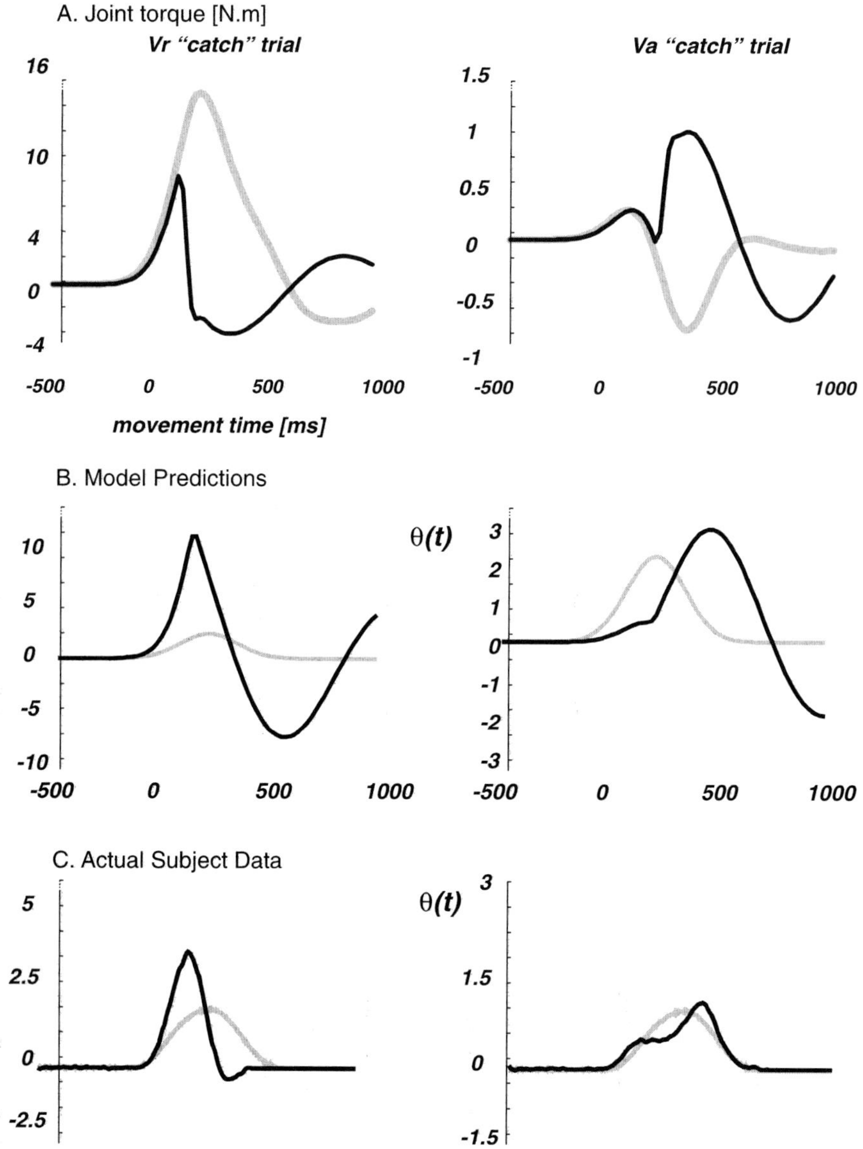

Fig. 4. Model predictions for 'Catch' trials. (A) Simulated torque profiles: temporal evolution of torque is shown in gray for the ideal case of an adequate IM. In any trial the two possible modes of operation described in the text proceed in parallel: the feedforward command from the IM, and sliding-mode feedback command. The evolution of torque actually produced by elbow muscles (black) starts by following the torque, resulting from applying an IM as indicated by task context. When the error criterion exceeds a threshold, muscle force output is dictated by the computed sliding-mode torque. (B) Simulated human performance: velocity is shown in thick gray for the ideal profile and in black for a 'catch' trial. The control system (Fig. 3) detects model mismatch and switches to sliding-mode control with comparable performance and delays to that experimentally observed in human subjects. (C) Actual human performance in the same scenarios as in B: here for a well-adapted subject, the average velocity (thick gray traces) from all perturbed trials is the equivalent of the ideal profile. Velocity traces recorded during 'catch' trials are shown in black. Externally applied force perturbations (see the *Behavioral data* paragraph above). *Vr*, resistive viscous; *Va*, assistive viscous.

(thick gray traces) of the meta-network in the two fields closely matched the ideal torques required to perform the desired movements in the corresponding external force fields.

In 'catch' trials, the expected external force fields were not present. The torque signals to the plant (black traces) were initially determined by the output of the IM. However, those torques produced movement velocities that were either much faster (Fig. 4B, left, black line) or much slower (Fig. 4B, right, black line) than the desired movement (Fig. 4B, gray lines). The overall network (Fig. 3) detected the mismatch between expected and sensed performance and switched to sliding-mode control after a criterion error threshold was crossed. At that point, the torque signal to the plant was determined by the output from the sliding-control feedback module. The resulting motor output performance of the network was strikingly similar to the kinematics of human subjects in 'catch' trials (Fig. 4C). Here the sliding-mode control is just one choice of a robust feedback mechanism that would encompass cases where the internal model embedded in the feedforward neural circuit would be inadequate. Notice that in each simulation the network shown on Fig. 3 had two possible modes of operation:

1. The external forces were as anticipated and the embedded (learned) IM was adequate.
2. The external forces were novel or *not* as anticipated (e.g., in "catch" trials).

In the latter case the feedback circuit took over after an initial accumulation of position and velocity errors, sufficient to trigger switching from pre-programmed (feedforward) to sliding-mode control. The nature of the sliding-variable makes it appropriate as a criterion for switching since it is a measure of the system's deviation from the desired movement.

Discussion

Deneve et al. (2001) demonstrated that reciprocally connected recurrent NNs that use population codes and compute functions of their input signals have many advantageous computational features. They possess multi-dimensional attractors that can extract the optimal estimate of the true nature of noisy signals. They can perform function approximation of linear and non-linear transformations, i.e. they can extract a third ('output') signal from two input signals. They can also perform cue integration, i.e. they can simultaneously find the optimal estimate of the true values of several noisy signals. The networks perform the computations efficiently (i.e., as reliably as possible), and rapidly, by relaxing to an attractor minimum that is near to the maximum likelihood estimate within three iterations.

Those features of DANs are ideally suited for many of the computational challenges faced by the motor system. Most motor control models assume that the motor system must make a series of linear and non-linear transformations to convert a central motor plan into a motor output command that will produce the required time-varying muscle activation patterns. It must integrate sensory signals from different modalities (e.g., proprioception, cutaneous receptors, vision) encoded in different sensory coordinate frameworks. Sensory feedback signals from these various sources are available at different delays. Furthermore, the reliability of those signals may vary from moment to moment due to the properties of peripheral sensory receptors and stochastic neural noise at every level in the system.

We show here that a meta-network comprised of a series of DAN modules similar to those of Deneve et al. (2001) can perform the computations needed to convert a kinematic plan into a dynamic output signal adequate to drive a motor plant. It accomplished this task by converting the motor control process from a continuous dynamic computation into a time series of quasi-static intermediate positions between the start and endpoint of the movement. At each time step, the network received as an input an estimate of the position error between its current position and the next desired position along the trajectory. Successive modules rapidly relaxed to equilibrium states that signaled the best estimate of the change in velocity and acceleration, and a final 'internal model' of task dynamics estimated the torque change necessary to drive the plant model to the next desired intermediate position (Richardson et al., 2005; Slotine and Lohmiller, 2001; Lohmiller and Slotine, 1998).

The meta-network was able to generate movements of the plant model that corresponded reasonably well to the desired ideal movement despite the fact that noise was deliberately added to each signal at each time step. This stability of performance was assured by the reciprocal connectivity between the visible layers and the hidden layer of each module and the resulting attractor properties which allowed each module to find the near optimal estimate of the signals on all visible layers at each time step in the movement. As a result, the net direction of computation in each module was not always fixed and unidirectional throughout the entire time course of the movement. Instead, it was influenced by the relative reliability (amplitude and noise level) of each of the visible-layer signals at each time step. Nevertheless, the introduction of a new position error at the start of each time step continued to drive the entire network toward the desired movement endpoint.

When the internal model of the motor plant and environment was correct, the meta-network could simulate movements in a null field and in resistive and assistive viscous fields, each of which required different temporal patterns of output torques. When the internal model was incorrect, such as in catch trials in which the expected viscous force field was not present, a parallel sliding-mode control module monitored the estimated errors in position and velocity and intervened when the accumulated error exceeded a certain criterion threshold. This permitted feedback error-mediated corrections for large motor output errors resulting from an inadequate internal model of the task dynamics.

One interesting feature of the network is that it incorporates a feedback-mediated position error signal at the beginning of the network cascade. This feedback signal is delayed by 40 ms in this particular model instantiation. This is consistent with the known latency of motor cortex responses to perturbations of the limb. The implication of this arrangement is that a movement is initiated by a largely feedforward computational process until such time as feedback signals about actual movement kinematics become available. Once the feedback arrives after the transmission delay, the network continues to function primarily in a forward computational direction to drive the limb to the next desired state but feedback about the actual current position is used to continually update the estimate of current position error.

The proposed computational organization is a modular neural implementation of environmental laws (IMs). By enforcing relations between unreliably encoded signals, the network dynamics drives estimates toward a coherent representation of the world. The network architecture thus has Kalman-filter-like properties (Kalman, 1960), offering a modular and coherent representation of the environment. Such properties are highly desirable for neural decoding.

Just as in a classical Kalman filter the output precision depends on several factors. First, how trustworthy is an estimate of sensory input? This is reflected in the amplitude and stochastic variability of the population activity in the input layers. Second, what is the network's own inherent capacity for encoding precision (e.g., number of neurons in the assembly)? For example, a discrepancy in coding precision between the input and hidden layers (as for instance in Deneve et al., 2001) may lead to an important bias in the model output. Finally, as in the Kalman filter philosophy we face the choice to use the model with any attractor (fixed-point) solution if the sensors are too noisy, and conversely we may enforce a given estimate based on reliable knowledge of the environment.

The DAN architecture exploits the nested-dynamics properties of neural assemblies built by interconnecting several basic modules, in a biologically plausible way of combining modalities, which does not lead to combinatorial explosion. The formalism is simple and coherent with many experimental studies which suggest the ubiquity of its main underlying computation — population-level summation. The recurrent structure of the network is also consistent with the recurrent nature of cortico-cortical connections.

The current model has two advantages when compared to previous work using conventional NNs to capture non-linear dynamics. First, a typical NN does not explicitly implement knowledge of natural laws, e.g. how to form velocity from position. This leads to higher dimensionality and complexity. Second, in the architecture here computation can proceed in either direction. The model is truly bi-directional and can be used to estimate or refine population codes for related external signals.

The current model is very simple and serves mainly as a proof of concept. Future developments will address a number of issues. Currently, it only simulates a one DOF single-joint movement. Furthermore, it encodes only joint-centered parameters of kinematics (position, velocity, acceleration) and dynamics (torques). We will extend the network to deal with the multi-muscle, multi-articular movements, surplus degrees of freedom and the complex non-linear dynamics of multi-articular movements. It will also be extended to resolve coordinate transformations between movement representations in different sensory and motor coordinate frameworks, and between movement representations in extrinsic versus intrinsic parameter spaces. It currently generates a movement by tracking a complete a priori kinematic plan. This is simplistic. We plan to develop dynamic error signals based on some cost function that would cause the kinematics of the movement to evolve in real time rather than by tracking a complete a priori kinematic plan. Finally, the current network does not learn. All network connections were optimized to provide a close approximation of the maximum likelihood estimate of each visible-layer signal (Deneve et al., 2001) and the IM module was parameterized to simulate a skilled performer. Error signals can be derived from the peripheral feedback signals to adjust connection weights in different modules to allow the network to deal with changes in task dynamics or other sources of performance errors.

Acknowledgments

The preparation of this chapter was supported by the Canadian Institutes of Health Research "New Emerging Team Grant in Computational Neuroscience" (NET-54000). We thank Paul Cisek, Trevor Drew and Andrea Green for many helpful suggestions.

References

Andersen, R.A., Snyder, L.H., Bradley, D.C. and Xing, J. (1997) Multimodal representation of space in the posterior parietal cortex and its use in planning movements. Annu. Rev. Neurosci., 20: 303–330.

Bengio, Y. and Frasconi, P. (1994a) Credit assignment through time: alternatives to backpropagation. In: Cowan, J.D., Tesauro, G. and Alspector, J. (Eds.), Advances in Neural Information Processing Systems, Vol. 6, Morgan Kaufmann, San Mateo, CA, pp. 75–82.

Bengio, Y. and Frasconi, P. (1994b) Learning long-term dependencies with gradient descent is difficult. IEEE Trans. Neural Netw., 5(2): 157–166.

Boucheny, C., Brunel, N. and Arleo, A. (2005) A continuous attractor network model without recurrent excitation: maintenance and integration in the head direction cell system. J. Comput. Neurosci., 18: 205–227.

Brunel, N. (1996) Hebbian learning of context in recurrent neural networks. Neural Comput., 8: 1677–1710.

Brunel, N. and Latham, P. (2003) Firing rate of noisy quadratic integrate-and-fire neurons. Neural Comput., 15: 2281–2306.

Brunel, N. and Nadal, P. (1998) Mutual information, fisher information and population coding. Neural Comput., 10: 1731–1757.

Deneve, S., Latham, P.E. and Pouget, A. (2001) Efficient computation and cue integration with noisy population codes. Nat. Neurosci., 4(8): 826–831.

Ditterich, J., Mazurek, M.E. and Shadlen, M.N. (2003) Microstimulation of visual cortex affects the speed of perceptual decisions. Nat. Neurosci., 6(8): 891–898.

Georgopoulos, A.P., Caminiti, R., Kalaska, J.F. and Massey, J.T. (1983) Spatial coding of movement: a hypothesis concerning the coding of movement direction by motor cortical populations. Exp. Brain Res., Suppl. 7: 327–336.

Gold, J.I. and Shadlen, M.N. (2002) Banburismus and the brain: decoding the relationship between sensory stimuli, decisions, and reward. Neuron, 36(2): 299–308.

Hanks, T.D., Ditterich, J. and Shadlen, M.N. (2006) Microstimulation of macaque area LIP affects decision-making in a motion discrimination task. Nat. Neurosci., 9(5): 682–689.

Hanneton, S., Berthoz, A., Droulez, J. and Slotine, J.J. (1997) Does the brain use sliding variables for the control of movements? Biol. Cybern., 77(6): 381–393.

Hertz, J., Krogh, A. and Palmer, R.G. (1991) Introduction to the Theory of Neural Computation. Addison-Wesley, Redwood City, CA.

Hinton, G.E. and Salakhutdinov, R. (2006) Reducing the dimensionality of data with neural networks. Science, 313(5786): 504–507.

Hopfield, J.J. (1982) Neural networks and physical systems with emergent collective computational abilities. Proc. Natl. Acad. Sci. U.S.A., 79(8): 2554–2558.

Kalman, R.E. (1960) A new approach to linear filtering and prediction. Trans. ASME — J. Basic Eng., 82: 35–45.
Latham, P.E., Deneve, S. and Pouget, A. (2003) Optimal computation with attractor networks. J. Physiol. (Paris), 97(4–6): 683–694.
Lo, C.C. and Wang, X.J. (2006) Cortico-basal ganglia circuit mechanism for a decision threshold in reaction time tasks. Nat. Neurosci., 9(7): 956–963.
Lohmiller, W. and Slotine, J.-J.E. (1998) Contraction analysis for nonlinear systems. Automatica, 34(6): 1–27.
Mazurek, M.E., Roitman, J.D., Ditterich, J. and Shadlen, M.N. (2003) A role for neural integrators in perceptual decision making. Cereb. Cortex, 13(11): 1257–1269.
Redish, A.D. and Touretzky, D.S. (1994) The reaching task: evidence for vector arithmetic in the motor system? Biol. Cybern., 71(4): 307.
Richardson, A., Tresch, M., Bizzi, E. and Slotine, J.J. (2005) Stability analysis of nonlinear muscle dynamics using contraction theory. Conf. Proc. IEEE Eng. Med. Biol. Soc., 5: 4986–4989.
Salinas, E. and Thier, P. (2000) Gain modulation: a major computational principle of the central nervous system. Neuron., 27(1): 15–21.
Schwartz, A.B. and Moran, D.W. (1999) Motor cortical activity during drawing movements: population representation during lemniscate drawing. J. Neurophysiol., 82: 2705–2718.
Shadlen, M.N. and Newsome, W.T. (2001) Neural basis of a perceptual decision in the parietal cortex (area LIP) of the rhesus monkey. J. Neurophysiol., 86(4): 1916–1936.
Slotine, J.J. and Lohmiller, W. (2001) Modularity, evolution, and the binding problem: a view from stability theory. Neural Netw., 14(2): 137–145.
Toulouse, G. (1987) Neural networks. Understanding physicists' brains. Nature, 327(6124): 662.
Wolpert, D.M. and Kawato, M. (1998) Multiple paired forward and inverse models for motor control. Neural Netw., 11(7–8): 1317–1329.
Xie, X., Hahnloser, R.H.R. and Seung, H.S. (2002) Double-ring network model of the head direction system. Phys. Rev. E, 66 041902-1-9.

P. Cisek, T. Drew & J.F. Kalaska (Eds.)
Progress in Brain Research, Vol. 165
ISSN 0079-6123
Copyright © 2007 Elsevier B.V. All rights reserved

CHAPTER 26

Computing movement geometry: a step in sensory-motor transformations

David Zipser[1,*] and Elizabeth Torres[2]

[1]Department of Cognitive Science, UCSD 0515, 9500 Gilman Drive, San Diego, CA 92093, USA
[2]Division of Biology, California Institute of Technology, Pasadena, CA 91125, USA

Abstract: The generation of goal-directed movements requires the solution of many difficult computational problems. In addition to generating the forces needed for movement there are a number of essentially geometric problems. Among these are transformations from extrinsic to intrinsic reference frames, removing under-specification due to excess degrees of freedom and path multiplicity, and error correction. There are no current motor control computational models that address these issues in the context of realistic arm movement with redundant degrees of freedom. In this chapter, we describe a geometric stage between sensory input and physical execution. The geometric stage determines movement paths without reference to forces. It is implemented with a gradient technique that can generate movement paths online. The model is demonstrated by simulating a seven degree of freedom arm that moves in three-dimensional space. Simulated orientation-matching movements generated by the model are compared with human experimental movement data to assess the validity of several of the model's behavioral predictions.

Keywords: arm model; geometric stage; motor control; gradient descent; movement speed; co-articulation; experimental results

Introduction

The movements we make to accomplish our cognitive goals are diverse, rapid and involve so much complex computation that it has proven daunting to explain how they are done. Indeed, we do not even fully know how to do the computations needed to make such remarkably versatile behavior possible. When faced with such massive complexity it is only reasonable to try to break the problem down onto manageable size chunks.

In this chapter we describe a computational model of a "chunk" or step in the sensory-motor transformation that we call the "geometric stage" (Torres and Zipser, 2002). The geometric stage translates from the extrinsic sensory description of the goal, such as its location and orientation, into some intrinsic representation, like joint angle changes, that can be used by more distal stages of the motor system to generate actual movements. The model can also be used to simulate movement paths. These simulated paths can be compared with experimental observations to test the consistency of the model.

To carry out the required transformation the model of the geometric stage solves the redundancy problem arising from the difference in the number of degrees of freedom in the sensory space where goals are specified and the motor space where they

*Corresponding author. Tel.: +1 858-775-1972;
E-mail: dzipser@ucsd.edu

DOI: 10.1016/S0079-6123(06)65026-7

are implemented, and it also chooses one path among the infinity of possible paths. In addition our model is able to do online error correction and constraint satisfaction. The geometric stage does not provide any information about the forces needed to generate movement and the paths it outputs do not depend on the speed of movement. This means that there is still plenty of work for other parts of the motor system. However, the geometric stage can generate movement paths that can be compared with experiments, so any predictions it makes can be tested directly, without reference to other aspects of the motor system.

The motivation for introducing the geometrical stage is to simplify sensory to motor transformation. Eliminating the need to deal with forces and speed makes it much simpler to transform sensory information into a form that specifies postural paths. It is important to keep in mind that a postural path is a complete description of the configuration of the arm and hand as they move to the location and orientation of the target.

There is some experimental justification for introducing a geometrical stage in sensory-motor transformations. For example, neurophysiological data from posterior parietal cortex (PPC) suggest a dynamics-free representation of movement (Kalaska et al., 1990) and other correlates of coordinate transformations (Andersen and Buneo, 2002).

We illustrate our model by simulating an arm with seven degrees of freedom making reach to grasp movements in which the arm moves to the location of a target while the hand rotates to match the target orientation. Both the translation and orientation movements use all seven degrees of freedom simultaneously, i.e., they are co-articulated. We also present experimental results that test some of the predictions of the model.

Background

Most of the theoretical work so far consists of either powerful general concepts, or fairly detailed models of highly simplified systems. Both of these lines of work have provided us with valuable insights into how movements can be generated. Perhaps the most influential general theory is optimal control that originated in engineering and robotics and has been adapted for use in biological systems [see Todorov (2004) for a review of optimal control and its sensory-motor applications]. Some form of optimization is generally needed to deal with the massive redundancy that occurs in sensory-motor transformations.

The most common computational approach in biological motor control has been to consider the problem as analogous to one in classical mechanics. Such models attempt to account simultaneously for all the parameters of the movement such as the path, speed, and force. In this formulation, one tries to find a function that gives the trajectory along which an energy-like quantity is minimized. Except in the simplest cases, this function cannot be found in an explicit form, so the minimizing trajectory is found by numerically integrating along many trajectories. These integrals generally require that the final posture and the elapsed time be known prior to movement, making the computations daunting when the tasks are underdetermined. Because of this, almost all the work has been done with simplified systems having two degrees of freedom and moving in a plane, i.e., with no excess degrees of freedom. Models of this general type such as minimum jerk (Flash and Hogan, 1985), minimum torque (Uno et al., 1989), minimum metabolic energy (Alexander, 1997), and minimum variance (Harris and Wolpert, 1998) have given good results in the study of planar arm movement. It seems odd that so many different optimization criteria give good results. It has been pointed out that all of these criteria resemble each other in that they are reduced by smooth motion that avoids rapid large changes in muscle activations that consume more energy. Extending these promising initial results to realistic, freely moving arms has proved very difficult so it seems that an attempt to break the problem down into manageable size pieces is warranted. This would be consistent with the observation that the posterior parietal areas seem to play an intermediate role in going from sensory input to motor output.

Implementing the model

Our problem now is to implement the geometric stage in such a way that it takes input in an extrinsic form from the sensory system and translates it into variables that can be controlled by the motor implementation stages. It is not yet known just what postural variables are controlled by the motor system. Joint angles and muscle lengths are possibilities. The computational strategy we propose here for the geometric stage will work with any set of postural variables. There are two important features of the geometric stage to keep in mind. First, it computes the geometrical aspects of movement, but not the forces required to bring these changes about. And second, it specifies movement paths, independent of the speed of movement. If our conjecture about the geometric stage is correct then observed movement paths should be independent of speed, and they should be described by a model of the geometric stage without reference to motor implementation.

The transformation from an extrinsic description of a goal for movement to variables that can be used directly by the motor system is underdetermined. This suggests that an optimization technique should be used. Since we do not know the actual variables used as input to the motor system, we have to pick some set that uniquely describes posture. It should then be possible to parameterize the transformation, using experimental data, so that veridical paths are generated. We have found that a modified form of a gradient descent strategy, originally developed for computer graphics (Witkin et al., 1987), has all these properties so we have used it to model the geometric stage.

Gradient descent is an optimization technique so the question arises what quantity should be optimized? The natural choice for the quantity to be reduced is the distance between the current hand configuration and the target configuration. If this distance is continuously reduced the hand configuration will eventually match that of the target and the goal will be achieved. The idea of incrementally reducing this distance has been used previously (Bullock and Grossberg, 1988; Bullock et al., 1993, 1998). These models use a neural network trained on randomly generated movements to solve the underdetermined extrinsic to intrinsic transformation needed for this movement, rather than the more general analytical approach described here.

Implementing gradient descent requires that we have a function that maps all the variables specifying both the current posture and the target for movement to a positive scalar representing the remaining distance. Various disciplines use different names for this kind of function such as *cost*, *error*, *principle*, and *energy* function. We will use *cost* function. So, in our model the cost is the distance remaining between the current hand configuration and its target.

Once we have specified the cost function mathematics tells us that the negative of the gradient of this function is a vector whose components specify the relative amount to change each postural variable to get the hand closer to the target configuration. The gradient descent procedure for the reach to grasp task consists of repeatedly computing the gradient of the cost function and changing the posture a small amount each time till the hand is at the target in the correct orientation. In this way it generates a postural path from the initial to the target posture.

We illustrate the gradient technique using an arm with seven degrees of freedom that does reach-to-grasp movements in three dimensions. For simplicity we first use a posture configuration consisting of a particular arbitrarily chosen set of joint angles. There is no a priori reason to think that the paths generated by this choice of joint angles will give realistic movement paths, and, in fact, it does not. However, a basic property of the gradient is that its value depends on the parameterization of its variables. Latter we show how to take advantage of this to generalize the gradient method using parameters obtained from experimental observation to generate realistic paths.

Computing the gradient

The model deals with two phases of grasping: *Transport* — moving the hand to the object — and *Orientation* — rotating the hand to match the object axis. Therefore, the cost function has two parts that are combined to map joint angles to a single value that represents both the remaining distance to the target and the remaining difference in orientation between hand and target. We have not attempted to deal with the finger movements involved in the grasp because this would add a great deal of complexity without further clarifying the basic concept of the gradient method.

We first describe using gradient descent for transport for an arm with seven degrees of freedom. The arm has three constant length segments linked end to end by seven flexible joints, three at the shoulder, two at the elbow, and two at the wrist. The shoulder end is attached to a fixed point in a 3-D reference frame and the other end, i.e., the hand, remains unconnected. It is convenient to think in terms of two spaces; the workspace, X, consisting of all the points that can be reached by the hand in 3-D space, and the 7D joint angle space, Q, each point of which specifies a set of joint angles that configure the arm so that the hand is at one of the points in the workspace X.

Given a point in Q there is a function that lets us compute where the hand is in X. However, the reverse is not true. Given a point in X there is no function to compute a point in Q specifying an arm configuration that puts the hand at that point in X, even though there are generally an infinity of such points in Q. Unfortunately, the target for reaching is typically given in X by sensory input. Since there is no function to compute an arm configuration that would put the hand at the target, we have to use some more complex procedure to solve this problem. Actually what we really need is a continuous path of arm configurations that will transport the hand from its current location to the target. There are an infinity of such paths and we have to choose one. This is where gradient descent comes in.

The gradient method

To use gradient descent we need a cost function that depends on all the joint angles and is equal to the distance between the current hand position and the target in 3-D space. This distance, r, is given by:

$$r = \sqrt{\sum_{i=1}^{i=3}(x_i^t - x_i)^2}$$

where $x = (x_1,x_2,x_3)$ is the current hand location and $x^t = (x_1^t,x_2^t,x_3^t)$ the location of the target, both in extrinsic space. To make r depend on the joint angles, $\boldsymbol{q}$, we have to use the vector-valued function, $\boldsymbol{f}$, that maps Q onto X. The components of $\boldsymbol{f}(\boldsymbol{q})$ are:

$$\boldsymbol{f}_i(\boldsymbol{q}) = x_i$$

So the cost function for translation is:

$$r = \sqrt{\sum_{i=1}^{i=3}(x_i^t - \boldsymbol{f}_i(\boldsymbol{q}))^2} \tag{1}$$

The function $\boldsymbol{f}$ for a seven-jointed arm can be found in the literature (Benati et al., 1980).

The gradient of r, $\nabla r(\boldsymbol{x^t},\boldsymbol{q})$, is a seven-component vector in Q. The negative of this vector points in a direction guaranteed to reduce r. This means that if we change all the joint angles a small amout in proportion to their corresponding components in $-\nabla r(\boldsymbol{x^t},\boldsymbol{q})$ we will move the hand closer to $\boldsymbol{x^t}$.

To solve the transport problem we move, in small steps, along the path given by the negative of the gradient of the cost function. Each step consists of computing the gradient and then changing the values of $\boldsymbol{q}$. Since $\nabla(\boldsymbol{x^t},\boldsymbol{q})$ is given in closed form its computation is straight forward and requires only that we know the current vaues of $\boldsymbol{q}$ and $\boldsymbol{x^t}$. The length of the gradient vector depends on the local details of the space Q and does not give any information about the speed of movement.

Next we consider using gradient descent to reduce the difference in orientation between the hand and the target object. We need to find a single value that decreases with this difference in orientation and is a function of $\boldsymbol{q}$ and the orientation of the target. Our strategy is to first find the matrix that represents a rotation that will align the target with the hand. This rotation can then be represented by a single angle ϕ about the principal axis $\vec{n}$ (Altman, 1986). This angle can be used in the cost function. The general scheme is given below.

Consider three reference frames: one fixed at the shoulder S, one on the hand H, and one on the object O, Fig. 1. The relative orientation of these frames can be represented with orthogonal rotation matrices: H for the rotation from shoulder to hand, O from shoulder to object, and $R = O^{-1}H$, from object to hand, where $O^{-1} = O^T$ is the rotation from object to shoulder. All of these reference frames are in 3-D space X as was target and hand

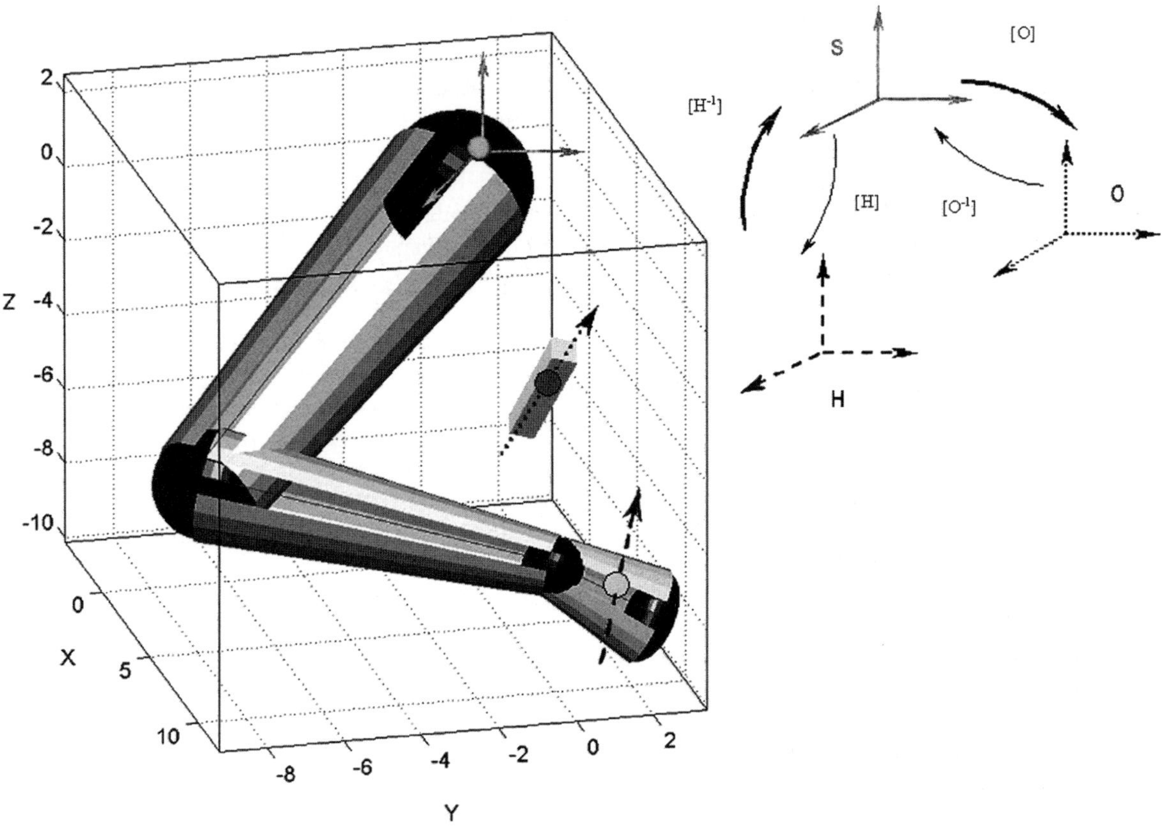

Fig. 1. The orientation part of the task requires rotating the hand, with orientation indicated by a dashed line, to that of the target, indicated by the dotted line (see the drawing on the left side of the figure). The relationship between the various coordinate systems is shown on the right side of the figure. All orientations are given relative to a coordinate system fixed to the shoulder, S (solid gray lines). H is a coordinate system fixed to the hand (dashed lines) and O is a system fixed to the target (dotted lines). The transformations between systems are indicated by curved arrows labeled by the matrix that does the transformation, i.e., [H] for shoulder to hand, [O] for shoulder to target, and [H^{-1}] and [O^{-1}] for the inverse transformations. The current target orientation, O, is assumed to be known from sensory input. The hand orientation, H, is a known function of the current posture. The rotation that will align the target with the hand is given by $R = O^{-1} \cdot H$, where O, H, and R are matrices. The rotation encoded in R can be represented by a single angle ϕ about the principal axis. This angle ϕ is used in the cost function. When $\phi = 0$ the hand matches the orientation of the object.

location. The orientation of the hand, H, depends, like hand location, on all seven joint angles of the posture, $\boldsymbol{q}$. As with hand position, hand orientation, H, is given by a known function (Benati et al., 1980) that is different from the one for hand location. The orientation of the target, O, comes from some sensory system, typically vision, and is an input to the geometric stage. The matrix R represents the orientation difference we want to reduce and depends on both the current posture and the object's orientation.

To get a single scalar value for the orientation difference we use the fact that for any rotation matrix, such as R, there is a principal axis $\vec{n}$ about which a single rotation ϕ has the same effect as the rotation matrix. A cost function for reaching-to-grasp can be obtained by combining ϕ, given by $\cos(\phi) = 1/2\text{Trace}(R) - 1$, with the distance term for transportation derived above:

$$r = \sqrt{\sum_{i=1}^{3} (x_i^t - \boldsymbol{f}_i(\boldsymbol{q}))^2 + \alpha(k\phi(\boldsymbol{q}, o))^2} \qquad (2)$$

where k is a scale factor between distance and rotation, and α a parameter that can be adjusted so that arm movement and hand rotation overlap realistically in time as the arm moves toward the target. In our simulations we set α to 1.0 so that the transport and orientation distance decrease proportionally. Subsequent experiments have shown this is what actually happens. More about this co-articulation later.

It is important to note that each term in Eq. (2) depends on all seven joint angles. This means that the wrist and upper arm are *not* treated as separate models. Posture changes are made in a way that simultaneously changes hand position and orientation. This is important since rotation of any joint angle can potentially change either the location or the orientation of the hand (Jeannerod, 1988).

The gradient descent algorithm starts at $\boldsymbol{q}^{\text{initial}}$, takes a small step in the direction of the negative gradient to a new point $\boldsymbol{q}^{\text{current}}$. This procedure is repeated using successive values of $\boldsymbol{q}^{\text{current}}$ until the value of r goes to zero, at which point $\boldsymbol{x}^{\text{current}}$ equals $\boldsymbol{x}^{\text{target}}$, and the hand orientation matches the object's. In practice, computation is stopped when r is less than some small tolerance. Ideally the steps should be infinitesimal, but the gradient is approximated quite well using moderately large steps. A feature of the gradient paradigm is that the paths obtained in Q provide a complete description of all arm configurations as movement proceeds from initial to final posture.

An important property of Eq. (2) is that r goes to zero if and only if the hand is at the target in the correct orientation. This enables the system to avoid the problem of local optima that can plague gradient descent. If the gradient of r goes to zero and r is not zero then a perturbation can be made in the trajectory and gradient descent continued until r is zero. This is not a trivial feature of the exact form of the cost function. Rather it is a consequence of the fact that distance and orientation difference are geometrical properties of the system that go to zero only when the reach-to-grasp task is complete.

Generalization of the model

In our initial description of the gradient method we used a posture configuration space consisting of joint angles as the axis of a 7D Cartesian space. This representation was chosen for illustrative simplicity, but it does not produce the experimentally observed paths. This is not surprising since there is no reason to think that the actual internal representation of posture configuration space would be the same as our example. We can get more realistic paths by transforming the original joint angle space into some new appropriate space. This works because the postural path generated by the gradient method depends on the representation of configuration space. However, this has to be done in a way that allows us to recover some joint angle representation so that we can compare the paths generated with experimental observations.

The formal way to do this is to find a transformation, $\boldsymbol{q}' = \boldsymbol{G}' \cdot \boldsymbol{q}$, from our original joint angles, $\boldsymbol{q}$, to new coordinates, $\boldsymbol{q}'$, such that the gradient in the new system, when transformed back to the original coordinates gives the paths we want. This transformation of the gradient back to the original Q system is obtained by pre-multiplying the original gradient by the inverse of the transformation matrix:

$$\nabla' r(\boldsymbol{x}^t, \boldsymbol{q}) = \boldsymbol{G}'^{-1} \cdot \nabla r(\boldsymbol{x}^t, \boldsymbol{q})$$

To illustrate how this method works we use it to generate two different paths for the tip of a simple arm with only two joint angles moving in a plane. In this system there are no excess degrees of freedom in the joint angles, however, the paths are underdetermined. We first find the path using a posture configuration space with each joint angle as an axis of a Cartesian coordinate system. This gives curved paths. Then we transform posture configuration space to get paths that are straight lines.

The arm is modeled with two unit-length segments and two joints, q_1 at the shoulder and q_2 at the elbow. The forward function $\boldsymbol{x} = \boldsymbol{f}(\boldsymbol{q})$ in Cartesian configuration space is given by:

$$\boldsymbol{f}_{x_1}(\boldsymbol{q}) = \cos(q_1) + \cos(q_1 + q_2)$$
$$\boldsymbol{f}_{x_2}(\boldsymbol{q}) = \sin(q_1) + \sin(q_1 + q_2)$$

where $[\boldsymbol{f}_{x_1}(\boldsymbol{q}), \boldsymbol{f}_{x_2}(\boldsymbol{q})]$ represent the coordinates of the arm's endpoint.

The translation cost function in this case is:

$$r(\boldsymbol{x}^t, \boldsymbol{q}) = \sqrt{(x_1^t - \boldsymbol{f}_{x_1}(\boldsymbol{q}))^2 + (x_2^t - f_{x_2}(\boldsymbol{q}))^2}$$

A typical curved arm tip path produced by the gradient of r projected onto X is shown in Fig. 2 (top right). The gradient path on the cost surface is shown in Fig. 2 (top left).

The value of $\boldsymbol{G}'$ that produces straight lines for this simple arm can be found analytically (Gray, 1998) and is given by:

$$\boldsymbol{G}'(\boldsymbol{q}) = \begin{bmatrix} 2(1+\cos(q_2)) & 1+\cos(q_2) \\ 1+\cos(q_2) & 1 \end{bmatrix}$$

When this matrix is inverted and applied to the original gradient the arm moves in straight lines. An example of straight-line movement in X and the corresponding gradient path on the new cost surface are shown in Fig. 2 (bottom).

Fitting the model to experimental data

It would be nice to be able to derive $\boldsymbol{G}'^{-1}$ from basic facts about the physical nature of the arm and the mathematical properties of posture space. We have made some progress toward this goal, (Torres and Zipser, 2002, 2004). In particular the constraints on joint movement can be accounted for by placing limiting functions along the diagonal of $\boldsymbol{G}'^{-1}$. Also, more realistic paths can also be obtained by assuming that movements follow the shortest paths in the posture space. The details of both of these techniques are rather involved and will not be discussed here.

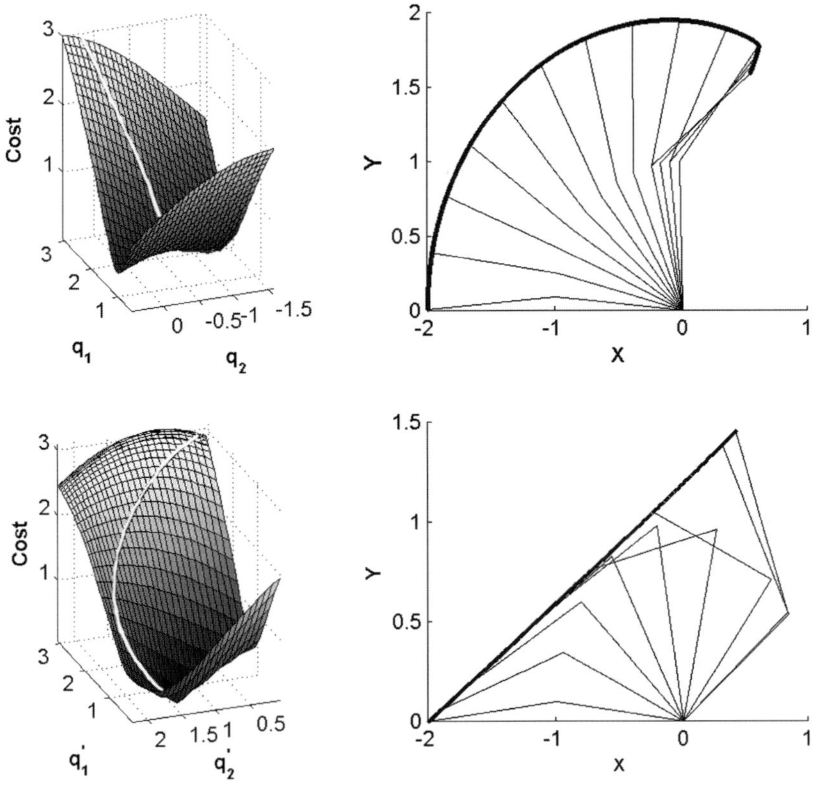

Fig. 2. Change of coordinates: (Top right) Hand path of two segment arm resulting from using the gradient in Cartesian joint angle space. (Top left) Corresponding cost surface with the gradient path shown on the surface. (Bottom right) Hand path of arm resulting from using the gradient in a transformed space that gives straight hand paths. (Bottom left) Corresponding cost surface with the gradient path shown on the surface.

Another approach to getting realistic postural paths is to use experimental data to find a set of constants for the matrix G'^{-1} so that it can be used as a linear transformation. It is not obvious that a good fit that works for all subjects can be represented by such a linear transformation.

The data we needed to do this were not available in the literature so we carried out an extensive series of experiments to generate it. The task we used involved transporting and rotating a hand-held cylinder to match the position and orientation of a similar target cylinder. The seven joint angle postural path was recorded using the Polhemus Fastrak motion tracking system (Torres and Zipser, 2002). We used six targets positioned and oriented differently, six subjects, and six repetitions per target. Then using all this data we found a set of constant components for $\tilde{G}^{-1}$ by optimization (Torres and Zipser, 2004).

The results of simulating paths with the model using the experimentally derived G'^{-1} are superimposed on the actual translation paths from the hand sensor and plotted in Fig. 3. The results for rotation, which are not shown, are similar. The close correspondence between model and observed paths shows that the model can be parameterized to fit the data fairly well with a constant linear transformation of the gradient vector represented by an arbitrary set of joint angles. This observation does not tell us exactly what posture space is like, but it demonstrates that it is related to joint angle space by a linear transformation.

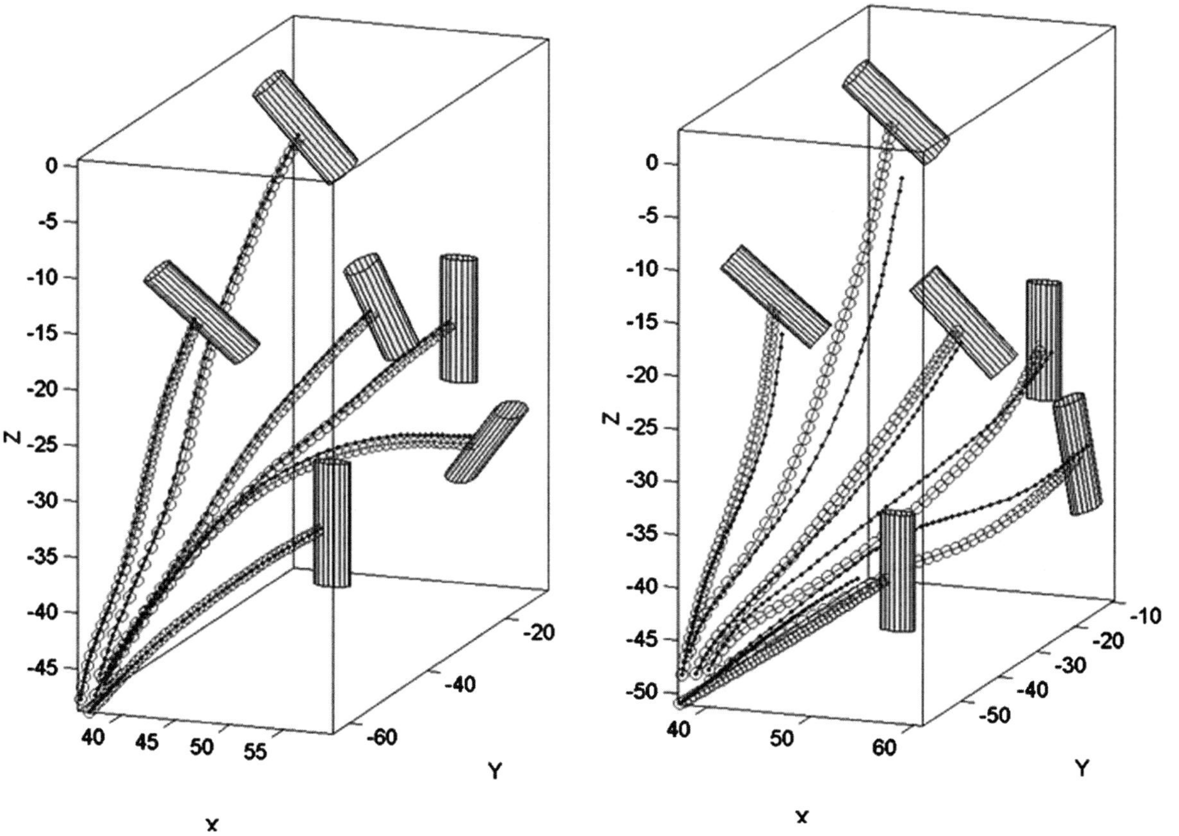

Fig. 3. Experimental example of metric correction using experimental data: Unconstrained arm movement paths were recorded from six subjects who performed orientation-matching motions to six targets located throughout the reachable workspace and oriented differently. Circled paths are the subjects' average across six repetitions. Superimposed on these are the model's simulated paths obtained using the experimentally derived value of G'^{-1}. (Left) Best case, (Right) worst case of six subjects.

This result is not trivial because the experimentally observed paths were generated by the motor system as a whole so that they have all the features imposed on the path by the physical constraints of the system. This demonstrates that a purely geometric posture configuration space can generate aspects of movement that would generally be considered the result of optimizing some physical quantity. For example, real movements tend to minimize the amount that the heavy upper arm moves relative to forearm and wrist, presumably to use less energy.

Error correction

The gradient descent approach is particularly well suited to error correction. Errors are a ubiquitous feature of movement generation. They arise from many causes such as noise generated by muscles and other parts of the neural apparatus, mismatches between sensory and motor representations, fatigue, etc. In addition the target may be perturbed during the course of the movement.

The gradient method does not have to detect error as such to correct for its effects. The reason for this is that both the target and the current posture are continuously supplied as input to the computation of the current value of the gradient. It is these current values that are used to determine the ongoing course of movement. If either the target or the posture is perturbed the gradient uses the new values. It is not necessarily stuck with a predetermined plan. But there is a problem of timing because there is a delay in the feedback of information about posture. For slow movements this is not a serious problem, but for fast movements it can lead to significant error. What is actually observed is that there is a speed-error tradeoff in movements. The faster the movement the greater the error. Part of this may reflect feedback delay. Movements also have a biphasic character, particularly when accuracy is important. The initial part of the movement is fast and brings the hand into the neighborhood of the target. Then there is a slower phase to accurately complete the task. This kind of behavior is consistent with the gradient method that can use internal integration of efference copy, which produces no significant delay, for fast movement, but is not able to correct errors. For slower movements feedback of new target and posture information can arrive in time to compensate for movement error.

Errors that come from the misalignment of the sensory and posture reference frames cannot be corrected by the gradient procedure we have described so far. The reason for this is that the configuration of the target is given by external sensory input while that of the hand is given by the internal somatic sensory system, and these are unlikely to be perfectly aligned.

To be more precise, for the case of vision and translation, consider three points: x^{target}, the true spatial location of the target; x^{vis}, the location of the target as measured by the visual system and input to the gradient computation; and x^{hand}, the final location where the arm moving system places the hand when given x^{vis} as input. In general, x^{target} will not coincide with x^{hand} because of misalignment errors in going from x^{target} to x^{vis} and then to x^{hand}. This sort of error can be corrected by visual feedback. The gradient gives a simple way to do this with a slight modification in the way we do the computation. We can separate the external variables, $\boldsymbol{x}$, from the intrinsic ones, $\boldsymbol{q}$, by using the chain rule, which gives us:

$$\nabla r(\boldsymbol{x}^t, \boldsymbol{q}) = \nabla r(\boldsymbol{x}, \boldsymbol{x}^t) J(\boldsymbol{f}(\boldsymbol{q})), (J = \textit{Jacobian}) \quad (3)$$

The *Jacobian* is a 3×7 matrix made up of the gradients of the three components of $\boldsymbol{f}$, the function that maps from Q to X described above. The term to its left is a three-component vector that gives the *offset* between the hand and the target. This offset can, in principle, be determined by the visual system using only the retinal image irrespective of the location of the eyes. This is not a trivial task because it involves recognizing the hand and the target and measuring the offset between them. However, this is the kind of task typically assigned to vision. Using just the retinal image is a great advantage because no intermediate transformations to body coordinates are needed and the eyes are free to move at will.

When the offset is used as input to the gradient computation movement continues until the visual system detects that the offset is zero, i.e., the hand is at the target. The hand will eventually reach the target even with the inevitable misalignment of the visual and postural systems because each step in the computation will bring the hand closer to the target as long as the direction of movement does not differ from the true gradient by more than $\pm\pi/2$, which gives a large margin for error. The computation for orientation is analogous, but the orientation offset is used instead of the distance.

Experimental tests

A geometric stage implemented with the gradient descent procedure has a number of testable consequences. Perhaps the most important is that movement paths are nearly speed independent. This follows from paths being described only to the degree of geometry. Some difference in paths may occur at high speeds due to slow feedback. Another consequence is co-articulation that arises from the fact that the cost function has two terms, one for transport and the other for orientation, both being optimized together, so that rotation and translation can overlap in time and in the joints they use. As we will see natural movements are completely co-articulated which can be difficult to compute by most of the standard methods used. Later we will dissect the cost function and show the possible biological significance of this observation.

Several aspects of the kinematics of movement are known to be invariant to changes in speed for pointing movements to targets constrained to the saggital plane (Atkeson and Hollerbach, 1985), and to targets distributed across the 3-D space (Nishikawa et al., 1999) where the wrist was constrained. It is unknown how these results extend to more complex orientation-matching motions toward targets positioned and oriented differently across the workspace, where the arm is unconstrained. It is also unknown how the postural paths behave under these conditions. We have done a set of experiments to investigate the effect of speed on the translation and orientation movements we simulated with our model. The data from the speed tests were also used to study co-articulation of translation and orientation.

The same task was used to test speed independence as was used to generate the data to calculate $\boldsymbol{G}'^{-1}$ described above, modified by the addition of three different speed of movement conditions. The experimental session began with a recorded instruction indicating the target number. This command was immediately followed by a tone, which served both as a cue to tell the subject the desired movement duration and to indicate that at the end of the tone the motion was to begin. The length of the tone was proportional to the desired movement time: fast (100 ms), normal (400 ms), and slow (700 ms). However, subjects adjusted the duration of the movement to their own comfortable pace for each speed. There was enough time between the recorded instruction and the tone so that subjects could visually locate the target. The target number–speed combinations were called at random by a computer. At the end of the movement, subjects waited at the target for a short beep that indicated the recording was over and they could move their arm back to the starting position. All movements were performed within 110 cm of the stationary system's origin. There was only one practice trial so that subjects could become familiar with the instructions, but not learn to make the task automatic. This was to avoid training the subjects to make habitual movements. We used six targets positioned and oriented differently, six subjects, three different speeds and six repetitions per speed–target combination. For statistical analysis, we divided the data set for each target (108 paths per target) into three sample groups corresponding to the three speed groups. Each sample group has 36 trajectories. Each target was analyzed independently.

The position-orientation hand paths in the three different speed conditions are quite similar (Fig. 4). We used a ratio from standard multivariate analysis of variance, the Wilk's lambda test to compare them (Rencher, 1995). In all the analyses we performed we failed to reject the null hypothesis, i.e., the means for all speeds are similar.

In the reach and orient task the hand changes location and rotates. How are these two actions related in time. Our cost function has a free parameter, α, that adjusts the relative rates of these two actions. With $\alpha = 1.0$ the two actions occur at

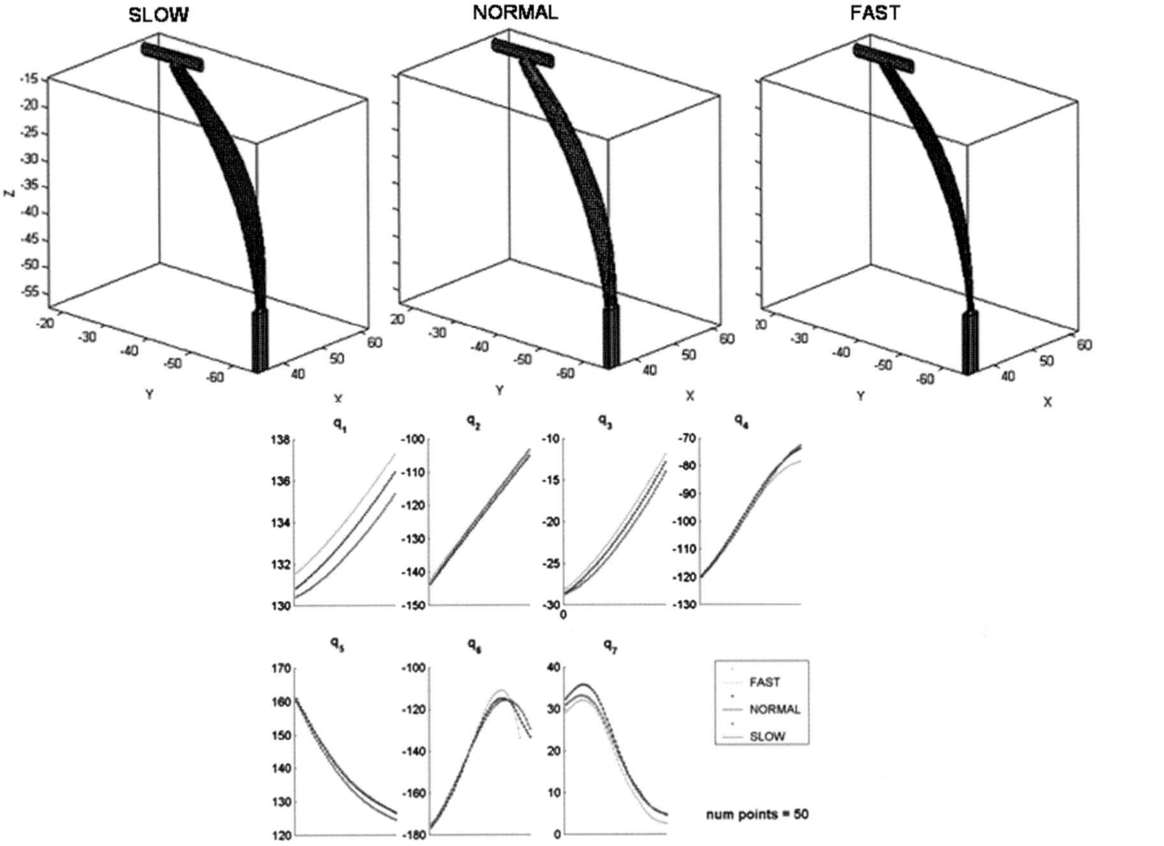

Fig. 4. (Top) Graphical representation of the speed data for a selected target. Cylinders at the beginning and at the end of the paths indicate the starting and target position-orientation respectively. The 95%-confidence region around the mean hand position path is given for each speed group. The averaged paths for the other two speed groups are plotted inside the confidence region corresponding to a given speed group. The idea is to identify the areas along the path where the groups differ significantly. The statistical details are given in Torres and Zipser (2004). As the figure shows, for each one of the groups, the paths from the other two speed groups fall inside the confidence region. (Bottom) Each of the panels show the rotation of one of the seven joint angles at three speed conditions to one target. All speed conditions are virtually the same. Note that two joint rotations change direction during the movement.

the same relative rate, i.e., they exactly overlap in time. There is no a priori way to know what will actually happen. To find out we used the data collected in the previously described experiments and plotted the normalized rate of translation against the rate of rotation, both squared to eliminate sign problems. The result is shown in Fig. 5. The result is a straight line with a slope of one. This means that rotation and translation are co-articulated in such a way that they not only start and end at the same time, but also they go at the same rate. Note that what is being plotted is the fraction of the remaining distance to the target vs. the fraction of the remaining rotation of the hand to match the orientation of the target. This struck us as interesting since the paths are not straight lines and it is true at all speeds.

The model and the brain

We now address the issue of how the computations described here might be used in the course ongoing movement behavior. This is not a simple problem. One of the complicating factors is that much motor behavior has been previously learned and has

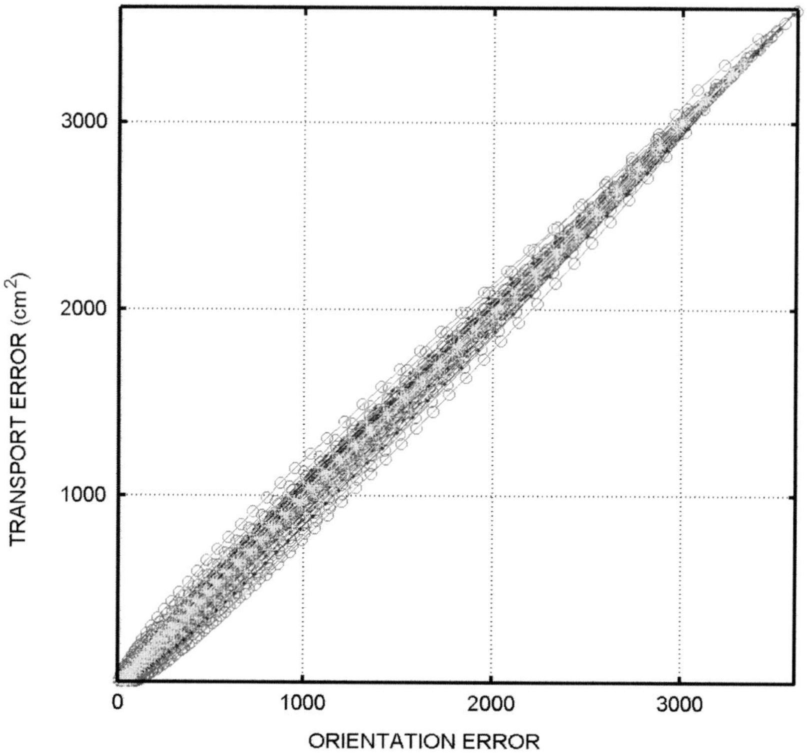

Fig. 5. The alpha ratio or co-articulation parameter plotted across the path for re-sampled motion to a target at three speeds. Each point represents the remaining orientation squared distance (horizontal axis) vs. the remaining squared positional distance (vertical axis). We use the square of the distances in each axis. At 0 the hand is on the target at the correct orientation. The slope of 1 in each plot is the co-articulation α-parameter.

become at least partly "automatic" in the sense that the information needed to make the movement is stored and is available without extensive computation. This means that there are few if any completely "unique" voluntary movements. On the other hand movements cannot be completely preprogrammed because there are generally variations in the details of even the most routine tasks. For example, eating is a very frequently performed stereotyped task, but there are always variations in the exact location of the food. This gives each movement some element of novelty. In addition there is the ever-present need to correct for inevitable errors arising from "noise" in a general sense. Other issues of how the geometric stage might be used arise when we consider learning new movements, planning movement and imagining consequences of a movement before doing it.

All of these considerations lead us to believe that a system that can compute movement geometry is required at one or more phases of all voluntary movements. During the learning of new movements or performance of relatively unique movements the geometric stage would be active before any motor activity and continue to be active as a supervisor able to correct errors during movement. This might involve different brain areas running the gradient computation at different times or one area repeating it. For well-learned movements that only required monitoring or slight modifications on line, the gradient might be computed only during movement and could easily lag a bit behind execution during most of a movements duration.

There is evidence that the different aspects of reaching to grasp, i.e., transport, orientation, hand shaping, and grasping, are processed in different brain regions (Jeannerod et al., 1995). Sometimes, as in the experiments described here, the different aspects are coupled and completely co-articulated, while under other experimental conditions they can be decoupled. The gradient method can account for both the physical separation of the brain areas computing different aspects of movement and also be used to simulate either co-articulation, which requires controlling all joint angles together, or for independent actions.

To separate transport from orientation we modify the original version of the cost function, Eq. (2), by removing the square root to give a sum of squares. This will give the same paths as Eq. (2) because the square root only contributes to the magnitude not the direction of the gradient. We get:

$$r^2 = \sum_{i=1}^{3}(x_i^t - \boldsymbol{f}_i(\boldsymbol{q}))^2 + \alpha(k\phi(\boldsymbol{q}, O))^2$$

Since the gradient is a linear operator we can compute the gradient of the terms separately and add the gradients and still get the same path. This shows that the various components of a movement can be computed in different brain areas and combined to give a single movement. Physical separation of translation and rotation computations could also help to explain our observation that when subjects are given explicit instructions to separate translation and orientation they are able to alter the pattern of co-articulation (Torres and Zipser, 2004). In simulations using the gradient method it is possible to manipulate the degree of co-articulation by changing the value of the α. When α equals one we get the observed complete co-articulation, when it is higher or lower orientation occurs either earlier or later during translation.

Conclusions

We have proposed that between sensory input and motor output there are brain areas that compute the geometrical aspects of movement paths independent of forces and speed. We further propose that gradient descent is used to implement these computations. The gradient is a fast, online computation that we show solves the excess degree of freedom problem and also the underdetermined path selection problem. It automatically corrects for many sources of error and can quickly update movement paths to respond to changes in target position and orientation. Simulations using the gradient method with a seven joint angle arm generated realistic reach to grasp postural paths. The gradient can be used for visual feedback from retinal images without regard to eye position. It also is compatible with different aspects of motion such as translation and orientation being computed in separate brain areas. It allows the coordination of these different aspects of motion that involve all joint angles. A nice feature of the model presented here is that even though it does not give speed or force information it can still account for observables such as movement paths, co-articulation of translation and rotation, and speed independence. Whether a gradient-like mechanism is actually used by the brain to compute movement parameters remains to be seen.

References

Alexander, R.M. (1997) A minimum energy cost hypothesis for human arm trajectories. Biol. Cybern., 76: 97–105.

Altman, S.L. (1986) Rotations, Quaternions and Double Groups. Oxford University Press, Oxford, pp. 65–79.

Andersen, R.A. and Buneo, C.A. (2002) Intentional maps in posterior parietal cortex. Annu. Rev. Neurosci., 25: 189–220.

Atkeson, C.G. and Hollerbach, J.M. (1985) Kinematic features of unrestrained vertical arm movements. J. Neurosci., 5: 2318–2330.

Benati, M., Gaglio, S., Morasso, P., Tagliasco, V. and Zaccaria, R. (1980) Anthropomorphic robotics I: representing mechanical complexity. Biol. Cybern., 38: 125–140.

Bullock, D., Cisek, P. and Grossberg, S. (1998) Cortical networks for control of voluntary arm movements under variable force conditions. Cereb. Cortex, 8: 48–62.

Bullock, D. and Grossberg, S. (1988) Neural dynamics of planned arm movements: emergent invariants and speed-accuracy properties during trajectory formation. Psychol. Rev., 95: 49–90.

Bullock, D., Grossberg, S. and Guenther, F.H. (1993) A self-organizing neural model of motor equivalent reaching and tool use by a multijoint arm. J. Cogn. Neurosci., 4: 408–435.

Flash, T. and Hogan, N. (1985) The coordination of arm movements: an experimentally confirmed mathematical model. J. Neurosci., 5(7): 1688–1703.

Gray, A. (1998) Modern Differential Geometry of Curves and Surfaces with Mathematica (2nd ed.). CRC Press, Boca Raton, FL.

Harris, C.M. and Wolpert, D.M. (1998) Signal-dependent noise determines motor planning. Nature, 394(20): 780–784.

Jeannerod, M. (1988) The Neural and Behavioural Organization of Goal-Directed Movements. Clarendon Press, Oxford, p. 55.

Jeannerod, M., Arbib, M.A., Rizzolatti, G. and Sakata, H. (1995) Grasping objects: the cortical mechanisms of visuomotor transformation. TINS, 18(7): 313–320.

Kalaska, J.F., Cohen, D.A.D., Hyde, M.L. and Prud'homme, M. (1990) Parietal area 5 neuronal activity encodes movement kinematics, not movement dynamics. Exp. Brain Res., 80: 351–364.

Nishikawa, K.C., Murray, S.T. and Flanders, M. (1999) Do arm postures vary with the speed of reaching? J. Neurophysiol., 81: 2582–2586.

Rencher, A.C. (1995) Methods of Multivariate Analysis. Wiley, New York.

Todorov, E. (2004) Optimality principles in sensorimotor control. Nat. Neurosci., 7: 907–915.

Torres, E.B. and Zipser, D. (2002) Reaching to grasp with a multi-jointed arm I: a computational model. J. Neurophysiol., 88: 1–13.

Torres, E.B. and Zipser, D. (2004) Simultaneous control of hand displacements and rotations in orientation-matching experiments. J. Appl. Physiol., 96: 1978–1987.

Uno, Y., Kawato, M. and Suzuki, R. (1989) Formation and control of optimal trajectory in human multijoint arm movement: minimum torque-change model. Biol. Cybern., 61: 89–101.

Witkin, A., Fleischer, K. and Barr, A. (1987) Energy constraints on parameterized models. Comput. Graph., 21(4): 225–232.

P. Cisek, T. Drew & J.F. Kalaska (Eds.)
Progress in Brain Research, Vol. 165
ISSN 0079-6123
Copyright © 2007 Elsevier B.V. All rights reserved

CHAPTER 27

Dynamics systems vs. optimal control — a unifying view

Stefan Schaal[1,2,*], Peyman Mohajerian[1] and Auke Ijspeert[1,3]

[1]*Computer Science & Neuroscience, University of Southern California, Los Angeles, CA 90089-2905, USA*
[2]*ATR Computational Neuroscience Laboratory, 2-2-2 Hikaridai Seika-cho Sorako-gun, Kyoto 619-02, Japan*
[3]*School of Computer and Communication Sciences, Ecole Polytechnique Fédérale de Lausanne (EPFL), Station 14, CH-1015 Lausanne, Switzerland*

Abstract: In the past, computational motor control has been approached from at least two major frameworks: the dynamic systems approach and the viewpoint of optimal control. The dynamic system approach emphasizes motor control as a process of self-organization between an animal and its environment. Nonlinear differential equations that can model entrainment and synchronization behavior are among the most favorable tools of dynamic systems modelers. In contrast, optimal control approaches view motor control as the evolutionary or development result of a nervous system that tries to optimize rather general organizational principles, e.g., energy consumption or accurate task achievement. Optimal control theory is usually employed to develop appropriate theories. Interestingly, there is rather little interaction between dynamic systems and optimal control modelers as the two approaches follow rather different philosophies and are often viewed as diametrically opposing. In this paper, we develop a computational approach to motor control that offers a unifying modeling framework for both dynamic systems and optimal control approaches. In discussions of several behavioral experiments and some theoretical and robotics studies, we demonstrate how our computational ideas allow both the representation of self-organizing processes and the optimization of movement based on reward criteria. Our modeling framework is rather simple and general, and opens opportunities to revisit many previous modeling results from this novel unifying view.

Keywords: discrete movement; rhythmic movement; movement primitives; dynamic systems; optimization; computational motor control

Introduction

Before entering a more detailed discussion on computational approaches to motor control, it is useful to start at a rather abstract level of modeling that can serve as a general basis for many theories. Following the classical control literature from around the 1950s and 1960s (Bellman, 1957; Dyer and McReynolds, 1970), the goal of motor control and motor learning can generally be formalized in terms of finding a task-specific control policy:

$$\mathbf{u} = \pi(\mathbf{x}, t, \alpha) \tag{1}$$

that maps the continuous state vector $\mathbf{x}$ of a control system and its environment, possibly in a time t dependent way, to a continuous control vector $\mathbf{u}$.

*Corresponding author. Tel.: +1 213 740 9418; Fax: +1 213 740 1510; E-mail: sschaal@usc.edu

DOI: 10.1016/S0079-6123(06)65027-9

The parameter vector α denotes the problem-specific adjustable parameters in the policy π, e.g., the weights in neural network or a generic statistical function approximator.[1] In simple words, all motor commands for all actuators (e.g., muscles or torque motors) at every moment of time depend (potentially) on all sensory and perceptual information available at this moment of time, and possibly even past information. We can think of different motor behaviors as different control policies π_i, such that motor control can be conceived of as a library of such control policies that are used in isolation, but potentially also in sequence and superposition in order to create more complex sensory-motor behaviors.

From a computational viewpoint, one can now examine how such control policies can be represented and acquired. Optimization theory offers one possible approach. Given some cost criterion $r(\mathbf{x}, \mathbf{u}, t)$ that can evaluate the quality of an action $\mathbf{u}$ in a particular state x (in a potentially time t dependent way), dynamic programming (DP), and especially its modern relative, reinforcement learning (RL), provide a well-founded set of algorithms of how to compute the policy π for complex nonlinear control problems. In essence, both RL and DP derive an optimal policy by optimizing the accumulated reward (in statistical expectation $E\{\}$) over a (potentially $\gamma \in [0, 1]$-discounted[2]) long-term horizon (Sutton and Barto, 1998):

$$J = E\left\{ \sum_{i=0}^{T} \gamma^i r(\mathbf{x}, \mathbf{u}, t) \right\} \tag{2}$$

Unfortunately, as already noted in Bellman's original work (Bellman, 1957), learning of π becomes computationally intractable for even moderately high-dimensional state-action spaces, e.g., starting from ~6 to 10 continuous dimensions, as the search space for an optimal policy becomes too large or too nonlinear to explore empirically. Although recent developments in RL increased the range of complexity that can be dealt with (e.g., Tesauro, 1992; Bertsekas and Tsitsiklis, 1996; Sutton and Barto, 1998), it still seems that there is a long way to go before general policy learning can be applied to complex control problems like human movement.

In many theories of biological motor control and most robotics applications, the full complexity of learning a control policy is strongly reduced by assuming prior information about the policy. The most common priors are that the control policy can be reduced to a desired trajectory, $[\mathbf{x}_d(t), \dot{\mathbf{x}}_d(t)]$. Optimal control or RL approaches for trajectory learning are computationally significantly more tractable (Kawato and Wolpert, 1998; Peters et al., 2005). For instance, by using a tracking error-driven feedback controller (e.g., proportional-derivative, PD), a (explicitly time dependent) control policy can be written as:

$$\begin{aligned} \mathbf{u} &= \pi(\mathbf{x}, \alpha(t), t) = \pi(\mathbf{x}, [\mathbf{x}_d(t), \dot{\mathbf{x}}_d(t)], t) \\ &= \mathbf{K}_x(\mathbf{x}_d(t) - \mathbf{x}) + \mathbf{K}_{\dot{x}}(\dot{\mathbf{x}}_d(t) - \dot{\mathbf{x}}) \end{aligned} \tag{3}$$

For problems in which the desired trajectory is easily generated and in which the environment is static or fully predictable, such a shortcut through the problem of policy generation is highly successful. However, since policies like those in Eq. (3) are usually valid only in a local vicinity of the time course of the desired trajectory $(\mathbf{x}_d(t), \dot{\mathbf{x}}_d(t))$, they are not very flexible. A typical toy example for this problem is the tracking of the surface of a ball with the fingertip. Assume the fingertip movement was planned as a desired trajectory that moves every second 1 cm forward in tracing the surface. Now imagine that someone comes and holds the fingertip for 10 s, i.e., no movement can take place. In these 10 s, however, the trajectory plan has progressed 10 cm, and upon the release of your finger, the error-driven control law in Eq. (3) would create a strong motor command to catch up. The bad part, however, is that Eq. (3) will try to take the shortest path to catch up with the desired trajectory, which, due to the concave surface in our example, will actually try to traverse through the inside of the ball. Obviously, this behavior is

[1]Note that different parameters may actually have different functionality in the policy: some may be more low level and just store a learned pattern, while others may be higher level, e.g., as the position of a goal, that may change every time the policy is used. See, for instance, Barto and Mahadevan (2003) or the following sections of this paper.

[2]The discount factor causes rewards far in the future to be weighted down, as can be verified when expanding Eq. (2) over a few terms.

inappropriate and would hurt the human and potentially destroy the ball. Many daily life motor behaviors have similar properties. Thus, when dealing with a dynamically changing environment in which substantial and reactive modifications of control commands are required, one needs to adjust desired trajectories appropriately, or even generate entirely new trajectories by generalizing from previously learned knowledge. In certain cases, it is possible to apply scaling laws in time and space to desired trajectories (Hollerbach, 1984; Kawamura and Fukao, 1994), but those can provide only limited flexibility. For the time being, the "desired trajectory" approach seems to be too restricted for general-purpose motor control and planning in dynamically changing environments, as needed in every biological motor system, and some biological evidence has been accumulated that completely preplanned desired trajectories may not exist in human behavior[3] (Desmurget and Grafton, 2000).

Given that the concept of time-indexed desired trajectories has its problems, both from a computational and a biological plausibility point of view, one might want to look for other ways to generate control policies. From a behavioral point of view, a control policy is supposed to take the motor system from an arbitrary start point to the desired behavior. In most biological studies of arm movements, the desired behavior is simply a goal for pointing or grasping. But there is also the large class of cyclic movements, like walking, swimming, chewing, etc. Both behavioral classes can be thought of as attractor dynamics, i.e., either a point attractor as in reaching and pointing, or a limit cycle attractor as in periodic movement. Systems with attractor dynamics have been studied extensively in the nonlinear dynamic systems literature (Guckenheimer and Holmes, 1983; Strogatz, 1994). A dynamic system can generally be written as a differential equation:

$$\dot{\mathbf{x}} = f(\mathbf{x}, \alpha, t) \qquad (4)$$

which is almost identical to Eq. (1), except that the left-hand side denotes a change of state, not a motor command. Such a *kinematic* formulation is, however, quite suitable for motor control if we conceive of this dynamic system as a kinematic policy that creates kinematic target values (e.g., positions, velocities, accelerations), which subsequently are converted to motor commands by an appropriate controller (Wolpert, 1997). Planning in kinematic space is often more suitable for motor control since kinematic plans generalize over a large part of the workspace — nonlinearities due to gravity and inertial forces are taken care of by the controller at the motor execution stage (cf. Fig. 1). Kinematic plans can also theoretically be cleanly superimposed to form more complex behaviors, which is not possible if policies code motor commands directly. It should be noted, however, that a kinematic representation of movement is not necessarily independent of the dynamic properties of the limb. Proprioceptive feedback can be used on-line to modify the attractor landscape of the policy in the same way as perceptual information (Rizzi and Koditschek, 1994; Schaal and Sternad, 1998; Williamson, 1998; Nakanishi et al., 2004). Figure 1 indicates this property with the "perceptual coupling" arrow.

Most dynamic systems approaches also emphasize removing the explicit time dependency of π, such that the control policies become "autonomous dynamic systems":

$$\dot{\mathbf{x}} = f(\mathbf{x}, \alpha) \qquad (5)$$

Explicit timing is cumbersome, as it requires maintaining a clocking signal, e.g., a time counter that increments at very small time steps (as typically done in robotics). Besides that it is disputed whether biological system have access to such clocks (e.g., Keating and Thach, 1997; Roberts and Bell, 2000; Ivry et al., 2002), there is an additional level of complexity needed for aborting, halting, or resetting the clock when unforeseen disturbances happen during movement execution, as mentioned in the ball-tracing example above. The power of modeling motor control with autonomous nonlinear dynamic systems is further enhanced, as it is now theoretically rather easy to modulate the control policy by additional,

[3]It should be noted, however, that some approaches exist that can create time indexed desired trajectories in a reactive fashion (Hoff and Arbib, 1993), but these approaches only apply to a very restricted class of analytically tractable trajectories, e.g., polynomial trajectories (Flash and Hogan, 1985).

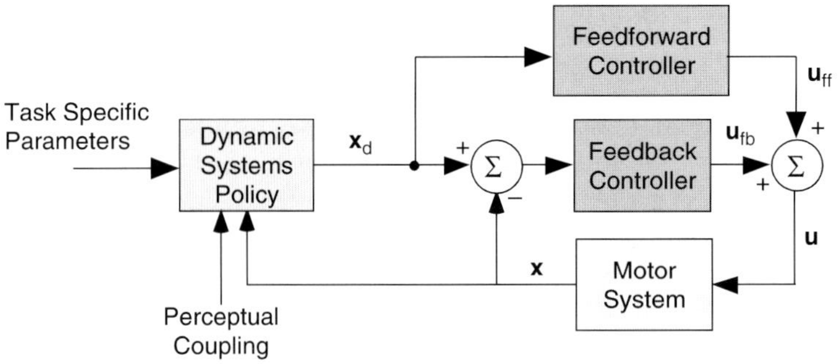

Fig. 1. Sketch of a control diagram with a dynamic systems kinematic policy, in particular how the policy is inserted into a controller with feedback (i.e., error-driven) and feedforward (i.e., anticipatory or model-based) components.

e.g., sensory or perceptual, variables, summarized in the coupling term C:

$$\dot{\mathbf{x}} = f(\mathbf{x}, \alpha) + C \tag{6}$$

We will return to such coupling ideas later in the paper.

Adopting the framework of dynamics systems theory for policy generation connects to a large body of previous work. For invertebrates and lower vertebrates, research on central pattern generators (Selverston, 1980; Getting, 1985; Kopell and Ermentrout, 1988; Marder, 2000; Righetti and Ijspeert, 2006; Ijspeert et al., 2007) has a long tradition of using coupled oscillator theories for modeling. From a behavioral point of view, many publications in the literature deal with coupled oscillator theories to explain perception–action coupling and other behavioral phenomena (Kugler et al., 1982; Turvey, 1990; Kelso, 1995). Thus, at the first glance, one might expect a straightforward and experimentally well-established framework to approach control policies as nonlinear dynamic systems. Unfortunately, this is not the case. First, modeling with nonlinear dynamics systems is mathematically quite difficult and requires usually very good intuition and deep knowledge in nonlinear systems theory — optimization approaches are often much easier to handle with well-established software tools. Second, with very few exceptions (Bullock and Grossberg, 1988; Schöner, 1990), dynamic systems approaches have only focused on periodic behavior, essentially assuming that discrete behavior is just an aborted limit cycle. In contrast, optimization approaches to motor control primarily have focused on discrete movement like reaching and pointing (e.g., Shadmehr and Wise, 2005), and rhythmic movement was frequently conceived of as cyclically concatenated discrete movements.

The goal of this paper is to demonstrate that a dynamic systems approach can offer a simple and powerful approach for both discrete and rhythmic movement phenomena, and that it can smoothly be combined with optimization approaches to address a large range of motor phenomena that have been observed in human behavior. For this purpose, first, we will review some behavioral and imaging studies that accumulated evidence that the distinction of discrete and rhythmic movement, as suggested by point and limit cycle attractors in dynamic systems theory, actually is also useful for classifying human movement. Second, we will suggest a modeling framework that can address both discrete and rhythmic movement in a simple and coherent dynamic systems framework. In contrast to any other dynamic systems approaches to motor control in the past, the suggested modeling approaches can easily be used from the viewpoint of optimization theory, too, and bridges thus the gap between dynamic systems and optimization approaches to motor control. We will demonstrate the power of our modeling approach in several synthetic and robotic studies.

Discrete and rhythmic movement — are they the same?

Since Morasso's and his coworkers' seminal work in the early 1980s (Morasso, 1981, 1983; Abend et al., 1982), a large amount of focus has been given to stroke-based trajectory formation. In the wake of this research, periodic movement was often regarded as a special case of discrete (i.e., stroke-based) movement generation, where two or more strokes are cyclically connected. In the following sections, we will review some of our own and other people's research that tried to emphasize periodic movement as an independent and equally important function of the nervous system, similar as point attractors and limit cycle attractors in dynamic systems theory require quite different treatment.

Dynamic manipulation as coupled dynamic systems

From the viewpoint of motor psychophysics, the task of bouncing a ball on a racket constitutes an interesting test bed to study trajectory planning and visuomotor coordination in humans. The bouncing ball has a strong stochastic component in its behavior and requires a continuous change of motor planning in response to the partially unpredictable behavior of the ball. In previous work (Schaal et al., 1996), we examined which principles were employed by human subjects to accomplish stable ball bouncing. Three alternative movement strategies were postulated. First, the point of impact could be planned with the goal of intersecting the ball with a well-chosen movement velocity such as to restore the correct amount of energy to accomplish a steady bouncing height (Aboaf et al., 1989); such a strategy is characterized by a constant velocity of the racket movement in the vicinity of the point of racket-ball impact. An alternative strategy was suggested by work in robotics: the racket movement was assumed to mirror the movement of the ball, thus impacting the ball within increasing velocity profile, i.e., positive acceleration (Rizzi and Koditschek, 1994). Both of these strategies are essentially stroke-based: a special trajectory is planned to hit the ball in its downward fall, and after the ball is hit, the movement is reset to redo this trajectory plan. A dynamic systems approach allows yet another way of accomplishing the ball bouncing task: an oscillatory racket movement creates a dynamically stable basin of attraction for ball bouncing, thus allowing even open-loop stable ball bouncing, i.e., ball bouncing with one's eyes closed. This movement strategy is characterized by a negative acceleration of the racket when impacting the ball (Schaal and Atkeson, 1993) — a quite nonintuitive solution: why would one break the movement before hitting the ball?

Examining the behavior of six subjects revealed the surprising result that dynamic systems captured the human behavior the best: all subjects reliably hit the ball with a negative acceleration at impact, as illustrated in Fig. 2 (note that some subjects, like Subject 5, displayed a learning process where early trials had positive acceleration at impact, but later trials switched to negative acceleration). Manipulations of bouncing amplitude also showed that the way the subjects accomplished such changes could easily be captured by a simple reparameterization of the oscillatory component of the movement, a principle that we will incorporate in our modeling approach below. Importantly, it was hard to imagine how the subjects could have achieved their behavioral characteristics with stroke-based movement generation scheme.

Apparent movement segmentation does not indicate segmented control

Invariants of human movement have been an important area of research for more than two decades. Here we will focus on two such invariants, the 2/3-power law and piecewise-planar movement segmentation, and how a parsimonious explanation of those effects can be obtained without the need of stroke-based movement planning.

Studying handwriting and 2D drawing movements, Viviani and Terzuolo (1980) were the first to identify a systematic relationship between angular velocity and curvature of the end-effector traces of human movement, an observation that was subsequently formalized in the "2/3-power law" (Lacquaniti et al., 1983):

$$a(t) = kc(t)^{2/3} \qquad (7)$$

$a(t)$ denotes the angular velocity of the endpoint trajectory, and $c(t)$ the corresponding curvature;

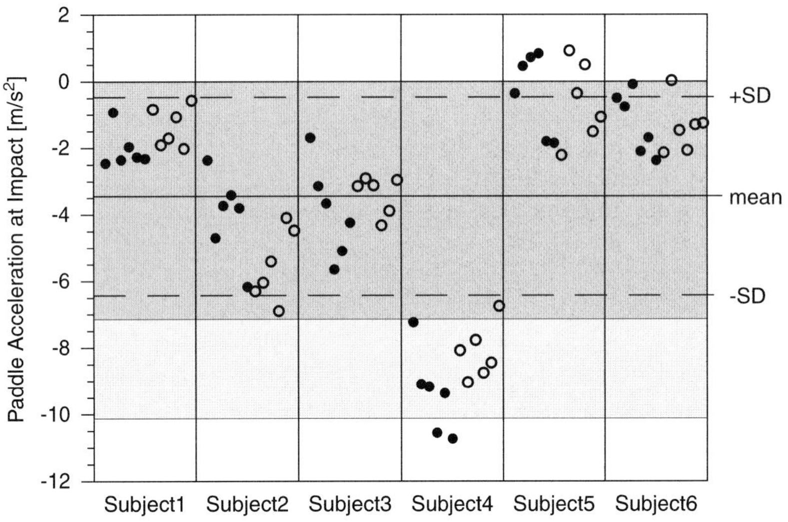

Fig. 2. Trial means of acceleration values at impact, $\ddot{x}_{P,n}$, for all six experimental conditions grouped by subject. The symbols differentiate the data for the two gravity conditions G. The dark shading covers the range of maximal local stability for G_{reduced} the light shading the range of maximal stability for G_{normal}. The overall mean and its standard deviation refers to the mean across all subjects and all conditions.

this relation can be equivalently expressed by a 1/3 power-law relating tangential velocity $v(t)$ with radius of curvature $r(t)$:

$$v(t) = kr(t)^{1/3} \tag{8}$$

Since there is no physical necessity for movement systems to satisfy this relation between kinematic and geometric properties, and since the relation has been reproduced in numerous experiments (for an overview, see Viviani and Flash, 1995), the 2/3-power law has been interpreted as an expression of a fundamental constraint of the CNS, although biomechanical properties may significantly contribute (Gribble and Ostry, 1996). Additionally, Viviani and Cenzato (1985) and Viviani (1986) investigated the role of the proportionality constant k as a means to reveal movement segmentation: as k is approximately constant during extended parts of the movement and only shifts abruptly at certain points of the trajectory, it was interpreted as an indicator for segmented control. Since the magnitude of k also appears to correlate with the average movement velocity in a movement segment, k was termed the "velocity gain factor." Viviani and Cenzato (1985) found that planar elliptical drawing patterns are characterized by a single k and, therefore, consist of one unit of action. However, in a fine-grained analysis of elliptic patterns of different eccentricities, Wann et al., 1988 demonstrated consistent deviations from this result. Such departures were detected from an increasing variability in the log-v to log-r-regressions for estimating k and the exponent β of Eq. (2), and ascribed to several movement segment each of which has a different velocity gain factor k.

The second movement segmentation hypothesis we want to address partially arose from research on the power law. Soechting and Terzuolo (1987a, b) provided qualitative demonstrations that 3D rhythmic endpoint trajectories are piecewise planar. Using a curvature criterion as the basis for segmentation, they confirmed and extended Morasso's (1983) results that rhythmic movements are segmented into piecewise planar strokes. After Pellizzer et al. (1992) demonstrated piecewise planarity even in an isometric task, movement segmentation into piecewise planar strokes has largely been accepted as one of the features of human and primate arm control.

We repeated some of the experiments that led to the derivation of the power law, movement segmentation based on the power law, and movement segmentation based on piecewise planarity. We tested six human subjects when drawing elliptical patterns and figure-8 patterns in 3D space freely in front of their bodies. Additionally, we used an anthropomorphic robot arm, a Sarcos Dexterous Arm, to create similar patterns as those performed by the subjects. Importantly, the robot generated the elliptical and figure-8 patterns solely out of joint-space oscillations, i.e., a nonsegmented movement control strategy. For both humans and the robot, we recorded the 3D position of the fingertip and the seven joint angles of the performing arm.

Figure 3 illustrates data traces of one human subject and the robot subject for elliptical drawing patterns of different sizes and different orientations. For every trajectory in this graph, we computed the tangential velocity of the fingertip of the arm and plotted it versus the radius of curvature raised to the power 1/3. If the power law were obeyed, all data points should lie on a straight line through the origin. Figure 3a, b clearly

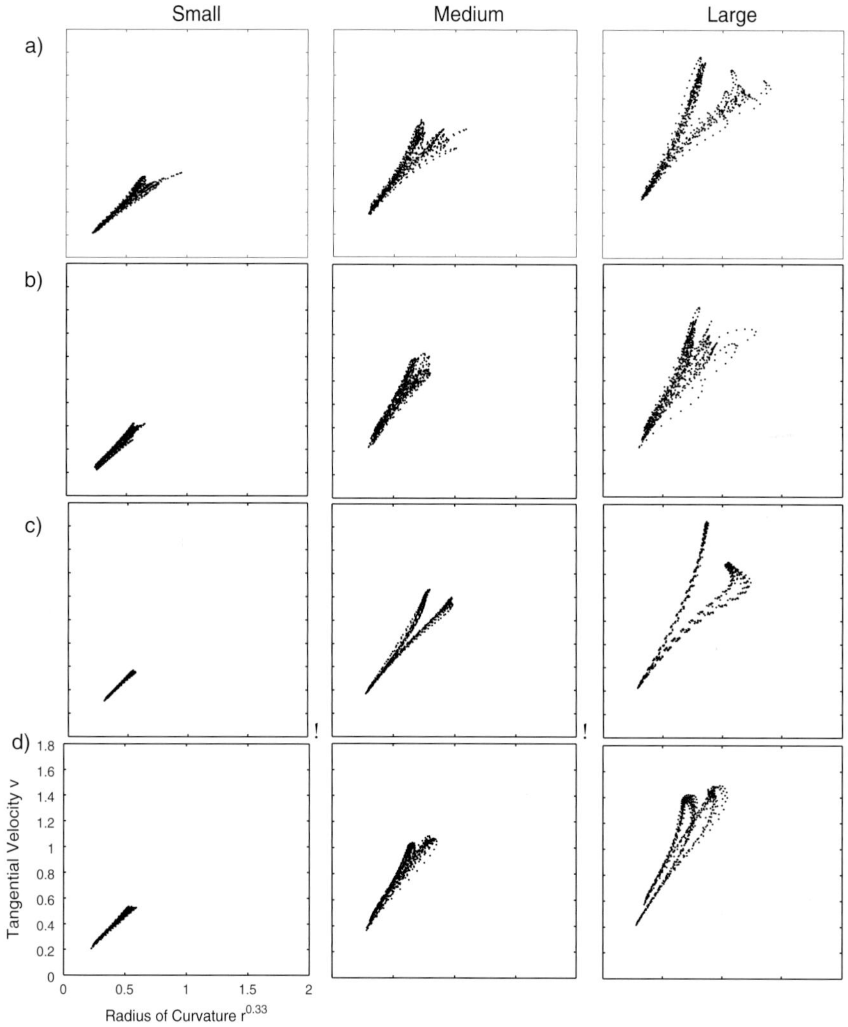

Fig. 3. Tangential velocity versus radius of curvature to the power 1/3 for ellipses of small, medium, and large size for elliptical pattern orientations in the frontal and oblique workspace plane: (a) human frontal; (b) human oblique; (c) robot frontal; (d) robot oblique.

demonstrates that for large size patterns, this is not the case, indicating that the power seems to be violated for large size patterns. However, the development of two branches for large elliptical patterns in Fig. 3a, b could be interpreted that large elliptical movement patterns are actually composed of two segments, each of which obeys the power law. The rejection of the latter point comes from the robot data in Fig. 3c, d. The robot produced strikingly similar features in the trajectory realizations as the human subjects.

However, the robot simply used oscillatory joint space movement to create these patterns, i.e., there was no segmented movement generation strategy. Some mathematical analysis of the power law and the kinematic structure of human arms could finally establish that the power law can be interpreted as an epiphenomenon of oscillatory movement generation: as long as movement patterns are small enough, the power law holds, while for large size patterns the law breaks down (Sternad and Schaal, 1999; Schaal and Sternad, 2001). Using figure-8 patterns instead of elliptical patterns, we were also able to illuminate the reason for apparent piecewise-planar movement segmentation in rhythmic drawing patterns. Figure 4 shows figure-8 patterns performed by human and robot subjects in a planar projection when looking at the figure-8 from the side. If realized with an appropriate width-to-height ratio, figure-8 patterns look indeed like piecewise planar trajectories in this projection and invite the hypothesis of movement segmentation at the node of the figure-8. However, as in the previous experiment, the robot subject produced the same features of movement segmentation despite the fact that it used solely joint space oscillations to create the patterns, i.e., no movement segmentation. Again, it was possible to explain the apparent piecewise planarity from a mathematical analysis of the kinematics of the human arm, rendering piecewise planarity to be an epiphenomenon of oscillatory joint space trajectories and the nonlinear kinematics of the human arm (Sternad and Schaal, 1999).

Superposition of discrete and rhythmic movement

In another experiment, we addressed the hypothesis that discrete and rhythmic movements are two separate movement regimes that can be used in superposition, sequence, or isolation. Subjects performed oscillatory movements around a given point in the workspace with one joint of the arm, and shifted the mean position of another joint of the same (or the other arm) at an auditory signal to another point. In previous work (Adamovich et al., 1994), it was argued that such a discrete shift terminates the oscillatory movement (generated by two cyclically connected movement strokes) and restarts it after the shift, i.e., the entire system of

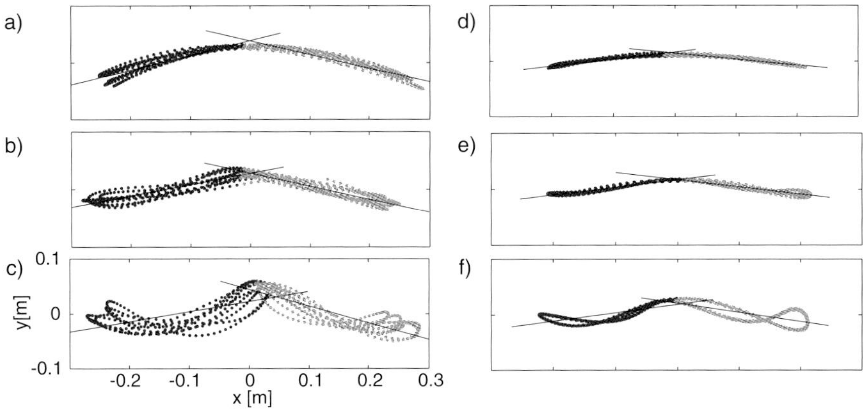

Fig. 4. Planar projection of one subject's figure-8 patterns of small, medium, and large width/height ratio: (a–c) human data; (d–f) corresponding robot data. The data on the left side of each plot belong to one lobe of the figure-8, and the data on the right side to the other figure-8 lobe.

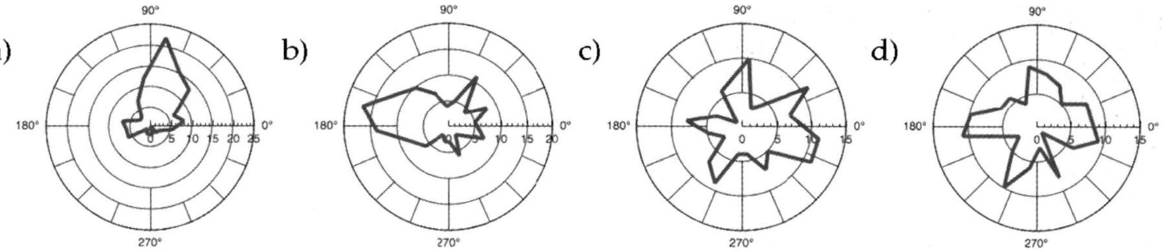

Fig. 5. Polar histograms of the phase of the discrete movement onset in various experimental conditions, averaged over six participants: (a) a rhythmic elbow movement is superimposed with a discrete elbow flexion; (b) a rhythmic elbow movement is superimposed with discrete wrist supination; (c) a rhythmic wrist flexion–extension movement with superimposed discrete shoulder flexion; (d) a right elbow flexion–extension movement superimposed with a discrete left elbow flexion movement. In (a) and (b), the onset of the discrete movement is confined to a phase window of the on-going rhythmic movement. In (c) and (d), no such phase window was found.

rhythmic and discrete movement was assumed to be generated by a sequence of discrete strokes.

Among the most interesting features of this experiment was that the initiation of the discrete movement superimposed onto ongoing rhythmic movement was constrained to a particular phase window of the ongoing rhythmic movement when both discrete and rhythmic movement used the same joint (Adamovich et al., 1994; Sternad et al., 2000, 2002; De Rugy and Sternad, 2003) (Fig. 5a) and even when the discrete and rhythmic movement used different joints (Fig. 5b) (Sternad and Dean, 2003). Furthermore, in both types of experiments the ongoing rhythmic movement was disrupted during the discrete initiation and showed phase resetting. Interestingly, in a bimanual task (Wei et al., 2003), where subjects performed rhythmic movement with their dominant arm and initiated a second discrete movement with their nondominant arm, there was no evidence of a preferred phase window for the discrete movement onset (Fig. 5d).

In Mohajerian et al. (2004), we repeated this experimental paradigm over a systematic set of combinations of discrete and rhythmic movement of different joints of the same arm, and also joints from the dominant and nondominant arm — some of the results are shown in Fig. 5. All observed phenomena of phase windows of the discrete movement onset and phase resetting of the rhythmic movement could be explained by superimposed rhythmic and discrete movement components and spinal reflexes. While the CNS executes the rhythmic movement, the discrete movement is triggered according to the auditory cue as a superimposed signal. If the rhythmic movement uses a muscle that is also needed for the discrete movement, and if this muscle is currently inhibited by the spinal interneuronal circuits due to reciprocal inhibition, the discrete movement onset is delayed. Such a superposition also leads to phase resetting of the rhythmic movement. Whenever the rhythmic and discrete joint did not share muscles for execution, no phase windows and phase resetting was observed (Fig. 5c, d). One more time, the hypothesis of independent circuits for discrete and rhythmic movement offered and elegant and simple explanation for observed behavioral phenomena.

Brain activation in discrete and rhythmic movement

Among the most compelling evidence in favor of the idea that discrete and rhythmic movement are independent functional circuits in the brain is a recent fMRI study that demonstrated that rhythmic and discrete movement activate different brain areas. Figure 6 illustrates the summary results from this experiment, where subjects performed either periodic wrist flexion–extension oscillations, or discrete flexion-to-extension or extension-to-flexion point-to-point movements with the same wrist. The major findings were that while rhythmic movement activated only a small number of unilateral primary motor areas (M1, S1, PMdc, SMA,

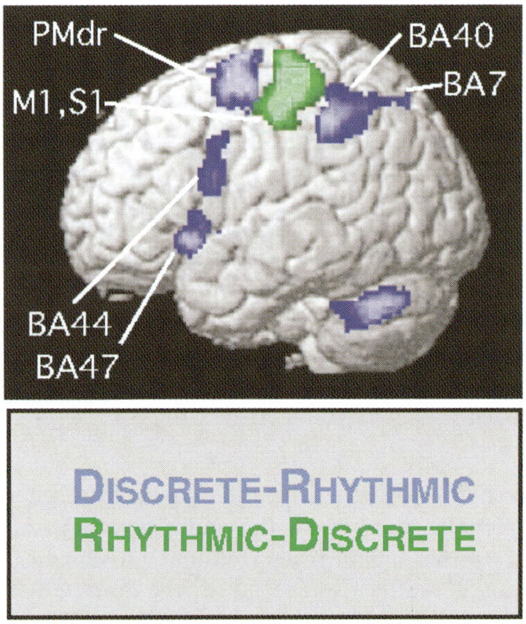

Fig. 6. Differences in brain activation between discrete and rhythmic wrist movements. Abbreviations are (Picard and Strick, 2001): CCZ: caudal cingulate zone; RCZ: rostral cingulate zone, divided in an anterior (RCZa) and posterior (RCZp) part; SMA: caudal portion of the supplementary motor area, corresponding to SMA proper; pre-SMA: rostral portion of the supplementary motor area; M1: primary motor cortex; S1: primary sensory cortex; PMdr: rostral part of the dorsal premotor cortex; PMdc: caudal part of the dorsal premotor cortex; BA: Brodman area; BA7: precuneus in parietal cortex; BA8: middle frontal gyrus; BA 9: middle frontal gyrus; BA10: anterior frontal lobe; BA47: inferior frontal gyrus; BA40: inferior parietal cortex; BA44: Broca's area.

pre-SMA, CCZ, RCZp, cerebellum), discrete movement activated a variety of additional contralateral nonprimary motor areas (BA7, BA40, BA44, BA47, PMdr, RCZa) and, moreover, showed very strong bilateral activity in both the cerebrum and cerebellum (Schaal et al., 2004). Figure 6 shows some of these results in as much as they can be visualized on the surface of the left hemisphere: most important are the Discrete-Rhythmic (blue) areas, which were unique to discrete movement. The Rhythmic-Discrete (green) area is actually active in both rhythmic and discrete movements, just to larger extend in rhythmic movement, which can be explained by the overall larger amount of movement in rhythmic trials. Control experiments examined whether such unbalanced amounts of movement in rhythmic movement, and, in discrete movement, the much more frequent movement initiation and termination and the associated cognitive effort could account for the observed differences. Only BA40, BA44, RCZa, and the cerebellum were potentially involved in such issues, leaving BA7, BA47, and PMdr as well as a large amount of bilateral activation a unique feature in discrete movement. Since rhythmic movement activates significantly fewer brain areas than discrete movement, it was concluded that it does not seem to be warranted to claim that rhythmic movement is generated on top of a discrete movement system, i.e., rhythmic arm movement is *not* composed of discrete strokes. The independence of discrete and rhythmic movement systems in the brain seemed to be the most plausible explanation of the imaging data, which is in concert with various other studies that demonstrated different behavioral phenomena in discrete and rhythmic movement (e.g., Smits-Engelsman et al., 2002; Buchanan et al., 2003; Spencer et al., 2003).

Discrete and rhythmic movement: a computational model

The previous section tried to establish that a large number of behavioral experiments support the idea that discrete and rhythmic movement should be treated as separate movement systems, and in particular, that there is strong evidence against the hypothesis that rhythmic movement is generated from discrete strokes. We will now turn to a unifying modeling framework for discrete and rhythmic movement, with the special focus to bridge dynamic systems approaches and optimization approaches to motor control. A useful start is to establish a list of properties that such a modeling framework should exhibit. In particular, we wish to model:

- point-to-point and periodic movements,
- multijoint movement that requires phase locking and arbitrary phase offsets between individual joints (e.g., as in biped locomotion),

- discrete and rhythmic movement that have rather complex trajectories (e.g., joint reversals, curved movement, a tennis forehand, etc.),
- learning and optimization of movement,
- coupling phenomena, in particular bimanual coupling phenomena and perception–action coupling,
- timing (without requiring an explicit time representation),
- generalization of learned movement to similar movement tasks,
- robustness of movements to disturbances and interactions with the environment.

As a starting point, we will use a dynamic systems model, as this approach seems to be the best suited for creating autonomous control policies that can accommodate coupling phenomena. Given that the modeling approach suggested below will be able to represent a library of different movements in the language of dynamic systems theory, we conceive of every member of this library as a movement primitive, and call our approach Dynamic Movement Primitives (DMPs) (Ijspeert et al., 2001, 2002a, b, 2003).

We assume that the variables of a DMP represent the desired kinematic state of a limb, i.e., desired positions, velocities, and accelerations for each joint. Alternatively, the DMP could also be defined in task space, and we would use appropriate task variables (e.g., the distance of the hand from an object to be grasped) as variables for the DMP — for the discussions in this paper, this distinction is, however, of subordinate importance, and, for the ease of presentation, we will focus on formulations in joint space. As shown in Fig. 1, kinematic variables are converted to motor commands through a feedforward controller — usually by employing an inverse dynamics model — and stabilized by low gain[4] feedback control. The example of Fig. 1 corresponds to a classical computed torque controller (Craig, 1986), which has also been suggested for biological motor control (Kawato, 1999), but any other control scheme could be inserted here. Thus, the motor execution of DMPs can incorporate any control technique that takes as input kinematic trajectory plans, and in particular, it is compatible with current theories of model-based control in computational motor control.

[4]The emphasis of low gain feedback control is motivated by the desire to have a movement system that is compliant when interacting with external objects or unforeseen perturbation, which is a hallmark of human motor control, but quite unlike traditional high gain control in most robotics applications.

Motor planning with DMPs

In order to accommodate discrete and rhythmic movement plans, two kinds of DMPs are needed: point attractive systems and limit-cycle systems. The key question of DMPs is how to formalize nonlinear dynamic equations such that they can be flexibly adjusted to represent complex motor behaviors without the need for manual parameter tuning and the danger of instability of the equations. We will sketch our approach in the example of a discrete dynamic system for reaching movements — an analogous development holds for rhythmic systems.

Assume we have a basic point attractive system, instantiated by the second order dynamics

$$\tau\dot{z} = \alpha_z(\beta_z(g - y) - z) + f, \quad \tau\dot{y} = z \qquad (9)$$

where g is a known goal state, α_z and β_z time constants, τ a temporal scaling factor (see below) and $y,\dot{y}$ correspond to the desired position and velocity generated by Eq. (9), interpreted as a movement plan as used in Fig. 1. For instance, $y,\dot{y}$ could be the desired states for a one degree-of-freedom motor system, e.g., the elbow flexion–extension. Without the function f, Eq. (9) is nothing but the first-order formulation of a linear spring-damper, and, after some reformulation, the time constants α_z and β_z have an interpretation in terms of spring stiffness and damping. For appropriate parameter settings and $f = 0$, these equations form a globally stable linear dynamic system with g as a unique point attractor, which means that for any start position the limb would reach g after a transient, just like a stretched spring, upon release, will return to its equilibrium point. Our key goal, however, is to instantiate the nonlinear function f in Eq. (9) to change the rather trivial exponential and monotonic convergence of y towards g to allow trajectories that are more complex on the way to the goal. As such a change of Eq. (9) enters the domain of nonlinear dynamics, an arbitrary complexity of the resulting equations might be expected. To the best of our knowledge, this problem has prevented research from employing nonlinear dynamic systems models on a larger scale so far. We will address this problem by first introducing a bit more formalism, and then by analyzing the resulting system equations.

The easiest way to force Eq. (9) to become more complex would be to create a function f as an explicit function of time. For instance, $f(t) = \sin(\omega t)$ would create an oscillating trajectory y, or $f(t) = \exp(-t)$ would create a speed up of the initial part of the trajectory y — such functions are called forcing functions in dynamic systems theory (Strogatz, 1994), and, after some reformulation, Eq. (9) could also be interpreted as PD controller that tracks a complex desired trajectory, expressed with the help f. But, as mentioned before, we would like to avoid explicit time dependencies. To achieve this goal, we need an additional dynamic system

$$\tau \dot{x} = -\alpha_x x \tag{10}$$

and the nonlinear function f in form of

$$f(x, g, y_0) = \frac{\sum_{i=1}^{N} \psi_i w_i x}{\sum_{i=1}^{N} \psi_i}(g - y_0),$$
$$\text{where } \psi_i = \exp(-h_i(x - c_i)^2) \tag{11}$$

Equation (10) is a simple first order "leaky-integrator" equation as used in many models of neural dynamics (e.g., Hodgkin and Huxley, 1952) — we will call this equation the *canonical* system from now on, as it is among the most basic dynamic systems available to create a point attractor. From any initial conditions, Eq. (10) can be guaranteed to converge monotonically to zero. This monotonic convergence of x becomes a substitute for time: all what time does is that it monotonically increases, similar to the time course of x. Of course, x behaves also a little bit different from time: it monotonically decreases (which, mathematically, is just a technically irrelevant detail), and it saturates exponentially at the value "0", which is appropriate as we expect that at this time the movement terminates. Equation (11) is a standard representation of a nonlinear function in terms of basis functions, as commonly employed in modeling population coding in the primate brain (e.g., Mussa-Ivaldi, 1988; Georgopoulos, 1991). Let us assume that the movement system is in an initial state $y = g = y_0$, $z = 0$, and $x = 0$. To trigger a movement, we change the goal g to a desired value and set $x = 1$ (where the value "1" is arbitrary and just chosen for convenience), similar as done with the "go" value in (Bullock and Grossberg, 1988). The duration of the movement is determined by the time constant τ. The value of x will now monotonically converge back to zero. Such a variable is called a "phase" variable as one can read out from its value in which phase of the movement we are, where "1" is the start, and "0" is the end. The nonlinear function f is generated by anchoring its Gaussian basis functions ψ_i (characterized by a center c_i and bandwidth h_i) in terms of the phase variable x. The phase x appears also multiplicative in Eq. (11) such that the influence of f vanishes at the end of the movement when x has converged to zero (see below). It can be shown that the combined system in Eqs. (9)–(11) asymptotically converge to the unique point attractor g.

The example in Fig. 7 clarifies the ingredients of the discrete DMP. The top row of Fig. 7 illustrates the position, velocity, and acceleration trajectories that serve as desired inputs to the motor command generation stage (cf. Fig. 1) — acceleration is equivalent to the time derivative of z, $\ddot{y} = \dot{z}$. In this example, the trajectories realize a minimum jerk trajectory (Hogan, 1984), a smooth trajectory as typically observed in human behavior (the ideal minimum jerk trajectory, which minimizes the integral of the squared jerk along the trajectory, is superimposed to the top three plots of Fig. 7, but the difference to the DMP output is hardly visible). The remaining plots of Fig. 7 show the time course of all internal variables of the DMP, as given by Eqs. (9)–(11). Note that the trajectory of x is just a strictly monotonically decreasing

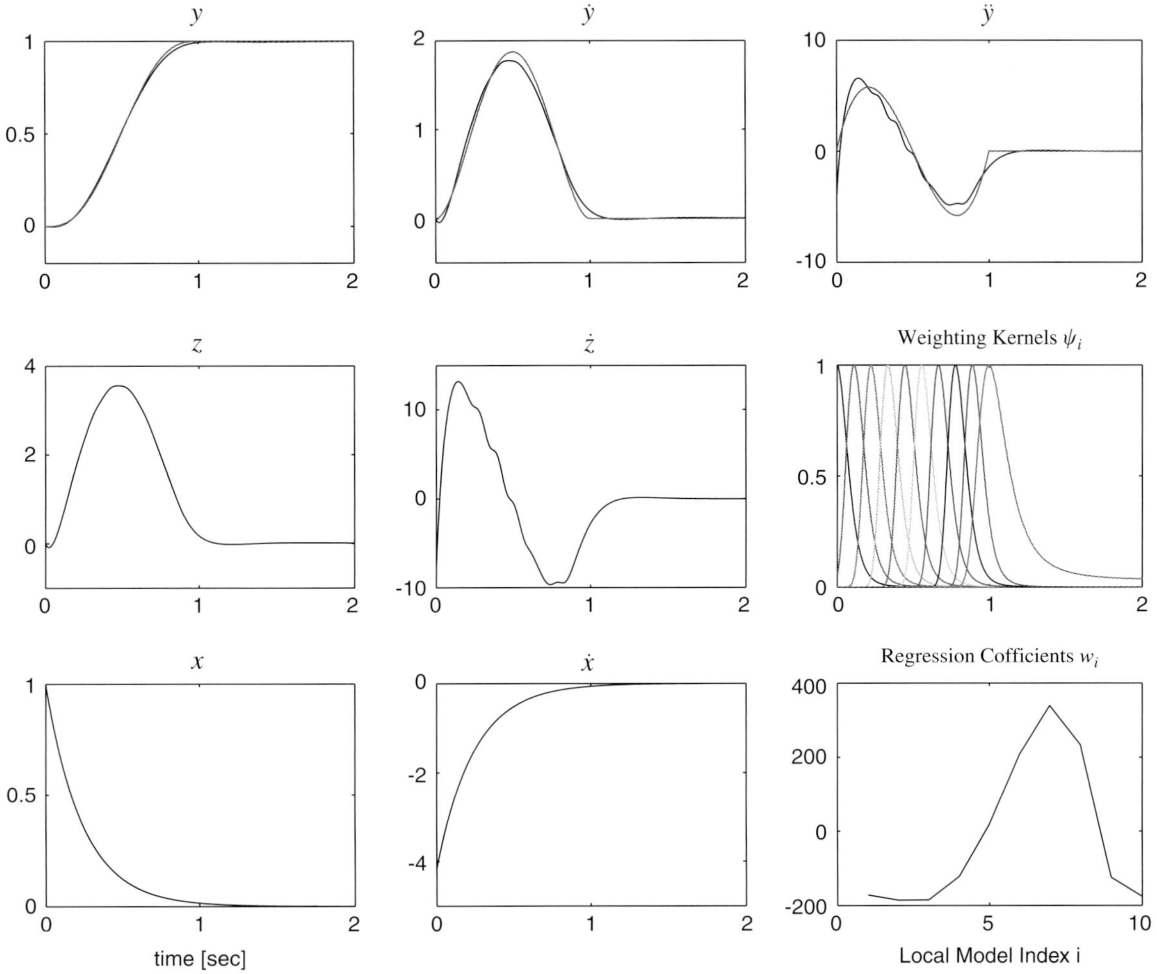

Fig. 7. Example of all variables of a discrete movement dynamic primitive as realized in a minimum jerk movement from zero initial conditions to goal state $g = 1$.

curve. As x multiplies the nonlinearity in Eq. (11), the nonlinearity only acts in a transient way, one of the main reasons that these nonlinear differential equations remain relatively easy to analyze. The basis function activations (ψ_i) are graphed as a function of time, and demonstrate how they essentially partition time into shorter intervals in which the function value of f can vary.

It is not the particular instantiation in Eqs. (9)–(11) that is the most important idea of DMPs, but rather it is the design principle that matters. A DMP consists of two sets of differential equations: a *canonical* system

$$\tau\dot{\mathbf{x}} = \mathbf{h}(\mathbf{x}, \theta) \tag{12}$$

and an *output* system

$$\tau\dot{\mathbf{y}} = \mathbf{g}(\mathbf{y}, f, \theta) \tag{13}$$

where we just inserted θ as a placeholder for all parameters of the these systems, like goal, time constants, etc. The canonical system needs to generate the phase variable x and is a substitute for time for anchoring our spatially localized basis functions Eq. (11). The appealing property of using a phase variable instead of an explicit time representation is that we can now manipulate the time evolution of phase, e.g., by speeding up or slowing down a movement as appropriate by means of additive coupling terms or phase resetting techniques (Nakanishi et al., 2004) — in contrast, an explicit time representation cannot be manipulated as easily. For instance, Eq. (10) could be augmented to be

$$\tau \dot{x} = \alpha_x x \frac{1}{1 + \alpha_c (y_{\text{actual}} - y)^2} \quad (14)$$

The term $(y_{\text{actual}} - y)$ is the tracking error of the motor system, if this error is large, the time development of the canonical system comes to a stop, until the error is reduced — this is exactly what one would want if a motor act got suddenly perturbed.

An especially useful feature of this general formalism is that it can be applied to rhythmic movements as well, simply by replacing the point attractor in the canonical system with a limit cycle oscillator (Ijspeert et al., 2003). Among the simplest oscillators is a phase representation, i.e., constant phase speed:

$$\tau \dot{\phi} = 1 \quad (15)$$

where r is the amplitude of the oscillator, A the desired amplitude, and ϕ its phase. For this case, Eq. (11) is modified to

$$f(\phi, A) = \frac{\sum_{i=1}^{N} \psi_i w_i}{\sum_{i=1}^{N} \psi_i} A,$$

$$\text{where } \psi_i = \exp(h_i(\cos(\phi - c_i) - 1)) \quad (16)$$

with A being the amplitude of the desired oscillation. The changes in Eq. (16) are motivated by the need to make the function f a function that lives on a circle, i.e., ψ_i are computed from a Gaussian function that lives on a circle (called von Mises function). The output system in Eq. (9) remains the same, except that we now identify the goal state g with a setpoint around which the oscillation takes place. Thus, by means of A, τ, and g, we can control amplitude, frequency, and setpoint of an oscillation independently.

DMPs for multidimensional motor systems

The previous section addressed only a one-dimensional motor system. If multiple dimensions are to be coordinated, e.g., as in the seven major degrees of freedom (DOFs) of a human arm, all that is required is to create a separate output system for every DOF (i.e., Eqs. (9) and (13)). The canonical system is shared across all DOFs. Thus, every DOF will have its own goal g (or amplitude A) and nonlinear function f. As all DOFs reference the same phase variable through the canonical system, it can be guaranteed that the DOFs remain properly coordinated throughout a movement, and in rhythmic movement, it is possible to create very complex stable phase relationship between the individual DOFs, e.g., as needed for biped locomotion. In comparison to previous work on modeling multidimensional oscillator systems for movement generation that required complex oscillator tuning to achieve phase locking and synchronization (e.g., Taga et al., 1991), our approach offers a drastic reduction of complexity.

Learning and optimization with DMPs

We can now address how the open parameters of DMPs are instantiated. We assume that goal g (or amplitude A) as well as the timing parameter τ is provided by some external behavioral constraints. Thus, all that is needed is to find the weights w_i in the nonlinear function f. Both supervised and reinforcement/optimization approaches are possible.

Supervised learning with DMPs

Given that f is a normalized basis function representation, linear in the coefficients of interest (i.e., w_i i) (e.g., Bishop, 1995), a variety of learning algorithms exist to find w_i. In supervised learning scenario, we can suppose that we are given a sample trajectory $y_{\text{demo}}(t)$, $\dot{y}_{\text{demo}}(t)$, $\ddot{y}_{\text{demo}}(t)$ with duration T, for instance, from the demonstration of a teacher. Based on this information, a supervised learning problem results with the following target for f:

$$f_{\text{target}} = \tau \ddot{y}_{\text{demo}} - \alpha_z(\beta_z(g - y_{\text{demo}}) - \tau \dot{y}_{\text{demo}}) \quad (17)$$

In order to obtain a matching input for f_{target}, the canonical system needs to be integrated. For this purpose, in Eq. (10), the initial state of the canonical system is set to $x = 1$ before integration. An analogous procedure is performed for the rhythmic DMPs. The time constant τ is chosen such that the DMP with $f = 0$ achieves 95% convergence at $t = T$. With this procedure, a clean supervised learning problem is obtained over the time course of the movement to be approximated with training samples (x, f_{target}).

For solving the function approximation problem, we chose a nonparametric regression technique from locally weighted learning (LWPR) (Vijayakumar and Schaal, 2000). This method allows us to determine the necessary number of basis functions N, their centers c_i, and bandwidth h_i automatically. In essence, every basis function ψ_i defines a small region in input space x, and point falling into this region are used to perform a linear regression analysis, which can be formalized as weighted regression (Atkeson et al., 1997). Predictions for a query point are generated by ψ_i-weighted average of the predictions of all local models. In simple words, we create a piecewise linear approximation of f_{target}, where each linear function piece belongs to one of the basis functions.

As evaluations of the suggested approach to movement primitives, in Ijspeert et al. (2002b), we demonstrated how a complex tennis forehand and tennis backhand swing can be learned from a human teacher, whose movements were captured at the joint level with an exoskeleton. Figure 8 illustrates imitation learning for a rhythmic trajectory using the phase oscillator DMP from Eqs. (15) and (16). The images in the top of Fig. 8 show four frames of the motion capture of a figure-8 pattern and its repetition on the humanoid robot after imitation learning of the trajectory. The plots in Fig. 9 demonstrate the motion captured and fitted trajectory of a bimanual drumming pattern, using 6 DOFs per arm. Note that rather complex phase relationships between the individual DOFs can be realized. For one joint angle, the right elbow joint (R_EB), Fig. 10 exemplifies the effect of various changes of parameter settings of the DMP (cf. also figure caption in Fig. 8). Here it is noteworthy how quickly the pattern converges to the new limit cycle attractor, and that parameter changes do not change the movement pattern qualitatively, an effect that can be predicted theoretically (Schaal et al., 2003). The nonlinear function of each DMP employed 15 basis functions.

Optimization of DMPs

The linear parameterization of DMPs allows any form of parameter optimization, not just supervised learning. As an illustrative example, we considered a 1 DOF movement linear movement system

$$m\ddot{y} + b\dot{y} + ky = u \tag{18}$$

with mass m, damping b, and spring stiffness k. For point-to-point movement, we optimized the following criteria:

$$\left.\begin{array}{ll} \text{Minimum Jerk} & J = \int_0^T \dddot{y}^2 dt \\ \text{Minimum Torque Change} & J = \int_0^T \dot{u}^2 dt \\ \text{Minimum Endpoint Variance} & \\ \quad \text{with signal dependent noise} & \\ J = \operatorname{var}(y(T) - g) + \operatorname{var}(\dot{y}(T)) & \end{array}\right\} \tag{19}$$

where in the case of the minimum-endpoint-variance criterion, we assumed signal dependent noise $u_{\text{noisy}} = (1 + \varepsilon)u$ and $\varepsilon \sim \text{Normal}(0, 0.04)$.

The results of these optimizations, using the Matlab optimization toolbox, are shown in Fig. 11. As a comparison, we superimposed the results of a minimum jerk trajectory in every plot. The velocity profiles obtained from the DMPs after optimization nicely coincide with what has been obtained in the original literature suggesting these optimization criteria (Flash and Hogan, 1985; Uno et al., 1989; Harris and Wolpert, 1998). What is the most important, however, was that it was essentially trivial to apply various optimization approaches to our dynamic systems representation of movement generation. Thus, we believe that these results are among the first that successfully combined dynamic systems representations to motor control and optimization approaches.

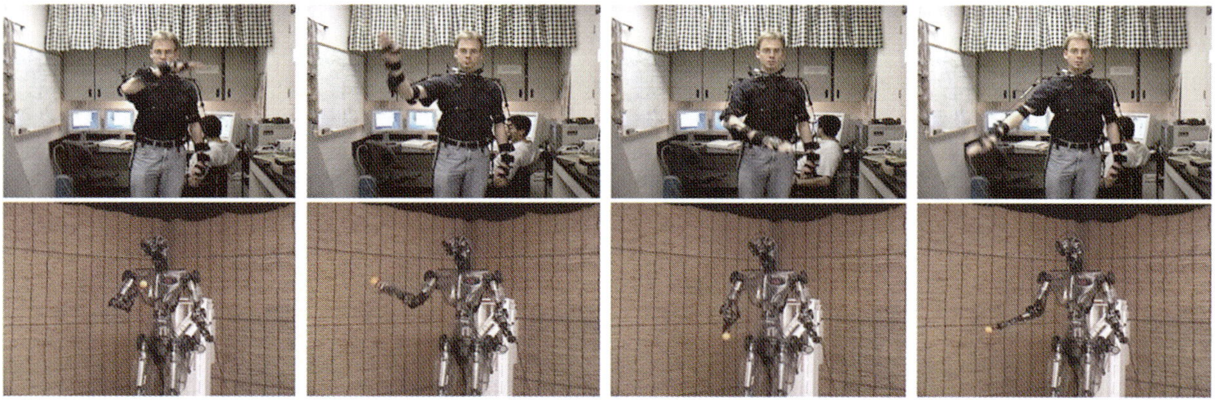

Fig. 8. Humanoid robot learning a figure-8 movement from a human demonstration.

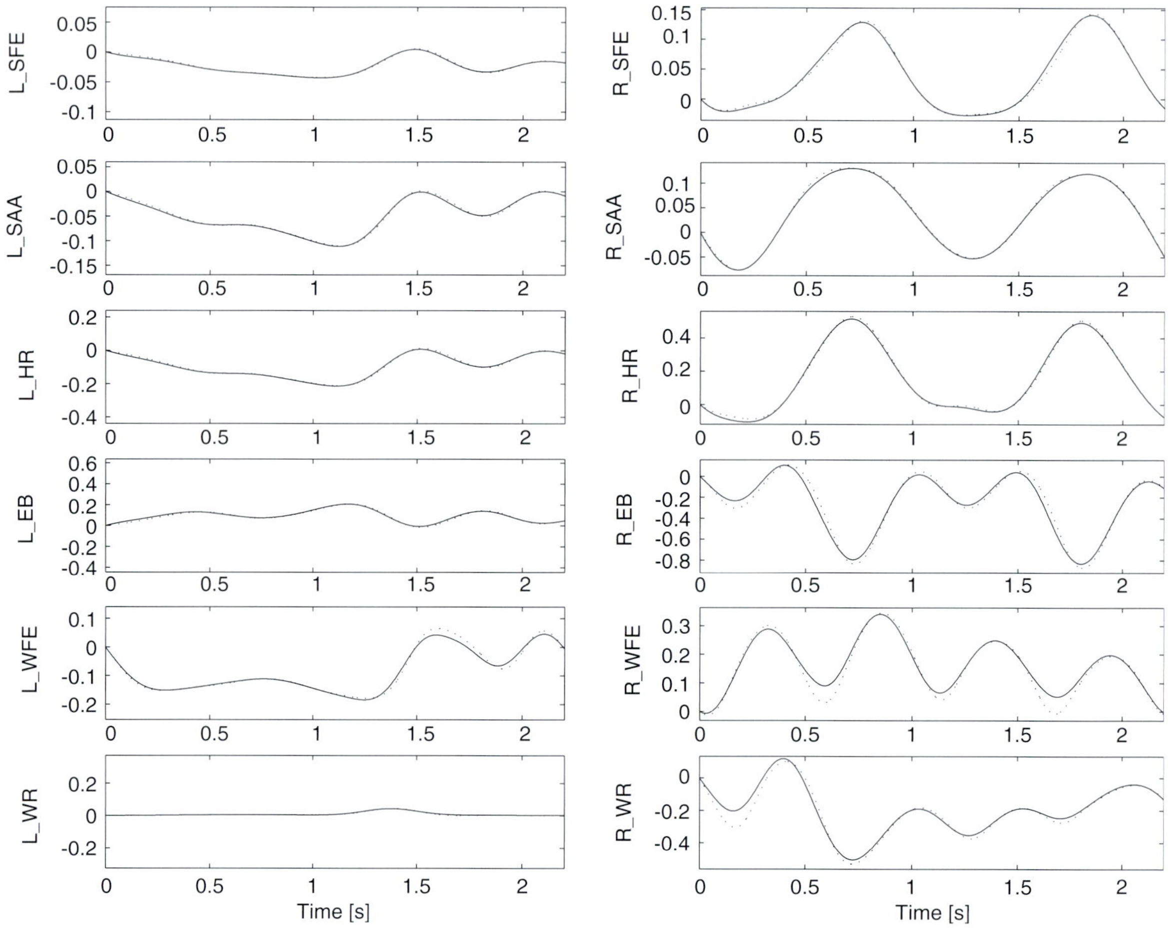

Fig. 9. Recorded drumming movement performed with both arms (6 DOFs per arm). The dotted lines and continuous lines correspond to one period of the demonstrated and learned trajectories, respectively — due to rather precise overlap, they are hardly distinguishable.

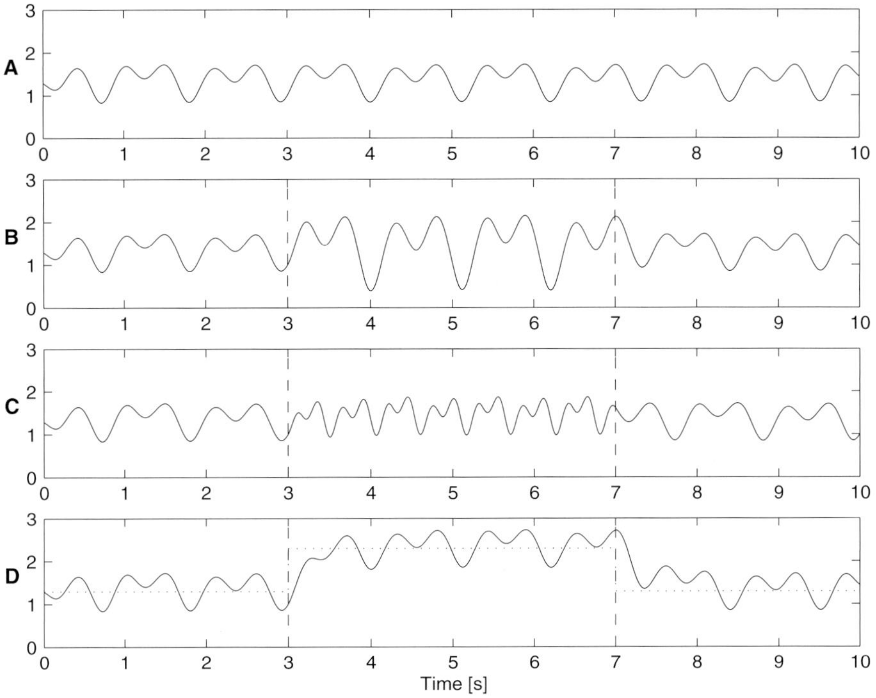

Fig. 10. Modification of the learned rhythmic drumming pattern (flexion/extension of the right elbow, R_EB). (A) Trajectory learned by the rhythmic DMP; (B) temporary modification with $A \leftarrow 2A$ in Eq. (16); (C): temporary modification with $\tau \leftarrow \tau/2$ in Eqs. (9) and (15); (D): temporary modification with $g \leftarrow g+1$ in Eq. (9) (dotted line). Modified parameters were applied between $t = 3$ s and $t = 7$ s. Note that in all modifications, the movement patterns do not change qualitatively, and convergence to the new attractor under changed parameters is very rapid.

Discussion

This paper addressed a computational model for movement generation in the framework of dynamic systems approaches, but with a novel formulation that also allows applying optimization and learning approaches to motor control. We started by reviewing some of our own work that established evidence that periodic and point-to-point movements need to be investigated as separate functionalities of motor control, similar to the fact that point attractors and limit cycle attractors require different theoretical treatment in dynamic systems theory. We also emphasized that models of movement generation should not have explicit time dependency, similar to autonomous dynamic systems, in order to accommodate coupling and perturbation effects in an easy way. While these requirements favor a dynamic systems formulation of motor control, there has been no acceptable computational framework so far that combines both the properties of dynamic systems approaches to motor control and the ease of applying learning and optimization approaches, which have played a dominant role in computational motor control over that last years (e.g., Shadmehr and Wise, 2005).

Our formulation of Dynamic Motor Primitives (DMP) offers a viable solution. Essentially DMPs are motivated by the VITE model of Bullock and Grossberg (1988) and other approaches that emphasized that movement should be driven by a difference vector between the current and the desired goal of a movement (for a review, see Shadmehr and Wise, 2005). DMPs create desired trajectories for a movement system out of the temporal evolution of autonomous nonlinear differential equations, i.e., the desired trajectory is created in real-time together with movement execution, and not as a preplanned entity. This

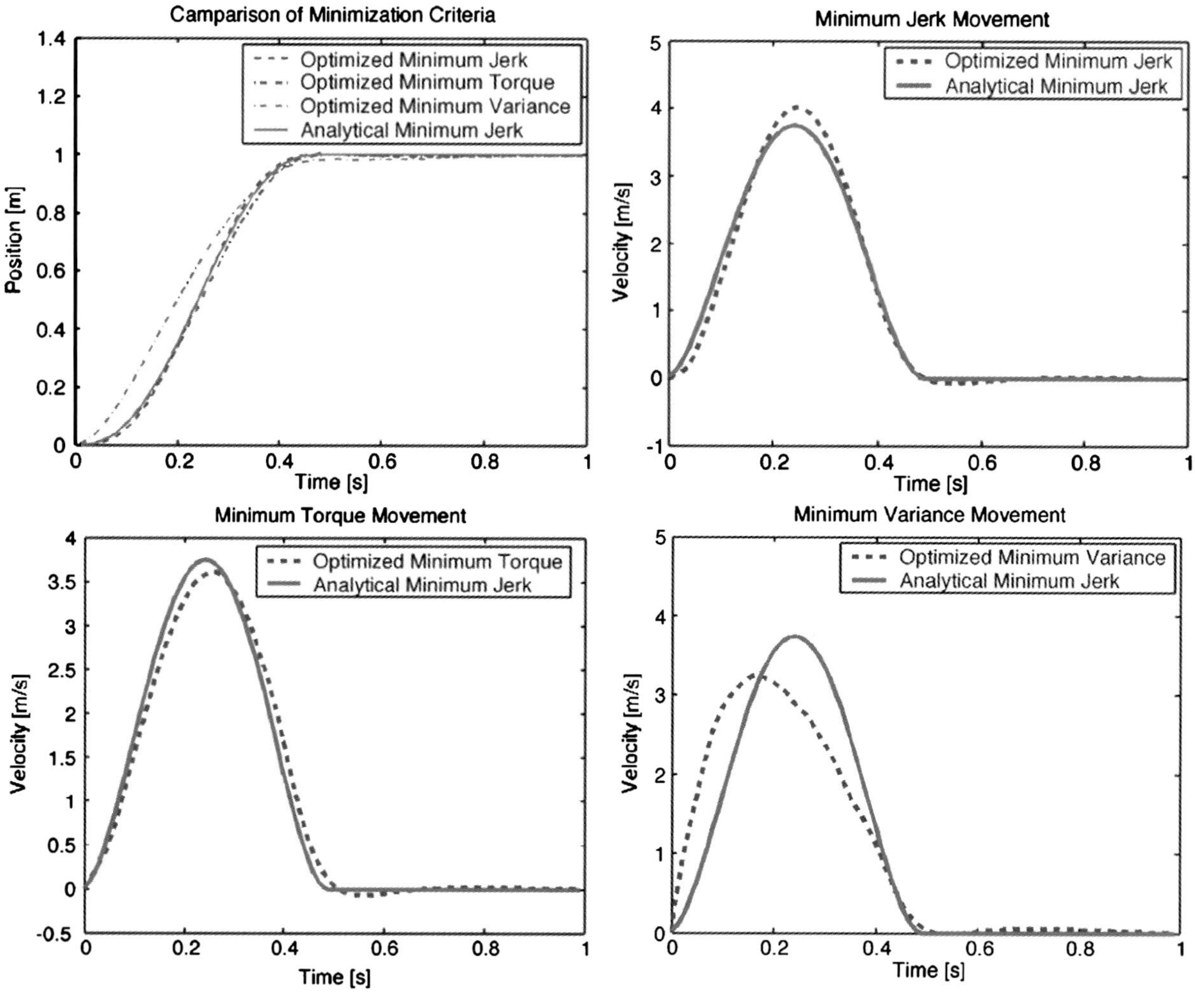

Fig. 11. Optimization results for DMPs for various criteria — see text for explanations.

real-time generation allows also real-time modification of the desired trajectory, a topic that we did not expand on in this paper, but which has been examined in previous work (Ijspeert et al., 2003). Such real-time modification is essential if one wishes to account for perception–action coupling or the reaction to perturbations during movement. Unlike other models of movement generation in the past, DMPs can represent rather complex movements in one simple coherent framework, e.g., a complete tennis forehand can be cast into one DMP. The complexity of a DMP is only limited by the number of basis functions that is provided to its core nonlinearity, a population-code basis function approximator that could be generated by many areas of the primate brain. This line of modeling opens the interesting question of where and when a complex movement needs to be segmented into smaller pieces, i.e., how complex a movement primitive can be in biology. Another point worth highlighting is that DMPs can represent both discrete and rhythmic movement. Complex multi-DOF periodic patterns can be generated, where all contributing DOFs are easily synchronized and phase locked in arbitrary relationships. This property is unlike traditional coupled-oscillator models for multi-DOF movement generation, which usually have major difficulties in modeling anything but synchronized in-phase and out-of-phase movement relationships. As a last point, DMPs can be scaled in time and space without losing the qualitative trajectory appearance that was originally coded in a DMP. For instance, a DMP coding a tennis forehand swing can easily create a very small and slow swing and a rather large and fast swing out of the exactly the same equations. We believe that this approach to modeling of movement could be a promising complement in many theories developed for human and primate motor control, and offers to revisit many previous movement models in one simple coherent framework.

Acknowledgments

This research was supported in part by National Science Foundation grants ECS-0325383, IIS-0312802, IIS-0082995, ECS-0326095, ANI-0224419, the DARPA program on Learning Locomotion, a NASA grant AC#98-516, an AFOSR grant on Intelligent Control, the ERATO Kawato Dynamic Brain Project funded by the Japanese Science and Technology Agency, and the ATR Computational Neuroscience Laboratories. We are very grateful for the insightful and thorough comments of the editors of this volume, which helped improving this article significantly.

References

Abend, W., Bizzi, E. and Morasso, P. (1982) Human arm trajectory formation. Brain, 105: 331–348.

Aboaf, E.W., Drucker, S.M. and Atkeson, C.G. (1989) Task-level robot learing: juggling a tennis ball more accurately. In: Proceedings of IEEE Interational Conference on Robotics and Automation. IEEE, Piscataway, NJ, May 14–19, Scottsdale, AZ, pp. 331–348.

Adamovich, S.V., Levin, M.F. and Feldman, A.G. (1994) Merging different motor patterns: coordination between rhythmical and discrete single-joint. Exp. Brain Res., 99: 325–337.

Atkeson, C.G., Moore, A.W. and Schaal, S. (1997) Locally weighted learning. Artif. Intell. Rev., 11: 11–73.

Barto, A.G. and Mahadevan, S. (2003) Recent advances in hierarchical reinforcement learning. Discrete Event Dyn. Syst., 13: 341–379.

Bellman, R. (1957) Dynamic Programming. Princeton University Press, Princeton, NJ.

Bertsekas, D.P. and Tsitsiklis, J.N. (1996) Neuro-Dynamic Programming. Athena Scientific, Bellmont, MA.

Bishop, C.M. (1995) Neural Networks for Pattern Recognition. Oxford University Press, New York.

Buchanan, J.J., Park, J.H., Ryu, Y.U. and Shea, C.H. (2003) Discrete and cyclical units of action in a mixed target pair aiming task. Exp. Brain Res., 150: 473–489.

Bullock, D. and Grossberg, S. (1988) Neural dynamics of planned arm movements: emergent invariants and speed-accuracy properties during trajectory formation. Psychol. Rev., 95: 49–90.

Craig, J.J. (1986) Introduction to Robotics. Addison-Wesley, Reading, MA.

De Rugy, A. and Sternad, D. (2003) Interaction between discrete and rhythmic movements: reaction time and phase of discrete movement initiation against oscillatory movement. Brain Res.

Desmurget, M. and Grafton, S. (2000) Forward modeling allows feedback control for fast reaching movements. Trends Cogn. Sci., 4: 423–431.

Dyer, P. and McReynolds, S.R. (1970) The Computation and Theory of Optimal Control. Academic Press, New York.

Flash, T. and Hogan, N. (1985) The coordination of arm movements: an experimentally confirmed mathematical model. J. Neurosci., 5: 1688–1703.

Georgopoulos, A.P. (1991) Higher order motor control. Annu. Rev. Neurosci., 14: 361–377.

Getting, P.A. (1985) Understanding central pattern generators: insights gained from the study of invertebrate systems. In: Neurobiology of Vertebrate Locomotion, Stockholm, pp. 361–377.

Gribble, P.L. and Ostry, D.J. (1996) Origins of the power law relation between movement velocity and curvature: modeling the effects of muscle mechanics and limb dynamics. J. Neurophysiol., 76: 2853–2860.

Guckenheimer, J. and Holmes, P. (1983) Nonlinear Oscillations, Dynamical Systems, and Bifurcations of Vector Fields. Springer, New York.

Harris, C.M. and Wolpert, D.M. (1998) Signal-dependent noise determines motor planning. Nature, 394: 780–784.

Hodgkin, A.L. and Huxley, A.F. (1952) A quantitative description of membrane current and its application to conduction and excitation in nerve. J. Physiol., 117: 500–544.

Hoff, B. and Arbib, M.A. (1993) Models of trajectory formation and temporal interaction of reach and grasp. J. Mot. Behav., 25: 175–192.

Hogan, N. (1984) An organizing principle for a class of voluntary movements. J. Neurosci., 4: 2745–2754.

Hollerbach, J.M. (1984) Dynamic scaling of manipulator trajectories. Trans. ASME, 106: 139–156.

Ijspeert, A., Nakanishi, J. and Schaal, S. (2001) Trajectory formation for imitation with nonlinear dynamical systems. In: IEEE International Conference on Intelligent Robots and Systems (IROS 2001). Weilea, HI, Oct. 29–Nov. 3, pp. 752–757.

Ijspeert, A., Nakanishi, J. and Schaal, S. (2003) Learning attractor landscapes for learning motor primitives. In: Becker S., Thrun S. and Obermayer K. (Eds.), Advances in Neural Information Processing Systems 15. MIT Press, Cambridge, MA, pp. 1547–1554.

Ijspeert, A.J., Crespi, A., Ryczko, D. and Cabelguen, J.M. (2007) From swimming to walking with a salamander robot driven by a spinal cord model. Science, 315: 1416–1420.

Ijspeert, J.A., Nakanishi, J. and Schaal, S. (2002a) Learning rhythmic movements by demonstration using nonlinear oscillators. In: IEEE International Conference on Intelligent Robots and Systems (IROS 2002). IEEE, Lausanne, Piscataway, NJ, Sept. 30–Oct. 4, pp. 958–963.

Ijspeert, J.A., Nakanishi, J. and Schaal, S. (2002b) Movement imitation with nonlinear dynamical systems in humanoid robots. In: International Conference on Robotics and Automation (ICRA2002). Washington, May 11–15.

Ivry, R.B., Spencer, R.M., Zelaznik, H.N. and Diedrichsen, J. (2002) The cerebellum and event timing. Ann. N.Y. Acad. Sci., 978: 302–317.

Kawamura, S. and Fukao, N. (1994) Interpolation for input torque patterns obtained through learning control. In: International Conference on Automation, Robotics and Computer Vision (ICARCV'94). Singapore, Nov. 8–11, pp. 183–191.

Kawato, M. (1999) Internal models for motor control and trajectory planning. Curr. Opin. Neurobiol., 9: 718–727.

Kawato, M. and Wolpert, D. (1998) Internal models for motor control. Novartis Found Symp., 218: 291–304.

Keating, J.G. and Thach, W.T. (1997) No clock signal in the discharge of neurons in the deep cerebellar nuclei. J. Neurophysiol., 77: 2232–2234.

Kelso, J.A.S. (1995) Dynamic Patterns: The Self-Organization of Brain and Behavior. MIT Press, Cambridge, MA.

Kopell, N. and Ermentrout, G.B. (1988) Coupled oscillators and the design of central pattern generators. Math. Biosci., 90: 87–109.

Kugler, P.N., Kelso, J.A.S. and Turvey, M.T. (1982) On control and co-ordination of naturally developing systems. In: Kelso J.A.S. and Clark J.E. (Eds.), The Development of Movement Control and Coordination. Wiley, New York, pp. 5–78.

Lacquaniti, F., Terzuolo, C. and Viviani, P. (1983) The law relating the kinematic and figural aspects of drawing movements. Acta Psychol., 54: 115–130.

Marder, E. (2000) Motor pattern generation. Curr. Opin. Neurobiol., 10: 691–698.

Mohajerian, P., Mistry, M. and Schaal, S. (2004) Neuronal or spinal level interaction between rhythmic and discrete motion during multi-joint arm movement. In: Abstracts of the 34th Meeting of the Society of Neuroscience. San Diego, CA, Oct. 23–27.

Morasso, P. (1981) Spatial control of arm movements. Exp. Brain Res., 42: 223–227.

Morasso, P. (1983) Three dimensional arm trajectories. Biol. Cybern., 48: 187–194.

Mussa-Ivaldi, F.A. (1988) Do neurons in the motor cortex encode movement direction? An alternative hypothesis. Neurosci. Lett., 91: 106–111.

Nakanishi, J., Morimoto, J., Endo, G., Cheng, G., Schaal, S. and Kawato, M. (2004) Learning from demonstration and adaptation of biped locomotion. Robot. Auton. Syst., 47: 79–91.

Pellizzer, G., Massey, J.T., Lurito, J.T. and Georgopoulos, A.P. (1992) Three-dimensional drawings in isometric conditions: planar segmentation of force trajectory. Exp. Brain Res., 92: 326–337.

Peters, J., Vijayakumar, S. and Schaal, S. (2005) Natural actor-critic. In: Gama J., Camacho R., Brazdil P., Jorge A. and Torgo L. (Eds.), Proceedings of the 16th European Conference on Machine Learning (ECML 2005), 3720. Springer, Porto, Portugal, pp. 280–291 Oct. 3–7.

Picard, N. and Strick, P.L. (2001) Imaging the premotor areas. Curr. Opin. Neurobiol., 11: 663–672.

Righetti, L. and Ijspeert, A. (2006) Design methodologies for central pattern generators: an application to crawling humanoids. In: Proceedings of Robotics: Science and Systems. MIT Press, Philadelphia, PA.

Rizzi, A.A. and Koditschek, D.E. (1994) Further progress in robot juggling: solvable mirror laws. In: IEEE International Conference on Robotics and Automation, Vol. 4. San Diego, CA, May 8–13, pp. 2935–2940.

Roberts, P.D. and Bell, C.C. (2000) Computational consequences of temporally asymmetric learning rules: II. Sensory image cancellation. J. Comput. Neurosci., 9: 67–83.

Schaal, S. and Atkeson, C.G. (1993) Open loop stable control strategies for robot juggling. In: IEEE International Conference on Robotics and Automation, Vol. 3. IEEE, Piscataway, NJ; Atlanta, GA, May 2–6, pp. 913–918.

Schaal, S., Peters, J., Nakanishi, J. and Ijspeert, A. (2003) Control, planning, learning, and imitation with dynamic movement primitives. In: Workshop on Bilateral Paradigms on Humans and Humanoids. IEEE International Conference on Intelligent Robots and Systems (IROS 2003). Las Vegas, NV, Oct. 27–31.

Schaal, S. and Sternad, D. (1998) Programmable pattern generators. In: 3rd International Conference on Computational Intelligence in Neuroscience. Research Triangle Park, NC, Oct. 24–28, pp. 48–51.

Schaal, S. and Sternad, D. (2001) Origins and violations of the 2/3 power law in rhythmic 3D movements. Exp. Brain Res., 136: 60–72.

Schaal, S., Sternad, D. and Atkeson, C.G. (1996) One-handed juggling: a dynamical approach to a rhythmic movement task. J. Mot. Behav., 28: 165–183.

Schaal, S., Sternad, D., Osu, R. and Kawato, M. (2004) Rhythmic movement is not discrete. Nat. Neurosci., 7: 1137–1144.

Schöner, G. (1990) A dynamic theory of coordination of discrete movement. Biol. Cybern., 63: 257–270.

Selverston, A.I. (1980) Are central pattern generators understandable? Behav. Brain Sci., 3: 555–571.

Shadmehr, R. and Wise, S.P. (2005) The computational neurobiology of reaching and pointing: a foundation for motor learning. MIT Press, Cambridge, MA.

Smits-Engelsman, B.C., Van Galen, G.P. and Duysens, J. (2002) The breakdown of Fitts' law in rapid, reciprocal aiming movements. Exp. Brain Res., 145: 222–230.

Soechting, J.F. and Terzuolo, C.A. (1987a) Organization of arm movements in three dimensional space. Wrist motion is piecewise planar. Neuroscience, 23: 53–61.

Soechting, J.F. and Terzuolo, C.A. (1987b) Organization of arm movements. Motion is segmented. Neuroscience, 23: 39–51.

Spencer, R.M., Zelaznik, H.N., Diedrichsen, J. and Ivry, R.B. (2003) Disrupted timing of discontinuous but not continuous movements by cerebellar lesions. Science, 300: 1437–1439.

Sternad, D., De Rugy, A., Pataky, T. and Dean, W.J. (2002) Interaction of discrete and rhythmic movements over a wide range of periods. Exp. Brain Res., 147: 162–174.

Sternad, D. and Dean, W.J. (2003) Rhythmic and discrete elements in multi-joint coordination. Brain Res.

Sternad, D., Dean, W.J. and Schaal, S. (2000) Interaction of rhythmic and discrete pattern generators in single joint movements. Hum. Mov. Sci., 19: 627–665.

Sternad, D. and Schaal, D. (1999) Segmentation of endpoint trajectories does not imply segmented control. Exp. Brain Res., 124: 118–136.

Strogatz, S.H. (1994) Nonlinear Dynamics and Chaos: With Applications to Physics, Biology, Chemistry, and Engineering. Addison-Wesley, Reading, MA.

Sutton, R.S. and Barto, A.G. (1998) Reinforcement Learning: An Introduction. MIT Press, Cambridge.

Taga, G., Yamaguchi, Y. and Shimizu, H. (1991) Self-organized control of bipedal locomotion by neural oscillators in unpredictable environment. Biol. Cybern., 65: 147–159.

Tesauro, G. (1992) Temporal difference learning of backgammon strategy. In: Sleeman D. and Edwards P. (Eds.), Proceedings of the Ninth International Workshop on Machine Learning. Morgan Kaufmann, Aberdeen, Scotland, UK, July 1–3, pp. 451–457.

Turvey, M.T. (1990) The challenge of a physical account of action: A personal view. In: Whiting, H.T.A., Meijer, O.G. and van Wieringen, P.C.W. (Eds.), The Natural Physical Approach to Movement Control. Amsterdam: Free University Press, Amsterdam, pp. 57–94.

Vijayakumar, S. and Schaal, S. (2000) Locally weighted projection regression: an O(n) algorithm for incremental real time learning in high dimensional spaces. In: Proceedings of the 17th International Conference on Machine Learning (ICML 2000), Vol. 1. Stanford, CA, pp. 288–293.

Viviani, P. (1986) Do units of motor action really exist? In: Experimental Brain Research Series 15. Springer, Berlin, pp. 828–845.

Viviani, P. and Cenzato, M. (1985) Segmentation and coupling in complex movements. J. Exp. Psychol. Hum. Percept. Perform., 11: 828–845.

Viviani, P. and Flash, T. (1995) Minimum-jerk, two-thirds power law, and isochrony: converging approaches to movement planning. J. Exp. Psychol. Hum. Percept. Perform., 21: 32–53.

Viviani, P. and Terzuolo, C. (1980) Space-time invariance in learned motor skills. In: Stelmach G.E. and Requin J. (Eds.), Tutorials in Motor Behavior. North-Holland, Amsterdam, pp. 525–533.

Wann, J., Nimmo-Smith, I. and Wing, A.M. (1988) Relation between velocity and curvature in movement: equivalence and divergence between a power law and a minimum jerk model. J. Exp. Psychol. Hum. Percept. Perform., 14: 622–637.

Wei, K., Wertman, G. and Sternad, D. (2003) Interactions between rhythmic and discrete components in a bimanual task. Motor Control, 7: 134–155.

Williamson, M. (1998) Neural control of rhythmic arm movements. Neural Netw., 11: 1379–1394.

Wolpert, D.M. (1997) Computational approaches to motor control. Trends Cogn. Sci., 1: 209–216.

P. Cisek, T. Drew & J.F. Kalaska (Eds.)
Progress in Brain Research, Vol. 165
ISSN 0079-6123
Copyright © 2007 Elsevier B.V. All rights reserved

CHAPTER 28

The place of 'codes' in nonlinear neurodynamics

Walter J. Freeman*

Department of Molecular and Cell Biology, University of California, Berkeley, CA 94720-3206, USA

Abstract: A key problem in cognitive science is to explain the neural mechanisms of the rapid transposition between stimulus energy and abstract concept — between the specific and the generic — in both material and conceptual aspects, not between neural and psychic aspects. Three approaches by researchers to a solution in terms of neural codes are considered. Materialists seek rate and frequency codes in the interspike intervals of trains of action potentials induced by stimuli and carried by topologically organized axonal lines. Cognitivists refer to the symbol grounding problem and search for symbolic codes in firings of hierarchically organized feature-detector neurons of phonemes, lines, odorants, pressures, etc., that object-detector neurons bind into representations of probabilities of stimulus occurrence. Dynamicists seek neural correlates of stimuli and associated behaviors in spatial patterns of oscillatory fields of dendritic activity that self-organize and evolve as trajectories through high-dimensional brain state space; the codes are landscapes of chaotic attractors. Unlike codes in DNA and the periodic table, these codes have neither alphabet nor syntax. They are epistemological metaphors required by experimentalists to measure neural activity and by engineers to model brain functions. Here I review the central neural mechanisms of olfaction as a paradigm for use of codes to explain how brains create cortical activities that mediate sensation, perception, comprehension, prediction, decision, and action or inaction.

Keywords: action–perception cycle; electroencephalogram; intentional arc; mesoscopic brain dynamics; neural code; phenomenology; reflex arc; scale-free cortical dynamics; wave packet

Introduction

Everyone knows the experience of smelling the scent of a rose. How does this happen? How do we interact with a material object and then know what it is and what it means for us? A neurobiologist says that we extract information from the chemicals and process it into a form suitable for comparison with information stored in memory; a cognitivist says that we make a representation and operate on the symbol according to certain rules; a dynamicist says that we intend the rose. These words denote complex concepts that we use to describe an elementary process. We need to simplify. We know that we share the process with animals, which often have better acuity than we do, though not our depth of comprehension, so we can study the process in animals with brains less complex than our own. The same elementary process occurs in all our senses, not just the traditional five of sight, sound, touch, taste, and smell, but also gravity, muscle tension, muscle length, joint angle, and countless senses for chemicals concentrations, pressures, temperatures, and volumes throughout our bodies and brains. Olfaction is the most versatile and universal, rivaled only by the immune system, yet also the simplest

*Corresponding author. Tel.: +1 510 642 4220;
Fax: +1 510 643 9290; E-mail: dfreeman@berkeley.edu

DOI: 10.1016/S0079-6123(06)65028-0

and most ancient. It is the prototype for all other perceptual systems.

For these reasons olfaction in rabbits is a paradigm of choice for study to understand the elementary process (Freeman, 2001) in order to compare the biological, cognitive, dynamic, and philosophical descriptions of brain/mind function and find commonalities. We seek answers to the question: How can we so simply and elegantly cross the border between odorant and odor, between the material and the perceptual: in one direction to perceive the smell of a substance, in the other direction to create a chemical with a desired fragrance (Burr, 2002; Turin, 2006)? A concept of critical utility in this quest is "intentionality" (Freeman, 2007b); we must "intend" to perceive and create. By this we mean our minds using our bodies to thrust out into the world and in part change it and in part accommodate and assimilate to it by learning from the experience. The concept had its origin in the work of Thomas Aquinas describing the functions of mind and body; derivative meanings are the psychologists' "intent" meaning purpose and the analytic philosophers' "aboutness", the relation of a mental symbol to that which it represents (Searle, 1983). Aquinas further distinguished between "first intention", which is the perception of objects that need not be conscious, and "second intention", which includes awareness of the self that does the perceiving.

In this review an answer is sought in neurodynamics by analyzing patterns of neural activity that self-organize in the brain. This neural activity is hierarchically organized. Sensory inflows from receptors and motor outflows to muscles are by myriad pulses on axons at the microscopic level, the level of the "phantasms" of Thomas Aquinas and the inaccessible raw sense data of phenomenologists. These data require rate and frequency codes. Below is the flux of molecules at submicroscopic and quantum levels. Above is the self-organization of local fields into pulse and wave activity in spatiotemporal patterns at the mesoscopic level, the first and incomplete stage of perception where abstraction and generalization take place. Next is the organization of widespread fields of coordinated neural activity at the macroscopic level. The fields are large enough to include many areas of the brain, perhaps at times in synchronized oscillations involving the entire forebrain. At this level the perceptual contents in patterned activity include the locations in time and space of perceptions of objects and events. These patterns are not representations of stimuli, actions, thoughts, beliefs, etc.; they are expressions of knowledge in active support of perception, recollection, and decision. They do not result from computations in any literal mathematical sense. They are dynamic entities akin to vortices in hurricanes, unlike numbers in computers. We use symbolic codes to represent them and to model them with statistics and differential equations.

These levels are not intrinsic to brains; they are imposed by the scales required by the techniques used for measurements using chemistry, electrical recording, and brain imaging. The fallout for the synthesis by brain modelers is the necessity for bridging across these levels. For heuristic purposes I find that these three levels suffice: micro-meso-macro.

I postulate that these macroscopic self-organized goal states, through recursive self-similarity, include perceptions of present states, projections of future states, plans at the mesoscopic level for action to achieve them, and trajectories of microscopic pulses that direct muscular activity in goal-directed actions modulated by sensory feedback. This hierarchy gives the behaviorists' reflex arc, the pragmatists' action–perception cycle, and the phenomenologists' intentional arc. My proposed explanation in terms of central action through field neurodynamics is consistent with the nonrepresentational systems of Aquinas, Heidegger, and Merleau-Ponty that avoid the Cartesian subject–object split. I will conclude that, at present, neurodynamics can explain first intention — understanding perception as direct grasp of objects and events by animals and prelingual children — but lacks the experimental data on brain activity that will be needed to explain second intention whereby the self comprehends the immanent action of understanding itself. Field studies open a pathway to remedy this deficiency.

The neurodynamical paradigm

Experimental neurobiologists are privileged in the search for understanding the process of transposition, because we have been granted the opportunity to record and measure the activity of neurons in the nose and in the many parts throughout the brain where the ongoing neural activity is modified by the elementary act that intends a rose, and represents it, and processes its information into knowledge. My group has recorded electrical activity from electrodes we fixed in the brains of rabbits trained to respond by sniffing or chewing after they learned the significance of simple odorant chemicals. By their actions we proved that they could identify the specific odorants that we presented to them. The rabbits acted the way they did because each time we presented an odorant we accompanied it by a reward or punishment that made the odorant meaningful for them. Without this reinforcement the odorants were meaningless for the rabbits, and they quickly learned to ignore them. With reinforcement they learned actions by which to get rewards and avoid punishments. They also learned to predict that any of several odorants would come in the near future, and they prepared their bodies to detect them and take appropriate action in response to whatever might occur, including the unexpected or unknown events, which in their limited and uncertain world could occur at any moment.

All these properties we derive from classical behaviorism. Psychologists describe and control these behaviors in terms of schedules of reinforcement (Ferster and Skinner, 1957); neurodynamicists describe them in terms of hierarchies of reflexes (Sherrington, 1906); philosophers describe them in terms of intentionality (Searle, 1983). Researchers comprehend the neural activity by recording and measuring the electric potential differences in and around the brains of the animals (e.g., scalp EEG, the magnetoencephalogram, MEG) as they anticipate, detect, and respond appropriately to the odorants in their learned repertoires. There is a notable reciprocity between the intention of a rabbit to perceive a signal of import and the intention of a researcher to perceive the neural activity. The animal prepares its body by orienting its nasal sensory receptors and sniffing; the researcher prepares and places electrode arrays, rigs electronics to amplify, filter and measure signals, and creates displays to bring the measurements to the observer's senses. The designs of the arrays, the filters, and the methods for measurements to extract information depend on the expectations of the researcher and the properties of the subjects. The details are complex and of interest only to specialists, but in principle the process is the same in man and rabbit. From our respective experiences we and our rabbit predict what the future holds; we plan appropriate tests of our predictions; we make the tests and detect the changes in our sensory input that are caused by our actions in making the test; we classify the results of our test by whether or not what happens conforms with what we expect to happen; and we modify our expectations accordingly. We and they are not observers; we are participants in a circumscribed relationship.

Of course, the rabbit is much simpler, and therein lies its utility. From its training it expects to receive any one of two or three odorants at some time in the near future, and it samples the air each time it breathes in. When an odorant comes, the rabbit detects it with its nose, determines with its brain which expected event has occurred, and with its body takes appropriate action such as sniffing or chewing or relaxing. The crux of the problem lies in the neural events by which the determination occurs in the brain of the odor from the odorant. We divide the neurobiological process into stages. In the first stage we observe the effect of the odorant on the receptor cells in the nose. In the second stage we observe the effect of the activated receptors on the olfactory brain. In the third stage we observe the effect of the olfactory system on the whole brain. Fourth, we observe the neural activities in the motor systems. Lastly we observe the effect of the brain on the body, as the rabbit responds to the odorant. The transposition from odorant to odor occurs in the second and third stages. We observe the process in these stages with electrodes in the brain by which to record, measure, and model neural activity, first in the olfactory system, then in the neocortices serving the limbic system and the other distance receptors in the eye, ear, and skin.

The network approach: information processing and linear causality

Consider again that we are contrasting the reflex arc with the intentional arc. The intentional arc begins with emergence from the present brain state of an extrapolation into the future that will require some appropriate action to direct the self into successful assimilation with an altering world. That foresight includes prior specification of what information might be needed through acts of observation and perception to achieve success. The details are formulated in the attractor landscapes emerging through preafference. The reflex arc is widely thought to begin with the stimulus that activates the sensory systems of a subject and to end with the response. To the contrary the reflex arc begins with the intentional action of the observer to explore the properties of the subject. The "features" of the stimulus emerge in the mind of the experimenter and are embodied in the selection and delivery of the stimulus. Neural correlates of the "features" are clearly detectable in the evoked activity of the brain, but whether and how the brains of subjects transform these evoked patterns of activity into percepts are matters for investigation. The aim of electrophysiological investigation is proposed here as challenging the "feature detector" concept and offering the "attractor landscape" concept as an alternative.

Each electrode inserted into the nose or the brain yields two forms of electrical activity. We see one form in trains of electric pulses (spikes, action potentials, units) from individual neurons. We see the other form in continuous waves of electric current (dendritic potentials, local field potentials, electrocorticograms — ECoG, scalp electroencephalograms — EEG) from populations of neurons. The study of pulses is based on the view of the organization of olfactory receptors and brain areas as networks of spiking neurons. The study of waves is based on the view of the same neurons generating continuous space-time fields, in which the identities of the neurons are submerged in the populations. The differences in views resemble those between the psychological analyses of individuals in families contrasted with sociological analyses of the organizations of cities and nations. At the start of the neurobiological experiments the electrodes are shaped and placed to maximize the detection of either pulses or waves, and the recordings of electrical activity containing both forms are filtered to separate the pulses and the waves for analyses. The data from each stream are used to construct hypotheses about the functions of the olfactory brain, on one hand as discrete networks of neurons that are connected by junctions, the synapses, and on the other hand as tissues that contain such high densities of neurons and synapses that the tissue can be described as a continuum, analogous to ways in which molecules can be described as forming a liquid or gas, and supporting both synaptic and nonsynaptic communication and modulation (Freeman, 2005c).

A selective synthesis of both views is essential for understanding brain function. This is because brains work at many levels of organization. An act of perception involves all levels of activity, ranging from the attachment of individual molecules of an odorant to the molecular structures on the surfaces of olfactory receptor cells to the initiation by the rabbit of sequences of social behaviors intended to enhance the likelihood of its species to survive. The guiding principles of experimental neurobiology are that we record activities of both kinds as the neural correlates of the process by which an odorant is comprehended as an odor, and that we use our observations of the correlates to construct explanations in the form of dynamic models of the brain systems that perform the process. Notably these numerical correlates interrelate patterns of neural activity with patterns of goal-directed behavior, not with consciousness or verbal descriptions of phenomenological states. We have no measure of what rabbits feel or what they are conscious of. We deal here with the process of inductive category formation in the accumulation and intentional utilization of knowledge, for which emotion is an integral part (Freeman, 1999), not with the 'hard problem' at the core of consciousness studies (Chalmers, 1996).

The network model is commonly assumed to begin with the reflex arc (but see "The continuity of circular causality across all levels"), in which the stimulus has the form of molecules of odorant that bind to receptor cells at the molecular and quantum levels. The binding releases a wave of electric current, the generator potential, that initiates and sustains firing of pulse trains from just those receptor neurons that can selectively bind the molecules. According to various authors (Lettvin and Gesteland, 1965; Lancet and Ben-Arie, 1993; Freeman, 2001; Burr, 2002; Buck and Axel, 2004) the microscopic neurons encode sensory information in their pulses and transmit it by axons into the olfactory brain, where it is directed by switching networks to selected neurons that by filtering or resonance act as feature detectors. The cortical neurons send the processed information to associational areas of the brain.

The steps beyond are conjectured from properties of artificial neural networks: higher areas are thought to compare the input information with previously stored information retrieved from memory by symbolic dynamics. Studies of perception in humans report the firing of neurons with remarkable specificity to stimuli such as photographs of famous persons (e.g., Quiroga et al., 2005), suggesting that their spike trains serve as symbols. Cognitivists propose that the best matching symbol is selected by competitive inhibition among such neurons and sent to the motor cortex, where an appropriate response is selected by winner-take-all for transmission into the motor systems of the brain stem and spinal cord. All this must occur in time frames lasting on the order of half a second.

The field approach: the action–perception cycle

The field theoretic model using the action–perception cycle begins not with the stimulus but instead with the formation in the forebrain of a macroscopic pattern that embodies anticipation of a desired future state of the brain and body, such as finding food or avoiding danger. We conjecture that within this macroscopic pattern the brain constructs mesoscopic activity patterns, which organize the local sensory and motor populations that control the actions intended to achieve the goal. Within each mesoscopic population the microscopic neurons are directed ("ordered") to fire pulses in prescribed sequences. These individual neurons also receive proprioceptive feedback from sensory receptors in the muscles and joints through the cerebellum and basal ganglia that is needed to continuously adapt the intended movement of the body to the intended goal. Knowledge about the neurobiology of these two downward steps is insufficient to detect and measure the mesoscopic patterns. They can be conceived in engineering terms as predictive systems such as those for controlling the flight of an airplane, which have an over-arching level in which the goal is selected by choosing a flight plan, outer loops that set the control surfaces to direct the aircraft to its goal, and inner loops that regulate the control surfaces in the wings and tail to compensate for air turbulence. In these terms the macroscopic pattern establishes a context embedding the mesoscopic patterns that self-organize in multiple populations comprising 'modules' (Houk and Wise, 1995; Houk, 2005), and the modules establish the local contexts in which microscopic neural networks perform the intended tasks.

The movements of the body in every intended overt action modify the positions with respect to the environment of the receptor cells in all sensory systems. The modifications change the sensory input. These self-induced changes are anticipated and predicted from past experience. The predictions have been described as communicated from the motor modules to the sensory modules of the brain by copies of the motor outflow known as "corollary discharges" (Sperry, 1950) and "efference copies" (Von Holst and Mittelstädt, 1950) in the process of "preafference" (Kay and Freeman, 1998), which is the basis for focused attention. The corollary discharges prime the sensory areas by making them selectively sensitive to each of the expected stimuli in the search for odorants signifying food or danger, be they from carrot or fox, cabbage, or man.

Studies of neural fields (Freeman, 2004a, b, 2005a, 2006a) show that the impact of the pulses from the receptors on the sensory areas of the brain is not at all the processing of spikes on a few hundred axons. In olfaction the millions of pulses with each inhalation cause a major change in function, which is equivalent to the change in state from a gas to a liquid (Freeman and Vitiello, 2006).

The nearly random activity before the impact is increased in amplitude, and at some point it condenses much as would water molecules forming a raindrop. In physical terms the impact induces a phase transition in the olfactory brain, which forces it out of its receiving state that is maintained at a pseudoequilibrium (Freeman, 2005b) into a transmitting state into which the bulbar dynamics converges. The transition period leading to convergence is a brief metastable state (Bressler and Kelso, 2001) of search through the selective classes of sensitivities stored by modifications of synaptic strengths from prior learning in an attractor landscape. We conceive each cortical dynamical system as having a state space through which the system travels as a point moving along a path (trajectory) through the state space (Kozma and Freeman, 2003).

A simple analogy is a spaceship flying over a landscape with valleys resembling the craters on the moon. An expected stimulus contained in the omnipresent background input selects a crater into which the ship descends. The convergent region in each crater defines the attractor to which the system trajectory goes, and the set of craters are the basins of attraction in the attractor landscape. There is a different attractor for each class of stimulus that has been learned and that preafference has primed the system to anticipate, each surrounded by its basin. The landscape is surrounded by a catch basin that signals unknown stimuli (Skarda and Freeman, 1987) that might be important. These output patterns trigger a fixed "auto-shaped" behavioral action known as the "orienting response". The animal receiving an unexpected stimulus freezes and directs its senses in search of something unknown and possibly threatening. If the unknown stimulus is accompanied by reinforcement, then a new attractor forms by Hebbian learning, which changes all of the other basins in deforming the landscape by attractor crowding. If there is no reinforcement, the system automatically adapts by habituation to block cortical responsiveness to that input in the future. These processes of Hebbian linkage and non-Hebbian habituation are the essence of associative memory. There is an exclusion principle at work in that only one attractor can be selected at a time (Freeman and Vitiello, 2006), though rapid rotation among two or more attractors may occur. Sequences of patterns indicate that "itinerant trajectories" (Tsuda, 2001) form through successions of attractors in the landscape, each attractor dissolving as soon as it is accessed and giving way to the next.

The dynamics in each sensory cortex (not just for olfaction but also vision, hearing, and touch) converges within milliseconds to an attractor, which transmits a modality-specific burst of neural activity that I call a "wave packet" (Freeman, 1975/2004, 2000) This is a spatially coherent oscillation of dendritic potentials typically in the gamma range (30–80 Hz) with relatively fixed spatial patterns of amplitude and phase modulation (AM, PM) of the shared wave form (Freeman, 2004b, 2007a). The perceptual contents of the AM patterns are determined by the previously learned synaptic connections in the sensory cortices, which constitute the integrated record of knowledge constructed during prior experience with the stimulus. That synaptic network determines the attractor and its basin in the landscape sustained by each cortex for each learned class of stimulus. A Hebbian network spans the basin of each class. The stimulus-evoked action potentials that are triggered by an expected stimulus select a basin by activating the network; this is the process of generalization to the class of the detected stimulus as the trajectory converges to the attractor, irrespective of where the cortex was placed within the basin by the particular receptors that the stimulus attached to, which vary from trial to trial. With each trial the process of learning continues to refine and update the Hebbian synaptic network. As the system converges to the attractor in the basin, it deletes the extraneous information about which particular receptors receive the stimulus; this is the process of abstraction. The attractor determines the transmitted wave packet, not the stimulus, which merely selects and refines the transmitted AM pattern, which is an expression of its knowledge by the rabbit that in terms of coding can be modeled as a symbol of its contents.

Owing to the large surface area of sensory cortex that is integrated by the attractor (Freeman, 2004b) and the divergent–convergent topology of the transmitting bundles of axons, the patterns are broadcast through the brain. Those cortical transmission pathways that have divergent–convergent projections and not topographic mapping perform a spatial integral transformation on the output. Transmitted activity having dispersed phase and frequency values is attenuated by cancellation and smoothing; activity that is spatially coherent (same frequency and phase) is relatively enhanced. The most salient among the targets of transmission is the limbic system. This is the core structure of every vertebrate brain that is identified with the expression of emotion. Its key structure, the hippocampus, was the first cortex to appear as laminated neuropil in the phylogenetic evolution of the brain (Maclean, 1969), and it well deserves its appellation, archicortex ("ancient cortex"). The hippocampus sustains the neural machinery by which sensory events and objects are assigned environmental spatial locations and times of occurrence in the stream of life history (Freeman, 2001). In mammalian brains the wave packets of all sensory cortices are received either directly from the olfactory bulb or by relays from other modalities by the hippocampal vestibule, the outer layer of the entorhinal cortex. Time and place are linked to each other and to the contents of multimodal stimuli (Gestalts) in the hippocampus. There the multiple sensory cortical wave packets are integrated into a multisensory pattern as they pass through the hippocampus back to the deep layers of the entorhinal cortex, whence it is disseminated back to the cortices of origin. Every event must make this passage, if it is to be assigned a space-time location in the stream of personal history.

These properties are commonly referred to as the spatial "cognitive map" and the temporal "short term memory" provided by the hippocampus (O'Keefe and Nadel, 1978; Buzsaki, 2002). The collective and incremental modification is the basis for self-assimilation by which the animal continuously updates its tenancy in the environment. The combined spatiotemporal pattern that is assembled in the hippocampus is re-transmitted by stages to all sensory areas by preafference. The result is that within half a second of the original event there emerges in the brain a global pattern of cortical activity that is participated in by every sensory area (Freeman and Burke, 2003: Freeman and Rogers, 2003). I postulate that this global pattern updates the contents of attractor landscapes, implement the prediction of new sensory inputs, and issue fresh motor commands. Preafference operates not as copies of motor commands for error correction (Von Holst and Mittelstädt, 1950) but by participation in a macroscopic, spatially coherent AM pattern. This emergence of the macroscopic pattern completes the action–perception cycle with assimilation (Freeman, 1995), literally within the time frame required for the blink of an eye.

Circular causality

One may ask where in the brain does one see the macroscopic pattern and in what form? My answer is that it appears in synchronized oscillations over broad ranges of beta frequencies in ECoG (Freeman and Burke, 2003; Freeman and Rogers, 2003) and EEG (Freeman et al., 2003a, b) and ECoG (Freeman and Burke, 2003), and that the underlying activity organizes all parts of brain and body that are simultaneously engaged with the material, formal and social environments. To focus again on olfaction, the molecular structures of the receptor cells in the nose are active in binding odorant molecules from the air stream. So also are the myriad synapses in the sensory and motor areas of cortex and at the neuromuscular synapses on muscle cells, which bind neurotransmitter molecules at the submicroscopic level. The networks of neurons in the olfactory brain are active in preprocessing the information delivered by pulses from receptor cells into cortical networks, executing the essentially engineering operations of amplification, range compression, normalization, filtering, and selective enhancement of the information (Freeman, 2001). The entire olfactory brain is reorganized in a phase transition by which the stimulus selects the class to which it belongs, and the entire olfactory system transmits a wave packet throughout the basal forebrain including the limbic system. The subsequent formation of a macroscopic pattern integrates the activity of the entire forebrain including the limbic, motor, and olfactory systems (Freeman and Burke, 2003). I conjecture that the pattern provides the context in which the appropriate behavior self-organizes, containing the trajectories of microscopic neural activity and mesoscopic limb movements that are required to achieve emergent goal states. Molecular, cellular, and mesoscopic assemblies are modulated and directed at all times, everywhere, and at all levels. How might this orchestration take place?

One might ask a similar question about any large-scale, self-organized physical process such as a hurricane or a tree. How does each molecule of air or water conform its trajectory into the gigantic vortex that feeds on solar energy? How does each pore on every leaf in the sunlight coordinate with every hair on every root branch in the ground? Correspondingly, how does each molecule of neurotransmitter substance and each neuron and each local assembly conform to the global organization that we observe in animal and human behavior? These questions we can answer now by combining neurobiological observation and experimentation with theory from physics, chemistry, and mathematics (Prigogine, 1980; Haken, 1983). But hurricanes and trees cannot intend, whereas brains can and do intend. The difference is twofold: hurricanes and trees cannot remember and utilize their past, and neither trees nor hurricanes can direct the movement of their bodies through their environments. They have no brains. Only animals with brains have the machinery for anticipating future states, planning for deployment of their bodies in pursuit of satisfaction of perceived needs, predicting the consequences on sensory inflow of their own actions, and above all for self-assimilating by which they bring their brains and bodies into conformance with their environments. In short, hurricanes and trees lack the mechanisms required for intentionality (Freeman, 1995).

It is immediately apparent that intention spans the entire range of material, psychological, and social behaviors, from the most distant conception of survival and procreation to the molecular changes in nerve and receptor cells that enable sensation, learning, and muscle contraction. The material basis at each level and its teleological relations to levels immediately below and above are well described by the particular science that is directed to the level. Of particular concern is the relation *between* levels that is described with the concept of circular causality (Haken, 1983): in self-organization the higher level order forms by the interactions of lower order parts. The now classic example in physics and engineering comes from the dynamics of a laser. The parts are the atoms in a gas that oscillate at frequencies in a wide distribution about some mean value, when they are in a state of low energy. When energy is pumped into the atoms, they oscillate more strongly and interact with each other more strongly. At some threshold they undergo a state transition and oscillate all at a shared frequency. The high-energy oscillation is called an "order parameter", because the atoms that generate the oscillation are "ordered" ("enslaved" according to Haken) by the whole to oscillate at one frequency. The reason this process is described as "circular causality" is that the particles (neurons) create the field (the wave packet) and the field imposes order onto the particles. Similarly in the olfactory brain at low energy before a stimulus input arrives, the neurons emit pulses seemingly at random with a distribution of pulse frequencies. When their energy level is increased by excitation from olfactory receptors, their pulse frequencies increase. At some threshold the whole population interacts so strongly that all the neuronal potentials oscillate at the same instantaneous frequency, though with different amplitudes and different levels of participation. The population signal seen in the AM pattern of the wave packet in that frequency range is an order parameter that brings all of the neurons into varying degrees of synchronous oscillation (Freeman and Vitiello, 2006).

The analogy is limited, because atoms are all indistinguishable, whereas neurons all differ from one another, no two being identical. Whereas all the atoms are locked into the one order parameter, the neurons in a population have varying degrees of sharing in the common signal; the order parameter is vectorial. Owing to their individual differences the classical descriptions from statistical mechanics are not adequate to describe population neurodynamics. Descriptions using concepts from classical thermodynamics certainly apply in terms of the requirements for disposal by brains of waste heat and entropy, as well as essential constraints on brain temperature, pressure, mass, and volume that are self-regulated. The analogy does have great value, because it expresses a fundamental property of brains in a simple way: populations of neurons interact by recurrent excitation and inhibition through synaptic transmission and create order parameters that regulate the same neurons. This is "circular causality". It differs from "entrainment", which denotes the reciprocal interaction of two entities at the same level, such as clocks or neurons. As introduced by Haken (1983) it denotes the conformance of the individual with the group, which requires field effects as seen in mobs and vortices. We observe the individual neural activity in pulse trains on axons; we observe the order parameter in waves of dendritic currents. The relation between pulses and waves is bidirectional. We predict the wave densities from pulse densities by averaging over the parts that form the whole. We deduce the effects of the waves on the pulse densities by calculating differences in wave densities. Integration carries us to the higher level; differentiation carries us to the lower level. These processes of summation and differencing occur simultaneously in all areas of cortex (Freeman, 2006b). The predominant direction of information flow through these processes in sensory areas is upward from individual neural activity to population densities; the predominant direction in motor areas is downward from population wave densities to more individually structured trajectories of pulse densities.

The continuity of circular causality across all levels

Looking downwardly, neurons are microscopic parts of mesoscopic populations, yet each neuron is a semi-autonomous whole that develops and maintains complex relations among its parts. It devotes most of its lifespan to its own janitorial functions; the typical cortical neuron fires a pulse lasting 1 ms at an average rate of 1/s, which would scale to 1 full day every 3 years. Yet it is ceaselessly active at all times in responding to input from an average 10,000 other neurons (Braitenberg and Schüz, 1998) by which it is modulated through multiple order parameters. Each of its parts is a subwhole, which is organized by assemblies of macromolecules that provide the energy for generating electric fields, opening and closing ion channels, and maintaining chemical balances. Each macromolecule is an organized assembly of atoms that performs a designated task that depends on collective, patterned action expressing an order parameter. Looking upwardly, mesoscopic neural populations are components of ongoing macroscopic fields comprising organized actions of the whole brain. The brain is one organ among many in the body that cooperate continually in directed actions. The body is embedded in organic relations with the material and social worlds, and so on. Each of these levels generates order parameters at differing scales of time and space, and operates with entities, states, and state variables that are unique to the point of view taken by scientists engaged in systematic study at each level. Yet brain wave dynamics is scale-free (Freeman, 2005b, 2007c), meaning that its wave patterns of electrical activity are self-similar (Barabási, 2002) across wide scales of time and space, as shown by measurements of distributions of its dynamic properties, most obviously those of the neocortex (Freeman, 2006b). It is the scale-free dynamics that appears to enable mammalian brains varying in mass 10^4 from mouse to whale to participate in and organize all levels of function simultaneously by transactions that extend seamlessly across the entire range, yet which can be abstracted for measurement and analysis at each desired level with its pertinent scales of measurement. These measurements give the numbers that are translated into information, and the numbers support the analyses by modeling based in symbolic codes.

The reflex arc actually begins not with a stimulus but with the intention of the investigator, who selects and delivers a stimulus to the subject with the goal of constructing a useful code. The stimulus is a pattern of chemical energy that impacts on individual receptors at the atomic level with binding of molecules of scent to the surfaces of receptor cells, initiating cascades of biochemical reactions resulting in microscopic pulses transmitted to the brain. The impact of myriads of pulses with inhalation destabilizes the olfactory brain and changes the order parameter to an intracortical search mode. Convergence to an attractor means that the collective population of neurons enters into an ordered state that modulates the pulse trains of the entire olfactory brain, sending an AM pattern that is carried by the patterns of myriad microscopic pulses to other parts of the brain. The pattern of the wave packet, being mesoscopic, is not detectable by observing the pulse trains of any small number of neurons; it is only seen in large averages. The integration of multiple wave packets supports the emergence of a global brain state that provides an order parameter that includes the motor areas simultaneously with the sensory areas. This macroscopic context modulates the mesoscopic populations that organize the motor areas into controlled sequences of oscillations and shape the sensitivities of the sensory areas by selection of attractor landscapes in preafference. The reflex arc is completed by the modular organization of microscopic pulse trains of motor neurons, which release the neurochemical synaptic transmitter molecules that are required for muscle contraction.

Whatever the intent of the investigator, the intentional arc of the animal begins with its intention as expressed in its macroscopic goal state and extends through mesoscopic patterns to the microscopic level of muscle contraction. Actions cause changes in the microscopic binding of chemicals to chemoreceptors, photons to visual receptors, and so on, which are transduced into rates and frequencies of firing, followed by mesoscopic phase transitions and, eventually by closure of the arc on assimilation and updating of the perceptual wave packets and the conceptual macroscopic state. This is the action–perception cycle.

From sensation to perception to conception; from goal to plan to action

The above descriptions of the neural correlates of intentional action and perception, when viewed in terms of scale-free brain dynamics across the broad range of scientific disciplines, leads to the view that engagement of the individual with the environment is simultaneous at all levels. Even though there are no "atomic propositions" (Barlow, 1972), the metaphors for coding are invaluable for communication among researchers. The material engagement takes place in the immersion of body and receptors in gases, liquids, and solids governed at the atomic and molecular levels by quantum field theory, and at macroscopic levels by Newtonian physics through forces that modulate the firings of stretch receptors in muscles, pressure receptors in skin, joints internal organs, and vestibular receptors for gravity and acceleration of the head. These chemical and physical forces permeate brains and bodies with continuous presentation of information to the brain, followed by its selective distillation into knowledge. At the mesoscopic level there is preconscious apprehension of the influx of new relationships between body and environment that go far beyond information processing in the emergence of wave packets, which can be interpreted as symbols of generalizations representing confirmations or disclaimers of anticipations regarding the continuity of the fabric of the world and the place claimed by the individual, the "horizons" of Merleau-Ponty (1942/1963). These surmises about the impending future accompany the preparations for rest or for incipient action to deal with predicted or unexpected contingencies in the surround, the arena of perception. Yet this is not all. Embedding the perceptual and premotor activities of body and brain is the guiding matrix of goals, ranging in scope and complexity from what to do in the next few seconds in the face of opportunity or danger to lifelong ambition to flourish and prevail. It is this self-structured dynamic edifice of anticipations rooted in the accumulated self-assimilations of a lifetime of knowledge that modulates, enriches, and integrates the experience so immediately reflected in mesoscopic and macroscopic patterns of brain activity. We have also discovered their traces in electrical fields at the surface of the human scalp (Freeman et al., 2003a), but we cannot yet read them, because we do not yet know how to encode their patterns in terms of information and symbols adequately to correlate them with behavioral measurements that include verbal communications.

This description of intentional brain dynamics was pioneered seven centuries ago by Aquinas (1272), who dismissed the passivity of the Platonic soul by conceiving intention as taking action (*intendere*) and coming to know the world by self-assimilation (*adequatio*), which is conforming the body and brain with the environment and not the Aristotelian processing and storing of forms (information). In the view of the intentional arc the goal pre-exists the action, whereas in the view of the reflex arc the goal exists as an achievement after completion of the action. According to Aquinas (Q 85, A 2) there are two kinds of intentional action. One is transitive action in mechanistically thrusting the body into the world in the manner of a robot or other machine. The other is immanent action by understanding, which distinguishes the actions of animals and humans from those of machines that act without comprehending what they are doing. Understanding includes contemplative withholding of action but still has reference to or engagement in the world that provides knowledge through self-assimilation through learning from the senses, herein differing from idealist conceptions that understanding is derived solely through reference to innate codes in the brain. Understanding does not occur at the microscopic level of single neural activity of pulses, which is unique and ephemeral and directly related to the particular stimulus that drives it. These Aquinian phantasms are likenesses of a thing and not the thing, in the manner that trains of action potentials that bear information to the brain are the likeness of a stimulus but not the stimulus. Being unique events, the phantasms (the patterns of the microscopic pulses, the raw sense data) are unknowable.

The mesoscopic level is that of the intelligible species, which forms by abstraction and generalization over multiple sequential phantasms. Here is the first step of crossing from the realm of the material to the realm of the perceptual, from the concrete to the abstract. The transposition begins in sensory areas with modality-specific wave packets, which embody a selection of all stored experience that is immediately relevant to the intended inputs (The information-bearing stimuli that are sought by intentional observations). The wave packets are not fully intelligible, because they lack multisensory integration and orientation in time and space from convergence and passage through the limbic system. Aquinas wrote (Q 79, A 4): "Therefore we must say that in the soul is some power derived from a higher intellect, whereby it is able to light up the phantasmata. And we know this by experience, since we perceive that we abstract universal forms from their particular conditions, which is to make them actually intelligible." His "light up" appears to correspond to the stage of self-assimilation when a macroscopic state emerges following the limbic integration of mesoscopic wave packets and preafference (Freeman and Burke, 2003; Freeman and Rogers, 2003). That macroscopic order parameter modulates all sensory cortices and includes the motor areas, which must be engaged in the process of deciding what to do in the light of new integrated input stemming from the senses.

The new state of knowledge is an engagement with the situation of brain and body in the world that by self-similarity contains mesoscopic preparatory states in both sensory and motor areas for planning action and predicting its sensory consequences. By virtue of scale-free dynamics the engagement occurs at all levels simultaneously, they may be material, formal, or social. Through mesoscopic and macroscopic constructions the brain conceives, grasps, and approaches by sequential actions with the body what Merleau-Ponty called "maximum grip" immediately and directly in the way that an aircraft pilot, a car driver, and a tennis player experience the instruments as extensions of the body, not as inner manipulation of symbols and representations or exercise of codes in computational logic. This elemental process does not posit consciousness; there is no need at this level for that hypothesis. Self-awareness in these actions is by neural mechanisms not yet adequately examined in humans to provide the experimental field data required to build the appropriate theory, but it readily appears that the recursive embedding provided by circular causality in macroscopic patterns of transient global synchrony will be identified as crucial in the process of consciousness.

First intention and second intention

This description of the neurodynamics of intentionality has been made possible only in the past few years, equally by advances in technology that enabled simultaneous EEG recording from large electrode arrays implanted onto the surface of the brain or on the scalp of humans, and by advances in theory that enabled modeling the EEG patterns using concepts from nonlinear dynamical theory (Freeman and Vitiello, 2006), neuropercolation theory (Kozma et al., 2005), and scale-free dynamics (Barabási, 2002; Freeman, 2006a, 2007c). These developments open the way to reconsider long-standing differences between cognitivists and phenomenologists in their interpretations of intentionality. Descartes abandoned the Thomist concept of intentionality in his dualist, subject–object description of the soul operating the brain like a pilot controlling machine functions using representational logic and mathematics. Intention was re-introduced by Brentano (1889/1969) as the basis for distinguishing the representations and operations on them of humans who know what they are doing from those of machines that do not know. The usages by his successors have led to Searle's (1983) characterization of intentionality as "aboutness", because a thought or a perception is "about" something. This interpretation suffers the intractable difficulty of grounding coding symbols in machines and brains to the entities they represent. For example, what is the relation between a word in a computer memory and the real person it represents? Similarly, how does the firing of neurons in the cortex of the fusiform gyrus signify the perception of a face, and how does that firing "cause" one to classify the person whose face it is?

Heidegger (1975/1988) reintroduced what he called "the enigmatic phenomenon of intentionality" in a form that is indistinguishable from that of Aquinas, despite his denial of any indebtedness to the "Scholastics". The only reason for citing his turgid, obfuscatory, quasimystical work for neuroscientists is that he addressed what he rightly called "the central problem of philosophy", the same as that with which this review began: in his terms, "... the 'transposition' [transcendence] of the Dasein over to things", and that he led other phenomenologists, principally Merleau-Ponty, back to this forgotten insight. By "Dasein" he simply meant the underlying, largely unconscious, intentional self and not the egoistic awareness of self. He usefully distinguished two widespread "misinterpretations" of "intentionality". First was the "common sense" assignment of intentionality to the subject; Searle (1983) wrote that the firing of neurons caused perception of an object, thus maintaining the Cartesian subject–object separation that is inherent in representationalism. Heidegger wrote that this view characterized "... intentionality as an extant relation between two things extant, a psychological subject and a physical object. The nature as well as the mode of being of intentionality is completely missed (pp. 60–61)." The second misconception was the "erroneous subjectivization of intentionality. ... Intentionality is neither objective nor subjective in the usual sense, although it is certainly both (pp. 63–65). This misconception is common among psychologists who conceive intention as purpose, a mental state of goal-directedness.

Here again is the core problem: understanding the relation between the abstractions and generalizations in the structures of brain dynamics and the material involvements that are understood, and how they are understood through and beyond "likenesses": the action potentials of neurodynamicists, the phantasms of Aquinas, and the raw sense data of psychologists. The dynamical view proposes that a self-similar hierarchy of patterns, emerging from the structures of knowledge that are stored in the synaptic tissues of the brain, is continually modified by interactions with the multiple environments of the body and brain. In some deep sense this patterned activity expresses the being that Heidegger conceived as the Dasein, but at present with a significant limitation that constrains intentional neurodynamics to describing only first intention that animals share with children still too young to remember their lives or to distinguish themselves from any other intentional being (Dasein). Operationally the capability is defined by the mirror test: toddlers in front of a mirror look behind it to see who is there; a few months later they watch themselves touching themselves. At present the evidence for macroscopic neurodynamics comes only from animals that cannot pass the test. Second intention in which the self reflects on the process of comprehending the likenesses provided by sensory processing early in first intention is barely touched by neurodynamicists, despite major efforts to explore consciousness and awareness. This is the domain of phenomenology. Dreyfus (2007) has described remarkably close correspondences between nonlinear brain dynamics and the basic conceptions of the dynamics of intentional behaviors as conceived by Heidegger and Merleau-Ponty, subject to the limitation that phenomenology can only begin with consciousness of concepts that emerge far above the raw sense data and wave packets. Owing to their entry at this high-level phenomenologists cannot reach down to the level of sensation so as to distinguish between sensation and perception, as neurophysiologists distinguish them, as shown by this exchange between Merleau-Ponty (1966) and a conference organizer:

> *M. Parodi*: Could you tell us what is your most important contribution on this question of fact. You began with very clear examples: we think we perceive things which we really only see in part, or more or less. What, according to you, is the essential element in this operation?
>
> *M. Merleau-Ponty*: To perceive is to render oneself present to something through the body. All the while the thing keeps its place within the horizon of the world, and the structurization consists in putting each detail in the perceptual horizons which belong to it. But such formulas are just so many enigmas unless we relate them to the concrete developments which they summarize.
>
> *M. Parodi*: I would be tempted to say that the body is much more essential for sensation than it is for perception.
>
> *M. Merleau-Ponty*: Can they be distinguished? ... (p. 42)"

Clearly M. Parodi did not grasp Merleau-Ponty's position, which was that sensation did not exist as a mental process, hence "the primacy of phenomenology".

Conclusions

Contemporary approaches used by researchers to understand and model both human and machine intelligence are commonly based in search for computational and representational codes. One reason is the clarity and simplicity of logical positivist concepts describing brain activity in terms of information and symbols, compared with the relative obscurity and impenetrability of the descriptors by dynamicists and phenomenologists. For nonscientists the arcane descriptions by brain dynamicists may appear just as opaque as Heidegger's and Merleau-Ponty's prose in translation appears to scientists, but scientists have the advantage of experimental grounding in brain physiology, the interpretation of which may be facilitated by translating concepts between fields. Alternative approaches to incorporate intentionality into neurobiology include those of pragmatists such as Dewey (1914): "Actions are not reactions to stimuli; they are actions into the stimuli"; Piaget (1930) in the study of child development; Köhler (1940) using field theory; Koffka (1935) using Gestalt theory; its extension by Gibson (1979) into ecological psychology; and situated cognition (Slezak, 1995). As shown by Dreyfus (2007) these and related cognitivist approaches are still shot through with strong reliance on information theory and representationalism for construction of explanatory codes. Indeed the inventor and chief architect of the programmable serial digital computer, the backbone of artificial intelligence for manipulation of symbols in coding systems, von Neumann (1958), realized early the limitations of the computer model:

> "Thus the outward forms of *our* mathematics are not absolutely relevant from the point of view of evaluating what the mathematical or logical language *truly* used by the central nervous system is. ... It is characterized by less logical and arithmetical depth than what we are normally used to. ... We require exquisite

> numerical precision over many logical steps to achieve what brains accomplish in very few short steps." (pp. 81–82)

Those few short steps can now be seen through the lens of nonlinear field neurodynamics.

Brain imaging also shows great promise as a source of new experimental data on global brain dynamics, but currently it is in a phase of empirical casuistry that in many ways resembles 19th century phrenology, owing to lack of adequate brain theory. Psychiatrists likewise rely heavily on empirical taxonomy following the failure of Freudian theory. Numerous proposals for theory have come from neurophilosophers on one hand and from mathematicians and physical scientists on the other, but with inadequate experimental support and with derivations often too strongly Cartesian to meet the challenge. Therefore, the new techniques for acquiring macroscopic data and interpreting them on the light of updated field theory and neuropercolation theory can provide the solid conceptual structure that is necessary to solve the core problem of philosophy. There is more. Thomist-Heideggerian philosophy will likely lead to constructing a totally new class of machine, the intentional robot, which is based in neurodynamics instead of digital logic (Kozma and Freeman, 2003; Kozma et al., 2003; Dreyfus, 2007). This possibility is as relevant to philosophers as it is to engineers. If an intelligent machine can comprehend and remember only the sensory consequences of its own intended actions, then it must be equipped with appropriate sensors, effectors, sources of reward, and the autonomy to explore its environment with learning by trial and error under reinforcement. Demonstration of a solution to the core problem of cognitive science and philosophy by such modeling of first intention must precede an approach to second intention, for which there is no realistic possibility at present.

From detailed measurements of the electric fields of the brain it is possible to infer that the essential operation in the sensory cortices is to replace (transpose) stimulus input with constructs by the brain of conceptions that stem from anticipation based in memory. These constructs emerge by cooperative neurodynamics operating over a continuum of scales in time and space that can be divided into levels corresponding to the techniques of observation and measurement of brain activity and behavior. The constructs are states of knowledge that support predictions by multisensory projections from the present into the future of desired rewards through patterns of sensory input from the body and the environment. The anticipations exist as macroscopic patterns of neural activity that order ("enslave") the mesoscopic populations of neurons comprising the sensory and motor areas. In the sensory cortical areas the local attractor landscapes embody the specific predictions. The motor cortical areas embed the tactical trajectories of neural activity that control the movements of the body and with proprioception shape the actions in the context of the changing environment. The changes in sensory inflow resulting from movements are transmitted to sensory cortical areas, where they encounter the attractor landscapes formulated through preafference as internal model-building. The sensory and motor mesoscopic activity patterns that exist in the forms and trajectories of the material substrate of neural activity are the abstract concepts that govern the engagement of the Dasein with the world by anticipating, acting, sensing, generalizing, and assimilating, encompassing first intention in animals and in preconscious states of humans.

In neurodynamics the process can be studied at the multiple levels of its material substrate in brain, body, and environment and the forms pertaining thereto. In physics the process can best be described by models that combine the agent of action with that part of the environment that is engaged, creating a mirror image or 'double' in order to balance the energy flows in the unified system (Vitiello, 2001). In philosophy the concepts referred to as phenomena constitute the mind, which directly enters into the world on its own terms, achieving closure and "maximum grip" without intermediation by representations of raw sense data (Dreyfus, 2007). What is still inaccessible to analysis with respect to neurodynamics is an explanation of second intention, the awareness of experiencing of the world. There is no physiological test for consciousness even at the elemental level of that which is obtunded by anesthesia or sleep. There is only the phenomenological test of asking a subject, "What do you remember?" and comparing the answer with objective records. In the lack of such a test the only acceptable conclusion is that we do not now understand the process of self-awareness. The aim of this essay is to describe a pathway in brain dynamics toward understanding by experimental observation and measurement of the macroscopic fields of the brains of normal subjects, which will require devising and applying new and advanced EEG technology supplemented in parallel with related techniques of noninvasive brain imaging.

References

Aquinas, St. Thomas. (1272) The Summa Theologica. Translated by Fathers of the English Dominican Province. Revised by Daniel J. Sullivan. Published by William Benton as Volume 19 in the Great Books Series. Encyclopedia Britannica, Inc., Chicago, IL, 1952.

Barabási, A.-L. (2002) Linked. The New Science of Networks. Perseus, Cambridge, MA.

Barlow, H.B. (1972) Single units and sensation: a neuron doctrine for perceptual psychology? Perception, 1: 371–394.

Braitenberg, V. and Schüz, A. (1998) Cortex: Statistics and Geometry of Neuronal Connectivity (2nd ed.). Springer-Verlag, Berlin.

Brentano, F.C. (1889/1969) The Origin of Our Knowledge of Right and Wrong. Chisolm R.M. and Schneewind E.H. (trans.). Humanities Press, New York.

Bressler, S.L. and Kelso, J.A.S. (2001) Cortical coordination dynamics and cognition. Trends Cogn. Sci., 5: 2–36.

Buck, L. and Axel, R. (2004) A novel multigene family may encode odorant receptors: a molecular basis for odor recognition. Cell, S116(2): 175–187.

Burr, C. (2002) The Emperor of Scent: A Story of Perfume, Obsession, and the Last Mystery of the Senses. Random House, New York.

Buzsaki, G. (2002) Theta oscillations in the hippocampus. Neuron, 33(3): 325–340.

Chalmers, D.J. (1996) The Conscious Mind: In Search of a Fundamental Theory. Oxford University Press, New York.

Dewey, J. (1914) Psychological doctrine in philosophical teaching. J. Philos., 11: 505–512.

Dreyfus, H. (2007) Why Heideggerian AI failed and how fixing it would require making it more Heideggerian. Philos. Psychol., 20(2): 247–268.

Ferster, C.B. and Skinner, B.F. (1957) Schedules of Reinforcement. Prentice-Hall, Englewood Cliffs, NJ.

Freeman, W.J. (1975/2004) Mass Action in the Nervous System. Academic Press, New York. http://sulcus.berkeley.edu/MANSWWW/MANSWWW.html

Freeman, W.J. (1995) Societies of Brains. A Study in the Neuroscience of Love and Hate. Lawrence Erlbaum, Mahwah, NJ.

Freeman, W.J. (1999) How Brains Make Up Their Minds. Weidenfeld & Nicolson, London, UK.

Freeman, W.J. (2000) Neurodynamics. An Exploration in Mesoscopic Brain Dynamics. Springer-Verlag, London, UK.

Freeman, W.J. (2001) The olfactory system: odor detection and classification. In: Frontiers in Biology, Vol. 3. Intelligent Systems. Part II. Brain Components as Elements of Intelligent Function. Academic Press, New York, pp. 509–526.

Freeman, W.J. (2004a) Origin, structure, and role of background EEG activity. Part 1. Analytic amplitude. Clin. Neurophysiol., 115: 2077–2088.

Freeman, W.J. (2004b) Origin, structure, and role of background EEG activity. Part 2. Analytic phase. Clin. Neurophysiol., 115: 2089–2107.

Freeman, W.J. (2005a) Origin, structure, and role of background EEG activity. Part 3. Neural frame classification. Clin. Neurophysiol., 116(5): 1118–1129.

Freeman, W.J. (2005b) A field-theoretic approach to understanding scale-free neocortical dynamics. Biol. Cybern., 92(6): 350–359.

Freeman, W.J. (2005c) NDN, volume transmission, and self-organization in brain dynamics. J. Integr. Neurosci., 4(4): 407–421.

Freeman, W.J. (2006a) Origin, structure, and role of background EEG activity. Part 4. Neural frame simulation. Clin. Neurophysiol., 117(3): 572–589.

Freeman, W.J. (2006b) Definitions of state variables and state space for brain-computer interface. Part 1. Multiple hierarchical levels of brain function. Cogn. Neurodyn., 1(1): 1871-3080 (print); 1871-4099 (online). http://dx.doi.org/10.1007/s11571-006-9001-x

Freeman, W.J. (2007a) Hilbert transform for brain waves. Scholarpedia. p. 7514. http://www.scholarpedia.org/article/Hilbert_transform_for_brain_waves

Freeman, W.J. (2007b) Intentionality. Scholarpedia. p. 8616. http://www.scholarpedia.org/article/Intentionality

Freeman, W.J. (2007c) Scale-free neocortical dynamics. Scholarpedia. p. 8780. http://www.scholarpedia.org/article/Scale-Free_Neocortical_Dynamics

Freeman, W.J. and Burke, B.C. (2003) A neurobiological theory of meaning in perception. Part 4. Multicortical patterns of amplitude modulation in gamma EEG. Int. J. Bifurc. Chaos, 13: 2857–2866.

Freeman, W.J., Burke, B.C. and Holmes, M.D. (2003a) Aperiodic phase re-setting in scalp EEG of beta-gamma oscillations by state transitions at alpha-theta rates. Hum. Brain Mapp., 19(4): 248–272.

Freeman, W.J., Burke, B.C., Holmes, M.D. and Vanhatalo, S. (2003b) Spatial spectra of scalp EEG and EMG from awake humans. Clin. Neurophysiol., 114: 1055–1060. http://repositories.cdlib.org/postprints/989

Freeman, W.J. and Rogers, L.J. (2003) A neurobiological theory of meaning in perception. Part 5. Multicortical patterns of phase modulation in gamma EEG. Int. J. Bifurc. Chaos, 13: 2867–2887.

Freeman, W.J. and Vitiello, G. (2006) Nonlinear brain dynamics as macroscopic manifestation of underlying many-body field dynamics. Phys. Life Rev., 3: 93–118.

Gibson, J.J. (1979) The Ecological Approach to Visual Perception. Houghton Mifflin, Boston, MA.

Haken, H. (1983) Synergetics: An Introduction. Springer, Berlin.

Heidegger, M. (1975/1988) The Basic Problems of Phenomenology (rev. ed.). Hofstadter A. (trans.). Indiana University Press, Bloomington, IN.

Houk, J.C. (2005) Agents of the mind. Biol. Cybern., 92: 427–437.

Houk, J.C. and Wise, S.P. (1995) Distributed modular architectures linking basal ganglia, cerebellum, and cerebral cortex: their role in planning and controlling action. Cereb. Cortex, 5: 95–110.

Kay, L.M. and Freeman, W.J. (1998) Bidirectional processing in the olfactory-limbic axis during olfactory behavior. Behav. Neurosci., 112: 541–553.

O'Keefe, J.M. and Nadel, L. (1978) The Hippocampus as a Cognitive Map. Oxford University Press, New York.

Koffka, K. (1935) Principles of Gestalt Psychology. Harcourt Brace, New York.

Köhler, W. (1940) Dynamics in Psychology. Grove Press, New York.

Kozma, R. and Freeman, W.J. (2003) Basic principles of the KIV model and its application to the navigation problem. J. Integr. Neurosci., 2: 125–145.

Kozma, R., Freeman, W.J. and Erdí, P. (2003) The KIV model — nonlinear spatio-temporal dynamics of the primordial vertebrate forebrain. Neurocomputing, 52: 819–826. http://repositories.cdlib.org/postprints/1049

Kozma, R., Puljic, M., Balister, P., Bollabás, B. and Freeman, W.J. (2005) Phase transitions in the neuropercolation model of neural populations with mixed local and non-local interactions. Biol. Cybern., 92: 367–379. http://repositories.cdlib.org/postprints/999

Lancet, D. and Ben-Arie, N. (1993) Olfactory receptors. Curr. Biol., 3: 6768–6774.

Lettvin, J.Y. and Gesteland, R.C. (1965) Speculations on smell. Cold Spring Harbor Symp. Quant. Biol., 30: 217–225.

Maclean, P.D. (1969) The Triune Brain. Plenum, New York.

Merleau-Ponty, M. (1942/1963) The Structure of Behavior. Fischer A.L. (trans.). Beacon, Boston, MA.

Merleau-Ponty, M. (1966) The Primacy of Perception. In: Edie J.M. (Ed.). Northwestern University Press, Evanston, IL.

von Neumann, J. (1958) The Computer and the Brain. Yale University Press, New Haven, CT.

Piaget, J. (1930) The child's conception of physical causality. Harcourt, Brace, New York.

Prigogine, I. (1980) From Being to Becoming: Time and Complexity in the Physical Sciences. WH Freeman, San Francisco, CA.

Quiroga, Q.R., Reddy, L., Kreiman, G., Koch, C. and Fried, I. (2005) Invariant visual representation by single-neurons in the human brain. Nature, 435: 1102–1107.

Searle, J.R. (1983) Intentionality. Cambridge University Press, Cambridge, UK.

Sherrington, C.S. (1906) The Integrative Action of the Nervous System. Yale University Press, New Haven, CT.

Skarda, C.A. and Freeman, W.J. (1987) How brains make chaos in order to make sense of the world. Behav. Brain Sci., 10: 161–195.

Slezak, P. (1995) The 'philosophical' case against visual imagery. In: Slezak P., Caelli T. and Clark R. (Eds.), Perspectives on Cognitive Science: Theories, Experiments and Foundations. Ablex Publ., Greenwich, CT, pp. 237–271.

Sperry, R.W. (1950) Neural basis of the spontaneous optokinetic response. J. Comp. Physiol., 43: 482–489.

Tsuda, I. (2001) Toward an interpretation of dynamics neural activity in terms of chaotic dynamical systems. Behav. Brain Sci., 24: 793–847.

Turin, L. (2006) The Secret of Scent. HarperCollins, New York.

Vitiello, G. (2001) My Double Unveiled. John Benjamins, Amsterdam.

Von Holst, E. and Mittelstädt, H. (1950) Das Reafferenz Prinzip. Wechselwirkung zwischen Zentralnervensystem und Peripherie. Naturewissenschaften, 37: 464–476.

P. Cisek, T. Drew & J.F. Kalaska (Eds.)
Progress in Brain Research, Vol. 165
ISSN 0079-6123
Copyright © 2007 Elsevier B.V. All rights reserved

CHAPTER 29

From a representation of behavior to the concept of cognitive syntax: a theoretical framework

Thomas Gisiger[1] and Michel Kerszberg[2,*]

[1]*Récepteurs et Cognition, Institut Pasteur, 25 rue du Docteur Roux, 75015 Paris Cedex 15, France*
[2]*Université Pierre et Marie Curie, Modélisation dynamique des systèmes intégrés UMR CNRS 7138 — Systématique, Adaptation, évolution, 7 quai Saint Bernard, 75252 Paris Cedex 05, France*

Abstract: Before we do anything, our brain must first construct a neural correlate of the various mental operations needed. Imaging and recording techniques have vastly improved our understanding of this process by providing detailed insight into how different regions of the brain contribute to behavior. However, exactly how these regions collaborate with each other to form the brain-scale activity necessary to generate even the simplest task remains elusive. Here we present a neural network model based on the hypothesis of a modular organization of brain activity, where basic neural functions useful to the current task are recruited and integrated into actual behavior. At the heart of this mechanism are regulating structures that restrain activity from flowing freely between the different cortical areas involved, releasing it instead in a controlled fashion designed to produce the different mental operations required by the task at hand. The resulting dynamics enables the network to perform the delayed-matching to sample and delayed-pair association tasks. The model suggests that brain activity coding for elementary tasks might be organized in modular fashion, simple neural functions becoming integrated into more complex behavior by executive structures harbored in prefrontal cortex and/or basal ganglia. We also argue that such an integration process might take place through an iterative process, by piecing together previously validated behavioral chunks, while creating new ones under the guidance of a partially innate cognitive syntax.

Keywords: computational model; task representation; syntax; electrophysiological data; memory; executive function; cognitive tasks

As we learn more about the structure and function of the various areas of the brain, it becomes increasingly important to understand how these different areas communicate and interact with each other to build the brain-scale neural activity necessary to implement behavior.

When action is needed, relevant neural circuits in the brain are somehow selected and allowed to interact, thereby creating the streams of neural computations necessary for the task at hand. How exactly this takes place, however, remains largely unclear. Mechanisms such as inhibition and excitatory reentry are often proposed as central to this complex process. Although they undoubtedly play an important role in neural processing, e.g., by removing unwanted exterior or interior influences and allowing information to be sustained over time, these mechanisms are more suited to stabilizing and refining ongoing neural activity than to

*Corresponding author. Tel.: +33 (0) 1 44 27 37 22;
Fax: +33 (0) 1 44 27 52 50; E-mail: mkersz@ccr.jussieu.fr

DOI: 10.1016/S0079-6123(06)65029-2

producing new patterns. Other mechanisms are clearly needed to account for the patterns of neural activity necessary to code even the simplest behaviors.

In a recent paper (Gisiger and Kerszberg, 2006), we have suggested that an important role in this process might be played by structures specifically devoted to controlling the flow of information circulating between the different parts of the cortex involved in a given behavior. According to this proposal, global brain activity would be constructed by mobilizing a set of neural circuits localized in separate cortical areas, with each of these circuits implementing one of the various components necessary to the task at hand. We proposed in addition that, aside from the circuitry coding for the different components necessary to build behavior, the brain is also equipped with specific circuits devoted to regulating, in a manner suitable for the desired behavior, the flow of information circulating between these components (Gisiger and Kerszberg, 2006). Thus, these control units, by resetting some circuits appropriately, or by opening neural pathways between certain circuits while keeping others closed, would allow meaningful neural computations to be performed, thereby producing the intended behavior.

To back these hypotheses, we built a neural network model which could be trained to perform the mixed-delayed response (MDR) task detailed below, while also qualitatively reproducing neural firing patterns gathered from the inferior temporal and prefrontal cortex of monkeys performing these same tasks [see Gisiger and Kerszberg (2006) for details].

Here, we take a closer look at the relationship between the flow of activity taking place in the neural circuits mobilized by a given task, and the behavior that results from it. We will do this by first describing the MDR task and listing the elementary processing actions it requires. We then summarize the main features of the model we introduced to perform this task. We next look at the works published by Chadderdon and Sporns (2006) and the group of Frank and O'Reilly (Frank et al., 2001; O'Reilly and Frank, 2006), which illustrate that managing the information accessing working memory sustained in it is critical to perform even simple memory tasks. We close this article by discussing the notion of task representation in our model, and how the latter suggests a link between the process leading to the emergence of task representations and the concept of cognitive syntax.

Task description

Figure 1 describes the MDR task, which was introduced to study memory retrieval in the monkey using visual associations (Sakai and Miyashita, 1991; Naya et al., 1996; Rainer et al., 1999). This task consists of randomly mixed delayed-matching to sample (DMS) and delayed-pair association (DPA) trials (Sakai and Miyashita, 1991; Rainer et al., 1999).

The DMS task is a simple memory task, which requires that the subject maintain the memory of an image during a delay. Therefore, to solve it, the subject must perform the following minimal set of actions: (1) looking at the sample image during the cue period and committing it to memory, (2) sustaining that memory throughout the delay, (3) looking at the target and distractor images displayed during the choice period, and (4) matching the target with the sample image stored in memory, and selecting it.

The DPA task unfolds in a manner identical to DMS. The only difference is that, to be successful, the subject needs to pick as target not the image presented before the delay, but rather the image which was associated with it during training (see legend of Fig. 1). DPA trials therefore require the same actions as DMS trials, except that, in addition, the subject must discard the memory of the sample image, and replace it with the memory of its paired-associate. Studies on the monkey (Naya et al., 1996; Rainer et al., 1999) have shown that this retrieval process already takes place during the delay, i.e., before the subject is presented with the target and distractor images.

We chose the MDR task because it imposes stringent constraints on the model. First, it requires the ability of either sustaining the representation of the sample image over the delay, or replacing it with that of the image associated with the sample. Second, it is a dual task, therefore requiring that all

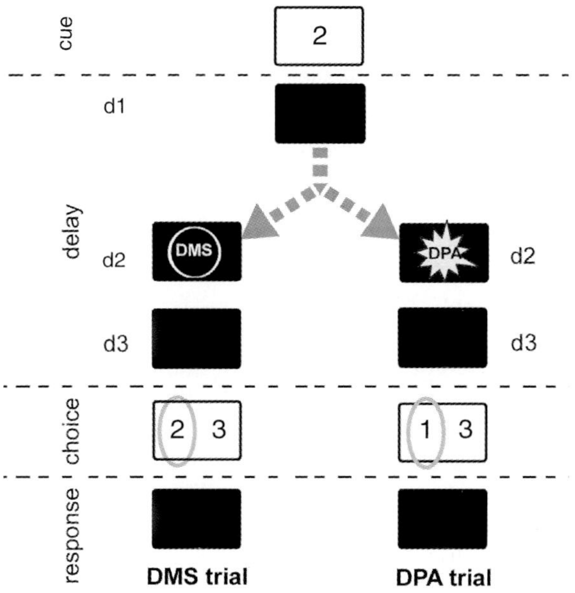

Fig. 1. Diagram of the MDR task. The MDR task consists of randomly mixed DMS and DPA trials (Naya et al., 1996). The stimuli used in the simulations are represented by numbers 1, 2, 3, and 4. A sample image is presented during the cue period and then hidden during the delay. A task signal specifying which task the subject is expected to perform is briefly displayed during the delay (subdelay d2). Two images are then presented during the choice period: a target (circled in gray) and a distractor. The target image depends on the trial type: it is identical to the sample image in DMS trials, and it is the sample's paired-associate image in DPA trials. In DPA trials, images have been associated in the arbitrarily chosen pairs {1,2} and {3,4}. If the subject chooses the target image, it will receive positive reward during the subsequent response period. Otherwise, negative reward is dispensed to the subject. Note that the task signal is not essential to trial success: the subject can figure out after the delay which task it is required to perform by inspecting the proposed stimuli. The signal however gives the subject the opportunity to act prospectively and to anticipate the target during the delay. Length of trial periods used: cue, 0.5 s; delay [divided in subdelays d1: 0.3 s, d2: 0.4 s, d3: 1.0 s]; choice, 0.5 s; and response, 0.5 s. (Adapted with permission from Gisiger and Kerszberg, 2006.)

these computations unfold as part of the same dynamics and in a unique neural architecture. Third, this task is well documented by several studies on the monkey, both at the behavioral and at the electrophysiological levels. The fact that these results should be reproduced by the network, at least qualitatively, provides further constraints on the behavior of the model, and on the activity adopted by its neurons.

Model description

Figure 2 presents a hierarchical view of the structure of the network. Successively higher cognitive areas, going from sensory areas to task processors, are displayed as layers tiled one over the other from the bottom of the graph to its top.

At the lowest level of the hierarchy is the Input layer, which is the primary visual area of the model, and through which visual information enters the network. Displayed over it is the visual representation layer VR. It models higher visual areas and contains neurons whose activity is both image-specific and unable to sustain itself. Higher still is the working memory layer WM which contains circuits capable of image-specific, sustained firing. The planning layer P placed above it houses similar circuits although their activity is image- as well as task-specific. Layers VR and WM, as well as layers WM and P, are interconnected by a dense array of diffuse, long-range connections running in both directions between them (see Fig. 2). These three layers provide the network with the various basic neural functions necessary to solve the MDR task. They are complemented by the following set of executive units, whose role is to control the information contained in, or traveling between, these layers.

There are first four gating units G_u, G_d, I_u, and I_d that regulate the flow of information along the vertical connections running from layer VR to WM and from layer WM to VR, as well as from layer WM to P and from layer P to WM, respectively (see Fig. 2). These gating units are defined in the model as two-state units: when open, information flowing along the gated connections is allowed to go through; when closed, no information is allowed to travel along the connections.

In addition, the network is equipped with a task layer T, which contains *sustain* and *recall* units. Their activity at a given time codes for the kind of processing action the network is currently engaged in: either sustaining the memory of the sample image, or recalling its paired-associate.

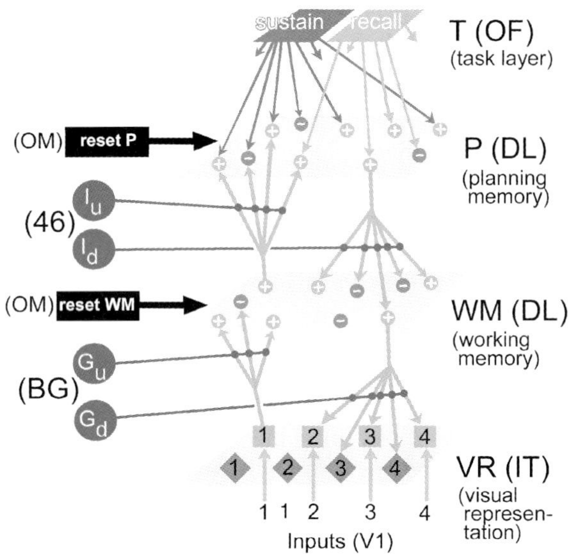

Fig. 2. Diagram of the structure of the network. Letters between parentheses indicate tentative assignment of network components to cortical or subcortical areas: OF = orbitofrontal, OM = orbitomedial, DL = dorsolateral, 46 = area 46, BG = basal ganglia, IT = inferotemporal, and V1 = primary visual cortex. Network areas and layers: T = task, P = planning, WM = working memory, and VR = visual representation. Excitatory and inhibitory neurons (represented by dots in light gray with a "+" sign, and by dark gray with a "−" sign, respectively) are arranged in two-dimensional layers P and WM and interconnected by short-distance connections (not shown). There are 900 of these neurons, which each represent a single cortical cell, in both layers WM and P. Layer VR is composed of four excitatory units (light gray squares), each representing a group of cortical cells coding for a single image, and four inhibitory units (dark gray diamonds), which implement lateral inhibition on excitatory VR units. Layers P, WM, and VR are connected via diffuse and homogenously distributed vertical excitatory projections (vertical arrow originating from excitatory neurons). All connections in the network are fitted with standard Hebbian learning algorithm, while downward connections have in addition reinforcement learning. Each P neuron also receives a single projection from one of the two task units of layer T. Layers WM and P are the targets of reset units (*reset* WM and *reset* P) which, when active, reinitialize to zero the membrane potential and output of all neurons in the layer. Units G_u, G_d, I_u, and I_d gate activity that travels from VR to WM, WM to VR, WM to P, and P to WM, respectively (dark horizontal lines). This gives the network the freedom to either transfer information from one layer to another, or to isolate layers so that they can work separately. Visual information from the exterior world enters the network via the *Inputs* variables, which feed stimulus-specific activity into layer VR (vertical arrows). (Adapted with permission from Gisiger and Kerszberg, 2006.)

Finally, the network is also equipped with two reset units, *reset* WM and *reset* P, which, when active, suppress the activity contained in layers WM and P, respectively.

These eight control units give considerable control over the activity occurring within the network. Indeed, by opening certain gatings while keeping others closed, it is possible to precisely dictate how information travels from one region of the network to the next. Also, behaviorally irrelevant activity can be removed from layers WM and P by activating the corresponding reset units. Further, the dynamics of the network can be tilted toward sustaining the representation of the sample image, or recalling its paired-associate, by activating the corresponding task unit in layer T.

It follows then that by activating executive units at the proper time, it becomes possible to induce elementary neural computations in the network. Such neural operations can then be linked together to produce various simple behaviors. Figure 3 shows how, by successively opening and closing the I and G gatings, a stable neural representation of image 2 can be created in the network.

This process takes place mainly through the collaboration between circuits harbored in layers VR, WM, and P. Indeed, the visual layer VR, although it contains image-specific neurons, is unable to sustain by itself its activity for more than about half a second. Similarly, the working memory layer, although it can harbor self-sustained, image-specific activity, does not have direct access to visual information from the primary visual area *Input*. However, proper activation of gatings G and I leads to a stable representation of image 2 spanning both the visual representation and the working memory layers (see Fig. 3). Moreover, layer P, when properly assisted by the task layer T, *reset* WM units, and gatings I_u and I_d, can perform neural computations on the content of working memory. During DPA trials, these units therefore implement the retrieval process through which the neural representation of the image associated with the sample is generated. Figure 2 displays a tentative assignment of the network's components to brain structures [see Gisiger and Kerszberg (2006) for details and short videos of the network performing DMS and DPA trials].

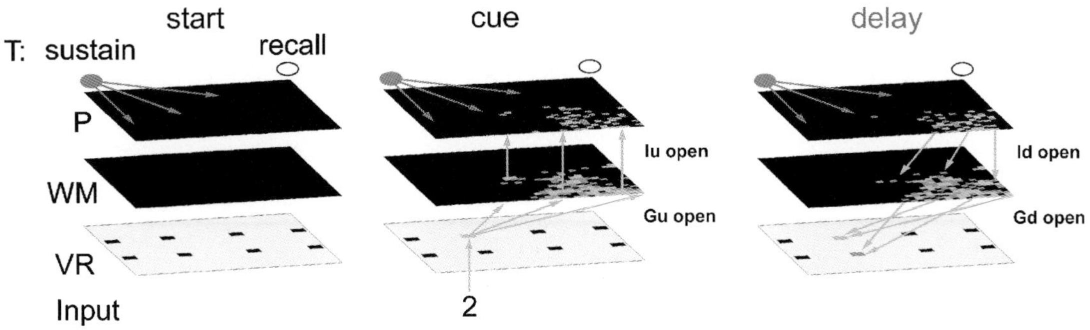

Fig. 3. Creating a stable neural representation of an image (here, image 2). The layers of the network are represented in three dimensions. Light gray squares represent firing neurons. Black square represent silent neurons. Light gray arrows between layers VR, WM, and P represent vertical, long-range connections that are gated into an open state. Notations are identical to those on Fig. 2. (Start) The network has readied itself for the upcoming trial by briefly triggering its *reset* WM and *reset* P units, thereby emptying layers WM and P from any activity left over from the preceding trial. (Cue) Image 2 is presented to the network by turning on *Input* unit 2. Gatings G_u and I_u on the vertical connections linking layer WM to layer VR, and layer P to layer WM are both open (G_d and I_d are both closed). As a result, visual information rises through the visual representation layer VR, where it triggers the neuron VR(2) specific to image 2, and upward to the working memory WM and planning P layers, where it creates self-sustained activity. (Delay) Image 2 is now hidden (*Input* unit 2 has now been turned off), and gatings G_d and I_d on the downward connections are now open (G_u and I_u are both closed). This allows the stable activity contained in layer WM to sustain the firing of neuron VR(2), therefore creating a stable neural representation of image 2 in the network. Activity in the planning layer P further stabilizes the component of this representation held in working memory.

Figure 4 shows the firing patterns of the control units, which enable the network to perform fixation trials (Fig. 4A), DMS trials (Fig. 4B), and DPA trials (Fig. 4C).

These firing patterns, which were hardwired from the start in the model, allow the network to be trained at first, and then to perform these tasks. Each simulation took place as follows. First, an immature network, i.e., one fitted with random connectivity between all its cells, is created. It is then submitted to ~20 fixation trials (Fig. 4A) during which it is only required to "observe" the sample image presented over the cue period. The network is then submitted to DMS training (Fig. 4B) and its performance sanctioned by a reward at the end of each trial. After DMS training has been successfully completed, the network is submitted to DPA training.

We found that, in ~50% of the simulations, the network is able to learn all tasks, which it then performs with a success rate of 90% or more. Further, the activities of neurons from the mature network reproduce qualitatively essential features of electrophysiological recordings gathered from inferior temporal (Naya et al., 1996) and prefrontal cortex (Rainer et al., 1999) of monkeys performing the same tasks.

Figure 5A presents the activity of excitatory unit 2 of the visual representation layer VR during DMS and DPA trials. This unit plays the role of the visual correlate of image 2 in the model: it fires when this image is seen (cue period of trials 2→1 and 2→2), retained (subdelays d1 and d2 of trial 2→1, delay period of trial 2→2), or recalled (subdelay d3 of trial 1→2) by the network. This unit is however silent when another image, e.g., image 1, is seen and retained (cue and subdelays d1 and d2 of trial 1→2, cue and delay of trial 1→1) or recalled (subdelay d3 of trial 2→1). As can be seen, these activities reproduce qualitatively the firing of monkey inferior temporal cells (Fig. 5B) measured by Naya et al. (1996). As discussed in Gisiger et al. (2005), the activity of VR units also reproduces data gathered from monkey inferior cortex during the choice period of DMS trials by Chelazzi et al. (1993).

Figure 5C displays the activity of three cells of layers WM and P during DMS and DPA trials. Cells α and β both belong to the working memory layer WM where they participate in the

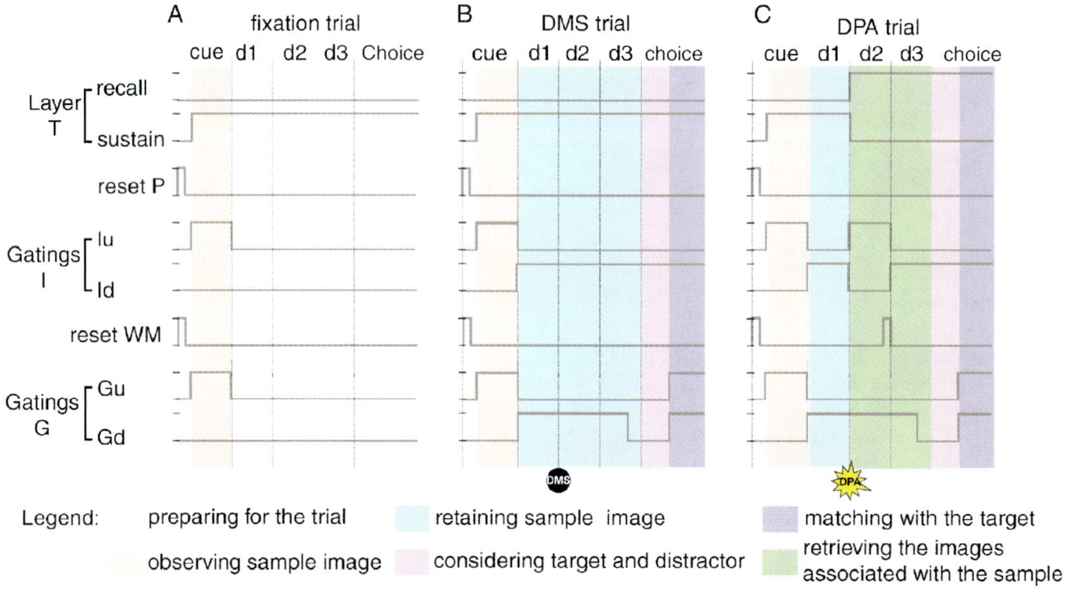

Fig. 4. Firing patterns of task, reset, and gating units coding for fixation, DMS, and DPA trials. (Adapted with permission from Gisiger and Kerszberg, 2006.) Each line represents the activity (either "on" or "off") of a single control unit as a function of time during the trial. The units' firing patterns are grouped in three sets, each corresponding to a different task: one for fixation trials (A), one for DMS trials (B), and one for DPA trials (C). These firing patterns specify the neural computations performed by the network to perform each task. For clarity, colored zones illustrate the series of actions necessary to solve each trial. Task parameters and control unit notations are as in Figs. 1 and 2, respectively. (A) The fixation task only requires that the network observes the sample image presented at the beginning of each trial. To do this, the network first clears from its WM and P layers any activity left over from the preceding trial (gray zone). It then allows visual information from the presented sample to rise into these layers (orange zone). (B) The DMS tasks generalizes the fixation task, requiring that the network retains the observed sample image during a delay (light blue zone) to then match it against target and distractor images during the choice period (pink zone), and finally selecting the target (dark blue zone). These operations are implemented by the above additional activities in the firing patterns of gatings I_d, G_u, and G_d. (C) The DPA task is identical to DMS except that the network needs to retrieve during subdelays d2 and d3 the image associated with the sample (green zone). This recall process is implemented during these periods by additional activities for gatings I_u and I_d, and the reset WM unit.

representations of images 3 and 2, respectively. Cell α therefore fires when image 3 is seen and retained (cue period and subdelays d1 and 2 of trial $3 \rightarrow 4$). Cell β has a more tonic activity, firing when image 2 is seen and retained (cue and delay periods of trial $2 \rightarrow 2$) or recalled (subdelay d3 of trial $1 \rightarrow 2$). Cell γ, on the other hand, fires only briefly during the delay of trials with sample image 1 and target 2. It belongs to the planning layer P where it participates in the creation of the recalled representation of image 2 during DPA trials. Although not a perfect match, the simulated activities of these three neurons reproduce important features of the firing of monkey prefrontal cells (Fig. 5D) gathered by Rainer et al. (1999).

All these results support the proposal that simple behavior might consist of a succession of elementary neural operations, which are orchestrated by well-defined neural structures directing the activity of the various cortical regions involved.

Possible correspondence between executive units and brain structures

As discussed in Gisiger and Kerszberg (2006), electrophysiological and neurophysiological studies have already suggested the existence in the brain of neural structures that might play a role similar to the executive units of the model.

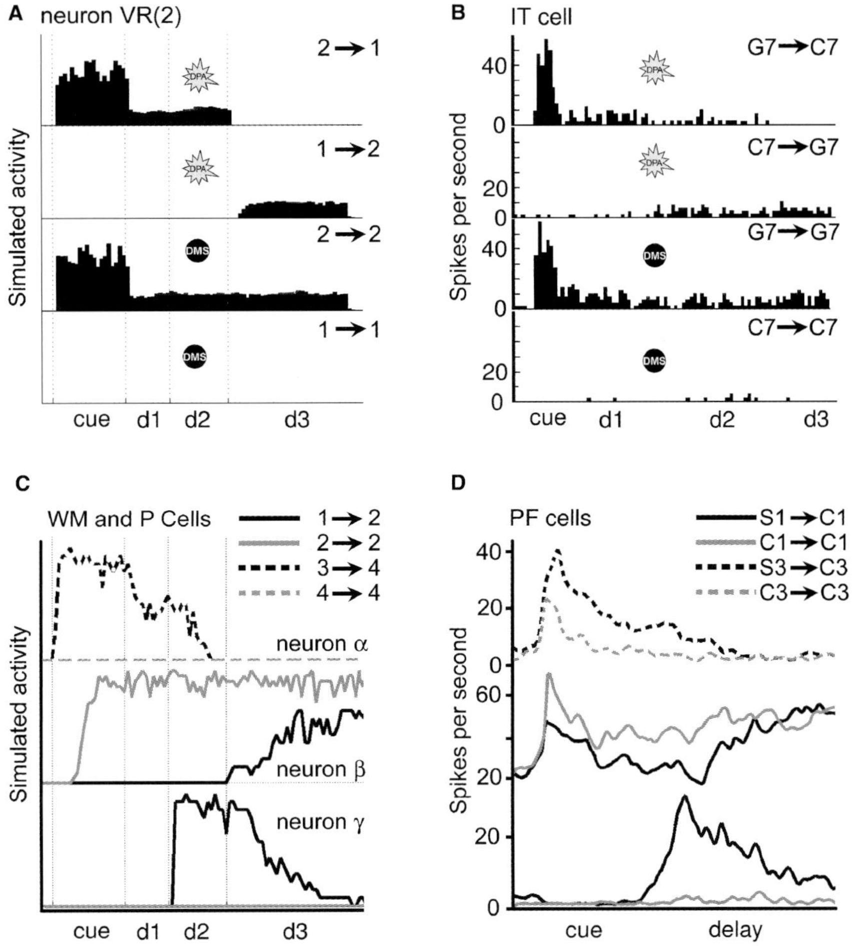

Fig. 5. Comparison of the simulated activity of neural network cells (A and C) with actual activity of monkey cortical neurons (B and D). Notation: $x \to x$ = DMS trial with sample and target x. $x \to y$ = DPA trial with sample x and target y. The black dots and bright stars on panels A and B represent the task signal that indicates the current trial type to the subject (i.e., the network or the monkey). Other notations are as in Figs. 1 and 2. (A) Activity of excitatory unit 2 of visual representation layer VR. Task parameters: cue (0.5 s), d1 (0.3 s), d2 (0.4 s), and d3 (1 s). (B) Recordings of a single IT neuron during four PACS trials, a task analogous to the MDR task that was introduced by Naya et al. (1996). Images G7 and C7 formed a pair in DPA trials. Task parameters: cue (0.5 s), d1 (2 s), d2 (3 s), and d3 (1 s). (C) Average firing patterns of two cells of layer WM (neurons α and β) and one cell of layer P (neuron γ) for DMS and DPA trials. Each curve represents the response of a single cell during trials featuring a given pair of sample and target images. Task parameters: cue (0.5 s), d1 (0.3 s), d2 (0.4 s), and d3 (1 s). (D) Recordings of three cells gathered from monkey dorsolateral PF cortex while the animal performed DMS and DPA trials. Each curve represents the response of a single cell during trials featuring a given pair of sample and target images. Images were associated in the pairs {S1,C1} and {S3,C3} during DPA trials. Task parameters: cue (0.5 s), delay (1 s). (Modified with permission from Figs. 5 and 13 from Gisiger and Kerszberg, 2006.)

The first such candidate is the prefrontal cortex, where neurons with task-specific firing have been detected (Asaad et al., 2000), and whose area 46 has been shown to participate in the shielding of information held in working memory from unrelated neural processes (Sakai et al., 2002). The neural circuits implementing these mechanisms present an interesting resemblance with the task units and gatings of our model, respectively.

The second candidate are the basal ganglia (BG) where projection neurons of the striatum have been proposed by Graybiel (Graybiel et al., 1994) to enhance or suppress movement through the BG pathways which run from the neocortex, through the striatum and other parts of the BG, and finally converge to frontal, or temporal (Middleton and Strick, 1996), cortex. This mechanism is especially interesting considering evidence that the BG might contain the neural substrate of genetically predetermined rules regulating certain behaviors [Cromwell and Berridge (1996) — see also discussion below].

Comparison with other models

Two models illustrating the need for managing the information entering working memory in simple delayed task, and possible mechanisms which implement such a function, have been recently published.

The first was proposed by Chadderdon and Sporns (2006). Its goal is to investigate the functional role of dopamine on working memory during simple memory tasks, such as DMS, and the delayed-nonmatch to sample (DNMS) task. The DNMS task is a straightforward generalization of DMS where the subject is required to select as target, not the sample image presented before the delay, but instead the distractor image. The model presents a set of highly detailed visual, motor response and working memory areas, each containing several interconnected subcircuits. At the core of its dynamics, and as a part of the working memory area, is a set of three task-specific units with mutually exclusive activity. Their firing reflects the task the network is currently engaged in: being idle at the beginning and the end of each trial, performing a DMS trial, or performing a DNMS trial, respectively. These units therefore act as a task memory in the network, very much like the task neurons of our own model. Their main role in the network is to stimulate the release of dopamine (presumably from ventral tegmental area terminals) in quantities which are correlated with the task the network is currently engaged in: this level will be low when the network is idle, and high when the network is involved in either the DMS or the DNMS task. In turn, the dopamine released controls the access to working memory by modifying the membrane gain of other prefrontal cells. Low dopamine levels reduce the membrane gain and the strength of reentrant connections, therefore allowing old activity to be eliminated from working memory and new activity to enter it. High dopamine levels, on the other hand, raise this gain and increase the strength of reentry, therefore simultaneously sustaining and protecting the activity pattern already present in working memory. Through these processes, the variation in the concentration of a single neurotransmitter lets the network reset the content of its working memory, allows new activity in it, and then sustains it until the end of the trial. This mechanism in fact plays a role very similar to the *reset* WM and G_u units of our model. The model is also equipped with a kind of "go" unit which withholds the network's response until after the delay period, possibly implementing an inhibitory effect attributed to the anterior cingulate cortex. The resulting network performs both the DMS and DNMS tasks, reproduces many neuroimaging and electrophysiological results, and makes several quantitative predictions.

The second model of task-directed access to working memory was proposed by O'Reilly, Frank and colleagues (Frank et al., 2001; O'Reilly and Frank, 2006). Here, the authors present a possible theoretical framework for understanding how the frontal cortex and the BG interact in providing the mechanisms necessary for selective working memory updating and robust maintenance. They argue that, to be behaviorally efficient, the brain must be equipped with mechanisms that control, depending on the requirement of the task at hand, how sensory information should access working memory. Building on a detailed analysis of cortical and sub-cortical connectivity, they suggest that mechanisms linking both sensory and prefrontal cortex to the striatum, thalamus, globus pallidus, and the substantia nigra might allow such a differential access to different parts of working memory. This gating structure would then be able to identify different stimuli, evaluate their relative behavioral significance and their role in the task, and then direct them into separate locations in working memory fitting the requirements of the task. It therefore implements functions similar to the G_u and *reset* WM units of our model, although here by contrast these mechanisms gate differently different stimuli instead of being purely task-specific. To test their hypotheses, the authors built a neural network model and submitted it to a complex delayed-response task where the subject is required to look for certain sequences of letters among a stream of letters and digits. Which sequence needs to be located (e.g., A–X or B–Y) changes sporadically, and is indicated to the subject by the presentation of a cue digit (e.g., 1 or 2). To perform the task, the subject must therefore simultaneously remember which cue digit was presented last (e.g., 1), while at the same time looking for the current correct sequence of letters it specifies (e.g., A–X). The network solves the task by storing the cue digits and letters in different areas of working memory. This way, memories of digits are protected from the computations performed on letters while the network looks for sequences. Simulations show that this differential access to working memory can be successfully implemented by the model with partial hardwiring of the rule of the task as part of the hypotheses of the model (Frank et al., 2001). In a more recent theoretical effort, the authors have shown that, however, it is possible for the network to learn the rule of the task only through the reinforcement dispensed by the experimenter (O'Reilly and Frank, 2006).

Both these models illustrate that, for working memory to be useful to its owner, it must be equipped with mechanisms which allow its content to be updated, or protected, following current behavior requirements. These studies further propose, with the support of current knowledge about the brain, biologically realistic implementations of these mechanisms. We agree with their conclusions and we put forward that they be extended to cortical circuits implementing functions other than working memory: working memory would then only be an example, among many others, of elementary cognitive functions that need to be assisted to participate efficiently to behavior. Indeed, in our view, it seems reasonable to assume that in the course of evolution, as brain areas specialized and diversified in architecture and function, mechanisms simultaneously emerged to allow these areas to work efficiently and in concert. To reflect this, we have equipped our model with control structures which manage the circulation of activity between all parts of the network, instead of just the two units G_u and *reset* WM that control the content and influx of information into the working memory layer WM. This gives the network considerable behavioral flexibility and extensive control over the movement of activity within the network. As we argue below, the precise manner, in which activity moves among the set of neural circuits mobilized for a given task, might be as important in defining behavior as the identity of the circuits present in that set.

Task representation and its emergence

In order to perform the DMS and DPA tasks, live subjects must have an understanding of what they are required to do. In other words, they must hold in their mental space a neural activity that represents these tasks, or at the very least, which codes for the different operations required to perform them. Animals are able to create this neural task representation in a few training sessions, using the reward dispensed by the experimenter as the only feedback on their actions from the exterior world.

The model presented here, by contrast, does not learn the DMS and DPA tasks since all the computational aspects of the tasks are already precoded in the firing patterns of the network's control units (see Fig. 4). During the training phase, this abstract task representation is then applied to the images chosen for the task, as the network modifies its connectivity to become able to perform MDR trials with that particular set of images. However, despite these simplifications, the model is able to suggest some insights into the neural activities that implement behaviors.

One important question it addresses is how much the behavior of a live subject is sensitive to the details of the neural activity that codes for it, or in other words, whether a given task has only one unique neural representation. The present model seems to suggest that it is not the case. Indeed, as was shown in Gisiger and Kerszberg (2006), the network can function with an acceptable degree of success even if random perturbations are imposed to the firing patterns of its control units. Therefore, there is a large spectrum of firing patterns that the control units could adopt that would still lead to task success. This compliant character of the model stems directly from its modular organization and its ability to sustain and protect information. This conclusion therefore raises the possibility that the same is true for the brain of primates.

Our model further suggests a concrete framework for studying how neural representations of behavior might emerge. A particularly attractive mechanism in this regard is the cognitive pattern generator proposed by Graybiel (1997, 1998). The author, generalizing studies on motor functions, suggests that cognitive behavior might be constructed in an iterative manner by piecing together small chunks or bits of elementary behavior. She also proposes the BG as central to this mechanism, based on their connectivity to cortical and subcortical structures, and also because of the presence in this region of neurons which code for innate complex series of movements such as the grooming actions of the rat (Cromwell and Berridge, 1996; Greer and Capecchi, 2002; Aldridge et al., 2004). These features should allow the gathering of information drawn from previously learned or innate behaviors, which would then be combined with new activities to produce possible solutions to the problems at hand. Related to this process is the notion of motor and cognitive syntax (Lashley, 1951; Lieberman, 2001), which would guide how these chunks of behavior might be combined together in order to converge toward neural representations coding for relevant behavior.

Indeed, most types of motor behavior are produced by a large number of neural operations that take place in the right order and at the proper time. This is also probably true for cognitive behavior, as has been argued for language, which is serial in nature and involves both motor and cognitive processes (Fuster, 2003). However, such a view leads to the following well-known problem: considering the huge amount of neural operations available and the seemingly infinite number of possible combinations between them, how does an organism manage to learn complex motor skills or even language so rapidly (e.g., language is mastered in only a few years of training)? A possible solution to this problem is the concept of "syntax," i.e., a set of predominantly genetically predetermined rules specifying the way in which interdependent neural operations should be articulated with each other. This notion, which was first introduced in the context of motor processing, has been more recently extended to the cognitive domain and the syntax of common language (Lashley, 1951; Fuster, 2003). The existence of such rules has already been demonstrated experimentally in the rat (Cromwell and Berridge, 1996; Aldridge et al., 2004) and the monkey (Fuji and Graybiel, 2003) for motor tasks.

The model presented here provides a formal framework where these hypotheses can be expressed and refined. Since each neural function is implemented by a distinct executive unit, constructing behavior in the model boils down to deciding whether and when each of these functions take effect. In the case of the three tasks, i.e., fixation, DMS, and DPA, these decisions are crystallized in the firing patterns displayed in Fig. 4. We now argue that these patterns, which have been predetermined here, could also be generated by a mechanism analog to that proposed above (Lashley, 1951; Graybiel, 1997, 1998; Lieberman, 2001).

For instance, as can be seen on Fig. 4, these activity patterns have a simple structure that can be divided in segments where only few control units are simultaneously active over a short time period. Each of these segments implements a different basic cognitive "action" in the model, and therefore represents an elementary behavioral chunk in this framework. Furthermore, the structure of the representations of the three tasks (fixation, DMS, and DPA) in the model, makes it likely that the firing patterns could be built through a process of successive generalizations (see Fig. 4 and its legend). We see for instance that the representations of the DMS and the fixation tasks share the firing patterns that code for the sample image perception taking place during the cue period (orange zone — Fig. 4A, B). The DMS task representation however generalizes that of the fixation task, by the addition of novel processings implementing sample retention and target selection (light blue, pink, and dark blue zones — Fig. 4B, delay and choice periods). The same is true for the DPA task representation, which generalizes that of DMS with the introduction of the target recall process (green zone — Fig. 4C, subdelays d2 and d3). The structures of these task representations, where novel processings have been added to previously successful ones to generate behavior of higher complexity, could clearly be produced by an iterative mechanism which has access to innate, learned, as well as new chunks of behavior (Graybiel, 1997, 1998). Such a generative process is expected to be guided by rules, i.e., a syntax, which would put constraints on the firing patterns that the control units might adopt. The implementation of this cognitive syntax would considerably accelerate the convergence toward relevant behavior.

Abbreviations

46	cortical area 46
BG	basal ganglia
d1	subdelay 1
d2	subdelay2
d3	subdelay 3
DL	dorsolateral cortex
DMS	delayed-matching to sample
DNMS	delayed-nonmatch to sample
DPA	delayed-pair association
IT	inferior temporal cortex
MDR	mixed delayed-response
OF	orbitofrontal cortex
OM	orbitomedial cortex
PF	prefrontal cortex
V1	primary visual cortex

Acknowledgments

We thank J.-P. Changeux for his support, for many stimulating discussions, and especially for directing our attention to the notion of syntax in the context of the executive units of our model. T.G. gratefully acknowledges a Châteaubriand post-doctoral scholarship from the Ministère Français des Affaires Etrangères, a fellowship of the Canadian Institutes of Health Research, as well as the support of V. Gisiger and R.-M. Gisiger throughout this project.

References

Aldridge, J.W., Berridge, K.C. and Rosen, A.R. (2004) Basal ganglia neural mechanism of natural movement sequences. Can. J. Physiol. Pharmacol., 82: 732–739.

Asaad, W.F., Rainer, G. and Miller, E.K. (2000) Task-specific neural activity in the primate prefrontal cortex. J. Neurophysiol., 84: 451–459.

Chadderdon, G.L. and Sporns, O. (2006) A large-scale neurocomputational model of task-oriented behavior selection and working memory in prefrontal cortex. J. Cogn. Neurosci., 18: 242–257.

Chelazzi, L., Miller, E.K., Duncan, J. and Desimone, R. (1993) A neural basis for visual search in inferior temporal cortex. Nature, 363: 345–347.

Cromwell, H.C. and Berridge, K.C. (1996) Implementation of action sequences by a neostriatal site: a lesion mapping study of grooming syntax. J. Neurosci., 16: 3444–3458.

Frank, M.J., Loughry, B. and O'Reilly, R.C. (2001) Interactions between frontal cortex and basal ganglia in working memory: a computational model. Cogn. Affect. Behav. Neurosci., 1: 137–160.

Fuji, N. and Graybiel, A.M. (2003) Representation of action sequence boundaries by macaque prefrontal cortical neurons. Science, 301: 1246–1249.

Fuster, J.M. (2003) Cortex and Mind: Unifying Cognition. Oxford University Press, New York, pp. 206–212.

Gisiger, T. and Kerszberg, M. (2006) A model for integrating elementary neural functions into delayed-response behavior. PLoS Comput. Biol., 2(4): e25.

Gisiger, T., Kerszberg, M. and Changeux, J.-P. (2005) Acquisition and performance of delayed-response tasks: a neural network model. Cereb. Cortex, 15: 489–506.

Graybiel, A.M. (1997) The basal ganglia and cognitive pattern generators. Schizophr. Bull., 23: 459–469.

Graybiel, A.M. (1998) The basal ganglia and chunking of action repertoires. Neurobiol. Learn. Mem., 70: 119–136.

Graybiel, A.M., Aosaki, T., Flaherty, A.W. and Kimura, M. (1994) The basal ganglia and adaptive motor control. Science, 265: 1826–1831.

Greer, J.M. and Capecchi, M.R. (2002) Hoxb8 is required for normal grooming behavior in mice. Neuron, 33: 23–24.

Lashley, K.S. (1951) The problem of serial order in behavior. In: Jeffress L.A. (Ed.), Cerebral Mechanisms in Behavior. Wiley, New York, pp. 112–146.

Lieberman, P. (2001) Human language and our reptilian brain: the subcortical bases of speech, syntax, and thought. Perspect. Biol. Med., 44: 32–51.

Middleton, F.A. and Strick, P.L. (1996) The temporal lobe is a target of output from the basal ganglia. Proc. Natl. Acad. Sci., 93: 8683–8687.
Naya, Y., Sakai, K. and Miyashita, Y. (1996) Activity of primate inferotemporal neurons related to a sought target in pair-association task. Proc. Natl. Acad. Sci. U.S.A., 93: 2664–2669.
Rainer, G., Rao, S.C. and Miller, E.K. (1999) Prospective coding for objects in primate prefrontal cortex. J. Neurosci., 19: 5493–5505.
O'Reilly, R.C. and Frank, M.J. (2006) Making working memory work: a computational model of learning in the prefrontal cortex and basal ganglia. Neural Comput., 18: 283–328.
Sakai, K. and Miyashita, Y. (1991) Neural organization for the long-term memory of paired associates. Nature, 354: 152–155.
Sakai, K., Rowe, J.B. and Passingham, R.E. (2002) Active maintenance in prefrontal area 46 creates distractor-resistant memory. Nat. Neurosci., 5: 479–484.

P. Cisek, T. Drew & J.F. Kalaska (Eds.)
Progress in Brain Research, Vol. 165
ISSN 0079-6123
Copyright © 2007 Elsevier B.V. All rights reserved

CHAPTER 30

A parallel framework for interactive behavior

Paul Cisek*

Groupe de Recherche sur le Système Nerveux Central, Département de Physiologie, Université de Montréal, Montréal, QC H3C 3J7, Canada

Abstract: Although theoretical models often assume that the basic organization of the nervous system involves separate systems for perception, cognition, and action, neural data often does not fit into any of these conceptual categories. Here, an alternative framework is described, which focuses on interactive behavior and treats it as a continuous competition between representations of currently available potential actions. This suggests a neural organization consisting of two parallel systems: a system for *action specification*, which uses sensory information to represent currently available potential actions, and a system for *action selection*, which involves attentional and decisional mechanisms which determine the action that will be performed. It is proposed that neural processing occurs through two waves of activation: an early wave which specifies several potential actions and a later wave of biasing influences which selects one action for execution. A computational model of decision making is described within the context of this proposal, and simulations of neural and behavioral phenomena are presented.

Keywords: action selection; decision making; affordances; lateral inhibition; population coding

Do neural systems classify into perception, cognition, and action?

Theoretical models of large-scale neural systems often subscribe, either implicitly or explicitly, to the view that animal behavior results from the operation of distinct systems responsible for sensing the world, thinking about it, and acting upon it (Fig. 1a). Within the context of overt behavior, their processing often follows a primarily serial scheme: First, the perceptual system collects sensory information to build an internal descriptive representation of objects in the external world (Marr, 1982). Next, this information is used along with representations of current needs and memories of past experience to make judgments and decide upon a course of action (Newell and Simon, 1972; Johnson-Laird, 1988; Shafir and Tversky, 1995). The resulting plan is then used to generate a desired movement, which is finally realized through muscular contraction (Miller et al., 1960; Keele, 1968). In other words, the brain first builds knowledge about the world using representations which are independent of actions, and this knowledge is later used to make decisions, compute an action plan, and finally execute a movement.

The architecture depicted in Fig. 1a addresses behavioral situations similar to a typical psychological experiment, in which a subject is shown a stimulus and asked to make an appropriate response. The framework is based on an analogy with digital computers: whereby perception is like input, action is like output, and cognition is like the information processing performed by computers

*Corresponding author. Tel.: +1 514-343-6111 x4355; Fax: +1 514-343-2111; E-mail: paul.cisek@umontreal.ca

DOI: 10.1016/S0079-6123(06)65030-9

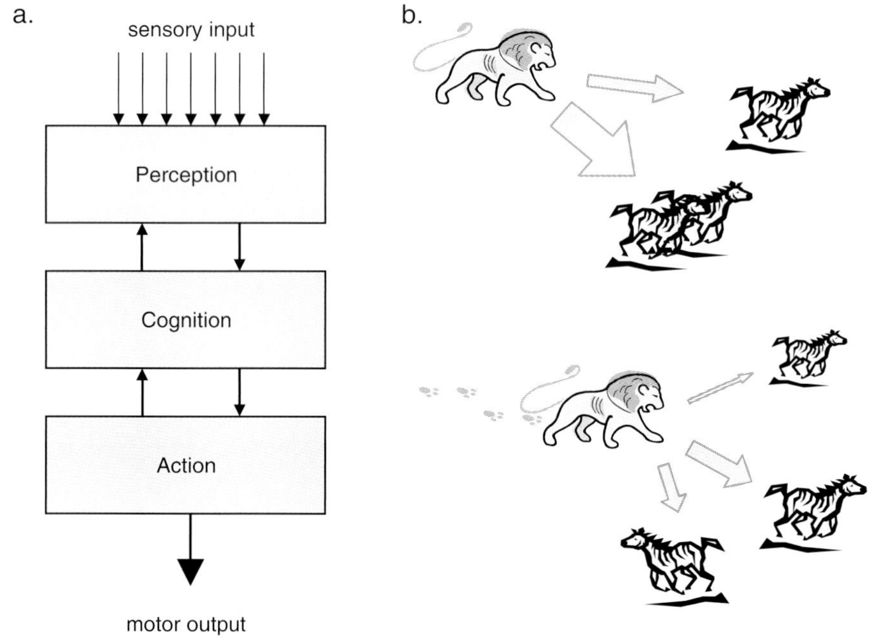

Fig. 1. (a) The traditional architecture that emphasizes distinctions between separate systems for perception, cognition, and action. (b) A schematic of a behavioral situation motivating continuous specification and selection of actions. Top: The lion is faced with two potential actions of running toward food sources (gray arrows). Bottom: As the lion moves, the potential actions change, sometimes splitting into distinct responses between which a decision must be made.

(Johnson-Laird, 1988; Block, 1995). This analogy has been at the foundation of the "cognitive revolution" in psychology (McCarthy et al., 1955), and it has now been inherited by cognitive neuroscience (Albright et al., 2000; Gazzaniga, 2000) as the basic blueprint for bridging neural and behavioral data. While not all theoretical accounts of behavior are strictly serial, many nevertheless subscribe to the systems-level description implied by cognitive psychology. The assumption of an internal representation of the world which cognition requires as input demarcates perception from cognition, while the concept of a motor plan delineates where cognition ends and motor control begins. These distinctions are very pervasive, and are even used to classify scientists themselves by the domain which they study and the conferences which they attend. Consequently, most models and theories either address perception, or motor control, or a variety of cognitive operations, which are defined to be distinct from either perception or motor control.

However, it is not entirely clear whether the distinctions between perception, cognition, and action are in fact supported by neural data. For example, it has often been assumed that the internal representation of the external world *unifies* diverse information into a centrally available form, which reflects the *stable* invariants present in the physical world (Marr, 1982). However, the most studied sensory system, vision, appears to be neither unified nor stable. The visual system diverges into a host of parallel streams which contain multiple representations of external space (Stein, 1992; Colby and Goldberg, 1999) and separate systems for processing color, shape, and motion (Felleman and Van Essen, 1991). Within each of these regions, neural activity is modulated by a variety of factors including attention (Treue, 2001) and decision variables (Platt and Glimcher, 1999), and fluctuates wildly even while viewing a motionless scene.

The distinction between cognitive and motor processes also does not find strong support in neural data. For example, neural activity related to decisions is distributed throughout the brain, including the same regions responsible for planning and performing movements (Schall and Bichot, 1998; Platt and Glimcher, 1999; Gold and Shadlen, 2000; Hoshi et al., 2000; Coe et al., 2002; Carello and Krauzlis, 2004; Horwitz et al., 2004; Romo et al., 2004; Cisek and Kalaska, 2005). Likewise, motor planning has traditionally been assumed to be separate from and antecedent to movement execution (Miller et al., 1960), but neurophysiological studies have found evidence that both of these processes occur within the very same neurons (Shen and Alexander, 1997a, b; Crammond and Kalaska, 2000; Cisek, 2005). Regions such as the posterior parietal cortex (PPC) have eluded classification into the traditional categories of sensory, motor, or cognitive systems, leading to persisting debates about their functional roles (Snyder et al., 1997; Platt and Glimcher, 1999; Kusunoki et al., 2000). For example, a recent review has suggested that "current hypotheses concerning parietal function may not be the actual dimensions along which the parietal lobes are functionally organized; on this view, what we are lacking is a conceptual advance that leads us to test better hypotheses" (Culham and Kanwisher, 2001, pp. 159–160). Perhaps the concepts of separate perceptual, cognitive, and motor systems, which theoretical neuroscience has inherited from cognitive psychology, are not appropriate for bridging neural data with behavior.

To summarize, neural data do not appear to support some of the most basic tenets of cognitive psychology. Processes that should be unified (e.g., visual processing) are divergent, while processes assumed to be distinct (e.g., decision making versus sensorimotor control) are strongly overlapping. This raises questions as to whether the classical distinctions between perception, cognition, and action systems truly provide the proper foundation for interpreting neural activity in terms of its contribution to complex mental and behavioral functions.

Even stronger concerns regarding cognitive psychology's suitability as a bridging framework are raised by considerations of evolutionary history (Sterelny, 1989; Hendriks-Jansen, 1996). Brain evolution is strikingly conservative and major features of modern neural organization can be seen in the humble *Haikouichthys*, a primitive jawless fish that lived during the early Cambrian epoch over 520 million years ago (Shu et al., 2003). Since the development of the telencephalon, the basic outline of the vertebrate nervous system has been strongly conserved (Butler and Hodos, 1996; Holland and Holland, 1999; Katz and Harris-Warrick, 1999), and even recently elaborated structures such as the mammalian neocortex have homologues among non-mammalian species (Medina and Reiner, 2000). Thus, the basic anatomical and functional organization of the primate brain reflects an ancient architecture which was well established by the time of the earliest terrestrial tetrapods. This architecture could not have been designed to serve the needs of higher cognitive abilities, such as those normally examined during psychological experiments, because those abilities simply did not exist. Instead, it was laid down so as to best address the needs of simple, interactive behavior.

A parallel framework for interactive behavior

The needs of interactive behavior are very different than the needs of an information processing system such as a digital computer or a subject in a classical psychological experiment. The natural environment at each moment presents animals with a multitude of opportunities and demands for action. Because all of these behaviors cannot be performed at the same time, one fundamental issue faced by every behaving creature is the question of action *selection*. That question must be resolved, in part, by using external sensory information about objects in the world, and in part, by using internal information about current behavioral needs. The parameters of each action must be *specified* to deal with the immediate situation. In particular, this requires information about the current spatial relationships among objects and surfaces in the world, represented in a coordinate frame relative to the orientation and configuration of the animal's body. Furthermore, ongoing actions are continuously modifying the landscape of possibilities. Sometimes, previously available actions become irrelevant while new actions become possible. Consider for example the situation schematized in Fig. 1b. At first, the lion is faced with two potential directions for running toward sources of food. As the action unfolds, the situation changes such that the directions for running are continuously updated and a single selected action can split into two, between which a decision must now be made.

The traditional view (Fig. 1a) suggests that we *select* what to do before *specifying* how to do it. However, continuous interaction with the world often does not allow one to stop to think or to collect information and build a complete knowledge of one's surroundings. To survive in a hostile environment, one must be ready to act at short notice, releasing into execution actions which are only partially prepared. These are the fundamental demands which shaped brain evolution. They motivate animals to process sensory information in an action-dependent manner, to build representations of the potential actions which the environment currently affords. In other words, the sensory processing of a given natural setting may involve not only representations that capture information about the identity of objects in the setting, but also representations which specify the parameters of possible actions that can be taken (Gibson, 1979; Kalaska et al., 1998; Fadiga et al., 2000; Cisek, 2001). With a set of such potential actions partially specified, the animal is ready to quickly perform actions if circumstances demand. In essence, it is possible that the nervous system addresses the questions of specification (how to do it) *before* performing selection (what to do). Indeed, for continuous interactive behavior, it may be best to perform both specification and selection processes at all times to enable continuous adjustment to the changing world (see Fig. 1b).

The proposal discussed here is that *the processes of action selection and specification occur simultaneously* and continue even during overt performance of movements. That is, sensory information arriving from the world is continuously used to specify several currently available potential actions, while other kinds of information are collected to select from among these the one that will be released into overt execution at a given moment (Kalaska et al. 1998; Kim and Shadlen, 1999; Glimcher, 2001; Gold and Shadlen, 2001; Platt, 2002; Cisek and Kalaska, 2005). From this perspective, behavior is viewed as a constant competition between internal representations of conflicting demands and opportunities, of the potential actions that Gibson (1979) termed "affordances." Hence, the framework described here is called the "affordance competition" hypothesis (Cisek, 2007a, b).

Figure 2 depicts a schematic representation of how the "affordance competition" framework may be used to interpret neural data on visually guided behavior. According to this hypothesis, visual information is processed at the level of cortex through at least two waves of activity. First, a wave of visually driven activation quickly sweeps through thalamocortical projections and through the occipitoparietal "dorsal stream" (Ungerleider and Mishkin, 1982; Milner and Goodale, 1995; Pisella et al., 1998), activating neurons in occipital, parietal, and frontal cortical areas within 40–60 ms of stimulus onset (Thompson et al., 1996; Schmolesky et al., 1998; Ledberg et al., 2007). This early wave of activity represents the immediate environment in terms of information about potential actions that are currently available. For example, visual targets for saccadic eye movements are represented in a retinotopic map in the lateral intraparietal area (LIP) and the frontal eye fields (FEF) (Snyder et al., 1997; Schall and Bichot, 1998; Snyder et al., 2000), while directions of reaching movements from the current hand location to graspable objects within reach are represented in medial intraparietal cortex (MIP) and the dorsal premotor cortex (PMd) (Wise, 1985; Alexander and Crutcher, 1990; Ferraina and Bianchi, 1994; Kalaska and Crammond, 1995; Hoshi and Tanji, 2000; Buneo et al., 2002). This initial very fast wave of activity causes multiple potential actions to be simultaneously encoded within each of these effector-specific fronto-parietal systems as distinct groups of active neurons within each local population (Cisek and Kalaska, 2005).

According to the model presented here, these potential actions compete for release into overt execution through mutual inhibition. In other words, if two different directions for reaching are simultaneously activated, they will each try to suppress the other. This competition plays out across

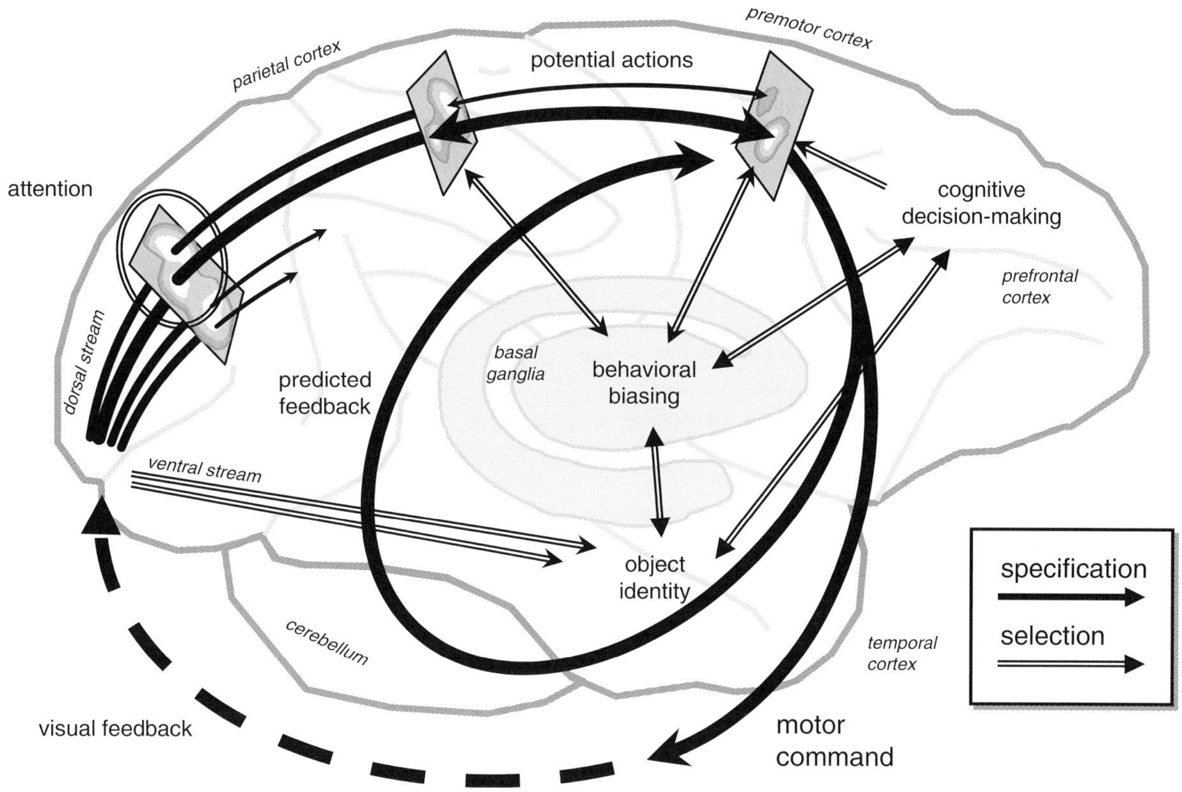

Fig. 2. Sketch of the proposed neural substrates of the affordance competition hypothesis, in the context of visually guided movement. The primate brain is shown, emphasizing the cerebral cortex, cerebellum, and basal ganglia. Filled dark arrows represent processes of action specification, which begin in the visual cortex and proceed rightward across the parietal lobe, transforming visual information into representations of potential actions. Polygons represent three neural populations along this route: (1) The leftmost represents the encoding of potential visual targets, modulated by attentional selection; (2) The middle represents potential actions encoded in parietal cortex; (3) The rightmost represents activity in premotor regions. Each population is depicted as a map of neural activity, with activity peaks corresponding to the lightest regions. As the action specification occurs across the fronto-parietal cortex, distinct potential actions compete for further processing. This competition is biased by input from the basal ganglia and prefrontal cortical regions, which collect information for action selection (double-line arrows). This biasing influences the competition in a number of loci, and because of reciprocal connectivity, these influences are reflected over a large portion of the cerebral cortex. The final selected action is released into execution and causes both overt feedback through the environment (dashed black arrow) and internal predictive feedback through the cerebellum. (Adapted with permission from Cisek, 2007a.)

a distributed set of cortical areas, and it is biased by a variety of influences arriving from basal ganglia, prefrontal cortex (PFC), and the limbic system. Within ~50–100 ms after the initial wave of activation, these biases become strong enough to cause a winning action to emerge and other potential actions to be suppressed, leaving activity throughout the fronto-parietal system to reflect a decision which is reached and an action which is about to be produced (Thompson et al., 1996; Cisek and Kalaska, 2005; Ledberg et al., 2007).

To summarize, the affordance competition hypothesis suggests that visual information leads to the very rapid *specification* of potential actions across a diverse set of regions distributed throughout the cerebral cortex, followed by a later biasing of this activity to *select* a winning action, which will be released into execution. These processes appear as two distinct waves of activation only in tightly controlled experiments in which stimuli are abruptly presented and single responses produced. During continuous interaction with a natural environment, they are entirely overlapping, with new actions being specified as others are being executed and selection biases shifting from one behavioral act to another. In contrast with traditional serial views, the hypothesis suggests that transformations of information from visual to motor coordinates occur in parallel within a diverse range of parieto-frontal circuits, and the selection of action occurs through a "distributed consensus" which is reached simultaneously across a wide cortical area.

Although this hypothesis is quite simplified, it nevertheless can be useful for interpreting a very diverse set of results on the neural correlates of sensory, cognitive, and motor processes (Cisek, 2007a, b). Here, I will focus on a specific behavioral domain: the processes for making simple decisions about voluntary reaching actions. In order to formalize ideas that are otherwise too abstract to translate into specific neural predictions, I will describe a computational model of reach decisions which is inspired by and framed within the framework of the affordance competition hypothesis.

A computational model of simple decisions

Figure 3a illustrates the circuit model, which is described in more detail in Cisek (2006). Because it focuses on visually guided reaching actions, the model includes some of the main cortical regions involved in reaching behavior, such as the PPC, PMd, primary motor cortex (M1), and PFC. Other relevant regions not currently modeled are the supplementary motor areas, somatosensory cortex, and many subcortical structures including the basal ganglia, red nucleus, etc. The input to the model consists of visual information about target direction and a signal triggering movement onset (GO signal), and the output is the direction of movement. The control of the overt movement is not simulated here (for compatible models of execution, see Bullock and Grossberg, 1988; Houk et al., 1993; Kettner et al., 1993; McIntyre and Bizzi, 1993; Bullock et al., 1998; Cisek et al., 1998).

In the model, each neural population was implemented as a set of 90 mean-rate leaky-integrator neurons each of which is broadly tuned to a particular direction of movement. Each neuron's behavior obeyed a differential equation of the following form:

$$\frac{dX}{dt} = -\alpha X + (\beta - X)\gamma E - XI + \Theta \qquad (1)$$

where X represents the mean firing rate of a neuron, dX/dt the change in the firing rate over time, E the excitatory input, I the inhibitory input, and Θ Gaussian noise. The parameter α controls the decay rate, β is the maximum activity, and γ is the excitatory gain. Excitatory input E consisted of topographic projections from upstream (and sometimes downstream) layers as well as local excitation from neurons in the same layer with similar tuning. Inhibitory input consisted of lateral inhibition from neurons in the same layer with different directional preferences. Thus, within each population, neurons with similar tuning excite each other while neurons with dissimilar tuning inhibit each other. Between populations, neurons with similar tuning excite each other through reciprocal topological connections. Noise is added to all neural activities. All of the weights are randomized around a mean value meant to resemble the known anatomical connections between the modeled regions. See Cisek (2006) for details of the model's implementation.

In the model, neural populations do not encode a unique value of a movement parameter (such as a single direction in space) but can represent an entire distribution of potential values of movement parameters (e.g., many possible directions represented simultaneously), as shown in Fig. 3b. This proposal is related to the attention model of Tipper et al. (2000), the "decision field" theory of Erlhagen and Schöner (2002), and the "Bayesian coding" hypothesis (Dayan and Abbott, 2001; Sanger, 2003; Knill and Pouget, 2004). It suggests that given a population of cells, each with a preferred value of a particular movement parameter, one can interpret the activity across the population as something akin to a probability density function of potential values of that parameter. Sometimes, the population may encode a range of contiguous values defining a single action, and at other times, several distinct and mutually exclusive potential actions can be represented simultaneously as

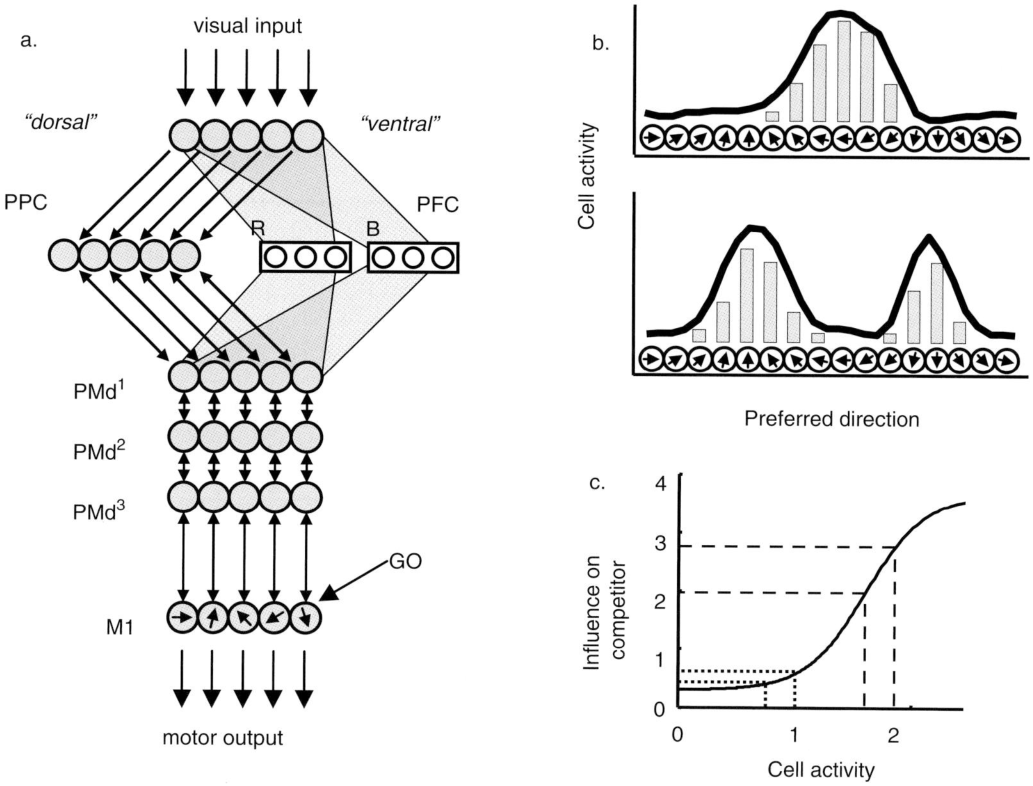

Fig. 3. Computational model. (a) Model circuit. Each neural population is depicted by a set of tuned neurons (circles), which are topographically arranged according to their spatial preferences. Visual information diverges into a dorsal stream which projects to the posterior parietal cortex (PPC) and a ventral stream that projects to the temporal lobes and to prefrontal cortex (PFC). Potential actions represented in PPC are recapitulated in dorsal premotor cortex (PMd; depicted by three identical layers). Within the PPC and PMd layers, simultaneously specified potential actions compete against each other, and this competition is biased by influences from PFC. When a GO signal is given, the resulting pattern of activity projects into M1 to initiate a voluntary movement. See Cisek (2006) for details. (b) Each population consists of cells with different preferred directions, and their pattern of activity can represent one potential reach direction (top) or several potential directions simultaneously (bottom). (c) The function governing the competitive influence of a given cell upon its neighbors within the same cortical layer. When two cells have low activities (dotted lines), the difference between their activities does not give either of them a strong advantage. When activities are higher (dashed lines), the same difference is exaggerated allowing the stronger activity to run away and win the competition.

distinct peaks of activity in the population. The strength of the activity associated with a particular value of the parameter reflects the likelihood that the final action will have that value, and it is influenced by a variety of factors including salience, expected reward, estimates of probability, etc. This hypothesis predicts that activity in the population is correlated with many decision variables, as observed in frontal (Kim and Shadlen, 1999; Gold and Shadlen, 2000; Hoshi et al., 2000; Coe et al., 2002; Roesch and Olson, 2004; Romo et al., 2004) and parietal cortex (Platt and Glimcher, 1999; Shadlen and Newsome, 2001; Coe et al., 2002; Glimcher, 2003; Dorris and Glimcher, 2004; Sugrue et al., 2004; Janssen and Shadlen, 2005).

The model suggests that sensory information in the dorsal visual stream is used to specify the spatial parameters of several currently available potential actions in parallel. These potential actions are represented simultaneously in frontal and parietal cortical regions, appearing as distinct peaks of activity in the neural populations involved in sensorimotor processing (Platt and Glimcher, 1997; Cisek et al., 2004; Cisek and Kalaska, 2005).

Whenever multiple peaks appear simultaneously within a specific frontal or parietal cortical region, they compete against each other through mutual inhibition. This is related to the biased competition mechanism in theories of visual attention (Desimone, 1998; Boynton, 2005). To state it briefly, cells with similar parameter preferences excite each other while cells with different preferences inhibit each other. This basic mechanism can explain a variety of neural phenomena, such as the inverse relationship between the number of options and neural activity associated with each (Basso and Wurtz, 1998; Cisek and Kalaska, 2005), narrowing of tuning functions with multiple options (Cisek and Kalaska, 2005), and relative coding of decision variables (Roesch and Olson, 2004).

The nature of the interaction among cells within individual layers of the model is critical to its behavior. Because neural activities are noisy, competition between distinct peaks of activity cannot follow a simple "winner-take-all" rule or random fluctuations would determine the winner each time, rendering informed decision making impossible. To prevent this, small differences in levels of activity should be ignored by the system. However, if activity associated with a given choice becomes sufficiently strong, it should be allowed to suppress its opponents and conclusively win the competition. In other words, there should be a threshold of activity above which a particular peak is selected as the final response choice. As described by Grossberg (1973), implementing such resistance to noise as well as a decision threshold within a competitive network can be done using a non-linear function defining interactions between neighboring cells, as shown in Fig. 3. When two or more peaks are present in the population and have low levels of activity, their influence on each other de-emphasizes differences between their activities. Thus, neither peak exerts inhibitory influence on the other strongly enough to overcome the positive feedback which sustains each peak. However, once one activity peak increases, it begins to exert stronger and stronger suppression upon its opponents, thus winning the competition. The point at which a given peak becomes the winner is called a "quenching threshold" (Grossberg, 1973), and it effectively acts as a threshold for committing to a particular decision. Unlike classical models of decision thresholds and reaction time (Carpenter and Williams, 1995; Mazurek et al., 2003; Smith and Ratcliff, 2004), the quenching threshold is not a preset constant but an emergent threshold which depends upon the number of choices, their relative and absolute strengths, and even the angular distance between them, as demonstrated below.

Finally, the model suggests that the competition that occurs between potential actions represented in the fronto-parietal system is biased by a variety of influences from other regions, including the basal ganglia (Redgrave et al., 1999) and PFC (Miller, 2000; Tanji and Hoshi, 2001), which accumulate evidence for each particular choice (Fig. 2). In the mathematical system described by Cisek (2006), only the influence of PFC is modeled, although it is likely that basal ganglia projections play a significant role in action selection (Redgrave et al., 1999). Several studies have shown that some cells in lateral PFC are sensitive to conjunctions of relevant sensory and cognitive information (Rainer et al., 1998; White and Wise, 1999; Miller, 2000; Tanji and Hoshi, 2001), and that they gradually accumulate evidence over time (Kim and Shadlen, 1999). Many studies have suggested that orbitofrontal cortex and the basal ganglia provide signals that predict the reward associated with a given response (Schultz et al., 2000), which could also serve as input to bias the fronto-parietal competition.

The formation of a decision is not perforce expressed as an overt action. There is a separate source of input to the model: a GO signal (Cisek et al., 1998), which modulates the strength of the PMd projection into M1. If the GO signal is nonzero, the pattern of activity in PMd flows into M1. However, lateral interactions within M1 are more strongly competitive than those in PMd, and multiple peaks of activity cannot survive in M1. Thus, activation of M1 forces the system to make a choice, regardless of the possible presence of multiple PMd peaks.

Model simulations

The operation of the model can be most easily understood in the context of a particular task.

For example, Fig. 4a shows a "two-target task" task in which the correct target for a reaching movement was indicated through a sequence of cues: during the spatial-cue period (SC), two possible targets were presented, and during a subsequent color-cue period (CC), one of these was designated as the correct target. In the model, the appearance of the spatial cue causes activity in two groups of cells in PPC, each tuned to one of the targets. Mutual excitation between nearby cells creates distinct peaks of activity that compete against each other through the inhibitory interactions between cells with different preferred directions. Because of the topographic projections between PPC and PMd (Fig. 3a), two peaks appear in PMd as well, although they are weaker in the PMd layers further downstream (compare layers PMd^1 and PMd^3). These two peaks continue to be active and to compete against each other even after the targets vanish, due to the positive feedback between layers (i.e., the system exhibits "working memory"). At the same time, activity accumulates in the PFC cells selective for the particular location-color conjunctions. The color cue is simulated as uniform excitation to all PFC cells preferring the given color (in this case, PFC^R), and it pushes that group of PFC cells toward stronger activity than the other group (in PFC^B). This causes the competition in PMd to become unbalanced, and one peak increases its activity until it crosses the quenching threshold and suppresses its competitor. In the model, this is equivalent to a decision. Finally, once the GO signal is given, activity is allowed to flow from PMd^3 into M1, and the peak of the M1 activity is taken to define the initial direction of the movement.

The simulation reproduces many features of neural activity recorded from the PMd and M1 of a monkey performing the same reach-decision task (Cisek and Kalaska, 2005). As shown in Fig. 4a (top), two groups of PMd neurons were active during the SC, and then during the CC one of these became more strongly active (predicting the monkey's choice) while the other was suppressed. Note how the activity was weaker while both options were present, consistent with the hypothesis that the two groups of cells exert an inhibitory influence on each other. As in the model, these phenomena were seen more strongly in the rostral part of PMd than in the caudal part.

Figure 4b shows a variation of the task in which the color cue is presented before the spatial cue. In this case, no directionally tuned activity appears in PMd during the color-cue period, and after the spatial targets are presented there is sustained activity corresponding only to the correct target. Thus, the neural activity during the SC is determined not by the sensory properties of the stimulus (which are the same as in Fig. 4a), but by the movement information specified by the stimulus (Wise et al., 1992; Boussaoud and Wise, 1993; di Pellegrino and Wise, 1993).

The simulation shown in Fig. 4b brings out an important point. Immediately after the appearance of the SC, there is a brief burst toward the incorrect target, in both the neurons recorded in rostral PMd (Cisek and Kalaska, 2005) and in the model PMd^1 and PPC populations (Fig. 4b). In other words, the earliest activities within the model PMd and PPC regions reflect the spatial locations of relevant sensory stimuli, just like the early wave of "stimulus-specific" activity observed in response to visual stimuli across a large area of the cerebral cortex (Schmolesky et al., 1998; Ledberg et al., 2007). Slightly later, this initial pattern of activity changes to reflect decision-making processes which select out the correct target from the two alternatives. This is similar to the "response-specific" activation reported by Ledberg et al. (2007), which begins $\sim$150 ms after stimulus onset in nearly all fronto-parietal regions including even striate cortex. Importantly, both the initial stimulus-specific and the later response-specific patterns occur nearly simultaneously in both frontal and parietal regions, both in the data and in the model. This may be taken to imply that "sensory" and "motor" processes overlap in time, but in the context of the affordance competition hypothesis a more accurate description is as follows: specification occurs quickly throughout nearly all of the cerebral cortex, involving representations related to both sensory and motor aspects of sensorimotor control, and is followed shortly thereafter by action selection processes which include both attentional and decisional modulation.

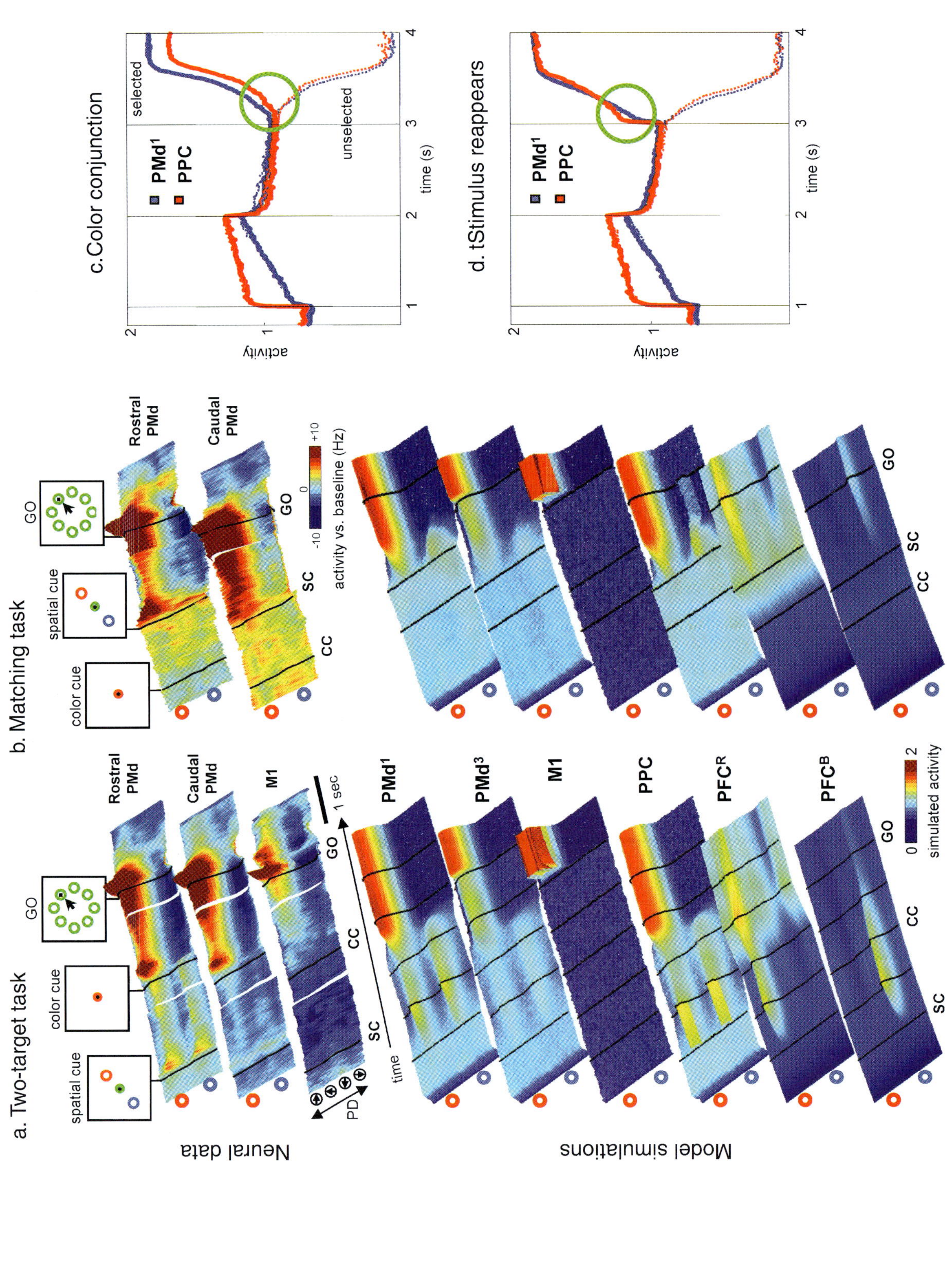

a. Two-target task
spatial cue
color cue
GO
Rostral PMd
Caudal PMd
M1
Neural data
PD
time
SC
CC
GO
1 sec
PMd1
PMd3
M1
PPC
PFCR
PFCB
Model simulations
0
2
simulated activity
b. Matching task
color cue
spatial cue
GO
Rostral PMd
Caudal PMd
CC
SC
GO
-10
0
+10
activity vs. baseline (Hz)
c.Color conjunction
PMd1
PPC
selected
unselected
activity
time (s)
d. tStimulus reappears
PMd1
PPC
activity
time (s)

While the distributions of latencies of stimulus-specific and response-specific activation are very similar across cortical regions (Ledberg et al., 2007), the present model predicts that they are not identical, and that they will follow a specific context-dependent trend. In particular, consider the case when a decision is made on the basis of cognitive information, such as a learned color cue (as in the two-target task described above). Because such cues are collected by prefrontal regions that project into rostral PMd, the bias introduced by the cue will begin to unbalance the PMd competition directly, which will then in turn (through fronto-parietal connections) cause the PPC competition to become unbalanced. Therefore, a decision made on the basis of such cognitive cues will first be expressed in frontal cortex and then, a very short time later, in parietal regions. This is indeed what was observed during neural recordings in the two-target task, which showed that PMd neurons tended to reflect the decision ~80 ms before PPC neurons (Cisek et al., 2004). This phenomenon is simulated in Fig. 4c. In particular, note that just after the color cue is presented (green circle) the neural activity tuned to the selected target begins to diverge from the activity tuned to the unselected target first in PMd (blue lines), and then shortly afterwards in PPC (red lines).

In contrast, consider a situation in which the decision is made on the basis of a more direct sensory signal, such as the reappearance of one of the targets. This information will first be available in parietal cortex, and will cause the PPC competition to become unbalanced, which will then in turn unbalance the PMd competition. Therefore, a decision made on the basis of the reappearance of a stimulus will first be expressed in parietal cortex and then very soon after appear in PMd. Figure 4d simulates this phenomenon. Note that just after the target reappears (green circle), the activity in PPC (red lines) reflects this event slightly before the activity in PMd (blue lines). Although such conditions have not been directly tested in neurophysiological recording experiments, the model predicts that the sequence by which decision-related activity spreads across the cerebral cortex is dependent upon the nature of the information which guides the choice that is made.

In addition to reproducing qualitative features of neural activity during the reach-decision tasks of Cisek and Kalaska (2005), the model can simulate important psychophysical results on the spatial and temporal characteristics of human motor decisions. For example, it reproduces the important finding that reducing the quality of evidence for a given choice makes reaction times longer and more broadly distributed. The model produces this (see Fig. 5a), through the same mechanism proposed by other models which involve a gradual accumulation to threshold: that with weaker evidence, the rate of accumulation is slower and the threshold is reached later in time, and therefore variability in accumulation rate produces broader distributions of reaction times (Carpenter and Williams, 1995; Ratcliff et al., 2003; Smith and Ratcliff, 2004).

It is also well known that reaction times in choice-tasks increase with the number of possible

Fig. 4. Comparison between neural activity and model simulations. (a) Two-target task. During the SC, two possible targets are presented, one red and one blue. During the CC, the center indicates which of these is the correct target. The GO signal instructs the monkey to begin the movement. Neural data (Cisek and Kalaska, 2005) is shown from three sets of neurons: rostral PMd, caudal PMd, and M1. In each, neural activity is depicted as a 3D colored surface in which time runs from left to right and cells are sorted by their preferred direction along the left edge. Colored circles indicate the locations of the two targets. Simulated model activities are depicted in the same format, where black lines indicate behavioral events (spatial cue on, spatial cue off, color cue on, color cue off, GO). (b) Matching task, same format. In this task, the color cue is presented prior to the spatial cue. (c) Simulation of the two-target task (in which the decision is made on the basis of a color conjunction rule). Blue lines show the time course of the average activity of two groups of PMd[1] cells — cells tuned to the selected target (thick line) and cells tuned to the unselected target (thin dotted line). Red lines show the activity of PPC cells tuned to the selected (thick line) and unselected targets (thin dotted line). Vertical lines indicate the time of SC onset, SC offset, and CC onset. The green circle emphasizes the first activity which reflects the decision made be the network, which appears in PMd prior to PPC. (d) Simulation of a task in which instead of a color cue, the decision is made when one of the target stimuli reappears at time $t = 3$. In this situation, the decision is reflected first in PPC before it appears in PMd (note activity emphasized by the green circle).

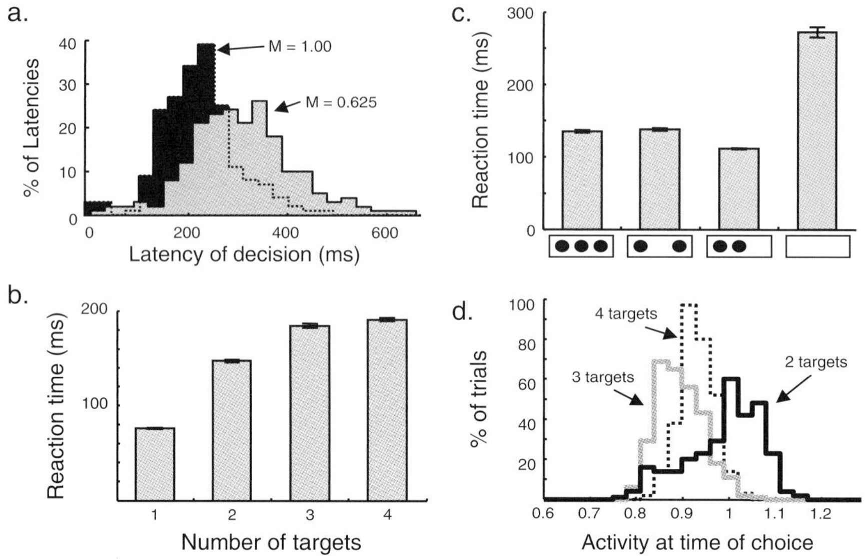

Fig. 5. (a) Distributions of decision latency computed during simulations (each with two targets) using a CC of different magnitudes (M). The decision latency was calculated as the time between the CC and the first time any PMd[3] cell activity exceeded 0.75. $N = 200$ for each condition. (b) Simulated reaction time during tasks with one, two, three, or four targets presented for 1.3 s, followed by a single correct target for 0.1 s, followed by the GO signal. Reaction times were calculated as first time after the GO signal that any neuron in the M1 population exceeded an activity threshold of 1.5. The mean and standard error are shown for $N = 300$ replications in each condition. (c) Simulated reaction time when cues are presented for 0.8 s followed by a single target for 0.3 s prior to the GO. The bars show mean$\pm$s.e. of reaction time in four conditions: when three cues are presented 80° apart, two cues 160° apart, two cues 80° apart, or no cue at all. $N = 100$ in each condition. (Adapted with permission from Cisek, 2007a.) (d) The effect of target number on the quenching threshold. Distributions are shown of the activity in the winning PMd[1] cell at the time that the decision is made, under conditions in which either 2, 3, or 4 targets are presented as options.

choices. This can be explained by the model (see Fig. 5b), because the activity associated with each option is reduced as the number of options is increased, and it therefore takes longer for the activity to reach the quenching threshold. This well-known result is reproduced by many models of decision making.

However, in addition to these results the model also addresses some phenomena which cannot be addressed by most models of decision making. For example, it has been shown that reaction time is not only determined by the number of targets presented to a subject, but also by their spatial configuration. For example, Bock and Eversheim (2000) showed that reaction time in a reaching task is similar with two or five targets as long as they subtend the same spatial angle, but shorter if two targets are closer together. This finding cannot be simulated by most decision-making models (Roe et al., 2001; Usher and McClelland, 2001; Mazurek et al., 2003; Reddi et al., 2003; Smith and Ratcliff, 2004) because the mechanism of decision is separate from the mechanisms of spatial planning. However, it falls out of the present model without need for any additional changes (see Fig. 5c).

A novel prediction of the present model is that the level of activity at which the decision is made will not be a constant across all trials, but will be dependent upon the number of targets. As shown in Fig. 5d, the relationship between the number of targets and the quenching threshold is non-monotonic: with two targets, the quenching threshold is highest, but it is lower with three targets than with four. This is caused by the non-linear effects upon the quenching threshold of the number of targets as well as their spatial separation.

The model also explains several observations on the spatial features of movements made in the presence of multiple choices. For example, Ghez et al. (1997) studied movements made in a "timed-response paradigm," in which subjects are trained to begin a movement at the end of a set of count-down tones. The time at which they are told which of two possible directions for movement is correct is then gradually brought closer and closer to the time of initiation, thus reducing the available time to perform response selection and planning. The results showed that when subjects are forced to make choices quickly (less that 80 ms between choice cue onset and movement), they move to targets randomly if they are spaced further than 60° apart, and in-between them if the targets are close together, as shown in Fig. 6a. This was interpreted as two modes of processing: a "discrete mode" for targets far apart and a "continuous mode" for nearby targets. However, as shown in Fig. 6b, the model reproduces all of these results with a single mechanism. When two targets are far apart, they create multiple competing peaks of activity in the PMd–PPC populations, and the decision is determined by which peak happens to fluctuate higher when the signal to move is given. However, if the targets are close together, then their two corresponding peaks merge into one because of the positive feedback between cells with similar parameter preferences (A similar explanation has been proposed by Erlhagen and Schöner, 2002).

In a related experiment, Favilla (1997) demonstrated that the discrete and continuous modes can occur at the same time when four targets are grouped into two pairs that are far apart but each of which consists of two targets close together (see Fig. 6c). This phenomenon is also reproduced by the model (Fig. 6d) (except for an additional central bias exhibited by human subjects). With four targets, peaks corresponding to targets within each pair merge together and then the two resulting peaks compete and are selected discretely.

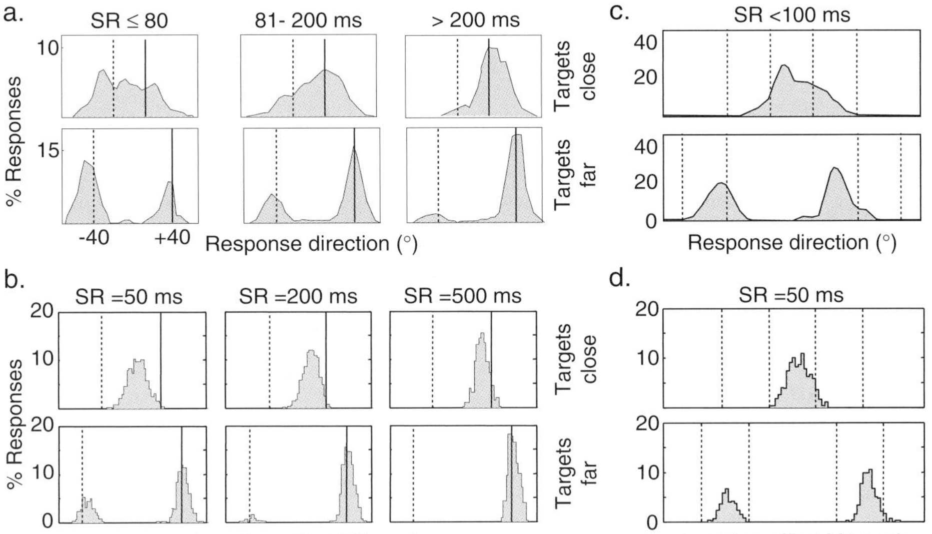

Fig. 6. Data and simulation of the timed response paradigms of Ghez et al. (1997) and Favilla (1997). (a) Behavioral data from the Ghez et al. (1997) task. Each panel shows the distribution of initial directions of force production with respect to two targets (vertical lines). Data are aligned such that the correct target (solid line) is on the right. Different distributions are reported for different delays (SR-intervals) between the onset of the choice cue stimulus and movement onset, and for different angular separations between the targets. (b) Simulations of the Ghez et al. (1997) task. Each panel shows the distribution of initial directions, calculated as the preferred direction of the first M1 cell whose activity exceeded a threshold of 1.75. (c) Behavioral data from the Favilla (1997) task, in which four targets are shown either all 30° apart or grouped into two pairs that are far apart. Same format as (a). (d) Simulations of the Favilla (1997) task, same format as (b). (Adapted with permission from Cisek, 2007a.)

Discussion

The computational model described above is essentially a model of decision making, and it shares many features with related models. For example, the accumulation of information until a decision threshold is reached has been central to a class of models called "sequential sampling models" (Roe et al., 2001; Usher and McClelland, 2001; Mazurek et al., 2003; Reddi et al., 2003; Smith and Ratcliff, 2004). The mechanism of transition from biased competition to winner-take-all dynamics is related to models of phase transitions in recurrent neural populations (Grossberg, 1973; Wang, 2002; Machens et al., 2005). The emergence of distinct response choices within populations of tuned neurons is related to the dynamic field theory of Erlhagen and Schöner (2002) and to "biased competition" models of attention (Desimone and Duncan, 1995; Tipper et al., 2000; Boynton, 2005).

However, although the mathematical model presented here is similar in some ways to previous models of decision making, it is based on a somewhat unusual theoretical foundation. The affordance competition hypothesis, illustrated schematically in Fig. 2, differs in several fundamental respects from the cognitive neuroscience frameworks within which models of decision making are usually developed. Importantly, it lacks the traditional emphasis on explicit representations which capture knowledge about the world. For example, the activity in the dorsal stream and the fronto-parietal system is not proposed to encode representations of objects in space, or representations of motor plans, or cognitive variables such as expected value. Instead, these regions encode a particular, functionally motivated mixture of all of these variables. From a traditional perspective, such activity appears surprising because it doesn't have any of the expected properties of a sensory, cognitive, or motor representation. It doesn't capture knowledge about the world in the explicit descriptive sense expected from cognitive theories, and has proven difficult to interpret from that perspective (see above). However, from the perspective of affordance competition, mixtures of sensory information with motor variables and cognitive biases make perfect sense. Their functional role is not to describe the world but to mediate adaptive interaction with the world.

Instead of viewing the functional architecture of behavior as serial stages of representation (Fig. 1a), we view it as a set of competing sensorimotor loops (Fig. 2). These loops are continuously processing sensory information to rapidly specify potential actions, which compete against each other both through lateral inhibition within local cortical areas and through a central switching mechanism within the basal ganglia (Prescott et al., 1999). Action selection occurs when the biases for one action become strong enough to suppress other competing actions, and this consensus spreads across a large distributed area of the cerebral cortex. When an action is released into execution, it continues to be specified through the sensorimotor loop involving the dorsal visual system, the fronto-parietal network, feedback through the environment, and predicted sensory feedback through cerebellar forward models.

This proposal is related to several theories which describe behavior as a competition between actions (Kornblum et al., 1990; Hendriks-Jansen, 1996; Toates, 1998; Prescott et al., 1999; Ewert et al., 2001), and the present discussion is an attempt to unify these and related ideas with a growing body of neurophysiological data. It is suggested that a great deal of neural activity in the cerebral cortex can be interpreted from the perspective of a competition between potential movements more easily than in terms of traditional distinctions between perception, cognition, and action (Cisek, 2001; Cisek, 2007a, b). It is not suggested that distinctions between perceptual, cognitive, and motor processes be discarded entirely, but only that other conceptual distinctions may be better suited to understanding central regions involved in interactive behavior.

In summary, the affordance competition hypothesis is a very general proposal on the basic functional organization of the nervous system. It is based on the assumption that the nervous system evolved to address the needs of simple, interactive behavior, and not the needs of complex cognitive abilities such as knowledge acquisition or deductive logic, as emphasized by cognitive psychology. The framework presented here clearly cannot address such cognitive abilities. However, it does suggest a place for them within the structure of sensorimotor interaction. It suggests that higher cognition evolved as an elaboration of action selection mechanisms, allowing increasingly sophisticated ways of making decisions on the basis of abstractions. In other words, it suggests that the basic biasing influences of basal ganglia and prefrontal regions upon a fronto-parietal sensorimotor competition have been retained as the frontal cortex expanded in primate phylogeny, allowing more and more complex criteria for resolving the ongoing competition. This proposal is consistent with the anatomical organization of projections from basal ganglia and cerebellum to frontal cortex (Middleton and Strick, 2000), suggesting that cognition itself evolved from a foundation of action selection (Cisek and Kalaska, 2001).

Acknowledgments

Supported by an NSERC Discovery Grant and the CIHR New Emerging Team Grant in Computational Neuroscience.

References

Albright, T.D., Kandel, E.R. and Posner, M.I. (2000) Cognitive neuroscience. Curr. Opin. Neurobiol., 10(5): 612–624.

Alexander, G.E. and Crutcher, M.D. (1990) Neural representations of the target (goal) of visually guided arm movements in three motor areas of the monkey. J. Neurophysiol., 64(1): 164–178.

Basso, M.A. and Wurtz, R.H. (1998) Modulation of neuronal activity in superior colliculus by changes in target probability. J. Neurosci., 18(18): 7519–7534.

Block, N. (1995) The mind as the software of the brain. In: Smith E.E. and Osherson D.N. (Eds.), Thinking: An Invitation to Cognitive Science. MIT Press, Cambridge, MA, pp. 377–425.

Bock, O. and Eversheim, U. (2000) The mechanisms of movement preparation: a precuing study. Behav. Brain Res., 108(1): 85–90.

Boussaoud, D. and Wise, S.P. (1993) Primate frontal cortex: effects of stimulus and movement. Exp. Brain Res., 95(1): 28–40.

Boynton, G.M. (2005) Attention and visual perception. Curr. Opin. Neurobiol., 15(4): 465–469.

Bullock, D., Cisek, P. and Grossberg, S. (1998) Cortical networks for control of voluntary arm movements under variable force conditions. Cereb. Cortex, 8: 48–62.

Bullock, D. and Grossberg, S. (1988) Neural dynamics of planned arm movements: emergent invariants and speed-accuracy properties during trajectory formation. Psychol. Rev., 95(1): 49–90.

Buneo, C.A., Jarvis, M.R., Batista, A.P. and Andersen, R.A. (2002) Direct visuomotor transformations for reaching. Nature, 416(6881): 632–636.

Butler, A.B. and Hodos, W. (1996) Comparative Vertebrate Neuroanatomy: Evolution and Adaptation. Wiley-Liss, New York.

Carello, C.D. and Krauzlis, R.J. (2004) Manipulating intent: evidence for a causal role of the superior colliculus in target selection. Neuron, 43(4): 575–583.

Carpenter, R.H. and Williams, M.L. (1995) Neural computation of log likelihood in control of saccadic eye movements. Nature, 377(6544): 59–62.

Cisek, P. (2001) Embodiment is all in the head. Behav. Brain Sci., 24(1): 36–38.

Cisek, P. (2005) Neural representations of motor plans, desired trajectories, and controlled objects. Cogn. Process., 6: 15–24.

Cisek, P. (2006) Integrated neural processes for defining potential actions and deciding between them: a computational model. J. Neurosci., 26(38): 9761–9770.

Cisek, P. (2007a) Cortical mechanisms of action selection: the affordance competition hypothesis. Philos. Trans. R. Soc. B Biol. Sci., 362 (http://www.journals.royalsoc.ac.uk/, doi:10.1098/rstb.2007.2054).

Cisek, P. (2007b) The affordance competition hypothesis: a framework for embodied behavior. In: Klatzky R., Behrmann M. and MacWhinney B. (Eds.), Embodiment, Ego-Space and Action. Lawrence Erlbaum Associates, Mahwah, NJ.

Cisek, P., Grossberg, S. and Bullock, D. (1998) A cortico-spinal model of reaching and proprioception under multiple task constraints. J. Cogn. Neurosci., 10(4): 425–444.

Cisek, P. and Kalaska, J.F. (2001) Common codes for situated interaction. Behav. Brain Sci., 24(5): 883–884.

Cisek, P. and Kalaska, J.F. (2005) Neural correlates of reaching decisions in dorsal premotor cortex: specification of multiple direction choices and final selection of action. Neuron, 45(5): 801–814.

Cisek, P., Michaud, N. and Kalaska, J.F. (2004) Integration of motor planning and sensory feedback in area 5. Soc. Neurosci. Abstr., 30.

Coe, B., Tomihara, K., Matsuzawa, M. and Hikosaka, O. (2002) Visual and anticipatory bias in three cortical eye fields of the monkey during an adaptive decision-making task. J. Neurosci., 22(12): 5081–5090.

Colby, C.L. and Goldberg, M.E. (1999) Space and attention in parietal cortex. Annu. Rev. Neurosci., 22: 319–349.

Crammond, D.J. and Kalaska, J.F. (2000) Prior information in motor and premotor cortex: activity during the delay period and effect on pre-movement activity. J. Neurophysiol., 84(2): 986–1005.

Culham, J.C. and Kanwisher, N.G. (2001) Neuroimaging of cognitive functions in human parietal cortex. Curr. Opin. Neurobiol., 11(2): 157–163.

Dayan, P. and Abbott, L.F. (2001) Theoretical Neuroscience. MIT Press, Cambridge, MA.
Desimone, R. (1998) Visual attention mediated by biased competition in extrastriate visual cortex. Philos. Trans. R. Soc. Lond. B Biol. Sci., 353(1373): 1245–1255.
Desimone, R. and Duncan, J. (1995) Neural mechanisms of selective visual attention. Annu. Rev. Neurosci., 18: 193–222.
Dorris, M.C. and Glimcher, P.W. (2004) Activity in posterior parietal cortex is correlated with the relative subjective desirability of action. Neuron, 44(2): 365–378.
Erlhagen, W. and Schöner, G. (2002) Dynamic field theory of movement preparation. Psychol. Rev., 109(3): 545–572.
Ewert, J.-P., Buxbaum-Conradi, H., Dreisvogt, F., Glagow, M., Merkel-Harff, C., Rottgen, A., Schurg-Pfeiffer, E. and Schwippert, W.W. (2001) Neural modulation of visuomotor functions underlying prey-catching behaviour in anurans: perception, attention, motor performance, learning. Comp. Biochem. Physiol. Part A, 128(3): 417–460.
Fadiga, L., Fogassi, L., Gallese, V. and Rizzolatti, G. (2000) Visuomotor neurons: ambiguity of the discharge or 'motor' perception? Int. J. Psychophysiol., 35(2–3): 165–177.
Favilla, M. (1997) Reaching movements: concurrency of continuous and discrete programming. Neuroreport, 8(18): 3973–3977.
Felleman, D.J. and Van Essen, D.C. (1991) Distributed hierarchical processing in the primate cerebral cortex. Cereb. Cortex, 1(1): 1–47.
Ferraina, S. and Bianchi, L. (1994) Posterior parietal cortex: functional properties of neurons in area 5 during an instructed-delay reaching task within different parts of space. Exp. Brain Res., 99(1): 175–178.
Gazzaniga, M.S. (2000) The New Cognitive Neurosciences (2nd ed.). The MIT Press, Cambridge, MA.
Ghez, C., Favilla, M., Ghilardi, M.F., Gordon, J., Bermejo, R. and Pullman, S. (1997) Discrete and continuous planning of hand movements and isometric force trajectories. Exp. Brain Res., 115(2): 217–233.
Gibson, J.J. (1979) The Ecological Approach to Visual Perception. Houghton Mifflin, Boston, MA.
Glimcher, P.W. (2001) Making choices: the neurophysiology of visual-saccadic decision making. Trends Neurosci., 24(11): 654–659.
Glimcher, P.W. (2003) The neurobiology of visual-saccadic decision making. Annu. Rev. Neurosci., 26: 133–179.
Gold, J.I. and Shadlen, M.N. (2000) Representation of a perceptual decision in developing oculomotor commands. Nature, 404(6776): 390–394.
Gold, J.I. and Shadlen, M.N. (2001) Neural computations that underlie decisions about sensory stimuli. Trends Cogn. Sci., 5(1): 10–16.
Grossberg, S. (1973) Contour enhancement, short term memory, and constancies in reverberating neural networks. Stud. Appl. Math., 52: 213–257.
Hendriks-Jansen, H. (1996) Catching Ourselves in the Act: Situated Activity, Interactive Emergence, Evolution, and Human Thought. MIT Press, Cambridge, MA.
Holland, L.Z. and Holland, N.D. (1999) Chordate origins of the vertebrate central nervous system. Curr. Opin. Neurobiol., 9(5): 596–602.
Horwitz, G.D., Batista, A.P. and Newsome, W.T. (2004) Representation of an abstract perceptual decision in macaque superior colliculus. J. Neurophysiol., 91(5): 2281–2296.
Hoshi, E., Shima, K. and Tanji, J. (2000) Neuronal activity in the primate prefrontal cortex in the process of motor selection based on two behavioral rules. J. Neurophysiol., 83(4): 2355–2373.
Hoshi, E. and Tanji, J. (2000) Integration of target and body-part information in the premotor cortex when planning action. Nature, 408(6811): 466–470.
Houk, J.C., Keifer, J. and Barto, A.G. (1993) Distributed motor commands in the limb premotor network. Trends Neurosci., 16(1): 27–33.
Janssen, P. and Shadlen, M.N. (2005) A representation of the hazard rate of elapsed time in macaque area LIP. Nat. Neurosci., 8(2): 234–241.
Johnson-Laird, P.N. (1988) The Computer and the Mind: An Introduction to Cognitive Science. Harvard University Press, Cambridge, MA.
Kalaska, J.F. and Crammond, D.J. (1995) Deciding not to GO: neuronal correlates of response selection in a GO/NOGO task in primate premotor and parietal cortex. Cereb. Cortex, 5: 410–428.
Kalaska, J.F., Sergio, L.E. and Cisek, P. (1998) Cortical control of whole-arm motor tasks. In: Glickstein M. (Ed.), Sensory Guidance of Movement, Novartis Foundation Symposium #218. Wiley, Chichester, UK, pp. 176–201.
Katz, P.S. and Harris-Warrick, R.M. (1999) The evolution of neuronal circuits underlying species-specific behavior. Curr. Opin. Neurobiol., 9(5): 628–633.
Keele, S.W. (1968) Movement control in skilled motor performance. Psychol. Bull., 70: 387–403.
Kettner, R.E., Marcario, J.K. and Port, N.L. (1993) A neural network model of cortical activity during reaching. J. Cogn. Neurosci., 5(1): 14–33.
Kim, J.-N. and Shadlen, M.N. (1999) Neural correlates of a decision in the dorsolateral prefrontal cortex of the macaque. Nat. Neurosci., 2(2): 176–185.
Knill, D.C. and Pouget, A. (2004) The Bayesian brain: the role of uncertainty in neural coding and computation. Trends Neurosci., 27(12): 712–719.
Kornblum, S., Hasbroucq, T. and Osman, A. (1990) Dimensional overlap: cognitive basis for stimulus-response compatibility — a model and taxonomy. Psychol. Rev., 97(2): 253–270.
Kusunoki, M., Gottlieb, J. and Goldberg, M.E. (2000) The lateral intraparietal area as a salience map: the representation of abrupt onset, stimulus motion, and task relevance. Vision Res., 40(10–12): 1459–1468.
Ledberg, A., Bressler, S.L., Ding, M., Coppola, R. and Nakamura, R. (2007) Large-scale visuomotor integration in the cerebral cortex. Cereb. Cortex, 17(1): 44–62.
Machens, C.K., Romo, R. and Brody, C.D. (2005) Flexible control of mutual inhibition: a neural model of two-interval discrimination. Science, 307(5712): 1121–1124.
Marr, D.C. (1982) Vision. W. H. Freeman, San Francisco, CA.

Mazurek, M.E., Roitman, J.D., Ditterich, J. and Shadlen, M.N. (2003) A role for neural integrators in perceptual decision making. Cereb. Cortex, 13(11): 1257–1269.

McCarthy, J., Minsky, M., Rochester, N. and Shannon, C.E. (1955) A Proposal for the Dartmouth Summer Research Project on Artificial Intelligence, http://www-formal.stanford.edu/jmc/history/dartmouth/dartmouth.html

McIntyre, J. and Bizzi, E. (1993) Servo hypotheses for the biological control of movement. J. Mot. Behav., 25(3): 193–202.

Medina, L. and Reiner, A. (2000) Do birds possess homologues of mammalian primary visual, somatosensory and motor cortices? Trends Neurosci., 23(1): 1–12.

Middleton, F.A. and Strick, P.L. (2000) Basal ganglia and cerebellar loops: motor and cognitive circuits. Brain Res. Rev., 31(2–3): 236–250.

Miller, E.K. (2000) The prefrontal cortex and cognitive control. Nat. Rev. Neurosci., 1(1): 59–65.

Miller, G.A., Galanter, E. and Pribram, K.H. (1960) Plans and the Structure of Behavior. Holt, Rinehart and Winston, Inc., New York.

Milner, A.D. and Goodale, M.A. (1995) The Visual Brain in Action. Oxford University Press, Oxford, UK.

Newell, A. and Simon, H.A. (1972) Human Problem Solving. Prentice-Hall, Englewood Cliffs, NJ.

di Pellegrino, G. and Wise, S.P. (1993) Visuospatial versus visuomotor activity in the premotor and prefrontal cortex of a primate. J. Neurosci., 13(3): 1227–1243.

Pisella, L., Arzi, M. and Rossetti, Y. (1998) The timing of color and location processing in the motor context. Exp. Brain Res., 121: 270–276.

Platt, M.L. (2002) Neural correlates of decisions. Curr. Opin. Neurobiol., 12(2): 141–148.

Platt, M.L. and Glimcher, P.W. (1997) Responses of intraparietal neurons to saccadic targets and visual distractors. J. Neurophysiol., 78(3): 1574–1589.

Platt, M.L. and Glimcher, P.W. (1999) Neural correlates of decision variables in parietal cortex. Nature, 400(6741): 233–238.

Prescott, T.J., Redgrave, P. and Gurney, K. (1999) Layered control architectures in robots and vertebrates. Adapt. Behav., 7: 99–127.

Rainer, G., Asaad, W.F. and Miller, E.K. (1998) Selective representation of relevant information by neurons in the primate prefrontal cortex. Nature, 363(6885): 577–579.

Ratcliff, R., Cherian, A. and Segraves, M. (2003) A comparison of macaque behavior and superior colliculus neuronal activity to predictions from models of two-choice decisions. J. Neurophysiol., 90(3): 1392–1407.

Reddi, B.A.J., Asrress, K.N. and Carpenter, R.H.S. (2003) Accuracy, information, and response time in a saccadic decision task. J. Neurophysiol., 90(5): 3538–3546.

Redgrave, P., Prescott, T.J. and Gurney, K. (1999) The basal ganglia: a vertebrate solution to the selection problem? Neuroscience, 89(4): 1009–1023.

Roe, R.M., Busemeyer, J.R. and Townsend, J.T. (2001) Multialternative decision field theory: a dynamic connectionist model of decision making. Psychol. Rev., 108(2): 370–392.

Roesch, M.R. and Olson, C.R. (2004) Neuronal activity related to reward value and motivation in primate frontal cortex. Science, 304(5668): 307–310.

Romo, R., Hernandez, A. and Zainos, A. (2004) Neuronal correlates of a perceptual decision in ventral premotor cortex. Neuron, 41(1): 165–173.

Sanger, T.D. (2003) Neural population codes. Curr. Opin. Neurobiol., 13(2): 238–249.

Schall, J.D. and Bichot, N.P. (1998) Neural correlates of visual and motor decision processes. Curr. Opin. Neurobiol., 8: 211–217.

Schmolesky, M.T., Wang, Y., Hanes, D.P., Thompson, K.G., Leutgeb, S., Schall, J.D. and Leventhal, A.G. (1998) Signal timing across the macaque visual system. J. Neurophysiol., 79(6): 3272–3278.

Schultz, W., Tremblay, L. and Hollerman, J.R. (2000) Reward processing in primate orbitofrontal cortex and basal ganglia. Cereb. Cortex, 10(3): 272–284.

Shadlen, M.N. and Newsome, W.T. (2001) Neural basis of a perceptual decision in the parietal cortex (area lip) of the rhesus monkey. J. Neurophysiol., 86(4): 1916–1936.

Shafir, E. and Tversky, A. (1995) Decision making. In: Smith E.E. and Osherson D.N. (Eds.), Thinking: An Invitation to Cognitive Science. MIT Press, Cambridge, MA, pp. 77–100.

Shen, L. and Alexander, G.E. (1997a) Preferential representation of instructed target location versus limb trajectory in dorsal premotor area. J. Neurophysiol., 77(3): 1195–1212.

Shen, L. and Alexander, G.E. (1997b) Neural correlates of a spatial sensory-to-motor transformation in primary motor cortex. J. Neurophysiol., 77: 1171–1194.

Shu, D.G., Morris, S.C., Han, J., Zhang, Z.F., Yasui, K., Janvier, P., Chen, L., Zhang, X.L., Liu, J.N., Li, Y. and Liu, H.Q. (2003) Head and backbone of the early Cambrian vertebrate Haikouichthys. Nature, 421(6922): 526–529.

Smith, P.L. and Ratcliff, R. (2004) Psychology and neurobiology of simple decisions. Trends Neurosci., 27(3): 161–168.

Snyder, L.H., Batista, A.P. and Andersen, R.A. (1997) Coding of intention in the posterior parietal cortex. Nature, 386: 167–170.

Snyder, L.H., Batista, A.P. and Andersen, R.A. (2000) Intention-related activity in the posterior parietal cortex: a review. Vision Res., 40(10–12): 1433–1441.

Stein, J.F. (1992) The representation of egocentric space in the posterior parietal cortex. Behav. Brain Sci., 15: 691–700.

Sterelny, K. (1989) Computational functional psychology: problems and prospects. In: Slezak P. and Albury W.R. (Eds.), Computers, Brains, and Minds. Kluwer Academic Publishers, Dordrecht, pp. 71–93.

Sugrue, L.P., Corrado, G.S. and Newsome, W.T. (2004) Matching behavior and the representation of value in the parietal cortex. Science, 304(5678): 1782–1787.

Tanji, J. and Hoshi, E. (2001) Behavioral planning in the prefrontal cortex. Curr. Opin. Neurobiol., 11(2): 164–170.

Thompson, K.G., Hanes, D.P., Bichot, N.P. and Schall, J.D. (1996) Perceptual and motor processing stages identified in the activity of macaque frontal eye field neurons during visual search. J. Neurophysiol., 76(6): 4040–4055.

Tipper, S.P., Howard, L.A. and Houghton, G. (2000) Behavioural consequences of selection from neural population codes. In: Monsell S. and Driver J. (Eds.), Control of Cognitive Processes: Attention and Performance XVIII. MIT Press, Cambridge, MA.

Toates, F. (1998) The interaction of cognitive and stimulus-response processes in the control of behaviour. Neurosci. Biobehav. Rev., 22(1): 59–83.

Treue, S. (2001) Neural correlates of attention in primate visual cortex. Trends Neurosci., 24(5): 295–300.

Ungerleider, L.G. and Mishkin, M. (1982) Two cortical visual systems. In: Ingle D.J., Goodale M.A. and Mansfield R.J.W. (Eds.), Analysis of Visual Behavior. MIT Press, Cambridge, MA, pp. 549–586.

Usher, M. and McClelland, J.L. (2001) The time course of perceptual choice: the leaky, competing accumulator model. Psychol. Rev., 108(3): 550–592.

Wang, X.J. (2002) Probabilistic decision making by slow reverberation in cortical circuits. Neuron, 36(5): 955–968.

White, I.M. and Wise, S.P. (1999) Rule-dependent neuronal activity in the prefrontal cortex. Exp. Brain Res., 126(3): 315–335.

Wise, S.P. (1985) The primate premotor cortex: past, present, and preparatory. Annu. Rev. Neurosci., 8: 1–19.

Wise, S.P., di Pellegrino, G. and Boussaoud, D. (1992) Primate premotor cortex: dissociation of visuomotor from sensory signals. J. Neurophysiol., 68(3): 969–972.

P. Cisek, T. Drew & J.F. Kalaska (Eds.)
Progress in Brain Research, Vol. 165
ISSN 0079-6123
Copyright © 2007 Elsevier B.V. All rights reserved

CHAPTER 31

Statistical models for neural encoding, decoding, and optimal stimulus design

Liam Paninski[1,*], Jonathan Pillow[2] and Jeremy Lewi[3]

[1]*Department of Statistics and Center for Theoretical Neuroscience, Columbia University, New York, NY, USA*
[2]*Gatsby Computational Neuroscience Unit, University College of London, London, UK*
[3]*Bioengineering Program, Georgia Institute of Technology, Atlanta, GA, USA*

Abstract: There are two basic problems in the statistical analysis of neural data. The "encoding" problem concerns how information is encoded in neural spike trains: can we predict the spike trains of a neuron (or population of neurons), given an arbitrary stimulus or observed motor response? Conversely, the "decoding" problem concerns how much information is in a spike train, in particular, how well can we estimate the stimulus that gave rise to the spike train? This chapter describes statistical model-based techniques that in some cases provide a unified solution to these two coding problems. These models can capture stimulus dependencies as well as spike history and interneuronal interaction effects in population spike trains, and are intimately related to biophysically based models of integrate-and-fire type. We describe flexible, powerful likelihood-based methods for fitting these encoding models and then for using the models to perform optimal decoding. Each of these (apparently quite difficult) tasks turn out to be highly computationally tractable, due to a key concavity property of the model likelihood. Finally, we return to the encoding problem to describe how to use these models to adaptively optimize the stimuli presented to the cell on a trial-by-trial basis, in order that we may infer the optimal model parameters as efficiently as possible.

Keywords: neural coding; decoding; optimal experimental design

Introduction

The neural coding problem is a fundamental question in systems neuroscience (Rieke et al., 1997): given some stimulus or movement, what is the probability of a neural response? For example, can we predict the activity of a population of neurons in response to a given visual stimulus? Conversely, can we decode this neural activity — e.g., can we reconstruct the image that the eye is actually seeing at any given moment, given only a few observed spike trains? What information is discarded in the neural code, and what features are most important? These questions are central both for our basic understanding of neural information processing and for engineering "neural prosthetic" devices that can interact with the brain directly (Donoghue, 2002). The problem is difficult both because neural responses are stochastic and because we want to identify these response properties given any possible stimulus in some very large set (e.g., all images that might occur in the world), and there are typically many more such stimuli than we

*Corresponding author. Tel.: +1 212-851-2166;
Fax: +1 212-851-2164; E-mail: liam@stat.columbia.edu

DOI: 10.1016/S0079-6123(06)65031-0

can hope to sample by brute force. Thus the neural coding problem is fundamentally *statistical*: given a finite number of samples of noisy physiological data, how do we estimate, in a global sense, the neural codebook?

This basic question has taken on a new urgency as neurophysiological recordings allow us to peer into the brain with ever greater facility; with the development of fast computers, cheap memory, and large-scale multineuronal recording and high-resolution imaging techniques, it has become feasible to directly observe and analyze neural activity at a level of detail that was impossible even 10 years ago. Experimentalists now routinely record from hundreds of neurons simultaneously, providing great challenges and opportunities for computational neuroscientists and statisticians. Indeed, it has become clear that, just as in systems biology more generally, sophisticated statistical techniques are necessary to understand the neural code, many of the key questions quite simply cannot be answered without powerful statistical tools. Conversely, many classical statistical methods are unenlightening when applied blindly to neural data; the most successful statistical models of the neural code incorporate increasingly detailed knowledge of biophysics and functional neuroanatomy.

This chapter summarizes some recent advances in model-based techniques for the analysis of spike train data. We begin ("Neural encoding models") by describing models of spike trains which have been developed to solve the neural "encoding" problem — the prediction of spike train responses (from one or multiple neurons simultaneously) given novel stimuli. In "Optimal model-based spike train decoding" we show how to use these encoding models to optimally decode population spike trains. The emphasis in each case is to employ well-justified, flexible, likelihood-based methods for model fitting and inference. Finally, "Optimal model-based closed-loop stimulus design" brings us back to the encoding problem, describing how we can apply the ideas developed in the first two sections to adaptively choose the optimal stimuli for characterizing the response function.

Neural encoding models

A neural "encoding model" is a model that assigns a conditional probability, $p(D|\vec{x})$, to any possible neural response D (for our purposes, D will represent an instantiation of a spike train, or of a population of spike trains), given an observed stimulus $\vec{x}$. The vector $\vec{x}(t)$ could include the stimulus presented at time t, or more generally the concatenated spatiotemporal stimulus history up to time t. As emphasized above, it is not feasible to directly estimate this probability $p(D|\vec{x})$ for all stimulus–response pairs $(\vec{x}, D)$; instead, therefore, we hypothesize some model, $p(D|\vec{x}, \theta)$, and then fit the model parameters θ to the observed data. Once θ is in hand we may compute the desired response probabilities as $p(D|\vec{x}) \approx p(D|\vec{x}, \theta)$; that is, knowing θ in some sense allows us to interpolate between the observed (noisy) stimulus–response pairs, in order to predict the response probabilities for novel stimuli $\vec{x}$ for which we have not yet observed any responses.

In choosing an encoding model, we must satisfy three (competing) desiderata: first, the model must be flexible and powerful enough to fit the observed data. Second, the model must be tractable: we need to be able to fit the model given the modest amount of data available in a physiological recording (preferably using modest computational resources as well); moreover, the model must not be so complex that we cannot interpret the form of the inferred parameters. Finally, the model must respect what is already known about the underlying physiology and anatomy of the system; ideally, we should be able to interpret the model parameters and predictions not only in statistical terms (e.g., confidence intervals, significance tests) but also in biophysical terms (membrane noise, dendritic filtering, etc.).

While in general there are many varieties of encoding models that could conceivably satisfy these three conditions, in this chapter we will focus on a particular model class known as the "generalized linear" model (GLM) (Paninski, 2004; Truccolo et al., 2005). This model class is a natural mathematical representation of the basic physiological concept of a "receptive field" and has proven useful in a wide variety of experimental preparations. In its simplest linear-nonlinear-Poisson (LNP) form, the model hypothesizes that spike trains are produced by an inhomogeneous Poisson process with rate

$$\lambda(t) = f(\vec{k} \cdot \vec{x}(t)) \quad (1)$$

given by a cascade of two simple steps (Fig. 1A). The linear stage, $\vec{k} \cdot \vec{x}(t)$, is a linear projection of $\vec{x}(t)$, the (vector) stimulus at time t, onto the receptive field $\vec{k}$; this linear stage is then followed by a simple scalar nonlinearity $f(.)$ which shapes the output [and in particular enforces the nonnegativity of the output firing rate $\lambda(t)$]. A great deal of the systems neuroscience literature concerns the quantification of these receptive field parameters $\vec{k}$.

How do we estimate the model parameters $\theta = \vec{k}$ here? We simply need to write down the likelihood $p(D|\vec{k}, \vec{x})$ of the observed spike data D given the model parameter $\vec{k}$ and the observed stimulus $\vec{x}$, and then we may employ standard likelihood optimization methods to obtain the maximum likelihood (ML) or maximum a posteriori (MAP) solutions for $\vec{k}$. It may be helpful to draw an analogy to standard linear regression here: imagine that we want to fit the standard linear regression model to our data. This model hypothesizes that each bin of observed spike train data n_t of width dt is generated by the formula $n_t = \vec{k} \cdot \vec{x}(t)dt + \varepsilon_t$, where ε_t is discrete Gaussian white noise. If we write down the likelihood of this model using the log of the Gaussian probability density function we have

$$\log p(D|X, \vec{k}) = c_1 - c_2 \sum_t (n_t - (\vec{k} \cdot \vec{x}(t))dt)^2,$$

where c_1, c_2 are constants that do not depend on the parameter $\vec{k}$, X abbreviates the matrix of observed stimuli (the t-th row of X is given by $X_t = \vec{x}(t)$), and the sum in t is over all observed time bins. Maximizing this likelihood leads to the usual least-squares regression solution $\vec{k}_{LS} = (X^T X)^{-1}(\Sigma_t n_t \vec{x}(t)/dt)$. Here $X^T X$ is a scaled

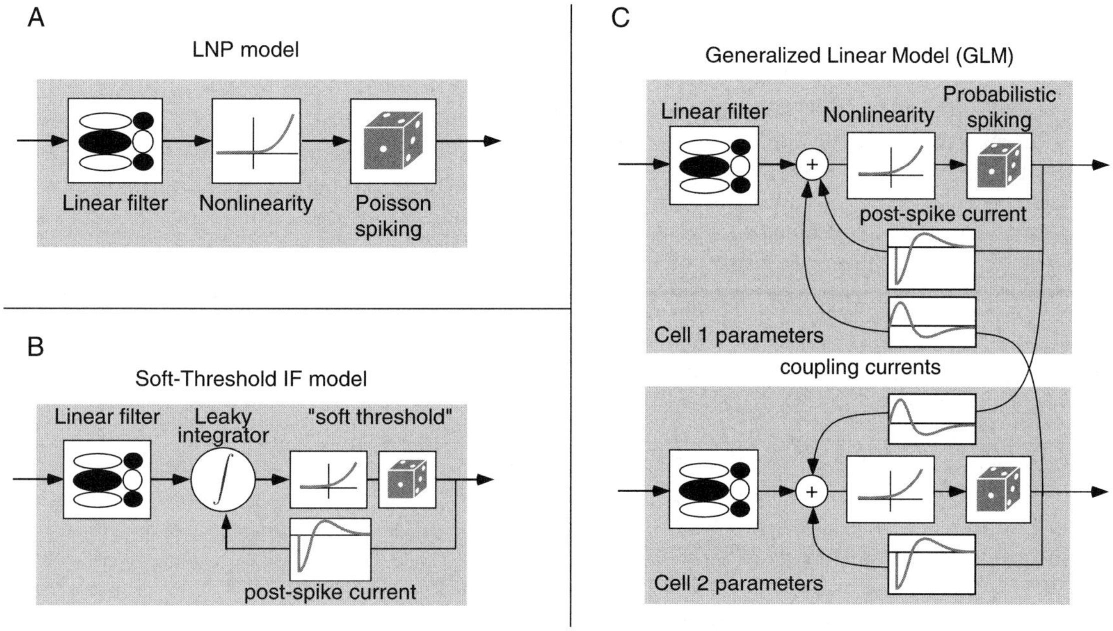

Fig. 1. Schematic diagrams of some of the encoding models discussed here. (A) The linear-nonlinear-Poisson (LNP) model is strictly feedforward, with no spike-history terms. (B) Illustration of the connection between the GLM with spike-history terms and the integrate-and-fire cell with a probabilistic ("soft") threshold. (C) GLM incorporating both spike-history and interneuronal coupling terms $h(.)$.

estimate of the covariance of $\vec{x}$, and the term on the right is proportional to the classical spike-triggered average.

Now, if we repeat the same exercise under the more plausible assumption that spike counts per bin follow a Poisson instead of Gaussian distribution [with the rate parameter of the Poisson distribution given by Eq. (1), $\lambda(t)\mathrm{d}t = f(\vec{k}\cdot\vec{x}(t))\mathrm{d}t$], we have $n_t \sim Poiss[f(\vec{k}\cdot\vec{x}(t))\mathrm{d}t]$, implying

$$\log p(D|X,\vec{k}) = c + \sum_t \left(n_t \log f(\vec{k}\cdot\vec{x}(t)) - f(\vec{x}(t)\cdot\vec{k})\mathrm{d}t\right).$$

This likelihood no longer has a simple analytical maximizer, as in the linear regression case, but nonetheless we can numerically optimize this function quite easily if we are willing to make two assumptions about the nonlinear rectification function $f(.)$: if we assume (1) $f(u)$ is a convex (upward-curving) function of its scalar argument u, and (2) $\log f(u)$ is concave (downward-curving) in u, then the loglikelihood above is guaranteed to be a concave function of the parameter $\vec{k}$, since in this case the loglikelihood is just a sum of concave functions of $\vec{k}$ (Paninski, 2004). This implies that the likelihood has no nonglobal local maxima, and therefore the ML parameter $\hat{k}_{ML}$ may be found by numerical ascent techniques. Functions $f(.)$ satisfying these two constraints are easy to think of, e.g., the standard linear rectifier and the exponential function both work. Interestingly, in the exponential case $f(.) = \exp(.)$, the ML estimate $\hat{\vec{k}}_{ML}$ turns out to be closely related to the least-squares estimate $\vec{k}_{LS}$ (Paninski, 2004), and therefore $\hat{k}_{ML}$ may be considered a generalization of classical spike-triggered average-based techniques for estimating the receptive field. See Paninski (2004) for further discussion.

Regularization: maximum penalized likelihood

In the linear regression case it is well-known that estimates of the receptive field $\vec{k}$ based on spike-triggered averaging can be quite noisy when $\vec{k}$ has many parameters (Sahani and Linden, 2003; Smyth et al., 2003); the noisiness of the estimate $\vec{k}_{LS}$ is roughly proportional to the dimensionality of $\vec{k}$ (the number of parameters in $\vec{k}$ that we need to estimate from data) divided by the total number of observed samples (Paninski, 2003). The same "overfitting" phenomenon (noisiness increasing with number of parameters) occurs in the GLM context. A variety of methods have been introduced to "regularize" the estimated $\vec{k}$, to incorporate prior knowledge about the shape and/or magnitude of the true $\vec{k}$ to reduce the noise in $\vec{k}_{LS}$. One basic idea is to restrict $\vec{k}$ to lie within a lower dimensional subspace; we then employ the same fitting procedure to estimate the coefficients of $\vec{k}$ within this lower dimensional basis [model selection procedures may be employed to choose the dimensionality of this subspace (Truccolo et al., 2005)].

A slightly less restrictive approach is to maximize the posterior $p(\vec{k}|X,D) = (1/Z)p(D|X,\vec{k})p(\vec{k})$ (with $\vec{k}$ allowed to take values in the full original basis, and Z is a normalizing constant independent of $\vec{k}$), instead of the likelihood $p(D|X,\vec{k})$; here $p(\vec{k})$ encodes our a priori beliefs about the true underlying $\vec{k}$. In the linear regression case, the computationally simplest prior is a zero-mean Gaussian, $\log p(\vec{k}) = c - \vec{k}^T A\vec{k}/2$, where A is a positive definite matrix (the inverse covariance matrix); maximizing the corresponding posterior analytically leads directly to the regularized least-square estimator $\vec{k}_{RLS} = (X^TX + A)^{-1}(\Sigma_t\, n_t\vec{x}(t)/\mathrm{d}t)$ (Sahani and Linden, 2003; Smyth et al., 2003). It is easy to incorporate this MAP idea in the GLM context as well (Paninski, 2004) (though once again we lose the nice analytic form of the optimal solution): we simply maximize

$$\begin{aligned}\log p(\vec{k}|X,D) &= c + \log p(D|X,\vec{k}) + \log p(\vec{k}) \\ &= c + \sum_t (n_t \log f(\vec{x}(t)\cdot\vec{k}) - f(\vec{x}(t)\cdot\vec{k})\mathrm{d}t) \\ &\quad + \log p(\vec{k}).\end{aligned}$$

Whenever $\log p(\vec{k})$ is a concave function of $\vec{k}$ (as in the Gaussian case described above), this "penalized" likelihood (where $\log p(\vec{k})$ acts to penalize improbable values of $\vec{k}$) is a concave function of $\vec{k}$, and ascent-based maximization may proceed (with no local maxima) as before.

Incorporating spike-history effects and interneuronal interactions

Above we have described how to adapt standard spike-triggered averaging techniques for the GL model. However, it is not immediately obvious, from a physiological point of view, what we have gained by this exercise. More importantly, it is clear that this simple model suffers from a number of basic deficiencies, e.g., the fact that we have assumed that the nonlinearity $f(.)$ is a convex function implies that the firing rate of our basic LNP model does not saturate: as we increase the magnitude of the stimulus $\vec{x}$, the rate must continue to increase at least linearly, whereas the firing rate of a real neuron will invariably saturate, leveling off after some peak discharge rate is attained. Moreover, neurons display a number of other related strongly nonlinear effects that are not captured by the model: e.g., refractory effects, burstiness and bistability of responses, and firing-rate adaptation. In other words, it seems the LNP model does not satisfy our first desideratum: model (1) is insufficiently flexible to accurately model real spiking responses.

Luckily, it turns out that we may simultaneously fix these problems and greatly enhance the GLM's flexibility, by the simple trick of enlarging our input matrix X. Recall that in the discussion above, the t-th row of this matrix consisted of the stimulus $\vec{x}(t)$. However, there is no mathematical reason why we cannot incorporate other observable variables into this matrix as well. For example, appending a column of ones to X corresponds to incorporating a constant "offset" parameter b in our model, $\lambda(t) = f(\vec{k} \cdot \vec{x}(t) + b)$, which provides an important degree of flexibility in setting the threshold and baseline firing rate of the model.

More importantly, we may incorporate terms corresponding to the neuron's observed past activity $n_s, s<t$, to obtain models of the form $\lambda(t) = f(b + \vec{k} \cdot \vec{x}(t) + \Sigma_{j=1}^{J} h_j n_{t-j})$ (Fig. 1C). Depending on the shape of the "spike-history filter" $\vec{h}$, the model can display all of the effects described above (Paninski et al., 2004b); e.g., a negative but sharply time-limited $\vec{h}$ corresponds to a refractory period (and firing rate saturation: increasing the firing rate will just increase the "hyperpolarizing" effect of the spike-history terms $\Sigma_J h_j n_{t-j}$), while a biphasic $\vec{h}$ induces burst effects in the spike train, and a slower negative $\vec{h}$ component can enforce spike-rate adaptation. Fitting these new model parameters proceeds exactly as above: we form the (augmented) matrix X (where now $X_t = \{1 \ \vec{x}(t) \ n_{t-J} \ n_{t-J+1} \ \ldots \ n_{t-1}\}$), then calculate the loglikelihood $\log p(D|X, \theta) = \Sigma_t (n_t \log f(X_t \cdot \theta) - f(X_t \cdot \theta) dt)$, and compute the ML or MAP solution for the model parameters $\theta = \{b, \vec{k}, \vec{h}\}$ by a concave optimization algorithm. Note that, while we still assume that the spike count n_t within a given short time bin is drawn from a one-dimensional Poiss($\lambda(t)dt$) distribution, the resulting model displays strong history effects and therefore the output of the model, considered as a vector of counts $D = \{n_t\}$, is no longer a Poisson process, unless $\vec{h} = 0$; see Fig. 2 for an example application.

As emphasized above, while this additive form of spike-history dependence fits conveniently within our GLM framework, this choice is by no means unique. A related approach for incorporating spike-history effects is multiplicative instead (Miller and Mark, 1992; Berry and Meister, 1998; Kass and Ventura, 2001): in this multiplicative model, the firing rate at time t is given by $\lambda(t) = f(\vec{k} \cdot \vec{x}(t)) r(t - t^*)$, where t^* denotes the time of the last observed spike and the function $r(.)$ encodes the spike-history effects. In general the additive (GL) and multiplicative models are distinct; e.g., the basic multiplicative model only incorporates spike-history information over the most recent interspike interval, whereas the additive model may incorporate information over many preceding spikes. Maximum-likelihood methods for estimating the parameters $\{\vec{k}, r(.)\}$ in this multiplicative model are discussed in Paninski (2004) but are beyond the scope of this chapter; however, it is worth noting that in the case of an exponential nonlinearity $f(.) = \exp(.)$ (Paninski et al., 2004a; Truccolo et al., 2005) the multiplicative model may be written as a one-spike modification of the GLM: $\lambda(t) = \exp(\vec{k} \cdot \vec{x}(t)) r(t - t^*) = \exp[\vec{k} \cdot \vec{x}(t) + \log r(t - t^*)]$, and we may estimate $h(.) = \log r(.)$ using the likelihood optimization techniques described above, once X is defined suitably.

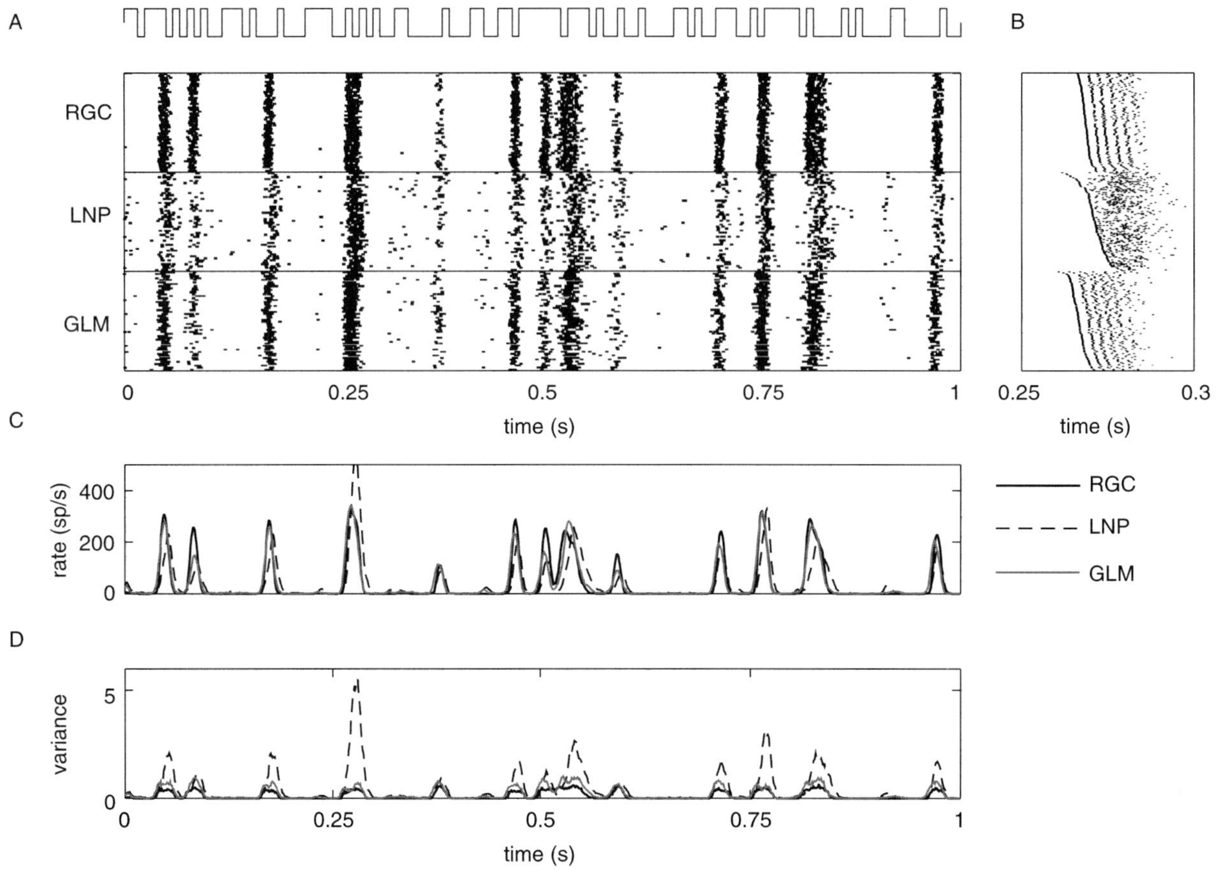

Fig. 2. Example predictions of retinal ganglion ON-cell activity using the GL encoding model with and without spike-history terms. Conventions are as in Figs. 3 and 4 in (Pillow et al., 2005b); physiological recording details as in (Uzzell and Chichilnisky, 2004; Pillow et al., 2005b). (A) Recorded responses to repeated full-field light stimulus (top) of true ON-cell ("RGC"), simulated LNP model (no spike-history terms; "LNP"), and GL model including spike-history terms ("GLM"). Each row corresponds to the response during a single stimulus presentation. Peristimulus rate and variance histograms are shown in panels C and D, respectively. (B) Magnified sections of rasters, with rows sorted in order of first spike time within the window in order to show spike timing details. Note that the predictions of the model including spike-history terms are in each case more accurate than those of the Poisson (LNP) model. The predictions of the GLM with spike-history terms are comparable in accuracy to those of the noisy integrate-and-fire model presented in Pillow et al. (2005b) (PSTH variance accounted for: 91% in each case, compared to 39% for the LNP model; IF data not shown here); predictions of the multiplicative model (Berry and Meister, 1998; Paninski, 2004) are significantly less accurate (78% v.a.f.). All data shown here are cross-validated "test" data (i.e., the estimated model parameters $\hat{\theta}_{ML}$ were in each case computed based on a nonoverlapping "training" data set not shown here).

Finally, we may expand the definition of X to include observations of other spike trains, and therefore develop GL models not just of single spike trains, but network models of how populations of neurons encode information jointly (Chornoboy et al., 1988; Paninski et al., 2004a; Pillow et al., 2005a; Truccolo et al., 2005). The resulting model is summarized (Fig. 1C):

$$n_{i,t} \sim Poiss(\lambda_i(t)\mathrm{d}t);$$
$$\lambda_i(t) = f(b + \vec{k}_i \cdot \vec{x}(t) + \sum_{i',j} h_{i',j} n_{i',t-j}),$$

with $\lambda_i(t)$ denoting the instantaneous firing rate of the i-th cell at time t, $\vec{k}_i$ the cell's linear receptive field, and $h_{i',j}$ a post-spike effect from the i'-th observed neuron in the population of cells from which we are recording simultaneously; these terms are summed over all past spike activity $n_{i',t-j}$. The $h_{i,j}$ terms (corresponding to the i-th cell's own past activity) plays the role of $\vec{h}$ above; the $h_{i',j}$ terms from the other cells in the population correspond to interneuronal interaction effects.

Connection to biophysical models: soft-threshold integrate-and-fire models

As emphasized above, one of our key goals in constructing an encoding model is to connect the model parameters to the underlying biophysics and known physiology. Thus it is worthwhile to consider the relationship between the GLM and the more biophysically motivated models employed in studies of intracellular dynamics. One connection is provided by the following model (Fig. 1B): consider the inhomogeneous Poisson process with rate given by $f(V(t)+b)$, where f is a convex, increasing, log-concave scalar function, b is a scalar, and $V(t)$ is the solution of the "leaky-integrator" differential equation $\mathrm{d}V/\mathrm{d}t = -gV(t) + \vec{k}\cdot\vec{x}(t) + \Sigma_j h_j n_{t-j}$, starting at the initial value V_{reset} after every spike. Here g denotes the membrane leak conductance; as usual, in the absence of input, V decays back to 0 with time constant $1/g$. This model is conceptually identical to a simple version of the "escape-rate" approximation to the noisy integrate-and-fire (IF) model described in (Gerstner and Kistler, 2002). Since this differential equation is linear, $V(t)$ here may be written in the form $\vec{k}_g\cdot\vec{x}(t) + \vec{h}_g\cdot\vec{n}(t)$, where $\vec{k}_g$ and $\vec{h}_g$ correspond to the original parameters $\vec{k}$ and $\vec{h}$ temporally convolved with the exponential function e^{-gt}; that is, this soft-threshold IF model is just a version of the GLM, and therefore the GLM parameters may be indirectly interpreted in biophysical terms. [It is worth noting, but beyond the scope of this article, that many of the same helpful concavity properties apply in the hard-threshold IF case, where noise is induced by an additive, unobserved intracellular noise current (Paninski et al., 2004b).]

Extensions

It should be clear that the GLM encoding framework described here can be extended in a number of important directions. We briefly describe two such directions here. First, as we have described the GLM above, it may appear that the model is restricted to including only linear dependencies on the stimulus $\vec{x}(t)$, through the $\vec{k}\cdot\vec{x}(t)$ term. However, if we modify our input matrix X once again, to include nonlinear transformations $F_j(\vec{x})$ of the stimulus $\vec{x}$, it is clear that we may fit nonlinear models of the form $\lambda(t) = f(\Sigma_j a_j F_j(\vec{x}))$ easily by maximizing the loglikelihood $\log p(D|X,\vec{a})$ with respect to the weight parameter $\vec{a}$ (Ahrens et al., 2006; Wu et al., 2006). Mathematically, the nonlinearities $F_j(\vec{x})$ may take essentially arbitrary form; physiologically speaking, it is clearly wise to choose $F_j(\vec{x})$ to reflect known facts about the anatomy and physiology of the system [e.g., $F_j(\vec{x})$ might model inputs from a presynaptic layer whose responses are better characterized than are those of the neuron of interest (Rust et al., 2006)].

Second, in many cases it is reasonable to include terms in X that we may not be able to observe or calculate directly (e.g., intracellular noise, or the dynamical state of the network); fitting the model parameters in this case requires that we perform a kind of average over these "latent," unobserved variables, e.g., via the expectation maximization algorithm (Smith and Brown, 2003; Paninski et al., 2004b; Kulkarni and Paninski, 2006). While inference in the presence of these hidden parameters is beyond the scope of this chapter, it is worth noting that this type of model may be fit tractably using generalizations of the methods described here, at the cost of increased computational complexity, but the benefit of enhanced model flexibility and realism.

Optimal model-based spike train decoding

"Decoding" refers to the problem of how to "read out" the information contained in a set of neural spike trains, and has both theoretical and practical implications for the study of neural coding (Rieke et al., 1997; Donoghue, 2002). A variety of statistical techniques have been applied to this problem (Brown et al., 1998); in this section, we focus specifically on decoding methods that rely on Bayesian "inversion" of the GL encoding model discussed above. That is, we apply Bayes' rule to obtain the posterior probability of the stimulus, conditional on the observed response: $p(\vec{x}|D) = (1/Z)p(D|\vec{x})p(\vec{x})$, where $p(\vec{x})$ is the prior stimulus probability. (We used a similar idea above when we incorporated prior knowledge to regularize our estimates of the encoding model parameter θ; here we are assuming that θ, or equivalently $p(D|\vec{x})$, has already been estimated to a reasonable degree of precision, and now we want to incorporate our prior knowledge of the stimulus $\vec{x}$.) The primary appeal of such Bayesian decoding methods is that they are optimal if we assume that the encoding model $p(D|\vec{x})$ is correct. Decoding therefore serves as a means for probing which aspects of the stimulus are preserved by the response, and also as a tool for comparing different encoding models. For example, we can decode a spike train using different models (e.g., including vs. ignoring spike-history effects) and examine which encoding model allows us to best decode the true stimulus (Pillow et al., 2005b). Such a test may in principle give a different outcome than a comparison that focuses the encoding model's ability to predict spike train statistics. In the following, we illustrate how to decode using the stimulus which maximizes the posterior distribution $p(\vec{x}|D)$, and show how a simple approximation to this posterior allows us to estimate how much information the spike train response carries about the stimulus.

Maximum a posteriori decoding

The MAP estimate is the stimulus $\vec{x}$ that is most probable given the observed spike response D, i.e., the $\vec{x}$ that maximizes $p(\vec{x}|D)$. Computing the MAP estimate for $\vec{x}$ once again requires that we search a high-dimensional space (the space of all possible stimuli $\vec{x}$) to find the maximizer of a nonlinear function, $p(\vec{x}|D)$. Luckily, in the GLM, the stimulus $\vec{x}$ interacts linearly with the model parameters θ, implying that concavity of the loglikelihood with respect to $\vec{x}$ holds under exactly the same conditions as does concavity in θ (Paninski, 2004). Moreover, the sum of two concave functions is concave, so the log-posterior, $\log p(\vec{x}|D) = \log p(D|\vec{x}) + \log p(\vec{x}) + c$, is concave as long as the stimulus log-prior $\log p(\vec{x})$ is itself a concave function of $\vec{x}$ (e.g., Gaussian). In this case, again, we may easily compute $\hat{x}_{\mathrm{MAP}}$ by numerically ascending the function $\log p(\vec{x}|D)$.

As an empirical test of the MAP estimate, we can compare its performance with that of the optimal linear estimate (OLE), the best linear estimate of the stimulus as a function of the observed spiking data D (Rieke et al., 1997). (Note that the MAP estimate, on the other hand, is in general a *nonlinear* function of D.) Parameters of the OLE can be obtained using standard least-squares regression of the spiking data onto the stimulus $\vec{x}$ (recall, conversely, that we discussed regressing the stimulus onto the spiking data in "Neural encoding models"). Figure 3 shows a comparison of the two decoding techniques, given responses D generated by a GLM encoding model with known parameters, as a function of stimulus contrast (variance) and size of the neuronal population. The MAP clearly outperforms the OLE at high contrasts or large population sizes.

Assessing decoding uncertainty

In addition to providing a reliable estimate of the stimulus underlying a set of spike responses, computing the MAP estimate $\hat{x}_{\mathrm{MAP}}$ gives us easy access to several important quantities for analyzing the neural code. In particular, the variance of the posterior distribution around $\hat{x}_{\mathrm{MAP}}$ tells us something about which stimulus features are best encoded by the response D. For example, along stimulus axes where the posterior has small variance (i.e., the posterior declines rapidly as we move away from $\hat{x}_{\mathrm{MAP}}$), we have relatively high certainty that the true $\vec{x}$ is close to $\hat{x}_{\mathrm{MAP}}$. Conversely, we have relatively low certainty about any feature axis along which the posterior variance is large.

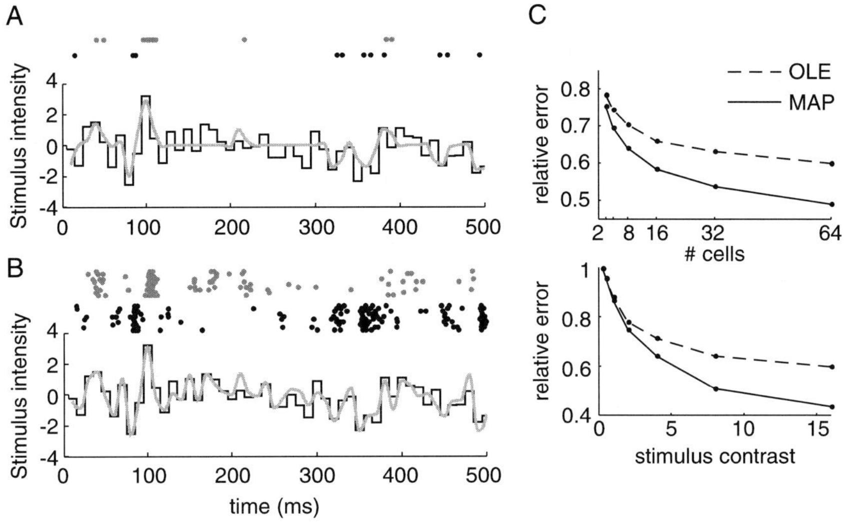

Fig. 3. Illustration of MAP decoding. (A) Simulated spike trains from a single pair of simulated ON and OFF retinal ganglion cells (above, gray and black dots) were used to compute the MAP estimate (gray) of a 500-ms Gaussian white noise stimulus (black), sampled at 100 Hz. (B) Spike trains from 10 identical, independent ON and OFF cells in response to the same stimulus, with the associated MAP estimate of the stimulus, illustrating convergence to the true stimulus as the responses of more cells are observed. (C) Comparison of the optimal linear estimate (OLE) and MAP estimate on simulated data, as a function of the number of observed cells (top) and stimulus contrast (variance; bottom). For each data point, the parameters of the OLE were estimated using a long run of simulated data. "Relative error" denotes the average RMS error between the true and estimated stimulus, averaged over 100 trials, divided by the RMS amplitude of the true stimulus.

We can measure the scale of the posterior distribution along an arbitrary axis in a fairly simple manner: since we know (by the above concavity arguments) that the posterior is characterized by a single "bump," and the position of the peak of this bump is already characterized by $\hat{x}_{\mathrm{MAP}}$, it is enough to measure the curvature of this bump at the peak $\hat{x}_{\mathrm{MAP}}$. Mathematically, we measure this curvature by computing the "Hessian" matrix A of second-derivatives of the log-posterior, $A_{ij} = -\partial^2 \log p(x_i|D)/\partial x_i \partial x_j$. Moreover, the eigendecomposition of this matrix A tells us exactly which axes of stimulus space correspond to the "best" and "worst" encoded features of the neural response: small eigenvalues of A correspond to directions of small curvature, where the observed data D poorly constrains the posterior distribution $p(\vec{x}|D)$ (and therefore the posterior variance will be relatively large in this direction), while conversely large eigenvalues in A imply relatively precise knowledge of $\vec{x}$, i.e., small posterior variance (Huys et al., 2006) (for this reason A is referred to as the "observed Fisher information matrix" in the statistics literature). In principle, this posterior uncertainty analysis can potentially clarify what features of the stimulus a "downstream" neuron might care most about.

We can furthermore use this Hessian to construct a useful approximation to the posterior $p(\vec{x}|D)$. The idea is simply to approximate this log-concave bump with a Gaussian function, where the parameters of the Gaussian are chosen to exactly match the peak and curvature of the true posterior; this Gaussian approximation makes it much easier to compute various quantities that are quite difficult to compute for general distributions $p(\vec{x}|D)$ [this approximation is quite common in the physics and statistics literature (Brown et al., 1998; Rieke et al., 1997)]. Specifically,

$$p(\vec{x}|D) \approx \left(\frac{1}{Z}\right) e^{-(\vec{x}-\hat{x}_{\mathrm{MAP}})^T A (\vec{x}-\hat{x}_{\mathrm{MAP}})/2}. \quad (2)$$

Figure 4 shows a comparison of the true posterior with the Gaussian approximation, for the examples illustrated in Fig. 3; the approximation is quite accurate here.

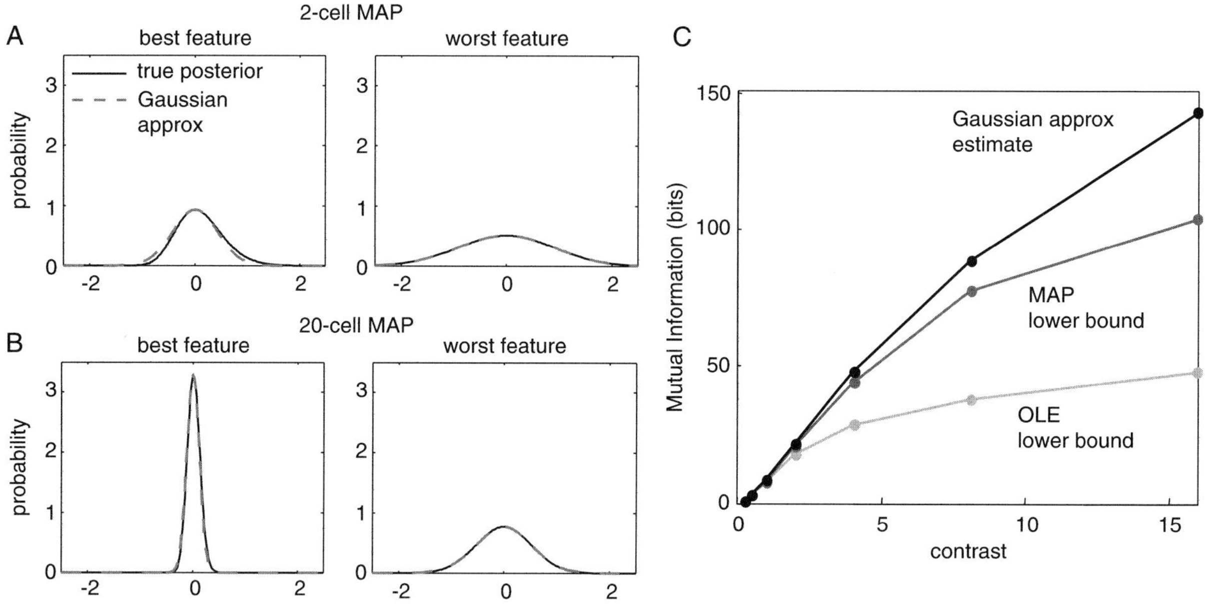

Fig. 4. Comparison of the Gaussian approximation and true posterior. (A) Slices through the true posterior $P(\vec{x}|D)$ (solid) and Gaussian approximation to the posterior (dotted), centered around the MAP estimate computed with two neural spike trains (Fig. 3a). Slices were taken along the principal axes of the posterior distribution with lowest (left) and highest (right) variance, which correspond to the "best" and "worst" encoded stimulus features (largest and smallest eigenvalues of the Hessian A, respectively). (B) Similar plots for the true posterior and Gaussian approximation around the MAP estimate computed with 20 cells (Fig. 3b). Note that the uncertainty (i.e., variance) of the distribution is greatly reduced relative to the 2-cell MAP estimate; the Gaussian approximation is quite accurate in each case. (C) Comparison of mutual information lower bounds computed with OLE and MAP vs. the information estimated directly from the Gaussian approximation, as a function of stimulus contrast (responses generated from a 20-cell GLM). The lower bounds appear to be tight for low stimulus contrast, but significantly underestimate information at higher contrasts.

Computing mutual information

A number of previous authors have drawn attention to the connections between the decoding problem and the problem of estimating how much information a population spike train carries about the stimulus (Rieke et al., 1997; Barbieri et al., 2004). Computing mutual information is quite difficult in general, as (roughly speaking) this requires estimating joint probability distributions over $\vec{x}$ and D, which is intractable in high dimensions. However, in the case that our forward model of $p(D|\vec{x})$ is sufficiently accurate, several model-based methods make the problem tractable. We express the mutual information as:

$$I(\vec{x}; D) = H(\vec{x}) - \langle H(\vec{x}|D) \rangle, \tag{3}$$

where $H(\vec{x})$ is the entropy of the raw stimulus and $H(\vec{x}|D)$ is the conditional entropy of the stimulus given the observed spiking data D, and $\langle . \rangle$ denotes averaging over D. The first term depends only on the prior stimulus distribution $p(\vec{x})$, and represents our uncertainty about the stimulus before any responses have been observed. The second term represents the average residual uncertainty about the stimulus *after* an observation of response data D. Mutual information therefore can be seen as the average amount we learn about $\vec{x}$ given D.

The most common approach to this information–estimation problem is not to estimate the true information at all, but rather to estimate a lower bound on this quantity (Rieke et al., 1997). The idea is to take a large number of stimulus–response pairs, $\{\vec{x}_i, D_i\}$, and compute the OLE $\hat{x}_{\mathrm{OLE}_i}$ from each response. We then approximate the distribution of the residuals, $p(\vec{x} - \vec{x}_{\mathrm{OLE}}|D)$, as Gaussian with mean zero and whose covariance $\hat{C}_{\mathrm{OLE}}$ we estimate with the covariance of the OLE residuals,

$\hat{C}_{\mathrm{OLE}} = \mathrm{cov}(\vec{x}_i - \hat{x}_{\mathrm{OLE}_i})$. This Gaussian distribution (with entropy $(1/2)\log|\hat{C}_{\mathrm{OLE}}|$, where $|.|$ is the matrix determinant) gives a maximum-entropy approximation to $p(\vec{x} - \hat{x}_{\mathrm{OLE}}|D)$: its entropy provides an upper bound on $\langle H(\vec{x}|D)\rangle$, and therefore a *lower* bound on $I(\vec{x}; D)$ via Eq. (3).

Our encoding model and MAP decoding framework suggest two approaches to improving this estimate of mutual information. First, we can simply substitute the MAP for the OLE in the above procedure. Since the MAP provides a better estimate of the stimulus (recall Fig. 3C), we can obtain a tighter upper bound on the conditional entropy with $\langle H(\vec{x}|D)\rangle = (1/2)\log|\hat{C}_{\mathrm{MAP}}|$, where $\hat{C}_{\mathrm{MAP}}$ is the covariance estimated from the residuals $\{x_i - \hat{x}_{\mathrm{MAP}_i}\}$. Second, we can directly estimate the mutual information (as opposed to lower bounding it), by making use of our Gaussian approximation to $p(\vec{x}|D_i)$ for every observed response D_i (Barbieri et al., 2004) (recall that this approximation is quite accurate here; Fig. 4). That is, instead of estimating information with the assumption that the posterior is independent of D (i.e., all the residuals come from the same Gaussian distribution), we can estimate the conditional entropy for every D_i, as $H(\vec{x}|D_i) = (1/2)\log|A_{D_i}^{-1}| = -(1/2)\log|A_{D_i}|$, where A_{D_i} is the Hessian of the log-posterior computed with data D_i [note that A plays the role of the inverse covariance matrix here, as in Eq. (2)]. We can then average this over trials i to obtain an estimate for the information, $\hat{I}(\vec{x}; D) = H(\vec{x}) + \langle(1/2)\log|A_{D_i}|\rangle$. Figure 4C shows a comparison of these three information estimates as a function of stimulus contrast. At low contrasts, the lower bounds derived from $\hat{x}_{\mathrm{OLE}}$ and $\hat{x}_{\mathrm{MAP}}$ are tight, while at higher contrasts the estimate derived from the Gaussian approximation suggests that the lower bounds significantly underestimate the true information.

Thus, the GLM encoding model enables tractable model-based decoding by simple MAP computations, and a straightforward Gaussian approximation has applications both for assessing uncertainty and estimating information–theoretic quantities. Although beyond the scope of this chapter, this approximation is also useful for testing hypotheses about changes in stimulus statistics (a.k.a. "change point detection," such as detecting a change in stimulus contrast), which requires computing integrals over the posterior $p(\vec{x}|D)$; see (Pillow and Paninski, 2007) for details. In addition, we can extend these techniques to employ more general "fully Bayesian" methods, in which we compute these integrals exactly by Monte Carlo techniques (Robert and Casella, 2005) instead of using the (computationally cheaper) Gaussian approximation.

Optimal model-based closed-loop stimulus design

In the previous sections we have developed robust and tractable approaches to understand neural encoding and decoding based on GL models. The framework we have developed is ultimately data-driven; both our encoding and decoding methods fail if the observed data do not sufficiently constrain our encoding model parameters θ. Therefore we will close by describing how to take advantage of the properties of the GLM to optimize our experiments: the objective is to select, in an online, closed-loop manner, the stimuli that will most efficiently characterize the neuron's response properties (Fig. 5A).

An important property of GL models is that not all stimuli will provide the same amount of information about the unknown coefficients $\vec{k}$. As a concrete example, we can typically learn much more about a visual neuron's response properties if we place stimulus energy within the receptive field, rather than "wasting" stimulus energy outside the receptive field. To make this idea more rigorous and generally applicable, we need a well-defined objective function that will rank any given stimulus according to its potential informativeness. Numerous objective functions have been proposed for quantifying the utility of different stimuli (Mackay, 1992; Nelken et al., 1994; Machens, 2002). When the goal is estimating the unknown parameters of a model, it makes sense to choose stimuli $\vec{x}(t)$ which will on average reduce the uncertainty in the parameters θ as quickly as possible (as in the game of 20 questions), given $D = \{\vec{x}(s), n_s\}_{s<t}$, the observed data up to the current trial. If we use the entropy of the posterior distribution on the model parameters $p(\theta|\vec{x}(t), D)$

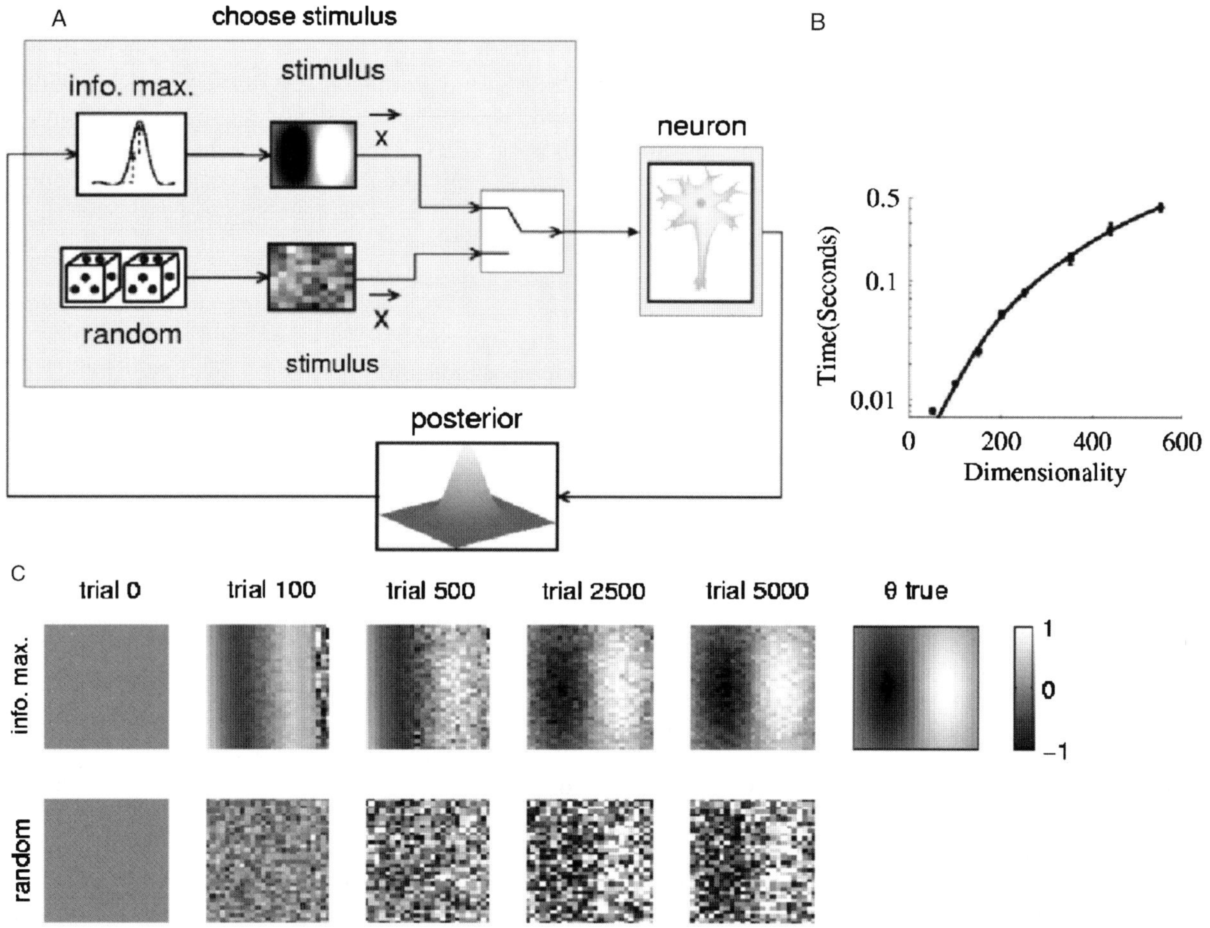

Fig. 5. (A) Closed-loop vs. open-loop stimulus design. (B) Plot of the total running time on a desktop computer for each iteration of the model-based stimulus optimization algorithm, as a function of the dimensionality of the stimulus $\vec{x}$. A quadratic polynomial $[O(\dim(\vec{x})^2)]$ fits the data quite well; note that <15 ms are necessary to optimize a 100-dimensional stimulus. (C) Plots of the estimated receptive field for a simulated visual neuron whose responses were generated by a GLM. The neuron's true receptive field $\vec{\theta}$ has the Gabor structure shown in the last panel; the nonlinearity $f(.)$ was assumed known a priori and the spike-history terms were assumed to be zero, for simplicity. Individual panels show $\vec{k}_{\mathrm{MAP}}$ after observing t stimulus–response pairs (the prior $p(\vec{k})$ was taken to be Gaussian with mean zero), comparing the accuracy of the estimates using information-maximizing vs. random stimuli (all stimuli were constrained to have unit norm, $||\vec{x}||_2 = 1$ here); the closed-loop approach is an order of magnitude more efficient in this case.

to quantify this uncertainty, we arrive at the objective function $I(\theta; n_t|\vec{x}(t), D)$, the mutual information between the response n_t and the model parameters θ given the stimulus and past data (Mackay, 1992; Paninski, 2005).

Therefore, to choose the optimal stimulus $\vec{x}(t)$ at time t, we need to do two things. First, we need to compute the objective function $I(\theta; n_t|\vec{x}(t), D)$, and then we need to optimize this function with respect to the stimulus $\vec{x}(t)$. In general, both of these problems are quite difficult: computing $I(\theta; n_t|\vec{x}(t), D)$ requires an integration over the parameter space, a task whose complexity in general scales exponentially with the number of parameters, $\dim(\theta)$. Then we need to compute this difficult objective function repeatedly as we search for the optimal $\vec{x}(t)$ [a search whose difficulty, again, scales exponentially with $\dim(\vec{x})$].

Here the special structure of the GLM comes into play. We saw in the last section how a

Gaussian approximation of a posterior distribution can greatly simplify the computation of the information; we use the same trick here, approximating $p(\theta|\vec{x}(t), D)$ by a Gaussian in this case instead of $p(\vec{x}|D, \theta)$ as before. [This Gaussian approximation may be justified by the same log-concavity arguments as before; moreover, asymptotic theory guarantees that this Gaussian approximation will be accurate — and moreover the MAP estimate $\vec{k}_{\text{MAP}}$ will converge to the true underlying parameter $\vec{k}$ — given a sufficiently long observation time t (Paninski, 2005).] Computing the entropy of this posterior has therefore been reduced from an intractable integration problem to the much more tractable computation of an average log-determinant of a Hessian matrix. (The geometric interpretation is that we want to minimize the volume of the confidence ellipsoid corresponding to this posterior Gaussian.)

While much simpler than the original integration problem, the determinant computation is in general still too slow for our goal of online, closed-loop stimulus optimization. Thus we make use of one more key feature of the GLM: the loglikelihood $\log p(n_t|\vec{k}, \vec{x}(t)) = c + n_t \log f(\vec{k} \cdot \vec{x}(t)) - f(\vec{k} \cdot \vec{x}(t))\text{d}t$ depends on the dim $(\vec{x})$-dimensional vector $\vec{k}$ only through the projection $\vec{k} \cdot \vec{x}(t)$. This effectively one-dimensional nature of the loglikelihood implies that the Hessian A_t of the log-posterior distribution given t observations is simply a rank-one perturbation of the Hessian A_{t-1} after $t-1$ observations:

$$
\begin{aligned}
A_t &= -\partial_\theta^2 \log p(\theta|D_t) \\
&= -\partial_\theta^2[\log p(\theta|D_{t-1}) + \log p(n_t|\theta, \vec{x}(t))] \\
&= A_{t-1} - \partial_\theta^2 \log p(n_t|\theta, \vec{x}(t)),
\end{aligned}
$$

where the last term is a matrix of rank one. (The equalities above are simple manipulations with Bayes rule and the definition of the Hessian.) This one-dimensional structure makes possible a very efficient recursive computation of the posterior log determinant; after making a few more simple approximations it turns out to be possible to reduce the full dim $(\vec{x})$-dimensional problem to a simple one-dimensional optimization, and this one-dimensional problem can be solved numerically rapidly enough to be used online. [See Lewi et al., 2006, for a full derivation; in addition Paninski, 2005, shows that, in a sense, the procedure is guaranteed not to get trapped in any "local optima," in a certain asymptotic sense.] The entire process — updating the posterior distribution, solving the one-dimensional optimization, and choosing the corresponding optimal stimulus — is quite fast (Fig. 5B), with the running time growing only as $O(\text{dim }(\vec{x})^2)$ (as opposed to the exponential growth in the general, nonmodel-based case). Figure 5C shows that the closed-loop optimization procedure leads to much more efficient experiments than does the standard open-loop approach of stimulating the cell with randomly chosen stimuli that are not optimized adaptively for the neuron under study.

A common argument against online stimulus optimization is that neurons are highly adaptive: a stimulus which might be optimal for a given neuron in a quiescent state may quickly become suboptimal due to adaptation (in the form of short- and long-term synaptic plasticity, slow network dynamics, etc.). Including spike-history terms in the GLM allows us to incorporate some forms of adaptation (particularly those due to intrinsic processes including, e.g., sodium channel inactivation and calcium-activated potassium channels), and these spike-history effects may be easily incorporated into the derivation of the optimal stimulus (Lewi et al., 2006). However, extending our results to models with more profound sources of adaptation is an important open research direction.

In addition to fast changes due to adaptation and spike-history effects, spiking properties typically change slowly and nonsystematically over the course of an experiment due to changes in the health, arousal, or attentive state of the preparation. We may handle these nonstationarities fairly easily using a Kalman filter-based approach: the idea is to incorporate the additional uncertainty due to these nonstationarities into our recursive updates of the posterior $\log p(\theta|D)$; again, see Lewi et al. (2006) for details. Future work will continue to improve the robustness and efficiency of these GLM-based methods, with the goal of applications to real-time optimized, adaptive neurophysiology experiments.

Conclusion

We have described a unified framework for attacking three key problems in neuroscience: (1) predicting a neuron's response (or that of a population of neurons) to a novel stimulus; (2) decoding the stimulus that led to an observed population spike train; and (3) designing a stimulus that will be as informative as possible about the neuron's properties. These techniques are flexible, well-justified statistically, and highly computationally tractable. We hope that the readers of this book will find these ideas and methods useful in their own explorations of the neural code.

Acknowledgments

JP is supported by a Royal Society International Research Fellowship and JL by a DOE Computational Science Graduate Fellowship. We thank R. Butera, D. Butts, J. Kulkarni, and E. Simoncelli for many interesting conversations, and especially V. Uzzell and E.J. Chichilnisky for permission to use data from the example cell shown in Fig. 2.

References

Ahrens, M., Paninski, L., Petersen, R. and Sahani, M. (2006) Input nonlinearity models of barrel cortex responses. Computational Neuroscience Meeting, Edinburgh.

Barbieri, R., Frank, L., Nguyen, D., Quirk, M., Solo, V., Wilson, M. and Brown, E. (2004) Dynamic analyses of information encoding in neural ensembles. Neural Comput., 16: 277–307.

Berry, M. and Meister, M. (1998) Refractoriness and neural precision. J. Neurosci., 18: 2200–2211.

Brown, E., Frank, L., Tang, D., Quirk, M. and Wilson, M. (1998) A statistical paradigm for neural spike train decoding applied to position prediction from ensemble firing patterns of rat hippocampal place cells. J. Neurosci., 18: 7411–7425.

Chornoboy, E., Schramm, L. and Karr, A. (1988) Maximum likelihood identification of neural point process systems. Biol. Cybern., 59: 265–275.

Donoghue, J. (2002) Connecting cortex to machines: recent advances in brain interfaces. Nat. Neurosci., 5: 1085–1088.

Gerstner, W. and Kistler, W. (2002) Spiking Neuron Models: Single Neurons, Populations, Plasticity. Cambridge University Press, Cambridge.

Huys, Q., Ahrens, M. and Paninski, L. (2006) Efficient estimation of detailed single-neuron models. J. Neurophysiol., 96: 872–890.

Kass, R. and Ventura, V. (2001) A spike-train probability model. Neural Comput., 13: 1713–1720.

Kulkarni, J. and Paninski, L. (2006) Common-input models for multiple neural spike-train data. COSYNE'06.

Lewi, J., Butera, R. and Paninski, L. (2006) Real-time adaptive information-theoretic optimization of neurophysiological experiments. NIPS.

Machens, C. (2002) Adaptive sampling by information maximization. Phys. Rev. Lett., 88: 228104–228107.

Mackay, D. (1992) Information-based objective functions for active data selection. Neural Comput., 4: 589–603.

Miller, M. and Mark, K. (1992) A statistical study of cochlear nerve discharge patterns in response to complex speech stimuli. J. Acoust. Soc. Am., 92: 202–209.

Nelken, I., Prut, Y., Vaadia, E. and Abeles, M. (1994) In search of the best stimulus: an optimization procedure for finding efficient stimuli in the cat auditory cortex. Hear. Res., 72: 237–253.

Paninski, L. (2003) Convergence properties of some spike-triggered analysis techniques. Netw. Comput. Neural Syst., 14: 437–464.

Paninski, L. (2004) Maximum likelihood estimation of cascade point-process neural encoding models. Netw. Comput. Neural Syst., 15: 243–262.

Paninski, L. (2005) Asymptotic theory of information-theoretic experimental design. Neural Comput., 17: 1480–1507.

Paninski, L., Fellows, M., Shoham, S., Hatsopoulos, N. and Donoghue, J. (2004a) Superlinear population encoding of dynamic hand trajectory in primary motor cortex. J. Neurosci., 24: 8551–8561.

Paninski, L., Pillow, J. and Simoncelli, E. (2004b) Maximum likelihood estimation of a stochastic integrate-and-fire neural model. Neural Comput., 16: 2533–2561.

Pillow, J. and Paninski, L. (2007) Model-based decoding, information estimation, and change-point detection in multineuron spike trains. Under review, Neural Computation.

Pillow, J., Paninski, L., Shlens, J., Simoncelli, E. and Chichilnisky, E. (2005a) Modeling multi-neuronal responses in primate retinal ganglion cells. Comp. Sys. Neur. '05.

Pillow, J., Paninski, L., Uzzell, V., Simoncelli, E. and Chichilnisky, E. (2005b) Prediction and decoding of retinal ganglion cell responses with a probabilistic spiking model. J. Neurosci., 25: 11003–11013.

Rieke, F., Warland, D., de Ruyter van Steveninck, R. and Bialek, W. (1997) Spikes: Exploring the Neural Code. MIT Press, Cambridge.

Robert, C. and Casella, G. (2005) Monte Carlo Statistical Methods. Springer, New York.

Rust, N., Mante, V., Simoncelli, E. and Movshon, J. (2006) How MT cells analyze the motion of visual patterns. Nat. Neurosci., 11: 1421–1431.

Sahani, M. and Linden, J. (2003) Evidence optimization techniques for estimating stimulus-response functions. NIPS, 15.

Smith, A. and Brown, E. (2003) Estimating a state-space model from point process observations. Neural Comput., 15: 965–991.
Smyth, D., Willmore, B., Baker, G., Thompson, I. and Tolhurst, D. (2003) The receptive-field organization of simple cells in primary visual cortex of ferrets under natural scene stimulation. J. Neurosci., 23: 4746–4759.
Truccolo, W., Eden, U., Fellows, M., Donoghue, J. and Brown, E. (2005) A point process framework for relating neural spiking activity to spiking history, neural ensemble and extrinsic covariate effects. J. Neurophysiol., 93: 1074–1089.
Uzzell, V. and Chichilnisky, E. (2004) Precision of spike trains in primate retinal ganglion cells. J. Neurophysiol., 92: 780–789.
Wu, M., David, S. and Gallant, J. (2006) Complete functional characterization of sensory neurons by system identification. Ann. Rev. Neurosci., 29(1): 477–505.

P. Cisek, T. Drew & J.F. Kalaska (Eds.)
Progress in Brain Research, Vol. 165
ISSN 0079-6123
Copyright © 2007 Elsevier B.V. All rights reserved

CHAPTER 32

Probabilistic population codes and the exponential family of distributions

J. Beck[1], W.J. Ma[1], P.E. Latham[2] and A. Pouget[1,*]

[1]*Department of Brain and Cognitive Sciences, University of Rochester, Rochester, NY, USA*
[2]*Gatsby Computational Neuroscience Unit, London WC1N 3AR, UK*

Abstract: Many experiments have shown that human behavior is nearly Bayes optimal in a variety of tasks. This implies that neural activity is capable of representing both the value and uncertainty of a stimulus, if not an entire probability distribution, and can also combine such representations in an optimal manner. Moreover, this computation can be performed optimally despite the fact that observed neural activity is highly variable (noisy) on a trial-by-trial basis. Here, we argue that this observed variability is actually expected in a neural system which represents uncertainty. Specifically, we note that Bayes' rule implies that a variable pattern of activity provides a natural representation of a probability distribution, and that the specific form of neural variability can be structured so that optimal inference can be executed using simple operations available to neural circuits.

Keywords: Bayes; neural coding; inference; noise

Introduction

The information available to our senses regarding the external world is ambiguous and often corrupted by noise. Despite such uncertainty, humans not only function successfully in the world, but seem capable of doing so in a manner that is optimal in a Bayesian sense. This has been observed in a variety of cue combination tasks, including visual and haptic cue combination (Ernst and Banks, 2002; Kording and Wolpert, 2004), visual and auditory cue combination (Gepshtein and Banks, 2003), and visual–visual cue combination (Knill and Richards, 1996; Saunders and Knill, 2003; Landy and Kojima, 2001; Hillis et al., 2004). Since cue combination processes lie at the heart of nearly every aspect of human perception, it is important to understand how cue combination tasks can be performed optimally, both in principle and in cortex.

Cue combination can be illustrated with the following example: consider a cat seeking a mouse using both visual and auditory cues. If it is dark and the mouse is partially occluded by its surroundings, there is a high degree of uncertainty in the visual cues available. The mouse may even be hiding in a field of gray mouse-sized rocks and facing in any number of directions, increasing this uncertainty. In such a context, when visual information is highly uncertain, the Bayesian cat would base an estimate of the position of the mouse primarily upon auditory cues. In contrast, when light is abundant, the mouse is easily visible, and auditory input becomes the less reliable of the two cues. In this second case, the cat should rely

*Corresponding author. Tel.: +1 (585) 275 0760;
Fax: +1 (585) 442 9216; E-mail: alex@bcs.rochester.edu

DOI: 10.1016/S0079-6123(06)65032-2

primarily on visual cues to locate the mouse. If the cat's auditory and visual cue-based estimates of the mouse's position (s_a and s_v respectively) are independent, unbiased and Gaussian distributed with standard deviations of σ_a and σ_v, then the optimal estimate of the position of the mouse is given by

$$s_{a+v} = \frac{s_a/\sigma_a^2 + s_v/\sigma_v^2}{1/\sigma_a^2 + 1/\sigma_v^2} \quad (1)$$

Cue combination studies have shown Eq. (1) to be compatible with behavior even when σ_a and σ_v are adjusted on a trial-by-trial basis. Thus one may conclude that the cortex utilizes a neural code that represents the uncertainty, if not an entire probability distribution, for each cue in a way that is amenable to the optimal cue combination as described by Eq. (1).

At first glance, it may seem that cortical neurons are not well-suited to the task of representing probability distributions, as they have been observed to exhibit a highly variable response when presented with identical stimuli. This variability is often thought of as noise, which makes neural decoding difficult in that estimates of various task-relevant parameters become somewhat unreliable. Here, however, we will argue that it is critical to realize that neural variability and the representation of uncertainty go hand-in-hand. For example, suppose that each time our hypothetical cat observes a mouse, a unique pattern of activity reliably occurs in some region of cortex. If this were the case, then observation of that pattern of activity would indicate with certainty that the mouse is in the cat's visual field. Thus, only when the pattern of activity is variable in such a way that it overlaps with patterns of activity for which the mouse is *not* present, can the pattern indicate that the mouse is present only with "some probability." In reality, absolute knowledge is an impractical goal. This is not just because sensory organs are unreliable, but also because many problems faced by biological organisms are both ill-posed (there are an infinite number of three-dimensional configurations that lead to the same two-dimensional image on the retina) and data limited (the signal reaching the brain is too noisy to determine precisely what two-dimensional image produced it). Regardless, the above example indicates that neural variability is not only compatible with the representation of probability distributions in cortex, but is, in fact, expected in this context.

Ultimately, this insight is simply an acknowledgement of Bayes' rule, which states that when the presentation of a given stimulus, s, yields a variable neural response vector $\mathbf{r}$, then for any particular response $\mathbf{r}$, the distribution of the stimulus is given by

$$p(s|\mathbf{r}) = \frac{p(\mathbf{r}|s)p(s)}{p(\mathbf{r})} \quad (2)$$

The construction of a posterior distribution, $p(s|\mathbf{r})$, from a likelihood function, $p(\mathbf{r}|s)$, in this manner corresponds to an ideal observer analysis and is, by definition, optimal. We are not suggesting that a Bayesian decoder is explicitly implemented in cortex, but rather that the existence of Bayes' rule renders such a computation unnecessary, since a variable population pattern of activity already provides a natural representation of the posterior distribution (Foldiak, 1993; Anderson, 1994; Sanger, 1996; Zemel et al., 1998). This view stands in contrast to previous work (Rao, 2004; Deneve, 2005) advocating the construction of a network that represents the posterior distribution by directly identifying neural activity with either the probability of a specific value of the stimulus, the log of that probability, or convolutions thereof.

It is not immediately clear whether or not optimal cue combination, or other desirable operations, can be performed. We will address this issue through the construction of a Probabilistic Population Code (PPC) that is capable of performing optimal cue combination via linear operations. We will then show that when distributions over neuronal activity, i.e., the stimulus-conditioned neuronal responses, $p(\mathbf{r}|s)$, belong to the exponential family of distributions with linear sufficient statistics, then optimal cue combination (and other type of Bayesian inference, such as integration over time) can be performed through simple linear combinations. Members of this family of likelihood functions, $p(\mathbf{r}|s)$, will then be shown to be compatible with populations of neurons that have arbitrarily shaped tuning curves, arbitrary covariance matrices, and can represent arbitrary posterior distributions.

Probabilistic Population Codes

We define a PPC as any code that uses Bayes' rule to optimally and *accessibly* encode a probability distribution. Here, we say that a code is accessible to a given neural circuit when that circuit is capable of performing the operations necessary to perform Bayesian inference and computation. For instance, in this work, we will be assuming that neural circuits are, at the very least, capable of performing linear operations and will seek the population code for which cue combination can be performed with some linear operation. To understand PPCs in a simplified setting, consider a Poisson distributed populations of neurons for which the tuning curve of neuron indexed by i is $f_i(s)$. In this case

$$p(\mathbf{r}|s,g) = \prod_i \frac{e^{-gf_i(s)}(gf_i(s))^{r_i}}{r_i!} \tag{3}$$

where r_i is the response or spike count of neuron i and g the amplitude, or gain, of population. When the prior is flat, i.e., $p(s)$ does not depend on s, application of Bayes' rule yields a posterior distribution that takes the form

$$\begin{aligned} p(s|\mathbf{r},g) &= \frac{p(\mathbf{r}|s,g)p(s|g)}{p(\mathbf{r}|g)} \\ &= \frac{1}{Lp(\mathbf{r}|g)}\prod_i \frac{e^{-gf_i(s)}(gf_i(s))^{r_i}}{r_i!} \\ &= \frac{1}{Lp(\mathbf{r}|g)}\left(\prod_i \frac{1}{r_i!}\right)\exp\left(\sum_i r_i \log g - gf_i(s)\right) \\ &\quad \exp\left(\sum_i r_i \log f_i(s)\right) \\ &= \frac{1}{Lp(\mathbf{r}|g)}\left(\prod_i \frac{1}{r_i!}\right)\exp\left(\sum_i r_i \log g - gc\right) \\ &\quad \exp\left(\sum_i r_i \log f_i(s)\right) \\ &\propto \exp\left(\sum_i r_i \log(f_i(s))\right) \end{aligned} \tag{4}$$

where $1/L = p(s)$, and we have assumed that tuning curves are sufficiently dense so that $\sum_i f_i(s) = c$. Because this last line of the equation represents an unnormalized probability distribution over s, we may conclude that the constant of proportionality depends only on $\mathbf{r}$ and is thus also independent of the gain g.

From Eq. (4), we can conclude that, if we knew the shape of the tuning curves $f_i(s)$, then for any given pattern of activity $\mathbf{r}$ (and any gain, g), we could simply plot this equation as a function of s to obtain the posterior distribution (see Fig. 1). In the language of a PPC we say that knowledge of the likelihood function, $p(\mathbf{r}|s)$, automatically implies knowledge of the posterior distribution $p(s|\mathbf{r})$. In this simple case of independent Poisson neurons, knowledge of the likelihood function means knowing the shape of the tuning curves. As we will now demonstrate, this knowledge is also sufficient for the identification of an optimal cue combination computation.

To this end, suppose that we have two such populations, $\mathbf{r}_1$ and $\mathbf{r}_2$, each of which encodes some piece of independent information about the same stimulus. In the context of our introductory example, $\mathbf{r}_1$ might encode the position of a mouse given visual information while $\mathbf{r}_2$ might encode the position of the mouse given auditory information. When the two populations are conditionally independent given the stimulus and the prior is flat, the posterior distribution of the stimulus given both population patterns of activity is simply given by the product of the posterior distributions given each population independently

$$\begin{aligned} p(s|\mathbf{r}_1,\mathbf{r}_2) &\propto p(\mathbf{r}_1,\mathbf{r}_2|s) \\ &\propto p(\mathbf{r}_1|s)p(\mathbf{r}_2|s) \\ &\propto p(s|\mathbf{r}_1)p(s|\mathbf{r}_2) \end{aligned} \tag{5}$$

Thus, in this case optimal cue combination corresponds to the multiplication of posteriors and subsequent normalization.

As illustrated in Fig. (2), a two layer network which performs the optimal cue combination operation would combine the two population patterns of activity, $\mathbf{r}_1$ and $\mathbf{r}_2$, into a third population, $\mathbf{r}_3$, so that

$$p(s|\mathbf{r}_3) = p(s|\mathbf{r}_1,\mathbf{r}_2) \tag{6}$$

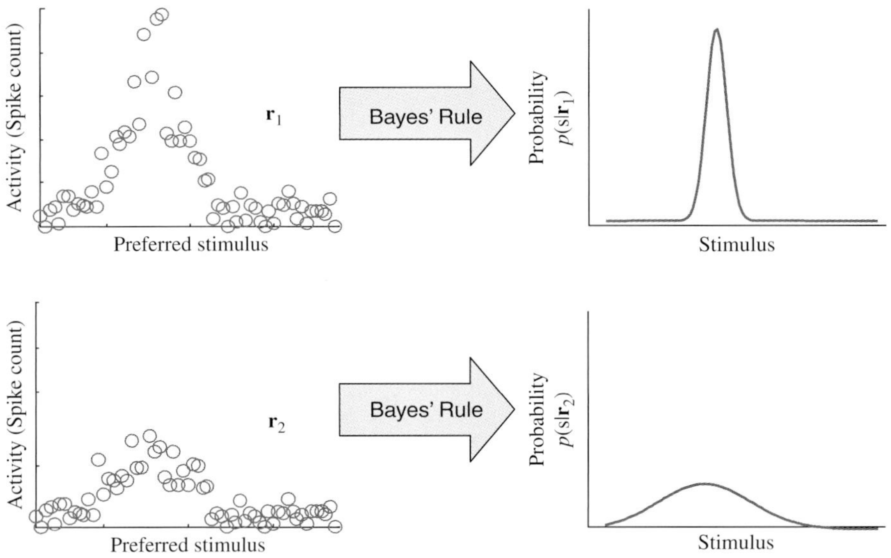

Fig. 1. Two population patterns of activity, $\mathbf{r}_1$ and $\mathbf{r}_2$, which were drawn from the distribution given by Eq. (3) with Gaussian-shaped tuning curves. Since the tuning curve shape is known, we can compute posterior $p(s|\mathbf{r})$ for any given $\mathbf{r}$, either by simply plotting the likelihood, $p(\mathbf{r}|s)$, as a function of the stimulus s or, equivalently, by using Eq. (4).

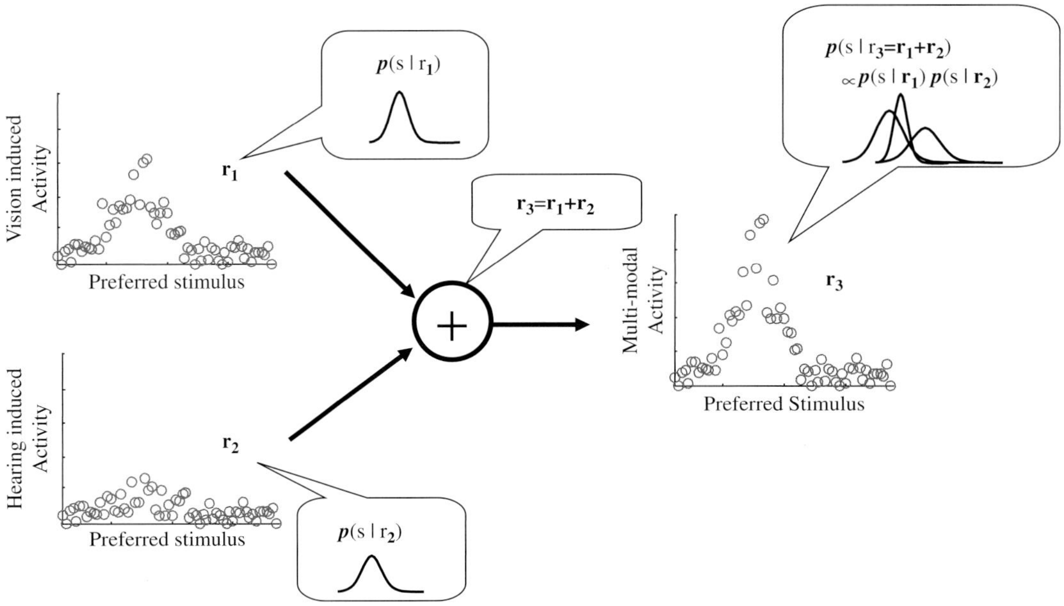

Fig. 2. On the left, the activities of two independent populations, $\mathbf{r}_1$ and $\mathbf{r}_2$, are added to yield a third population pattern $\mathbf{r}_3$. In the insets, we plot the posterior distributions associated with each of the three activity patterns are shown. In this case, optimal cue combination corresponds to a multiplication of the posteriors associated with the two independent populations.

For identically tuned populations, $\mathbf{r}_3$ is simply the sum, $\mathbf{r}_1+\mathbf{r}_2$. Since $\mathbf{r}_3$ is the sum of two Poisson random variables with identically shaped tuning curves, it too is a Poisson random variable with the same shaped tuning curve (but with a larger amplitude). As such, the Bayesian decoder applied to $\mathbf{r}_3$ takes the same form as the Bayesian decoder for $\mathbf{r}_1$ and $\mathbf{r}_2$. This implies

$$\begin{aligned} p(s|\mathbf{r}_3) &\propto \exp\left(\sum_i r_{3i}\log(f_i(s))\right) \\ &\propto \exp\left(\sum_i (r_{1i}+r_{2i})\log(f_i(s))\right) \\ &\propto \exp\left(\sum_i r_{1i}\log(f_i(s))\right)\exp\left(\sum_i r_{2i}\log(f_i(s))\right) \\ &\propto p(s|\mathbf{r}_1)p(s|\mathbf{r}_2) \\ &\propto p(s|\mathbf{r}_1,\mathbf{r}_2) \end{aligned} \quad (7)$$

and we conclude that multiplication of the two associated posterior distributions corresponds to the addition of two population codes. When the tuning curves are Gaussian, this result can also be obtained through a consideration of the variance of the maximum likelihood estimate obtained from the posterior distribution associated with the population $\mathbf{r}_3$. This results from the fact that, for Gaussian tuning curves, the log of $f_i(s)$ is quadratic in s and thus the posterior distribution is also Gaussian with maximum likelihood estimate, $\hat{s}(\mathbf{r})$, and estimate variance, $\sigma^2(\mathbf{r})$. These quantities are related to the population pattern activity $\mathbf{r}$, via the expressions

$$\begin{aligned} \hat{s}(\mathbf{r}) &= \frac{\sum_i s_i r_i}{\sum_i r_i} \\ \frac{1}{\sigma^2(\mathbf{r})} &= \frac{1}{\sigma_{tc}^2}\sum_i r_i \end{aligned} \quad (8)$$

where s_i is the preferred stimulus of the ith and σ_{tc} gives the width of the tuning curve. The estimate, $\hat{s}(\mathbf{r})$, is the well-known population vector decoder, which is known to be optimal in this case (Snippe, 1996). Now, we use the fact that the expression for the mean and variance of the posterior associated with $\mathbf{r}_3$ is the same as the expression associated with $\mathbf{r}_1$ and $\mathbf{r}_2$. This implies that

$$\begin{aligned} \frac{1}{\sigma_3^2(\mathbf{r}_3)} &= \frac{1}{\sigma_{tc}^2}\sum_i r_{3i} = \frac{1}{\sigma_{tc}^2}\sum_i r_{1i}+r_{2i} \\ &= \frac{1}{\sigma_1^2(\mathbf{r}_1)}+\frac{1}{\sigma_2^2(\mathbf{r}_2)} \end{aligned} \quad (9)$$

and

$$\begin{aligned} \hat{s}_3(\mathbf{r}_3) &= \frac{\sum_i r_{3i}s_i}{\sum_i r_{3i}} = \frac{\sum_i r_{1i}s_i+\sum_i r_{2i}s_i}{\sum_i r_{1i}+\sum_i r_{2i}} \\ &= \frac{\hat{s}_1(\mathbf{r}_1)/\sigma_1^2(\mathbf{r}_1)+\hat{s}_2(\mathbf{r}_2)/\sigma_2^2(\mathbf{r}_2)}{1/\sigma_1^2(\mathbf{r}_1)+1/\sigma_2^2(\mathbf{r}_2)} \end{aligned} \quad (10)$$

Comparison with Eq. (1) demonstrates optimality. Moreover, optimality is achieved on a trial-by-trial basis, since the estimate of each population is weighted by a variance which is computed from the actual population pattern of activity.

It is also important to note that, in the computation above, we did not explicitly compute the posterior distributions, $p(s|\mathbf{r}_i)$, and then multiply them together. Rather we operated on the population patterns of activity (by adding them together) and then simply remarked (Fig. 2) that we could have applied Bayes' rule to optimally decode these population patterns and noted the optimality of the computation. This is the essence of a PPC. Specifically, a PPC consists of three things: (1) a set of operations on neural responses $\mathbf{r}$; (2) a desired set of operations in posterior space, $p(s|\mathbf{r})$; and (3) the family of likelihood functions, $p(\mathbf{r}|s)$, for which this operation pair is optimal in a Bayesian sense. In the cue combination example above, the operation of addition of population patterns of activity (list item 1) was shown to correspond to the operation of operation of multiplication (and renormalization) of posterior distributions (list item 2), when the population patterns of activity were drawn from likelihood functions which corresponded to an independent Poisson spiking populations with identically shaped tuning curves (list item 3). Moreover, simple addition was shown to be optimal regardless of the variability of each cue, i.e., unlike other proposed cue combination schemes (Rao, 2004; Navalpakkam and Itti, 2005), this scheme does not require that the weights of the linear combination be adjusted on a trial-by-trial basis.

Generalization to the exponential family with linear sufficient statistics

So far we have relied on the assumption that populations consist of independent and Poisson spiking neurons with identically shaped tuning curves. However, this is not a limitation of this approach. As it turns out, constant coefficient linear operations can be found that correspond to optimal cue combination for a broad class of Poisson-like likelihood functions described by the so-called exponential family with linear sufficient statistics. This family includes members that can have any tuning curve shape, any correlation structure, and can represent any shape of the posterior, i.e., not just Gaussian posteriors. Below, we show that, in the case of contrast-invariant turning curves, the requirement that optimal cue combination occur via linear combination of population patterns of activity with fixed coefficients only limits us to likelihood functions for which the variance is proportional to the mean.

Optimal cue combination via linear combination of population codes

Consider two population codes, $\mathbf{r}_1$ and $\mathbf{r}_2$, encoding visual and auditory location cues, which are jointly distributed according to $p(\mathbf{r}_1, \mathbf{r}_2|s)$. The goal of an optimal cue combination computation is to combine these two populations into a third population pattern of activity $\mathbf{r}_3 = \mathbf{F}(\mathbf{r}_1, \mathbf{r}_2)$, so that an application of the Bayes rule yields

$$p(s|\mathbf{r}_3) = p(s|\mathbf{r}_1, \mathbf{r}_2) \tag{11}$$

Note that optimal cue combination can be trivially achieved by selecting any invertible function $\mathbf{F}(\mathbf{r}_1, \mathbf{r}_2)$. To avoid this degenerate case, we assume the lengths of the vectors $\mathbf{r}_1$, $\mathbf{r}_2$, and $\mathbf{r}_3$ are the same. Thus the function $\mathbf{F}$ cannot be invertible. The likelihood of $\mathbf{r}_3$ is related to the likelihood of $\mathbf{r}_1$ and $\mathbf{r}_2$ via the equation

$$p(\mathbf{r}_3|s) = \int p(\mathbf{r}_1, \mathbf{r}_2|s)\delta(\mathbf{r}_3 - F(\mathbf{r}_1, \mathbf{r}_2))d\mathbf{r}_1 d\mathbf{r}_2 \tag{12}$$

Application of Bayes' rule and the condition of optimality [Eq. (11)], indicates that an optimal cue combination operation, $\mathbf{F}(\mathbf{r}_1, \mathbf{r}_2)$, depends on the likelihood, $p(\mathbf{r}_1, \mathbf{r}_2|s)$.

Interestingly, if the likelihood is in the exponential family with linear sufficient statistics, a linear function $\mathbf{F}(\mathbf{r}_1, \mathbf{r}_2)$ can be found such that Eq. (11) is satisfied. Members of this family take the functional form

$$p(\mathbf{r}_1, \mathbf{r}_2|s) = \frac{\phi_{12}(\mathbf{r}_1, \mathbf{r}_2)}{\eta_{12}(s)} \exp(\mathbf{h}_1^{\mathrm{T}}(s)\mathbf{r}_1 + \mathbf{h}_2^{\mathrm{T}}(s)\mathbf{r}_2) \tag{13}$$

where the superscript "T" denotes transpose, $\phi_{12}(\mathbf{r}_1, \mathbf{r}_2)$ is the so-called measure function, and $\eta_{12}(s)$ the normalization factor, often called the partition function. Here $\mathbf{h}_1(s)$ and $\mathbf{h}_2(s)$ are vector functions of s, which are called the stimulus-dependent kernels associated with each population. We make the additional assumption that $\mathbf{h}_1(s)$ and $\mathbf{h}_2(s)$ share a common basis $\mathbf{b}(s)$ which can also be represented as vector of functions of s. This implies that we may write $\mathbf{h}_i(s) = \mathbf{A}_i\mathbf{b}(s)$ for some stimulus independent matrix $\mathbf{A}_i$ ($i = 1, 2$). We will now show that, when this is the case, optimal combination is performed by the linear function

$$\mathbf{r}_3 = \mathbf{F}(\mathbf{r}_1, \mathbf{r}_2) = \mathbf{A}_1^{\mathrm{T}}\mathbf{r}_1 + \mathbf{A}_2^{\mathrm{T}}\mathbf{r}_2 \tag{14}$$

Moreover, this we will show that the likelihood function $p(\mathbf{r}_3|s)$ is also in the same family of distributions as $p(\mathbf{r}_1, \mathbf{r}_2|s)$. This is important, as it demonstrates that this approach — taking linear combinations of firing rates to perform optimal Bayesian inference — can be either repeated iteratively over time or cascaded from one population to the next.

Optimality of this operation is most easily demonstrated by computing the likelihood, $p(\mathbf{r}_3|s)$, applying Bayes' rule to obtain $p(s|\mathbf{r}_3)$ and then showing that $p(s|\mathbf{r}_3) = p(s|\mathbf{r}_1, \mathbf{r}_2)$. Combining

Eqs. (12–14) above indicates that

$$
\begin{aligned}
p(\mathbf{r}_3|s) &= \int \frac{\phi_{12}(\mathbf{r}_1,\mathbf{r}_2)}{\eta_{12}(s)} \exp(\mathbf{h}_1^{\mathrm{T}}(s)\mathbf{r}_1 \\
&\quad + \mathbf{h}_2^{\mathrm{T}}(s)\mathbf{r}_2)\delta(\mathbf{r}_3 - \mathbf{A}_1^{\mathrm{T}}\mathbf{r}_1 - \mathbf{A}_2^{\mathrm{T}}\mathbf{r}_2)d\mathbf{r}_1 d\mathbf{r}_2 \\
&= \int \frac{\phi_{12}(\mathbf{r}_1,\mathbf{r}_2)}{\eta_{12}(s)} \exp(\mathbf{b}^{\mathrm{T}}(s)\mathbf{A}_1^{\mathrm{T}}\mathbf{r}_1 \\
&\quad - \mathbf{b}^{\mathrm{T}}(s)\mathbf{A}_2^{\mathrm{T}}\mathbf{r}_2)\delta\big(\mathbf{r}_3 - \mathbf{A}_1^{\mathrm{T}}\mathbf{r}_1 - \mathbf{A}_2^{\mathrm{T}}\mathbf{r}_2\big)d\mathbf{r}_1 d\mathbf{r}_2 \\
&= \frac{\exp(\mathbf{b}^{\mathrm{T}}(s)\mathbf{r}_3)}{\eta_{12}(s)} \int \phi_{12}(\mathbf{r}_1,\mathbf{r}_2) \\
&\quad \delta(\mathbf{r}_3 - \mathbf{A}_1^{\mathrm{T}}\mathbf{r}_1 - \mathbf{A}_2^{\mathrm{T}}\mathbf{r}_2)d\mathbf{r}_1 d\mathbf{r}_2 \\
&= \frac{\phi_3(\mathbf{r}_3)}{\eta_{12}(s)} \exp(\mathbf{b}^{\mathrm{T}}(s)\mathbf{r}_3) \qquad (15)
\end{aligned}
$$

where $\phi_3(\mathbf{r}_3)$ is a new measure function that is independent of s. This demonstrates that $\mathbf{r}_3$ is also a member of this family of distributions with a stimulus-dependent kernel drawn from the common basis associated with $\mathbf{h}_1(s)$ and $\mathbf{h}_2(s)$. As in the independent Poisson case, the Bayesian decoder applied to a member of this family of distributions takes the form

$$
p(s|\mathbf{r}_1,\mathbf{r}_2) \propto \frac{\exp(\mathbf{h}_1^{\mathrm{T}}(s)\mathbf{r}_1 + \mathbf{h}_2^{\mathrm{T}}(s)\mathbf{r}_2)}{\eta_{12}(s)} \qquad (16)
$$

and we may conclude that optimal cue combination has been performed by this linear operation, since

$$
\begin{aligned}
p(s|\mathbf{r}_3) &\propto \exp\frac{(\mathbf{b}^{\mathrm{T}}(s)\mathbf{r}_3)}{\eta_{12}(s)} \\
&\propto \exp\frac{(\mathbf{b}^{\mathrm{T}}(s)\mathbf{A}_1^{\mathrm{T}}\mathbf{r}_1 + \mathbf{b}^{\mathrm{T}}(s)\mathbf{A}_2^{\mathrm{T}}\mathbf{r}_2)}{\eta_{12}(s)} \\
&\propto \exp\frac{(\mathbf{h}_1^{\mathrm{T}}(s)\mathbf{r}_1 + \mathbf{h}_2^{\mathrm{T}}(s)\mathbf{r}_2)}{\eta_{12}(s)} \\
&\propto p(s|\mathbf{r}_1,\mathbf{r}_2) \qquad (17)
\end{aligned}
$$

Note that the measure function $\phi_{12}(\mathbf{r}_1,\mathbf{r}_2)$, which was defined in Eq. (13), is completely arbitrary so long as it does not depend on the stimulus.

Nuisance parameters and gain

In the calculation above, we assumed that the likelihood function, $p(\mathbf{r}|s)$, is a function *only* of the stimulus. In fact, the likelihood often depends on what are commonly called *nuisance parameters*. These are quantities that affect the response distributions of the individual neural populations, but that the brain would like to ignore when performing inference. For example, it is well-known that contrast and attention strongly affect the gain and information content of a population, as was the case in the independent Poisson example above. For members of the exponential family of distributions, this direct relationship between gain and information content is, in fact, expected. Recalling that the posterior distribution takes the form

$$
p(s|\mathbf{r}) \propto \exp(\mathbf{h}^{\mathrm{T}}(s)\mathbf{r}) \qquad (18)
$$

it is easy to see that a large amplitude population pattern of activity is associated with a significantly sharper distribution than a low amplitude pattern of activity (see Fig. 1). From this we can conclude that amplitude or gain is an example of a nuisance parameter of particular interest as it is directly related to the variance of the posterior distribution.

We can model this gain dependence by writing the likelihood function for populations 1 and 2 as $p(\mathbf{r}_1,\mathbf{r}_2|s,g_1,g_2)$ where g_k is the gain parameter for population k ($k = 1,2$). Although we could apply this Bayesian formalism and treat g_1 and g_2 as part of the stimulus, if we did that the likelihood for $\mathbf{r}_3$ would contain the term $\exp(\mathbf{b}^{\mathrm{T}}(s,g_1,g_2)\ \mathbf{r}_3)$ [see Eq. (16)]. This is clearly inconvenient, as it means we would have to either know g_1 and g_2, or integrate these quantities out of the posterior distribution to extract the a posterior distribution for the stimulus alone, i.e., we would have to find a neural operation which effectively performs the integral

$$
p(s|\mathbf{r}) = \int p(s,g|\mathbf{r})dg \qquad (19)
$$

Fortunately, it is easy to show that this problem can be avoided if the nuisance parameter does not appear in the stimulus-dependent kernel, i.e., when the likelihood for a given population takes the form

$$
p(\mathbf{r}|s,g) = \phi(\mathbf{r},g)\exp(\mathbf{h}^{\mathrm{T}}(s)\mathbf{r}) \qquad (20)
$$

When this is the case, the posterior distribution is given by

$$p(s|\mathbf{r}, g) \propto \exp(\mathbf{h}^{\mathrm{T}}(s)\mathbf{r}) \tag{21}$$

and the value of g does not affect posterior distribution over s and thus does not affect the optimal combination operation. If $\mathbf{h}(s)$ were a function of g, this would not necessarily be true. Note that the normalization factor, $\eta(s, g)$, from Eq. (16) is not present in Eq. (20). This is because the posterior is only unaffected by g when $\eta(s, g)$ factorizes into a term that is dependent only on s and a term that is dependent only on g and this occurs only when $\eta(s, g)$ is independent of s. Fortunately, this seemingly strict condition is satisfied in many biologically relevant scenarios, and seems to be intricately related to the very notion of a tuning curve. Specifically, when silence is uninformative ($\mathbf{r} = 0$ gives no information about the stimulus), it can be shown that the normalization factor, $\eta(s, g)$, is dependent only on g

$$\begin{aligned} p(s) &= p(s|\mathbf{r} = \mathbf{0}, g) \\ &= \frac{\phi(\mathbf{0}, g)p(s)}{\eta(s, g)} \left(\int \frac{\phi(\mathbf{0}, g)p(s')}{\eta(s', g)} ds' \right)^{-1} \\ &= \frac{p(s)}{\eta(s, g)} \left(\int \frac{p(s')}{\eta(s', g)} ds' \right)^{-1} \end{aligned} \tag{22}$$

Since the second term in the product on the right hand side is a function only of g, equality holds only when $\eta(s, g)$ is independent of s.

Relationship between the tuning curves, the covariance matrix and the stimulus-dependent kernel h(s)

At this point we have demonstrated that the family of likelihood function described above is compatible with the identification of linear operations on neural activity with the optimal cue combination of associated posterior distribution. What remains unclear is whether or not this family of likelihood functions is capable of describing the statistics of neural populations. In this section, we show that this family of distribution is applicable to a very wide range of tuning curves and covariance matrices, i.e., members of this family of distributions can model the s-dependence of the first and second order statistics of any population. We will then show that when the shape of the tuning curve is gain invariant, we expect to observe that the covariance matrix should also be proportional to gain. This is an important result, as it is a widely observed property of the statistics of cortical neurons.

We begin by showing that the tuning curve and covariance matrix of a population pattern of interest are related to the stimulus-dependent kernel by a simple relationship obtained via consideration of the derivative of the mean of the population pattern of activity, $\mathbf{f}(s, g)$, with respect to the stimulus as follows:

$$\begin{aligned} \mathbf{f}'(s, g) &= \frac{d}{ds} \int \mathbf{r}\phi(\mathbf{r}, g) \exp(\mathbf{h}^{\mathrm{T}}(s)\mathbf{r}) d\mathbf{r} \\ &= \int \mathbf{r}\mathbf{r}^{\mathrm{T}}\mathbf{h}'(s)\phi(\mathbf{r}, g) \exp(\mathbf{h}^{\mathrm{T}}(s)\mathbf{r}) d\mathbf{r} \\ &= \int \mathbf{r}\mathbf{r}^{\mathrm{T}}\mathbf{h}'(s)p(\mathbf{r}|s, g) d\mathbf{r} \\ &= \langle \mathbf{r}\mathbf{r}^{\mathrm{T}} \rangle_{s,g}\mathbf{h}'(s) - \mathbf{f}(s, g)\mathbf{f}^{\mathrm{T}}(s, g)\mathbf{h}'(s) \\ &= \mathbf{\Gamma}(s, g)\mathbf{h}'(s) \end{aligned} \tag{23}$$

Here $\langle\rangle_{s,g}$ is an expected value conditioned on s and g, $\mathbf{\Gamma}(s,g)$ the covariance matrix and we have used the fact that $\mathbf{f}^{\mathrm{T}}(s, g)\mathbf{h}'(s) = 0$ for all distributions given by Eq. (20), which follows from the assumption that silence is uninformative. Next, we rewrite Eq. (23) as

$$\mathbf{h}'(s) = \mathbf{\Gamma}^{-1}(s, g)\mathbf{f}'(s, g) \tag{24}$$

and observe that, in the absence of nuisance parameters, a stimulus-dependent kernel can be found for any tuning curve and any covariance matrix, regardless of their stimulus-dependence. Thus this family of distributions is as general as the Gaussian family in terms of its ability to model the first and second order statistics of any experimentally observed population pattern of activity. However, when nuisance parameters, such as gain, are present the requirement that the stimulus-dependent kernel, $\mathbf{h}(s)$, be independent of these parameters restricts the set of tuning curves and covariance matrices that are compatible with this family of distributions. For example, when the tuning curve shape is gain invariant (i.e., $\mathbf{f}'(s,g) = g\bar{\mathbf{f}}'(s)$ where $\bar{\mathbf{f}}(s)$ is independent of gain), $\mathbf{h}$(s) is independent of the gain if the covariance matrix is proportional to the gain. Since variance and mean are both proportional to the gain, their ratio, known as the Fano factor, is also constant. Thus we conclude that constant Fano factors are associated with neurons that implement a linear PPC using tuning curves which have gain invariant shape. Hereafter, likelihood functions with these properties will be referred to as "Poisson-like" likelihoods.

Constraint on the posterior distribution over s

In addition to being compatible with a wide variety of tuning curves and covariance matrices, Poisson-like likelihoods can be used to represent many types of posterior distributions, including non-Gaussian ones. Specifically, as in the independent Poisson case, when the prior $p(s)$ is flat, application of Bayes rule yields

$$p(s|\mathbf{r}) \propto \exp(\mathbf{h}^{\mathrm{T}}(s)\mathbf{r}) \tag{25}$$

Thus, the log of the posterior is a linear combination of the functions that make up the vector $\mathbf{h}(s)$, and we may conclude any posterior distribution may be well approximated when this set of functions is "sufficiently rich." Of course, it is also possible to restrict the set of posterior distributions by an appropriate choice for $\mathbf{h}(s)$. For instance, if a Gaussian posterior is required, we can simply restrict the basis of $\mathbf{h}(s)$ to the set quadratic functions of s.

An example: combining population codes

To illustrate this point, in Fig. 3 we show a simulation in which there are three input layers in which the tuning curves are Gaussian sigmoidal with a positive slope, and sigmoidal with a negative slope (Fig. 3a). The parameters of the individual tuning curves, such as the widths, slopes, amplitude, and baseline activity, are randomly selected. Associated with each population is a stimulus-dependent kernel, $\mathbf{h}_k(s$k $= 1, 2, 3$. Since the set of Gaussian functions of s form a basis, $\mathbf{b}(s)$, each of these stimulus-dependent kernels can be represented as a linear combination of these functions, i.e., $\mathbf{h}_k(s) = \mathbf{A}_k\mathbf{b}(s)$. Thus the linear combination of the input activities, $\mathbf{A}_1^{\mathrm{T}}\mathbf{r}_1 + \mathbf{A}_2^{\mathrm{T}}\mathbf{r}_2 + \mathbf{A}_3^{\mathrm{T}}\mathbf{r}_3$, corresponds to the product of the three posterior distributions. To ensure that all responses are positive, a baseline is removed, and the output population pattern of activity is given by

$$\mathbf{r}_4 = \mathbf{A}_1^{\mathrm{T}}\mathbf{r}_1 + \mathbf{A}_2^{\mathrm{T}}\mathbf{r}_2 + \mathbf{A}_3^{\mathrm{T}}\mathbf{r}_3 - \min(\mathbf{A}_1^{\mathrm{T}}\mathbf{r}_1 + \mathbf{A}_2^{\mathrm{T}}\mathbf{r}_2 + \mathbf{A}_3^{\mathrm{T}}\mathbf{r}_3) \tag{26}$$

Note that the choice of a Gaussian basis, $\mathbf{b}(s)$, yields more or less Gaussian-shaped tuning curves (Fig. 3c) in the output population and that the resulting population pattern of activity is highly correlated. Additionally, for this basis, the removal of the baseline can be shown to have no affect of the resulting posterior, regardless of how that baseline is chosen.

Figure 3b–d shows the network activity patterns and corresponding probability distributions on a given trial. As can be seen in Fig. 3d, the probability distribution encoded by the output layer is identical to the distribution obtained from multiplying the input distributions.

Discussion

We have shown that when the neuronal variability is Poisson-like, i.e., it belongs to the exponential family with linear sufficient statistics, Bayesian inference such as the one required for optimal cue combination reduces to simple linear combinations of neural activities.

In the case in which probability distributions are all Gaussian, reducing Bayesian inference to linear combination may not appear to be so remarkable, since Bayesian inferences are linear in this case. Indeed, as can be seen in Eq. (1), the visual–auditory estimate is obtained through a linear combination of the visual estimate and auditory estimate. However, in this equation, the weights of the linear combination are related to the variance of each estimate, in such a way that the combination favors the cue with the smallest variance, i.e., the most reliable cue. This is problematic, because it implies that the weights must be adjusted every time the reliability of the cues

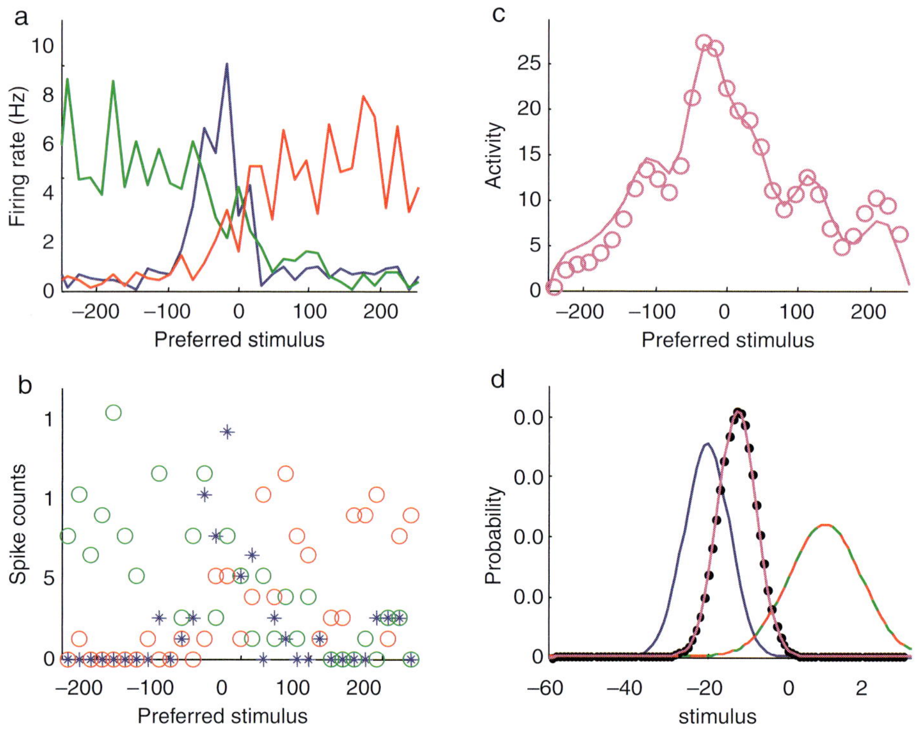

Fig. 3. Inference with non-translation invariant Gaussian and sigmoidal tuning curves. (a) Mean activity in the three input layers when $s = 0$. Blue curves: input layer with Gaussian tuning curves. Red curves: input layers with sigmoidal tuning curves with positive slopes. Green curves: input layers with sigmoidal tuning curves with negative slopes. The noise in the curves is due to variability in the baseline, widths, slopes, and amplitudes of the tuning curves, and to the fact that the tuning curves are not equally spaced along the stimulus axis. (b) Activity in the three input layers on a given trial. These activities were sampled from Poisson distributions with means as in a. Color legend as in a. (c) Solid lines: mean activity in the output layer. Circles: output activity on a given trial, obtained by a linear combination of the input activities shown in b. (d) Blue curves: probability distribution encoded by the blue stars in b (input layer with Gaussian tuning curves). Red-green curve: probability distribution encoded by the red and green circles in b (the two input layers with sigmoidal tuning curves). Magenta curve: probability distribution encoded by the activity shown in c (magenta circles). Black dots: probability distribution obtained with Bayes rule (i.e., the product of the blue and red-green curves appropriately normalized). The fact that the black dots are perfectly lined up with the magenta curve demonstrates that the output activity shown in **c** encodes the probability distribution expected from Bayes rule.

changes. By contrast, with the approach described in this chapter, there is no need for such weight adjustments. If the noise is Poisson-like, a linear combination with fixed coefficients works across any range of cue reliability. This is the main advantage of our approach. Moreover, it explains how humans remain optimal even when the reliability of the cue changes from trial to trial, without having to invoke any trial-by-trial adjustment of synaptic weights.

Another appealing feature of our theory is that it suggests an explanation for why all cortical neurons exhibit Poisson-like noise. In early stages of sensory processing, the statistics of spike trains are not necessarily Poisson-like, and differ across sensory systems. In the subcortical stages of the auditory system, spike timing is very precise and spike trains are oscillatory, reflecting the oscillatory nature of sound waves. By contrast, in the LGN (the thalamic relay of the visual system), the spike trains in response to static stimuli are neither very precise nor oscillatory. Performing optimal multisensory integration with such differences in spike statistics is a difficult problem. If our theory is correct, the cortex solves the problem by first reformatting all spike trains into the Poisson-like family, so as to reduce optimal integration to simple sums of spikes. This transformation is particularly apparent in the auditory system. In the auditory cortex of awake animals, the response of most neurons show Poisson-like statistics, in sharp contrast with the oscillatory spike train seen in early subcortical stages.

The idea that the cortex uses a format that reduced Bayesian inference to linear combination is certainly appealing, but one should not forget that many aspects of neural computation are highly nonlinear. In its present form, our theory does not require those nonlinearities. However, we have only considered one type of Bayesian inference, namely, products of distributions. There are other types of Bayesian inference, such as marginalization, that are just as important. In fact, marginalization is needed to perform the optimal nonlinear computations which are needed for most sensorimotor transformations. We suspect that nonlinearities will be needed to implement marginalization optimally in neural hardware when the noise is Poisson-like, and may also be necessary to implement optimal cue combination when noise is not Poisson-like. We intend to investigate these and related issues in future studies. It is also important to note that, in its most general formulation, a PPC does not necessarily assign equality of the operations of addition of neural responses to optimal cue combination of related posteriors. Rather this is just a particular example of a PPC which seems to be compatible with neural statistics.

References

Anderson, C. (1994) Computational Intelligence Imitating Life. IEEE Press, New York, pp. 213–222.

Deneve, S. (2005) Neural Information Processing System. MIT Press, Cambridge, MA.

Ernst, M.O. and Banks, M.S. (2002) Humans integrate visual and haptic information in a statistically optimal fashion. Nature, 415: 429–433.

Foldiak, P. (1993) In: Eeckman F. and Bower J. (Eds.), Computation and Neural Systems. Kluwer Academic Publishers, Norwell, MA, pp. 55–60.

Hillis, J.M., Watt, S.J., Landy, M.S. and Banks, M.S. (2004) Slant from texture and disparity cues: optimal cue combination. J. Vis., 4(12): 967–992.

Knill, D.C. and Richards, W. (1996) Perception as Bayesian Inference. Cambridge University Press, New York.

Kording, K.P. and Wolpert, D.M. (2004) Bayesian integration in sensorimotor learning. Nature, 427: 244–247.

Landy, M.S. and Kojima, H. (2001) Ideal cue combination for localizing texture-defined edges. J. Opt. Soc. Am. A., 18(9): 2307–2320.

Navalpakkam, V. and Itti, L. (2005) Optimal cue selection strategy. In: Neural Information Processing System. MIT Press, Cambridge, MA.

Rao, R.P. (2004) Bayesian computation in recurrent neural circuits. Neural Comput., 16: 1–38.

Sanger, T. (1996) Probability density estimation for the interpretation of neural population codes. J. Neurophysiol., 76: 2790–2793.

Saunders, J.A. and Knill, D. (2003) Perception of 3D surface orientation from skew symmetry. Vision Res., 41(24): 3163–3183.

Snippe, H.P. (1996) Parameter extraction from population codes: a critical assessment. Neural Comput., 8: 511–529.

Zemel, R., Dayan, P. and Pouget, A. (1998) Probabilistic interpretation of population code. Neural Comput., 10: 403–430.

P. Cisek, T. Drew & J.F. Kalaska (Eds.)
Progress in Brain Research, Vol. 165
ISSN 0079-6123
Copyright © 2007 Elsevier B.V. All rights reserved

CHAPTER 33

On the challenge of learning complex functions

Yoshua Bengio*

Department IRO, Université de Montréal, P.O. Box 6128, Downtown Branch, Montreal, QC, H3C 3J7, Canada

Abstract: A common goal of computational neuroscience and of artificial intelligence research based on statistical learning algorithms is the discovery and understanding of computational principles that could explain what we consider adaptive intelligence, in animals as well as in machines. This chapter focuses on what is required for the learning of complex behaviors. We believe it involves the learning of highly varying functions, in a mathematical sense. We bring forward two types of arguments which convey the message that many currently popular machine learning approaches to learning flexible functions have fundamental limitations that render them inappropriate for learning highly varying functions. The first issue concerns the representation of such functions with what we call shallow model architectures. We discuss limitations of *shallow architectures*, such as so-called kernel machines, boosting algorithms, and one-hidden-layer artificial neural networks. The second issue is more focused and concerns kernel machines with a *local kernel* (the type used most often in practice) that act like a collection of template-matching units. We present mathematical results on such computational architectures showing that they have a limitation similar to those already proved for older non-parametric methods, and connected to the so-called curse of dimensionality. Though it has long been believed that efficient learning in deep architectures is difficult, recently proposed computational principles for learning in deep architectures may offer a breakthrough.

Keywords: theory of learning algorithms; artificial intelligence; template matching; deep neural networks; kernel machines; deep belief networks; multi-layer neural networks; learning abstractions

Introduction

Much research in artificial intelligence (AI) and in computational neuroscience has focused on *how to perform a "function"*.[1] This is tedious work that is being done in both AI and computational neuroscience because of the very large number of tasks, sub-tasks, and concepts that need to be considered to explain the rich array of behaviors that are observed or desired. Research in learning algorithms started from the premise that much more robust and adaptive behaviors would result if we focused on how to *learn* "function", i.e., the development of procedures that apply more general-purpose knowledge (how to learn to perform a task) to the specific examples encountered by the machine or the animal in order to yield behaviors tuned to the specifics of the environment. However, as argued here, we believe that something important has been missing from most accounts of how brains or machines learn. Expressing complex behaviors requires highly varying mathematical functions, or high-level abstractions, i.e., mathematical functions that are highly non-linear, in

*Corresponding author. Tel.: +1 514-343-6804;
Fax: +1 514-343-5834; E-mail: bengioy@iro.umontreal.ca

[1]For example, object recognition in vision, motor control for grasping, etc.: we quote "function" to talk about its biological sense, and otherwise use the word in its mathematical sense.

DOI: 10.1016/S0079-6123(06)65033-4

complex ways, in terms of the raw sensory inputs of the machine or animal. Consider, e.g., the task of interpreting an input image such as the one in Fig. 1. A high-level abstraction such as the ones illustrated in that figure has the property that it corresponds to a very large set of possible raw inputs, which may be very different from each other from the point of view of simple Euclidean distance in pixel space. Hence that set (e.g., the set of images to which the label "man" could be attributed) forms a highly convoluted region in pixel space, which we call a *manifold.*

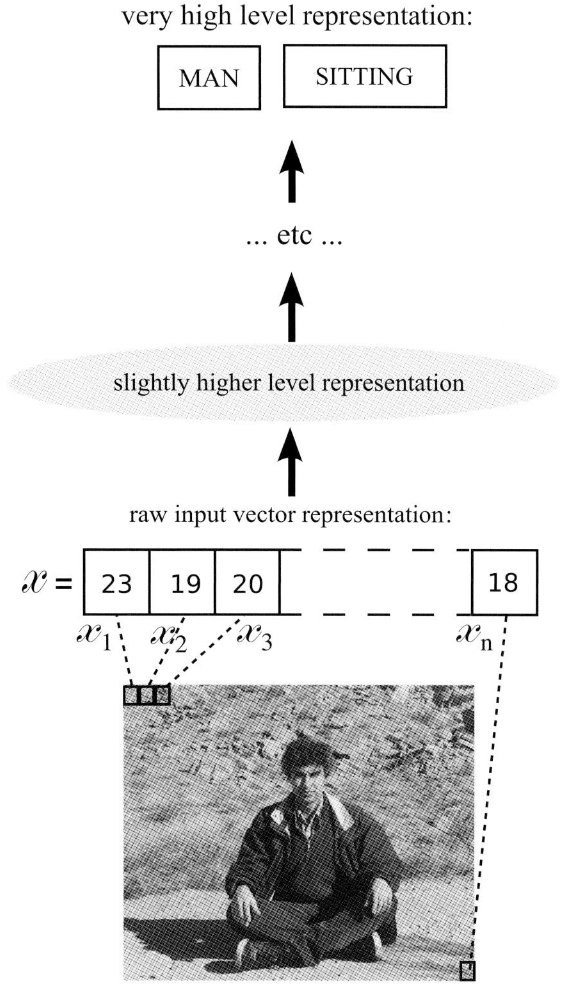

Fig. 1. The raw input image is transformed into gradually higher levels of representation, representing more and more abstract functions of the raw input. In practice, we do not know ahead of time what the "right" representation (in the AI or in the biological senses) should be for all these levels of abstractions, although linguistic concepts sometimes help us to imagine what the higher levels might implicitly represent. We need learning algorithms that can discover most of these abstractions, from mostly unlabeled data.

The raw input to the learning system is a high-dimensional entity, made of many observed variables, related by unknown intricate statistical relationships. Over the years, research in the field of statistical machine learning has revealed fundamental principles for learning predictive models from data, in addition to a plethora of useful machine learning algorithms (Duda et al., 2001; Hastie et al., 2001; Bishop, 2006). The last quarter century has given us a set of flexible (i.e., non-parametric) statistical learning algorithms that can, at least in principle, learn any continuous input–output mapping, if provided with enough computing resources and training data. In practice, traditional machine learning and statistics research has focused on relatively simple problems (in comparison to full-fledged AI). In these simple cases, the data distribution was explicitly *or implicitly* assumed to have a rather simple form: it was either assumed known ahead of time up to a few parameters, or smooth, or to involve only a few interactions between variables.

One long-term goal of machine learning research is to produce methods that are applicable to highly complex tasks, such a perception, reasoning, intelligent control, and other artificially intelligent behaviors. However, despite impressive progress on both the academic and technological sides, these long-term goals remain elusive. What makes these learning tasks challenging and difficult for the computer and presumably for animals as well is that not enough is known ahead of time about the generating distribution (or process) from which the data come. There are currently two major modes of performing statistical machine learning: one using rich explicit prior knowledge about the generating process (such as using so-called probabilistic graphical models (Jordan, 1998), Bayes nets, etc.); and another in which prior knowledge is implicit in the choice of a metric or similarity function that compares examples (e.g., kernel-based models — see www.kernel-machines.org — and other non-parametric models such as artificial neural networks (Rumelhart et al., 1986) and boosting (Freund and Schapire, 1997)). Non-parametric models also implicitly assume a notion of smoothness of the function to be learned, i.e., if x is close to y then we favor a function f such that $f(x)$ is close to $f(y)$. These principles work quite well and have been employed in sophisticated ways in the last few years, but they also have limits. This chapter aims at studying them in order to orient future research towards learning algorithms that can plausibly learn the kind of intelligent behaviors that animals and humans can learn.

An example of a complex task in which the smoothness prior is insufficient is the recognition of multiple objects in an image, when each of the objects in the image can come in many variations of shape, geometry (location, scale, angle), and when backgrounds can also vary. This task involves a number of factors that can co-vary, creating a combinatorial explosion of possible variations. This setting involves highly varying functions in the sense that if we were to parameterize all the instances of a particular object, we would find that many pixel values would oscillate between the same colors, e.g., as we translated the object from left to right across the scene. This setting has been studied in Bengio and Le Cun (2007), which draws many comparisons between different learning algorithms with deep vs. shallow architectures, and finds that shallow architectures are fundamentally limited in their capacity to learn efficient representations of this sort of function.

Most of the current non-parametric statistical learning techniques, studied by AI researchers and computational neuroscientists, rely almost exclusively, implicitly, or explicitly, on the smoothness prior. We discuss two fundamental limitations of these computational architectures. The first limitation, discussed in "The problem with shallow architectures", is more general and concerns the representation of a function with an architecture of one or two levels of adaptive elements (elements which one can think of as neurons, or groups of neurons, and that we sometimes call units, here). We call such architectures *shallow*. We give several examples suggesting that such architectures can be very inefficient, in the sense that many more units are required to represent some functions than in a *deep architecture*. An example of a deep architecture is that of a multi-layer neural network with many layers (e.g., 5 to 10 layers). The second limitation is more specific, and concerns architectures of *kernel machines*. Most learning algorithms for kernel machines compute linear combinations of the outputs of template matchers, units that produce a large output (activation) when the raw input matches a specific template associated with the unit, and a small output otherwise. The mathematical form of the Gaussian kernel permits the theoretical analysis summarized in "The problem with template matching and local kernels", which gives us insight into the limitations of more general forms of template matcher, which we call local kernels. This second limitation might also plague feed-forward artificial neural networks with one-hidden-layer, which become kernel machines in the mathematical limit as the number of neurons becomes very large (Bengio et al., 2006b).

These limitations both lead us to the same conclusion: in order to learn and represent highly varying functions (such as those we believe are required in the computations involved with complex behaviors) with a shallow architecture, one would need a very large number of units. This number may even grow exponentially with the number of factors influencing the variations that can occur in the sensory inputs. With a shallow architecture, the number of examples required to learn a task to a given degree of accuracy would also be very large (in contradiction with what is observed with animals and humans) with the consequence that it would be impractical to implement an AI that can learn truly complex behaviors.

One way around the limitations of shallow or local architectures is to embed a lot of prior knowledge in the architecture. For example, if the hard-wired form of the first layer of units in such an architecture already computes an appropriate representation for the task at hand, then the number of units required may remain small: in the right space, any learning problem becomes easy. However, such hard-wired non-linear transformations would have to be *designed* by human engineers in the case of machines, and evolved in the case of biological brains. It would seem much more efficient, for human designers, as well as for evolution, to take advantage of broad priors about a large set of tasks, such as those that humans solve easily. Therefore, we set our goals towards learning algorithms that do not require very detailed priors on each task, but yet, can learn the kind of complex functions required for intelligent behavior.

In "What is needed", we put the requirement for deep architectures in a broader context: we present a number of computational principles which we believe should be present, either in an attempt to explain biological learning of complex behaviors or in an attempt to achieve AI through machine learning. These include: a deep architecture, on-line learning (not having to store all examples and return to them many times), semi-supervised learning (dealing with mostly unlabeled examples), multi-task learning (capitalizing on the common processing involved in a large number of tasks), reinforcement learning (learning using reinforcement signals rather than supervised signals, which may be delayed in time), and probably active learning (choosing actions to acquire more information about the data-generating process) as well.

The problem with shallow architectures

Here we consider a limitation of the general class of architectures that are *shallow*, i.e., have one or two levels of adaptive elements. In addition to kernel machines, this includes ordinary feed-forward artificial neural networks with one-hidden-layer.

Any specific function can be implemented by a suitably designed shallow architecture or by a deep architecture. Furthermore, when parameterizing a family of functions, we have the choice between shallow or deep architectures. The important questions are: (1) how large is the corresponding architecture (with how many parameters, how many neurons, how much computation to compute the output); (2) how much manual labor/evolutionary time is required in specializing the architecture to the task.

Using a number of examples, we shall demonstrate that deep architectures are often more efficient (more compact) for representing many functions. Let us first consider the task of adding two N-bit binary numbers. The most natural digital circuit for doing so involves adding the bits pair by pair and propagating the carry. The carry propagation takes a number of steps and circuit elements proportional to N. Hence a natural architecture for binary addition is a deep one, with N layers and on the order of N elements in total. It is also possible to implement it efficiently with a less deep circuit, with log N layers, with less than N elements per layer, for a total of about N log N computations and elements. On the other hand, a shallow architecture with two layers can implement any Boolean formula expressed in disjunctive normal form (DNF), by computing the min-terms (applying AND functions on the inputs) in the first layer, and applying an OR on the second layer. Unfortunately, even for primitive Boolean operations such as binary addition and multiplication, the number of terms in the intermediate layer can be extremely large (up to 2^N for N-bit inputs in the worst case). In fact, most functions in the set of all possible Boolean functions of N bits require an exponential number of min-terms (i.e., an exponentially large number of basis functions) (Bengio and Le Cun, 2007). Only an exponentially small fraction of possible Boolean functions require a less than exponential number of min-terms. The computer industry has devoted a considerable amount of effort to optimize the implementation of two-level Boolean functions with an exponential number of terms, but the largest it can put on a single chip has only about 32 input bits (a 4 Gbit RAM chip). This is why practical digital circuits, e.g., for adding or multiplying two numbers are built with deep circuits, i.e., with multiple layers of logic gates: their two-layer implementation would be prohibitively expensive. There are many theoretical results from circuits complexity analysis which clearly indicate that circuits with a small number of layers can be extremely inefficient at representing functions that can otherwise be represented compactly with a deep circuit (Hastad, 1987; Allender, 1996). See Utgoff and Stracuzzi (2002), Bengio and Le Cun (2007) for discussions of this question in the context of learning architectures.

Another interesting example is the Boolean parity function. The N-bit Boolean parity function can be implemented in at least these three ways:

1. with N daisy-chained XOR gates (an N-layer architecture or a recurrent circuit with one XOR gate and N time steps);
2. with $N-1$ XOR gates arranged in a tree (a $\log_2 N$ layer architecture);
3. a DNF formula (i.e., two layers) with a number of min-terms proportional to 2^N.

In Theorem 2 ("The problem with template matching and local kernels") we state a similar result for learning architectures: an exponential number of terms is required with a Gaussian kernel machine in order to represent the parity function. In many instances, space can be traded for time (or layers) with considerable advantage. In the case of Boolean circuits, a similar theorem shows that an exponential number of elements (logical AND, OR, or NOT gates) are required to implement parity with a two-level circuit (Ajtai, 1983).

Another interesting example in which adding layers is beneficial is the fast Fourier transform algorithm (FFT). Since the discrete Fourier transform is a linear operation, it can be performed by a matrix multiplication with N^2 scalar multiplications, which can all be performed in parallel, followed by on the order of N^2 additions to collect the sums. However the FFT algorithm can reduce the total cost to $1/2N \log_2 N$, multiplications, with the trade-off of requiring $\log_2 N$ sequential steps involving $N/2$ multiplications each. This example shows that, even with linear functions, adding layers allows us to take advantage of the intrinsic regularities in the task.

These examples and the complexity theory of circuits highlight a severe limitation of shallow architectures, i.e., of kernel machines (grandmother cells followed by linear predictors) and one-hidden-layer neural networks. They may need to be exponentially large to represent functions that may otherwise be represented compactly with a deep architecture, e.g., a deep neural network.

The problem with template matching and local kernels

This section focuses more mathematically than the previous one on specific shallow architectures, kernel machines with *local kernels*, corresponding to a layer of template-matching units followed by linear aggregation. It shows that when the function to be learned has many variations (twists and turns), which can come about because of the interaction of many explanatory factors, such architectures may require a very large number of training examples and computational elements. This section can be skipped without losing the main thread of the chapter.

Kernel machines compute a function of their input that has the following structure:

$$f(x) = b + \sum_i \alpha_i K(x, x_i) \qquad (1)$$

where $K(\cdot, \cdot)$ can be understood as a matching function. It is chosen a priori and represents a notion of similarity between two input objects. A typical kernel function is the Gaussian kernel

$$K_\sigma(u, v) = e^{-\frac{1}{\sigma^2}||u-v||^2} \qquad (2)$$

in which the width σ controls the locality of the kernel. In biological terms, this approach corresponds to two levels of processing: first the estimation of similarity of the current sensory pattern x with many previously encountered patterns, and then some form of voting between the matching patterns, in order to come up with a decision or a prediction $f(x)$. When $K(x, x_i)$ is substantially greater than its base output level, this would be like a grandmother cell for the "grandmother image" x_i firing in response to input x. This architecture can be summarized as follows: at the bottom, a matching level and at the top, a linear classification or linear regression level that aggregates all the matches into one prediction or decision. Only the top level is fully tuned to the task (the bottom one is learned in a simple and unsupervised way by copying raw inputs). In an artificial neural network with a single-hidden-layer, trained in a supervised way or by reinforcement learning, there are also two levels but both can be fully tuned to the task. Both are shallow architectures. A more shallow architecture is the linear classifier or linear regressor, corresponding to a single layer of neurons, as in Rosenblatt's Perceptron (Rosenblatt, 1957). The limits of the Perceptron are well understood

(Minsky and Papert, 1969). In this chapter, we emphasize the less obvious but nonetheless serious limits of shallow architectures such as a kernel machine. One reason why these limitations may have been overlooked is that unlike the Perceptron, such architectures are universal approximators; with enough training data, they can approximate any continuous function arbitrarily closely. However, the number of required elements (i.e., the number of training examples, or grandmother cells) could be very large. As we show below, in many cases the requirement may be exponential with respect to the size of the input pattern.

To establish the intuition motivating the mathematical results below, consider the apparently simple problem of pattern recognition and classification when the input images are handwritten characters with various backgrounds. One of the fundamental problems in pattern recognition is how to handle intra-class variability. For example, we can picture the set of all the possible '4' images as a continuous manifold in the pixel space. Any pair of '4' images on this manifold can be connected by a path, along which every point corresponds to some image of a '4'. The dimensionality of this manifold at one location corresponds to the number of independent distortions that can be applied to a shape while preserving its category. For handwritten character categories, the manifold has a high dimension: characters can be distorted using affine transforms (six parameters), distorted using an elastic sheet deformation (high dimension), or modified so as to cover the range of possible writing styles (with or without a loop, with tick or thin stroke,...), and the backgrounds can change (high dimension). Even for simple character images, the manifold is highly non-linear, with high curvature. Moreover, manifolds for different categories are closely intertwined. Consider the shape of a capital U and an O at the same location. They have many pixels in common, many more pixels in fact than with a shifted version of the same U. Another insight about the high curvature of these manifolds can be obtained from the example of translating a high-contrast image. The tangent of a manifold is a locally linear approximation of the manifold, appropriate around a particular point of the manifold, as illustrated in Fig. 2. Consider the one-dimensional manifold that is the set of images generated by taking a particular image and allowing it to be rotated or translated left or right by different amounts. Analyzing such a manifold shows that the tangent vector of this manifold (which can be seen as an image itself) changes abruptly as we translate the image only one pixel to the right, indicating high curvature of the manifold. As discussed earlier, many kernel algorithms make an implicit assumption of a locally smooth function around each training example x_i. They approximate the manifold with a locally linear patch around each x_i. For support vector machines or SVMs (Boser et al., 1992; Cortes and Vapnik, 1995; Vapnik, 1998), as discussed in "Smoothness vs. locality", the function is locally linear. Hence a high curvature implies the necessity of a large number of training examples in order to cover all the desired turns with locally constant or locally linear patches. The basic problem is that we have many factors of variation that can be combined in order to give rise to a rich set of possible patterns. With pattern matching architectures such as kernel machines with a local kernel (e.g., the Gaussian), one must cover the space of these variations, with at least one grandmother cell for each combination of values, especially if a change in values can give rise to a change in the desired response. Even more variations than those outlined above could be obtained by simply combining multiple objects in each image. The number of possible variations then grows exponentially with the number of objects and with the number of dimensions of variation per object. An even more dire situation occurs if the background is not uniformly white, but can contain random clutter. To cover all the important combinations of variations the kernel machine will need many different templates containing each motif with a wide variety of different backgrounds.

Minimum number of bases required

The following theorem informs us about the number of sign changes that a Gaussian kernel machine can achieve, when it has k bases (i.e., k support vectors, or at least k training examples).

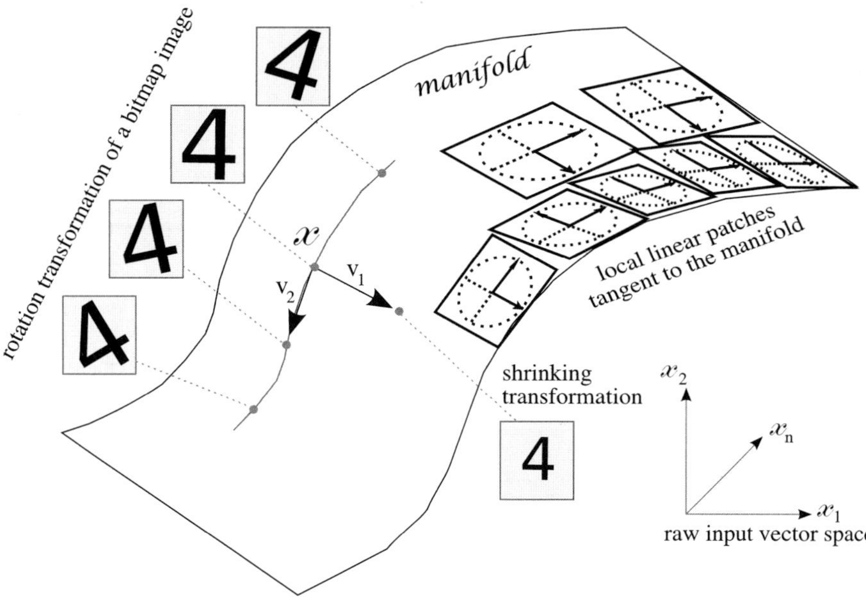

Fig. 2. The set of images associated with the same object class forms a manifold, i.e., a region of lower dimension than the original space of images. By rotating, translating, or shrinking the image we get other images of the same class, i.e., on the same manifold. Since the manifold is locally smooth, it can in principle be approximated locally by linear patches. However, if the manifold is highly curved, many such patches will be needed.

Theorem 1. *Theorem 2 of* Schmitt (2002). *Let* f: $\mathbb{R} \to \mathbb{R}$ *computed by a Gaussian kernel machine [Eq. (1)] with k bases (non-zero* α_i*'s). Then f has at most 2k zeros.*

We would like to say something about kernel machines in $\mathbb{R}^d$, and we can do this simply by considering a straight line in $\mathbb{R}^d$ and the number of sign changes that the solution function f can achieve along that line.

Corollary 1. *Suppose that the learning problem is such that in order to achieve a given error level for samples from a distribution P with a Gaussian kernel machine [Eq. (1)], f must change sign at least 2k times along some straight line (i.e., in the case of a classifier, a good decision surface must be crossed at least 2k times by that straight line). Then the kernel machine must have at least k bases (non-zero* α_i*'s).*

A proof can be found in Bengio et al. (2006a).

Example 1. Consider the decision surface shown in Fig. 3, which is a sinusoidal function. One may take advantage of the global regularity to learn it with few parameters (thus requiring few examples). By contrast, with an affine combination of Gaussians, Corollary 1 implies one would need at least $\lceil m/2 \rceil = 10$ Gaussians. For more complex tasks in higher dimension, the complexity of the decision surface could quickly make learning impractical when using such a local kernel method.

Of course, one only seeks to approximate the decision surface S, and does not necessarily need to learn it perfectly: Corollary 1 says nothing about the existence of an easier-to-learn decision surface approximating S. For instance, in the example of Fig. 3, the dotted line could turn out to be a good enough estimated decision surface if most samples were far from the true decision surface, and this line can be obtained with only two Gaussians.

The above theorem tells us that if we are trying to represent a function that locally varies a lot (in the sense that its sign along a straight line changes many times), then we need many training examples to do so with a Gaussian kernel machine. Note that it says nothing about the dimensionality of the space, but we might expect to have to learn functions that vary more when the data is high-dimensional. The next

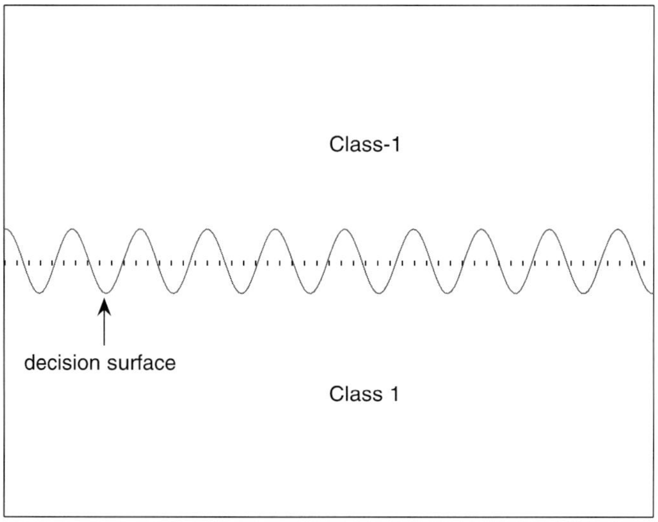

Fig. 3. The dotted line crosses the decision surface 19 times: one thus needs at least 10 Gaussians to learn it with an affine combination of Gaussians with same width.

theorem confirms this suspicion in the case of a highly varying function, the d-bits parity function, which changes value whenever any one of its input bits is flipped:

$$\text{parity}: \quad (b_1, \ldots, b_d) \in \{0,1\}^d \mapsto \begin{cases} 1 \text{ if } \sum_{i=1}^{d} b_i \text{ is even} \\ -1 \text{ otherwise} \end{cases}$$

This function is interesting because it varies a lot as we move around in input space. Learning this apparently simple function with Gaussians centered on any of the possible input bit patterns: it requires a number of Gaussians exponential in d (for a fixed Gaussian width). Note that our earlier Corollary 1 does not apply to this function, so it represents another type of local variation (not along a line). However, it is also possible to prove a strong result about that case.

Theorem 2. *Let $f(x) = b + \sum_{i=1}^{2^d} \alpha_i K_\sigma(x_i, x)$ be an affine combination of Gaussians with same width σ centered on points $x_i \in X_d$. If f solves the parity problem, then there are at least 2^{d-1} non-zero coefficients α_i.*

A proof can be found in Bengio et al. (2006a).

The bound in Theorem 2 is tight, since it is possible to solve the parity problem with exactly 2^{d-1} Gaussians and a bias, for instance by using a negative bias and putting a positive weight on each example satisfying parity $(x_i) = 1$.

One may argue that parity is a simple discrete toy problem of little interest. But even if we have to restrict the analysis to discrete samples in $\{0,1\}^d$ for mathematical reasons, the parity function can be extended to a smooth function on the $[0,1]^d$ hypercube depending only on the continuous sum $b_1 + \ldots + b_d$. Theorem 2 is thus a basis to argue that the number of Gaussians needed to learn a function with many variations in a continuous space may scale linearly with these variations, and thus possibly exponentially in the dimension.

Smoothness vs. locality

Consider a Gaussian SVM and how that estimator changes as one varies the width σ of the Gaussian kernel. For large σ one would expect the estimated function to be very smooth, whereas for small σ one would expect the estimated function to be more local, in the sense discussed earlier: the near neighbors of x have dominating influence in the shape of the predictor at x.

The following proposition tells us what happens when σ is large, or when we consider what happens in a ball of training examples with small radius (compared with σ). It considers the space of possible input examples, and a sphere in that space, that contains all the training examples. The proposition focuses on what happens when the Gaussian width σ becomes large compared to that sphere.

Proposition 1. *For the Gaussian kernel classifier, as σ increases and becomes large compared with the diameter of the data, within the smallest sphere containing the data the decision surface becomes linear if $\sum_i \alpha_i = 0$ (e.g., for SVMs), or else the normal vector of the decision surface becomes a linear combination of two sphere surface normal vectors, with each sphere centered on a weighted average of the examples of the corresponding class.*

A proof can be found in Bengio et al. (2006a).

This proposition shows that when σ becomes large, a kernel classifier becomes non-local (it approaches a linear classifier). However, this non-locality is at the price of constraining the decision surface to be very smooth, making it difficult to model highly varying decision surfaces. This is the essence of the trade-off between smoothness and locality in many similar non-parametric models (including the classical ones such as *k*-nearest-neighbor and Parzen windows algorithms).

Now consider in what senses a Gaussian kernel machine is local (thinking about σ small). Consider a test point x that is near the decision surface. We claim that the orientation of the decision surface is dominated by the Euclidean neighbors x_i of x in the training set, making the predictor *local in its derivative. It means that variations around an input x of the decision surface represented by the kernel machine f are mostly determined by the training examples in the neighborhood of x.* If we consider the coefficients α_i fixed (i.e., ignoring their dependence on the training x_i's), then it is obvious that the prediction $f(x)$ is dominated by the near neighbors x_i of x, since $K(x, x_i) \rightarrow 0$ quickly when $||x-x_i||/\sigma$ becomes large. However, the α_i can be influenced by all the x_j's. The following proposition skirts that issue by looking at the first derivative of f.

Proposition 2. *For the Gaussian kernel classifier, the normal of the tangent of the decision surface at x, with x on the decision surface, is constrained to approximately lie in the span of the vectors $(x-x_i)$ with $||x-x_i||$ not large compared to σ and x_i in the training set.*

See Bengio and Le Cun (2007) for a proof.

The constraint of $\partial f(x)/\partial x$ being in the span of the vectors $x-x_i$ for neighbors x_i of x is not strong if the region of the decision surface where data concentrate (a manifold of possibly lower dimension than the decision surface itself) has low dimensionality. Indeed if that dimensionality is smaller or equal to the number of dominating neighbors, then the kernel machine decision surface is not strongly constrained by the neighboring examples. However, when modeling complex dependencies involving many factors of variation, the region of interest may have high dimension (e.g., consider the effect of variations that have arbitrarily large dimension, such as changes in clutter, background, etc. in images). For such a complex highly varying target function, we also need a local predictor (σ small) in order to accurately represent all the desired variations. With a small σ, the number of dominating neighbors will be small compared to the dimension of the manifold of interest, making this locality in the derivative a strong constraint, and allowing the following curse of dimensionality argument.

This notion of locality in the sense of the derivative allows us to define a ball around each test point x, containing neighbors that have a dominating influence on $\partial f(x)/\partial x$. Smoothness within that ball constrains the decision surface to be approximately either linear (case of SVMs) or a particular quadratic form. Let N be the number of such balls necessary to cover the region Ω where the value of the predictor is desired (e.g., near the target decision surface, in the case of classification problems). Let k be the smallest number such that one needs at least k examples in each ball to reach error level ε. The number of examples thus required is kN. To see that N can be exponential in some dimension, consider the maximum radius r of all these balls and the radius R of Ω. If Ω has intrinsic dimension d, then N could be as large as the number of radius r balls that can tile a d-dimensional manifold of radius R, which is on the order of $(R/r)^d$. This means that *in order to cover the possibly variations in the data that matter to define the decision surface, one may need a number of examples that grows as* $(R/r)^d$.

Learning abstractions one on top of the other

The analyses of the previous sections point to the difficulty of learning *highly-varying* functions, i.e., functions that have a large number of *variations* in the domain of interest. These analyses focus on shallow architectures and learning algorithms that generalize only locally, such as kernel machines with Gaussian kernels, i.e., grandmother cells. This problem also plagues many non-parametric unsupervised learning algorithms, which attempt to cover the manifold near which the data concentrate. Such models can in principle cover the space of variations of the input examples by a large number of locally linear patches. Since the number of locally linear patches can be made to grow exponentially with the number of input variables, this problem is directly connected with the well-known curse of dimensionality for classical non-parametric learning algorithms (for regression, classification, and density estimation). If the shapes of all these linear patches are unrelated, one needs enough examples for each patch in order to generalize properly. However, if these shapes are related and can be predicted from each other, *non-local learning algorithms* have the potential to generalize to regions not covered by the patches arising from the training set. Such an ability would seem necessary for learning in complex domains such as those involved in intelligent behavior.

One way to represent a highly varying function compactly (with few parameters) is through the composition of many non-linearities. Such multiple composition of non-linearities appear to grant non-local properties to the estimator, in the sense that the value of $f(x)$ or $f'(x)$ can be strongly dependent on training examples far from x_i while at the same time allowing $f(\cdot)$ to capture a large number of variations. We have already discussed the example of parity in the previous two sections. Other arguments can be brought to bear to strongly suggest that learning of more abstract functions is much more efficient when it is done sequentially, composing previously learned concepts in order to represent and learn more abstract concepts (Utgoff and Stracuzzi, 2002).

The raw input object, e.g., the vector of gray-scale values of all the pixels in an image constitutes an initial *low-level representation* of the input. This is to be contrasted with *high-level representations* such as a set of symbolic categories for that input object (e.g., in an image: man, sitting...). Many intermediate levels of representation can exist in between these two extremes. Indeed, deep multi-layered systems, with each layer extracting a slightly higher level representation of its input patterns, have long been believed to be the key to building the ultimate intelligent learning systems (Utgoff and Stracuzzi, 2002). Unfortunately, we generally do not know a set of intermediate and high-level concepts which would be appropriate to explain the data. It would therefore be important to have algorithms that can discover such abstractions. However this vision has proved difficult to actualize; learning deep-layered architecture is known to be problematic (Tesauro, 1992) because it is a difficult *optimization* problem. For this reason, many recent successful approaches in machine learning (Schökopf et al., 1999) seem to have given up on the notion of multiple levels of transformations altogether, in favor of analytical simplicity and theoretical guarantees. Recently, however, Hinton et al. (2006) have demonstrated algorithms that suggest that the difficulties of learning with many layers can be overcome.

Deep architectures are perhaps best exemplified by multi-layer neural networks with several hidden layers. In general terms, deep architectures form a *composition* of non-linear modules, each of which can be adapted. Deep architectures rarely appear in the literature. Indeed, the vast majority of neural network research has focused on shallow architectures with a single-hidden-layer, because of the difficulty of training networks with more than two or three layers (Tesauro, 1992). Notable exceptions include work on convolutional networks (LeCun et al., 1989, 1998), and recent work on deep belief networks (Hinton et al., 2006; Bengio et al., 2007) or DBNs. The latter are probabilistic generative models of the data, along with a fast approximation of the posterior (determining what higher level causes or abstractions are likely to be involved in explaining the current input), and a greedy layer-wise training algorithm that works by training one layer at a time in an unsupervised fashion. Each layer is trained to model the distribution of its input, which is the output of the previous layer. Upper layers of a DBN are supposed to represent more abstract concepts that explain the input observation x, whereas lower layers extract low-level features from x. They learn simpler concepts first, and more abstract concepts are learned by composing them. Although DBNs have been introduced only recently, it has already been shown that they can learn efficient high-level representations. They have also been shown to beat state-of-the-art methods by a comfortable margin (Hinton et al., 2006; Bengio and Le Cun, 2007) on MNIST (a well-known benchmark task involving classification of digit images), when no prior knowledge on images is allowed. DBNs are described in more detail in this book, in Hinton's chapter.

However, the idea that learning occurs in stages, with different levels of concepts, dates back at least to Piaget (1952). Humans do not learn very abstract mathematical concepts until the end of adolescence. For example, they start by learning the notion of objects in the world, they learn to count and doing simple mathematical operations on them, and gradually build on these early concepts in order to learn to represent more abstract concepts. This strategy makes a lot of sense from a mathematical point of view: the optimization problem of learning the more abstract concepts directly would appear too difficult (e.g., training a deep neural network gets stuck in poor local minima), whereas sequentially breaking the problem into simpler ones (e.g., learning less abstract concepts first, and gradually more abstract ones on top) is a common type of optimization heuristic. Biological evidence for maturation one stage after the other is less clear (Guillery, 2005), but some observations support a hierarchical sequence of maturations.

What is needed

To design learning algorithms that handle more complex data distributions such as those presumably involved in artificial and natural intelligence, we believe that bold steps are required, not just fine-tuning of existing algorithms. We hypothesize that significant progress can be achieved through a small set of computational and mathematical general-purpose principles, by opposition to a large set of engineered special-purpose tricks. By general-purpose or special-purpose we want to distinguish methods that apply to a large class vs. a tiny class of tasks. Keeping in mind that there exists no *completely universal* statistical learning algorithm (Wolpert, 1996), it suffices that such broadly applicable generalization principles be relevant to the type of learning tasks that we care about, such as those solved by humans and animals.

Another hypothesis on which this chapter has focused, is that we are better off using *deep architectures* than *shallow architectures* in order to learn complex highly varying functions such as those involved in intelligent behaviors. In particular one of our objectives is to investigate deep architectures that are obtained by the principle of composition: more abstract and more non-linear functions are represented as the composition of simpler ones, and this is done at multiple levels. The number of such levels is the depth of the architecture, and corresponds to depth of a circuit if that function is represented as a circuit. To achieve generalization on examples that are truly novel, and to generalize across tasks, it is important that the components get to be re-used in different places and different tasks.

In order to be able to learn highly varying complex functions, we believe that one needs learning algorithms that can cope with very large datasets, that can take advantage of a large number of inter-related tasks, that have a deep architecture obtained by composition of simpler components, and that can learn even when most of the examples are unlabeled, and when the inputs are very high-dimensional.

We put together below a set of requirements which we believe are necessary in learning algorithms for intelligent behaviors.

- *Learning complex abstractions through composition of simpler ones.* Our work already strongly suggests that high-level abstractions require deep architectures, but there remains the question of optimizing these architectures. Learning complex abstractions (highly varying functions) is likely to involve a difficult optimization problem, but a promising strategy is to break this problem into simpler (even possibly convex) sub-problems. Successful examples of this strategy are found in recent work, following the work on DBNs (Hinton et al., 2006), i.e., with greedy layer-wise training of deep networks (Hinton and Salakhutdinov, 2006; Ranzato et al., 2006; Bengio et al., 2007).

- *Unsupervised and semi-supervised learning are key*. To learn complex abstractions requires a lot of data, and tagging it would be prohibitively expensive. Most current state-of-the-art unsupervised and semi-supervised learning algorithms are of the memorizing type (local non-parametric models) and the mathematical results outlined here show that they will suffer from the curse of dimensionality, i.e., they are unlikely to have the ability to learn abstractions that actually have a simple expression but appear complex because they correspond to a great variety of examples or give rise to highly variable functions.
- *The more variables and tasks, the better?* Although a common belief (that has some justifications) is that learning is harder when there are more variables involved (high-dimensional spaces, curse of dimensionality, etc.), we hypothesize that one can take advantage of the presence of many variables, as long as their relations are not arbitrary but relate to shared underlying realities. If we consider predicting one variable from the others to be one of a series of tasks, and we apply the principles of *multi-task learning*, there should be an advantage to working with more random variables, as long as they are meaningfully related. This idea also means that instead of separately tackling each task, we devote a great part of our effort on learning concepts that are relevant to a large number of tasks, e.g., concepts that help to make sense of the world around us.
- *Great quantities of data call for on-line learning*. If a large number of examples are required to learn complex concepts then we should strive to develop learning algorithms whose computational requirements scale linearly with the number of examples. On-line learning algorithms visit each example only once. Other variants are possible, but the overall training time should not scale much worse than linearly in the number of examples.
- *Predictive, reinforcement, and active learning*. Although most data are unlabeled, the task of predicting what comes next can be achieved with supervised learning algorithms, as components in the unsupervised learning task. Because what we are trying to learn is complex, passively observing it for a lifetime may not be sufficient to collect and process enough data. Active learning algorithms (Cohn et al., 1995; Fukumizu, 1996) suggest actions that influence what examples are seen next, i.e., in which direction to explore in order to acquire data that brings more information. They can potentially give rise to exponentially faster learning. In this context, the learning algorithms must consider the optimization of a *sequence of decisions*, as in reinforcement learning.
- *Learning to represent context at multiple levels*. Another challenge for learning algorithms is that many of the statistical dependencies that matter in the performance of intelligent tasks involve events at different times and are greatly influenced by temporal context. This means that learning algorithms must involve models of the dynamics in the data, and that an unobserved state representation must be learned, which necessarily involves *long-term dependencies*, that are unfortunately hard to learn with currently known techniques (Bengio et al., 1994; Bengio and Frasconi, 1995). We believe that one of the keys to achieving this goal is to represent context at different levels of abstraction and different time scales (ElHihi and Bengio, 1996).

Conclusion

Because the brain may be seen as a deep network, computational neuroscience research on the learning mechanisms that involve many layers could serve as very useful inspiration for AI research. Conversely, the algorithmic and mathematical development of ideas in statistical machine learning geared towards training deep networks could also provide hypotheses to inspire computational neuroscience research into learning mechanisms.

The main messages of this chapter are the following:

- Shallow architectures such as those of linear predictors, kernel machines, (grandmother cells and template matching) and single-hidden-layer neural networks are not efficient enough in terms of representation to address the learning of complex functions such as those involved in intelligent behavior. Computational research should pay particular attention to possible learning mechanisms involving many layers of processing together.
- Local estimators such as kernel machines with a local kernel (e.g., the Gaussian kernel, template matching) are similarly limited, because they cannot discover regularities in the data that are both fine-grained (many local variations) but span a large region of data space (globally coherent structures, principles applicable to many different types of possible inputs). Computational neuroscience models that are limited to template matching followed by linear prediction or linear classification are insufficient to explain the richness of human or animal learning.
- Deep architectures (e.g., neural networks with many layers) appear to be the only way to avoid these limitations, and although they were until recently thought to be too difficult to train, new algorithms strongly suggest that they can be trained efficiently using a greedy layer-wise unsupervised strategy.
- Deep architectures and the greedy layer-wise strategy exploits the principle, apparently also exploited by humans (Piaget, 1952), that one can more easily learn high-level abstractions if these are defined by the composition of lower level abstractions, with the property that these lower level abstractions are useful by themselves to describe the data, and can thus be learned before the higher level abstractions are learned. How such a training by stages could occur in brain remains a question. However, our work (Bengio et al., 2007) suggests that *all the levels could be learning simultaneously*, even though lower- levels would presumably converge near their final state earlier.
- To achieve the learning of intelligent behaviors, this multi-level learning of abstractions should be combined with several other characteristics of learning algorithms: ability to exploit unlabeled data (unsupervised and semi-supervised learning), ability to exploit commonalities between a large number of tasks (multi-task learning) and a large number of inputs (multi-modal learning), on-line learning, learning to represent context at multiple levels, active learning, predictive learning, and reinforcement learning. Most of these have been considered separately in the machine learning community, but it is time to start putting them together in one system.

Acknowledgments

The author wants to acknowledge the intellectual influence and contributions to the results and ideas discussed in this chapter, primarily from Yann Le Cun, but also from Geoffrey Hinton, Pascal Lamblin, François Rivest, Olivier Delalleau, Pascal Vincent, Nicolas Le Roux, and Hugo Larochelle. The following funding agencies have also contributed to this work: NSERC, the NCE (MITACS), and the Canada Research Chairs.

References

Ajtai, M. (1983) $\sum_1^1$- formulae on finite structures. Ann. Pure Appl. Logic, 24(1): 48.

Allender, E. (1996) Circuit complexity before the dawn of the new millennium. In: 16th Annual Conference on Foundations of Software Technology and Theoretical Computer Science, Hyderabad, India, pp. 1–18. Lecture Notes in Computer Science 1180.

Bengio, Y., Delalleau, O. and Le Roux, N. (2006a) The curse of highly variable functions for local kernel machines. In: Weiss Y., Schölkopf B. and Platt J. (Eds.), Advances in Neural

Information Processing Systems 18. MIT Press, Cambridge, MA, pp. 107–114.
Bengio, Y. and Frasconi, P. (1995) Diffusion of context and credit information in Markovian models. J. Artif. Intell. Res., 3: 223–244.
Bengio, Y., Lamblin, P., Popovici, D. and Larochelle, H. (2007) Greedy layer-wise training of deep networks. In: Advances in Neural Information Processing Systems 19. MIT Press, Cambridge, MA.
Bengio, Y. and Le Cun, Y. (2007) Scaling learning algorithms towards AI. In: Bottou L., Chapelle O., DeCoste D. and Weston J. (Eds.), Large Scale Kernel Machines. MIT Press, Cambridge, MA.
Bengio, Y., Le Roux, N., Vincent, P., Delalleau, O. and Marcotte, P. (2006b) Convex neural networks. In: Weiss Y., Schölkopf B. and Platt J. (Eds.), Advances in Neural Information Processing Systems 18. MIT Press, Cambridge, MA, pp. 123–130.
Bengio, Y., Simard, P. and Frasconi, P. (1994) Learning long-term dependencies with gradient descent is difficult. IEEE Trans. Neural Netw., 5(2): 157–166.
Bishop, C. (2006) Pattern Recognition and Machine Learning. Springer.
Boser, B., Guyon, I. and Vapnik, V. (1992) A training algorithm for optimal margin classifiers. In: Fifth Annual Workshop on Computational Learning Theory, Pittsburgh, pp. 144–152 .
Cohn, D., Ghahramani, Z. and Jordan, M.I. (1995) Active learning with statistical models. In: Tesauro G., Touretzky D. and Leen T. (Eds.), Advances in Neural Information Processing Systems 7. MIT Press, Cambridge, MA.
Cortes, C. and Vapnik, V. (1995) Support vector networks. Machine Learn., 20: 273–297.
Duda, R., Hart, P. and Stork, D. (2001) Pattern Classification (2nd ed.). Wiley, New York.
ElHihi, S. and Bengio, Y. (1996) Hierarchical recurrent neural networks for long-term dependencies. In: Touretzky D., Mozer M. and Hasselmo M. (Eds.), Advances in Neural Information Processing Systems 8. MIT Press, Cambridge, MA, pp. 493–499.
Freund, Y. and Schapire, R.E. (1997) A decision theoretic generalization of on-line learning and an application to boosting. J. Comput. Syst. Sci., 55(1): 119–139.
Fukumizu, K. (1996) Active learning in multilayer perceptrons. In: Touretzky D., Mozer M. and Hasselmo M. (Eds.), Advances in Neural Information Processing Systems 8. MIT Press, Cambridge, MA.
Guillery, R. (2005) Is postnatal neocortical maturation hierarchical? Trends Neurosci., 28(10): 512–517.
Hastad, J.T. (1987) Computational Limitations for Small Depth Circuits. MIT Press, Cambridge, MA.
Hastie, T., Tibshirani, R. and Friedman, J. (2001) The elements of statistical learning: data mining, inference and prediction. Springer Verlag Springer Series in Statistics.
Hinton, G. and Salakhutdinov, R. (2006) Reducing the dimensionality of data with neural networks. Science, 313(5786): 504–507.
Hinton, G.E., Osindero, S. and Teh, Y. (2006) A fast learning algorithm for deep belief nets. Neural Comput., 18: 1527–1554.
Jordan, M. (1998) Learning in Graphical Models. Kluwer, Dordrecht, Netherlands.
LeCun, Y., Boser, B., Denker, J., Henderson, D., Howard, R., Hubbard, W. and Jackel, L. (1989) Backpropagation applied to handwritten zip code recognition. Neural Comput., 1(4): 541–551.
LeCun, Y., Bottou, L., Bengio, Y. and Haffner, P. (1998) Gradient based learning applied to document recognition. Proc. IEEE, 86(11): 2278–2324.
Minsky, M. and Papert, S. (1969) Perceptrons. MIT Press, Cambridge.
Piaget, J.-P. (1952) The Origins of Intelligence in Children. International Universities Press, New York.
Ranzato, M., Poultney, C., Chopra, S. and LeCun, Y. (2006) Efficient learning of sparse representations with an energy-based model. In: Scholkopf B., Platt J. and Hoffman T. (Eds.), Advances in Neural Information Processing Systems (NIPS 2006). MIT Press, Cambridge, MA.
Rosenblatt, F. (1957) The perceptron: a perceiving and recognizing automaton. Tech. rep. 85-460-1, Cornell Aeronautical Laboratory, Ithaca, NY.
Rumelhart, D., Hinton, G. and Williams, R. (1986) Learning representations by backpropagating errors. Nature, 323: 533–536.
Schmitt, M. (2002) Descartes' rule of signs for radial basis function neural networks. Neural Comput., 14(12): 2997–3011.
Schökopf, B., Burges, C.J.C. and Smola, A.J. (1999) Advances in Kernel Methods. Support Vector Learning. MIT Press, Cambridge, MA.
Tesauro, G. (1992) Practical issues in temporal difference learning. Machine Learn., 8: 257–277.
Utgoff, P. and Stracuzzi, D. (2002) Many-layered learning. Neural Comput., 14: 2497–2539.
Vapnik, V. (1998) Statistical Learning Theory, Vol. 454. Wiley Lecture Notes in Economics and Mathematical Systems.
Wolpert, D. (1996) The lack of a priori distinction between learning algorithms. Neural Comput., 8(7): 1341–1390.

P. Cisek, T. Drew & J.F. Kalaska (Eds.)
Progress in Brain Research, Vol. 165
ISSN 0079-6123
Copyright © 2007 Elsevier B.V. All rights reserved

CHAPTER 34

To recognize shapes, first learn to generate images

Geoffrey E. Hinton*

Department of Computer Science, University of Toronto, 10 Kings College Road, Toronto, M5S 3G4 Canada

Abstract: The uniformity of the cortical architecture and the ability of functions to move to different areas of cortex following early damage strongly suggest that there is a single basic learning algorithm for extracting underlying structure from richly structured, high-dimensional sensory data. There have been many attempts to design such an algorithm, but until recently they all suffered from serious computational weaknesses. This chapter describes several of the proposed algorithms and shows how they can be combined to produce hybrid methods that work efficiently in networks with many layers and millions of adaptive connections.

Keywords: learning algorithms; multilayer neural networks; unsupervised learning; Boltzmann machines; wake–sleep algorithm; contrastive divergence; feature discovery; shape recognition; generative models

Five strategies for learning multilayer networks

Half a century ago, Oliver Selfridge (1958) proposed a pattern recognition system called "Pandemonium" consisting of multiple layers of feature detectors. His actual proposal contained a considerable amount of detail about what each of the layers would compute but the basic idea was that each feature detector would be activated by some familiar pattern of firing among the feature detectors in the layer below. This would allow the layers to extract more and more complicated features culminating in a top layer of detectors that would fire if and only if a familiar object was present in the visual input. Over the next 25 years, many attempts were made to find a learning algorithm that would be capable of discovering appropriate connection strengths (weights) for the feature detectors in every layer. Learning the weights of a single feature detector is quite easy if we are given both the inputs to the feature detector and its desired firing behavior, but learning is much more difficult if we are not told how the intermediate layers of feature detectors ought to behave. These intermediate feature detectors are called "hidden units" because their desired states cannot be observed. There are several strategies for learning the incoming weights of the hidden units.

The first strategy is denial. If we assume that there is only one layer of hidden units, it is often possible to set their incoming weights by hand using domain-specific knowledge. So the problem of learning hidden units does not exist. Within neuroscience, the equivalent of using hand-coded features is to assume that the receptive fields of feature detectors are specified innately — a view that is increasingly untenable (Merzenich et al., 1983; Karni et al., 1994; Sharma et al., 2000). Most of the work on perceptrons (Rosenblatt, 1962; Minsky and Papert, 1969) used hand-coded feature detectors, so learning only occurred for the weights from the feature detectors to the final decision units whose desired states were known. To

*Corresponding author. Tel.: +1 416-978-7564;
Fax: +1 416-978-1455; E-mail: hinton@cs.toronto.edu

DOI: 10.1016/S0079-6123(06)65034-6

be fair to Rosenblatt, he was well aware of the limitations of this approach — he just did not know how to learn multiple layers of features efficiently. The current version of denial is called "support vector machines" (Vapnik, 2000). These come with a fixed, non-adaptive recipe for converting a whole training image into a feature and a clever optimization technique for deciding which training cases should be turned into features and how these features should be weighted. Their main attraction is that the optimization method is guaranteed to find the global minimum. They are inadequate for tasks like 3-D object recognition that cannot be solved efficiently using a single layer of feature detectors (LeCun et al., 2004) but they work undeniably well for many of the simpler tasks that are used to evaluate machine learning algorithms (Decoste and Schoelkopf, 2002).

The second strategy is based on an analogy with evolution — randomly jitter the weights and save the changes that cause the performance of the whole network to improve. This is attractive because it is easy to understand, easy to implement in hardware (Jabri and Flower, 1992) and it works for almost any type of network. But it is hopelessly inefficient when the number of weights is large. Even if we only change one weight at a time, we still have to classify a large number of images to see if that single change helps or hurts. Changing many weights at the same time is no more efficient because the changes in other weights create noise that prevents each weight from detecting what effect it has on the overall performance. The evolutionary strategy can be significantly improved by applying the jitter to the activities of the feature detectors rather than to the weights (Mazzoni et al., 1991; Seung, 2003), but it is still an extremely inefficient way to discover gradients. Even blind evolution must have stumbled across a better strategy than this.

The third strategy is procrastination. Instead of learning feature detectors that are designed to be helpful in solving the classification problem, we can learn a layer of feature detectors that capture interesting regularities in the input images and put off the classification problem until later. This strategy can be applied recursively: we can learn a second layer of feature detectors that capture interesting regularities in the patterns of activation of the first layer detectors, and so on for as many layers as we like. The hope is that the features in the higher layers will be much more useful for classification than the raw inputs or the features in lower layers. As we shall see, this is not just wishful thinking. The main difficulties with the layer-by-layer strategy are that we need a quantitative definition of what it means for a regularity to be "interesting" and we need a way of ensuring that different feature detectors within a layer learn to detect different regularities even if they receive the same inputs.

The fourth strategy is to use calculus. To apply this strategy we need the output of each hidden unit to be a smooth function of the inputs it receives from the layer below. We also need a cost function that measures how poorly the network is performing. This cost function must change smoothly with the weights, so the number of classification errors is not the right function. For classification tasks, we can interpret the outputs of the top-level units as class probabilities and an appropriate cost function is then the cross-entropy, which is the negative log probability that the network assigns to the correct class. Given appropriate hidden units and an appropriate cost function, the chain rule can be used to compute how the cross-entropy changes as each weight in the network is changed. This computation can be made very efficient by first computing, for each hidden unit, how the cross-entropy changes as the activity of that hidden unit is changed. This is known as backpropagation because the computation starts at the output layer and proceeds backwards through the network one layer at a time. Once we know how the activity of a hidden unit affects the cross-entropy on each training case we have a surrogate for the desired state of the hidden unit and it is easy to change the incoming weights to decrease the sum of the cross-entropies on all the training cases. Compared with random jittering of the weights or feature activations, backpropagation is more efficient by a factor of the number of weights or features in the network.

Backpropagation was discovered independently by several different researchers (Werbos, 1974; Bryson and Ho, 1975; LeCun, 1985; Parker, 1985;

Rumelhart et al., 1986) and it was the first effective way to learn neural networks that had one or more layers of adaptive hidden units. It works very well for tasks such as the recognition of handwritten digits (LeCun et al., 1998; Simard et al., 2003), but it has two serious computational problems that will be addressed in this chapter. First, it is necessary to choose initial random values for all the weights. If these values are small, it is very difficult to learn deep networks because the gradients decrease multiplicatively as we backpropagate through each hidden layer. If the initial values are large, we have randomly chosen a particular region of the weight-space and we may well become trapped in a poor local optimum within this region. Second, the amount of information that each training case provides about the mapping between images and classes is at most the log of the number of possible classes. This means that large networks require a large amount of labeled training data if they are to learn weights that generalize well to novel test cases.

Many neuroscientists treat backpropagation with deep suspicion because it is not at all obvious how to implement it in cortex. In particular, it is hard to see how a single neuron can communicate both its activity and the deriviative of the cost function with respect to its activity. It seems very unlikely, however, that hundreds of millions of years of evolution have failed to find an effective way of tuning lower level feature detectors so that they provide the outputs that higher level detectors need in order to make the right decision.

The fifth and last strategy in this survey was designed to allow higher level feature detectors to communicate their needs to lower level ones whilst also being easy to implement in layered networks of stochastic, binary neurons that have activation states of 1 or 0 and turn on with a probability that is a smooth non-linear function of the total input they receive:

$$p(s_j = 1) = \frac{1}{1 + \exp(-b_j - \sum_i s_i w_{ij})} \tag{1}$$

where s_i and s_j are the binary activities of units i and j, w_{ij} the weight on the connection from i to j and b_j the bias of unit j. Imagine that the training data was generated top-down by a multilayer "graphics" model of the type shown in Fig. 1. The binary state of a hidden unit that was actually used to generate an image top-down could then be used as its desired state when learning the bottom-up "recognition" weights. At first sight, this idea of using top-down "generative" connections to provide desired states for the hidden units does not appear to help because we now have to learn a graphics model that can generate the training data. If, however, we already had some good recognition connections we could use a bottom-up pass from the real training data to activate the units in every layer and then we could learn the generative weights by trying to reconstruct the activities in each layer from the activities in the layer above. So we have a chicken-and-egg problem: given the generative weights we can learn the recognition weights and given the recognition weights we can learn the generative weights. It turns out that we can learn both sets of weights by starting with small random values and alternating between two phases of learning. In the "wake" phase, the recognition weights are used to drive the units bottom-up, and the binary states of units in adjacent layers can then be used to train the generative

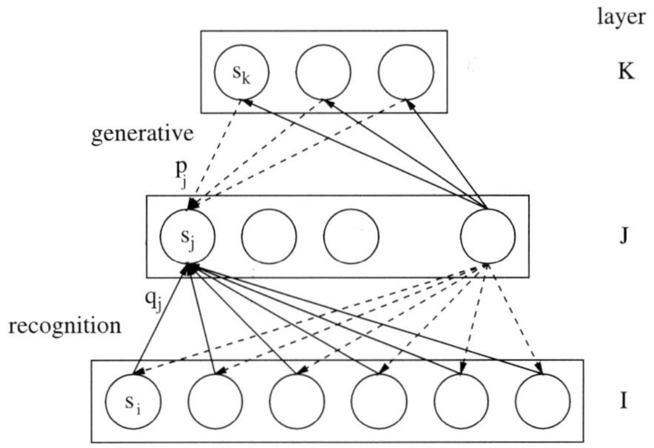

Fig. 1. This shows a three-layer neural network. Activities in the bottom layer represent the sensory input and activities in higher layers learn to represent the causes of the sensory input. The bottom-up "recognition" connections convert the sensory input into an internal representation. They can be trained by assuming that they are trying to invert a generative process (like computer graphics) that converts hidden causes into sensory data. The assumed generative process is represented in the top-down "generative" connections and it too is learned just by observing sensory data.

weights. In the "sleep" phase, the top-down generative connections are used to drive the network, so it produces fantasies from its generative model. The binary states of units in adjacent layers can then be used to learn the bottom-up recognition connections (Hinton et al., 1995). The learning rules are very simple. During the wake phase, a generative weight, g_{kj}, is changed by

$$\Delta g_{kj} = \varepsilon s_k (s_j - p_j) \tag{2}$$

where unit k is in the layer above unit j, ε a learning rate and p_j the probability that unit j would turn on if it were being driven by the current states of the units in the layer above using the current generative weights. During the sleep phase, a recognition weight, w_{ij}, is changed by

$$\Delta w_{ij} = \varepsilon s_i (s_j - q_j) \tag{3}$$

where q_j is the probability that unit j would turn on if it were being driven by the current states of the units in the layer below using the current recognition weights.

The rest of this chapter shows that the performance of both backpropagation (strategy four) and the "wake–sleep" algorithm (strategy five) can be greatly improved by using a "pretraining" phase in which unsupervised layer-by-layer learning is used to make the hidden units in each layer represent regularities in the patterns of activity of units in the layer below (strategy three). With this type of pretraining, it is finally possible to learn deep, multilayer networks efficiently and to demonstrate that they are better at classification than shallow methods.

Learning feature detectors with no supervision

Classification of isolated, normalized shapes like those shown in Fig. 2 has been one of the paradigm tasks for demonstrating the pattern recognition abilities of artificial neural networks. The connection weights in a multilayer network are typically initialized by using small random values, which are then iteratively adjusted by backpropagating the difference between the desired output of the network on each training case and the output that it actually produces on that training case. To prevent the network from modeling accidental regularities that arise from the random selection of training examples, it is common to stop the training early or to impose a penalty on large

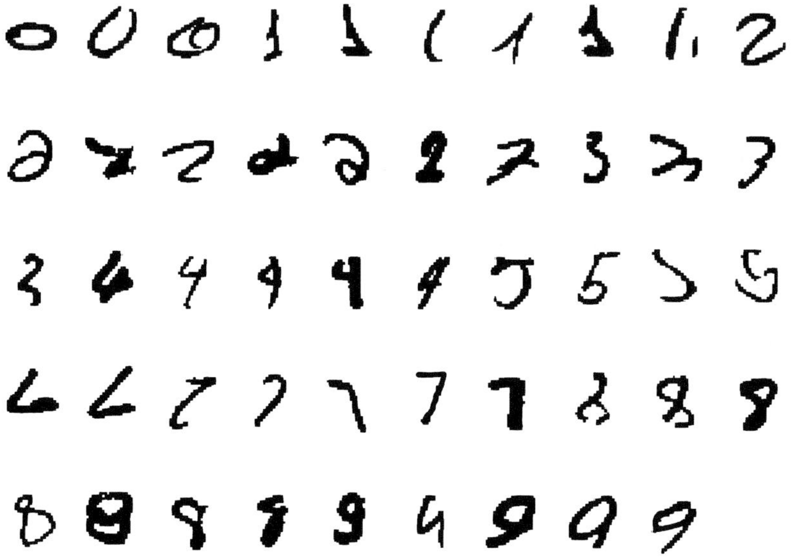

Fig. 2. Some examples of real handwritten digits from the MNIST test set that are hard to recognize. A neural network described at the end of this chapter gets all of these examples right, even though it has never seen them before. However, it is not confident about its classification for any of these examples. The true classes are arranged in standard scan order.

connection weights. This improves the final performance of the network on a test set, but it is not nearly as effective as using a more intelligent strategy for initializing the weights.

A discriminative training procedure like backpropagation ignores the structure in the input and only tries to model the way in which the output depends on the input. This is a bad idea if the input contains a lot of structure that can be modeled by latent variables and the output is a class label that is more simply related to these latent variables than it is to the raw input. Consider, for example a set of images of a dog. Latent variables such as the position, size, shape and color of the dog are a good way of explaining the complicated, higher order correlations between the individual pixel intensities, and some of these latent variables are very good predictors of the class label. In cases like this, it makes sense to start by using unsupervised learning to discover latent variables (i.e. features) that model the structure in the ensemble of training images. Once a good set of features has been found using unsupervised learning, discriminative learning can then be used to model the dependence of the class label on the features and to fine-tune the features so that they work better for discrimination. The features are then determined mainly by the input images, which contain a lot of information, and only slightly by the labels which typically contain much less information.

Learning one layer of feature detectors

Images composed of binary pixels can be modeled by using a "Restricted Boltzmann machine" (RBM) that uses a layer of binary feature detectors to model the higher order correlations between pixels. If there are no direct interactions between the feature detectors and no direct interactions between the pixels, there is a simple and efficient way to learn a good set of feature detectors from a set of training images (Hinton, 2002). We start with zero weights on the symmetric connections between each pixel i and each feature detector j. Then we repeatedly update each weight, w_{ij}, using the difference between two measured, pairwise correlations

$$\Delta w_{ij} = \varepsilon(\langle s_i s_j \rangle_{\text{data}} - \langle s_i s_j \rangle_{\text{recon}}) \quad (4)$$

where ε is a learning rate, $\langle s_i s_j \rangle_{\text{data}}$ the frequency with which pixel i and feature detector j are on together when the feature detectors are being driven by images from the training set and $\langle s_i s_j \rangle_{\text{recon}}$ the corresponding frequency when the feature detectors are being driven by reconstructed images. A similar learning rule can be used for the biases.

Given a training image, we set the binary state, s_j, of each feature detector to be 1 with probability

$$p(s_j = 1) = \frac{1}{1 + \exp(-b_j - \sum_{i \in \text{pixels}} s_i w_{ij})} \quad (5)$$

where b_j is the bias of j and s_i the binary state of pixel i. Once binary states have been chosen for the hidden units we produce a "reconstruction" of the training image by setting the state of each pixel to be 1 with probability

$$p(s_i = 1) = \frac{1}{1 + \exp(-b_i - \sum_{j \in \text{features}} s_j w_{ij})} \quad (6)$$

On 28×28 pixel images of handwritten digits like those shown in Fig. 2, good features can be found by using 100 passes through a training set of 50,000 images, with the weights being updated after every 100 images using the pixel-feature correlations measured on those 100 images and their reconstructions. Figure 3 shows a randomly selected subset of the features that are learned. We will use the letters A–F to refer to the rows of this figure and the numbers 1–10 to refer to the columns. Some features have an on-center off-surround structure (e.g. B3) or the reverse (A5). These features are a good way to model the simple fact that if a pixel is on, nearby pixels tend to be on. Some features detect parts of strokes (A9), and they typically inhibit the region of the image that is further from the center than the stroke fragment. Some features, which look more like fingerprints (D2), encode the phase and amplitude of high-frequency Fourier components of a large part of the whole image. These features tend to turn on about half the time and can be eliminated by forcing features to only turn on rarely (Ranzato et al., 2007). The three features with unnaturally sharp black lines (A4, D9, E4) capture the fact that if a pixel is on, pixels that are more than 20 rows above or below it cannot be on because of the way the data was normalized.

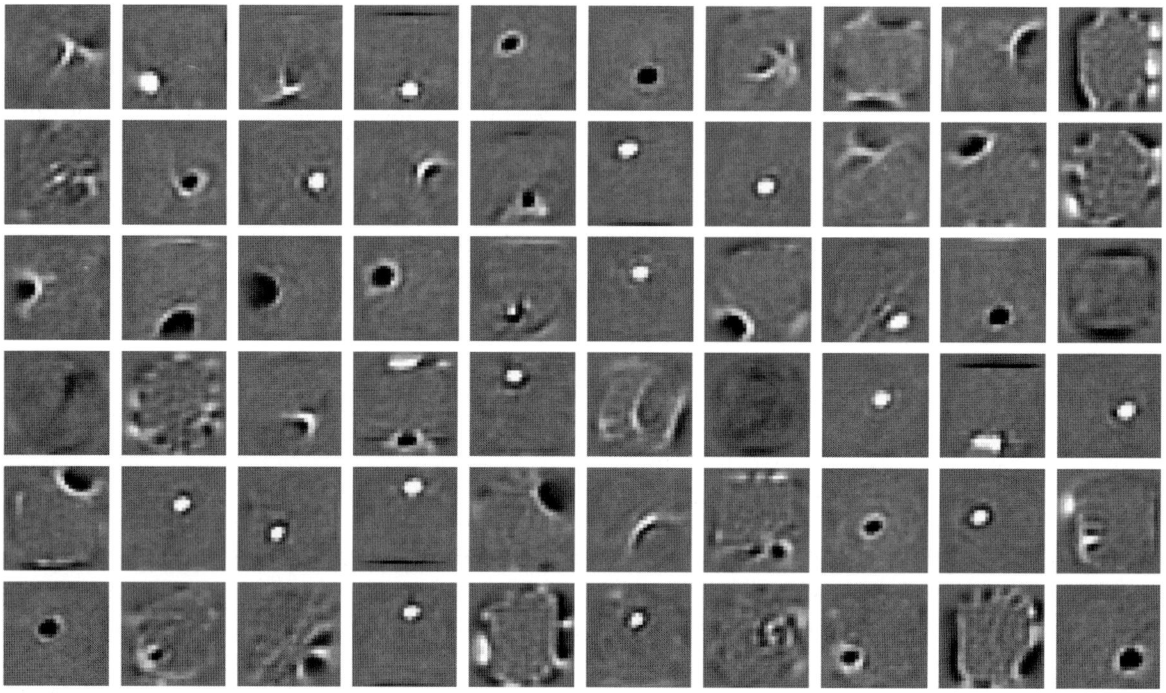

Fig. 3. The receptive fields of some feature detectors. Each gray square shows the incoming weights to one feature detector from all the pixels. Pure white means a positive weight of at least 3 and pure black means a negative weight of at least −3. Most of the feature detectors learn highly localized receptive fields.

The learned weights and biases of the features implicitly define a probability distribution over all possible binary images. Sampling from this distribution is difficult, but it can be done by using "alternating Gibbs sampling." This starts with a random image and then alternates between updating all of the features in parallel using Eq. (5) and updating all of the pixels in parallel using Eq. (6). After Gibbs sampling for sufficiently long, the network reaches "thermal equilibrium." The states of pixels and features detectors still change, but the probability of finding the system in any particular binary configuration does not.

A greedy learning algorithm for multiple hidden layers

A single layer of binary features is not the best way to model the structure in a set of images. After learning the first layer of feature detectors, a second layer can be learned in just the same way by treating the existing feature detectors, when they are being driven by training images, as if they were data. To reduce noise in the learning signal, the binary states of feature detectors (or pixels) in the "data" layer are replaced by their real-valued probabilities of activation when learning the next layer of feature detectors, but the new feature detectors have binary states to limit the amount of information they can convey. This greedy, layer-by-layer learning can be repeated as many times as desired. To justify this layer-by-layer approach, it would be good to show that adding an extra layer of feature detectors always increases the probability that the overall generative model would generate the training data. This is almost true: provided the number of feature detectors does not decrease and their weights are initialized correctly, adding an extra layer is guaranteed to raise a lower bound on the log probability of the training data (Hinton et al., 2006). So after learning several layers there is good reason to believe that the feature detectors will have captured many of the statistical regularities in the set of training images and we can now test the hypothesis that these feature detectors will be useful for classification.

Using backpropagation for discriminative fine-tuning

After greedily learning layers of 500, 500 and 2000 feature detectors without using any information about the class labels, gentle backpropagation was used to fine-tune the weights for discrimination. This produced much better classification performance on test data than using backpropagation without the initial, unsupervised phase of learning. The MNIST dataset used for these experiments has been used as a benchmark for many years and many different researchers have tried using many different learning methods, including variations of backpropagation in nets with different numbers of hidden layers and different numbers of hidden units per layer.

There are several different versions of the MNIST learning task. In the most difficult version, the learning algorithm is not given any prior knowledge of the geometry of images and it is forbidden to increase the size of the training set by using small affine or elastic distortions of the training images. Consequently, if the same random permutation is applied to the pixels of every training and test image, the performance of the learning algorithm will be unaffected. For this reason, this is called the "permutation invariant" version of the task. So far as the learning algorithm is concerned, each 28×28 pixel image is just a vector of 784 numbers that has to be given one of 10 labels. The best published backpropagation error rate for this version of the task is 1.6% (Simard et al., 2003). Support vector machines can achieve 1.4% (Decoste and Schoelkopf, 2002). Table 1 shows that the error rate of backpropagation can be reduced to about 1.12% if it is only used for fine-tuning features that are originally discovered by layer-by-layer pretraining.

Details of the discriminative fine-tuning procedure

Using three different splits of the 60,000 image training set into 50,000 training examples and 10,000 validation examples, the greedy learning algorithm was used to initialize the weights and biases and gentle backpropagation was then used to fine-tune the weights. After each sweep through the training set (which is called an "epoch"), the classification error rate was measured on the validation set. Training was continued until two conditions were satisfied. The first condition involved the average cross-entropy error on the validation set. This is the quantity that is being minimized by the learning algorithm so it always falls on the training data. On the validation data, however, it starts rising as soon as overfitting occurs. There is a strong tendency for the number of classification errors to continue to fall after the cross-entropy has bottomed-out on the validation data, so the

Table 1. Neta, Netb and Netc were greedily pretrained on different, unlabeled, subsets of the training data that were obtained by removing disjoint validation sets of 10,000 images

Pretrained Network	Backpropagation training Set size	Train Epochs	Train cost Per 100	Train Errors	Valid. cost Per 100	Valid. errors In 10^4	Test cost Per 100	Test errors In 10^4
Neta	50,000	33	0.12	1	6.49	129	6.22	122
Netb	50,000	56	0.04	0	7.81	118	6.21	116
Netc	50,000	63	0.03	0	8.12	118	6.73	124
Combined							5.75	110
Neta	60,000	33+16	<0.12	1			5.81	113
Netb	60,000	56+28	<0.04	0			5.90	106
Netc	60,000	63+31	<0.03	0			5.93	118
Combined							5.40	106
Not pretrained	60,000	119	<0.063	0			18.43	227

Note: After pretraining, they were trained on those same subsets using backpropagation. Then the training was continued on the full training set until the cross-entropy error reached the criterion explained in the text.

first condition is that the learning must have already gone past the minimum of the cross-entropy on the validation set. It is easy to detect when this condition is satisfied because the cross-entropy changes very smoothly during the learning. The second condition involved the number of errors on the validation set. This quantity fluctuates unpredictably, so the criterion was that the minimum value observed so far should have occurred at least 10 epochs ago. Once both conditions were satisfied, the weights and biases were restored to the values they had when the number of validation set errors was at its minimum, and performance on the 10,000 test cases was measured. As shown in Table 1, this gave test error rates of 1.22, 1.16 and 1.24% on the three different splits. The fourth line of the table shows that these error rates can be reduced to 1.10% by multiplying together the three probabilities that the three nets predict for each digit class and picking the class with the maximum product.

Once the performance on the validation set has been used to find a good set of weights, the cross-entropy error on the *training* set is recorded. Performance on the test data can then be further improved by adding the validation set to the training set and continuing the training until the cross-entropy error on the expanded training set has fallen to the value it had on the original training set for the weights selected by the validation procedure. As shown in Table 1 this eliminates about 8% of the errors. Combining the predictions of all three models produces less improvement than before because each model has now seen all of the training data. The final line of Table 1 shows that backpropagation in this relatively large network gives much worse results if no pretraining is used. For this last experiment, the stopping criterion was set to be the average of the stopping criteria from the previous experiments.

To avoid making large changes to the weights found by the pretraining, the backpropagation stage of learning used a very small learning rate which made it very slow, so a new trick was introduced which sped up the learning by about a factor of three. Most of the computational effort is expended computing the almost non-existent gradients for "easy" training cases that the network can already classify confidently and correctly. It is tempting to make a note of these easy cases and then just ignore them, checking every few epochs to see if the cross-entropy error on any of the ignored cases has become significant. This can be done without changing the expected value of the overall gradient by using a method called importance sampling. Instead of being completely ignored, easy cases are selected with a probability of 0.1, but when they are selected, the computed gradients are multiplied by 10. Using more extreme values like 0.01 and 100 is dangerous because a case that used to be easy might have developed a large gradient while it was being ignored, and multiplying this gradient by 100 could give the network a shock. When using importance sampling, an "epoch" was redefined to be the time it takes to sample as many training cases as the total number in the training set. So an epoch typically involves several sweeps through the whole set of training examples, but it is the same amount of computation as one sweep without importance sampling.

After the results in Table 1 were obtained using the rather complicated version of backpropagation described above, Ruslan Salakhutdinov discovered that similar results can be obtained using a standard method called "conjugate gradient" which takes the gradients delivered by backpropagation and uses them in a more intelligent way than simply changing each weight in proportion to its gradient (Hinton and Salakhutdinov, 2006). The MNIST data together with the Matlab code required for pretraining and fine-tuning the network are available at http://www.cs.toronto.edu/~hinton/MatlabForSciencePaper.html.

Using extra unlabeled data

Since the greedy pretraining algorithm does not require any labeled data, it should be a very effective way to make use of unlabeled examples to improve performance on a small labeled dataset. Learning with only a few labeled examples is much more characteristic of human learning. We see many instances of many different types of object, but we are very rarely told the name of an object.

Preliminary experiments confirm that pretraining on unlabeled data helps a lot, but for a proper comparison it will be necessary to use networks of the appropriate size. When the number of labeled examples is small, it is unfair to compare the performance of a large network that makes use of unlabeled examples with a network of the same size that does not make use of the unlabeled examples.

Using geometric prior knowledge

The greedy pretraining improves the error rate of backpropagation by about the same amount as methods that make use of prior knowledge about the geometry of images, such as weight-sharing (LeCun et al., 1998) or enlarging the training set by using small affine or elastic distortions of the training images. But pretraining can also be combined with these other methods. If translations of up to two pixels are used to create 12 extra versions of each training image, the error rate of the best support vector machine falls from 1.4% to 0.56% (Decoste and Schoelkopf, 2002). The average error rate of the pretrained neural net falls from 1.12% to 0.65%. The translated data is presumably less helpful to the multilayer neural net because the pretraining can already capture some of the geometrical structure even without the translations. The best published result for a single method is currently 0.4%, which was obtained using backpropagation in a multilayer neural net that uses *both* weight-sharing and sophisticated, elastic distortions (Simard et al., 2003). The idea of using unsupervised pretraining to improve the performance of backpropagation has recently been applied to networks that use weight-sharing and it consistently reduces the error rate by about 0.1% even when the error rate is already very low (Ranzato et al., 2007).

Using contrastive wake–sleep for generative fine-tuning

Figure 4 shows a multilayer generative model in which the top two layers interact via undirected connections and form an associative memory. At the start of learning, all configurations of this top-level associative memory have roughly equal energy. Learning sculpts the energy landscape and after learning, the associative memory will settle into low-energy states that represent images of digits. Valleys in the high-dimensional energy-landscape represent digit classes. Directions along the valley floor represent the allowable variations of a digit and directions up the side of a valley represent implausible variations that make the image surprising to the network. Turning on one of the 10 label units lowers one whole valley and raises the other valleys. The number of valleys and the dimensionality of each valley floor are determined by the set of training examples.

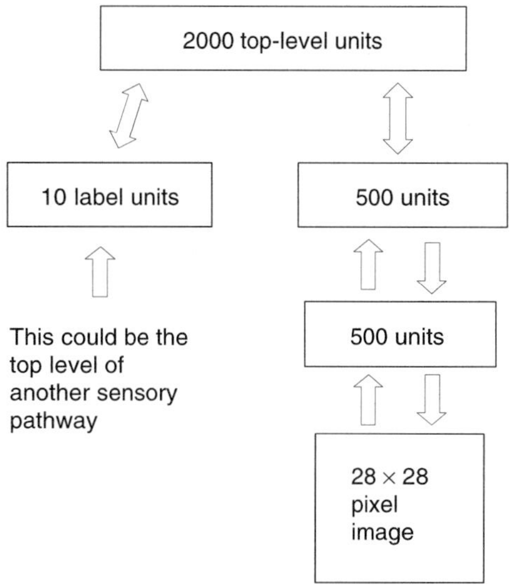

Fig. 4. A multilayer neural network that learns to model the joint distribution of digit images and digit labels. The top two layers have symmetric connections and form an associative memory. The layers below have directed, top-down, generative connections that can be used to map a state of the associative memory to an image. There are also directed, bottom-up, recognition connections that are used to infer a factorial representation in one layer from the binary activities in the layer below.

The states of the associative memory are just binary activity vectors that look nothing like the images they represent, but it is easy to see what the associative memory has in mind. First, the 500 hidden units that form part of the associative memory are used to stochastically activate some of the units in the layer below via the top-down, generative connections. Then these activated units are used to provide top-down input to the pixels. Figure 5 shows some fantasies produced by the trained network when the top-level associative memory is allowed to wander stochastically between low-energy states, but with one of the label units clamped so that it tends to stay in the same valley. The fact that it can generate a wide variety of slightly implausible versions of each type of digit makes it very good at recognizing poorly written digits. A demonstration that shows the network generating and recognizing digit images is available at http://www.cs.toronto.edu/hinton/digits.html. In this chapter, each training case consists of an image and an explicit class label, but the same learning algorithm can be used if the "labels" are replaced by a multilayer pathway whose inputs are spectrograms from multiple different speakers saying isolated digits (Kaganov et al., 2007). The network then learns to generate pairs that consist of an image and a spectrogram of the same digit class.

The network was trained in two stages — pretraining and fine-tuning. The layer-by-layer pretraining was the same as in the previous section, except that when training the top layer of 2000 feature detectors, each "data" vector had 510 components. The first 500 were the activation probabilities of the 500 feature detectors in the penultimate layer and the last 10 were the label values. The value of the correct label was set to 1 and the remainder were set to 0. So the top layer of feature detectors learns to model the joint distribution of the 500 penultimate features and the 10 labels.

At the end of the layer-by-layer pretraining, the weight between any two units in adjacent layers is the same in both directions and we can view the result of the pretraining as a set of three different RBMs whose only interaction is that the data for the higher RBMs is provided by the feature activations of the lower RBMs. It is possible, however, to take a very different view of exactly the same system (Hinton et al., 2006). We can view it as a single generative model that generates data by first letting the top-level RBM settle to thermal equilibrium, which may take a very long time, and then performing a single top-down pass to convert the 500 binary feature activations in the penultimate layer into an image. When it is viewed as a single generative model, the weights between the top two

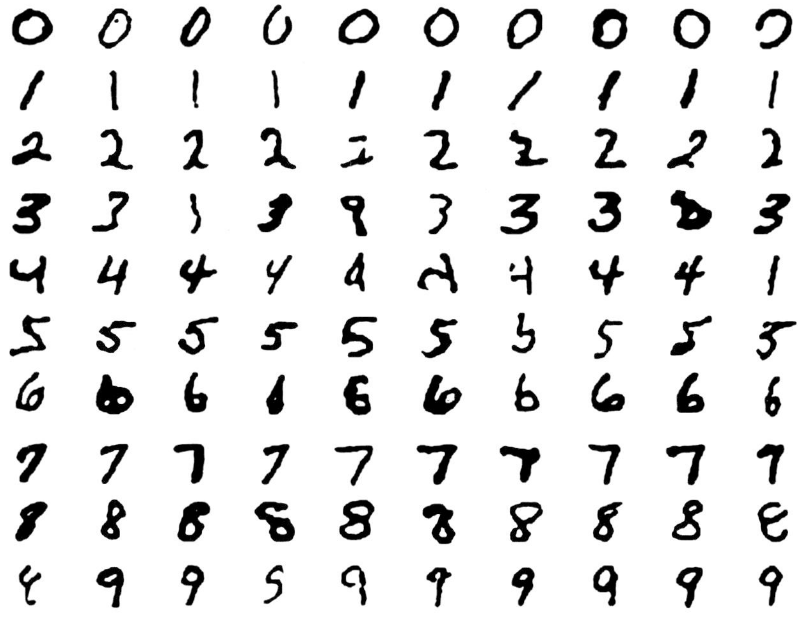

Fig. 5. Each row shows 10 samples from the generative model with a particular label clamped on. The top-level associative memory is run for 1000 iterations of alternating Gibbs sampling between samples.

layers need to be symmetric, but the weights between lower layers do not. In the top-down, generative direction, these weights form part of the overall generative model, but in the bottom-up, recognition direction they are not part of the model. They are merely an efficient way of inferring what hidden states probably caused the observed image. If the whole system is viewed as a single generative model, we can ask whether it is possible to fine-tune the weights produced by the pretraining to make the overall generative model more likely to generate the set of image-label pairs in the training data. The answer is that the generative model can be significantly improved by using a contrastive form of the wake–sleep algorithm. In the lower layers, this makes the recognition weights differ from the generative weights. In addition to improving the overall generative model, the generative fine-tuning makes the model much better at assigning labels to test images using a method, which will be described later.

In the standard wake–sleep algorithm, the network generates fantasies by starting with a pattern of activation of the top-level units that is chosen stochastically using only the generative bias of each top-level unit to influence its probability of being on. This way of initiating fantasies cannot be used if the top two layers of the generative model form an associative memory because it will not produce samples from the generative model. The obvious alternative is to use prolonged Gibbs sampling in the top two layers to sample from the energy landscape defined by the associative memory, but this is much too slow. A very effective alternative is to use the bottom-up recognition connections to convert a image-label pair from the training set into a state of the associative memory and then to perform brief alternating Gibbs sampling which allows the associative memory to produce a "confabulation" that it prefers to its initial representation of the training pair. The top-level associative memory is then trained as an RBM by using Eq. (4) to lower the energy of the initial representation of the training pair and raise the energy of the confabulation. The confabulation in the associative memory is also used to drive the system top-down, and the states of all the hidden units that are produced by this generative, top-down pass are used as targets to train the bottom-up recognition connections. The "wake" phase is just the same as in the standard wake–sleep algorithm: After the initial bottom-up pass, the top-down, generative connections in the bottom two layers are trained, using Eq. (2), to reconstruct the activities in the layer below from the activities in the layer above. The details are given in Hinton et al. (2006).

Fine-tuning with the contrastive wake–sleep algorithm is about an order of magnitude slower than fine-tuning with backpropagation, partly because it has a more ambitious goal. The network shown in Fig. 4 takes a week to train on a 3 GHz machine. The examples shown in Fig. 2 were all classified correctly by this network which gets a test error rate of 1.25%. This is slightly worse than pretrained networks with the same architecture that are fine-tuned with backpropagation, but it is better than the 1.4% achieved by the best support vector machine on the permutation-invariant version of the MNIST task. It is rare for a generative model to outperform a good discriminative model *at discrimination*.

There are several different ways of using the generative model for discrimination. If time was not an issue, it would be possible to use sampling methods to measure the relative probabilities of generating each of the ten image-label pairs that are obtained by pairing the test image with each of the 10 possible labels. A fast and accurate approximation can be obtained by first performing a bottom-up pass in which the activation probabilities of the first layer of hidden units are used to compute activation probabilities for the penultimate hidden layer. Using probabilities rather than stochastic binary states suppresses the noise due to sampling. Then the vector of activation probabilities of the feature detectors in the penultimate layer is paired with each of the 10 labels in turn and the "free energy" of the associative memory is computed. Each of the top-level units contributes additively to this free energy, so it is easy to calculate exactly (Hinton et al., 2006). The label that gives the lowest free-energy is the network's guess.

Fitting a generative model constrains the weights of the network far more strongly than fitting a discriminative model, but if the ultimate objective is discrimination, it also wastes a lot of the discriminative capacity. This waste shows up in the fact that after fine-tuning the generative model, its discriminative performance on the training data is about the same as its discriminative performance on the test data — there is almost no overfitting. This suggests one final experiment. After first using contrastive wake–sleep for fine-tuning, further fine-tuning can be performed using a weighted average of the gradients computed by backpropagation and by contrastive wake–sleep. Using a validation set, the coefficient controlling the contribution of the backpropagation gradient to the weighted average was gradually increased to find the coefficient value at which the error rate on the validation set was minimized. Using this value of the coefficient, the test error rate was 0.97%, which is the current record for the permutation-invariant MNIST task. It is also possible to combine the gradient from backpropagation with the gradient computed by the pretraining (Bengio et al., 2007). This is much less computational effort than using contrastive wake–sleep, but does not perform as well.

Acknowledgments

I thank Yoshua Bengio, Yann LeCun, Peter Dayan, David MacKay, Sam Roweis, Terry Sejnowski, Max Welling and my past and present graduate students for their numerous contributions to these ideas. The research was supported by NSERC, CFI and OIT. GEH is a fellow of the Canadian Institute for Advanced Research and holds a Canada Research Chair in Machine Learning.

References

Bengio, Y., Lamblin, P., Popovici, D. and Larochelle, H. (2007) Greedy layer-wise training of deep networks. In: Advances in Neural Information Processing Systems, 19. MIT Press, Cambridge, MA, pp. 153–160.

Bryson, A. and Ho, Y. (1975) Applied optimal control. Wiley, New York.

Decoste, D. and Schoelkopf, B. (2002) Training invariant support vector machines. Machine Learn., 46: 161–190.

Hinton, G.E. (2002) Training products of experts by minimizing contrastive divergence. Neural Comput., 14: 1771–1800.

Hinton, G.E., Dayan, P., Frey, B.J. and Neal, R. (1995) The wake-sleep algorithm for self-organizing neural networks. Science, 268: 1158–1161.

Hinton, G.E., Osindero, S. and Teh, Y.W. (2006) A fast learning algorithm for deep belief nets. Neural Comput., 18: 1527–1554.

Hinton, G.E. and Salakhutdinov, R.R. (2006) Reducing the dimensionality of data with neural networks. Science, 313: 504–507.

Jabri, M. and Flower, B. (1992) Weight perturbation: an optimal architecture and learning technique for analog VLSI feedforward and recurrent multilayer networks. IEEE Trans. Neural Netw., 3(1): 154–157.

Kaganov, A., Osindero, S. and Hinton, G.E. (2007) Learning the relationship between spoken digits and digit images. In: Technical Report, Department of Computer Science, University of Toronto.

Karni, A., Tanne, D., Rubenstein, B., Askenasy, J. and Sagi, D. (1994) Dependence on REM sleep of overnight improvement of a perceptual skill. Science, 265(5172): 679.

LeCun, Y. (1985) Une procédure d'apprentissage pour réseau a seuil asymmetrique (a learning scheme for asymmetric threshold networks). In: Proceedings of Cognitiva 85, Paris, France, pp. 599–604.

LeCun, Y., Bottou, L., Bengio, Y. and Haffner, P. (1998) Gradient based learning applied to document recognition. Proc. IEEE, 86(11): 2278–2324.

LeCun, Y., Huang, F.-J. and Bottou, L. (2004) Learning methods for generic object recognition with invariance to pose and lighting. In: Proceedings of CVPR'04. IEEE Press, New York.

Mazzoni, P., Andersen, R. and Jordan, M. (1991) A more biologically plausible learning rule for neural networks. Proc. Natl. Acad. Sci., 88(10): 4433–4437.

Merzenich, M., Kaas, J., Wall, J., Nelson, R., Sur, M. and Felleman, D. (1983) Topographic reorganization of somatosensory cortical areas 3b and 1 in adult monkeys following restricted deafferentation. Neuroscience, 8(1): 33–55.

Minsky, M. and Papert, S. (1969) Perceptrons: an introduction to computational geometry. MIT Press, Cambridge, MA.

Parker, D. (1985) Learning logic. In: Technical Report TR-47. Center for Computational Research in Economics and Management Science. Massachusetts Institute of Technology, Cambridge, MA.

Ranzato, M., Poultney, C., Chopra, S. and LeCun, Y. (2007) Efficient learning of sparse representations with an energy-based model. In: Advances in Neural Information Processing Systems 17. MIT Press, Cambridge, MA.

Rosenblatt, F. (1962) Principles of Neurodynamics. Spartan Books, New York.

Rumelhart, D.E., Hinton, G.E. and Williams, R.J. (1986) Learning representations by back-propagating errors. Nature, 323: 533–536.

Selfridge, O.G. (1958) Pandemonium: a paradigm for learning. In: Mechanisation of thought processes: Proceedings of a symposium held at the National Physical Laboratory. HMSO, London.

Seung, H. (2003) Learning in spiking neural networks by reinforcement of stochastic synaptic transmission. Neuron, 40(6): 1063–1073.

Sharma, J., Angelucci, A. and Sur, M. (2000) Induction of visual orientation modules in auditory cortex. Nature, 404(6780): 841–847.

Simard, P.Y., Steinkraus, D. and Platt, J. (2003) Best practice for convolutional neural networks applied to visual document analysis. In: International Conference on Document Analysis and Recogntion (ICDAR). IEEE Computer Society, Los Alamitos, pp. 958–962.

Vapnik, V.N. (2000) The Nature of Statistical Learning Theory. Springer, New York.

Werbos, P. (1974) Beyond regression: new tools for prediction and analysis in the behavioral sciences. PhD thesis, Harvard University.

Subject Index